T0181731

Lecture Notes in Computer Science   **14229**

Founding Editors

Gerhard Goos
Juris Hartmanis

The series Lecture Notes in Computer Science (LNCS), including its subseries Lecture Notes in Artificial Intelligence (LNAI) and Lecture Notes in Bioinformatics (LNBI), has established itself as a medium for the publication of new developments in computer science and information technology research, teaching, and education.

LNCS enjoys close cooperation with the computer science R & D community, the series counts many renowned academics among its volume editors and paper authors, and collaborates with prestigious societies. Its mission is to serve this international community by providing an invaluable service, mainly focused on the publication of conference and workshop proceedings and postproceedings. LNCS commenced publication in 1973.

Hayit Greenspan · Anant Madabhushi ·
Parvin Mousavi · Septimiu Salcudean ·
James Duncan · Tanveer Syeda-Mahmood ·
Russell Taylor
Editors

Medical Image Computing and Computer Assisted Intervention – MICCAI 2023

26th International Conference
Vancouver, BC, Canada, October 8–12, 2023
Proceedings, Part X

 Springer

Editors
Hayit Greenspan
Icahn School of Medicine, Mount Sinai,
NYC, NY, USA

Tel Aviv University
Tel Aviv, Israel

Parvin Mousavi
Queen's University
Kingston, ON, Canada

James Duncan ⓘ
Yale University
New Haven, CT, USA

Russell Taylor ⓘ
Johns Hopkins University
Baltimore, MD, USA

Anant Madabhushi ⓘ
Emory University
Atlanta, GA, USA

Septimiu Salcudean ⓘ
The University of British Columbia
Vancouver, BC, Canada

Tanveer Syeda-Mahmood ⓘ
IBM Research
San Jose, CA, USA

ISSN 0302-9743   ISSN 1611-3349 (electronic)
Lecture Notes in Computer Science
ISBN 978-3-031-43998-8   ISBN 978-3-031-43999-5 (eBook)
https://doi.org/10.1007/978-3-031-43999-5

This Springer imprint is published by the registered company Springer Nature Switzerland AG
The registered company address is: Gewerbestrasse 11, 6330 Cham, Switzerland

Paper in this product is recyclable.

Preface

We are pleased to present the proceedings for the 26th International Conference on Medical Image Computing and Computer-Assisted Intervention (MICCAI). After several difficult years of virtual conferences, this edition was held in a mainly in-person format with a hybrid component at the Vancouver Convention Centre, in Vancouver, BC, Canada October 8–12, 2023. The conference featured 33 physical workshops, 15 online workshops, 15 tutorials, and 29 challenges held on October 8 and October 12. Co-located with the conference was also the 3rd Conference on Clinical Translation on Medical Image Computing and Computer-Assisted Intervention (CLINICCAI) on October 10.

MICCAI 2023 received the largest number of submissions so far, with an approximately 30% increase compared to 2022. We received 2365 full submissions of which 2250 were subjected to full review. To keep the acceptance ratios around 32% as in previous years, there was a corresponding increase in accepted papers leading to 730 papers accepted, with 68 orals and the remaining presented in poster form. These papers comprise ten volumes of Lecture Notes in Computer Science (LNCS) proceedings as follows:

- Part I, LNCS Volume 14220: Machine Learning with Limited Supervision and Machine Learning – Transfer Learning
- Part II, LNCS Volume 14221: Machine Learning – Learning Strategies and Machine Learning – Explainability, Bias, and Uncertainty I
- Part III, LNCS Volume 14222: Machine Learning – Explainability, Bias, and Uncertainty II and Image Segmentation I
- Part IV, LNCS Volume 14223: Image Segmentation II
- Part V, LNCS Volume 14224: Computer-Aided Diagnosis I
- Part VI, LNCS Volume 14225: Computer-Aided Diagnosis II and Computational Pathology
- Part VII, LNCS Volume 14226: Clinical Applications – Abdomen, Clinical Applications – Breast, Clinical Applications – Cardiac, Clinical Applications – Dermatology, Clinical Applications – Fetal Imaging, Clinical Applications – Lung, Clinical Applications – Musculoskeletal, Clinical Applications – Oncology, Clinical Applications – Ophthalmology, and Clinical Applications – Vascular
- Part VIII, LNCS Volume 14227: Clinical Applications – Neuroimaging and Microscopy
- Part IX, LNCS Volume 14228: Image-Guided Intervention, Surgical Planning, and Data Science
- Part X, LNCS Volume 14229: Image Reconstruction and Image Registration

The papers for the proceedings were selected after a rigorous double-blind peer-review process. The MICCAI 2023 Program Committee consisted of 133 area chairs and over 1600 reviewers, with representation from several countries across all major continents. It also maintained a gender balance with 31% of scientists who self-identified as women. With an increase in the number of area chairs and reviewers, the reviewer load on the experts was reduced this year, keeping to 16–18 papers per area chair and about 4–6 papers per reviewer. Based on the double-blinded reviews, area chairs' recommendations, and program chairs' global adjustments, 308 papers (14%) were provisionally accepted, 1196 papers (53%) were provisionally rejected, and 746 papers (33%) proceeded to the rebuttal stage. As in previous years, Microsoft's Conference Management Toolkit (CMT) was used for paper management and organizing the overall review process. Similarly, the Toronto paper matching system (TPMS) was employed to ensure knowledgeable experts were assigned to review appropriate papers. Area chairs and reviewers were selected following public calls to the community, and were vetted by the program chairs.

Among the new features this year was the emphasis on clinical translation, moving Medical Image Computing (MIC) and Computer-Assisted Interventions (CAI) research from theory to practice by featuring two clinical translational sessions reflecting the real-world impact of the field in the clinical workflows and clinical evaluations. For the first time, clinicians were appointed as Clinical Chairs to select papers for the clinical translational sessions. The philosophy behind the dedicated clinical translational sessions was to maintain the high scientific and technical standard of MICCAI papers in terms of methodology development, while at the same time showcasing the strong focus on clinical applications. This was an opportunity to expose the MICCAI community to the clinical challenges and for ideation of novel solutions to address these unmet needs. Consequently, during paper submission, in addition to MIC and CAI a new category of "Clinical Applications" was introduced for authors to self-declare.

MICCAI 2023 for the first time in its history also featured dual parallel tracks that allowed the conference to keep the same proportion of oral presentations as in previous years, despite the 30% increase in submitted and accepted papers.

We also introduced two new sessions this year focusing on young and emerging scientists through their Ph.D. thesis presentations, and another with experienced researchers commenting on the state of the field through a fireside chat format.

The organization of the final program by grouping the papers into topics and sessions was aided by the latest advancements in generative AI models. Specifically, Open AI's GPT-4 large language model was used to group the papers into initial topics which were then manually curated and organized. This resulted in fresh titles for sessions that are more reflective of the technical advancements of our field.

Although not reflected in the proceedings, the conference also benefited from keynote talks from experts in their respective fields including Turing Award winner Yann LeCun and leading experts Jocelyne Troccaz and Mihaela van der Schaar.

We extend our sincere gratitude to everyone who contributed to the success of MIC-CAI 2023 and the quality of its proceedings. In particular, we would like to express our profound thanks to the MICCAI Submission System Manager Kitty Wong whose meticulous support throughout the paper submission, review, program planning, and proceeding preparation process was invaluable. We are especially appreciative of the effort and dedication of our Satellite Events Chair, Bennett Landman, who tirelessly coordinated the organization of over 90 satellite events consisting of workshops, challenges and tutorials. Our workshop chairs Hongzhi Wang, Alistair Young, tutorial chairs Islem

Rekik, Guoyan Zheng, and challenge chairs, Lena Maier-Hein, Jayashree Kalpathy-Kramer, Alexander Seitel, worked hard to assemble a strong program for the satellite events. Special mention this year also goes to our first-time Clinical Chairs, Drs. Curtis Langlotz, Charles Kahn, and Masaru Ishii who helped us select papers for the clinical sessions and organized the clinical sessions.

We acknowledge the contributions of our Keynote Chairs, William Wells and Alejandro Frangi, who secured our keynote speakers. Our publication chairs, Kevin Zhou and Ron Summers, helped in our efforts to get the MICCAI papers indexed in PubMed. It was a challenging year for fundraising for the conference due to the recovery of the economy after the COVID pandemic. Despite this situation, our industrial sponsorship chairs, Mohammad Yaqub, Le Lu and Yanwu Xu, along with Dekon's Mehmet Eldegez, worked tirelessly to secure sponsors in innovative ways, for which we are grateful.

An active body of the MICCAI Student Board led by Camila Gonzalez and our 2023 student representatives Nathaniel Braman and Vaishnavi Subramanian helped put together student-run networking and social events including a novel Ph.D. thesis 3-minute madness event to spotlight new graduates for their careers. Similarly, Women in MICCAI chairs Xiaoxiao Li and Jayanthi Sivaswamy and RISE chairs, Islem Rekik, Pingkun Yan, and Andrea Lara further strengthened the quality of our technical program through their organized events. Local arrangements logistics including the recruiting of University of British Columbia students and invitation letters to attendees, was ably looked after by our local arrangement chairs Purang Abolmaesumi and Mehdi Moradi. They also helped coordinate the visits to the local sites in Vancouver both during the selection of the site and organization of our local activities during the conference. Our Young Investigator chairs Marius Linguraru, Archana Venkataraman, Antonio Porras Perez put forward the startup village and helped secure funding from NIH for early career scientist participation in the conference. Our communications chair, Ehsan Adeli, and Diana Cunningham were active in making the conference visible on social media platforms and circulating the newsletters. Niharika D'Souza was our cross-committee liaison providing note-taking support for all our meetings. We are grateful to all these organization committee members for their active contributions that made the conference successful.

We would like to thank the MICCAI society chair, Caroline Essert, and the MICCAI board for their approvals, support and feedback, which provided clarity on various aspects of running the conference. Behind the scenes, we acknowledge the contributions of the MICCAI secretariat personnel, Janette Wallace, and Johanne Langford, who kept a close eye on logistics and budgets, and Diana Cunningham and Anna Van Vliet for including our conference announcements in a timely manner in the MICCAI society newsletters. This year, when the existing virtual platform provider indicated that they would discontinue their service, a new virtual platform provider Conference Catalysts was chosen after due diligence by John Baxter. John also handled the setup and coordination with CMT and consultation with program chairs on features, for which we are very grateful. The physical organization of the conference at the site, budget financials, fund-raising, and the smooth running of events would not have been possible without our Professional Conference Organization team from Dekon Congress & Tourism led by Mehmet Eldegez. The model of having a PCO run the conference, which we used at

MICCAI, significantly reduces the work of general chairs for which we are particularly grateful.

Finally, we are especially grateful to all members of the Program Committee for their diligent work in the reviewer assignments and final paper selection, as well as the reviewers for their support during the entire process. Lastly, and most importantly, we thank all authors, co-authors, students/postdocs, and supervisors for submitting and presenting their high-quality work, which played a pivotal role in making MICCAI 2023 a resounding success.

With a successful MICCAI 2023, we now look forward to seeing you next year in Marrakesh, Morocco when MICCAI 2024 goes to the African continent for the first time.

October 2023

Tanveer Syeda-Mahmood
James Duncan
Russ Taylor
General Chairs

Hayit Greenspan
Anant Madabhushi
Parvin Mousavi
Septimiu Salcudean
Program Chairs

Organization

General Chairs

Tanveer Syeda-Mahmood   IBM Research, USA
James Duncan   Yale University, USA
Russ Taylor   Johns Hopkins University, USA

Program Committee Chairs

Hayit Greenspan   Tel-Aviv University, Israel and Icahn School of Medicine at Mount Sinai, USA
Anant Madabhushi   Emory University, USA
Parvin Mousavi   Queen's University, Canada
Septimiu Salcudean   University of British Columbia, Canada

Satellite Events Chair

Bennett Landman   Vanderbilt University, USA

Workshop Chairs

Hongzhi Wang   IBM Research, USA
Alistair Young   King's College, London, UK

Challenges Chairs

Jayashree Kalpathy-Kramer   Harvard University, USA
Alexander Seitel   German Cancer Research Center, Germany
Lena Maier-Hein   German Cancer Research Center, Germany

Tutorial Chairs

Islem Rekik   Imperial College London, UK
Guoyan Zheng   Shanghai Jiao Tong University, China

Clinical Chairs

Curtis Langlotz   Stanford University, USA
Charles Kahn   University of Pennsylvania, USA
Masaru Ishii   Johns Hopkins University, USA

Local Arrangements Chairs

Purang Abolmaesumi   University of British Columbia, Canada
Mehdi Moradi   McMaster University, Canada

Keynote Chairs

William Wells   Harvard University, USA
Alejandro Frangi   University of Manchester, UK

Industrial Sponsorship Chairs

Mohammad Yaqub   MBZ University of Artificial Intelligence,
                                   Abu Dhabi
Le Lu   DAMO Academy, Alibaba Group, USA
Yanwu Xu   Baidu, China

Communication Chair

Ehsan Adeli   Stanford University, USA

Publication Chairs

Ron Summers	National Institutes of Health, USA
Kevin Zhou	University of Science and Technology of China, China

Young Investigator Chairs

Marius Linguraru	Children's National Institute, USA
Archana Venkataraman	Boston University, USA
Antonio Porras	University of Colorado Anschutz Medical Campus, USA

Student Activities Chairs

Nathaniel Braman	Picture Health, USA
Vaishnavi Subramanian	EPFL, France

Women in MICCAI Chairs

Jayanthi Sivaswamy	IIIT, Hyderabad, India
Xiaoxiao Li	University of British Columbia, Canada

RISE Committee Chairs

Islem Rekik	Imperial College London, UK
Pingkun Yan	Rensselaer Polytechnic Institute, USA
Andrea Lara	Universidad Galileo, Guatemala

Submission Platform Manager

Kitty Wong	The MICCAI Society, Canada

Virtual Platform Manager

John Baxter   INSERM, Université de Rennes 1, France

Cross-Committee Liaison

Niharika D'Souza   IBM Research, USA

Program Committee

Sahar Ahmad	University of North Carolina at Chapel Hill, USA
Shadi Albarqouni	University of Bonn and Helmholtz Munich, Germany
Angelica Aviles-Rivero	University of Cambridge, UK
Shekoofeh Azizi	Google, Google Brain, USA
Ulas Bagci	Northwestern University, USA
Wenjia Bai	Imperial College London, UK
Sophia Bano	University College London, UK
Kayhan Batmanghelich	University of Pittsburgh and Boston University, USA
Ismail Ben Ayed	ETS Montreal, Canada
Katharina Breininger	Friedrich-Alexander-Universität Erlangen-Nürnberg, Germany
Weidong Cai	University of Sydney, Australia
Geng Chen	Northwestern Polytechnical University, China
Hao Chen	Hong Kong University of Science and Technology, China
Jun Cheng	Institute for Infocomm Research, A*STAR, Singapore
Li Cheng	University of Alberta, Canada
Albert C. S. Chung	University of Exeter, UK
Toby Collins	Ircad, France
Adrian Dalca	Massachusetts Institute of Technology and Harvard Medical School, USA
Jose Dolz	ETS Montreal, Canada
Qi Dou	Chinese University of Hong Kong, China
Nicha Dvornek	Yale University, USA
Shireen Elhabian	University of Utah, USA
Sandy Engelhardt	Heidelberg University Hospital, Germany
Ruogu Fang	University of Florida, USA

Aasa Feragen	Technical University of Denmark, Denmark
Moti Freiman	Technion - Israel Institute of Technology, Israel
Huazhu Fu	IHPC, A*STAR, Singapore
Adrian Galdran	Universitat Pompeu Fabra, Barcelona, Spain
Zhifan Gao	Sun Yat-sen University, China
Zongyuan Ge	Monash University, Australia
Stamatia Giannarou	Imperial College London, UK
Yun Gu	Shanghai Jiao Tong University, China
Hu Han	Institute of Computing Technology, Chinese Academy of Sciences, China
Daniel Hashimoto	University of Pennsylvania, USA
Mattias Heinrich	University of Lübeck, Germany
Heng Huang	University of Pittsburgh, USA
Yuankai Huo	Vanderbilt University, USA
Mobarakol Islam	University College London, UK
Jayender Jagadeesan	Harvard Medical School, USA
Won-Ki Jeong	Korea University, South Korea
Xi Jiang	University of Electronic Science and Technology of China, China
Yueming Jin	National University of Singapore, Singapore
Anand Joshi	University of Southern California, USA
Shantanu Joshi	UCLA, USA
Leo Joskowicz	Hebrew University of Jerusalem, Israel
Samuel Kadoury	Polytechnique Montreal, Canada
Bernhard Kainz	Friedrich-Alexander-Universität Erlangen-Nürnberg, Germany and Imperial College London, UK
Davood Karimi	Harvard University, USA
Anees Kazi	Massachusetts General Hospital, USA
Marta Kersten-Oertel	Concordia University, Canada
Fahmi Khalifa	Mansoura University, Egypt
Minjeong Kim	University of North Carolina, Greensboro, USA
Seong Tae Kim	Kyung Hee University, South Korea
Pavitra Krishnaswamy	Institute for Infocomm Research, Agency for Science Technology and Research (A*STAR), Singapore
Jin Tae Kwak	Korea University, South Korea
Baiying Lei	Shenzhen University, China
Xiang Li	Massachusetts General Hospital, USA
Xiaoxiao Li	University of British Columbia, Canada
Yuexiang Li	Tencent Jarvis Lab, China
Chunfeng Lian	Xi'an Jiaotong University, China

Jianming Liang	Arizona State University, USA
Jianfei Liu	National Institutes of Health Clinical Center, USA
Mingxia Liu	University of North Carolina at Chapel Hill, USA
Xiaofeng Liu	Harvard Medical School and MGH, USA
Herve Lombaert	École de technologie supérieure, Canada
Ismini Lourentzou	Virginia Tech, USA
Le Lu	Damo Academy USA, Alibaba Group, USA
Dwarikanath Mahapatra	Inception Institute of Artificial Intelligence, United Arab Emirates
Saad Nadeem	Memorial Sloan Kettering Cancer Center, USA
Dong Nie	Alibaba (US), USA
Yoshito Otake	Nara Institute of Science and Technology, Japan
Sang Hyun Park	Daegu Gyeongbuk Institute of Science and Technology, South Korea
Magdalini Paschali	Stanford University, USA
Tingying Peng	Helmholtz Munich, Germany
Caroline Petitjean	LITIS Université de Rouen Normandie, France
Esther Puyol Anton	King's College London, UK
Chen Qin	Imperial College London, UK
Daniel Racoceanu	Sorbonne Université, France
Hedyeh Rafii-Tari	Auris Health, USA
Hongliang Ren	Chinese University of Hong Kong, China and National University of Singapore, Singapore
Tammy Riklin Raviv	Ben-Gurion University, Israel
Hassan Rivaz	Concordia University, Canada
Mirabela Rusu	Stanford University, USA
Thomas Schultz	University of Bonn, Germany
Feng Shi	Shanghai United Imaging Intelligence, China
Yang Song	University of New South Wales, Australia
Aristeidis Sotiras	Washington University in St. Louis, USA
Rachel Sparks	King's College London, UK
Yao Sui	Peking University, China
Kenji Suzuki	Tokyo Institute of Technology, Japan
Qian Tao	Delft University of Technology, Netherlands
Mathias Unberath	Johns Hopkins University, USA
Martin Urschler	Medical University Graz, Austria
Maria Vakalopoulou	CentraleSupelec, University Paris Saclay, France
Erdem Varol	New York University, USA
Francisco Vasconcelos	University College London, UK
Harini Veeraraghavan	Memorial Sloan Kettering Cancer Center, USA
Satish Viswanath	Case Western Reserve University, USA
Christian Wachinger	Technical University of Munich, Germany

Hua Wang	Colorado School of Mines, USA
Qian Wang	ShanghaiTech University, China
Shanshan Wang	Paul C. Lauterbur Research Center, SIAT, China
Yalin Wang	Arizona State University, USA
Bryan Williams	Lancaster University, UK
Matthias Wilms	University of Calgary, Canada
Jelmer Wolterink	University of Twente, Netherlands
Ken C. L. Wong	IBM Research Almaden, USA
Jonghye Woo	Massachusetts General Hospital and Harvard Medical School, USA
Shandong Wu	University of Pittsburgh, USA
Yutong Xie	University of Adelaide, Australia
Fuyong Xing	University of Colorado, Denver, USA
Daguang Xu	NVIDIA, USA
Yan Xu	Beihang University, China
Yanwu Xu	Baidu, China
Pingkun Yan	Rensselaer Polytechnic Institute, USA
Guang Yang	Imperial College London, UK
Jianhua Yao	Tencent, China
Chuyang Ye	Beijing Institute of Technology, China
Lequan Yu	University of Hong Kong, China
Ghada Zamzmi	National Institutes of Health, USA
Liang Zhan	University of Pittsburgh, USA
Fan Zhang	Harvard Medical School, USA
Ling Zhang	Alibaba Group, China
Miaomiao Zhang	University of Virginia, USA
Shu Zhang	Northwestern Polytechnical University, China
Rongchang Zhao	Central South University, China
Yitian Zhao	Chinese Academy of Sciences, China
Tao Zhou	Nanjing University of Science and Technology, USA
Yuyin Zhou	UC Santa Cruz, USA
Dajiang Zhu	University of Texas at Arlington, USA
Lei Zhu	ROAS Thrust HKUST (GZ), and ECE HKUST, China
Xiahai Zhuang	Fudan University, China
Veronika Zimmer	Technical University of Munich, Germany

Reviewers

Alaa Eldin Abdelaal
John Abel
Kumar Abhishek
Shahira Abousamra
Mazdak Abulnaga
Burak Acar
Abdoljalil Addeh
Ehsan Adeli
Sukesh Adiga Vasudeva
Seyed-Ahmad Ahmadi
Euijoon Ahn
Faranak Akbarifar
Alireza Akhondi-asl
Saad Ullah Akram
Daniel Alexander
Hanan Alghamdi
Hassan Alhajj
Omar Al-Kadi
Max Allan
Andre Altmann
Pablo Alvarez
Charlems Alvarez-Jimenez
Jennifer Alvén
Lidia Al-Zogbi
Kimberly Amador
Tamaz Amiranashvili
Amine Amyar
Wangpeng An
Vincent Andrearczyk
Manon Ansart
Sameer Antani
Jacob Antunes
Michel Antunes
Guilherme Aresta
Mohammad Ali Armin
Kasra Arnavaz
Corey Arnold
Janan Arslan
Marius Arvinte
Muhammad Asad
John Ashburner
Md Ashikuzzaman
Shahab Aslani

Mehdi Astaraki
Angélica Atehortúa
Benjamin Aubert
Marc Aubreville
Paolo Avesani
Sana Ayromlou
Reza Azad
Mohammad Farid
    Azampour
Qinle Ba
Meritxell Bach Cuadra
Hyeon-Min Bae
Matheus Baffa
Cagla Bahadir
Fan Bai
Jun Bai
Long Bai
Pradeep Bajracharya
Shafa Balaram
Yaël Balbastre
Yutong Ban
Abhirup Banerjee
Soumyanil Banerjee
Sreya Banerjee
Shunxing Bao
Omri Bar
Adrian Barbu
Joao Barreto
Adrian Basarab
Berke Basaran
Michael Baumgartner
Siming Bayer
Roza Bayrak
Aicha BenTaieb
Guy Ben-Yosef
Sutanu Bera
Cosmin Bercea
Jorge Bernal
Jose Bernal
Gabriel Bernardino
Riddhish Bhalodia
Jignesh Bhatt
Indrani Bhattacharya

Binod Bhattarai
Lei Bi
Qi Bi
Cheng Bian
Gui-Bin Bian
Carlo Biffi
Alexander Bigalke
Benjamin Billot
Manuel Birlo
Ryoma Bise
Daniel Blezek
Stefano Blumberg
Sebastian Bodenstedt
Federico Bolelli
Bhushan Borotikar
Ilaria Boscolo Galazzo
Alexandre Bousse
Nicolas Boutry
Joseph Boyd
Behzad Bozorgtabar
Nadia Brancati
Clara Brémond Martin
Stéphanie Bricq
Christopher Bridge
Coleman Broaddus
Rupert Brooks
Tom Brosch
Mikael Brudfors
Ninon Burgos
Nikolay Burlutskiy
Michal Byra
Ryan Cabeen
Mariano Cabezas
Hongmin Cai
Tongan Cai
Zongyou Cai
Liane Canas
Bing Cao
Guogang Cao
Weiguo Cao
Xu Cao
Yankun Cao
Zhenjie Cao

Jaime Cardoso
M. Jorge Cardoso
Owen Carmichael
Jacob Carse
Adrià Casamitjana
Alessandro Casella
Angela Castillo
Kate Cevora
Krishna Chaitanya
Satrajit Chakrabarty
Yi Hao Chan
Shekhar Chandra
Ming-Ching Chang
Peng Chang
Qi Chang
Yuchou Chang
Hanqing Chao
Simon Chatelin
Soumick Chatterjee
Sudhanya Chatterjee
Muhammad Faizyab Ali
  Chaudhary
Antong Chen
Bingzhi Chen
Chen Chen
Cheng Chen
Chengkuan Chen
Eric Chen
Fang Chen
Haomin Chen
Jianan Chen
Jianxu Chen
Jiazhou Chen
Jie Chen
Jintai Chen
Jun Chen
Junxiang Chen
Junyu Chen
Li Chen
Liyun Chen
Nenglun Chen
Pingjun Chen
Pingyi Chen
Qi Chen
Qiang Chen

Runnan Chen
Shengcong Chen
Sihao Chen
Tingting Chen
Wenting Chen
Xi Chen
Xiang Chen
Xiaoran Chen
Xin Chen
Xiongchao Chen
Yanxi Chen
Yixiong Chen
Yixuan Chen
Yuanyuan Chen
Yuqian Chen
Zhaolin Chen
Zhen Chen
Zhenghao Chen
Zhennong Chen
Zhihao Chen
Zhineng Chen
Zhixiang Chen
Chang-Chieh Cheng
Jiale Cheng
Jianhong Cheng
Jun Cheng
Xuelian Cheng
Yupeng Cheng
Mark Chiew
Philip Chikontwe
Eleni Chiou
Jungchan Cho
Jang-Hwan Choi
Min-Kook Choi
Wookjin Choi
Jaegul Choo
Yu-Cheng Chou
Daan Christiaens
Argyrios Christodoulidis
Stergios Christodoulidis
Kai-Cheng Chuang
Hyungjin Chung
Matthew Clarkson
Michaël Clément
Dana Cobzas

Jaume Coll-Font
Olivier Colliot
Runmin Cong
Yulai Cong
Laura Connolly
William Consagra
Pierre-Henri Conze
Tim Cootes
Teresa Correia
Baris Coskunuzer
Alex Crimi
Can Cui
Hejie Cui
Hui Cui
Lei Cui
Wenhui Cui
Tolga Cukur
Tobias Czempiel
Javid Dadashkarimi
Haixing Dai
Tingting Dan
Kang Dang
Salman Ul Hassan Dar
Eleonora D'Arnese
Dhritiman Das
Neda Davoudi
Tareen Dawood
Sandro De Zanet
Farah Deeba
Charles Delahunt
Herve Delingette
Ugur Demir
Liang-Jian Deng
Ruining Deng
Wenlong Deng
Felix Denzinger
Adrien Depeursinge
Mohammad Mahdi
  Derakhshani
Hrishikesh Deshpande
Adrien Desjardins
Christian Desrosiers
Blake Dewey
Neel Dey
Rohan Dhamdhere

Maxime Di Folco
Songhui Diao
Alina Dima
Hao Ding
Li Ding
Ying Ding
Zhipeng Ding
Nicola Dinsdale
Konstantin Dmitriev
Ines Domingues
Bo Dong
Liang Dong
Nanqing Dong
Siyuan Dong
Reuben Dorent
Gianfranco Doretto
Sven Dorkenwald
Haoran Dou
Mitchell Doughty
Jason Dowling
Niharika D'Souza
Guodong Du
Jie Du
Shiyi Du
Hongyi Duanmu
Benoit Dufumier
James Duncan
Joshua Durso-Finley
Dmitry V. Dylov
Oleh Dzyubachyk
Mahdi (Elias) Ebnali
Philip Edwards
Jan Egger
Gudmundur Einarsson
Mostafa El Habib Daho
Ahmed Elazab
Idris El-Feghi
David Ellis
Mohammed Elmogy
Amr Elsawy
Okyaz Eminaga
Ertunc Erdil
Lauren Erdman
Marius Erdt
Maria Escobar

Hooman Esfandiari
Nazila Esmaeili
Ivan Ezhov
Alessio Fagioli
Deng-Ping Fan
Lei Fan
Xin Fan
Yubo Fan
Huihui Fang
Jiansheng Fang
Xi Fang
Zhenghan Fang
Mohammad Farazi
Azade Farshad
Mohsen Farzi
Hamid Fehri
Lina Felsner
Chaolu Feng
Chun-Mei Feng
Jianjiang Feng
Mengling Feng
Ruibin Feng
Zishun Feng
Alvaro Fernandez-Quilez
Ricardo Ferrari
Lucas Fidon
Lukas Fischer
Madalina Fiterau
Antonio
    Foncubierta-Rodríguez
Fahimeh Fooladgar
Germain Forestier
Nils Daniel Forkert
Jean-Rassaire Fouefack
Kevin François-Bouaou
Wolfgang Freysinger
Bianca Freytag
Guanghui Fu
Kexue Fu
Lan Fu
Yunguan Fu
Pedro Furtado
Ryo Furukawa
Jin Kyu Gahm
Mélanie Gaillochet

Francesca Galassi
Jiangzhang Gan
Yu Gan
Yulu Gan
Alireza Ganjdanesh
Chang Gao
Cong Gao
Linlin Gao
Zeyu Gao
Zhongpai Gao
Sara Garbarino
Alain Garcia
Beatriz Garcia Santa Cruz
Rongjun Ge
Shiv Gehlot
Manuela Geiss
Salah Ghamizi
Negin Ghamsarian
Ramtin Gharleghi
Ghazal Ghazaei
Florin Ghesu
Sayan Ghosal
Syed Zulqarnain Gilani
Mahdi Gilany
Yannik Glaser
Ben Glocker
Bharti Goel
Jacob Goldberger
Polina Golland
Alberto Gomez
Catalina Gomez
Estibaliz
    Gómez-de-Mariscal
Haifan Gong
Kuang Gong
Xun Gong
Ricardo Gonzales
Camila Gonzalez
German Gonzalez
Vanessa Gonzalez Duque
Sharath Gopal
Karthik Gopinath
Pietro Gori
Michael Götz
Shuiping Gou

Maged Goubran
Sobhan Goudarzi
Mark Graham
Alejandro Granados
Mara Graziani
Thomas Grenier
Radu Grosu
Michal Grzeszczyk
Feng Gu
Pengfei Gu
Qiangqiang Gu
Ran Gu
Shi Gu
Wenhao Gu
Xianfeng Gu
Yiwen Gu
Zaiwang Gu
Hao Guan
Jayavardhana Gubbi
Houssem-Eddine Gueziri
Dazhou Guo
Hengtao Guo
Jixiang Guo
Jun Guo
Pengfei Guo
Wenzhangzhi Guo
Xiaoqing Guo
Xueqi Guo
Yi Guo
Vikash Gupta
Praveen Gurunath Bharathi
Prashnna Gyawali
Sung Min Ha
Mohamad Habes
Ilker Hacihaliloglu
Stathis Hadjidemetriou
Fatemeh Haghighi
Justin Haldar
Noura Hamze
Liang Han
Luyi Han
Seungjae Han
Tianyu Han
Zhongyi Han
Jonny Hancox

Lasse Hansen
Degan Hao
Huaying Hao
Jinkui Hao
Nazim Haouchine
Michael Hardisty
Stefan Harrer
Jeffry Hartanto
Charles Hatt
Huiguang He
Kelei He
Qi He
Shenghua He
Xinwei He
Stefan Heldmann
Nicholas Heller
Edward Henderson
Alessa Hering
Monica Hernandez
Kilian Hett
Amogh Hiremath
David Ho
Malte Hoffmann
Matthew Holden
Qingqi Hong
Yoonmi Hong
Mohammad Reza
    Hosseinzadeh Taher
William Hsu
Chuanfei Hu
Dan Hu
Kai Hu
Rongyao Hu
Shishuai Hu
Xiaoling Hu
Xinrong Hu
Yan Hu
Yang Hu
Chaoqin Huang
Junzhou Huang
Ling Huang
Luojie Huang
Qinwen Huang
Sharon Xiaolei Huang
Weijian Huang

Xiaoyang Huang
Yi-Jie Huang
Yongsong Huang
Yongxiang Huang
Yuhao Huang
Zhe Huang
Zhi-An Huang
Ziyi Huang
Arnaud Huaulmé
Henkjan Huisman
Alex Hung
Jiayu Huo
Andreas Husch
Mohammad Arafat
    Hussain
Sarfaraz Hussein
Jana Hutter
Khoi Huynh
Ilknur Icke
Kay Igwe
Abdullah Al Zubaer Imran
Muhammad Imran
Samra Irshad
Nahid Ul Islam
Koichi Ito
Hayato Itoh
Yuji Iwahori
Krithika Iyer
Mohammad Jafari
Srikrishna Jaganathan
Hassan Jahanandish
Andras Jakab
Amir Jamaludin
Amoon Jamzad
Ananya Jana
Se-In Jang
Pierre Jannin
Vincent Jaouen
Uditha Jarayathne
Ronnachai Jaroensri
Guillaume Jaume
Syed Ashar Javed
Rachid Jennane
Debesh Jha
Ge-Peng Ji

Luping Ji
Zexuan Ji
Zhanghexuan Ji
Haozhe Jia
Hongchao Jiang
Jue Jiang
Meirui Jiang
Tingting Jiang
Xiajun Jiang
Zekun Jiang
Zhifan Jiang
Ziyu Jiang
Jianbo Jiao
Zhicheng Jiao
Chen Jin
Dakai Jin
Qiangguo Jin
Qiuye Jin
Weina Jin
Baoyu Jing
Bin Jing
Yaqub Jonmohamadi
Lie Ju
Yohan Jun
Dinkar Juyal
Manjunath K N
Ali Kafaei Zad Tehrani
John Kalafut
Niveditha Kalavakonda
Megha Kalia
Anil Kamat
Qingbo Kang
Po-Yu Kao
Anuradha Kar
Neerav Karani
Turkay Kart
Satyananda Kashyap
Alexander Katzmann
Lisa Kausch
Maxime Kayser
Salome Kazeminia
Wenchi Ke
Youngwook Kee
Matthias Keicher
Erwan Kerrien

Afifa Khaled
Nadieh Khalili
Farzad Khalvati
Bidur Khanal
Bishesh Khanal
Pulkit Khandelwal
Maksim Kholiavchenko
Ron Kikinis
Benjamin Killeen
Daeseung Kim
Heejong Kim
Jaeil Kim
Jinhee Kim
Jinman Kim
Junsik Kim
Minkyung Kim
Namkug Kim
Sangwook Kim
Tae Soo Kim
Younghoon Kim
Young-Min Kim
Andrew King
Miranda Kirby
Gabriel Kiss
Andreas Kist
Yoshiro Kitamura
Stefan Klein
Tobias Klinder
Kazuma Kobayashi
Lisa Koch
Satoshi Kondo
Fanwei Kong
Tomasz Konopczynski
Ender Konukoglu
Aishik Konwer
Thijs Kooi
Ivica Kopriva
Avinash Kori
Kivanc Kose
Suraj Kothawade
Anna Kreshuk
AnithaPriya Krishnan
Florian Kromp
Frithjof Kruggel
Thomas Kuestner

Levin Kuhlmann
Abhay Kumar
Kuldeep Kumar
Sayantan Kumar
Manuela Kunz
Holger Kunze
Tahsin Kurc
Anvar Kurmukov
Yoshihiro Kuroda
Yusuke Kurose
Hyuksool Kwon
Aymen Laadhari
Jorma Laaksonen
Dmitrii Lachinov
Alain Lalande
Rodney LaLonde
Bennett Landman
Daniel Lang
Carole Lartizien
Shlomi Laufer
Max-Heinrich Laves
William Le
Loic Le Folgoc
Christian Ledig
Eung-Joo Lee
Ho Hin Lee
Hyekyoung Lee
John Lee
Kisuk Lee
Kyungsu Lee
Soochahn Lee
Woonghee Lee
Étienne Léger
Wen Hui Lei
Yiming Lei
George Leifman
Rogers Jeffrey Leo John
Juan Leon
Bo Li
Caizi Li
Chao Li
Chen Li
Cheng Li
Chenxin Li
Chnegyin Li

Dawei Li
Fuhai Li
Gang Li
Guang Li
Hao Li
Haofeng Li
Haojia Li
Heng Li
Hongming Li
Hongwei Li
Huiqi Li
Jian Li
Jieyu Li
Kang Li
Lin Li
Mengzhang Li
Ming Li
Qing Li
Quanzheng Li
Shaohua Li
Shulong Li
Tengfei Li
Weijian Li
Wen Li
Xiaomeng Li
Xingyu Li
Xinhui Li
Xuelu Li
Xueshen Li
Yamin Li
Yang Li
Yi Li
Yuemeng Li
Yunxiang Li
Zeju Li
Zhaoshuo Li
Zhe Li
Zhen Li
Zhenqiang Li
Zhiyuan Li
Zhjin Li
Zi Li
Hao Liang
Libin Liang
Peixian Liang

Yuan Liang
Yudong Liang
Haofu Liao
Hongen Liao
Wei Liao
Zehui Liao
Gilbert Lim
Hongxiang Lin
Li Lin
Manxi Lin
Mingquan Lin
Tiancheng Lin
Yi Lin
Zudi Lin
Claudia Lindner
Simone Lionetti
Chi Liu
Chuanbin Liu
Daochang Liu
Dongnan Liu
Feihong Liu
Fenglin Liu
Han Liu
Huiye Liu
Jiang Liu
Jie Liu
Jinduo Liu
Jing Liu
Jingya Liu
Jundong Liu
Lihao Liu
Mengting Liu
Mingyuan Liu
Peirong Liu
Peng Liu
Qin Liu
Quan Liu
Rui Liu
Shengfeng Liu
Shuangjun Liu
Sidong Liu
Siyuan Liu
Weide Liu
Xiao Liu
Xiaoyu Liu

Xingtong Liu
Xinwen Liu
Xinyang Liu
Xinyu Liu
Yan Liu
Yi Liu
Yihao Liu
Yikang Liu
Yilin Liu
Yilong Liu
Yiqiao Liu
Yong Liu
Yuhang Liu
Zelong Liu
Zhe Liu
Zhiyuan Liu
Zuozhu Liu
Lisette Lockhart
Andrea Loddo
Nicolas Loménie
Yonghao Long
Daniel Lopes
Ange Lou
Brian Lovell
Nicolas Loy Rodas
Charles Lu
Chun-Shien Lu
Donghuan Lu
Guangming Lu
Huanxiang Lu
Jingpei Lu
Yao Lu
Oeslle Lucena
Jie Luo
Luyang Luo
Ma Luo
Mingyuan Luo
Wenhan Luo
Xiangde Luo
Xinzhe Luo
Jinxin Lv
Tianxu Lv
Fei Lyu
Ilwoo Lyu
Mengye Lyu

Qing Lyu
Yanjun Lyu
Yuanyuan Lyu
Benteng Ma
Chunwei Ma
Hehuan Ma
Jun Ma
Junbo Ma
Wenao Ma
Yuhui Ma
Pedro Macias Gordaliza
Anant Madabhushi
Derek Magee
S. Sara Mahdavi
Andreas Maier
Klaus H. Maier-Hein
Sokratis Makrogiannis
Danial Maleki
Michail Mamalakis
Zhehua Mao
Jan Margeta
Brett Marinelli
Zdravko Marinov
Viktoria Markova
Carsten Marr
Yassine Marrakchi
Anne Martel
Martin Maška
Tejas Sudharshan Mathai
Petr Matula
Dimitrios Mavroeidis
Evangelos Mazomenos
Amarachi Mbakwe
Adam McCarthy
Stephen McKenna
Raghav Mehta
Xueyan Mei
Felix Meissen
Felix Meister
Afaque Memon
Mingyuan Meng
Qingjie Meng
Xiangzhu Meng
Yanda Meng
Zhu Meng

Martin Menten
Odyssée Merveille
Mikhail Milchenko
Leo Milecki
Fausto Milletari
Hyun-Seok Min
Zhe Min
Song Ming
Duy Minh Ho Nguyen
Deepak Mishra
Suraj Mishra
Virendra Mishra
Tadashi Miyamoto
Sara Moccia
Marc Modat
Omid Mohareri
Tony C. W. Mok
Javier Montoya
Rodrigo Moreno
Stefano Moriconi
Lia Morra
Ana Mota
Lei Mou
Dana Moukheiber
Lama Moukheiber
Daniel Moyer
Pritam Mukherjee
Anirban Mukhopadhyay
Henning Müller
Ana Murillo
Gowtham Krishnan
    Murugesan
Ahmed Naglah
Karthik Nandakumar
Venkatesh
    Narasimhamurthy
Raja Narayan
Dominik Narnhofer
Vishwesh Nath
Rodrigo Nava
Abdullah Nazib
Ahmed Nebli
Peter Neher
Amin Nejatbakhsh
Trong-Thuan Nguyen

Truong Nguyen
Dong Ni
Haomiao Ni
Xiuyan Ni
Hannes Nickisch
Weizhi Nie
Aditya Nigam
Lipeng Ning
Xia Ning
Kazuya Nishimura
Chuang Niu
Sijie Niu
Vincent Noblet
Narges Norouzi
Alexey Novikov
Jorge Novo
Gilberto Ochoa-Ruiz
Masahiro Oda
Benjamin Odry
Hugo Oliveira
Sara Oliveira
Arnau Oliver
Jimena Olveres
John Onofrey
Marcos Ortega
Mauricio Alberto
    Ortega-Ruíz
Yusuf Osmanlioglu
Chubin Ou
Cheng Ouyang
Jiahong Ouyang
Xi Ouyang
Cristina Oyarzun Laura
Utku Ozbulak
Ece Ozkan
Ege Özsoy
Batu Ozturkler
Harshith Padigela
Johannes Paetzold
José Blas Pagador
    Carrasco
Daniel Pak
Sourabh Palande
Chengwei Pan
Jiazhen Pan

Jin Pan
Yongsheng Pan
Egor Panfilov
Jiaxuan Pang
Joao Papa
Constantin Pape
Bartlomiej Papiez
Nripesh Parajuli
Hyunjin Park
Akash Parvatikar
Tiziano Passerini
Diego Patiño Cortés
Mayank Patwari
Angshuman Paul
Rasmus Paulsen
Yuchen Pei
Yuru Pei
Tao Peng
Wei Peng
Yige Peng
Yunsong Peng
Matteo Pennisi
Antonio Pepe
Oscar Perdomo
Sérgio Pereira
Jose-Antonio
    Pérez-Carrasco
Mehran Pesteie
Terry Peters
Eike Petersen
Jens Petersen
Micha Pfeiffer
Dzung Pham
Hieu Pham
Ashish Phophalia
Tomasz Pieciak
Antonio Pinheiro
Pramod Pisharady
Theodoros Pissas
Szymon Płotka
Kilian Pohl
Sebastian Pölsterl
Alison Pouch
Tim Prangemeier
Prateek Prasanna

Raphael Prevost
Juan Prieto
Federica Proietto Salanitri
Sergi Pujades
Elodie Puybareau
Talha Qaiser
Buyue Qian
Mengyun Qiao
Yuchuan Qiao
Zhi Qiao
Chenchen Qin
Fangbo Qin
Wenjian Qin
Yulei Qin
Jie Qiu
Jielin Qiu
Peijie Qiu
Shi Qiu
Wu Qiu
Liangqiong Qu
Linhao Qu
Quan Quan
Tran Minh Quan
Sandro Queirós
Prashanth R
Febrian Rachmadi
Daniel Racoceanu
Mehdi Rahim
Jagath Rajapakse
Kashif Rajpoot
Keerthi Ram
Dhanesh Ramachandram
João Ramalhinho
Xuming Ran
Aneesh Rangnekar
Hatem Rashwan
Keerthi Sravan Ravi
Daniele Ravì
Sadhana Ravikumar
Harish Raviprakash
Surreerat Reaungamornrat
Samuel Remedios
Mengwei Ren
Sucheng Ren
Elton Rexhepaj

Mauricio Reyes
Constantino
    Reyes-Aldasoro
Abel Reyes-Angulo
Hadrien Reynaud
Razieh Rezaei
Anne-Marie Rickmann
Laurent Risser
Dominik Rivoir
Emma Robinson
Robert Robinson
Jessica Rodgers
Ranga Rodrigo
Rafael Rodrigues
Robert Rohling
Margherita Rosnati
Łukasz Roszkowiak
Holger Roth
José Rouco
Dan Ruan
Jiacheng Ruan
Daniel Rueckert
Danny Ruijters
Kanghyun Ryu
Ario Sadafi
Numan Saeed
Monjoy Saha
Pramit Saha
Farhang Sahba
Pranjal Sahu
Simone Saitta
Md Sirajus Salekin
Abbas Samani
Pedro Sanchez
Luis Sanchez Giraldo
Yudi Sang
Gerard Sanroma-Guell
Rodrigo Santa Cruz
Alice Santilli
Rachana Sathish
Olivier Saut
Mattia Savardi
Nico Scherf
Alexander Schlaefer
Jerome Schmid

Adam Schmidt
Julia Schnabel
Lawrence Schobs
Julian Schön
Peter Schueffler
Andreas Schuh
Christina
 Schwarz-Gsaxner
Michaël Sdika
Suman Sedai
Lalithkumar Seenivasan
Matthias Seibold
Sourya Sengupta
Lama Seoud
Ana Sequeira
Sharmishtaa Seshamani
Ahmed Shaffie
Jay Shah
Keyur Shah
Ahmed Shahin
Mohammad Abuzar
 Shaikh
S. Shailja
Hongming Shan
Wei Shao
Mostafa Sharifzadeh
Anuja Sharma
Gregory Sharp
Hailan Shen
Li Shen
Linlin Shen
Mali Shen
Mingren Shen
Yiqing Shen
Zhengyang Shen
Jun Shi
Xiaoshuang Shi
Yiyu Shi
Yonggang Shi
Hoo-Chang Shin
Jitae Shin
Keewon Shin
Boris Shirokikh
Suzanne Shontz
Yucheng Shu

Hanna Siebert
Alberto Signoroni
Wilson Silva
Julio Silva-Rodríguez
Margarida Silveira
Walter Simson
Praveer Singh
Vivek Singh
Nitin Singhal
Elena Sizikova
Gregory Slabaugh
Dane Smith
Kevin Smith
Tiffany So
Rajath Soans
Roger Soberanis-Mukul
Hessam Sokooti
Jingwei Song
Weinan Song
Xinhang Song
Xinrui Song
Mazen Soufi
Georgia Sovatzidi
Bella Specktor Fadida
William Speier
Ziga Spiclin
Dominik Spinczyk
Jon Sporring
Pradeeba Sridar
Chetan L. Srinidhi
Abhishek Srivastava
Lawrence Staib
Marc Stamminger
Justin Strait
Hai Su
Ruisheng Su
Zhe Su
Vaishnavi Subramanian
Gérard Subsol
Carole Sudre
Dong Sui
Heung-Il Suk
Shipra Suman
He Sun
Hongfu Sun

Jian Sun
Li Sun
Liyan Sun
Shanlin Sun
Kyung Sung
Yannick Suter
Swapna T. R.
Amir Tahmasebi
Pablo Tahoces
Sirine Taleb
Bingyao Tan
Chaowei Tan
Wenjun Tan
Hao Tang
Siyi Tang
Xiaoying Tang
Yucheng Tang
Zihao Tang
Michael Tanzer
Austin Tapp
Elias Tappeiner
Mickael Tardy
Giacomo Tarroni
Athena Taymourtash
Kaveri Thakoor
Elina Thibeau-Sutre
Paul Thienphrapa
Sarina Thomas
Stephen Thompson
Karl Thurnhofer-Hemsi
Cristiana Tiago
Lin Tian
Lixia Tian
Yapeng Tian
Yu Tian
Yun Tian
Aleksei Tiulpin
Hamid Tizhoosh
Minh Nguyen Nhat To
Matthew Toews
Maryam Toloubidokhti
Minh Tran
Quoc-Huy Trinh
Jocelyne Troccaz
Roger Trullo

Chialing Tsai
Apostolia Tsirikoglou
Puxun Tu
Samyakh Tukra
Sudhakar Tummala
Georgios Tziritas
Vladimír Ulman
Tamas Ungi
Régis Vaillant
Jeya Maria Jose Valanarasu
Vanya Valindria
Juan Miguel Valverde
Fons van der Sommen
Maureen van Eijnatten
Tom van Sonsbeek
Gijs van Tulder
Yogatheesan Varatharajah
Madhurima Vardhan
Thomas Varsavsky
Hooman Vaseli
Serge Vasylechko
S. Swaroop Vedula
Sanketh Vedula
Gonzalo Vegas
    Sanchez-Ferrero
Matthew Velazquez
Archana Venkataraman
Sulaiman Vesal
Mitko Veta
Barbara Villarini
Athanasios Vlontzos
Wolf-Dieter Vogl
Ingmar Voigt
Sandrine Voros
Vibashan VS
Trinh Thi Le Vuong
An Wang
Bo Wang
Ce Wang
Changmiao Wang
Ching-Wei Wang
Dadong Wang
Dong Wang
Fakai Wang
Guotai Wang

Haifeng Wang
Haoran Wang
Hong Wang
Hongxiao Wang
Hongyu Wang
Jiacheng Wang
Jing Wang
Jue Wang
Kang Wang
Ke Wang
Lei Wang
Li Wang
Liansheng Wang
Lin Wang
Ling Wang
Linwei Wang
Manning Wang
Mingliang Wang
Puyang Wang
Qiuli Wang
Renzhen Wang
Ruixuan Wang
Shaoyu Wang
Sheng Wang
Shujun Wang
Shuo Wang
Shuqiang Wang
Tao Wang
Tianchen Wang
Tianyu Wang
Wenzhe Wang
Xi Wang
Xiangdong Wang
Xiaoqing Wang
Xiaosong Wang
Yan Wang
Yangang Wang
Yaping Wang
Yi Wang
Yirui Wang
Yixin Wang
Zeyi Wang
Zhao Wang
Zichen Wang
Ziqin Wang

Ziyi Wang
Zuhui Wang
Dong Wei
Donglai Wei
Hao Wei
Jia Wei
Leihao Wei
Ruofeng Wei
Shuwen Wei
Martin Weigert
Wolfgang Wein
Michael Wels
Cédric Wemmert
Thomas Wendler
Markus Wenzel
Rhydian Windsor
Adam Wittek
Marek Wodzinski
Ivo Wolf
Julia Wolleb
Ka-Chun Wong
Jonghye Woo
Chongruo Wu
Chunpeng Wu
Fuping Wu
Huaqian Wu
Ji Wu
Jiangjie Wu
Jiong Wu
Junde Wu
Linshan Wu
Qing Wu
Weiwen Wu
Wenjun Wu
Xiyin Wu
Yawen Wu
Ye Wu
Yicheng Wu
Yongfei Wu
Zhengwang Wu
Pengcheng Xi
Chao Xia
Siyu Xia
Wenjun Xia
Lei Xiang

Jiawei Zhang
Jingqing Zhang
Jingyang Zhang
Jinwei Zhang
Jiong Zhang
Jiping Zhang
Ke Zhang
Lefei Zhang
Lei Zhang
Li Zhang
Lichi Zhang
Lu Zhang
Minghui Zhang
Molin Zhang
Ning Zhang
Rongzhao Zhang
Ruipeng Zhang
Ruisi Zhang
Shichuan Zhang
Shihao Zhang
Shuai Zhang
Tuo Zhang
Wei Zhang
Weihang Zhang
Wen Zhang
Wenhua Zhang
Wenqiang Zhang
Xiaodan Zhang
Xiaoran Zhang
Xin Zhang
Xukun Zhang
Xuzhe Zhang
Ya Zhang
Yanbo Zhang
Yanfu Zhang
Yao Zhang
Yi Zhang
Yifan Zhang
Yixiao Zhang
Yongqin Zhang
You Zhang
Youshan Zhang

Yu Zhang
Yubo Zhang
Yue Zhang
Yuhan Zhang
Yulun Zhang
Yundong Zhang
Yunlong Zhang
Yuyao Zhang
Zheng Zhang
Zhenxi Zhang
Ziqi Zhang
Can Zhao
Chongyue Zhao
Fenqiang Zhao
Gangming Zhao
He Zhao
Jianfeng Zhao
Jun Zhao
Li Zhao
Liang Zhao
Lin Zhao
Mengliu Zhao
Mingbo Zhao
Qingyu Zhao
Shang Zhao
Shijie Zhao
Tengda Zhao
Tianyi Zhao
Wei Zhao
Yidong Zhao
Yiyuan Zhao
Yu Zhao
Zhihe Zhao
Ziyuan Zhao
Haiyong Zheng
Hao Zheng
Jiannan Zheng
Kang Zheng
Meng Zheng
Sisi Zheng
Tianshu Zheng
Yalin Zheng

Yefeng Zheng
Yinqiang Zheng
Yushan Zheng
Aoxiao Zhong
Jia-Xing Zhong
Tao Zhong
Zichun Zhong
Hong-Yu Zhou
Houliang Zhou
Huiyu Zhou
Kang Zhou
Qin Zhou
Ran Zhou
S. Kevin Zhou
Tianfei Zhou
Wei Zhou
Xiao-Hu Zhou
Xiao-Yun Zhou
Yi Zhou
Youjia Zhou
Yukun Zhou
Zongwei Zhou
Chenglu Zhu
Dongxiao Zhu
Heqin Zhu
Jiayi Zhu
Meilu Zhu
Wei Zhu
Wenhui Zhu
Xiaofeng Zhu
Xin Zhu
Yonghua Zhu
Yongpei Zhu
Yuemin Zhu
Yan Zhuang
David Zimmerer
Yongshuo Zong
Ke Zou
Yukai Zou
Lianrui Zuo
Gerald Zwettler

Outstanding Area Chairs

Mingxia Liu	University of North Carolina at Chapel Hill, USA
Matthias Wilms	University of Calgary, Canada
Veronika Zimmer	Technical University Munich, Germany

Outstanding Reviewers

Kimberly Amador	University of Calgary, Canada
Angela Castillo	Universidad de los Andes, Colombia
Chen Chen	Imperial College London, UK
Laura Connolly	Queen's University, Canada
Pierre-Henri Conze	IMT Atlantique, France
Niharika D'Souza	IBM Research, USA
Michael Götz	University Hospital Ulm, Germany
Meirui Jiang	Chinese University of Hong Kong, China
Manuela Kunz	National Research Council Canada, Canada
Zdravko Marinov	Karlsruhe Institute of Technology, Germany
Sérgio Pereira	Lunit, South Korea
Lalithkumar Seenivasan	National University of Singapore, Singapore

Honorable Mentions (Reviewers)

Kumar Abhishek	Simon Fraser University, Canada
Guilherme Aresta	Medical University of Vienna, Austria
Shahab Aslani	University College London, UK
Marc Aubreville	Technische Hochschule Ingolstadt, Germany
Yaël Balbastre	Massachusetts General Hospital, USA
Omri Bar	Theator, Israel
Aicha Ben Taieb	Simon Fraser University, Canada
Cosmin Bercea	Technical University Munich and Helmholtz AI and Helmholtz Center Munich, Germany
Benjamin Billot	Massachusetts Institute of Technology, USA
Michal Byra	RIKEN Center for Brain Science, Japan
Mariano Cabezas	University of Sydney, Australia
Alessandro Casella	Italian Institute of Technology and Politecnico di Milano, Italy
Junyu Chen	Johns Hopkins University, USA
Argyrios Christodoulidis	Pfizer, Greece
Olivier Colliot	CNRS, France

Wolf-Dieter Vogl   RetInSight GmbH, Austria
Vibashan VS   Johns Hopkins University, USA
Lin Wang   Harbin Engineering University, China
Yan Wang   Sichuan University, China
Rhydian Windsor   University of Oxford, UK
Ivo Wolf   University of Applied Sciences Mannheim,
                                    Germany
Linshan Wu   Hunan University, China
Xin Yang   Chinese University of Hong Kong, China

Contents – Part X

Image Reconstruction

Image Registration

Image Reconstruction

CDiffMR: Can We Replace the Gaussian Noise with K-Space Undersampling for Fast MRI?

Jiahao Huang[1,2]([⊠]), Angelica I. Aviles-Rivero[3], Carola-Bibiane Schönlieb[3], and Guang Yang[4,5,6,7]([⊠])

[1] National Heart and Lung Institute, Imperial College London, London, UK
j.huang21@imperial.ac.uk
[2] Cardiovascular Research Centre, Royal Brompton Hospital, London, UK
[3] Department of Applied Mathematics and Theoretical Physics, University of Cambridge, Cambridge, UK
[4] Bioengineering Department and Imperial-X, Imperial College London, London W12 7SL, UK
g.yang@imperial.ac.uk
[5] National Heart and Lung Institute, Imperial College London, London SW7 2AZ, UK
[6] Cardiovascular Research Centre, Royal Brompton Hospital, London SW3 6NP, UK
[7] School of Biomedical Engineering & Imaging Sciences, King's College London, London WC2R 2LS, UK

Abstract. Deep learning has shown the capability to substantially accelerate MRI reconstruction while acquiring fewer measurements. Recently, diffusion models have gained burgeoning interests as a novel group of deep learning-based generative methods. These methods seek to sample data points that belong to a target distribution from a Gaussian distribution, which has been successfully extended to MRI reconstruction. In this work, we proposed a Cold Diffusion-based MRI reconstruction method called CDiffMR. Different from conventional diffusion models, the degradation operation of our CDiffMR is based on k-space undersampling instead of adding Gaussian noise, and the restoration network is trained to harness a de-aliaseing function. We also design starting point and data consistency conditioning strategies to guide and accelerate the reverse process. More intriguingly, the pre-trained CDiffMR model can be reused for reconstruction tasks with different undersampling rates. We demonstrated, through extensive numerical and visual experiments, that the proposed CDiffMR can achieve comparable or even superior reconstruction results than state-of-the-art models. Compared to the diffusion model-based counterpart, CDiffMR reaches readily competing results using only 1.6–3.4% for inference time. The code is publicly available at https://github.com/ayanglab/CDiffMR.

Keywords: Diffusion Models · Fast MRI · Deep Learning

Supplementary Information The online version contains supplementary material available at https://doi.org/10.1007/978-3-031-43999-5_1.

1   Introduction

Magnetic Resonance Imaging (MRI) is an essential non-invasive technique that enables high-resolution and reproducible assessments of structural and functional information, for clinical diagnosis and prognosis, without exposing the patient to radiation. Despite its widely use in clinical practice, MRI still suffers from the intrinsically slow data acquisition process, which leads to uncomfortable patient experience and artefacts from voluntary and involuntary physiological movements [5].

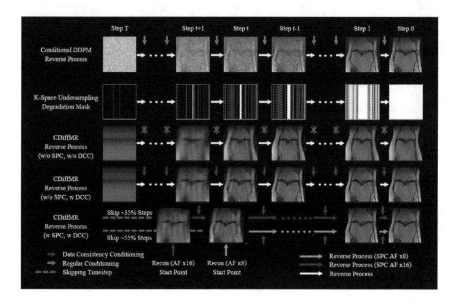

Fig. 1. Row 1: The reverse process of conditional DDPM [10]; Row 2: The K-Space Undersampling Degradation Mask for the proposed CDiffMR; Row 3: The reverse process of CDiffMR without DDC and SPC; Row 4: The reverse process of CDiffMR with DDC but without SPC; Row 5: The reverse process of CDiffMR with DDC and SPC; SPC: Starting Point Conditioning; DDC: Data Consistency Conditioning.

Deep learning has achieved considerable success across various research domains in recent years, including the ability to substantially accelerate MRI reconstruction while requiring fewer measurements. Various kinds of deep learning-based models, including Convolutional Neural Networks [15,21], Recurrent Neural Networks [4,9], Graph Neural Networks [11] or Transformers [12,13], have been explored for MRI reconstruction and achieved impressive success with a high accelerate factor (AF). However, most of these methods are based on a strong degradation prior, i.e., the undersampling mask, which entails a performance drop when the training and testing undersampling masks mismatch [14,16]. Therefore, additional training is required when applying different undersampling mask condition, leading to a waste of computational resources.

Diffusion models [10,18,20] represent a group of unconditional generative methods which sample data points that belong to target distribution from a

Gaussian distribution. The earliest diffusion models type was known as Denoising Diffusion Probabilistic Models (DDPMs) [10] and Score Matching with Langevin Dynamics (SMLD) [18], which were later unified into a framework by Score-based Stochastic Differential Equation (SDE) [20].

Diffusion models have been widely applied for inverse problems [6,19] including MRI Reconstruction [2,3,7,8,14]. Peng et al. [14] proposed a diffusion model-based MR reconstruction method, called DiffuseRecon, which did not require additional training on specific acceleration factors. Chung et al. [7] designed a score-based model for MRI reconstruction, which performed the reconstruction task iteratively using a numerical SDE solver and data consistency step. Cao et al. [3] proposed a complex diffusion probabilistic model for MRI reconstruction for better preservation of the MRI complex-value information. Gungor et al. [8] introduced an adaptive diffusion prior, namely AdaDiff, for enhancing reconstruction performance during the inference stage. Cao et al. [2] designed a modified high-frequency DDPM model for high-frequency information preservation of MRI data. However, they do share a commonality–the prolonged inference time due to the iterative nature of diffusion models. Chung et al. [6] proposed a new reverse process strategy for accelerating the sampling for the reverse problem, Come-Closer-Diffuse-Faster (CCDF), suggesting that *starting from Gaussian noise is necessary for diffusion models*. CCDF-MRI achieved outstanding reconstruction results with reduced reverse process steps.

Most existing diffusion models, including the original DDPM, SMLD and their variants, are strongly based on the use of Gaussian noise, which provides the 'random walk' for 'hot' diffusion. Cold Diffusion Model [1] rethought the role of the Gaussian noise, and generalised the diffusion models using different kinds of degradation strategies, e.g., blur, pixelate, mask-out, rather than the Gaussian noise applied on conventional diffusion models.

In this work, a novel Cold Diffusion-based MRI Reconstruction method (CDiffMR) is proposed (see Fig. 1). CDiffMR introduces a novel K-Space Undersampling Degradation (KSUD) module for the degradation, which means CDiffMR does not depend on the Gaussian noise. Instead of building an implicit transform to target distribution by Gaussian noise, CDiffMR explicitly learns the relationship between undersampled distribution and target distribution by KSUD.

We propose two novel k-space conditioning strategies to guide the reverse process and to reduce the required time steps. 1) Starting Point Conditioning (SPC). The k-space undersampled zero-filled images, which is usually regarded as the network input, can act as the reverse process starting point for conditioning. The number of reverse time steps therefore depends on the undersamping rate, i.e., the higher k-space undersampling rate (lower AF, easier task), the fewer reverse time steps required. 2) Data Consistency Conditioning (DCC). In every step of the reverse process, data consistency is applied to further guide the reverse process in the correct way.

It is note that our CDiffMR is a one-for-all model. This means that once CDiffMR is trained, it can be reused for all the reconstruction tasks, with any reasonable undersampling rates conditioning, as long as the undersampling rate is larger than the preset degraded images x_T at the end of forward process (e.g., 1% undersampling rate). Experiments were conducted on FastMRI dataset [22]. The proposed CDiffMR achieves comparable or superior reconstruction results with respect to state-of-the-art methods, and reaches a much faster reverse process compared with diffusion model-based counterparts. For the sake of clarity, we use 'sampling' or 'undersampling' to specify the k-space data acquisition for MRI, and use 'reverse process' to represent sampling from target data distribution in the inference stage of diffusion models. Our main contributions are summarised as follows:

- An innovative Cold Diffusion-based MRI Reconstruction methods is proposed. To best of our knowledge, CDiffMR is the first diffusion model-based MRI reconstruction method that exploits the k-space undersampling degradation.
- Two novel k-space conditioning strategies, namely SPC and DCC, are developed to guide and accelerate the reverse process.
- The pre-trained CDiffMR model can be reused for MRI reconstruction tasks with a reasonable range of undersampling rates.

2   Methodology

This section details two key parts of the proposed CDiffMR: 1) the optimisation and training schemes and 2) the k-space conditioning reverse process.

2.1   Model Components and Training

Diffusion models are generally composed of a degradation operator $D(\cdot, t)$ and a learnable restoration operator $R_\theta(\cdot, t)$ [1]. For standard diffusion models, the $D(\cdot, t)$ disturbs the images via Gaussian noise according to a preset noise schedule controlled by time step t. The $R_\theta(\cdot, t)$ is a denoising function controlled by t for various noise levels.

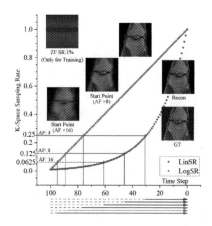

Fig. 2. Linear and Log k-space undersampling degradation schedules of CDiffMR.

Instead of applying Gaussian noise, CDiffMR provides a new option, namely KSUD operator $D(\cdot, t)$ and de-aliasing restoration operator $R_\theta(\cdot, t)$. With the support of KSUD, CDiffMR can explicitly learn the relationship between input distribution of k-space undersampled images and target distribution of fully-sampled images that is built implicitly by Gaussian noise in the conventional diffusion model. The KSUD operator $D(x,t) = \mathcal{F}^{-1}\mathcal{M}_t\mathcal{F}x$, undersamples input images via different k-space mask of which undersampling rate is controlled by the time step t.

Two k-space sampling rate SR_t schedule are designed in this study, linear (LinSR) and log (LogSR) sampling rate schedules. We set $SR_t = 1-(1-SR_{min})\frac{t}{T}$ for LinSR schedule, and $SR_t = SR_{min}^{\frac{t}{T}}$ for LogSR schedule (see Fig. 2). $D(x,t+1)$ contains less k-space information compared to $D(x,t)$, and when $t=0$ we have:

$$x = D(x,0) = \mathcal{F}^{-1}\mathcal{M}_0\mathcal{F}x = \mathcal{F}^{-1}\mathcal{I}\mathcal{F}x, \tag{1}$$

where $\mathcal{M}_t$ is the undersampling mask at step t corresponding to SR_t, and $\mathcal{M}_0 = \mathcal{I}$ is the fully-sampling mask (identity map). $\mathcal{F}$ and $\mathcal{F}^{-1}$ denote Fourier and inverse Fourier transform.

The restoration operator $R_\theta(\cdot,t)$ is an improved U-Net with time embedding module, following the official implementation[1] of Denoising Diffusion Implicit Models [17]. An ideal $R_\theta(\cdot,t)$ should satisfy $x_0 \approx R_\theta(x_t,t)$.

For the training of the restoration operator $R_\theta(\cdot,t)$, x_{true} is the fully-sampled images randomly sampled from target distribution $\mathcal{X}$. Practically, time step t is randomly chosen from $(1,T]$ during the training. The driven optimisation scheme reads:

$$\min_\theta \mathbb{E}_{x_{true}\sim\mathcal{X}}\| R_\theta(D(x_{true},t),t) - x\|, \qquad t = 1...T. \tag{2}$$

2.2   K-Space Conditioning Reverse Process

Two k-space conditioning strategies, SPC and DCC, are designed to guide and accelerate the reverse process.

Starting Point Conditioning enables the reverse process of CDiffMR to start from the *half way* step T' instead of step T (see Fig. 1 and Fig. 2). The starting point of the reverse process depends on the k-space undersampling rate of the reconstruction task. Specifically, for the reconstruction task with $\mathcal{M}$, T' can be checked by comparing the sampling rate of $\mathcal{M}$ in the degradation schedule, and the corresponding reverse process start step T' can be located, which is expressed as:

$$\text{Sampling Rate:}\quad \mathcal{M}_T < \mathcal{M}_{T'} < \mathcal{M} < \mathcal{M}_{T'-1} < \mathcal{M}_0 = \mathcal{I}. \tag{3}$$

With the start step T', the reverse process is conditioned by the reverse process of the initial image $x_{T'} \leftarrow \mathcal{F}^{-1}y$. The framework of the reverse process follows Algorithm 2 in [1], whereas we applied DCC strategy for further guiding (Eq. (5)). The result of the reverse process x_0 is the final reconstruction result. The whole reverse process is formulated as:

$$\hat{x}'_{0,t} = R_\theta(x_t,t), \quad \text{s.t. } t = T'...1. \tag{4}$$

$$\hat{x}_{0,t} = \mathcal{F}^{-1}(1-\mathcal{M})\mathcal{F}\hat{x}'_{0,t} + \mathcal{F}^{-1}\mathcal{M}\mathcal{F}x_{T'}, \quad \text{s.t. } t = T'...1. \tag{5}$$

$$x_{t-1} = x_t - D(\hat{x}_{0,t},t) + D(\hat{x}_{0,t},t-1), \quad \text{s.t. } t = T'...1. \tag{6}$$

---

[1] https://github.com/ermongroup/ddim.

3   Experimental Results

This section describes in detail the set of experiments conducted to validate the proposed CDiffMR.

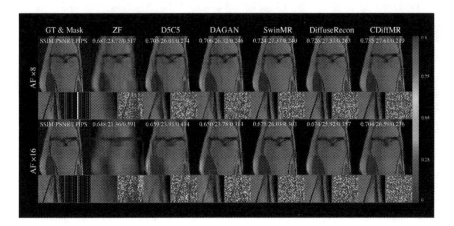

Fig. 3. Visual comparison against the state-of-the-art approaches with accelerate rate (AF) ×8 and ×16. The colour bar denotes the difference between the reconstructed result and ground truth. Zoom-in views displays selected fine detailed regions.

3.1   Implementation Details and Evaluation Methods

The experiments were conducted on the FastMRI dataset [22], which contains single-channel complex-value MRI data. For the FastMRI dataset, we applied 684 proton-density weighted knee MRI scans without fat suppression from the official training and validation sets, which were randomly divided into training set (420 cases), validation set (64 cases) and testing set (200 cases), approximately according to a ratio of 6:1:3. For each case, 20 coronal 2D single-channel complex-value slices near the centre were chosen, and all slices were centre-cropped to 320 × 320.

Undersampling mask $\mathcal{M}$ and $\mathcal{M}_t$ were generated by the fastMRI official implementation. We applied AF×8 and ×16 Cartesian mask for all experiments.

Our proposed CDiffMR was trained on two NVIDIA A100 (80 GB) GPUs, and tested on an NVIDIA RTX3090 (24 GB) GPU. CDiffMR was trained for 100,000 gradient steps, using the Adam optimiser with a learning rate 2e−5 and a batch size 24. We set the total diffusion time step to $T = 100$ for both LinSR and LogSR degradation schedules. The minimal sampling rate (when $t = 100$) was set to 1%.

For comparison, we used CNN-based methods D5C5 [15], DAGAN [21], Transformers-based method SwinMR [12], novel diffusion model-based method

DiffuseRecon [14]. We trained D5C5 and DiffuseRecon following the official setting, while we modified DAGAN and SwinMR for 2-channel input, output and loss function, as they were officially proposed for single-channel reconstruction.

In the quantitative experiments, Peak Signal-to-Noise Ratio (PSNR), Structural Similarity Index Measure (SSIM), Learned Perceptual Image Patch Similarity (LPIPS) [23] were applied to examine the reconstruction quality. Among them, LPIPS is a deep learning-based perceptual metric, which can match human visual perception well. The inference time was measured on a NVIDIA RTX3090 (24 GB) GPU with an input shape of $(1, 1, 320, 320)$ for ten times.

Table 1. Quantitative reconstruction results with accelerate rate (AF) $\times 8$ and $\times 16$. LinSR (LogSR): linear (log) sampling rate schedule; $^\star(^\dagger)$: significantly different from CDiffMR-LinSR(LogSR) by Mann-Whitney Test.

Model (AF $\times 8$)	SSIM ↑	PSNR ↑	LPIPS ↓	Inference Time ↓
ZF	0.678 (0.087)*†	22.74 (1.73)*†	0.504 (0.058)*†	–
D5C5 [15]	0.719 (0.104)*†	25.99 (2.13)*†	0.291 (0.039)*†	0.044
DAGAN [21]	0.709 (0.095)*†	25.19 (2.21)*†	0.262 (0.043)*†	0.004
SwinMR [12]	0.730 (0.107)*†	26.98 (2.46)*†	0.254 (0.042)*†	0.037
DiffuseRecon [14]	0.738 (0.108)*	**27.40 (2.40)**	0.286 (0.038)*†	183.770
CDiffMR-LinSR	**0.745 (0.108)**	27.35 (2.56)	0.240 (0.042)	5.862
CDiffMR-LogSR	0.744 (0.107)	27.26 (2.52)	**0.236 (0.041)**	3.030
Model (AF $\times 16$)	SSIM ↑	PSNR ↑	LPIPS ↓	Inference Time ↓
ZF	0.624 (0.080)*†	20.04 (1.59)*†	0.580 (0.049)*†	–
D5C5 [15]	0.667 (0.108)*†	23.35 (1.78)*†	0.412 (0.049)*†	0.044
DAGAN [21]	0.673 (0.102)*†	23.87 (1.84)*†	0.317 (0.044)*†	0.004
SwinMR [12]	0.673 (0.115)*†	24.85 (2.12)*†	0.327 (0.045)*†	0.037
DiffuseRecon [14]	0.688 (0.119)*†	25.75 (2.15)*†	0.362 (0.047)*†	183.770
CDiffMR-LinSR	**0.709 (0.117)**	**25.83 (2.27)**	0.297 (0.042)	6.263
CDiffMR-LogSR	0.707 (0.116)	25.77 (2.25)	**0.293 (0.042)**	4.017

3.2   Comparison and Ablation Studies

The quantitative results are reported in Table 1 and further supported by visual comparisons in Fig. 3. The proposed CDiffMR achieves promising results compared to the SOTA MRI reconstruction methods. Compared with the diffusion model-based method DiffuseRecon, CDiffMR achieves comparable or better results with only 1.6–3.4% inference time of DiffuseRecon. For ablation studies, we explored how DDC and SPC affect the speed of reverse process and reconstruction quality (see Fig. 4(A)(B)).

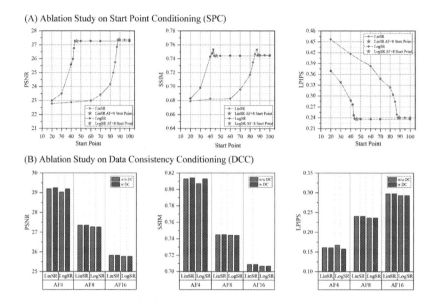

Fig. 4. (A) The relationship between reconstruction metrics and the reverse process starting point conditioning (SPC); (B) The relationship between reconstruction metrics and the reverse process data consistency conditioning (DDC).

4   Discussion and Conclusion

This work has exploited Cold Diffusion-based model for MRI reconstruction and proposed CDiffMR. We have designed the novel KSUD for degradation operator $D(\cdot, t)$ and trained a restoration function $R_\theta(\cdot, t)$ for de-aliasing under various undersampling rates. We pioneered the harmonisation of the degradation function in reverse problem (k-space undersampling in MRI reconstruction) and the degradation operator in diffusion model (KSUD). In doing so, CDiffMR is able to explicitly learn the k-space undersampling operation to further improve the reconstruction, while providing the basis for the reverse process acceleration. Two k-space conditioning strategies, SPC and DCC, have been designed to guide and accelerate the reverse process. Experiments have demonstrated that k-space undersampling can be successfully used as degradation in diffusion models for MRI reconstruction.

In this study, two KSUD schedules, i.e., have been designed for controlling the k-space sampling rate of every reverse time steps. According to Table 1, LogSR schedule achieves better perceptual score while LinSR has better fidelity score, where the difference is actually not significant. However, the required reverse process steps of LogSR is much fewer than LinSR's, which significantly accelerates the reverse process. This is because for LogSR schedule, a larger proportion steps are corresponding to lower sampling rate (high AF), therefore the starting point of LogSR is closer to step 0 than LinSR (see Fig. 2 and Fig. 4(A)). For AF×16 reconstruction task, CDiffMR-LinSR theoretically requires 95 reverse steps, while CDiffMR-LogSR only requires 61 reverse steps, and for AF×8 reconstruction task, CDiffMR-LinSR requires 89 reverse steps, while CDiffMR-LogSR only requires 46 reverse steps. The lower AF of the reconstruction task, the less reverse process steps required. Therefore, we recommend CDiffMR-LogSR as it achieves similiar results of CDiffMR-LinSR with much faster reverse process.

In the ablation studies, we have further examined the selection of reverse process starting point T'. Figure 4(A) has shown the reconstruction quality using different starting point. Reconstruction performance keeps stable with a range of reverse process starting points, but suddenly drops after a tuning point, which exactly matches the theoretical starting point. This experiment has proven the validity of our starting point selection method (Eq. (3)), and shown that our theoretical starting point keeps optimal balance between the reconstruction process and reverse process speed.

We also explored the validity of DDC in the ablation studies. Figure 4(B) has shown the reconstruction quality with or without DDC using different sampling rate schedule with different AF. The improvement by the DDC is significant with a lower AF (×4), but limited with a higher AF (×8, ×16). Therefore, CDiffMR keeps the DDC due to its insignificant computational cost.

The proposed CDiffMR heralds a new kind of diffusion models for solving inverse problems, i.e., applying the degradation model in reverse problem as the degradation module in diffusion model. CDiffMR has proven that this idea performs well for MRI reconstruction tasks. We can envision that our CDiffMR can serve as the basis for general inverse problems.

Acknowledgement. This study was supported in part by the ERC IMI (101005122), the H2020 (952172), the MRC (MC/PC/21013), the Royal Society (IEC\NSFC\211235), the NVIDIA Academic Hardware Grant Program, the SABER project supported by Boehringer Ingelheim Ltd, Wellcome Leap Dynamic Resilience, and the UKRI Future Leaders Fellowship (MR/V023799/1).

References

1. Bansal, A., et al.: Cold diffusion: inverting arbitrary image transforms without noise. arXiv e-prints p. arXiv:2208.09392 (2022)
2. Cao, C., Cui, Z.X., Liu, S., Zheng, H., Liang, D., Zhu, Y.: High-frequency space diffusion models for accelerated MRI. arXiv e-prints p. arXiv:2208.05481 (2022)
3. Cao, Y., Wang, L., Zhang, J., Xia, H., Yang, F., Zhu, Y.: Accelerating multi-echo MRI in k-space with complex-valued diffusion probabilistic model. In: 2022 16th IEEE International Conference on Signal Processing (ICSP), vol. 1, pp. 479–484 (2022)
4. Chen, E.Z., Wang, P., Chen, X., Chen, T., Sun, S.: Pyramid convolutional RNN for MRI image reconstruction. IEEE Trans. Med. Imaging **41**(8), 2033–2047 (2022)
5. Chen, Y., et al.: AI-based reconstruction for fast MRI-A systematic review and meta-analysis. Proc. IEEE **110**(2), 224–245 (2022)
6. Chung, H., Sim, B., Ye, J.C.: Come-closer-diffuse-faster: accelerating conditional diffusion models for inverse problems through stochastic contraction. In: Proceedings of the IEEE/CVF Conference on Computer Vision and Pattern Recognition (CVPR), pp. 12413–12422 (2022)

7. Chung, H., Ye, J.C.: Score-based diffusion models for accelerated MRI. Med. Image Anal. **80**, 102479 (2022)
8. Güngör, A., et al.: Adaptive diffusion priors for accelerated MRI reconstruction. arXiv e-prints p. arXiv:2207.05876 (2022)
9. Guo, P., Valanarasu, J.M.J., Wang, P., Zhou, J., Jiang, S., Patel, V.M.: Over-and-under complete convolutional RNN for MRI reconstruction. In: de Bruijne, M., et al. (eds.) MICCAI 2021. LNCS, vol. 12906, pp. 13–23. Springer, Cham (2021). https://doi.org/10.1007/978-3-030-87231-1_2
10. Ho, J., Jain, A., Abbeel, P.: Denoising diffusion probabilistic models. In: Advances in Neural Information Processing Systems, vol. 33. Curran Associates, Inc. (2020)
11. Huang, J., Aviles-Rivero, A., Schonlieb, C.B., Yang, G.: ViGU: vision GNN U-net for fast MRI. arXiv e-prints p. arXiv:2302.10273 (2023)
12. Huang, J., et al.: Swin transformer for fast MRI. Neurocomputing **493**, 281–304 (2022)
13. Korkmaz, Y., Dar, S.U.H., Yurt, M., Özbey, M., Çukur, T.: Unsupervised MRI reconstruction via zero-shot learned adversarial transformers. IEEE Trans. Med. Imaging **41**(7), 1747–1763 (2022)
14. Peng, C., Guo, P., Zhou, S.K., Patel, V.M., Chellappa, R.: Towards performant and reliable undersampled MR reconstruction via diffusion model sampling. In: Wang, L., Dou, Q., Fletcher, P.T., Speidel, S., Li, S. (eds.) MICCAI 2022. LNCS, vol. 13436, pp. 623–633. Springer, Cham (2022)
15. Schlemper, J., Caballero, J., Hajnal, J.V., Price, A., Rueckert, D.: A deep cascade of convolutional neural networks for MR image reconstruction. In: Niethammer, M., et al. (eds.) IPMI 2017. LNCS, vol. 10265, pp. 647–658. Springer, Cham (2017). https://doi.org/10.1007/978-3-319-59050-9_51
16. Shimron, E., Tamir, J.I., Wang, K., Lustig, M.: Implicit data crimes: machine learning bias arising from misuse of public data. Proc. Natl. Acad. Sci. **119**(13), e2117203119 (2022)
17. Song, J., Meng, C., Ermon, S.: Denoising diffusion implicit models. arXiv e-prints p. arXiv:2010.02502 (2020)
18. Song, Y., Ermon, S.: Generative modeling by estimating gradients of the data distribution. In: Advances in Neural Information Processing Systems, vol. 32. Curran Associates, Inc. (2019)
19. Song, Y., Shen, L., Xing, L., Ermon, S.: Solving inverse problems in medical imaging with score-based generative models. arXiv e-prints arXiv:2111.08005 (2021)
20. Song, Y., Sohl-Dickstein, J., Kingma, D.P., Kumar, A., Ermon, S., Poole, B.: Score-based generative modeling through stochastic differential equations. arXiv e-prints arXiv:2011.13456 (2020)
21. Yang, G., et al.: DAGAN: deep de-aliasing generative adversarial networks for fast compressed sensing MRI reconstruction. IEEE Trans. Med. Imaging **37**(6), 1310–1321 (2018)
22. Zbontar, J., et al.: fastMRI: an open dataset and benchmarks for accelerated MRI. arXiv e-prints p. arXiv:1811.08839 (2018)
23. Zhang, R., Isola, P., Efros, A.A., Shechtman, E., Wang, O.: The unreasonable effectiveness of deep features as a perceptual metric. In: 2018 IEEE/CVF Conference on Computer Vision and Pattern Recognition, pp. 586–595 (2018)

Learning Deep Intensity Field for Extremely Sparse-View CBCT Reconstruction

Yiqun Lin[1], Zhongjin Luo[2], Wei Zhao[3], and Xiaomeng Li[1(✉)]

[1] The Hong Kong University of Science and Technology, Kowloon, Hong Kong
eexmli@ust.hk
[2] The Chinese University of Hong Kong, Shenzhen, China
[3] Beihang University, Beijing, People's Republic of China

Abstract. Sparse-view cone-beam CT (CBCT) reconstruction is an important direction to reduce radiation dose and benefit clinical applications. Previous voxel-based generation methods represent the CT as discrete voxels, resulting in high memory requirements and limited spatial resolution due to the use of 3D decoders. In this paper, we formulate the CT volume as a continuous intensity field and develop a novel DIF-Net to perform high-quality CBCT reconstruction from extremely sparse (≤ 10) projection views at an ultrafast speed. The intensity field of a CT can be regarded as a continuous function of 3D spatial points. Therefore, the reconstruction can be reformulated as regressing the intensity value of an arbitrary 3D point from given sparse projections. Specifically, for a point, DIF-Net extracts its view-specific features from different 2D projection views. These features are subsequently aggregated by a fusion module for intensity estimation. Notably, thousands of points can be processed in parallel to improve efficiency during training and testing. In practice, we collect a knee CBCT dataset to train and evaluate DIF-Net. Extensive experiments show that our approach can reconstruct CBCT with high image quality and high spatial resolution from extremely sparse views within 1.6 s, significantly outperforming state-of-the-art methods. Our code will be available at https://github.com/xmed-lab/DIF-Net.

Keywords: CBCT Reconstruction · Implicit Neural Representation · Sparse View · Low Dose · Efficient Reconstruction

1 Introduction

Cone-beam computed tomography (CBCT) is a common 3D imaging technique used to examine the internal structure of an object with high spatial resolution and fast scanning speed [20]. During CBCT scanning, the scanner rotates around the object and emits cone-shaped beams, obtaining 2D projections in the detection panel to reconstruct 3D volume. In recent years, beyond dentistry, CBCT

Supplementary Information The online version contains supplementary material available at https://doi.org/10.1007/978-3-031-43999-5_2.

H. Greenspan et al. (Eds.): MICCAI 2023, LNCS 14229, pp. 13–23, 2023.
https://doi.org/10.1007/978-3-031-43999-5_2

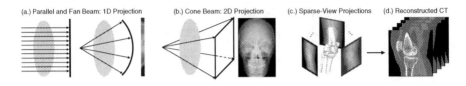

Fig. 1. (a-b): Comparison of conventional CT and cone-beam CT scanning. (c-d): CBCT reconstruction from a stack of sparse 2D projections.

has been widely used to acquire images of the human knee joint for applications such as total knee arthroplasty and postoperative pain management [3,4,9,15]. To maintain image quality, CBCT typically requires hundreds of projections involving high radiation doses from X-rays, which could be a concern in clinical practice. Sparse-view reconstruction is one of the ways to reduce radiation dose by reducing the number of scanning views (10× fewer). In this paper, we study a more challenging problem, extremely sparse-view CBCT reconstruction, aiming to reconstruct a high-quality CT volume from fewer than 10 projection views.

Compared to conventional CT (e.g., parallel beam, fan beam), CBCT reconstructs a 3D volume from 2D projections instead of a 2D slice from 1D projections, as comparison shown in Fig. 1, resulting in a significant increase in spatial dimensionality and computational complexity. Therefore, although sparse-view conventional CT reconstruction [2,23,25,26] has been developed for many years, these methods cannot be trivially extended to CBCT. CBCT reconstruction can be divided into dense-view ($\geq$100), sparse-view (20~50), extremely sparse-view ($\leq$10), and single/orthogonal-view reconstructions depending on the number of projection views required. A typical example of dense-view reconstruction is FDK [6], which is a filtered-backprojection (FBP) algorithm that accumulates intensities by backprojecting from 2D views, but requires hundreds of views to avoid streaking artifacts. To reduce required projection views, ART [7] and its extensions (e.g., SART [1], VW-ART [16]) formulate reconstruction as an iterative minimization process, which is useful when projections are limited. Nevertheless, such methods often take a long computational time to converge and cope poorly with extremely sparse projections; see results of SART in Table 1. With the development of deep learning techniques and computing devices, learning-based approaches are proposed for CBCT sparse-view reconstruction. Lahiri et al. [12] propose to reconstruct a coarse CT with FDK and use 2D CNNs to denoise each slice. However, the algorithm has not been validated on medical datasets, and the performance is still limited as FDK introduces extensive streaking artifacts with sparse views. Recently, neural rendering techniques [5,14,19,21,29] have been introduced to reconstruct CBCT volume by parameterizing the attenuation coefficient field as an implicit neural representation field (NeRF), but they require a long time for per-patient optimization and do not perform well with extremely sparse views due to lack of prior knowledge; see results of NAF in Table 2. For single/orthogonal-view reconstruction, voxel-based approaches [10,22,27] are proposed to build 2D-to-3D generation networks that consist of 2D encoders and 3D decoders with large training parameters, leading to high memory requirements and limited spatial resolution. These methods are special designs with the networks [10,27] or patient-specific training data [22], which are difficult to extend to general sparse-view reconstruction.

In this work, our goal is to reconstruct a CBCT of high image quality and high spatial resolution from extremely sparse (≤ 10) 2D projections, which is an important yet challenging and unstudied problem in sparse-view CBCT reconstruction. Unlike previous voxel-based methods that represent the CT as discrete voxels, we formulate the CT volume as a continuous intensity field, which can be regarded as a continuous function $g(\cdot)$ of 3D spatial points. The property of a point p in this field represents its intensity value v, i.e., $v = g(p)$. Therefore, the reconstruction problem can be reformulated as regressing the intensity value of an arbitrary 3D point from a stack of 2D projections $\mathcal{I}$, i.e., $v = g(\mathcal{I}, p)$. Based on the above formulation, we develop a novel reconstruction framework, namely DIF-Net (**D**eep **I**ntensity **F**ield **Net**work). Specifically, DIF-Net first extracts feature maps from K given 2D projections. Given a 3D point, we project the point onto the 2D imaging panel of each $view_i$ by corresponding imaging parameters (distance, angle, etc.) and query its view-specific features from the feature map of $view_i$. Then, K view-specific features from different views are aggregated by a cross-view fusion module for intensity regression. By introducing the continuous intensity field, it becomes possible to train DIF-Net with a set of sparsely sampled points to reduce memory requirement, and reconstruct the CT volume with any desired resolution during testing. Compared with NeRF-based methods [5,14,19,21,29], the design of DIF-Net shares the similar data representation (i.e., implicit neural representation) but additional training data can be introduced to help DIF-Net learn prior knowledge. Benefiting from this, DIF-Net can not only reconstruct high-resolution CT in a very short time since only inference is required for a new test sample (no retraining), but also performs much better than NeRF-based methods with extremely limited views.

To summarize, the main contributions of this work include 1.) we are the first to introduce the continuous intensity field for supervised CBCT reconstruction; 2.) we propose a novel reconstruction framework DIF-Net that reconstructs CBCT with high image quality (PSNR: 29.3 dB, SSIM: 0.92) and high spatial resolution ($\geq 256^3$) from extremely sparse (≤ 10) views within 1.6 s; 3.) we conduct extensive experiments to validate the effectiveness of the proposed sparse-view CBCT reconstruction method on a clinical knee CBCT dataset.

2 Method

2.1 Intensity Field

We formulate the CT volume as a continuous intensity field, where the property of a 3D point $p \in \mathbb{R}^3$ in this field represents its intensity value $v \in \mathbb{R}$. The intensity field can be defined as a continuous function $g : \mathbb{R}^3 \to \mathbb{R}$, such that $v = g(p)$. Hence, the reconstruction problem can be reformulated as regressing the intensity value of an arbitrary point p in the 3D space from K projections $\mathcal{I} = \{I_1, I_2, \ldots, I_K\}$, i.e., $v = g(\mathcal{I}, p)$. Based on the above formulation, we propose a novel reconstruction framework, namely DIF-Net, to perform efficient sparse-view CBCT reconstruction, as the overview shown in Fig. 2.

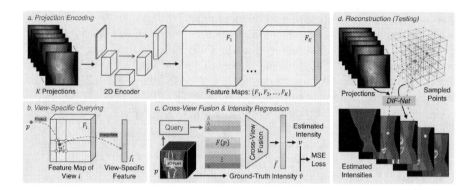

Fig. 2. Overview of DIF-Net. (a) Given K projections, a shared 2D encoder is used for feature extraction. (b) For a point p in the 3D space, its view-specific features are queried from feature maps of different views by projection and interpolation. (c) Queried features are aggregated to estimate the intensity value of p. (d) During testing, given input projections, DIF-Net predicts intensity values for points uniformly sampled in 3D space to reconstruct the target CT image.

2.2   DIF-Net: Deep Intensity Field Network

DIF-Net first extracts feature maps $\{F_1, F_2, \ldots, F_K\} \subset \mathbb{R}^{C \times H \times W}$ from projections $\mathcal{I}$ using a shared 2D encoder, where C is the number of feature channels and H/W are height/width. In practice, we choose U-Net [18] as the 2D encoder because of its good feature extraction ability and popular applications in medical image analysis [17]. Then, given a 3D point, DIF-Net gathers its view-specific features queried from feature maps of different views for intensity regression.

View-Specific Feature Querying. Considering a point $p \in \mathbb{R}^3$ in the 3D space, for a projection view_i with scanning angle α_i and other imaging parameters β (distance, spacing, etc.), we project p to the 2D imaging panel of view_i and obtain its 2D projection coordinates $p_i' = \varphi(p, \alpha_i, \beta) \in \mathbb{R}^2$, where $\varphi(\cdot)$ is the projection function. Projection coordinates p_i' are used for querying view-specific features $f_i \in \mathbb{R}^C$ from the 2D feature map F_i of view_i:

$$f_i = \pi(F_i, p_i') = \pi\left(F_i, \varphi(p, \alpha_i, \beta)\right), \tag{1}$$

where $\pi(\cdot)$ is bilinar interpolation. Similar to perspective projection, the CBCT projection function $\varphi(\cdot)$ can be formulated as

$$\varphi(p, \alpha_i, \beta) = H\left(A(\beta)R(\alpha_i)\begin{bmatrix} p \\ 1 \end{bmatrix}\right), \tag{2}$$

where $R(\alpha_i) \in \mathbb{R}^{4 \times 4}$ is a rotation matrix that transforms point p from the world coordinate system to the scanner coordinate system of view_i, $A(\beta) \in \mathbb{R}^{3 \times 4}$ is a projection matrix that projects the point onto the 2D imaging panel of view_i, and $H : \mathbb{R}^3 \to \mathbb{R}^2$ is the homogeneous division that maps the homogeneous coordinates of p_i' to its Cartesian coordinates. Due to page limitations, the detailed formulation of $\varphi(\cdot)$ is given in the supplementary material.

Cross-View Feature Fusion and Intensity Regression. Given K projection views, K view-specific features of the point p are queried from different views to form a feature list $F(p) = \{f_1, f_2, \ldots, f_K\} \subset \mathbb{R}^C$. Then, the cross-view feature fusion $\delta(\cdot)$ is introduced to gather features from $F(p)$ and generate a 1D vector $\bar{f} = \delta(F(p)) \in \mathbb{R}^C$ to represent the semantic features of p. In general, $F(p)$ is an unordered feature set, which means that $\delta(\cdot)$ should be a set function and can be implemented with a pooling layer (e.g., max/avg pooling). In our experiments, the projection angles of the training and test samples are the same, uniformly sampled from $0°$ to $180°$ (half rotation). Therefore, $F(p)$ can be regarded as an ordered list ($K \times C$ tensor), and $\delta(\cdot)$ can be implemented by a 2-layer MLP ($K \rightarrow \lfloor \frac{K}{2} \rfloor \rightarrow 1$) for feature aggregation. We will compare different implementations of $\delta(\cdot)$ in the ablation study. Finally, a 4-layer MLP ($C \rightarrow 2C \rightarrow \lfloor \frac{C}{2} \rfloor \rightarrow \lfloor \frac{C}{8} \rfloor \rightarrow 1$) is applied to $\bar{f}$ for the regression of intensity value $v \in \mathbb{R}$.

2.3 Network Training

Assume that the shape and spacing of the original CT volume are $H \times W \times D$ and (s_h, s_w, s_d) mm, respectively. During training, different from previous voxel-based methods that regard the entire 3D CT image as the supervision target, we randomly sample a set of N points $\{p_1, p_2, \ldots, p_N\}$ with coordinates ranging from $(0, 0, 0)$ to $(s_h H, s_w W, s_d D)$ in the world coordinate system (unit: mm) as the input. Then DIF-Net will estimate their intensity values $\mathcal{V} = \{v_1, v_2, \ldots, v_N\}$ from given projections $\mathcal{I}$. For supervision, ground-truth intensity values $\hat{\mathcal{V}} = \{\hat{v}_1, \hat{v}_2, \ldots, \hat{v}_N\}$ can be obtained from the ground-truth CT image based on the coordinates of points by trilinear interpolation. We choose mean-square-error (MSE) as the objective function, and the training loss can be formulated as

$$\mathcal{L}(\mathcal{V}, \hat{\mathcal{V}}) = \frac{1}{N} \sum_{i=1}^{N} (v_i - \hat{v}_i)^2. \tag{3}$$

Because background points (62%, e.g., air) occupy more space than foreground points (38%, e.g., bones, organs), uniform sampling will bring imbalanced prediction of intensities. We set an intensity threshold 10^{-5} to identify foreground and background areas by binary classification and sample $\frac{N}{2}$ points from each area for training.

2.4 Volume Reconstruction

During inference, a regular and dense point set to cover all CT voxels is sampled, i.e., to uniformly sample $H \times W \times D$ points from $(0, 0, 0)$ to $(s_h H, s_w W, s_d D)$. Then the network will take 2D projections and points as the input and generate intensity values of sampled points to form the target CT volume. Unlike previous voxel-based methods that are limited to generating fixed-resolution CT volumes, our method enables scalable output resolutions by introducing the representation of continuous intensity field. For example, we can uniformly sample $\lfloor \frac{H}{s} \rfloor \times \lfloor \frac{W}{s} \rfloor \times$

$\lfloor \frac{D}{s} \rfloor$ points to generate a coarse CT image but with a faster reconstruction speed, or sample $\lfloor sH \rfloor \times \lfloor sW \rfloor \times \lfloor sD \rfloor$ points to generate a CT image with higher resolution, where $s > 1$ is the scaling ratio.

3   Experiments

We conduct extensive experiments on a collected knee CBCT dataset to show the effectiveness of our proposed method on sparse-view CBCT reconstruction. Compared to previous works, our DIF-Net can reconstruct a CT volume with high image quality and high spatial resolution from extremely sparse (≤ 10) projections at an ultrafast speed.

3.1   Experimental Settings

Dataset and Preprocessing. We collect a knee CBCT dataset consisting of 614 CT scans. Of these, 464 are used for training, 50 for validation, and 100 for testing. We resample, interpolate, and crop (or pad) CT scans to have isotropic voxel spacing of $(0.8, 0.8, 0.8)$ mm and shape of $256 \times 256 \times 256$. 2D projections are generated by digitally reconstructed radiographs (DRRs) at a resolution of 256×256. Projection angles are uniformly selected in the range of $180°$.

Implementation. We implement DIF-Net using PyTorch with a single NVIDIA RTX 3090 GPU. The network parameters are optimized using stochastic gradient descent (SGD) with a momentum of 0.98 and an initial learning rate of 0.01. The learning rate is decreased by a factor of $0.001^{1/400} \approx 0.9829$ per epoch, and we train the model for 400 epochs with a batch size of 4. For each CT scan, $N = 10,000$ points are sampled as the input during one training iteration. For the full model, we employ U-Net [18] with $C = 128$ output feature channels as the 2D encoder, and cross-view feature fusion is implemented with MLP.

Baseline Methods. We compare four publicly available methods as our baselines, including traditional methods FDK [6] and SART [1], NeRF-based method NAF [29], and data-driven denoising method FBPConvNet [11]. Due to the increase in dimensionality (2D to 3D), denoising methods should be equipped with 3D conv/deconvs for a dense prediction when extended to CBCT reconstruction, which leads to extremely high computational costs and low resolution ($\leq 64^3$). For a fair comparison, we use FDK to obtain an initial result and apply the 2D network for slice-wise denoising.

Evaluation Metrics. We follow previous works [27–29] to evaluate the reconstructed CT volumes with two quantitative metrics, namely peak signal-to-noise ratio (PSNR) and structural similarity (SSIM) [24]. Higher PSNR/SSIM values represent superior reconstruction quality.

3.2   Results

Performance. As shown in Table 1, we compare DIF-Net with four previous methods [1,6,22,29] under the setting of reconstruction with different output

Table 1. Comparison of DIF-Net with previous methods under measurements of PSNR (dB) and SSIM. We evaluate reconstructions with different output resolutions (Res.) and from different numbers of projection views (K).

Method	Res. = 128^3			Res. = 256^3		
	$K = 6$	$K = 8$	$K = 10$	$K = 6$	$K = 8$	$K = 10$
FDK [6]	14.1/.18	15.7/.22	17.0/.25	14.1/.16	15.7/.20	16.9/.23
SART [1]	25.4/.81	26.6/.85	27.6/.88	24.7/.81	25.8/.84	26.7/.86
NAF [29]	20.8/.54	23.0/.64	25.0/.73	20.1/.58	22.4/.67	24.3/.75
FBPConvNet [11]	26.4/.84	27.0/.87	27.8/.88	25.1/.83	25.9/.83	26.7/.84
DIF-Net (Ours)	**28.3/.91**	**29.6/.92**	**30.7/.94**	**27.1/.89**	**28.3/.90**	**29.3/.92**

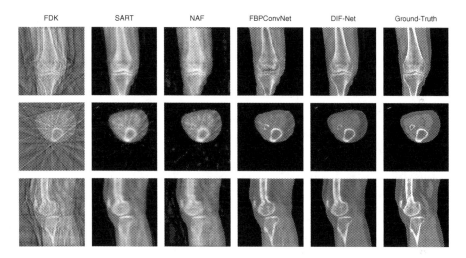

Fig. 3. Qualitative comparison of 10-view reconstruction.

resolutions (i.e., $128^3, 256^3$) and from different numbers of projection views (i.e., 6, 8, and 10). Experiments show that our proposed DIF-Net can reconstruct CBCT with high image quality even using only 6 projection views, which significantly outperforms previous works in terms of PSNR and SSIM values. More importantly, DIF-Net can be directly applied to reconstruct CT images with different output resolutions without the need for model retraining or modification. As visual results are shown in Fig. 3, FDK [6] produces results with many streaking artifacts due to lack of sufficient projection views; SART [1] and NAF [29] produce results with good shape contours but lack detailed internal information; FBPConvNet [11] reconstructs good shapes and moderate details, but there are still some streaking artifacts remaining; our proposed DIF-Net can reconstruct high-quality CT with better shape contour, clearer internal information, and fewer artifacts. More visual comparisons of the number of input views are given in the supplementary material.

Table 2. Comparison of different methods in terms of reconstruction quality (PSNR/S-SIM), reconstruction time, parameters, and training memory cost. Default setting: 10-view reconstruction with the output resolution of 256^3; training with a batch size of 1. †: evaluated with the output resolution of 128^3 due to the memory limitation.

Method	PSNR/SSIM	Time (s)	Parameters (M)	Memory Cost (MB)
FDK [6]	16.9/.23	0.3	-	-
SART [1]	26.7/.86	106	-	339
NAF [29]	24.3/.75	738	14.3	3,273
FBPConvNet [11]	26.7/.84	1.7	34.6	3,095
DIF-Net (Ours)	29.3/.92	1.6	31.1	7,617

Table 3. Ablation study (10-view) on different cross-view fusion strategies.

Cross-View Fusion	PSNR	SSIM
Avg pooling	27.6	0.88
Max pooling	28.9	0.92
MLP	**29.3**	**0.92**

Table 4. Ablation study (10-view) on different numbers of training points N.

# Points	PSNR	SSIM
5,000	28.8	0.91
10,000	**29.3**	**0.92**
20,000	29.3	0.92

Reconstruction Efficiency. As shown in Table 2, FDK [6] requires the least time for reconstruction, but has the worst image quality; SART [1] and NAF [29] require a lot of time for optimization or training; FBPConvNet [11] can reconstruct 3D volumes faster, but the quality is still limited. Our DIF-Net can reconstruct high-quality CT within 1.6 s, much faster than most compared methods. In addition, DIF-Net, which benefits from the intensity field representation, has fewer training parameters and requires less computational memory, enabling high-resolution reconstruction.

Ablation Study. Tables 3 and 4 show the ablative analysis of cross-view fusion strategy and the number of training points N. Experiments demonstrate that 1.) MLP performs best, but max pooling is also effective and would be a general solution when the view angles are not consistent across training/test data, as discussed in Sect. 2.2; 2.) fewer points (e.g., 5,000) may destabilize the loss and gradient during training, leading to performance degradation; 10,000 points are enough to achieve the best performance, and training with 10,000 points is much sparser than voxel-based methods that train with the entire CT volume (i.e., 256^3 or 128^3). We have tried to use a different encoder like pre-trained ResNet18 [8] with more model parameters than U-Net [18]. However, ResNet18 does not bring any improvement (PSNR/SSIM: 29.2/0.92), which means that U-Net is powerful enough for feature extraction in this task.

4   Conclusion

In this work, we formulate the CT volume as a continuous intensity field and present a novel DIF-Net for ultrafast CBCT reconstruction from extremely sparse (≤ 10) projection views. DIF-Net aims to estimate the intensity value of an arbitrary point in 3D space from input projections, which means 3D CNNs are not required for feature decoding, thereby reducing memory requirement and computational cost. Experiments show that DIF-Net can perform efficient and high-quality CT reconstruction, significantly outperforming previous state-of-the-art methods. More importantly, DIF-Net is a general sparse-view reconstruction framework, which can be trained on a large-scale dataset containing various body parts with different projection views and imaging parameters to achieve better generalization ability. This will be left as our future work.

Acknowledgement. This work was supported by the Hong Kong Innovation and Technology Fund under Projects PRP/041/22FX and ITS/030/21, as well as by grants from Foshan HKUST Projects under Grants FSUST21-HKUST10E and FSUST21-HKUST11E.

References

1. Andersen, A.H., Kak, A.C.: Simultaneous algebraic reconstruction technique (SART): a superior implementation of the art algorithm. Ultrason. Imaging **6**(1), 81–94 (1984)
2. Anirudh, R., Kim, H., Thiagarajan, J.J., Mohan, K.A., Champley, K., Bremer, T.: Lose the views: limited angle CT reconstruction via implicit sinogram completion. In: Proceedings of the IEEE Conference on Computer Vision and Pattern Recognition, pp. 6343–6352 (2018)
3. Bier, B., et al.: Range imaging for motion compensation in C-arm cone-beam CT of knees under weight-bearing conditions. J. Imaging **4**(1), 13 (2018)
4. Dartus, J., et al.: The advantages of cone-beam computerised tomography (CT) in pain management following total knee arthroplasty, in comparison with conventional multi-detector ct. Orthop. Traumatol. Surg. Res. **107**(3), 102874 (2021)
5. Fang, Y., et al.: SNAF: sparse-view CBCT reconstruction with neural attenuation fields. arXiv preprint arXiv:2211.17048 (2022)
6. Feldkamp, L.A., Davis, L.C., Kress, J.W.: Practical cone-beam algorithm. Josa a **1**(6), 612–619 (1984)
7. Gordon, R., Bender, R., Herman, G.T.: Algebraic reconstruction techniques (art) for three-dimensional electron microscopy and x-ray photography. J. Theor. Biol. **29**(3), 471–481 (1970)
8. He, K., Zhang, X., Ren, S., Sun, J.: Deep residual learning for image recognition. In: Proceedings of the IEEE Conference on Computer Vision and Pattern Recognition, pp. 770–778 (2016)
9. Jaroma, A., Suomalainen, J.S., Niemitukia, L., Soininvaara, T., Salo, J., Kröger, H.: Imaging of symptomatic total knee arthroplasty with cone beam computed tomography. Acta Radiol. **59**(12), 1500–1507 (2018)
10. Jiang, Y.: MFCT-GAN: multi-information network to reconstruct CT volumes for security screening. J. Intell. Manuf. Spec. Equipment **3**, 17–30 (2022)

11. Jin, K.H., McCann, M.T., Froustey, E., Unser, M.: Deep convolutional neural network for inverse problems in imaging. IEEE Trans. Image Process. **26**(9), 4509–4522 (2017)
12. Lahiri, A., Klasky, M., Fessler, J.A., Ravishankar, S.: Sparse-view cone beam CT reconstruction using data-consistent supervised and adversarial learning from scarce training data. arXiv preprint arXiv:2201.09318 (2022)
13. Lechuga, L., Weidlich, G.A.: Cone beam CT vs. fan beam CT: a comparison of image quality and dose delivered between two differing CT imaging modalities. Cureus **8**(9) (2016)
14. Mildenhall, B., Srinivasan, P.P., Tancik, M., Barron, J.T., Ramamoorthi, R., Ng, R.: Nerf: representing scenes as neural radiance fields for view synthesis. Commun. ACM **65**(1), 99–106 (2021)
15. Nardi, C., et al.: The role of cone beam CT in the study of symptomatic total knee arthroplasty (TKA): a 20 cases report. Br. J. Radiol. **90**(1074), 20160925 (2017)
16. Pan, J., Zhou, T., Han, Y., Jiang, M.: Variable weighted ordered subset image reconstruction algorithm. Int. J. Biomed. Imaging **2006** (2006)
17. Punn, N.S., Agarwal, S.: Modality specific u-net variants for biomedical image segmentation: a survey. Artif. Intell. Rev. **55**(7), 5845–5889 (2022)
18. Ronneberger, O., Fischer, P., Brox, T.: U-Net: convolutional networks for biomedical image segmentation. In: Navab, N., Hornegger, J., Wells, W.M., Frangi, A.F. (eds.) MICCAI 2015. LNCS, vol. 9351, pp. 234–241. Springer, Cham (2015). https://doi.org/10.1007/978-3-319-24574-4_28
19. Rückert, D., Wang, Y., Li, R., Idoughi, R., Heidrich, W.: Neat: neural adaptive tomography. ACM Trans. Graph. (TOG) **41**(4), 1–13 (2022)
20. Scarfe, W.C., Farman, A.G., Sukovic, P., et al.: Clinical applications of cone-beam computed tomography in dental practice. J. Can. Dent. Assoc. **72**(1), 75 (2006)
21. Shen, L., Pauly, J., Xing, L.: NeRP: implicit neural representation learning with prior embedding for sparsely sampled image reconstruction. IEEE Trans. Neural Netw. Learn. Syst. (2022)
22. Shen, L., Zhao, W., Xing, L.: Patient-specific reconstruction of volumetric computed tomography images from a single projection view via deep learning. Nat. Biomed. Eng. **3**(11), 880–888 (2019)
23. Tang, C., et al.: Projection super-resolution based on convolutional neural network for computed tomography. In: 15th International Meeting on Fully Three-Dimensional Image Reconstruction in Radiology and Nuclear Medicine, vol. 11072, pp. 537–541. SPIE (2019)
24. Wang, Z., Bovik, A.C., Sheikh, H.R., Simoncelli, E.P.: Image quality assessment: from error visibility to structural similarity. IEEE Trans. Image Process. **13**(4), 600–612 (2004)
25. Wu, W., Guo, X., Chen, Y., Wang, S., Chen, J.: Deep embedding-attention-refinement for sparse-view CT reconstruction. IEEE Trans. Instrum. Meas. **72**, 1–11 (2022)
26. Wu, W., Hu, D., Niu, C., Yu, H., Vardhanabhuti, V., Wang, G.: Drone: dual-domain residual-based optimization network for sparse-view CT reconstruction. IEEE Trans. Med. Imaging **40**(11), 3002–3014 (2021)
27. Ying, X., Guo, H., Ma, K., Wu, J., Weng, Z., Zheng, Y.: X2CT-GAN: reconstructing CT from biplanar x-rays with generative adversarial networks. In: Proceedings of the IEEE/CVF Conference on Computer Vision and Pattern Recognition, pp. 10619–10628 (2019)

28. Zang, G., Idoughi, R., Li, R., Wonka, P., Heidrich, W.: Intratomo: self-supervised learning-based tomography via sinogram synthesis and prediction. In: Proceedings of the IEEE/CVF International Conference on Computer Vision, pp. 1960–1970 (2021)
29. Zha, R., Zhang, Y., Li, H.: NAF: neural attenuation fields for sparse-view CBCT reconstruction. In: Wang, L., Dou, Q., Fletcher, P.T., Speidel, S., Li, S. (eds.) MICCAI 2022. LNCS, vol. 13436, pp. 442–452. Springer, Cham (2022). https:// doi.org/10.1007/978-3-031-16446-0_42

Revealing Anatomical Structures in PET to Generate CT for Attenuation Correction

Yongsheng Pan[1], Feihong Liu[1,2], Caiwen Jiang[1], Jiawei Huang[1], Yong Xia[3], and Dinggang Shen[1,4(✉)]

[1] School of Biomedical Engineering, ShanghaiTech University, Shanghai, China
{panysh,dgshen}@shanghaitech.edu.cn
[2] School of Information Science and Technology, Northwest University, Xi'an, China
[3] School of Computer Science, Northwestern Polytechnical University, Xi'an, China
yxia@nwpu.edu.cn
[4] Shanghai United Imaging Intelligence Co., Ltd., Shanghai, China

Abstract. Positron emission tomography (PET) is a molecular imaging technique relying on a step, namely attenuation correction (AC), to correct radionuclide distribution based on pre-determined attenuation coefficients. Conventional AC techniques require additionally-acquired computed tomography (CT) or magnetic resonance (MR) images to calculate attenuation coefficients, which increases imaging expenses, time costs, or radiation hazards to patients, especially for whole-body scanners. In this paper, considering technological advances in acquiring more anatomical information in raw PET images, we propose to conduct attenuation correction to PET by itself. To achieve this, we design a deep learning based framework, namely anatomical skeleton-enhanced generation (ASEG), to generate pseudo CT images from non-attenuation corrected PET images for attenuation correction. Specifically, ASEG contains two sequential modules, i.e., a skeleton prediction module and a tissue rendering module. The former module first delineates anatomical skeleton and the latter module then renders tissue details. Both modules are trained collaboratively with specific *anatomical-consistency constraint* to guarantee tissue generation fidelity. Experiments on four public PET/CT datasets demonstrate that our ASEG outperforms existing methods by achieving better consistency of anatomical structures in generated CT images, which are further employed to conduct PET attenuation correction with better similarity to real ones. This work verifies the feasibility of generating pseudo CT from raw PET for attenuation correction without acquising additional images. The associated implementation is available at https://github.com/YongshengPan/ASEG-for-PET2CT.

Keywords: PET · Attenuation correction · CT · Image generation

Supplementary Information The online version contains supplementary material available at https://doi.org/10.1007/978-3-031-43999-5_3.

1   Introduction

Positron emission tomography (PET) is a general nuclear imaging technique, which has been widely used to characterize tissue metabolism, protein deposition, etc. [9]. According to the PET imaging principle, radioactive tracers injected into the body involve in the metabolism and produce γ decay signals externally. However, due to photoelectric absorption and Compton scattering, the decay signals are attenuated when passing through human tissues to external receivers, resulting in incorrect tracer distribution reasoning (see non-attenuation corrected PET (NAC-PET) in Fig. 1(a)). To obtain correct tracer distribution (see AC-PET in Fig. 1(a)), attenuation correction (AC) on the received signals is required.

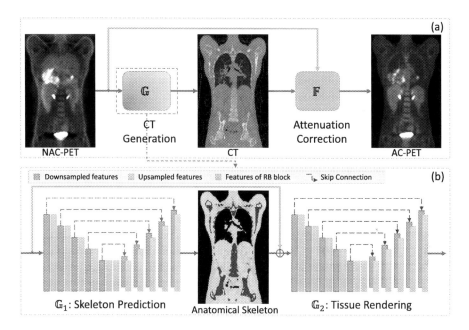

Fig. 1. ASEG framework. (a) The general process for reconstructing AC-PET by generating CT from NAC-PET. Rather than directly generating CT from NAC-PET, (b) our ASEG first delineates the anatomical skeleton and then renders the tissue details.

Traditional AC accompanies additional costs caused by the simultaneously obtained MR or CT images which are commonly useless for diagnosis. The additional costs are especially significant for advanced total-body PET/CT scanners [12,13], which have effective sensitivity and low radiation dose during PET scanning but accumulative radiation dose during CT scanning. In another word, CT becomes a non-negligible source of radiation hazards. To reduce the costs, including expense, time, and radiation hazards, some studies proposed to conduct AC by exploiting each PET image itself. Researchers have been motivated to generate pseudo CT images from NAC-PET images [2,7], or more directly, to generate AC-PET images from NAC-PET images [5,11]. Since pseudo CT is convenient to be integrated into conventional AC processes, generating pseudo CT images is feasible in clinics for AC.

The pseudo CT images should satisfy two-fold requests. Firstly, the pseudo CT images should be visually similar in anatomical structures to corresponding actual CT images. Secondly, PET images corrected by pseudo CT images should be consistent with that corrected by actual CT images. However, current techniques of image generation tend to produce statistical average values and patterns, which easily erase significant tissues (e.g., bones and lungs). As a result, for those tissues with relatively similar metabolism but large variances in attenuation coefficient, these methods could cause large errors as they are blind to the correct tissue distributions. Therefore, special techniques should be investigated to guarantee the fidelity of anatomical structures in these generated pseudo CT images.

In this paper, we propose a deep learning framework, named anatomical skeleton enhanced generation (ASEG), to generate pseudo CT from NAC-PET for attenuation correction. ASEG focuses more on the fidelity of tissue distribution, i.e., anatomical skeleton, in pseudo CT images. As shown in Fig. 1(b), this framework contains two sequential modules structure prediction module $\mathbb{G}_1$ and tissue rendering module $\mathbb{G}_2$}. $\mathbb{G}_1$ devotes to delineating the anatomical skeleton from a NAC-PET image, thus producing a prior tissue distribution map to $\mathbb{G}_2$, while $\mathbb{G}_2$ devotes to rendering the tissue details according to both the skeleton and NAC-PET image. We regard $\mathbb{G}_1$ as a segmentation network that is trained under the combination of cross-entropy loss and Dice loss and outputs the anatomical skeleton. For training the generative module $\mathbb{G}_2$, we further propose the *anatomical-consistency constraint* to guarantee the fidelity of tissue distribution besides general constraints in previous studies. Experiments on four publicly collected PET/CT datasets demonstrate that our ASEG outperforms existing methods by preserving better anatomical structures in generated pseudo CT images and achieving better visual similarity in corrected PET images.

2   Method

We propose the anatomical skeleton enhanced generation (ASEG, as illustrated in Fig. (1) framework that regards the CT generation as two sequential tasks, i.e., skeleton prediction and tissue rendering, instead of simply mapping pseudo CT from NAC-PET. ASEG composes of two sequential generative modules $\{\mathbb{G}_1, \mathbb{G}_2\}$ to deal with them, respectively. $\mathbb{G}_1$ devotes itself to decoupling the anatomical skeleton from NAC-PET to provide rough prior information of attenuation coefficients to $\mathbb{G}_2$, particularly for lungs and bones that have the most influential variances. $\mathbb{G}_2$ then devotes to rendering the tissue details in the CT pattern exploiting both the skeleton and NAC-PET images. In short, the skeleton decoupled by $\mathbb{G}_1$ is a prior guidance to $\mathbb{G}_2$, and in turn, $\mathbb{G}_2$ can serve as a target supervision for $\mathbb{G}_1$. These two modules are trained with different constraints according to the corresponding tasks. Specially, the general Dice loss and cross-entropy loss [16] are employed to guarantee $\mathbb{G}_1$ for the fidelity of tissue distributions while general *mean absolute error* and *feature matching* losses are utilized to guarantee $\mathbb{G}_2$ for potential coarse-to-fine semantic constraint. To improve the fidelity of anatomical structures, we further propose the *anatomical consistency loss* to encourage $\mathbb{G}_2$ to generate CT images that are consistent in tissue distributions with actual CT images in particular.

Network Architecture. As illustrated in Fig. 1(b), our ASEG has two generative modules for skeleton prediction and tissue rendering, respectively, where $\mathbb{G}_1$ and $\mathbb{G}_2$ share the same network structure but $\mathbb{G}_2$ is accompanied by an adversarial network $\mathbb{D}$ (not drawn, same structure in [8]). Each generative network consists of an input convolutional layer, four encoding blocks, two residual blocks (RBs) [8], four decoding blocks, and an output convolutional layer. Each encoding block contains a RB and a convolutional layer with strides of $2 \times 2 \times 2$ for downsampling while each decoding block contains an upsampling operation of $2 \times 2 \times 2$ and a convolutional layer. The kernel size for the input and output convolutional layers is $7 \times 7 \times 7$ while for others is $3 \times 3 \times 3$. Skip connections are further used locally in RBs and globally between corresponding layers to empower information transmission. Meanwhile, the adversarial network $\mathbb{D}$ consists of five $4 \times 4 \times 4$ convolutional layers with strides of $2 \times 2 \times 2$ for the first four layers and $1 \times 1 \times 1$ for the last layer.

Model Formulation. Let X_{nac} and X_{ac} denote the NAC-PET and AC-PET images, and Y be the actual CT image used for AC. Since CT image is highly crucial in conventional AC algorithms, they generally have a relationship as

$$X_{ac} = \mathbb{F}(X_{nac}, Y), \tag{1}$$

under an AC algorithm $\mathbb{F}$. To avoid scanning an additional CT image, we attempt to predict Y from X_{nac} as an alternative in AC algorithm. Namely, a mapping $\mathbb{G}$ is required to build the relationship between Y and X_{nac}, i.e., $\hat{Y} = \mathbb{G}(X_{nac})$. Then, X_{ac} can be acquired by

$$X_{ac} \approx \hat{X}_{ac} = \mathbb{F}(X_{nac}, \hat{Y}) = \mathbb{F}(X_{nac}, \mathbb{G}(X_{nac})). \tag{2}$$

This results in a pioneering AC algorithm that requires only a commonly reusable mapping function $\mathbb{G}$ for all PET images rather than a corresponding CT image Y for each PET image.

As verified in some previous studies [1,2,7], $\mathbb{G}$ can be assigned by some image generation techniques, e.g. GANs and CNNs. However, since these general techniques tend to produce statistical average values, directly applying them may lead to serious brightness deviation, for those tissues with large intensity ranges. To overcome this drawback, we propose ASEG as a specialized AC technique, which decouple the CT generation process $\mathbb{G}$ in two sequential parts, i.e., $\mathbb{G}_1$ for skeleton prediction and $\mathbb{G}_2$ for tissue rendering, as formulated as

$$\hat{Y} = \mathbb{G}(X_{nac}) = \mathbb{G}_2(Y_{as}, X_{nac}) \approx \mathbb{G}_2(\hat{Y}_{as}, X_{nac}) = \mathbb{G}_2(\mathbb{G}_1(X_{nac}), X_{nac}). \tag{3}$$

Herein, $\mathbb{G}_1$ devotes to delineating anatomical skeleton Y_{as} from X_{nac}, thus providing a prior tissue distribution to $\mathbb{G}_2$ while $\mathbb{G}_2$ devotes to rendering the tissue details from X_{nac} and $\hat{Y}_{as} = \mathbb{G}_1(X_{nac})$.

To avoid annotating the ground truth, Y_{as} can be derived from the actual CT image by a segmentation algorithm (denoted as $\mathbb{S} : Y_{as} = \mathbb{S}(Y)$). As different tissues have obvious differences in intensity ranges, we define $\mathbb{S}$ as a simple thresholding-based algorithm. Herein, we first smooth each non-normalized CT image with a small recursive Gaussian filter to suppress the impulse noise, and then threshold this CT image to four binary masks according to the Hounsfield scale of tissue density [6], including the air-lung mask (intensity ranges from -950HU to -125HU), the fluids-fat mask (ranges from -125HU to 10HU), the soft-tissue mask (ranges from 10HU to 100HU), and the bone mask (ranges from 100HU to 3000HU), as demonstrated by anatomical skeleton in Fig. 1(b). This binarization trick highlights the difference among different tissues, and thus is easier perceived.

General Constraints. As mentioned above, two generative modules $\{\mathbb{G}_1, \mathbb{G}_2\}$ work for two tasks, namely the skeleton prediction and tissue rendering, respectively. Thus, they are trained with different target-oriented constraints. In the training scheme, the loss function for $\mathbb{G}_1$ is the combination of Dice loss $\mathcal{L}_{dice}$ and cross-entropy loss $\mathcal{L}_{ce}$ [16], denoted as

$$\mathcal{L}_1(\hat{Y}_{as}, Y_{as}) = \mathcal{L}_{dice}(\hat{Y}_{as}, Y_{as}) + \mathcal{L}_{ce}(\hat{Y}_{as}, Y_{as}). \tag{4}$$

Meanwhile, the loss function for $\mathbb{G}_2$ combines the mean absolute error (MAE) $\mathcal{L}_{mae}$, perceptual feature matching loss $\mathcal{L}_{fm}$ [14], and anatomical-consistency loss $\mathcal{L}_{ac}$, denoted as

$$\mathcal{L}_2(\hat{Y}, Y) = \mathcal{L}_{mae}(\hat{Y}, Y) + \mathcal{L}_{fm}(\hat{Y}; \mathbb{D}) + \mathcal{L}_{ac}(\hat{Y}, Y). \tag{5}$$

where the anatomical consistency loss $\mathcal{L}_{st}$ is explained below.

Anatomical consistency. It is generally known that CT images can provide anatomical observation because different tissues have a distinctive appearance in Hounsfield scale (linear related to attenuation coefficients). Therefore, it is crucial to ensure the consistency of tissue distribution in the pseudo CT images, tracking which we propose to use the tissue distribution consistency to guide the network learning. Based on the segmentation algorithm $\mathbb{S}$, both the actual and generated CTs $\{Y, \hat{Y}\}$ can be segmented to anatomical structure/tissue distribution masks $\{\mathbb{S}(Y), \mathbb{S}(\hat{Y})\}$, and their consistency can then be measured by Dice coefficient. Accordingly, the anatomical-consistency loss $\mathcal{L}_{ac}$ is a Dice loss as

$$\mathcal{L}_{ac}(Y, \hat{Y}) = \mathcal{L}_{dice}(\mathbb{S}(\hat{Y}), \mathbb{S}(Y)). \tag{6}$$

During the inference phase, only the NAC-PET image of each input subject is required, where the pseudo CT image is derived by $\hat{Y} \approx \mathbb{G}_2(\mathbb{G}_1(X_{nac}), X_{nac})$.

3 Experiments

3.1 Materials

The data used in our experiments are collected from The Cancer Image Archive (TCIA) [4] (https://www.cancerimagingarchive.net/collections/), where a series of public datasets with different types of lesions, patients, and scanners are open-access. Among them, 401, 108, 46, and 20 samples are extracted from the Head and Neck Scamorous Cell Carcinoma (HNSCC), Non-Small Cell Lung Cancer (NSCLC), The Cancer Genome Atlas (TCGA) - Head-Neck Squamous Cell Carcinoma (TCGA-HNSC), and TCGA - Lung Adenocarcinoma (TCGA-LUAD), respectively. We use these samples in HNSCC for training and in other three datasets for evaluation.

Each sample contains co-registered (acquired with PET-CT scans) CT, PET, and NAC-PET whole-body scans. In our experiments, we re-sampled all of them to a voxel spacing of $2 \times 2 \times 2$ and re-scaled the intensities of NAC-PET/AC-PET images to a range of $[0, 1]$, of CT images by multiplying 0.001. The input and output of our ASEG framework are cropped patches with the size of $192 \times 192 \times 128$ voxels. To achieve full-FoV output, the consecutive outputs of each sample are composed into a single volume where the overlapped regions are averaged.

3.2 Comparison with Other Methods

We compared our ASEG with three state-of-the-art methods, including (i) a U-Net based method [3] that directly learns a mapping from NAC-PET to CT image with MAE loss (denoted as U-Net), (ii) a conventional GAN-based method [1,2] that uses the U-Net as the backbone and employ the style-content loss and adversarial loss as an extra constraint (denoted as CGAN), and (iii) an auxiliary GAN-based method [10] that uses the CT-based segmentation (i.e., the simple thresholding $\mathbb{S}$) as an auxiliary task for CT generation (denoted as AGAN). For a fair comparison, we implemented these methods by ourselves in a TensorFlow platform with an NVIDIA 3090 GPU. All methods share the same backbone structure as $\mathbb{G}_*$ in Fig. 1(b) and follow the same experimental settings. Particularly, the adversarial loss of methods (ii) and (iii) are replaced by the perceptual feature matching loss. These two methods could be considered as variants of our method without using predicted prior anatomic skeleton.

Quantitative Analysis of CT. As the most import application of CT that is to display the anatomical information, we propose to measure the anatomical consistency between the pseudo CT images and actual CT images, where the Dice coefficients on multiple anatomical regions that extracted from the pseudo/actual CT images are calculated. To avoid excessive self-referencing in evaluating anatomical consistency, instead of employing the simple thresholding segmentation (i.e., $\mathbb{S}$), we resort to the open-access TotalSegmentator [15] to finely segment the actual and pseudo CT images to multiple anatomical structures, and compose them to nine independent tissues for simplifying result

Table 1. Comparison of pseudo CT images generated by different methods.

(a) Anatomical consistency (%)

Method	SSIM	Dice_l	Dice_h	Dice_k	Dice_b	Dice_d	Dice_r	Dice_v	Dice_{il}
UNet	78.40	58.34	78.89	66.91	52.23	43.00	6.83	50.19	41.81
CGAN	**80.42**	80.97	79.25	67.38	52.86	43.91	18.95	52.40	41.19
AGAN	74.71	83.40	74.41	60.26	46.71	39.25	22.39	64.58	49.52
ASEG	76.31	**85.26**	**80.63**	**75.89**	**63.78**	**52.32**	**31.60**	**69.05**	**52.79**

(b) Effectiveness in PET attenuation correction (%)

CT Source	MAE (%)	PSNR (dB)	NCC (%)	SSIM (%)
No CT	1.55±0.56	34.83±2.58	97.13±2.04	94.08±3.44
UNet	1.47±0.50	34.76±2.45	97.20±1.75	95.18±2.49
CGAN	1.52±0.49	34.51±2.33	97.05±1.77	94.92±2.49
AGAN	1.45±0.51	35.09±2.38	97.11±1.80	95.22±2.61
ASEG	**1.36±0.49**	**35.89±2.48**	**97.66±1.72**	**95.67±2.49**
Actual CT	1.20±0.48	37.28±3.37	98.25±1.56	96.10±2.51

report, e.g., lung (Dice_l), heart (Dice_h), liver (Dice_{li}), kidneys (Dice_k), blood vessels (Dice_k), digestive system (Dice_d), ribs (Dice_r), vertebras (Dice_v), and iliac bones (Dice_{ib}). Additionally, the Structure Similarity Index Measure (SSIM) values are also reported to measure the global intensity similarity.

Results of various methods are provided in Table 1(a), where the following conclusions can be drawn. Firstly, U-Net and CGAN generate CT images with slightly better global intensity similarity but worse anatomical consistency in some tissues than AGAN and ASEG. This indicates that the general constraints (MAE and perceptual feature matching) cannot preserve the tissue distribution since they tend to produce statistical average values or patterns, particularly in these regions with large intensity variants. Secondly, AGAN achieves the worst intensity similarity and anatomical consistency for some organs. Such inconsistent metrics suggest that the global intensity similarity may have a competing relationship with anatomical consistency in the learning procedure, thus it is not advisable to balance them in a single network. Thirdly, CGAN achieves better anatomical consistency than U-Net, but worse than ASEG. It implies that the perceptual feature matching loss can also identify the variants between different tissues implicitly but cannot compare to our strategy to explicitly enhance the anatomical skeleton. Fourthly, our proposed ASEG achieves the best anatomical consistency for all tissues, indicating it is reasonable to enhance tissue variations. In brief, the above results supports the strategy to decouple the skeleton prediction as a preceding task is effective for CT generation.

Effectiveness in Attenuation Correction. As the pseudo CT images generated from NAC-PET are expected to be used in AC, it is necessary to further evaluate the effectiveness of pseudo CT images in PET AC. Because we cannot access the original scatters [5], inspired by [11], we propose to resort CGAN to simulate the AC process, denoted as ACGAN and trained on HNSCC dataset. The input of ACGAN is a concatenation of NAC-PET and actual CT, while the output is actual AC-PET. To evaluate the pseudo CT images, we simply use them to take place of the actual CT. Four metrics, including the Peak Signal to Noise Ratio (PSNR), Mean Absolute Error (MAE), Normalized Cross Correlation (NCC), and SSIM, are used to measure ACGAN with pseudo CT images on test datasets (NSCLC, TCGA-HNSC, and TCGA-LUDA). The results are reported in Table 1(b), where the fourth column list the ACGAN results with actual CT images. Meanwhile, we also report the results of direct mapping NAC-PET to AC-PET without CT images in the third column ("No CT"), which is trained from scratch and independent from ACGAN.

It can be observed from Table 1(b) that: (1) ACGAN with actual CT images can predict images very close to the actual AC-PET images, thus is qualified to simulate the AC process; (2) With actual or pseudo CT images, ACGAN can predict images closer to the actual AC-PET images than without CT, demonstrating the necessity of CT images in process of PET AC; (3) These pseudo CTs cannot compare to actual CTs, reflecting that there exist some relative information that can hardly be mined from NAC-PET; (4) The pseudo CTs generated by ASEG achieve the best in three metrics (MAE, PSNR, NCC) and second in the other metric (SSIM), demonstrating the advance of our ASEG. Figure 2 displayed the detailed diversity of the AC-PET corrected by different pseudo

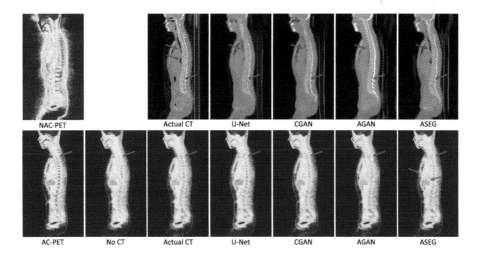

Fig. 2. Visualization of different pseudo CT images (top) and their AC effect (bottom). From left to right are actual NAC-PET/AC-PET and PET corrected without CT (no CT) or with actual or pseudo CT generated by U-Net, CGAN, AGAN, and our ASEG.

CTs. It can be found that the structures of AC-PET are highly dependent on CT, particularly the lung regions. However, errors in corners and shapes are relatively large (see these locations marked by red arrows), which indicates there are still some space in designing more advanced mapping methods. Nonetheless, compared to other pseudo CTs, these generated by ASEG result in more realistic AC-PET with fewer errors, demonstrating the AC usability of ASEG.

4   Conclusion

In this paper, we proposed the anatomical skeleton-enhance generation (ASEG) to generate pseudo CT images for PET attenuation correction (AC), with the goal of avoiding acquiring extra CT or MR images. ASEG divided the CT generation into the skeleton prediction and tissue rendering, two sequential tasks, addressed by two designed generative modules. The first module delineates the anatomical skeleton to explicitly enhance the tissue distribution which are vital for AC, while the second module renders the tissue details based on the anatomical skeleton and NAC-PET. Under the collaboration of two modules and specific *anatomical-consistency constraint,* our ASEG can generate more reasonable pseudo CT from NAC-PET. Experiments on a collection of public datasets demonstrate that our ASEG outperforms existing methods by achieving advanced performance in anatomical consistency. Our study support that ASEG could be a promising and lower-cost alternative of CT acquirement for AC. Our future work will extend our study to multiple PET tracers.

Acknowledgements. This work was supported in part by The China Postdoctoral Science Foundation (Nos. 2021M703340, BX2021333), National Natural Science Foundation of China (Nos. 62131015, 62203355), Science and Technology Commission of Shanghai Municipality (STCSM) (No. 21010502600), and The Key R&D Program of Guangdong Province, China (No. 2021B0101420006).

References

1. Armanious, K., et al.: Independent attenuation correction of whole body [18F] FDG-PET using a deep learning approach with generative adversarial networks. EJNMMI Res. **10**(1), 1–9 (2020)
2. Dong, X., et al.: Synthetic CT generation from non-attenuation corrected PET images for whole-body PET imaging. Phys. Med. Biol. **64**(21), 215016 (2019)
3. Çiçek, Ö., Abdulkadir, A., Lienkamp, S.S., Brox, T., Ronneberger, O.: 3D U-Net: learning dense volumetric segmentation from sparse annotation. In: Ourselin, S., Joskowicz, L., Sabuncu, M.R., Unal, G., Wells, W. (eds.) MICCAI 2016. LNCS, vol. 9901, pp. 424–432. Springer, Cham (2016). https://doi.org/10.1007/978-3-319-46723-8_49
4. Clark, K., Vendt, B., Smith, K., Freymann, J., Kirby, J., et al.: The cancer imaging archive (TCIA): maintaining and operating a public information repository. J. Digit. Imaging **26**(6), 1045–1057 (2013)

5. Guo, R., Xue, S., Hu, J., Sari, H., Mingels, C., et al.: Using domain knowledge for robust and generalizable deep learning-based CT-free PET attenuation and scatter correction. Nat. Commun. **13**, 5882 (2022)
6. Häggström, M.: Hounsfield units. https://radlines.org/Hounsfield_unit
7. Liu, F., Jang, H., Kijowski, R., Bradshaw, T., McMillan, A.B.: Deep learning MR imaging-based attenuation correction for PET/MR imaging. Radiology **286**(2), 676–684 (2018)
8. Pan, Y., Liu, M., Xia, Y., Shen, D.: Disease-image-specific learning for diagnosis-oriented neuroimage synthesis with incomplete multi-modality data. IEEE Trans. Pattern Anal. Mach. Intell. **44**, 6839–6853 (2021)
9. Rabinovici, G.D., Gatsonis, C., Apgar, C., Chaudhary, K., Gareen, I., et al.: Association of amyloid positron emission tomography with subsequent change in clinical management among medicare beneficiaries with mild cognitive impairment or dementia. JAMA **321**(13), 1286–1294 (2019)
10. Rodríguez Colmeiro, R., Verrastro, C., Minsky, D., Grosges, T.: Towards a whole body [18F] FDG positron emission tomography attenuation correction map synthesizing using deep neural networks. J. Comput. Sci. Technol. **21**, 29–41 (2021)
11. Shiri, I., et al.: Direct attenuation correction of brain PET images using only emission data via a deep convolutional encoder-decoder (Deep-DAC). Eur. Radiol. **29**(12), 6867–6879 (2019). https://doi.org/10.1007/s00330-019-06229-1
12. Spencer, B.A., Berg, E., Schmall, J.P., Omidvari, N., Leung, E.K., et al.: Performance evaluation of the uEXPLORER total-body PET/CT scanner based on NEMA NU 2-2018 with additional tests to characterize PET scanners with a long axial field of view. J. Nucl. Med. **62**(6), 861–870 (2021)
13. Tan, H., et al.: Total-body PET/CT: current applications and future perspectives. Am. J. Roentgenol. **215**(2), 325–337 (2020)
14. Wang, T.C., Liu, M.Y., Zhu, J.Y., Tao, A., Kautz, J., Catanzaro, B.: High-resolution image synthesis and semantic manipulation with conditional GANs. In: 2018 IEEE/CVF Conference on Computer Vision and Pattern Recognition, pp. 8798–8807 (2018)
15. Wasserthal, J., Meyer, M., Breit, H., Cyriac, J., Yang, S., Segeroth, M.: TotalSegmentator: robust segmentation of 104 anatomical structures in CT images. arXiv preprint arXiv:2208.05868 (2022)
16. Zhao, R., et al.: Rethinking dice loss for medical image segmentation. In: 2020 IEEE International Conference on Data Mining (ICDM), pp. 851–860. IEEE (2020)

LLCaps: Learning to Illuminate Low-Light Capsule Endoscopy with Curved Wavelet Attention and Reverse Diffusion

Long Bai[1], Tong Chen[2], Yanan Wu[1,3], An Wang[1], Mobarakol Islam[4], and Hongliang Ren[1,5(✉)]

[1] Department of Electronic Engineering, The Chinese University of Hong Kong (CUHK), Hong Kong SAR, China
{b.long,wa09}@link.cuhk.edu.hk
yananwu@cuhk.edu.hk
[2] The University of Sydney, Sydney, NSW, Australia
tche2095@uni.sydney.edu.au
[3] Northeastern University, Shenyang, China
[4] Wellcome/EPSRC Centre for Interventional and Surgical Sciences (WEISS), University College London, London, UK
mobarakol.islam@ucl.ac.uk
[5] Shun Hing Institute of Advanced Engineering, CUHK, Hong Kong SAR, China
hlren@ee.cuhk.edu.hk

Abstract. Wireless capsule endoscopy (WCE) is a painless and non-invasive diagnostic tool for gastrointestinal (GI) diseases. However, due to GI anatomical constraints and hardware manufacturing limitations, WCE vision signals may suffer from insufficient illumination, leading to a complicated screening and examination procedure. Deep learning-based low-light image enhancement (LLIE) in the medical field gradually attracts researchers. Given the exuberant development of the denoising diffusion probabilistic model (DDPM) in computer vision, we introduce a WCE LLIE framework based on the multi-scale convolutional neural network (CNN) and reverse diffusion process. The multi-scale design allows models to preserve high-resolution representation and context information from low-resolution, while the curved wavelet attention (CWA) block is proposed for high-frequency and local feature learning. Moreover, we combine the reverse diffusion procedure to optimize the shallow output further and generate images highly approximate to real ones. The proposed method is compared with eleven state-of-the-art (SOTA) LLIE methods and significantly outperforms quantitatively and qualitatively. The superior performance on GI disease segmentation further demon-

---

L. Bai and T. Chen—are co-first authors.

Supplementary Information The online version contains supplementary material available at https://doi.org/10.1007/978-3-031-43999-5_4.

strates the clinical potential of our proposed model. Our code is publicly accessible at github.com/longbai1006/LLCaps.

1 Introduction

Currently, the golden standard of gastrointestinal (GI) examination is endoscope screening, which can provide direct vision signals for diagnosis and analysis. Benefiting from its characteristics of being non-invasive, painless, and low physical burden, wireless capsule endoscopy (WCE) has the potential to overcome the shortcomings of conventional endoscopy [21]. However, due to the anatomical complexity, insufficient illumination, and limited performance of the camera, low-quality images may hinder the diagnosis process. Blood vessels and lesions with minor color changes in the early stages can be hard to be screened out [15]. Figure 1 shows WCE images with low illumination and contrast. The disease features clearly visible in the normal image become challenging to be found in the low-light images. Therefore, it is necessary to develop a low-light image enhancement framework for WCE to assist clinical diagnosis.

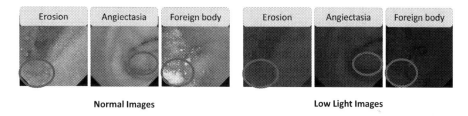

Fig. 1. Comparison of normal images with low-light images. Obvious lesions are visible on normal images, but the same lesions can hardly be distinguished by human eyes in the corresponding low-light images.

Many traditional algorithms (e.g., intensity transformation [7], histogram equalization [14], and Retinex theory [13]) have been proposed for low-light image enhancement (LLIE). For WCE, Long *et al.* [15] discussed adaptive fraction-power transformation for image enhancement. However, traditional methods usually require an ideal assumption or an effective prior, limiting their wider applications. Deep learning (DL) provides novel avenues to solve LLIE problems [6,10,16]. Some DL-based LLIE schemes for medical endoscopy have been proposed [5,18]. Gomez *et al.* [5] offered a solution for laryngoscope low-light enhancement, and Ma *et al.* [18] proposed a medical image enhancement model with unpaired training data.

Recently, denoising diffusion probabilistic model (DDPM) [8] is the most popular topic in image generation, and has achieved success in various applications. Due to its unique regression process, DDPM has a stable training process and excellent output results, but also suffers from its expensive sampling procedure and lack of low-dimensional representation [19]. It has been proved that DDPM can be combined with other existing DL techniques to speed up the sampling process [19]. In our work, we introduce the reverse diffusion process of DDPM into our end-to-end LLIE process, which can preserve image details without introducing excessive computational costs. Our contributions to this work can be summarized as three-fold:

- We design a **Low-Light** image enhancement framework for **Caps**ule endoscopy (**LLCaps**). Subsequent to the feature learning and preliminary shallow image reconstruction by the convolutional neural network (CNN), the reverse diffusion process is employed to further promote image reconstruction, preserve image details, and close in the optimization target.
- Our proposed curved wavelet attention (CWA) block can efficiently extract high-frequency detail features via wavelet transform, and conduct local representation learning with the curved attention layer.
- Extensive experiments on two publicly accessible datasets demonstrate the excellent performance of our proposed model and components. The high-level lesion segmentation tasks further show the potential power of LLCaps on clinical applications.

2   Methodology

2.1   Preliminaries

Multi-scale Residual Block. Multi-scale Residual Block (MSRB) [12] constructs a multi-scale neuronal receptive field, which allows the network to learn multi-scale spatial information in the same layer. Therefore, the network can acquire contextual information from the low-resolution features while preserving high-resolution representations. We establish our CNN branch with six stacked multi-scale residual blocks (MSRB), and every two MSRBs are followed by a 2D convolutional layer (Conv2D). Besides, each MSRB shall require feature learning and the multi-scale feature aggregation module. Specifically, we propose our curved wavelet attention (CWA) module to conduct multi-scale feature learning, and employ the selective kernel feature fusion (SKFF) [29] to combine multi-scale features, as shown in Fig. 2(a).

Denoising Diffusion Probabilistic Models. Denoising Diffusion Probabilistic Models (DDPMs) [8] can be summarised as a model consisting of a forward noise addition $q(i_{1:T}|i_0)$ and a reverse denoising process $p(i_{0:T})$, which are both parameterized Markov chains. The forward diffusion process gradually adds noise to the input image until the original input is destroyed. Correspondingly, the reverse process uses the neural network to model the Gaussian distribution and achieves image generation through gradual sampling and denoising.

2.2   Proposed Methodology

Curved Wavelet Attention. Curved Wavelet Attention (CWA) block is the core component of our CNN branch, which is constructed via a curved dual attention mechanism and wavelet transform, as shown in Fig. 2(b). Firstly, the input feature map F_{in} is divided into identity feature $F_{identity}$ and processing feature F_p. Medical LLIE shall require high image details. In this case, we transform the F_p into wavelet domain F_w to extract high-frequency detail information based on discrete wavelet transform. F_w is then propagated through the feature selector and dual attention module for deep representation learning. Finally, we conduct reverse wavelet transform (IWT) to get F'_p, and concatenate it with $F_{identity}$ before the final output convolution layer.

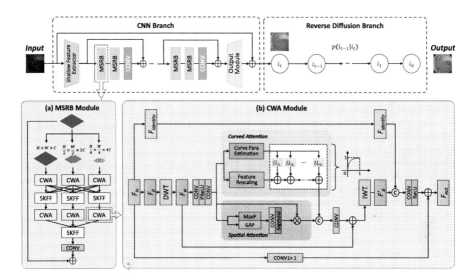

Fig. 2. The overview of our proposed LLCaps. The CNN branch shall extract the shallow image output while the DDPM branch further optimizes the image via Markov chain inference. (a) represents the multi-scale residual block (MSRB), which allows the model to learn representation on different resolutions. (b) denotes our curved wavelet attention (CWA) block for attention learning and feature restoration. In (b), DWT and IWT denote discrete wavelet transform and inverse wavelet transform, respectively. The PReLU with two convolutional layers constructs the feature selector. 'MaxP' denotes max pooling, and 'GAP' means global average pooling.

We construct our curved dual attention module with parallel spatial and curved attention blocks. The spatial attention (SA) layer exploits the inter-spatial dependencies of convolutional features [29]. The SA layer performs the global average pooling and max pooling on input features respectively, and concatenates the output $F_{w(mean)}$ and $F_{w(max)}$ to get F_{cat}. Then the feature map will be dimensionally reduced and passed through the activation function.

However, literature [6,33] has discussed the problem of local illumination in LLIE. If we simply use a global computing method such as the SA layer, the model may not be able to effectively understand the local illumination/lack of illumination. Therefore, in order to compensate for the SA layer, we design the Curved Attention (CurveA) layer, which is used to model the high-order curve of the input features. Let $\mathrm{IL}_{n(c)}$ denote the curve function, c denote the feature location coordinates, and $Curve_{(n-1)}$ denote the pixel-wise curve parameter, we can obtain the curve estimation equation as:

$$\frac{\mathrm{IL}_{n(c)}}{\mathrm{IL}_{n-1(c)}} = Curve_{n-1}(1 - \mathrm{IL}_{n-1(c)}) \tag{1}$$

The detailed CurveA layer is presented in the top of Fig. 2(b), and the Eq. (1) is related to the white area. The Curve Parameter Estimation module consists of a Sigmoid activation and several Conv2D layers, and shall estimate the pixel-wise curve parameter at each order. The Feature Rescaling module will rescale the input feature into [0, 1] to learn the concave down curves. By applying the CurveA layer to the channels of the feature map, the CWA block can better estimate local areas with different illumination.

Reverse Diffusion Process. Some works [19,26] have discussed combining diffusion models with other DL-based methods to reduce training costs and be used for downstream applications. In our work, we combine the reverse diffusion process of DDPM in a simple and ingenious way, and use it to optimize the shallow output by the CNN branch. Various experiments shall prove the effectiveness of our design in improving image quality and assisting clinical applications.

In our formulation, we assume that i_0 is the learning target Y^* and i_T is the output shallow image from the CNN branch. Therefore, we only need to engage the reverse process in our LLIE task. The reverse process is modeled using a Markov chain:

$$p_\theta (i_{0:T}) = p(i_T) \prod_{t=1}^{T} p_\theta (i_{t-1} \mid i_t)$$
$$p_\theta (i_{t-1} \mid i_t) = \mathcal{N} (i_{t-1}; \boldsymbol{\mu}_\theta (i_t, t), \boldsymbol{\Sigma}_\theta (i_t, t)) \tag{2}$$

$p_\theta (i_{t-1} \mid i_t)$ are parameterized Gaussian distributions whose mean $\boldsymbol{\mu}_\theta (i_t, t)$ and variance $\boldsymbol{\Sigma}_\theta (i_t, t)$ are given by the trained network. Meanwhile, we simplify the network and directly include the reverse diffusion process in the end-to-end training of the entire network. Shallow output is therefore optimized by the reverse diffusion branch to get the predicted image Y. We further simplify the optimization function and only employ a pixel-level loss on the final output image, which also improves the training and convergence efficiency.

Overall Network Architecture. An overview of our framework can be found in Fig. 2. Our LLCaps contains a CNN branch (including a shallow feature extractor (SFE), multi-scale residual blocks (MSRBs), an output module (OPM)), and the reverse diffusion process. The SFE is a Conv2D layer that maps the input image into the high-dimensional feature representation $F_{SFE} \in \mathbb{R}^{C \times W \times H}$ [27]. Stacked

MSRBs shall conduct deep feature extraction and learning. OPM is a Conv2D layer that recovers the feature space into image pixels. A residual connection is employed here to optimize the end-to-end training and converge process. Hence, given a low-light image $x \in \mathbb{R}^{3 \times W \times H}$, where W and H represent the width and height, the CNN branch can be formulated as:

$$F_{SFE} = H_{SFE}(x)$$
$$F_{MSRBs} = H_{MSRBs}(F_{SFE}), \tag{3}$$
$$F_{OPM} = H_{OPM}(F_{MB}) + x$$

The shallow output $F_{OPM} \in \mathbb{R}^{3 \times W \times H}$ shall further be propagated through the reverse diffusion process and achieve the final enhanced image $Y \in \mathbb{R}^{3 \times W \times H}$. The whole network is constructed in an end-to-end mode and optimized by Charbonnier loss [1]. The ε is set to 10^{-3} empirically.

$$\mathcal{L}(x, x^*) = \sqrt{\|Y - Y^*\|^2 + \varepsilon^2} \tag{4}$$

in which Y and Y^* denote the input and ground truth images, respectively.

3   Experiments

3.1   Dataset

We conduct our experiments on two publicly accessible WCE datasets, the Kvasir-Capsule [22] and the Red Lesion Endoscopy (RLE) dataset [3].

Kvasir-Capsule dataset [22] is a WCE classification dataset with three anatomy classes and eleven luminal finding classes. By following [2], we randomly select 2400 images from the Kvasir-Capsule dataset, of which 2000 are used for training and 400 for testing. To create low-light images, we adopt random Gamma correction and illumination reduction following [11,16]. Furthermore, to evaluate the performance on real data, we add an external validation on 100 real images selected from the Kvasir-Capsule dataset. These images are with low brightness and are not included in our original experiments.

Red Lesion Endoscopy dataset [3] (RLE) is a WCE dataset for red lesion segmentation tasks (e.g., angioectasias, angiodysplasias, and bleeding). We randomly choose 1283 images, of which 946 images are used for training and 337 for testing. We adopt the same method in the Kvasir-Capsule dataset to generate low-light images. Furthermore, we conduct a segmentation task on the RLE test set to investigate the effectiveness of the LLIE models in clinical applications.

3.2   Implementation Details

We compare the performance of our LLCaps against the following state-of-the-art (SOTA) LLIE methodologies: LIME [7], DUAL [31], Zero-DCE [6], Enlight-enGAN [10], LLFlow [24], HWMNet [4], MIRNet [29], SNR-Aware [28], Still-GAN [17], MIRNetv2 [30], and DDPM [8]. Our models are trained using Adam optimizer for 200 epochs with a batch size of 4 and a learning rate of 1×10^{-4}. For evaluation, we adopt three commonly used image quality assessment metrics: Peak Signal-to-Noise Ratio (PSNR) [9], Structural Similarity Index (SSIM) [25], and Learned Perceptual Image Patch Similarity (LPIPS) [32]. For the external validation set, we evaluate with no-reference metrics LPIPS [32] and Perception-based Image Quality Evaluator (PIQE) [23] due to the lack of ground truth images. To verify the usefulness of the LLIE methods for downstream medical tasks, we conduct red lesion segmentation on the RLE test set and evaluate the performance via mean Intersection over Union (mIoU), Dice similarity coefficient (Dice), and Hausdorff Distance (HD). We train UNet [20] using Adam optimizer for 20 epochs. The batch size and learning rate are set to 4 and 1×10^{-4}, respectively. All experiments are implemented by Python PyTorch and conducted on NVIDIA RTX 3090 GPU. Results are the average of 3-fold cross-validation.

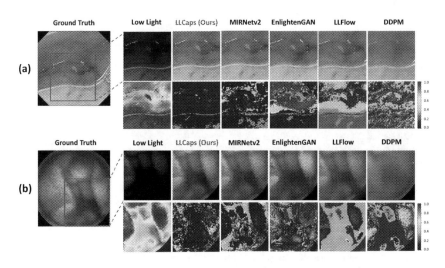

Fig. 3. The quantitative results for LLCaps compared with SOTA approaches on (a) Kvasir-Capsule dataset [22] and (b) RLE dataset [3]. The first row visualizes the enhanced images from different LLIE approaches, and the second row contains the reconstruction error heat maps. The blue and red represent low and high error, respectively. (Color figure online)

3.3   Results

We compare the performance of our LLCaps to the existing approaches, as demonstrated in Table 1 and Fig. 3 quantitatively and qualitatively. Compared with other methods, our proposed method achieves the best performance among all metrics. Specifically, our method surpasses MIRNetv2 [30] by 3.57 dB for the Kvasir-Capsule dataset and 0.33 dB for the RLE dataset. The SSIM of our method has improved to 96.34% in the Kvasir-Capsule dataset and 93.34% in the RLE dataset. Besides that, our method also performs the best in the no-reference metric LPIPS. The qualitative results of the comparison methods and

Table 1. Image quality comparison with existing methods on Kvasir-Capsule [22] and RLE dataset [3]. The 'External Val' denotes the external validation experiment conducted on 100 selected real low-light images from the Kvasir-Capsule dataset [22]. The red lesion segmentation experiment is also conducted on RLE test set [3].

Models	Kvasir-Capsule			RLE			External Val		RLE Segmentation		
	PSNR ↑	SSIM ↑	LPIPS ↓	PSNR ↑	SSIM ↑	LPIPS ↓	LPIPS ↓	PIQE ↓	mIoU ↑	Dice ↑	HD ↓
LIME [7]	12.07	29.66	0.4401	14.21	15.93	0.5144	0.3498	26.41	60.19	78.42	56.20
DUAL [31]	11.61	29.01	0.4532	14.64	16.11	0.4903	0.3305	25.47	61.89	78.15	55.70
Zero-DCE [6]	14.03	46.31	0.4917	14.86	34.18	0.4519	0.6723	21.47	54.77	71.46	56.24
EnlightenGAN [10]	27.15	85.03	0.1769	23.65	80.51	0.1864	0.4796	34.75	61.97	74.15	54.89
LLFlow [24]	29.69	92.57	0.0774	25.93	85.19	0.1340	0.3712	35.67	61.06	**78.55**	60.04
HWMNet [4]	27.62	92.09	0.1507	21.81	76.11	0.3624	0.5089	35.37	56.48	74.17	59.90
MIRNet [29]	31.23	95.77	0.0436	25.77	86.94	0.1519	0.3485	34.28	59.84	78.32	63.10
StillGAN [17]	28.28	91.30	0.1302	26.38	83.33	0.1860	0.3095	38.10	58.32	71.56	55.02
SNR-Aware [28]	30.32	94.92	0.0521	27.73	88.44	0.1094	0.3992	26.82	58.95	70.26	57.73
MIRNetv2 [30]	31.67	95.22	0.0486	32.85	92.69	0.0781	0.3341	41.24	63.14	75.07	53.71
DDPM [8]	25.17	73.16	0.4098	22.97	70.31	0.4198	0.5069	43.64	54.09	75.10	67.54
LLCaps (Ours)	**35.24**	**96.34**	**0.0374**	**33.18**	**93.34**	**0.0721**	**0.3082**	**20.67**	**66.47**	78.47	**44.37**

our method on the Kvasir-Capsule and RLE datasets are visualized in Fig. 3 with the corresponding heat maps. Firstly, we can see that directly performing LLIE training on DDPM [8] cannot obtain good image restoration, and the original structures of the DDPM images are largely damaged. EnlightenGAN [10] also does not perform satisfactorily in structure restoration. Our method successfully surpasses LLFlow [24] and MIRNetv2 [30] in illumination restoration. The error heat maps further reflect the superior performance of our method in recovering the illumination and structure from low-light images. Moreover, our solution yields the best on the real low-light dataset during the external validation, proving the superior performance of our solution in real-world applications.

Furthermore, a downstream red lesion segmentation task is conducted to investigate the usefulness of our LLCaps on clinical applications. As illustrated in Table 1, LLCaps achieve the best lesion segmentation results, manifesting the superior performance of our LLCaps model in lesion segmentation. Additionally, LLCaps surpasses all SOTA methods in HD, showing LLCaps images perform perfectly in processing the segmentation boundaries, suggesting that our method possesses better image reconstruction and edge retention ability.

Besides, an ablation study is conducted on the Kvasir-Capsule dataset to demonstrate the effectiveness of our design and network components, as shown in Table 2. To observe and compare the performance changes, we try to (i) remove the wavelet transform in CWA blocks, (ii) degenerate the curved attention (CurveA) layer in CWA block to a simple channel attention layer [29], and (iii) remove the reverse diffusion branch. Experimental results demonstrate that the absence of any component shall cause great performance degradation. The significant improvement in quantitative metrics is a further testament to the effectiveness of our design for each component.

Table 2. Ablation experiments of our LLCaps on the Kvasir-Capsule Dataset [22]. In order to observe the performance changes, we (i) remove the wavelet transform, (ii) degenerate the CurveA layer, and (iii) remove the reverse diffusion branch.

Wavelet Transform	Curve Attention	Reverse Diffusion	PSNR ↑	SSIM ↑	LPIPS ↓
✗	✗	✗	31.12	94.96	0.0793
✓	✗	✗	32.78	96.26	0.0394
✗	✓	✗	32.08	96.27	0.0415
✗	✗	✓	33.10	94.53	0.0709
✓	✓	✗	33.92	96.20	0.0381
✓	✗	✓	34.07	95.61	0.0518
✗	✓	✓	33.41	95.03	0.0579
✓	✓	✓	**35.24**	**96.34**	**0.0374**

4   Conclusion

We present LLCaps, an end-to-end capsule endoscopy LLIE framework with multi-scale CNN and reverse diffusion process. The CNN branch is constructed by stacked MSRB modules, in which the core CWA block extracts high-frequency detail information through wavelet transform, and learns the local representation of the image via the Curved Attention layer. The reverse diffusion process further optimizes the shallow output, achieving the closest approximation to the real image. Comparison and ablation studies prove that our method and design bring about superior performance improvement in image quality. Further medical image segmentation experiments demonstrate the reliability of our method in clinical applications. Potential future works include extending our model to various medical scenarios (e.g., surgical robotics, endoscopic navigation, augmented reality for surgery) and clinical deep learning model deployment.

Acknowledgements. This work was supported by Hong Kong RGC CRF C4063-18G, CRF C4026-21GF, RIF R4020-22, GRF 14216022, GRF 14211420, NSFC/RGC JRS N_CUHK420/22; Shenzhen-Hong Kong-Macau Technology Research Programme (Type C 202108233000303); GBABF #2021B1515120035.

References

1. Charbonnier, P., Blanc-Feraud, L., Aubert, G., Barlaud, M.: Two deterministic half-quadratic regularization algorithms for computed imaging. In: Proceedings of 1st International Conference on Image Processing, vol. 2, pp. 168–172. IEEE (1994)
2. Chen, W., Liu, Y., Hu, J., Yuan, Y.: Dynamic depth-aware network for endoscopy super-resolution. IEEE J. Biomed. Health Inform. **26**(10), 5189–5200 (2022)
3. Coelho, P., Pereira, A., Leite, A., Salgado, M., Cunha, A.: A deep learning approach for red lesions detection in video capsule endoscopies. In: Campilho, A., Karray, F.,

ter Haar Romeny, B. (eds.) ICIAR 2018. LNCS, vol. 10882, pp. 553–561. Springer, Cham (2018). https://doi.org/10.1007/978-3-319-93000-8_63

4. Fan, C.M., Liu, T.J., Liu, K.H.: Half wavelet attention on M-Net+ for low-light image enhancement. In: 2022 IEEE International Conference on Image Processing (ICIP), pp. 3878–3882. IEEE (2022)

5. Gómez, P., Semmler, M., Schützenberger, A., Bohr, C., Döllinger, M.: Low-light image enhancement of high-speed endoscopic videos using a convolutional neural network. Med. Biol. Eng. Comput. **57**(7), 1451–1463 (2019). https://doi.org/10.1007/s11517-019-01965-4

6. Guo, C., et al.: Zero-reference deep curve estimation for low-light image enhancement. In: Proceedings of the IEEE/CVF Conference on Computer Vision and Pattern Recognition, pp. 1780–1789 (2020)

7. Guo, X., Li, Y., Ling, H.: LIME: low-light image enhancement via illumination map estimation. IEEE Trans. Image Process. **26**(2), 982–993 (2016)

8. Ho, J., Jain, A., Abbeel, P.: Denoising diffusion probabilistic models. Adv. Neural. Inf. Process. Syst. **33**, 6840–6851 (2020)

9. Huynh-Thu, Q., Ghanbari, M.: Scope of validity of PSNR in image/video quality assessment. Electron. Lett. **44**(13), 800–801 (2008)

10. Jiang, Y., et al.: EnlightenGAN: deep light enhancement without paired supervision. IEEE Trans. Image Process. **30**, 2340–2349 (2021)

11. Li, C., et al.: Low-light image and video enhancement using deep learning: a survey. IEEE Trans. Pattern Anal. Mach. Intell. **44**(12), 9396–9416 (2021)

12. Li, J., Fang, F., Mei, K., Zhang, G.: Multi-scale residual network for image super-resolution. In: Proceedings of the European Conference on Computer Vision (ECCV), pp. 517–532 (2018)

13. Li, M., Liu, J., Yang, W., Sun, X., Guo, Z.: Structure-revealing low-light image enhancement via robust retinex model. IEEE Trans. Image Process. **27**(6), 2828–2841 (2018)

14. Liu, Y.F., Guo, J.M., Yu, J.C.: Contrast enhancement using stratified parametric-oriented histogram equalization. IEEE Trans. Circuits Syst. Video Technol. **27**(6), 1171–1181 (2016)

15. Long, M., Li, Z., Xie, X., Li, G., Wang, Z.: Adaptive image enhancement based on guide image and fraction-power transformation for wireless capsule endoscopy. IEEE Trans. Biomed. Circuits Syst. **12**(5), 993–1003 (2018)

16. Lore, K.G., Akintayo, A., Sarkar, S.: LLNet: a deep autoencoder approach to natural low-light image enhancement. Pattern Recogn. **61**, 650–662 (2017)

17. Ma, Y., et al.: Structure and illumination constrained GAN for medical image enhancement. IEEE Trans. Med. Imaging **40**(12), 3955–3967 (2021)

18. Ma, Y., et al.: Cycle structure and illumination constrained GAN for medical image enhancement. In: Martel, A.L., et al. (eds.) MICCAI 2020, Part II. LNCS, vol. 12262, pp. 667–677. Springer, Cham (2020). https://doi.org/10.1007/978-3-030-59713-9_64

19. Pandey, K., Mukherjee, A., Rai, P., Kumar, A.: DiffuseVAE: efficient, controllable and high-fidelity generation from low-dimensional latents. arXiv preprint arXiv:2201.00308 (2022)

20. Ronneberger, O., Fischer, P., Brox, T.: U-net: convolutional networks for biomedical image segmentation. In: Navab, N., Hornegger, J., Wells, W.M., Frangi, A.F. (eds.) MICCAI 2015, Part III. LNCS, vol. 9351, pp. 234–241. Springer, Cham (2015). https://doi.org/10.1007/978-3-319-24574-4_28

21. Sliker, L.J., Ciuti, G.: Flexible and capsule endoscopy for screening, diagnosis and treatment. Expert Rev. Med. Devices **11**(6), 649–666 (2014)

22. Smedsrud, P.H., et al.: Kvasir-capsule, a video capsule endoscopy dataset. Sci. Data **8**(1), 142 (2021)
23. Venkatanath, N., Praneeth, D., Bh, M.C., Channappayya, S.S., Medasani, S.S.: Blind image quality evaluation using perception based features. In: 2015 Twenty First National Conference on Communications (NCC), pp. 1–6. IEEE (2015)
24. Wang, Y., Wan, R., Yang, W., Li, H., Chau, L.P., Kot, A.: Low-light image enhancement with normalizing flow. In: Proceedings of the AAAI Conference on Artificial Intelligence, vol. 36, pp. 2604–2612 (2022)
25. Wang, Z., Bovik, A.C., Sheikh, H.R., Simoncelli, E.P.: Image quality assessment: from error visibility to structural similarity. IEEE Trans. Image Process. **13**(4), 600–612 (2004)
26. Wu, J., Fu, R., Fang, H., Zhang, Y., Xu, Y.: MedSegDiff-V2: diffusion based medical image segmentation with transformer. arXiv preprint arXiv:2301.11798 (2023)
27. Xiao, T., Singh, M., Mintun, E., Darrell, T., Dollár, P., Girshick, R.: Early convolutions help transformers see better. Adv. Neural. Inf. Process. Syst. **34**, 30392–30400 (2021)
28. Xu, X., Wang, R., Fu, C.W., Jia, J.: SNR-aware low-light image enhancement. In: Proceedings of the IEEE/CVF Conference on Computer Vision and Pattern Recognition, pp. 17714–17724 (2022)
29. Zamir, S.W., et al.: Learning enriched features for real image restoration and enhancement. In: Vedaldi, A., Bischof, H., Brox, T., Frahm, J.-M. (eds.) ECCV 2020, Part XXV. LNCS, vol. 12370, pp. 492–511. Springer, Cham (2020). https://doi.org/10.1007/978-3-030-58595-2_30
30. Zamir, S.W., et al.: Learning enriched features for fast image restoration and enhancement. IEEE Trans. Pattern Anal. Mach. Intell. **45**(2), 1934–1948 (2022)
31. Zhang, Q., Nie, Y., Zheng, W.S.: Dual illumination estimation for robust exposure correction. In: Computer Graphics Forum, vol. 38, pp. 243–252 (2019)
32. Zhang, R., Isola, P., Efros, A.A., Shechtman, E., Wang, O.: The unreasonable effectiveness of deep features as a perceptual metric. In: Proceedings of the IEEE Conference on Computer Vision and Pattern Recognition, pp. 586–595 (2018)
33. Zhou, S., Li, C., Change Loy, C.: LEDNet: joint low-light enhancement and deblurring in the dark. In: Avidan, S., Brostow, G., Cissé, M., Farinella, G.M., Hassner, T. (eds.) ECCV 2022. LNCS, vol. 13666, pp. 573–589. Springer, Cham (2022). https://doi.org/10.1007/978-3-031-20068-7_33

An Explainable Deep Framework: Towards Task-Specific Fusion for Multi-to-One MRI Synthesis

Luyi Han[1,2], Tianyu Zhang[1,2,3], Yunzhi Huang[5], Haoran Dou[6], Xin Wang[2,3], Yuan Gao[2,3], Chunyao Lu[1,2], Tao Tan[2,4(✉)], and Ritse Mann[1,2]

[1] Department of Radiology and Nuclear Medicine, Radboud University Medical Centre, Nijmegen, The Netherlands
[2] Department of Radiology, Netherlands Cancer Institute (NKI), Amsterdam, The Netherlands
taotanjs@gmail.com
[3] GROW School for Oncology and Development Biology, Maastricht University, P. O. Box 616, 6200 MD Maastricht, The Netherlands
[4] Faculty of Applied Sciences, Macao Polytechnic University, Macao 999078, China
[5] Institute for AI in Medicine, School of Artificial Intelligence, Nanjing University of Information Science and Technology, Nanjing, China
[6] Centre for Computational Imaging and Simulation Technologies in Biomedicine (CISTIB), School of Computing, University of Leeds, Leeds, UK

Abstract. Multi-sequence MRI is valuable in clinical settings for reliable diagnosis and treatment prognosis, but some sequences may be unusable or missing for various reasons. To address this issue, MRI synthesis is a potential solution. Recent deep learning-based methods have achieved good performance in combining multiple available sequences for missing sequence synthesis. Despite their success, these methods lack the ability to quantify the contributions of different input sequences and estimate region-specific quality in generated images, making it hard to be practical. Hence, we propose an explainable task-specific synthesis network, which adapts weights automatically for specific sequence generation tasks and provides interpretability and reliability from two sides: (1) visualize and quantify the contribution of each input sequence in the fusion stage by a trainable task-specific weighted average module; (2) highlight the area the network tried to refine during synthesizing by a task-specific attention module. We conduct experiments on the BraTS2021 dataset of 1251 subjects, and results on arbitrary sequence synthesis indicate that the proposed method achieves better performance than the state-of-the-art methods. Our code is available at https://github.com/fiy2W/mri_seq2seq.

Keywords: Missing-Sequence MRI Synthesis · Explainable Synthesis · Multi-Sequence Fusion · Task-Specific Attention

---

T. Zhang and L. Han—Contributed equally to this work.

Supplementary Information The online version contains supplementary material available at https://doi.org/10.1007/978-3-031-43999-5_5.

1   Introduction

Magnetic resonance imaging (MRI) consists of a series of pulse sequences, *e.g.* T1-weighted (T1), contrast-enhanced (T1Gd), T2-weighted (T2), and T2-fluid-attenuated inversion recovery (Flair), each showing various contrast of water and fat tissues. The intensity contrast combination of multi-sequence MRI provides clinicians with different characteristics of tissues, extensively used in disease diagnosis [16], lesion segmentation [17], treatment prognosis [7], *etc.* However, some acquired sequences are unusable or missing in clinical settings due to incorrect machine settings, imaging artifacts, high scanning costs, time constraints, contrast agents allergies, and different acquisition protocols between hospitals [5]. Without rescanning or affecting the downstream pipelines, the MRI synthesis technique can generate missing sequences by leveraging redundant shared information between multiple sequences [18].

Many studies have demonstrated the potential of deep learning methods for image-to-image synthesis in the field of both nature images [8,11,12] and medical images [2,13,19]. Most of these works introduce an autoencoder-like architecture for image-to-image translation and employ adversarial loss to generate more realistic images. Unlike these one-to-one approaches, MRI synthesis faces the challenge of fusing complementary information from multiple input sequences. Recent studies about multi-sequence fusion can specifically be divided into two groups: (1) image fusion and (2) feature fusion. The image fusion approach is to concatenate sequences as a multi-channel input. Sharma *et al.* [18] design a network with multi-channel input and output, which combines all the available sequences and reconstructs the complete sequences at once. Li *et al.* [14] add an availability condition branch to guide the model to adapt features for different input combinations. Dalmaz *et al.* [9] equip the synthesis model with residual transformer blocks to learn contextual features. Image-level fusion is simple and efficient but unstable – zero-padding inputs for missing sequences lead to training unstable and slight misalignment between images can easily cause artifacts. In contrast, efforts have been made on feature fusion, which can alleviate the discrepancy across multiple sequences, as high-level features focus on the semantic regions and are less affected by input misalignment compared to images. Zhou *et al.* [23] design operation-based (*e.g.* summation, product, maximization) fusion blocks to densely combine the hierarchical features. And Li *et al.* [15] employ self-attention modules to integrate multi-level features. The model architectures of these methods are not flexible and difficult to adapt to various sequence combinations. More importantly, recent studies only focus on proposing end-to-end models, lacking quantifying the contributions for different sequences and estimating the qualities of generated images.

In this work, we propose an explainable task-specific fusion sequence-to-sequence (TSF-Seq2Seq) network, which has adaptive weights for specific synthesis tasks with different input combinations and targets. Specially, this framework can be easily extended to other tasks, such as segmentation. Our primary contributions are as follows: (1) We propose a flexible network to synthesize the target MRI sequence from an arbitrary combination of inputs; (2) The network

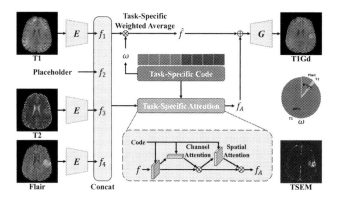

Fig. 1. Overview of the TSF-Seq2Seq network. By giving the task-specific code, TSF-Seq2Seq can synthesize a target sequence from existing sequences, and meanwhile, output the weight of input sequences ω and the task-specific enhanced map (TSEM).

shows interpretability for fusion by quantifying the contribution of each input sequence; (3) The network provides reliability for synthesis by highlighting the area the network tried to refine.

2   Methods

Figure 1 illustrates the overview of the proposed TSF-Seq2Seq network. Our network has an autoencoder-like architecture including an encoder $\mathbf{E}$, a multi-sequence fusion module, and a decoder $\mathbf{G}$. Available MRI sequences are first encoded to features by $\mathbf{E}$, respectively. Then features from multiple input sequences are fused by giving the task-specific code, which identifies sources and targets with a binary code. Finally, the fused features are decoded to the target sequence by $\mathbf{G}$. Furthermore, to explain the mechanism of multi-sequence fusion, our network can quantify the contributions of different input sequences with the task-specific weighted average module and visualize the TSEM with the task-specific attention module.

To leverage shared information between sequences, we use $\mathbf{E}$ and $\mathbf{G}$ from Seq2Seq [10], which is a one-to-one synthetic model that integrates arbitrary sequence synthesis into single $\mathbf{E}$ and $\mathbf{G}$. They can reduce the distance between different sequences at the feature level to help more stable fusion. Details of the multi-sequence fusion module and TSEM are described in the following sections.

2.1   Multi-sequence Fusion

Define a set of N sequences MRI: $\mathcal{X} = \{X_i | i = 1, ..., N\}$ and corresponding available indicator $\mathcal{A} \subset \{1, ..., N\}$ and $\mathcal{A} \neq \varnothing$. Our goal is to predict the target set $\mathcal{X}_T = \{X_i | i \notin \mathcal{A}\}$ by giving the available set $\mathcal{X}_A = \{X_i | i \in \mathcal{A}\}$ and the corresponding task-specific code $c = \{c_{\text{src}}, c_{\text{tgt}}\} \in \mathbb{Z}^{2N}$. As shown in Fig. 1, c_{src} and

c_{tgt} are zero-one codes for the source and the target set, respectively. To fuse multiple sequences at the feature level, we first encode images and concatenate the features as $\vec{f} = \{\mathbf{E}(X_i) | i = 1, ..., N\}$. Specifically, we use zero-filled placeholders with the same shape as $\mathbf{E}(X_i)$ to replace features of $i \notin \mathcal{A}$ to handle arbitrary input sequence combinations. The multi-sequence fusion module includes: (1) a task-specific weighted average module for the linear combination of available features; (2) a task-specific attention module to refine the fused features.

Task-Specific Weighted Average. The weighted average is an intuitive fusion strategy that can quantify the contribution of different sequences directly. To learn the weight automatically, we use a trainable fully connected (FC) layer to predict the initial weight $\omega_0 \in \mathbb{R}^N$ from c.

$$\omega_0 = \text{softmax}(c\mathbf{W} + \mathbf{b}) + \epsilon \tag{1}$$

where $\mathbf{W}$ and $\mathbf{b}$ are weights and bias for the FC layer, $\epsilon = 10^{-5}$ to avoid dividing 0 in the following equation. To eliminate distractions and accelerate training, we force the weights of missing sequences in ω_0 to be 0 and guarantee the output $\omega \in \mathbb{R}^N$ to sum to 1.

$$\omega = \frac{\omega_0 \cdot c_{\text{src}}}{\langle \omega_0, c_{\text{src}} \rangle} \tag{2}$$

where $\cdot$ refers to the element-wise product and $\langle \cdot, \cdot \rangle$ indicates the inner product. With the weights ω, we can fuse multi-sequence features as $\hat{f}$ by the linear combination.

$$\hat{f} = \langle \vec{f}, \omega \rangle \tag{3}$$

Specially, $\hat{f} \equiv \mathbf{E}(X_i)$ when only one sequence i is available, i.e. $\mathcal{A} = \{i\}$. It demonstrates that the designed ω can help the network excellently inherit the synthesis performance of pre-trained $\mathbf{E}$ and $\mathbf{G}$. In this work, we use ω to quantify the contribution of different input combinations.

Task-Specific Attention. Apart from the sequence-level fusion of $\hat{f}$, a task-specific attention module $\mathbf{G}_A$ is introduced to refine the fused features at the pixel level. The weights of $\mathbf{G}_A$ can adapt to the specific fusion task with the given target code. To build a conditional attention module, we replace convolutional layers in convolutional block attention module (CBAM) [20] with Hyper-Conv [10]. HyperConv is a dynamic filter whose kernel is mapped from a shared weight bank, and the mapping function is generated by the given target code. As shown in Fig. 1, channel attention and spatial attention can provide adaptive feature refinement guided by the task-specific code c to generate residual attentional fused features f_A.

$$f_A = \mathbf{G}_A(\vec{f} | c) \tag{4}$$

Loss Function. To force both $\hat{f}$ and $\hat{f} + f_A$ can be reconstructed to the target sequence by the conditional $\mathbf{G}$, a supervised reconstruction loss is given as,

$$
\begin{aligned}
\mathcal{L}_{rec} =&\lambda_r \cdot \|X' - X_{\text{tgt}}\|_1 + \lambda_p \cdot \mathcal{L}_p(X', X_{\text{tgt}}) \\
&+ \lambda_r \cdot \|X'_A - X_{\text{tgt}}\|_1 + \lambda_p \cdot \mathcal{L}_p(X'_A, X_{\text{tgt}})
\end{aligned}
\tag{5}
$$

where $X' = \mathbf{G}(\hat{f}|c_{\text{tgt}})$, $X'_A = \mathbf{G}(\hat{f} + f_A|c_{\text{tgt}})$, $X_{\text{tgt}} \in \mathcal{X}_{\mathcal{T}}$, $\|\cdot\|_1$ refers to a L_1 loss, and $\mathcal{L}_p$ indicates the perceptual loss based on pre-trained VGG19. λ_r and λ_p are weight terms and are experimentally set to be 10 and 0.01.

2.2   Task-Specific Enhanced Map

As f_A is a task-specific contextual refinement for fused features, analyzing it can help us understand more what the network tried to do. Many studies focus on visualizing the attention maps to interpret the principle of the network, especially for the transformer modules [1,6]. However, visualization of the attention map is limited by its low resolution and rough boundary. Thus, we proposed the TSEM by subtracting the reconstructed target sequences with and without f_A, which has the same resolution as the original images and clear interpretation for specific tasks.

$$
\text{TSEM} = |X'_A - X'|
\tag{6}
$$

3   Experiments

3.1   Dataset and Evaluation Metrics

We use brain MRI images of 1,251 subjects from Brain Tumor Segmentation 2021 (BraTS2021) [3,4,17], which includes four aligned sequences, T1, T1Gd, T2, and Flair, for each subject. We select 830 subjects for training, 93 for validation, and 328 for testing. All the images are intensity normalized to $[-1, 1]$ and central cropped to $128 \times 192 \times 192$. During training, for each subject, a random number of sequences are selected as inputs and the rest as targets. For validation and testing, we fixed the input combinations and the target for each subject.

The synthesis performance is quantified using the metrics of peak signal noise rate (PSNR), structural similarity index measure (SSIM), and learned perceptual image patch similarity (LPIPS) [21], which evaluate from intensity, structure, and perceptual aspects.

3.2   Implementation Details

The models are implemented with PyTorch and trained on the NVIDIA GeForce RTX 3090 Ti GPU. $\mathbf{E}$ comprises three convolutional layers and six residual blocks. The initial convolutional layer is responsible for encoding intensities to features, while the second and third convolutional layers downsample images by a factor of four. The residual blocks then extract the high-level representation. The

Table 1. Results for a set of sequences to a target sequence synthesis on BraTS2021.

Number of inputs	Methods	PSNR↑	SSIM↑	LPIPS↓
1	Pix2Pix [12]	25.6 ± 3.1	0.819 ± 0.086	15.85 ± 9.41
	MM-GAN [18]	27.3 ± 2.4	0.864 ± 0.039	11.47 ± 3.76
	DiamondGAN [14]	27.0 ± 2.3	0.857 ± 0.040	11.95 ± 3.65
	ResViT [9]	26.8 ± 2.1	0.857 ± 0.037	11.82 ± 3.54
	Seq2Seq [10]	27.7 ± 2.4	0.869 ± 0.038	10.49 ± 3.63
	TSF-Seq2Seq (w/o f_A)	27.8 ± 2.4	0.871 ± 0.039	$\mathbf{10.15 \pm 3.67}$
	TSF-Seq2Seq	$\mathbf{27.8 \pm 2.4}$	$\mathbf{0.872 \pm 0.039}$	10.16 ± 3.69
2	Pix2Pix [12] (Average)	26.2 ± 2.7	0.834 ± 0.054	15.84 ± 6.05
	MM-GAN [18]	28.0 ± 2.3	0.878 ± 0.037	10.33 ± 3.58
	DiamondGAN [14]	27.7 ± 2.3	0.872 ± 0.038	10.82 ± 3.36
	ResViT [9]	27.7 ± 2.2	0.875 ± 0.035	10.53 ± 3.26
	Hi-Net [23]	27.1 ± 2.3	0.866 ± 0.039	11.11 ± 3.76
	MMgSN-Net [15]	27.1 ± 2.7	0.865 ± 0.044	11.38 ± 4.37
	Seq2Seq [10] (Average)	28.2 ± 2.2	0.879 ± 0.035	11.11 ± 3.72
	TSF-Seq2Seq (w/o f_A)	28.0 ± 2.4	0.875 ± 0.039	9.89 ± 3.63
	TSF-Seq2Seq	$\mathbf{28.3 \pm 2.4}$	$\mathbf{0.882 \pm 0.038}$	$\mathbf{9.48 \pm 3.58}$
3	Pix2Pix [12] (Average)	26.6 ± 2.5	0.842 ± 0.041	15.77 ± 5.08
	MM-GAN [18]	28.5 ± 2.5	0.883 ± 0.040	9.65 ± 3.57
	DiamondGAN [14]	28.2 ± 2.5	0.877 ± 0.041	10.20 ± 3.33
	ResViT [9]	28.3 ± 2.4	0.882 ± 0.039	9.87 ± 3.30
	Seq2Seq [10] (Average)	28.5 ± 2.3	0.880 ± 0.038	11.61 ± 3.87
	TSF-Seq2Seq (w/o f_A)	28.3 ± 2.6	0.876 ± 0.044	9.61 ± 4.00
	TSF-Seq2Seq	$\mathbf{28.8 \pm 2.6}$	$\mathbf{0.887 \pm 0.042}$	$\mathbf{8.89 \pm 3.80}$

channels are 64, 128, 256, and 256, respectively. **G** has an inverse architecture with E, and all the convolutional layers are replaced with HyperConv. The **E** and **G** from Seq2Seq are pre-trained using the Adam optimizer with an initial learning rate of 2×10^{-4} and a batch size of 1 for 1,000,000 steps, taking about 60 h. Then we finetune the TSF-Seq2Seq with the frozen **E** using the Adam optimizer with an initial learning rate of 10^{-4} and a batch size of 1 for another 300,000 steps, taking about 40 h.

3.3   Quantitative Results

We compare our method with one-to-one translation, image-level fusion, and feature-level fusion methods. One-to-one translation methods include Pix2Pix [12] and Seq2Seq [10]. Image-level fusion methods consist of MM-GAN [18], DiamondGAN [14], and ResViT [9]. Feature-level fusion methods

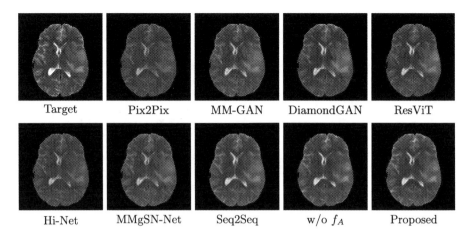

Fig. 2. Examples of synthetic T2 of comparison methods given the combination of T1Gd and Flair.

include Hi-Net [23] and MMgSN-Net [15]. Figure 2 shows the examples of synthetic T2 of comparison methods input with the combinations of T1Gd and Flair. Table 1 reports the sequence synthesis performance for comparison methods organized by the different numbers of input combinations. Note that, for multiple inputs, one-to-one translation methods synthesize multiple outputs separately and average them as one. And Hi-Net [23] and MMgSN-Net [15] only test on the subset with two inputs due to fixed network architectures. As shown in Table 1, the proposed method achieves the best performance in different input combinations.

3.4 Ablation Study

We compare two components of our method, including (1) task-specific weighted average and (2) task-specific attention, by conducting an ablation study between Seq2Seq, TSF-Seq2Seq (w/o f_A), and TSF-Seq2Seq. TSF-Seq2Seq (w/o f_A) refers to the model removing the task-specific attention module. As shown in Table 1, when only one sequence is available, our method can inherit the performance of Seq2Seq and achieve slight improvements. For multi-input situations, the task-specific weighted average can decrease LPIPS to achieve better perceptual performance. And task-specific attention can refine the fused features to achieve the best synthesis results.

3.5 Interpretability Visualization

The proposed method not only achieves superior synthesis performance but also has good interpretability. In this section, we will visualize the contribution of different input combinations and TSEM.

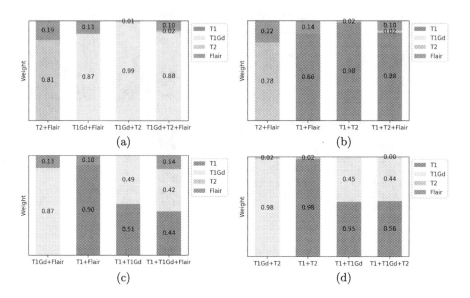

Fig. 3. Bar chart for the weights of the input set of sequences to synthesize different target sequences: (a) T1; (b) T1Gd; (c) T2; (d) Flair.

Table 2. Results of PSNR for regions highlighted or not highlighted by TSEM.

Number of inputs	TSEM > 99%	TSEM < 99%	Total
1	18.0 ± 3.2	28.3 ± 2.4	27.8 ± 2.4
2	18.8 ± 3.7	28.8 ± 2.4	28.3 ± 2.4
3	19.5 ± 4.0	29.3 ± 2.5	28.8 ± 2.6

Sequence Contribution. We use ω in Eq. 2 to quantify the contribution of different input combinations for synthesizing different target sequences. Figure 3 shows the bar chart for the sequence contribution weight ω with different task-specific code c. As shown in Fig. 3, both T1 and T1Gd contribute greatly to the sequence synthesis of each other, which is expected because T1Gd are T1-weighted scanning after contrast agent injection, and the enhancement between these two sequences is indispensable for cancer detection and diagnosis. The less contribution of T2, when combined with T1 and/or T1Gd, is consistent with the clinical findings [22, 23] that T2 can be well-synthesized by T1 and/or T1Gd.

TSEM vs. Attention Map. Figure 4 shows the proposed TSEM and the attention maps extracted by ResViT [9]. As shown in Fig. 4, TSEM has a higher resolution than the attention maps and can highlight the tumor area which is hard to be synthesized by the networks. Table 2 reports the results of PSNR for regions highlighted or not highlighted by TSEM with a threshold of the 99th percentile. To assist the synthesis models deploying in clinical settings, TSEM can be used as an attention and uncertainty map to remind clinicians of the possible unreliable synthesized area.

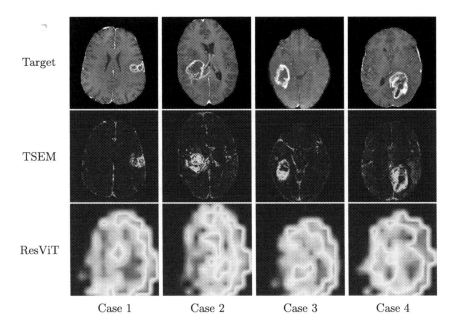

Fig. 4. Examples of the proposed TSEM and the attention maps extracted by ResViT [9] when generating T1Gd by given with T1, T2, and Flair.

4   Conclusion

In this work, we introduce an explainable network for multi-to-one synthesis with extensive experiments and interpretability visualization. Experimental results based on BraTS2021 demonstrate the superiority of our approach compared with the state-of-the-art methods. And we will explore the proposed method in assisting downstream applications for multi-sequence analysis in future works.

Acknowledgement. Luyi Han was funded by Chinese Scholarship Council (CSC) scholarship. Tianyu Zhang was supported by the Guangzhou Elite Project (TZ-JY201948).

References

1. Abnar, S., Zuidema, W.: Quantifying attention flow in transformers. arXiv preprint arXiv:2005.00928 (2020)
2. Armanious, K., et al.: Medgan: medical image translation using gans. Comput. Med. Imaging Graph. **79**, 101684 (2020)
3. Baid, U., et al.: The rsna-asnr-miccai brats 2021 benchmark on brain tumor segmentation and radiogenomic classification. arXiv preprint arXiv:2107.02314 (2021)
4. Bakas, S., et al.: Advancing the cancer genome atlas glioma MRI collections with expert segmentation labels and radiomic features. Sci. Data **4**(1), 1–13 (2017)

5. Chartsias, A., Joyce, T., Giuffrida, M.V., Tsaftaris, S.A.: Multimodal mr synthesis via modality-invariant latent representation. IEEE Trans. Med. Imaging **37**(3), 803–814 (2017)
6. Chefer, H., Gur, S., Wolf, L.: Transformer interpretability beyond attention visualization. In: Proceedings of the IEEE/CVF Conference on Computer Vision and Pattern Recognition, pp. 782–791 (2021)
7. Chen, J.H., Su, M.Y.: Clinical application of magnetic resonance imaging in management of breast cancer patients receiving neoadjuvant chemotherapy. BioMed Res. Int. **2013** (2013)
8. Choi, Y., Choi, M., Kim, M., Ha, J.W., Kim, S., Choo, J.: Stargan: unified generative adversarial networks for multi-domain image-to-image translation. In: Proceedings of the IEEE Conference on Computer Vision and Pattern Recognition, pp. 8789–8797 (2018)
9. Dalmaz, O., Yurt, M., Çukur, T.: Resvit: residual vision transformers for multimodal medical image synthesis. IEEE Trans. Med. Imaging **41**(10), 2598–2614 (2022)
10. Han, L., et al.: Synthesis-based imaging-differentiation representation learning for multi-sequence 3d/4d mri. arXiv preprint arXiv:2302.00517 (2023)
11. Huang, X., Liu, M.Y., Belongie, S., Kautz, J.: Multimodal unsupervised image-to-image translation. In: Proceedings of the European Conference on Computer Vision (ECCV), pp. 172–189 (2018)
12. Isola, P., Zhu, J.Y., Zhou, T., Efros, A.A.: Image-to-image translation with conditional adversarial networks. In: Proceedings of the IEEE Conference on Computer Vision and Pattern Recognition, pp. 1125–1134 (2017)
13. Jung, E., Luna, M., Park, S.H.: Conditional GAN with an attention-based generator and a 3D discriminator for 3D medical image generation. In: de Bruijne, M., et al. (eds.) MICCAI 2021. LNCS, vol. 12906, pp. 318–328. Springer, Cham (2021). https://doi.org/10.1007/978-3-030-87231-1_31
14. Li, H., et al.: DiamondGAN: unified multi-modal generative adversarial networks for MRI sequences synthesis. In: Shen, D., et al. (eds.) MICCAI 2019. LNCS, vol. 11767, pp. 795–803. Springer, Cham (2019). https://doi.org/10.1007/978-3-030-32251-9_87
15. Li, W., et al.: Virtual contrast-enhanced magnetic resonance images synthesis for patients with nasopharyngeal carcinoma using multimodality-guided synergistic neural network. Int. J. Radiat. Oncol.* Biol.* Phys. **112**(4), 1033–1044 (2022)
16. Mann, R.M., Cho, N., Moy, L.: Breast mri: state of the art. Radiology **292**(3), 520–536 (2019)
17. Menze, B.H., et al.: The multimodal brain tumor image segmentation benchmark (brats). IEEE Trans. Med. Imaging **34**(10), 1993–2024 (2014)
18. Sharma, A., Hamarneh, G.: Missing mri pulse sequence synthesis using multimodal generative adversarial network. IEEE Trans. Med. Imaging **39**(4), 1170–1183 (2019)
19. Uzunova, H., Ehrhardt, J., Handels, H.: Memory-efficient gan-based domain translation of high resolution 3d medical images. Comput. Med. Imaging Graph. **86**, 101801 (2020)
20. Woo, S., Park, J., Lee, J.Y., Kweon, I.S.: Cbam: convolutional block attention module. In: Proceedings of the European Conference on Computer Vision (ECCV), pp. 3–19 (2018)
21. Zhang, R., Isola, P., Efros, A.A., Shechtman, E., Wang, O.: The unreasonable effectiveness of deep features as a perceptual metric. In: Proceedings of the IEEE Conference on Computer Vision and Pattern Recognition, pp. 586–595 (2018)

22. Zhang, T., et al.: Important-net: integrated mri multi-parameter reinforcement fusion generator with attention network for synthesizing absent data. arXiv preprint arXiv:2302.01788 (2023)
23. Zhou, T., Fu, H., Chen, G., Shen, J., Shao, L.: Hi-net: hybrid-fusion network for multi-modal mr image synthesis. IEEE Trans. Med. Imaging **39**(9), 2772–2781 (2020)

Structure-Preserving Synthesis: MaskGAN for Unpaired MR-CT Translation

Vu Minh Hieu Phan[1](✉)(iD), Zhibin Liao[1](iD), Johan W. Verjans[1](iD),
and Minh-Son To[2](iD)

[1] Australian Institute for Machine Learning, University of Adelaide, Adelaide,
Australia
vuminhhieu.phan@adelaide.edu.au
[2] Flinders Health and Medical Research Institute, Flinders University, Adelaide,
Australia

Abstract. Medical image synthesis is a challenging task due to the scarcity of paired data. Several methods have applied CycleGAN to leverage unpaired data, but they often generate inaccurate mappings that shift the anatomy. This problem is further exacerbated when the images from the source and target modalities are heavily misaligned. Recently, current methods have aimed to address this issue by incorporating a supplementary segmentation network. Unfortunately, this strategy requires costly and time-consuming pixel-level annotations. To overcome this problem, this paper proposes MaskGAN, a novel and cost-effective framework that enforces structural consistency by utilizing automatically extracted coarse masks. Our approach employs a mask generator to outline anatomical structures and a content generator to synthesize CT contents that align with these structures. Extensive experiments demonstrate that MaskGAN outperforms state-of-the-art synthesis methods on a challenging pediatric dataset, where MR and CT scans are heavily misaligned due to rapid growth in children. Specifically, MaskGAN excels in preserving anatomical structures without the need for expert annotations. The code for this paper can be found at https://github.com/HieuPhan33/MaskGAN.

Keywords: Unpaired CT synthesis · Structural consistency

1 Introduction

Magnetic resonance imaging (MRI) and computed tomography (CT) are two commonly used cross-sectional medical imaging techniques. MRI and CT produce different tissue contrast and are often used in tandem to provide complementary information. While MRI is useful for visualizing soft tissues (e.g. muscle,

Acknowledgement: This study was supported by Channel 7 Children's Research Foundation of South Australia Incorporated (CRF).

Supplementary Information The online version contains supplementary material available at https://doi.org/10.1007/978-3-031-43999-5_6.

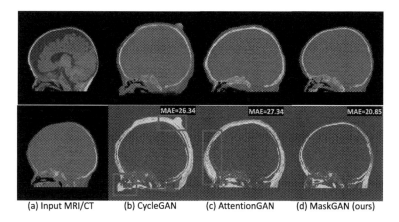

(a) Input MRI/CT   (b) CycleGAN   (c) AttentionGAN   (d) MaskGAN (ours)

Fig. 1. Visual results (Row 1) and the error map (Row 2) between the ground-truth and synthetic CT on pediatric dataset. (a) Input MRI and the paired CT. (b) Cycle-GAN [20] fails to preserve the smooth anatomy of the MRI. (c) AttentionGAN [12] inflates the head area in the synthetic CT, which is inconsistent with the original MRI. Quantitative evaluations in MAE (lower is better) are shown in yellow.

fat), CT is superior for visualizing bony structures. Some medical procedures, such as radiotherapy for brain tumors, craniosynostosis, and spinal surgery, typically require both MRI and CT for planning. Unfortunately, CT imaging exposes patients to ionizing radiation, which can damage DNA and increase cancer risk [9], especially in children and adolescents. Given these issues, there are clear advantages for synthesizing anatomically accurate CT data from MRI.

Most synthesis methods adopt supervised learning paradigms and train generative models to synthesize CT [1–3,6,17]. Despite the superior performance, supervised methods require a large amount of paired data, which is prohibitively expensive to acquire. Several unsupervised MRI-to-CT synthesis methods [4,6,14], leverage CycleGAN with cycle consistency supervision to eliminate the need for paired data. Unfortunately, the performance of unsupervised CT synthesis methods [4,14,15] is inferior to supervised counterparts. Due to the lack of direct constraints on the synthetic outputs, CycleGAN [20] struggles to preserve the anatomical structure when synthesizing CT images, as shown in Fig. 1(b). The structural distortion in synthetic results exacerbates when data from the two modalities are heavily misaligned, which usually occurs in pediatric scanning due to the rapid growth in children.

Recent unsupervised methods impose structural constraints on the synthesized CT through pixel-wise or shape-wise consistency. Pixel-wise consistency methods [8,14,15] capture and align pixel-wise correlations between MRI and synthesized CT. However, enforcing pixel-wise consistency may introduce undesirable artifacts in the synthetic results. This problem is particularly relevant in brain scanning, where both the pixel-wise correlation and noise statistics in MR and CT images are different, as a direct consequence of the signal acquisition technique. The alternative shape-wise consistency methods [3,4,19] aim to preserve the shapes of major body parts in the synthetic image. Notably, shape-CycleGAN [4] segments synthesized CT and enforces consistency with the ground-truth MRI segmentation. However, these methods rely on segmentation annotations, which are time-consuming, labor-intensive, and require expert radiological annotators. A recent natural image synthesis approach, called Attention-GAN [12], learns attention masks to identify discriminative structures. AttentionGAN implicitly learns prominent structures in the image without using the ground-truth shape. Unfortunately, the lack of explicit mask supervision can lead to imprecise attention masks and, in turn, produce inaccurate mappings of the anatomy, as shown in Fig. 1(c).

Table 1. Comparisons of different shape-aware image synthesis.

Method	Mask Supervision	Human Annotation	Structural Consistency
Shape-cycleGAN [4]	Precise mask	Yes	Yes
AttentionGAN [12]	Not required	No	No
MaskGAN (Ours)	**Coarse mask**	**No**	**Yes**

In this paper, we propose **MaskGAN**, a novel unsupervised MRI-to-CT synthesis method, that preserves the anatomy under the explicit supervision of coarse masks without using costly manual annotations. Unlike segmentation-based methods [4,18], MaskGAN bypasses the need for precise annotations, replacing them with standard (unsupervised) image processing techniques, which can produce coarse anatomical masks. Such masks, although imperfect, provide sufficient cues for MaskGAN to capture anatomical outlines and produce structurally consistent images. Table 1 highlights our differences compared with previous shape-aware methods [4,12]. Our major contributions are summarized as follows. **1)** We introduce **MaskGAN**, a novel unsupervised MRI-to-CT synthesis method. MaskGAN is the first framework that maintains shape consistency without relying on human-annotated segmentation. **2)** We present two new structural supervisions to enforce consistent extraction of anatomical structures across MRI and CT domains. **3)** Extensive experiments show that our method outperforms state-of-the-art methods by using automatically extracted coarse masks to effectively enhance structural consistency.

2   Proposed Method

In this section, we first introduce the **MaskGAN** architecture, shown in Fig. 2, and then describe the three supervision losses we use for optimization.

2.1   MaskGAN Architecture

The network comprises two generators, each learning an MRI-CT and a CT-MRI translation. Our generator design has two branches, one for generating masks

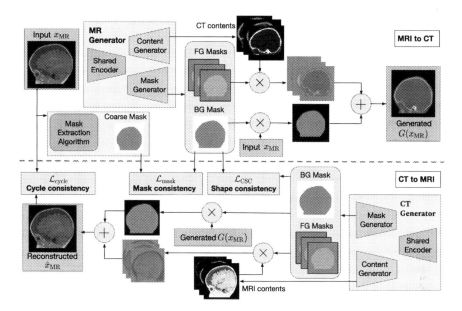

Fig. 2. Overview of our proposed **MaskGAN**. First, we automatically extract coarse masks from the input x_{MR}. MaskGAN then learns via two new objectives in addition to the standard CycleGAN loss. The mask loss $\mathcal{L}_{mask}$ minimizes the L1 distance between the extracted background mask and the coarse mask, ensuring accurate anatomical structure generation. The cycle shape consistency (CSC) loss $\mathcal{L}_{CSC}$ minimizes the L1 distance between the masks learned by the MRI and CT generators, promoting consistent anatomy segmentation across domains.

and the other for synthesizing the content in the masked regions. The mask branch learns N attention masks A_i, where the first $N-1$ masks capture foreground (FG) structures and the last mask represents the background (BG). The content branch synthesizes $N-1$ outputs for the foreground structures, denoted as C. Each output, C_i, represents the synthetic content for the corresponding foreground region that is masked by the attention mask A_i.

Intuitively, each channel A_i in the mask tensor A focuses on different anatomical structures in the medical image. For instance, one channel emphasizes on synthesizing the skull, while another focuses on the brain tissue. The last channel A_N in A corresponds to the background and is applied to the original input to preserve the background contents. The final output is the sum of masked foreground contents and masked background input. Formally, the synthetic CT output generated from the input MRI x is defined as

$$O_{CT} = G_{CT}(x) = A_{CT}^N x + \sum_{i=1}^{N-1} A_{CT}^i C_{CT}^i. \tag{1}$$

The synthetic MRI output from the CT scan y is defined similarly based on the attention masks and the contents from the MR generator. The proposed network is trained using three training objectives described in the next sections.

2.2   CycleGAN Supervision

The two generators, G_{MR} and G_{CT}, map images from MRI domain (X) and CT domain (Y), respectively. Two discriminators, D_{MR} and D_{CT}, are used to distinguish real from fake images in the MRI and CT domains. The adversarial loss for training the generators to produce synthetic CT images is defined as

$$\mathcal{L}_{\text{CT}}(G_{\text{MR}}, D_{\text{CT}}, x, y) = \mathbb{E}_{y \sim p_{\text{data}}(y)} \big[\log D_{\text{CT}}(y) \big] + \mathbb{E}_{x \sim p_{\text{data}}(x)} \big[\log(1 - D_{\text{CT}}(G_{\text{MR}}(x))) \big]. \tag{2}$$

The adversarial loss $\mathcal{L}_{\text{MR}}$ for generating MRI images is defined in a similar manner. For unsupervised training, CycleGAN imposes the cycle consistency loss, which is formulated as follows

$$\mathcal{L}_{\text{cycle}} = \mathbb{E}_{x \sim p_{\text{data}}}(x) \left\| x - G_{CT}(G_{MR}(x)) \right\| + \mathbb{E}_{y \sim p_{\text{data}}}(y) \left\| y - G_{MR}(G_{CT}(y)) \right\|. \tag{3}$$

The CycleGAN's objective $\mathcal{L}_{\text{GAN}}$ is the combination of adversarial and cycle consistency loss.

2.3   Mask and Cycle Shape Consistency Supervision

Mask Loss. To reduce spurious mappings in the background regions, MaskGAN explicitly guides the mask generator to differentiate the foreground objects from the background using mask supervision. We extract the coarse mask B using basic image processing operations. Specifically, we design a simple but robust algorithm that works on both MRI and CT scans, with a binarization stage followed by a refinement step. In the binarization stage, we normalize the intensity to the range [0, 1] and apply a binary threshold of 0.1, selected based on histogram inspection, to separate the foreground from the background. In the post-processing stage, we refine the binary image using morphological operations, specifically employing a binary opening operation to remove small artifacts. We perform connected component analysis [11] and keep the largest component as the foreground. Column 6 in Fig. 3 shows examples of extracted masks.

We introduce a novel mask supervision loss that penalizes the difference between the background mask A_N learned from the input image and the ground-truth background mask B in both MRI and CT domains. The mask loss for the attention generators is formulated as

$$\mathcal{L}_{\text{mask}} = \mathbb{E}_{x \sim p_{\text{data}}(x)} \left\| A_N^{\text{MR}} - B_N^{\text{MR}} \right\|_1 + \mathbb{E}_{y \sim p_{\text{data}}(y)} \left\| A_N^{\text{CT}} - B_N^{\text{CT}} \right\|_1. \tag{4}$$

Discussion. Previous shape-aware methods [4,18] use a pre-trained U-Net [10] segmentation network to enforce shape consistency on the generator. U-Net is pre-trained in a separate stage and frozen when the generator is trained. Hence, any errors produced by the segmentation network cannot be corrected. In contrast, we jointly train the shape extractor, *i.e.*, the mask generator, and the content generator end-to-end. Besides mask loss $\mathcal{L}_{\mathrm{mask}}$, the mask generator also receives supervision from adversarial loss $\mathcal{L}_{\mathrm{GAN}}$ to adjust the extracted shape and optimize the final synthetic results. Moreover, in contrast to previous methods that train a separate shape extractor, our MaskGAN uses a shared encoder for mask and content generators, as illustrated in Fig. 2. Our design embeds the extracted shape knowledge into the content generator, thus improving the structural consistency of the synthetic contents.

Cycle Shape Consistency Loss. Spurious mappings can occur when the anatomy is shifted during translation. To preserve structural consistency across domains, we introduce the cycle shape consistency (CSC) loss as our secondary contribution. Our loss penalizes the discrepancy between the background attention mask A_N^{MR} learned from the input MRI image and the mask $\tilde{A}_N^{\mathrm{CT}}$ learned from synthetic CT. Enforcing consistency in both domains, we formulate the shape consistency loss as

$$\mathcal{L}_{\mathrm{shape}} = \mathbb{E}_{x \sim p_{\mathrm{data}}(x)} \left\| A_N^{\mathrm{MR}} - \tilde{A}_N^{\mathrm{CT}} \right\|_1 + \mathbb{E}_{y \sim p_{\mathrm{data}}(y)} \left\| A_N^{\mathrm{CT}} - \tilde{A}_N^{\mathrm{MR}} \right\|_1. \tag{5}$$

The final loss for MaskGAN is the sum of three loss objectives weighted by the corresponding loss coefficients: $\mathcal{L} = \mathcal{L}_{\mathrm{GAN}} + \lambda_{\mathrm{mask}}\mathcal{L}_{\mathrm{mask}} + \lambda_{\mathrm{shape}}\mathcal{L}_{\mathrm{shape}}$.

3 Experimental Results

3.1 Experimental Settings

Data Collection. We collected 270 volumetric T1-weighted MRI and 267 thin-slice CT head scans with bony reconstruction performed in pediatric patients under routine scanning protocols[1]. We targeted the age group from 6–24 months since pediatric patients are more susceptible to ionizing radiation and experience a greater cancer risk (up to 24% increase) from radiation exposure [7]. Furthermore, surgery for craniosynostosis, a birth defect in which the skull bones fuse too early, typically occurs during this age [5,16]. The scans were acquired by Ingenia 3.0T MRI scanners and Philips Brilliance 64 CT scanners. We then resampled the volumetric scans to the same resolution of $1.0 \times 1.0 \times 1.0$ mm^3.

The dataset comprises brain MR and CT volumes from 262 subjects. 13 MRI-CT volumes from the same patients that were captured less than three months apart are registered using rigid registration algorithms. The dataset is divided into 249, 1 and 12 subjects for training, validating and testing set. Following [13], we conducted experiments on sagittal slices. Each MR and CT volume consists of 180 to 200 slices, which are resized and padded to the size of 224×224. The intensity range of CT is clipped into [-1000, 2000]. All models are trained using

---

[1] Ethics approval was granted by Southern Adelaide Clinical Human Research Ethics Committee.

the Adam optimizer for 100 epochs, with a learning rate of 0.0002 which linearly decays to zero over the last 50 epochs. We use a batch size of 16 and train on two NVIDIA RTX 3090 GPUs.

Evaluation Metrics. To provide a quantitative evaluation of methods, we compute the same standard performance metrics as in previous works [6,14] including mean absolute error (MAE), peak signal-to-noise ratio (PSNR), and structural similarity (SSIM) between ground-truth and synthesized CT. The scope of the paper centers on theoretical development; clinical evaluations such as dose calculation and treatment planning will be conducted in future work.

3.2   Results and Discussions

Comparisons with State-of-the-Art. We compare the performance of our proposed MaskGAN with existing state-of-the-art image synthesis methods, including CycleGAN [20], AttentionGAN [12], structure-constrained CycleGAN (sc-CycleGAN) [14] and shape-CycleGAN [4]. Shape-CycleGAN requires annotated segmentation to train a separate U-Net. For a fair comparison, we implement shape-CycleGAN using our extracted coarse masks based on the authors' official code. Note that CT-to-MRI synthesis is a secondary task supporting the primary MRI-to-CT synthesis task. As better MRI synthesis leads to improved CT synthesis, we also report the model's performance on MRI synthesis.

Table 2. Quantitative comparison of different methods on the primary MRI-CT task and the secondary CT-MRI task. The results of an ablated version of our proposed MaskGAN are also reported. ± standard deviation is reported over five evaluations. The paired t-test is conducted between MaskGAN and a compared method at $p = 0.05$. The improvement of MaskGAN over all compared methods is statistically significant.

Methods	Primary: MRI-to-CT			Secondary: CT-to-MRI		
	MAE ↓	PSNR ↑	SSIM (%) ↑	MAE ↓	PSNR ↑	SSIM (%) ↑
CycleGAN [20]	32.12 ± 0.31	31.57 ± 0.12	46.17 ± 0.20	34.21 ± 0.33	29.88 ± 0.24	45.73 ± 0.17
AttentionGAN [12]	28.25 ± 0.25	32.88 ± 0.09	53.57 ± 0.15	30.47 ± 0.22	30.15 ± 0.10	50.66 ± 0.14
sc-CycleGAN [14]	24.55 ± 0.24	32.97 ± 0.07	57.08 ± 0.11	26.13 ± 0.15	31.22 ± 0.07	54.14 ± 0.10
shape-CycleGAN [4]	24.30 ± 0.28	33.14 ± 0.05	57.73 ± 0.13	25.96 ± 0.19	31.69 ± 0.08	54.88 ± 0.09
MaskGAN (w/o Shape)	22.78 ± 0.19	34.02 ± 0.09	60.19 ± 0.06	23.58 ± 0.23	32.43 ± 0.07	57.35 ± 0.08
MaskGAN (Ours)	$\mathbf{21.56 \pm 0.18}$	$\mathbf{34.75 \pm 0.08}$	$\mathbf{61.25 \pm 0.10}$	$\mathbf{22.77 \pm 0.17}$	$\mathbf{32.55 \pm 0.06}$	$\mathbf{58.32 \pm 0.10}$

Table 2 demonstrates that our proposed MaskGAN outperforms existing methods for statistical significance of $p = 0.05$ in both tasks. The method reduces the MAE of CycleGAN and AttentionGAN by 29.07% and 19.36%, respectively. Furthermore, MaskGAN outperforms shape-CycleGAN, reducing its MAE by 11.28%. Unlike shape-CycleGAN, which underperforms when trained with coarse segmentations, our method obtains consistently higher results. Figure 3 shows the visual results of different methods. sc-CycleGAN produces artifacts (e.g., the eye socket in the first sample and the nasal cavity in the second sample),

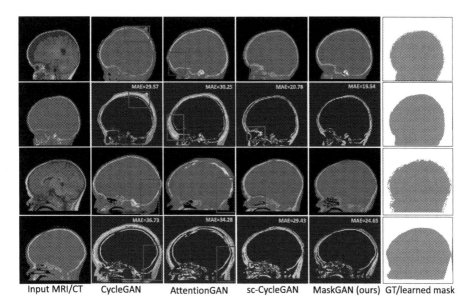

Input MRI/CT   CycleGAN   AttentionGAN   sc-CycleGAN   MaskGAN (ours)   GT/learned mask

Fig. 3. Visual comparison of synthesized CT images by different methods on two samples. Column 1: Input MRI (Row 1 and 3) and the corresponding paired CT scan (Row 2 and 4). Column 2–5: Synthesized CT results (Row 1 and 3) and the corresponding error maps (Row 2 and 4). Column 6: Extracted coarse background (ground-truth) masks (Row 1 and 3) and attention masks learned by our MaskGAN (Row 2 and 4).

as it preserves pixel-wise correlations. In contrast, our proposed MaskGAN preserves shape-wise consistency and produces the smoothest synthetic CT. Unlike adult datasets [4,14], pediatric datasets are easily misaligned due to children's rapid growth between scans. Under this challenging setting, unpaired image synthesis can have non-optimal visual results and SSIM scores. Yet, our MaskGAN achieves the highest quality, indicating its suitability for pediatric image synthesis.

We perform an ablation study by removing the cycle shape consistency loss (w/o Shape). Compared with shape-CycleGAN, MaskGAN using only a mask loss significantly reduces MAE by 6.26%. The combination of both mask and cycle shape consistency losses results in the largest improvement, demonstrating the complementary contributions of our two losses.

Robustness to Error-Prone Coarse Masks. We compare the performance of our approach with shape-CycleGAN [4] using deformed masks that simulate human errors during annotation. To alter object shapes, we employ random elastic deformation, a standard data augmentation technique [10] that applies random displacement vectors to objects. The level of distortion is controlled by the standard deviation of the normal distribution from which the vectors are sampled. Figure 4 (Left) shows MAE of the two methods under increasing levels of distortion. MAE of shape-CycleGAN drastically increases as the masks become more distorted. Figure 4 (Right) shows that our MaskGAN (d) better preserves the anatomy.

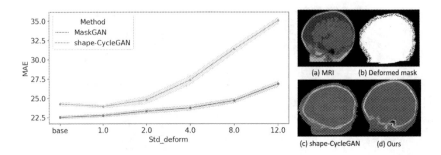

Fig. 4. *Left*: MAE of the two shape-aware methods using deformed masks. *Right*: Qualitative results of shape-CycleGAN (c) and our MaskGAN (d) when training using coarse masks deformed by the standard deviation of 2.0.

4   Conclusion

This paper proposes MaskGAN - a novel automated framework that maintains the shape consistency of prominent anatomical structures without relying on expert annotated segmentations. Our method generates a coarse mask outlining the shape of the main anatomy and synthesizes the contents for the masked foreground region. Experimental results on a clinical dataset show that MaskGAN significantly outperforms existing methods and produces synthetic CT with more consistent mappings of anatomical structures.

References

1. Armanious, K., et al.: MedGAN: medical image translation using GANs. Computer. Med. Imag. Graphic. **79**, 101684 (2020)
2. Dalmaz, O., Yurt, M., Çukur, T.: ResViT: residual vision transformers for multimodal medical image synthesis. IEEE Trans. Med. Imag. **41**(10), 2598–2614 (2022)
3. Emami, H., Dong, M., Nejad-Davarani, S.P., Glide-Hurst, C.K.: SA-GAN: structure-aware GAN for organ-preserving synthetic CT generation. In: de Bruijne, M., et al. (eds.) MICCAI 2021. LNCS, vol. 12906, pp. 471–481. Springer, Cham (2021). https://doi.org/10.1007/978-3-030-87231-1_46
4. Ge, Y., et al.: Unpaired MR to CT synthesis with explicit structural constrained adversarial learning. In: IEEE International Symposium Biomedical Imaging. IEEE (2019)
5. Governale, L.S.: Craniosynostosis. Pediatr. Neurol. **53**(5), 394–401 (2015)
6. Liu, Y., et al.: CT synthesis from MRI using multi-cycle GAN for head-and-neck radiation therapy. Computer. Med. Imag. Graphic. **91**, 101953 (2021)
7. Mathews, J.D., et al.: Cancer risk in 680000 people exposed to computed tomography scans in childhood or adolescence: data linkage study of 11 million AUSTRALIANS. BMJ **346** (2013)
8. Matsuo, H., et al.: Unsupervised-learning-based method for chest MRI-CT transformation using structure constrained unsupervised generative attention networks. Sci. Rep. **12**(1), 11090 (2022)

9. Richardson, D.B., et al.: Risk of cancer from occupational exposure to ionising radiation: retrospective cohort study of workers in France, the united kingdom, and the united states (INWORKS). BMJ **351** (2015)

10. Ronneberger, O., Fischer, P., Brox, T.: U-Net: convolutional networks for biomedical image segmentation. In: Navab, N., Hornegger, J., Wells, W.M., Frangi, A.F. (eds.) MICCAI 2015. LNCS, vol. 9351, pp. 234–241. Springer, Cham (2015). https://doi.org/10.1007/978-3-319-24574-4_28

11. Samet, H., Tamminen, M.: Efficient component labeling of images of arbitrary dimension represented by linear bintrees. IEEE Trans. Pattern Anal. Mach. Intell. **10**(4), 579–586 (1988)

12. Tang, H., Liu, H., Xu, D., Torr, P.H., Sebe, N.: AttentionGAN: unpaired image-to-image translation using attention-guided generative adversarial networks. IEEE Trans. Neu. Netw. Learn. Syst. **34**, 1972–1987 (2021)

13. Wolterink, J.M., Dinkla, A.M., Savenije, M.H.F., Seevinck, P.R., van den Berg, C.A.T., Išgum, I.: Deep MR to CT synthesis using unpaired data. In: Tsaftaris, S.A., Gooya, A., Frangi, A.F., Prince, J.L. (eds.) SASHIMI 2017. LNCS, vol. 10557, pp. 14–23. Springer, Cham (2017). https://doi.org/10.1007/978-3-319-68127-6_2

14. Yang, H., et al.: Unsupervised MR-to-CT synthesis using structure-constrained CycleGAN. IEEE Trans. Med. Imag. **39**(12), 4249–4261 (2020)

15. Yang, Heran, et al.: Unpaired brain MR-to-CT synthesis using a structure-constrained CycleGAN. In: Stoyanov, Danail, et al. (eds.) DLMIA/ML-CDS -2018. LNCS, vol. 11045, pp. 174–182. Springer, Cham (2018). https://doi.org/10.1007/978-3-030-00889-5_20

16. Zakhary, G.M., Montes, D.M., Woerner, J.E., Notarianni, C., Ghali, G.: Surgical correction of craniosynostosis: a review of 100 cases. J. Cranio-Maxillofac. Surg. **42**(8), 1684–1691 (2014)

17. Zhang, J., Cui, Z., Jiang, C., Zhang, J., Gao, F., Shen, D.: Mapping in cycles: dual-domain PET-CT synthesis framework with cycle-consistent constraints. In: Wang, L., Dou, Q., Fletcher, P.T., Speidel, S., Li, S. (eds.) MICCAI 2022. Lecture Notes in Computer Science, vol. 13436, pp. 758–767. Springer, Cham (2022). https://doi.org/10.1007/978-3-031-16446-0_72

18. Zhang, Z., Yang, L., Zheng, Y.: Translating and segmenting multimodal medical volumes with cycle-and shape-consistency generative adversarial network. In: Proceedings of the IEEE Conference on Computer Vision and Pattern Recognition, pp. 9242–9251 (2018)

19. Zhou, B., Augenfeld, Z., Chapiro, J., Zhou, S.K., Liu, C., Duncan, J.S.: Anatomy-guided multimodal registration by learning segmentation without ground truth: application to intraprocedural CBCT/MR liver segmentation and registration. Med. Image Anal. **71**, 102041 (2021)

20. Zhu, J.Y., Park, T., Isola, P., Efros, A.A.: Unpaired image-to-image translation using cycle-consistent adversarial networks. In: Proceedings of the IEEE International Conference on Computer Vision, pp. 2223–2232 (2017)

Alias-Free Co-modulated Network for Cross-Modality Synthesis and Super-Resolution of MR Images

Zhiyun Song[1], Xin Wang[1], Xiangyu Zhao[1], Sheng Wang[1,3], Zhenrong Shen[1], Zixu Zhuang[1,3], Mengjun Liu[1], Qian Wang[2], and Lichi Zhang[1(✉)]

[1] School of Biomedical Engineering, Shanghai Jiao Tong University, Shanghai, China
`lichizhang@sjtu.edu.cn`
[2] School of Biomedical Engineering, ShanghaiTech University, Shanghai, China
[3] Shanghai United Imaging Intelligence Co., Ltd., Shanghai, China

Abstract. Cross-modality synthesis (CMS) and super-resolution (SR) have both been extensively studied with learning-based methods, which aim to synthesize desired modality images and reduce slice thickness for magnetic resonance imaging (MRI), respectively. It is also desirable to build a network for simultaneous cross-modality and super-resolution (CMSR) so as to further bridge the gap between clinical scenarios and research studies. However, these works are limited to specific fields. None of them can flexibly adapt to various combinations of resolution and modality, and perform CMS, SR, and CMSR with a single network. Moreover, alias frequencies are often treated carelessly in these works, leading to inferior detail-restoration ability. In this paper, we propose Alias-Free Co-Modulated network ($AFCM$) to accomplish all the tasks with a single network design. To this end, we propose to perform CMS and SR consistently with co-modulation, which also provides the flexibility to reduce slice thickness to various, non-integer values for SR. Furthermore, the network is redesigned to be alias-free under the Shannon-Nyquist signal processing framework, ensuring efficient suppression of alias frequencies. Experiments on three datasets demonstrate that AFCM outperforms the alternatives in CMS, SR, and CMSR of MR images. Our codes are available at https://github.com/zhiyuns/AFCM.

Keywords: Cross-modality synthesis · Super-resolution · Magnetic resonance imaging

1 Introduction

Magnetic Resonance Imaging (MRI) is widely recognized as a pivotally important neuroimaging technique, which provides rich information about brain tissue

Supplementary Information The online version contains supplementary material available at https://doi.org/10.1007/978-3-031-43999-5_7.

anatomy in a non-invasive manner. Several studies have shown that multi-modal MR images offer complementary information on tissue morphology, which help conduct more comprehensive brain region analysis [4,8,20]. For instance, T1-weighted images distinguish grey and white matter tissues, while FLAIR images differentiate edematous regions from cerebrospinal fluid [19]. Furthermore, 3D MR imaging with isotropic voxels contains more details than anisotropic ones and facilitates the analyzing procedure [22], especially for automated algorithms whose performance degrades severely when dealing with anisotropic images [6,23]. However, the scanning protocols in the clinic are different from that in research studies. Due to the high costs of radiation examinations, physicians and radiologists prefer to scan a specific MR contrast, anisotropic MRI with 2D scanning protocols, or their combinations [2,10,17]. Such simplified protocols may hinder the potential possibility of utilizing them for further clinical studies. Therefore, there has been ever-growing interest in reconstructing images with the desired modality and resolution from the acquired images, which may have different modalities or resolutions.

Relevant literature in this area can be divided into cross-modality synthesis (CMS), super-resolution (SR), or cross-modality super-resolution (CMSR). Most of the works concerning CMS model modality-invariant and coherent high-frequency details to translate between modalities. For example, pGAN [5] extended conditional GAN with perceptual loss and leveraged neighboring cross-sections to translate between T1- and T2-weighted images. AutoGAN [7] explored NAS to automatically search for optimal generator architectures for MR image synthesis. ResViT [4] proposed a transformer-based generator to capture long-range spatial dependencies for multi-contrast synthesis. On the other hand, SR methods are usually designed to build 3D models for reconstruction. DeepResolve [2] employed a ResNet-based 3D CNN to reduce thickness for knee MR images. DCSRN [3] designed a DenseNet-based 3D model to restore HR features of brain MR images. In more special clinical scenarios where both the CMS and SR need to be performed, current works prefer to build a single model that simultaneously performs the task, rather than performing them sequentially. For instance, SynthSR [10] used a modified 3D-UNet to reconstruct MP-RAGE volumes with thickness of 1 mm from other MR images scanned with different protocols. WEENIE [9] proposed a weakly-supervised joint convolutional sparse coding method to acquire high-resolution modality images with low-resolution ones with other modalities.

However, these works have two main limitations. On the one hand, these works are limited to specified fields, and cannot be flexibly switched to other tasks or situations. For instance, CMS models described above will fail to produce consistent 3D results due to the lack of modeling inter-plane correlations. Moreover, CMSR models only produce MR images with fixed thickness, which may constrain their further applications. On the other hand, high-frequency details with alias frequencies are often treated carelessly in these generative networks. This may lead to unnatural details and even aliasing artifacts, especially for the ill-posed inverse tasks (SR or CMSR) that require restoring high-frequency details based on low-resolution images.

In this paper, we propose an Alias-Free Co-Modulated network (AFCM) to address the aforementioned issues. AFCM is inspired by the recent advances in foundation models [24], which achieve superior results in various tasks and even outperform the models for specific tasks. To flexibly accomplish 2D-CMS and 3D-SR with a single model, we propose to extend style-based modulation [15,16] with image-conditioned and position-embedded representations, termed as co-modulation, to ensure consistency between these tasks. The design also enables more flexible SR by reconstructing slices in arbitrary positions with continuous positional embeddings. Moreover, the discrete operators in the generator are carefully redesigned to be alias-free under the Shannon-Nyquist signal processing framework. Our main contributions are as follows: 1) We propose a co-modulated network that can accomplish CMS, SR, and CMSR with a single model. 2) We propose the continuous positional embedding strategy in AFCM, which can synthesize high-resolution MR images with non-integer thickness. 3) With the redesigned alias-free generator, AFCM is capable of restoring high-frequency details more naturally for the reconstructed images. 4) AFCM achieves state-of-the-art results for CMS, SR, and CMSR of MR images.

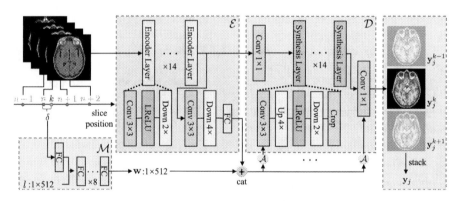

Fig. 1. Overall architecture of AFCM. The encoder $\mathcal{E}$ embeds multiple slices into image-conditioned representation, while the mapping network $\mathcal{M}$ maps the latent code l and embedded δ into position-embedded representation $\mathbf{w}$. The concatenated representations are transformed with affine transformation $\mathcal{A}$ and modulate the weights of the convolution kernels of the decoder $\mathcal{D}$. The target MR image $\mathbf{y}_j$ is generated by walking through all target positions and concatenating the generated slices. The resampling filter and nonlinearity in each layer are redesigned to suppress the corresponding alias frequencies and ensure alias-free generation.

2 Methodology

2.1 Co-modulated Network

Here we propose a novel style-based co-modulated network to accomplish CMS, SR, and CMSR consistently. As shown in Fig. 1, the architecture design of our network is based on the style-based unconditional generative architecture [15,16].

Different from original stochastic style representation, we design both position-embedded representation and image-conditioned representation as the modulations, which control the generation process consistently. Remind that we aim to translate low-resolution MR images $\mathbf{x}_i$ with modality i and thickness p into high-resolution MR images $\mathbf{y}_j$ with modality j and thickness q. The position-embedded representation $\mathbf{w}$ is transformed from the latent code $l \sim N(0, \mathbf{I})$ and the relative slice index $\delta = kq/p - \lfloor kq/p \rfloor$ that controls the position k for each output slice $\mathbf{y}_j^k$. This is accomplished with a mapping network $\mathcal{M}$ where the relative slice index is embedded using the class-conditional design [13]. The image-conditioned representation $\mathcal{E}(\mathbf{x}_i^{in})$ is obtained by encoding $2m$ adjacent slices (i.e., $\mathbf{x}_i^{in} = [\mathbf{x}_i^{n-(m-1)}, ..., \mathbf{x}_i^n, ..., \mathbf{x}_i^{n+m}]$ where $n = \lfloor kq/p \rfloor$) with the encoder $\mathcal{E}$. In this paper, m is experientially set to 2 following [12]. The two representations are then concatenated and produce a style vector s with the affine transform $\mathcal{A}$:

$$s = \mathcal{A}(\mathcal{E}(\mathbf{x}_i^{in}), \mathcal{M}(l, \delta)). \tag{1}$$

The style vector s is then used to modulate the weights of the convolution kernels in the synthesis network $\mathcal{D}$ [16]. Moreover, skip connections are implemented between $\mathcal{E}$ and $\mathcal{D}$ to preserve the structure of conditional images. The target volume $\mathbf{y}_j$ is created by stacking generated slices together. Note that our network is flexible to perform various tasks. It performs 2D-CMS or 3D-CMS alone when δ is fixed to 0, which is denoted as AFCM_{multi}. In this case, if the information of adjacent cross-sections is unknown, we perform the one-to-one translation with AFCM_{one} where only one slice is given as input (i.e., $\mathbf{x}_i^{in} = \mathbf{x}_i^k$). Moreover, AFCM_{sr} performs SR alone when $i = j$.

2.2 Alias-Free Generator

We notice that results produced by the vanilla generator are contaminated by "texture sticking", where several fine-grained details are fixed in pixel coordinates when δ changes, as illustrated in our ablation study. The phenomenon was found to originate from aliasing caused by carelessly designed operators (i.e., convolution, resampling, and nonlinearity) in the network [14]. To solve the problem, we consider the feature map Z in the generator as a regularly sampled signal, which can represent the continuous signal z with limited frequencies under the Shannon-Nyquist signal processing framework [26]. Therefore, the discrete operation $\mathbf{F}$ on Z has its continuous counterpart $\mathbf{f}(z)$:

$$\mathbf{F}(Z) = \text{III}_{r'} \odot \mathbf{f}(\phi_r * Z), \tag{2}$$

where $\text{III}_{r'}$ is the two-dimensional Dirac comb with sampling rate r', and ϕ_r is the interpolation filter with a bandlimit of $r/2$ so that z can be represented as $\phi_r * Z$. The operator $\odot$ denotes pointwise multiplication and $*$ denotes continuous convolution. The operation $\mathbf{F}$ is alias-free when any frequencies higher than $r'/2$ in the output signal, also known as alias frequencies, are efficiently suppressed. In this way, we re-design the generator by incorporating the alias-free mechanism based on the analysis above, which consists of the antialiasing decoder and encoder which are further illustrated as follows.

Antialiasing Decoder. Considering that any details with alias frequencies need to be suppressed, the resampling and nonlinearity operators in the decoder are carefully redesigned following Alias-Free GAN [14]. The convolution preserves the original form as it introduces no new frequencies. In low-resolution layers, the resampling filter is designed as non-critical sinc one whose cutoff frequency varies with the resolution of feature maps. Moreover, nonlinear operation (i.e., LeakyRelu here) is wrapped between upsampling and downsampling to ensure that any high-frequency content introduced by the operation is eliminated. Other minor modifications including the removal of noise, simplification of the network, and flexible layers are also implemented. Please refer to [14] for more details. Note that Fourier features are replaced by feature maps obtained by the encoder $\mathcal{E}$.

Antialiasing Encoder. Given that our network has a U-shaped architecture, the encoder also needs to be alias-free so that the skip connections would not introduce extra content with undesired frequency. The encoder consists of 14 layers, each of which is further composed of a convolution, a nonlinear operation, and a downsampling filter. The parameters of the filter are designed to vary with the resolution of the corresponding layer. Specifically, the cutoff frequency geometrically decreases from $f_c = r_N/2$ in the first non-critically sampled layer to $f_c = 2$ in the last layer, where r_N is the image resolution. The minimum acceptable stopband frequency starts at $f_t = 2^{0.3} \cdot r_N/2$ and geometrically decreases to $f_t = 2^{2.1}$, whose value determines the resolution $r = \min(\text{ceil}(2 \cdot f_t), r_N)$ and the transition band half-width $f_h = \max(r/2, f_t) - f_c$. The target feature map of each layer is surrounded by a 10-pixel margin, and the final feature map is resampled to 4×4 before formulating the image-conditioned representation.

2.3 Optimization

The overall loss is composed of an adversarial loss and a pixel-wise L_1 loss:

$$L_G = L_{adv} + \lambda L_{pix}, \tag{3}$$

where λ is used to balance the losses. The adversarial loss is defined as non-saturation loss with R_1 regularization, and the discriminator preserves the architecture of that in StyleGAN2 [16]. We combine $\mathbf{x}_i^{in}$ with real/fake $\mathbf{y}_j^k$ as the input for the discriminator, and embed δ with the projection discriminator strategy [21]. Following the stabilization trick [14], we blur target and reconstructed images using a Gaussian filter with $\sigma = 2$ pixels in the early training stage.

3 Experiments

3.1 Experimental Settings

Datasets. We evaluate the performance of AFCM on three brain MRI datasets:

Table 1. Quantitative results for cross-modality synthesis on the IXI dataset. Except for $AFCM_{multi}$, all experiments are performed under one-to-one translation protocol where only one slice is taken as the input.

	T1-T2		T2-PD		PD-T1	
	PSNR	SSIM	PSNR	SSIM	PSNR	SSIM
pix2pix [11]	23.35 ± 0.71	0.8754 ± 0.0126	30.09 ± 0.75	0.9558 ± 0.0063	25.92 ± 0.81	0.9258 ± 0.0094
pGAN [5]	26.09 ± 1.13	0.9205 ± 0.0130	32.87 ± 0.46	0.9737 ± 0.0023	27.38 ± 0.91	0.9424 ± 0.0089
AutoGAN [7]	26.33 ± 0.92	0.9193 ± 0.0174	33.01 ± 0.79	0.9669 ± 0.0115	27.49 ± 0.89	0.9384 ± 0.0088
ResViT [4]	26.17 ± 1.10	0.9209 ± 0.0131	33.15 ± 0.55	0.9753 ± 0.0024	27.73 ± 1.00	0.9443 ± 0.0091
$AFCM_{one}$	26.38 ± 1.08	0.9297 ± 0.0119	33.56 ± 0.51	$\mathbf{0.9779 \pm 0.0020}$	27.88 ± 0.94	0.9453 ± 0.0088
$AFCM_{multi}$	$\mathbf{26.92 \pm 1.06}$	$\mathbf{0.9355 \pm 0.0196}$	$\mathbf{33.58 \pm 0.49}$	$\mathbf{0.9779 \pm 0.0019}$	$\mathbf{28.76 \pm 1.05}$	$\mathbf{0.9525 \pm 0.0074}$

IXI[1] is a collection of multi-modality MRI images with spatial resolution of $0.94 \times 0.94 \times 1.2 \, mm^3$ obtained from three hospitals. It includes images from 309 randomly selected healthy subjects, which are further divided into three subsets: training (285 subjects), validation (12 subjects), and testing (12 subjects).

CSDH is an in-house dataset comprising 100 patients diagnosed with chronic subdural hematoma (CSDH). It includes T1-weighted images with spatial resolution of $0.75 \times 0.75 \times 1 \, mm^3$ and T2-weighted flair images with spatial resolution of $0.75 \times 0.75 \times 8 \, mm^3$. Pixel-wise annotations of liquefied blood clots made by an experienced radiologist are used for segmentation accuracy evaluation.

ADNI[2] is composed of 50 patients diagnosed with Alzheimer's disease and 50 elderly controls. Each subject has one near-isotropic T1 MP-RAGE scan with thickness of 1 mm and one axial FLAIR scan with thickness of 5 mm.

ADNI and CSDH are both further divided into 70, 10, and 20 subjects for training, validating, and testing. For all three datasets, MR images from the same subjects are co-registered using a rigid registration model with ANTs [1].

Implementation Details. We use the translation equivariant configuration of the Alias-Free GAN, and discard the rotational equivariance to avoid generating overly-symmetric images [25]. AFCM is trained with batch size 16 on an NVIDIA GeForce RTX 3090 GPU with 24G memory for 100 epochs, which takes about 36 GPU hours for each experiment. The learning rates are initialized as 0.0025/0.0020 for the generator and the discriminator respectively in the first 50 epochs and linearly decrease to 0.

3.2 Comparative Experiments

Cross-Modality Synthesis. In this section, we report three subtasks, namely CMS from T1 to T2, T2 to PD, and PD to T1-weighted MR images on the IXI dataset in Table 1. $AFCM_{one}$ achieves superior results for all tasks when performing one-to-one translation, where the corresponding slice for the target is

---

[1] https://brain-development.org/ixi-dataset/.
[2] https://adni.loni.usc.edu/.

taken as the input. The improvement in the generation quality can be attributed to the alias-free design of AFCM, which can restore high-frequency details more accurately, as shown in Fig. 2. Moreover, $AFCM_{multi}$ demonstrates higher generation quality and a significant improvement over other methods ($p < 0.05$) when adjacent slices are used as extra guidance. The inverted translation results provided in the supplementary file also indicate the superiority of our method.

| Input | Ground Truth | Pix2pix | PGAN | AutoGAN | ResViT | $AFCM_{one}$ | $AFCM_{multi}$ |

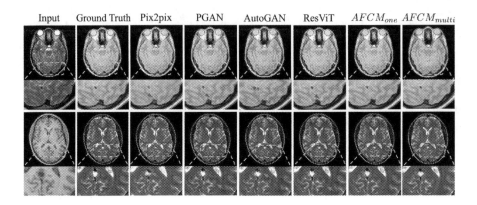

Fig. 2. Qualitative results for cross-modality synthesis on the IXI dataset.

Super-Resolution. The SR experiment is performed using the CSDH dataset with downsampling factor (DSF) of 8 and the ADNI dataset with DSF of 5. Following previous work [2], the low-resolution training data are simulated by filtering and downsampling thin-slice images. We compare with SR methods for MR images (Deepresolve [2] and SynthSR [10]), as well as the 3D version of an SR method for natural images (EDSR [18]). It may seem counterintuitive to generate 2D images slice-by-slice instead of reconstructing the entire volume directly. However, results in Fig. 3(right) suggest that $AFCM_{sr}$ yields higher quality than other 3D-based SR models due to the continuous position-embedded representation and the alias-free design for detail-restoration. Although the improvement is not significant compared with the SOTA methods, the qualitative results in the supplementary file demonstrate that $AFCM_{sr}$ produces coherent details compared to other methods whose results are blurred due to imperfect pre-resampling, especially when the DSF is high. We also evaluate whether the synthesized images can be used for downstream tasks. As observed from Fig. 3(right), when using a pre-trained segmentation model to segment the liquefied blood clots in the reconstructed images, $AFCM_s r$ can produce most reliable results, which also indicates the superiority of our method for clinical applications.

Cross-Modality and Super-Resolution. As previously addressed, it is more challenging to implement CMS and SR simultaneously. One possible approach is to perform CMS and SR in a two-stage manner, which is accomplished with a combination of ResViT and DeepResolve. We also compare our approach with SynthSR [10] (fully supervised version for fair comparison) that directly performs the task. As reported in Fig. 3(right), the performance of the baseline method is improved when either DeepResolve is replaced with $AFCM_{sr}$ for SR ($ResViT+AFCM_{sr}$) or ResViT is replaced with $AFCM_{multi}$ for CMS ($AFCM_{multi}+$DeepResolve). Additionally, although the two-stage performance is slightly lower than SynthSR, AFCM achieves the best results among all combinations ($p < 0.05$, SSIM). We also qualitatively evaluate whether our network can perform CMSR with flexible target thickness. To this end, we directly use the model trained on CSDH to generate isotropic images with thickness of 0.75mm, which means $10.67\times$ CMSR. As depicted in Fig. 3, when δ is set to values not encountered during training (fixed at [0, 0.125, 0.25, 0.375, 0.5, 0.625, 0.75, 0.875]), AFCM still generates reasonable results. Multiple views of the reconstructed MR image with reduced thickness of 0.75 mm also demonstrate the effectiveness of reducing the thickness to flexible values, which are even blind for training.

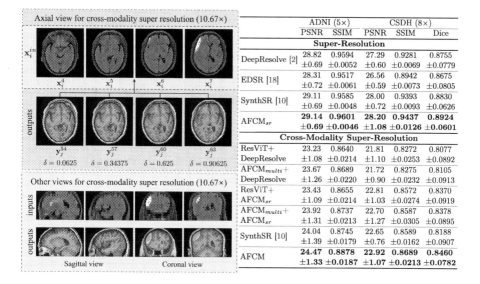

	ADNI (5×)		CSDH (8×)		
	PSNR	SSIM	PSNR	SSIM	Dice
Super-Resolution					
DeepResolve [2]	28.82	0.9594	27.29	0.9281	0.8755
	±0.69	±0.0052	±0.60	±0.0069	±0.0779
EDSR [18]	28.31	0.9517	26.56	0.8942	0.8675
	±0.72	±0.0061	±0.59	±0.0073	±0.0805
SynthSR [10]	29.11	0.9585	28.00	0.9393	0.8830
	±0.69	±0.0048	±0.72	±0.0093	±0.0626
$AFCM_{sr}$	**29.14**	**0.9601**	**28.20**	**0.9437**	**0.8924**
	±0.69	**±0.0046**	**±1.08**	**±0.0126**	**±0.0601**
Cross-Modality Super-Resolution					
ResViT+ DeepResolve	23.23	0.8640	21.81	0.8272	0.8077
	±1.08	±0.0214	±1.10	±0.0253	±0.0892
$AFCM_{multi}+$ DeepResolve	23.67	0.8689	21.72	0.8275	0.8105
	±1.26	±0.0220	±0.90	±0.0232	±0.0913
ResViT+ $AFCM_{sr}$	23.43	0.8655	22.81	0.8572	0.8370
	±1.09	±0.0214	±1.03	±0.0274	±0.0919
$AFCM_{multi}+$ $AFCM_{sr}$	23.92	0.8737	22.70	0.8587	0.8378
	±1.31	±0.0213	±1.27	±0.0305	±0.0895
SynthSR [10]	24.04	0.8745	22.65	0.8589	0.8188
	±1.39	±0.0179	±0.76	±0.0162	±0.0907
AFCM	**24.47**	**0.8878**	**22.92**	**0.8689**	**0.8460**
	±1.33	**±0.0187**	**±1.07**	**±0.0213**	**±0.0782**

Axial view for cross-modality super resolution (10.67×)

x_i^{in}

x_i^4 x_i^5 x_i^6 x_i^7

outputs

y_j^{54} y_j^{57} y_j^{60} y_j^{63}
$\delta = 0.0625$ $\delta = 0.34375$ $\delta = 0.625$ $\delta = 0.90625$

Other views for cross-modality super resolution (10.67×)

inputs

outputs

Sagittal view   Coronal view

Fig. 3. Left: Qualitative results for arbitrary-scale cross-modality super-resolution on the CSDH dataset (input: anisotropic Flair, output: isotropic T1) with flexible target thickness (training: 1 mm, testing: 0.75 mm). Right: Quantitative results for super-resolution and cross-modality super-resolution.

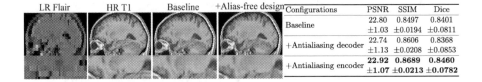

LR Flair	HR T1	Baseline	+Alias-free design	Configurations	PSNR	SSIM	Dice
				Baseline	22.80 ±1.03	0.8497 ±0.0194	0.8401 ±0.0811
				+Antialiasing decoder	22.74 ±1.13	0.8606 ±0.0208	0.8368 ±0.0853
				+Antialiasing encoder	**22.92** ±1.07	**0.8689** ±0.0213	**0.8460** ±0.0782

Fig. 4. Qualitative (left) and quantitative (right) results for ablation study.

Ablation Study. In this section, we evaluate the impact of our alias-free design by performing CMSR on the CSDH dataset. Note that we replace the encoder and decoder in AFCM with vanilla ones [27] as the baseline for comparison. Table in Fig. 4 indicates that although redesigning the decoder leads to the improvement of SSIM, it is when we also redesign the encoder that the dropped metrics (PSNR and Dice) recover, which highlights the importance of redesigning both the encoder and decoder to achieve optimal results. The qualitative results in Fig. 4 also demonstrate that our design successfully suppresses "texture sticking".

4   Conclusion

We propose a novel alias-free co-modulated network for CMS, SR, and CMSR of MR images. Our method addresses the problems of task inconsistency between CMS and SR with a novel co-modulated design, and suppresses aliasing artifacts by a redesigned alias-free generator. AFCM is also flexible enough to reconstruct MR images with various non-integer target thickness. The experiments on three independent datasets demonstrate our state-of-the-art performance in CMS, SR, and CMSR of MR images.

Acknowledgement. This work was supported by the National Natural Science Foundation of China (No. 62001292 and No. 82227807).

References

1. Avants, B.B., Tustison, N.J., Song, G., Cook, P.A., Klein, A., Gee, J.C.: A reproducible evaluation of ants similarity metric performance in brain image registration. Neuroimage **54**(3), 2033–2044 (2011)
2. Chaudhari, A.S., et al.: Super-resolution musculoskeletal MRI using deep learning. Magn. Reson. Med. **80**(5), 2139–2154 (2018)
3. Chen, Y., Xie, Y., Zhou, Z., Shi, F., Christodoulou, A.G., Li, D.: Brain MRI super resolution using 3D deep densely connected neural networks. In: 2018 IEEE 15th International Symposium on Biomedical Imaging (ISBI 2018), pp. 739–742 (2018)
4. Dalmaz, O., Yurt, M., Çukur, T.: ResViT: residual vision transformers for multimodal medical image synthesis. IEEE Trans. Med. Imaging **41**(10), 2598–2614 (2022)

5. Dar, S.U., Yurt, M., Karacan, L., Erdem, A., Erdem, E., Cukur, T.: Image synthesis in multi-contrast MRI with conditional generative adversarial networks. IEEE Trans. Med. Imaging **38**(10), 2375–2388 (2019)
6. Greve, D.N., Fischl, B.: Accurate and robust brain image alignment using boundary-based registration. Neuroimage **48**(1), 63–72 (2009)
7. Hu, X., Shen, R., Luo, D., Tai, Y., Wang, C., Menze, B.H.: AutoGAN-synthesizer: neural architecture search for cross-modality MRI synthesis. In: Wang, L., Dou, Q., Fletcher, P.T., Speidel, S., Li, S. (eds.) MICCAI 2022. LNCS, vol. 13436, pp. 397–409. Springer, Cham (2022). https://doi.org/10.1007/978-3-031-16446-0_38
8. Huang, J., Chen, C., Axel, L.: Fast multi-contrast MRI reconstruction. Magn. Reson. Imaging **32**(10), 1344–1352 (2014)
9. Huang, Y., Shao, L., Frangi, A.F.: Simultaneous super-resolution and cross-modality synthesis of 3D medical images using weakly-supervised joint convolutional sparse coding. In: Proceedings of the IEEE Conference on Computer Vision and Pattern Recognition (CVPR) (2017)
10. Iglesias, J.E., et al.: Joint super-resolution and synthesis of 1 mm isotropic MP-RAGE volumes from clinical MRI exams with scans of different orientation, resolution and contrast. Neuroimage **237**, 118206 (2021)
11. Isola, P., Zhu, J.Y., Zhou, T., Efros, A.A.: Image-to-image translation with conditional adversarial networks. In: Proceedings of the IEEE Conference on Computer Vision and Pattern Recognition (CVPR) (2017)
12. Kalluri, T., Pathak, D., Chandraker, M., Tran, D.: FLAVR: flow-agnostic video representations for fast frame interpolation. In: Proceedings of the IEEE/CVF Winter Conference on Applications of Computer Vision (WACV), pp. 2071–2082 (2023)
13. Karras, T., Aittala, M., Hellsten, J., Laine, S., Lehtinen, J., Aila, T.: Training generative adversarial networks with limited data. In: Advances in Neural Information Processing Systems, vol. 33, pp. 12104–12114 (2020)
14. Karras, T., et al.: Alias-free generative adversarial networks. In: Advances in Neural Information Processing Systems, vol. 34, pp. 852–863 (2021)
15. Karras, T., Laine, S., Aila, T.: A style-based generator architecture for generative adversarial networks. In: Proceedings of the IEEE/CVF Conference on Computer Vision and Pattern Recognition, pp. 4401–4410 (2019)
16. Karras, T., Laine, S., Aittala, M., Hellsten, J., Lehtinen, J., Aila, T.: Analyzing and improving the image quality of StyleGAN. In: Proceedings of the IEEE/CVF Conference on Computer Vision and Pattern Recognition (CVPR) (2020)
17. Krupa, K., Bekiesińska-Figatowska, M.: Artifacts in magnetic resonance imaging. Pol. J. Radiol. **80**, 93 (2015)
18. Lim, B., Son, S., Kim, H., Nah, S., Mu Lee, K.: Enhanced deep residual networks for single image super-resolution. In: Proceedings of the IEEE Conference on Computer Vision and Pattern Recognition (CVPR) Workshops (2017)
19. Abd-Ellah, M.K., Awad, A.I., Khalaf, A.A., Hamed, H.F.: A review on brain tumor diagnosis from MRI images: practical implications, key achievements, and lessons learned. Magn. Reson. Imaging **61**, 300–318 (2019)
20. Menze, B.H., et al.: The multimodal brain tumor image segmentation benchmark (BRATS). IEEE Trans. Med. Imaging **34**(10), 1993–2024 (2014)
21. Miyato, T., Koyama, M.: cGANs with projection discriminator. In: International Conference on Learning Representations (2018)
22. Moraal, B., et al.: Multi-contrast, isotropic, single-slab 3D MR imaging in multiple sclerosis. Neuroradiol. J. **22**(1_suppl), 33–42 (2009)

23. Patenaude, B., Smith, S.M., Kennedy, D.N., Jenkinson, M.: A Bayesian model of shape and appearance for subcortical brain segmentation. Neuroimage **56**(3), 907–922 (2011)
24. Rombach, R., Blattmann, A., Lorenz, D., Esser, P., Ommer, B.: High-resolution image synthesis with latent diffusion models. In: Proceedings of the IEEE/CVF Conference on Computer Vision and Pattern Recognition (CVPR), pp. 10684–10695 (2022)
25. Sauer, A., Schwarz, K., Geiger, A.: StylegAN-XL: scaling StyleGAN to large diverse datasets. In: ACM SIGGRAPH 2022 Conference Proceedings (2022)
26. Shannon, C.: Communities in the presence of noise. Proc. Instit. Radio Eng. **37**(1), 10–21 (1949)
27. Zhao, S., et al.: Large scale image completion via co-modulated generative adversarial networks. In: International Conference on Learning Representations (2021)

Multi-perspective Adaptive Iteration Network for Metal Artifact Reduction

Haiyang Mao[1], Yanyang Wang[1], Hengyong Yu[2], Weiwen Wu[1(✉)],
and Jianjia Zhang[1(✉)]

[1] School of Biomedical Engineering, Shenzhen Campus of Sun Yat-sen University,
Shenzhen 518107, China
{wuweiw7,zhangjj225}@mail.sysu.edu.cn
[2] University of Massachusetts Lowell, Lowell, MA 01854, USA

Abstract. Metal artifact reduction (MAR) is important to alleviate the impacts of metal implants on clinical diagnosis with CT images. However, enhancing the quality of metal-corrupted image remains a challenge. Although the deep learning-based MAR methods have achieved impressive success, their interpretability and generalizability need further improvement. It is found that metal artifacts mainly concentrate in high frequency, and their distributions in the wavelet domain are significantly different from those in the image domain. Decomposing metal artifacts into different frequency bands is conducive for us to characterize them. Based on these observations, a model is constructed with dual-domain constraints to encode artifacts by utilizing wavelet transform. To facilitate the optimization of the model and improve its interpretability, a novel multi-perspective adaptive iteration network (MAIN) is proposed. Our MAIN is constructed under the guidance of the proximal gradient technique. Moreover, with the usage of the adaptive wavelet module, the network gains better generalization performance. Compared with the representative state-of-the-art deep learning-based MAR methods, the results show that our MAIN significantly outperforms other methods on both of a synthetic and a clinical datasets.

Keywords: CT · metal artifact reduction · multi-perspective regularizations · iterative learning · wavelet transform

1 Introduction

Metal artifact could significantly affect the clinical diagnoses with computed tomography (CT) images, and how to effectively reduce it is a critical but challenging issue. Specifically, metal artifact is caused by metallic implants and often exhibits as bright and dark streaks in the reconstructed images [7]. These streaks could hinder the perception of the actual contents, posing a serious obstacle for radiologists in making an accurate diagnosis [26]. Making matters worse, with the increasing employment of metallic implants, metal artifacts in CT images have become more widespread. In this case, effectively reducing metal artifacts while maintaining the tissue details is of great clinical significance [20].

H. Greenspan et al. (Eds.): MICCAI 2023, LNCS 14229, pp. 77–87, 2023.
https://doi.org/10.1007/978-3-031-43999-5_8

Since the metal artifacts are structured and non-local, they are tough to be removed from images directly [8]. Most of the traditional methods propose to reduce metal artifacts in the sinogram domain [6,13,21]. For instance, linear interpolation (LI) [6] and normalization metal artifacts reduction (NMAR) [13] weakened metal artifacts by substituting metal trace regions with interpolated data. However, severe secondary artifacts are induced by interpolation errors. Another commonly used iterative reconstruction algorithm [12] is computationally expensive. Some other researchers also explore a combination of multiple traditional methods to leverage their advantages [5] and improve their performance.

In contrast to the traditional methods mentioned above, deep learning-based MAR methods are undergoing more intensive studies and become a dominant approach to MAR. Depending on the performing domain, they can be categorized into three types, i.e., sinogram domain-based, image domain-based, and dual-domain-based. Specifically, i). the sinogram domain-based methods leverage the advantage that the signals of metal artifacts are concentrated in form of metal trace(s) and can be easily separated from the informative image contents in the sinogram domain [3,15,17]. However, they are restricted by the availability of the physical scanning configurations, and slight disturbance in the sinogram could cause serious artifacts in the image domain; ii). the image domain-based methods directly reduce metal artifacts in the image domain by utilizing residual [22] or adversarial learning [10,24,25,29] techniques. However, the deep intertwinement of artifacts and the image contents in the image domain makes them arduous to be differentiated, limiting the network performance; iii). the dual-domain-based methods utilize both the sinogram and image domains to reduce metal artifacts [8,28]. They typically involve alternative reduction of metal artifacts in the sinogram domain and refinement of image contents in the image domain, e.g., dual-domain data consistent recurrent network [32] and deep unrolling dual-domain network [23]. However, they cannot completely resolve the problem of secondary artifact and still require physical scanning configurations. In addition, most of the three above types of deep learning-based methods lack interpretability since they perform MAR in a black-box mechanism.

To address the issues of the existing methods aforementioned, we propose a Multi-perspective Adaptive Iteration Network(MAIN), and our main contributions are as follows:

1) **Multi-perspective Regularizations:** Based on the insightful analysis on the limitations of using sinograms in MAR, this paper innovatively identifies that the desirable properties of wavelet transform could well address the issues of sinograms in MAR. i.e., the spatial distribution characteristics of metal artifacts under different domains and resolutions. Based on this, we integrate multi-domain, multi-frequency band, and multi-constraint into our scheme by exploiting wavelet transform. Therefore, we explicitly formulate such knowledge as a multi-perspective optimization model as shown in Fig. 1.

2) **Iterative Optimization Algorithm:** To solve the multi-perspective optimization model, we develop an iterative algorithm with the proximal gradient

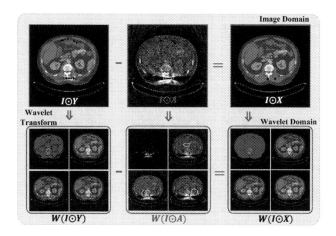

Fig. 1. The different distributions of metal artifacts. Here, Y, X, A and I donate the metal-corrupted CT image, clean CT image, metal artifacts and binary non-metal mask, respectively; $\odot$ is the element-wise multiplication; and W denotes the adaptive wavelet transform.

technique [1]. The network is designed according to the algorithm to keep it in accordance with the procedure of the theoretical MAR optimization, making the network more interpretable.

3) **Adaptive Wavelet Transform:** In order to increase the flexibility and adaptivity of the proposed model, the proposed model conducts wavelet transforms with neural technology rather than the traditional fixed wavelet transform.

2 Method

Mathematical Model. In the image domain, it is easy to segment metal regions with much higher CT values [22]. As the metal regions have no human tissues, the clean CT image X can be defined as:

$$I \odot X = I \odot Y - I \odot A, \tag{1}$$

where $Y \in \mathbf{R}^{H \times W}$ is a metal-corrupted image, A and I denote the metal artifact and binary non-metal mask, respectively. $\odot$ is the element-wise multiplication. To obtain a promising solution, various regularization terms representing prior constrains are introduced as:

$$\begin{aligned} \{X, A\} = \underset{\{X,A\}}{\arg\min} \{ &\frac{1}{2} \|I \odot (X + A - Y)\|_F^2 + \lambda_1 f_1(W(I \odot A)) \\ &+ \lambda_2 f_2(W(I \odot X)) + \lambda_3 f_3(I \odot X) \} \end{aligned} \tag{2}$$

where W denotes the wavelet transform, $f_1(.)$, $f_2(.)$ and $f_3(.)$ are regularization functions, and λ_1, λ_2 and λ_3 are regularization weights. Specifically, $f_1(.)$ and

$f_2(.)$ represent wavelet domain constraints of X and A respectively, and $f_3(.)$ introduces prior knowledge of X in the image domain. ε is an arbitrarily small number, and $\|.\|_F^2$ is the Frobenius norm.

When the wavelet components are transformed back to the image domain, the image will be blurred due to information loss. To recover a more precise image, let $U = I \odot X'$ and introduce an error feedback item [14] into Eq. (2), where X' represents the CT image in the image domain obtained after performing inverse wavelet transform. Then, the optimization problem becomes:

$$\{X, A, U\} = \arg\min_{\{X,A,U\}}\{\frac{1}{2}\left\|I \odot (X + A - Y)\right\|_F^2 + \frac{\beta}{2}\left\|I \odot X - U\right\|_F^2$$
$$+ \lambda_1 f_1(W(I \odot A)) + \lambda_2 f_2(W(I \odot X)) + \lambda_3 f_3(U)\}, \tag{3}$$

where β is an adaptive weight.

Optimization Algorithm. In this paper, an alternating minimization strategy is used to solve Eq. (3). At the $(k + 1)^{th}$ iteration, $A^{(k+1)}$, $X^{(k+1)}$ and $U^{(k+1)}$ are derived as the following three sub-problems:

$$\begin{cases} A^{(k+1)} = \arg\min_{A}\{\frac{1}{2}\left\|I \odot (X^{(k)} + A - Y)\right\|_F^2 + \lambda_1 f_1(W(I \odot A))\}, \\ X^{(k+1)} = \arg\min_{X}\{\frac{1}{2}\left\|I \odot (X + A^{(k+1)} - Y)\right\|_F^2 + \frac{\beta}{2}\left\|I \odot X - U^{(k)}\right\|_F^2 \\ \qquad + \lambda_2 f_2(W(I \odot X)), \\ U^{(k+1)} = \arg\min_{U}\{\frac{\beta}{2}\left\|(I \odot X^{(k+1)} - U)\right\|_F^2\} + \lambda_3 f_3(U). \end{cases} \tag{4}$$

The Proximal Gradient Descent algorithm(PGD) [1] is applied to solve each sub-problems above. Taking the first sub-problem as an example, the Taylor formula is utilized to introduce $A^{(k)}$ into the approximation of $A^{(k+1)}$. Taylor's second-order expansion can be expressed as $f(x) = \frac{f''(x_0)}{2!}(x - x_0)^2 + f'(x_0)(x - x_0) + f(x_0) + o^n$. Let $g(A) = \left\|I \odot (X^{(k)} + A - Y)\right\|_F^2$. At $A = A^{(k)}$, the quadratic approximation of $A^{(k+1)}$ can be expressed as:

$$A^{(k+1)} = \arg\min_{A}\{\frac{\nabla^2 g(A^{(k)})}{2}\left\|I \odot (A - A^{(k)})\right\|_F^2 + g(A^{(k)}) + o(A)$$
$$+ \nabla g(A^{(k)})(I \odot (A - A^{(k)})) + 2\lambda_1 f_1(W(I \odot A))\} \tag{5}$$

Assuming that $\nabla^2 g(A^{(k)})$ is a non-zero constant that is replaced by $\frac{1}{\eta_1}$. Besides, $g(A^{(k)})$ is replaced by another constant $(\eta_1 \nabla g(A^{(k)}))^2$ to form a perfect square trinomial since changing constant does not affect the minimization of our objective function. Therefore, Eq. (5) is reformatted as

$$A^{(k+1)} = \arg\min_{A}\{\frac{1}{2\eta_1}\left\|I \odot A - I \odot A^{(k)} + \eta_1 \nabla g(A^{(k)})\right\|_F^2 + 2\lambda_1 f_1(W(I \odot A))\}, \tag{6}$$

In code implementation, a trainable parameter is used to represent η_1. It is excepted that the application of adaptive parameter can assist network fitting A more preferably. In order to reveal the iterative procedure more apparently, an intermediate variable $A^{(k+0.5)}$ is introduced:

$$
\begin{aligned}
A^{(k+0.5)} &= I \odot A^{(k)} - \eta_1 \nabla g(A^{(k)}) \\
&= (1 - 2\eta_1)I \odot A^{(k)} + 2\eta_1 I \odot (Y - X^{(k)}).
\end{aligned}
\tag{7}
$$

Based on the above derivations, an iteration equation can be formulated as [22]:

$$
A^{(k+1)} = W^T prox_{\eta_1}(W A^{(k+0.5)}),
\tag{8}
$$

where $prox_{\eta_1}$ is the proximal operator [22] related to $f_1(.)$. Since $f_1(.)$ denotes the constraint on metal artifacts in wavelet domain, the iterative solution of A is carried out in wavelet domain. Since we eventually want to obtain artifacts in the image domain, W^T transforms the artifacts from wavelet domain to image domain. Similarly, $X^{(k+1)}$ and $U^{(k+1)}$ are derived as:

$$
\begin{cases}
X^{(k+1)} = W^T prox_{\eta_2}(W X^{(k+0.5)}), \\
U^{(k+1)} = prox_{\eta_3}(U^{(k+0.5)}).
\end{cases}
\tag{9}
$$

where $prox_{\eta_2}$ and $prox_{\eta_3}$ are the proximal operator related to $f_2(.)$ and $f_3(.)$ respectively.

Network Design. Figure 2 depicts the flowchart of the proposed MAIN network. The network contains T iteration blocks. Following the optimization algorithm, each block contains three key modules of $proxNet_A$, $proxNet_X$ and $proxNet_U$. The three network modules are designed under the guidance of Eqs. (8) and (9) to respectively emulate the proximal operators of $prox_{\eta_1}$, $prox_{\eta_2}$ and $prox_{\eta_3}$.

At the $(k+1)^{th}$ block, $A^{(k)}$, $X^{(k)}$ and $U^{(k)}$ are first decomposed by adaptive wavelet transform module [18]. In the wavelet domain, the $proxNet_A$ and $proxNet_X$ are built by following the DnCNN [30], subsequently, $A^{(k+1)}$ and $X^{(k+1)}$ are computed. Next, $A^{(k+1)}$ and $X^{(k+1)}$ are converted to the image domain. And a lightweight U-Net [14,19] is employed as $proxNet_U$.

3 Experiments

Synthetic Data. A synthetic dataset is generated by following the simulation procedure in [31]. Specifically, 1,200 clean CT images from Deeplesion [27] and 100 metal masks are collected for image synthesis. For network training, 1,000 CT images and 90 metal masks are randomly selected, creating 90,000 unique combinations. An additional 2,000 images are synthesized for test using the remaining 200 CT images and 10 metal masks. The pixel sizes of the 10 test masks are: 2054, 879, 878, 448, 242, 115, 115, 111, 53, and 32.

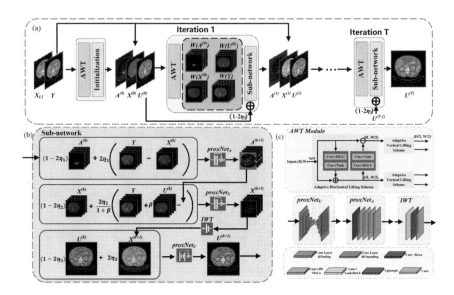

Fig. 2. The architecture of MAIN. (a) The overall architecture of our network with T iterations. (b) The procedure of sub-network at k^{th} iteration. (c) The detailed structure of adaptive wavelet transform and different proximal operator.

Clinical Data. In addition to the above synthetic data, the proposed model is also evaluated on the publicly available clinical dataset CLINIC-metal [9]. To keep consistent with Yu *et al.* [28], we segment the metal masks using a threshold of 2,000 HU. Then, the linear interpolated image is computed with the same procedure as synthetic data.

Implementation Details. The MAIN network is implemented using PyTorch [16] and trained with a single NVIDIA A6000 GPU for 120 epochs. We use the Adam optimizer with parameters $(\beta 1, \beta 2) = (0.5, 0.999)$, a learning rate of $5e^{-5}$, and a batch size of 16 to train the network. To enhance the stability of the model training, various image augmentations, such as image rotation and transposition, are applied.

Baseline and Evaluation Metric. Six state-of-the-art methods for metal artifact reduction are compared, including traditional method (NMAR [13]) and deep learning-based approaches (ACDNet [22], DuDoNet++ [11], DuDoNet [8] and CNNMAR [31]). DuDoNet and DuDoNet++ are reimplemented strictly adhering to the original methodology since they lacked open-source code. The Root Mean Square Error (RMSE) and Structural Similarity Index (SSIM) are used for quantitative assessment on the synthetic dataset. As reference images are unavailable on the clinical dataset, only visual comparison is conducted in terms of metal artifact reduction.

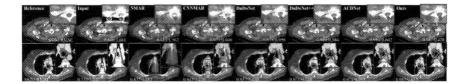

Fig. 3. Visual comparisons of MAR results on synthetic images. The display window is $[-224\ 634]$ HU.

Table 1. Quantitative comparison on synthetic testing dataset

Methods	NMAR	CNNMAR	DuDoNet	DuDoNet++	ACDNet	Ours
SSIM	0.910 ± 0.03	0.962 ± 0.02	0.974 ± 0.01	0.974 ± 0.01	0.980 ± 0.01	$\mathbf{0.984 \pm 0.01}$
RMSE	2.671 ± 0.82	1.077 ± 0.44	1.032 ± 0.41	0.716 ± 0.30	0.680 ± 0.21	$\mathbf{0.628 \pm 0.26}$

4   Results

Experimental Results on Synthetic Data. The visual comparison between our MAIN and other methods are shown in Fig. 3. Traditional method NMAR suffers from severe secondary artifacts due to interpolation errors, leading to blurred detailed structures. Although deep learning-based methods gets better results, the detailed structures around the metals are still blamed. In contrast to the above methods, our MAIN achieves the best result in metal artifact reduction and detail restoration. Table 1 presents the quantitative evaluation results of different methods, demonstrating that the MAIL network achieves the best performance with the highest values of SSIM (0.984) and RMSE (0.628).

Moreover, Fig. 4(a) displays the statistical results of different methods on the test set. Figure 4(b) shows the noise power spectrum (NPS) maps [2]. Figure 4(c) shows the intensity profiles [4] of different MAR results. These results demonstrate that the proposed method is highly stable and effective in dealing with different metallic implants.

Experimental Results on Clinical Data. Figure 5 shows the visual comparison on a clinical CT image with metal artifacts. It can be observed that the secondary artifacts caused by NMAR are severe, and other baselines would blur the tissue near the metallic implants. In comparison, our MAIN has achieved the best results in reducing metal artifacts and recovering tissue details.

Ablation Studies. We re-configure our network and retrain the model in image domain and wavelet domain, respectively. We also utilize 'db3' wavelet to re-configure the network and compare the corresponding model with the adaptive wavelet. It can be observed from Fig. 6 and Table 2 that our approach for reducing metal artifacts in both the wavelet and image domains is more effective than the single domain scheme. Moreover, the effectiveness of the adaptive wavelet transform is also confirmed.

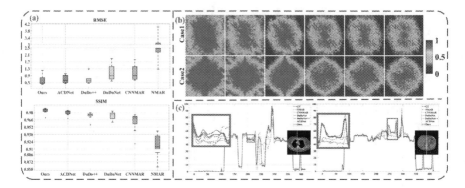

Fig. 4. (a) Visualization of statistical results on 2000 testing data. (b) NPS maps of MAR in cases 1 and 2. The 1st - 6th columns denote NMAR, CNNMAR, DuDoNet, DuDoNet++, ACDNet and our method. (c) The intensity profiles along the specified yellow line. (Color figure online)

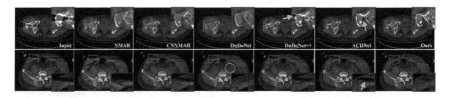

Fig. 5. Visual comparisons of MAR results on clinical images. The display window is [−142 532] HU.

Fig. 6. Effectiveness verification of dual-domain and adaptive wavelet scheme.

Table 2. Quantitative analysis results of our methods on synthetic dataset

Method	wavelet domain	image domain	normal wavelet	ours
SSIM	0.978 ± 0.009	0.983 ± 0.005	0.983 ± 0.006	$\mathbf{0.984 \pm 0.005}$
RMSE	0.823 ± 0.345	0.664 ± 0.271	0.663 ± 0.307	$\mathbf{0.628 \pm 0.263}$

5   Discussion and Conclusion

In this study, we propose a multi-perspective adaptive iteration learning network for metal artifact reduction. The multi-perspective regularizations introduces prior knowledge in wavelet and image domain to constrain the feasible region of the solution space. The employment of adaptive wavelet transform makes full use of the powerful learning ability of data-driven model, and it enhances the flexibility of the model. The comprehensive experiments on both of a synthetic and a clinical datasets have consistently verified the effectiveness of our method.

Acknowledgements. This work was supported in part by National Natural Science Foundation of China (grant numbers 62101611 and 62201628), National Key Research and Development Program of China (2022YFA1204200), Guangdong Basic and Applied Basic Research Foundation (grant number 2022A1515011375,2023A1515012278, 2023A1515011780) and Shenzhen Science and Technology Program (grant number JCYJ20220530145411027, JCYJ20220818102414031).

References

1. Beck, A., Teboulle, M.: A fast iterative Shrinkage-Thresholding algorithm for linear inverse problems. SIAM J. Imaging Sci. **2**(1), 183–202 (2009)
2. Diwakar, M., Kumar, M.: A review on CT image noise and its denoising. Biomed. Signal Process. Control **42**, 73–88 (2018)
3. Ghani, M.U., Karl, W.: Deep learning based sinogram correction for metal artifact reduction. Electron. Imaging **2018**, 4721–4728 (2018)
4. Ghose, S., Singh, N., Singh, P.: Image denoising using deep learning: convolutional neural network. In: 2020 10th International Conference on Cloud Computing, Data Science & Engineering (Confluence), pp. 511–517 (2020)
5. Gjesteby, L., et al.: Metal artifact reduction in CT: where are we after four decades? IEEE Access **4**, 5826–5849 (2016)
6. Kalender, W.A., Hebel, R., Ebersberger, J.: Reduction of CT artifacts caused by metallic implants. Radiology **164**(2), 576–577 (1987)
7. Katsura, M., Sato, J., Akahane, M., Kunimatsu, A., Abe, O.: Current and novel techniques for metal artifact reduction at CT: practical guide for radiologists. Radiographics **38**(2), 450–461 (2018)
8. Lin, W.A., et al.: DuDoNet: dual domain network for CT metal artifact reduction. In: Proceedings of the IEEE/CVF Conference on Computer Vision and Pattern Recognition (CVPR) (2019)
9. Liu, P., et al.: Deep learning to segment pelvic bones: large-scale CT datasets and baseline models. Int. J. Comput. Assist. Radiol. Surg. **16**(5), 749–756 (2021). https://doi.org/10.1007/s11548-021-02363-8
10. Luo, Y., et al.: Adaptive rectification based adversarial network with spectrum constraint for high-quality PET image synthesis. Med. Image Anal. **77**, 102335 (2022)
11. Lyu, Y., Lin, W.-A., Liao, H., Lu, J., Zhou, S.K.: Encoding metal mask projection for metal artifact reduction in computed tomography. In: Martel, A.L., et al. (eds.) MICCAI 2020. LNCS, vol. 12262, pp. 147–157. Springer, Cham (2020). https://doi.org/10.1007/978-3-030-59713-9_15
12. Medoff, B.P., Brody, W.R., Nassi, M., Macovski, A.: Iterative convolution backprojection algorithms for image reconstruction from limited data. J. Opt. Soc. Am. **73**(11), 1493–1500 (1983)
13. Meyer, E., Raupach, R., Lell, M., Schmidt, B., Kachelrieß, M.: Normalized metal artifact reduction (NMAR) in computed tomography. Med. Phys. **37**(10), 5482–5493 (2010)

14. Pan, J., Zhang, H., Wu, W., Gao, Z., Wu, W.: Multi-domain integrative swin transformer network for sparse-view tomographic reconstruction. Patterns **3**(6), 100498 (2022)
15. Park, H.S., Lee, S.M., Kim, H.P., Seo, J.K., Chung, Y.E.: CT sinogram-consistency learning for metal-induced beam hardening correction. Med. Phys. **45**(12), 5376–5384 (2018)
16. Paszke, A., et al.: Automatic differentiation in PyTorch (2017)
17. Peng, C., et al.: An irregular metal trace inpainting network for X-ray CT metal artifact reduction. Med. Phys. **47**(9), 4087–4100 (2020)
18. Rodriguez, M.X.B., et al.: Deep adaptive wavelet network. In: Proceedings of the IEEE/CVF Winter Conference on Applications of Computer Vision (WACV) (2020)
19. Ronneberger, O., Fischer, P., Brox, T.: U-Net: convolutional networks for biomedical image segmentation. In: Navab, N., Hornegger, J., Wells, W.M., Frangi, A.F. (eds.) MICCAI 2015. LNCS, vol. 9351, pp. 234–241. Springer, Cham (2015). https://doi.org/10.1007/978-3-319-24574-4_28
20. Rousselle, A., et al.: Metallic implants and CT artefacts in the CTV area: where are we in 2020? Cancer/Radiothérapie **24**(6), 658–666 (2020). 31e Congrès national de la Société française de radiothérapie oncologique
21. Verburg, J.M., Seco, J.: CT metal artifact reduction method correcting for beam hardening and missing projections. Phys. Med. Biol. **57**(9), 2803 (2012)
22. Wang, H., Li, Y., Meng, D., Zheng, Y.: Adaptive convolutional dictionary network for CT metal artifact reduction. In: Raedt, L.D. (ed.) Proceedings of the Thirty-First International Joint Conference on Artificial Intelligence, IJCAI 2022, pp. 1401–1407. International Joint Conferences on Artificial Intelligence Organization (2022)
23. Wang, H., et al.: InDuDoNet: an interpretable dual domain network for CT metal artifact reduction. In: de Bruijne, M., et al. (eds.) MICCAI 2021. LNCS, vol. 12906, pp. 107–118. Springer, Cham (2021). https://doi.org/10.1007/978-3-030-87231-1_11
24. Wang, J., Zhao, Y., Noble, J.H., Dawant, B.M.: Conditional generative adversarial networks for metal artifact reduction in CT images of the ear. In: Frangi, A.F., Schnabel, J.A., Davatzikos, C., Alberola-López, C., Fichtinger, G. (eds.) MICCAI 2018. LNCS, vol. 11070, pp. 3–11. Springer, Cham (2018). https://doi.org/10.1007/978-3-030-00928-1_1
25. Wang, Y., et al.: 3D auto-context-based locality adaptive multi-modality GANs for PET synthesis. IEEE Trans. Med. Imaging **38**(6), 1328–1339 (2019)
26. Wellenberg, R., Hakvoort, E., Slump, C., Boomsma, M., Maas, M., Streekstra, G.: Metal artifact reduction techniques in musculoskeletal CT-imaging. Eur. J. Radiol. **107**, 60–69 (2018)
27. Yan, K., et al.: Deep lesion graphs in the wild: relationship learning and organization of significant radiology image findings in a diverse large-scale lesion database. In: Proceedings of the IEEE Conference on Computer Vision and Pattern Recognition (CVPR) (2018)
28. Yu, L., Zhang, Z., Li, X., Xing, L.: Deep sinogram completion with image prior for metal artifact reduction in CT images. IEEE Trans. Med. Imaging **40**(1), 228–238 (2021)
29. Zhan, B., et al.: Multi-constraint generative adversarial network for dose prediction in radiotherapy. Med. Image Anal. **77**, 102339 (2022)

30. Zhang, K., Zuo, W., Chen, Y., Meng, D., Zhang, L.: Beyond a Gaussian denoiser: residual learning of deep CNN for image denoising. IEEE Trans. Image Process. **26**(7), 3142–3155 (2017)
31. Zhang, Y., Yu, H.: Convolutional neural network based metal artifact reduction in X-ray computed tomography. IEEE Trans. Med. Imaging **37**(6), 1370–1381 (2018)
32. Zhou, B., Chen, X., Zhou, S.K., Duncan, J.S., Liu, C.: DuDoDR-Net: dual-domain data consistent recurrent network for simultaneous sparse view and metal artifact reduction in computed tomography. Med. Image Anal. **75**, 102289 (2022)

Noise Conditioned Weight Modulation for Robust and Generalizable Low Dose CT Denoising

Sutanu Bera$^{(\boxtimes)}$ and Prabir Kumar Biswas

Department of Electronics and Electrical Communication Engineering, Indian
Institute of Technology Kharagpur, Kharagpur, India
`sutanu.bera@iitkgp.ac.in`

Abstract. Deep neural networks have been extensively studied for
denoising low-dose computed tomography (LDCT) images, but some
challenges related to robustness and generalization still need to be
addressed. It is known that CNN-based denoising methods perform optimally when all the training and testing images have the same noise variance, but this assumption does not hold in the case of LDCT denoising.
As the variance of the CT noise varies depending on the tissue density
of the scanned organ, CNNs fails to perform at their full capacity. To
overcome this limitation, we propose a novel noise-conditioned feature
modulation layer that scales the weight matrix values of a particular
convolutional layer based on the noise level present in the input signal. This technique creates a neural network that is conditioned on the
input image and can adapt to varying noise levels. Our experiments on
two public benchmark datasets show that the proposed dynamic convolutional layer significantly improves the denoising performance of the
baseline network, as well as its robustness and generalization to previously unseen noise levels.

Keywords: LDCT denoising · Dynamic Convolution · CT noise
variance

1 Introduction

Convolutional neural networks (CNN) have emerged as one of the most popular methods for noise removal and restoration of LDCT images [1,2,5,6,14].
While CNNs can produce better image quality than manually designed functions, there are still some challenges that hinder their widespread adoption in
clinical settings. Convolutional denoisers are known to perform best when the
training and testing images have similar or identical noise variance [15,16]. On
the other hand, different anatomical sites of the human body have different tissue densities and compositions, which affects the amount of radiation that is
absorbed and scattered during CT scanning; as a result, noise variance in LDCT
images also varies significantly among different sites of the human body [13].

© The Author(s), under exclusive license to Springer Nature Switzerland AG 2023
H. Greenspan et al. (Eds.): MICCAI 2023, LNCS 14229, pp. 88–97, 2023.
https://doi.org/10.1007/978-3-031-43999-5_9

Furthermore, the noise variance is also influenced by the differences in patient size and shape, imaging protocol, etc. [11]. Because of this, CNN-based denoising networks fail to perform optimally in LDCT denoising. In this study, we have introduced a novel dynamic convolution layer to combat the issue of noise level variability in LDCT images. Dynamic convolution layer is a type of convolutional layer in which the convolutional kernel is generated dynamically at each layer based on the input data [3,4,8]. Unlike the conventional dynamic convolution layer, here we have proposed to use a modulating signal to scale the value of the weight vector(learned via conventional backpropagation) of a convolutional layer. The modulating signal is generated dynamically from the input image using an encoder network. The proposed method is very simple, and learning the network weight is a straightforward one-step process, making it manageable to deploy and train. We evaluated the proposed method on the recently released large-scale LDCT database of TCIA Low Dose CT Image and Projection Data [10] and the 2016 NIH-AAPM-Mayo Clinic low dose CT grand challenge database [9]. These databases contain low-dose CT data from three anatomical sites, i.e., head, chest, and abdomen. Extensive experiments on these databases validate the proposed method improves the baseline network's performance significantly. Furthermore, we have shown the generalization ability to the out-of-distribution data, and the robustness of the baseline network is also increased significantly via using the proposed weight-modulated dynamic convolutional layer.

2   Method

Motivation: Each convolutional layer in a neural network performs the sum of the product operation between the weight vector and input features. However, as tissue density changes in LDCT images, the noise intensity also changes, leading to a difference in the magnitude of intermediate feature values. If the variation in input noise intensity is significant, the magnitude of the output feature of the convolutional layer can also change substantially. This large variation in input feature values can make the CNN layer's response unstable, negatively impacting the denoising performance. To address this issue, we propose to modulate the weight vector values of the CNN layer based on the noise level of the input image. This approach ensures that the CNN layer's response remains consistent, even when the input noise variance changes drastically.

Weight Modulation: Figure 1 depicts our weight modulation technique, which involves the use of an additional anatomy encoder network, $\mathcal{E}_a$, along with the backbone denoising network, CNN_D. The output of the anatomy encoder, denoted as e_x, is a D-dimensional embedding, i.e., $e_x = \mathcal{E}_a(\nabla^2(x))$. Here, x is the input noisy image, and $\nabla^2(.)$ is a second-order Laplacian filter. This embedding e_x serves as a modulating signal for weight modulation in the main denoising network (CNN_D). Specifically, the lth weight-modulated convolutional layer, $\mathcal{F}_l$, of the backbone network, CNN_D, takes the embedding e_x as input. Then the embedding e_x is passed to a 2 Layer MLP, denoted as ϕ_l, which learns a non-linear mapping between the layer-specific code, denoted as $s_l \in \mathbb{R}^{N_l}$, and the

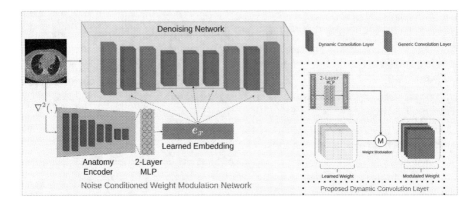

Fig. 1. Overview of the proposed noise conditioned weight modulation framework.

embedding e_x, i.e., $s_l = \phi_l(e_x)$. Here, N_l represents the number of feature maps in the layer $\mathcal{F}_l$. The embedding e_x can be considered as the high dimensional code containing the semantics information and noise characteristic of the input image. The non-linear mapping ϕ_l maps the embedding e_x to a layer-specific code s_l, so that different layers can be modulated differently depending on the depth and characteristic of the features. Let $w_l \in \mathbb{R}^{N_l \times N_{l-1} \times k \times k}$ be the weight vector of $\mathcal{F}_l$ learned via standard back-propagation learning. Here $(k \times k)$ is the size of the kernel, N_{l-1} is the number of feature map in the previous layer. Then the w_l is modulated using s_l as following,

$$\hat{w}_l = w_l \odot s_l \tag{1}$$

Here, $\hat{w}_l$ is the modulated weight value, and $\odot$ represents component wise multiplication. Next, the scaled weight vector is normalized by its L2 norm across channels as follows:

$$\tilde{w}_l = \hat{w}_l \Big/ \sqrt{\sum_{N_{l-1},k,k} \hat{w}_l^2 + \epsilon} \tag{2}$$

Normalizing the modulated weights takes care of any possible instability arise due to high or too low weight value and also ensures that the modulated weight has consistent scaling across channels, which is important for preserving the spatial coherence of the denoised image [7]. The normalized weight vectors, $\tilde{w}_l$ are then used for convolution, i.e., $f_l = \mathcal{F}_l\big(\tilde{w}_l * f_{l-1}\big)$. Here, f_l, and f_{l-1} are the output feature map of lth, $l-1$th layer, and $*$ is the convolution operation.

Relationship with Recent Methods: The proposed weight modulation technique leveraged the recent concept of style-based image synthesis proposed in StyleGAN2 [7]. However, StyleGAN2 controlled the structure and style of the generated image by modulating weight vectors using random noise and latent code. Whereas, we have used weight modulation for dynamic filter generation conditioned on input noisy image to generate a consistent output image.

Implementation Details: The proposed dynamic convolutional layer is very generic and can be integrated into various backbone networks. For our denoising task, we opted for the encoder-decoder-based UNet [12] architecture and replaced some of its generic convolutional layers with our weight-modulated dynamic convolution layer. To construct the anatomy encoder network, we employed ten convolutional blocks and downscaled the input feature map's spatial resolution by a factor of nine through two max-pooling operations inside the network. We fed the output of the last convolutional layer into a global average pooling layer to generate a 512-dimensional feature vector. This vector was then passed through a 2-layer MLP to produce the final embedding, $e_x \in \mathbb{R}^{512}$.

3 Experimental Setting

We used two publicly available data sets, namely, 1. TCIA Low Dose CT Image and Projection Data, 2. 2016 NIH-AAPM-Mayo Clinic low dose CT grand challenge database to validate the proposed method. The first dataset contains LDCT data of different patients of three anatomical sites, i.e., head, chest, and abdomen, and the second dataset contains LDCT images of the abdomen with two different slice thicknesses (3 mm, 1 mm). We choose 80% data from each anatomical site for training and the remaining 20% for testing. We used the Adam optimizer with a batch size of 16. The learning rate was initially set to $1e^{-4}$ and was assigned to decrease by a factor of 2 after every 6000 iterations.

Table 1. Objective and computational cost comparison between different methods. Objective metrics are reported by averaging the values for all the images present in the test set.

Model	Abdomen			Head			Chest			FLOPs
	PSNR	SSIM	RMSE	PSNR	SSIM	RMSE	PSNR	SSIM	RMSE	
M1	33.84	0.912	8.46	39.45	0.957	2.42	29.39	0.622	103.27	75.53G
M2	34.15	0.921	7.41	40.04	0.968	2.02	29.66	.689	89.23	98.47G

4 Result and Discussion

Comparison with Baseline: This section discusses the efficacy of the proposed weight modulation technique, comparing it with a baseline UNet network (M1) and the proposed weight-modulated convolutional network (M2). The networks were trained using LDCT images from a single anatomical region and tested on images from the same region. Table 1 provides an objective comparison between the two methods in terms of PSNR, SSIM, and RMSE for different anatomical regions. The results show that the proposed dynamic weight modulation technique significantly improved the denoising performance of the baseline UNet for all settings. For example, the PSNR for head images was improved by 0.59 dB, and similar improvements were observed for other anatomical regions. Additionally, Table 1 shows the floating point computational requirements of the different

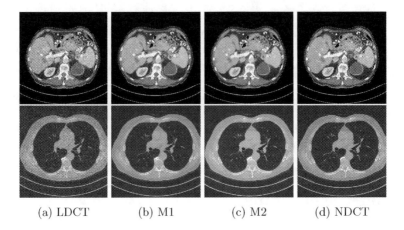

(a) LDCT   (b) M1   (c) M2   (d) NDCT

Fig. 2. Result of Denoising for comparison. The display window for the abdomen image (top row) is set to $[-140, 260]$, and $[-1200, 600]$ for the chest image.

methods. It can be seen that the number of FLOPs of the dynamic weight modulation technique is not considerably higher than the baseline network M1, yet the improvement in performance is much appreciable.

In Fig. 2, we provide a visual comparison of the denoised output produced by different networks. Two sample images from datasets D1 and D2, corresponding to the abdomen and chest regions, respectively, are shown. The comparison shows that the proposed network M2 outperforms the baseline model M1 in terms of noise reduction and details preservation. For instance, in the denoised image of the abdomen region, the surface of the liver in M1 appears rough and splotchy due to noise, while in M2, the image is crisp, and noise suppression is adequate. Similarly, in the chest LDCT images, noticeable streaking artifacts near the breast region are present in the M1 output, and the boundaries of different organs like the heart and shoulder blade are not well-defined. In contrast, M2 produces crisp and definite boundaries, and streaking artifacts are significantly reduced. Moreover, M1 erases finer details like tiny blood vessels in the lung region, leading to compromised visibility, while M2 preserves small details much better than M1, resulting in output that is comparable with the original NDCT image.

Robustness Analysis: In this section, we evaluate the performance of existing denoising networks in a challenging scenario where the networks are trained to remove noise from a mixture of LDCT images taken from different anatomical regions with varying noise variances and patterns. We compared two networks in this analysis: M3, which is a baseline UNet model trained using a mixture of LDCT images, and M4, which is the proposed weight-modulated network, trained using same training data. Table 2 provides an objective comparison between these two methods. We found that joint training has a negative impact on the performance of the baseline network, M3, by a significant margin. Specifically, M3 yielded 0.88 dB lower PSNR than model M1 for head images, which

Table 2. Objective comparison among different methods. Objective metrics are reported by averaging the values for all the images present in the test set.

Model	Abdomen			Head			Chest		
M3	33.64	0.895	8.54	38.67	0.937	3.45	29.28	0.612	105.2
M4	34.17	0.921	7.45	39.70	0.964	2.12	29.69	0.689	89.21

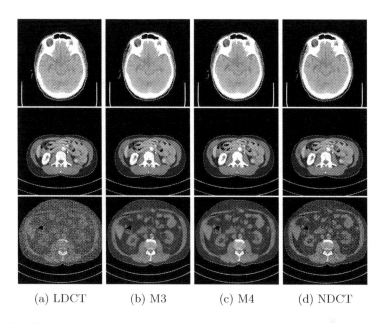

(a) LDCT   (b) M3   (c) M4   (d) NDCT

Fig. 3. Result of Denoising for comparison. The display window for the abdomen image is set to $[-140, 260]$, $[-175, 240]$ for the chest image, and $[-80, 100]$ for the head.

were trained using only head images. Similar observations were also noted for other anatomical regions like the abdomen and chest. The differences in noise characteristics among the different LDCT images make it difficult for a single model to denoise images efficiently from a mixture of anatomical regions. Furthermore, the class imbalance between small anatomical sites (e.g., head, knee, and prostate) and large anatomical locations (e.g., lung, abdomen) in a training set introduces a bias towards large anatomical sites, resulting in unacceptably lower performance for small anatomical sites. On the other hand, M4 showed robustness to these issues. Its performance was similar to M2 for all settings, and it achieved 0.69 dB higher PSNR than M3. Noise-conditioned weight modulation enables the network to adjust its weight based on the input images, allowing it to denoise every image with the same efficiency.

Figure 3 provides a visual comparison of the denoising performance of two methods on LDCT images from three anatomical regions. The adverse effects of joint training on images from different regions are apparent. Head LDCT images, which had the lowest noise, experienced a loss of structural and textural information in the denoising process by baseline M3. For example, the head lobes appeared distorted in the reconstructed image. Conversely, chest LDCT images, which were the noisiest, produced artefacts in the denoised image by M3, significantly altering the image's visual appearance. In contrast, M4 preserved all structural information and provided comparable noise reduction across all anatomical structures. CNN-based denoising networks act like a subtractive method, where the network learns to subtract the noise from the input signal by using a series of convolutional layers. A fixed set of subtracters is inefficient for removing noise from images with various noise levels. As a result, images with low noise are over smoothed and structural information is lost, whereas images with high noise generate residual noise and artefacts. In case of images containing a narrow range of noise levels, such as images from a single anatomical region, the above-mentioned limitation of naive CNN-based denoisers remains acceptable, but when a mixture of images with diverge noise levels is used in training and testing, it becomes problematic. The proposed noise conditioned weight modulation addresses this major limitation of CNN based denoising network, by designing an adjustable subtractor which is adjusted based on the input signal.

Figure 4 presents a two-dimensional projection of the learned embedding for all the test images using the TSNE transformation. The embedding has created three distinct clusters in the 2D feature space, each corresponding to images from one of three different anatomical regions. This observation validates our claim that the embedding learned by the anatomy encoder represents a meaningful representation of the input image. Notably, the noise level of low dose chest CT images differs significantly from those of the other two regions, resulting in a

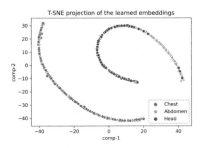

Fig. 4. 2 dimensional projection of learned embedding. The projection are learned using TSNE transformation.

Table 3. Objective comparison among different networks. Objective metrics are reported by averaging the values for all the images present in the test set of abdominal images taken with 1mm slice thickness.

Model	M5	M6	M7	M8
PSNR	22.23	22.55	22.80	22.96
SSIM	0.759	0.762	0.777	0.788
RMSE	32.13	30.13	29.37	29.14

separate cluster that is located at a slightly greater distance from the other two clusters.

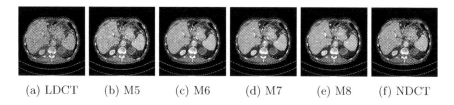

(a) LDCT   (b) M5   (c) M6   (d) M7   (e) M8   (f) NDCT

Fig. 5. Result of Denoising for comparison. The display window for the abdomen image is set to $[-140, 260]$

Generalization Analysis: In this section, we evaluate the generalization ability of different networks on out-of-distribution test data using LDCT abdomen images taken with a 1mm slice thickness from dataset D1. We consider four networks for this analysis: 1) M5, the baseline UNet trained on LDCT abdomen images with a 3mm slice thickness from dataset D1, 2) M6, the baseline UNet trained on a mixture of LDCT images from all anatomical regions except the abdomen with a 1mm slice thickness, 3) M7, the proposed weight-modulated network trained on the same training set as M6, and 4) M8, the baseline UNet trained on LDCT abdomen images with a 1mm slice thickness. Objective comparisons among these networks are presented in Table 3. The results show that the performance of M5 and M6 is poor on this dataset, indicating their poor ability to generalize to unseen data. In contrast, M7 performs similarly to the supervised model M8. Next, we compared the denoising performance of different methods visually in Fig. 5. It can be seen that M5 completely failed to remove noise from these images despite the fact the M5 was trained using the abdominal image. Now the output of M6 is better than the M5 in terms of noise removal, but a lot of over-smoothness and loss of structural information can be seen, for example, the over-smooth texture of the liver and removal of blood vessels. M6 benefits from being trained on diverse LDCT images, which allows it to learn robust features applicable to a range of inputs and generalize well to new images. However, the CNN networks' limited ability to handle diverse noise levels results in M6 failing to preserve all the structural information in some cases. In contrast, M7 uses a large training set and dynamic convolution to preserve all structural information and remove noise effectively, comparable to the baseline model M8.

5   Conclusion

This study proposes a novel noise-conditioned feature modulation layer to address the limitations of convolutional denoising networks in handling variability in noise levels in low-dose computed tomography (LDCT) images. The proposed technique modulates the weight matrix of a convolutional layer according to the noise present in the input signal, creating a slightly modified neural network. Experimental results on two public benchmark datasets demonstrate that this dynamic convolutional layer significantly improves denoising performance, as well as robustness and generalization to unseen noise levels. The proposed method has the potential to enhance the accuracy and reliability of LDCT image analysis in various clinical applications.

References

1. Bera, S., Biswas, P.K.: Noise conscious training of non local neural network powered by self attentive spectral normalized Markovian patch GAN for low dose CT denoising. IEEE Trans. Med. Imaging **40**(12), 3663–3673 (2021). https://doi.org/10.1109/TMI.2021.3094525
2. Chen, H., et al.: Low-dose CT with a residual encoder-decoder convolutional neural network. IEEE Trans. Med. Imaging **36**(12), 2524–2535 (2017)
3. He, T., Shen, C., Van Den Hengel, A.: DyCo3D: robust instance segmentation of 3D point clouds through dynamic convolution. In: Proceedings of the IEEE/CVF Conference on Computer Vision and Pattern Recognition, pp. 354–363 (2021)
4. Jia, X., De Brabandere, B., Tuytelaars, T., Gool, L.V.: Dynamic filter networks. In: Advances in Neural Information Processing Systems, vol. 29 (2016)
5. Kang, E., Chang, W., Yoo, J., Ye, J.C.: Deep convolutional framelet denosing for low-dose CT via wavelet residual network. IEEE Trans. Med. Imaging **37**(6), 1358–1369 (2018)
6. Kang, E., Min, J., Ye, J.C.: A deep convolutional neural network using directional wavelets for low-dose X-ray CT reconstruction. Med. Phys. **44**(10), e360–e375 (2017)
7. Karras, T., Laine, S., Aittala, M., Hellsten, J., Lehtinen, J., Aila, T.: Analyzing and improving the image quality of StyleGAN. In: Proceedings of the IEEE/CVF Conference on Computer Vision and Pattern Recognition, pp. 8110–8119 (2020)
8. Klein, B., Wolf, L., Afek, Y.: A dynamic convolutional layer for short range weather prediction. In: Proceedings of the IEEE Conference on Computer Vision and Pattern Recognition, pp. 4840–4848 (2015)
9. McCollough, C.H., et al.: Low-dose CT for the detection and classification of metastatic liver lesions: results of the 2016 low dose CT grand challenge. Med. Phys. **44**(10), e339–e352 (2017)
10. Moen, T.R., et al.: Low-dose CT image and projection dataset. Med. Phys. **48**(2), 902–911 (2021)
11. Murphy, A., Bell, D., Rock, P., et al.: Noise (CT). Reference article, Radiopaedia.org (2023). https://doi.org/10.53347/rID-51832. Accessed 08 Mar 2023
12. Ronneberger, O., Fischer, P., Brox, T.: U-Net: convolutional networks for biomedical image segmentation. In: Navab, N., Hornegger, J., Wells, W.M., Frangi, A.F. (eds.) MICCAI 2015, Part III. LNCS, vol. 9351, pp. 234–241. Springer, Cham (2015). https://doi.org/10.1007/978-3-319-24574-4_28
13. Sprawls, P.: AAPM tutorial. CT image detail and noise. Radiographics **12**(5), 1041–1046 (1992)
14. Yin, X., et al.: Domain progressive 3D residual convolution network to improve low-dose CT imaging. IEEE Trans. Med. Imaging **38**(12), 2903–2913 (2019)

15. Zhang, K., Zuo, W., Chen, Y., Meng, D., Zhang, L.: Beyond a Gaussian denoiser: residual learning of deep CNN for image denoising. IEEE Trans. Image Process. **26**(7), 3142–3155 (2017)
16. Zhang, K., Zuo, W., Zhang, L.: FFDNet: toward a fast and flexible solution for CNN-based image denoising. IEEE Trans. Image Process. **27**(9), 4608–4622 (2018)

Low-Dose CT Image Super-Resolution Network with Dual-Guidance Feature Distillation and Dual-Path Content Communication

Jianning Chi[(✉)], Zhiyi Sun, Tianli Zhao, Huan Wang, Xiaosheng Yu, and Chengdong Wu

Faculty of Robot Science and Engineering, Northeastern University, Shenyang, China
chijianning@mail.neu.edu.cn

Abstract. Low-dose computer tomography (LDCT) has been widely used in medical diagnosis yet suffered from spatial resolution loss and artifacts. Numerous methods have been proposed to deal with those issues, but there still exists drawbacks: (1) convolution without guidance causes essential information not highlighted; (2) features with fixed-resolution lose the attention to multi-scale information; (3) single super-resolution module fails to balance details reconstruction and noise removal. Therefore, we propose an LDCT image super-resolution network consisting of a dual-guidance feature distillation backbone for elaborate visual feature extraction, and a dual-path content communication head for artifacts-free and details-clear CT reconstruction. Specifically, the dual-guidance feature distillation backbone is composed of a dual-guidance fusion module (DGFM) and a sampling attention block (SAB). The DGFM guides the network to concentrate the feature representation of the 3D inter-slice information in the region of interest (ROI) by introducing the average CT image and segmentation mask as complements of the original LDCT input. Meanwhile, the elaborate SAB utilizes the essential multi-scale features to capture visual information more relative to edges. The dual-path reconstruction architecture introduces the denoising head before and after the super-resolution (SR) head in each path to suppress residual artifacts, respectively. Furthermore, the heads with the same function share the parameters so as to efficiently improve the reconstruction performance by reducing the amount of parameters. The experiments compared with 6 state-of-the-art methods on 2 public datasets prove the superiority of our method. The code is made available at https://github.com/neu-szy/dual-guidance_LDCT_SR.

Keywords: Low-dose computed tomography · Image denoising · Image super-resolution · Deep learning

H. Greenspan et al. (Eds.): MICCAI 2023, LNCS 14229, pp. 98–108, 2023.
https://doi.org/10.1007/978-3-031-43999-5_10

1 Introduction

Following the "as low as reasonably achievable" (ALARA) principle [22], low-dose computer tomography (LDCT) has been widely used in various medical applications, for example, clinical diagnosis [18] and cancer screening [28]. To balance the high image quality and low radiation damage compared to normal-dose CT (NDCT), numerous algorithms have been proposed for LDCT super-resolution [3,4].

In the past decades, image post-processing techniques attracted much attention from researchers because they did not rely on the vendor-specific parameters [2] like iterative reconstruction algorithms [1,23] and could be easily applied to current CT workflows [29]. Image post-processing super-resolution (SR) methods could be divided into 3 categories: interpolated-based methods [16,25], model-based methods [13,14,24,26] and learning-based methods [7–9,17]. Interpolated-based methods could recover clear results in those flattened regions but failed to reconstruct detailed textures because they equally recover information with different frequencies. And model-based methods often involved time-consuming optimization processes and degraded quickly when image statistics were biased from the image prior [6].

With the development of deep learning (DL), various learning-based methods have been proposed, such as EDSR [20], RCAN [31], and SwinIR [19]. Those methods optimized their trainable parameters by pre-degraded low-resolution (LR) and high-resolution (HR) pairs to build a robust model with generalization and finally reconstruct SR images. However, they were designed for known degradation (for example bicubic degradation) and failed to deal with more complex and unknown degradation processes (such as LDCT degradation). Facing more complex degradation processes, blind SR methods have attracted attention. Huang et al. [11] introduced a deep alternating network (DAN) which estimated the degradation kernels and corrected those kernels iteratively and reconstructed results following the inverse process of the estimated degradation. More recently, aiming at improving the quality of medical images further, Huang et al. [12] first composited degradation model proposed for radiographs and proposed attention denoising super-resolution generative adversarial network (AID-SRGAN) which could denoise and super-resolve radiographs simultaneously. To accurately reconstruct HR CT images from LR CT images, Hou et al. [10] proposed a dual-channel joint learning framework which could process the denoising reconstruction and SR reconstruction in parallel.

The aforementioned methods still have drawbacks: (1) They treated the regions of interest (ROI) and regions of uninterest equally, resulting in the extra cost in computing source and inefficient use for hierarchical features. (2) Most of them extracted the features with a fixed resolution, failing to effectively leverage multi-scale features which are essential to image restoration task [27,32]. (3) They connected the SR task and the LDCT denoising task stiffly, leading to smooth texture, residual artifacts and unclear edges.

To deal with those issues, as shown in Fig. 1(a), we propose an LDCT image SR network with dual-guidance feature distillation and dual-path content com-

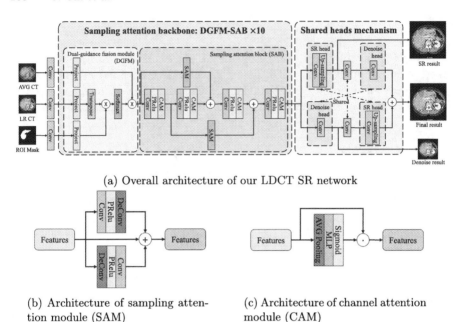

(a) Overall architecture of our LDCT SR network

(b) Architecture of sampling atten-   (c) Architecture of channel attention
tion module (SAM)   module (CAM)

Fig. 1. Architecture of our proposed method. SAM is sampling attention module. CAM
is channel attention module. AVG CT is the average image among adjacent CT slices
of each patient.

munication. Our contributions are as follows: (1) We design a dual-guidance
fusion module (DGFM) which could fuse the 3D CT information and ROI guidance by mutual attention to make full use of CT features and reconstruct clearer
textures and sharper edges. (2) We propose a sampling attention block (SAB)
which consists of sampling attention module (SAM), channel attention module
(CAM) and elaborate multi-depth residual connection aiming at the essential
multi-scale features by up-sampling and down-sampling to leverage the features
in CT images. (3) We design a multi-supervised mechanism based on shared
task heads, which introducing the denoising head into SR task to concentrate on
the connection between the SR task and the denoising task. Such design could
suppress more artifacts while decreasing the number of parameters.

2   Method

2.1   Overall Architecture

The pipeline of our proposed method is shown in Fig. 1(a). We first calculate the
average CT image of adjacent CT slices of each patient to provide the 3D spatial
structure information of CT volume. Meanwhile, the ROI mask is obtained by
a pre-trained segmentation network to guide the network to concentrate on the
focus area or tissue area. Then those guidance images and the input LDCT image

are fed to the dual-guidance feature distillation backbone to extract the deep features. Finally, the proposed dual-path architecture consisting of parameter-shared SR heads and denoising heads leverages the deep visual features obtained by our backbone to build the connection between the SR task and the denoising task, resulting in noise-free and detail-clear reconstructed results.

Dual-Guidance Feature Distillation Backbone. To decrease the redundant computation and make full use of the above-mentioned extra information, we design a dual-guidance feature distillation backbone consisting of a dual-guidance fusion module (DGFM) and sampling attention block(SAB).

Firstly, we use a 3×3 convolutional layer to extract the shallow features of the three input images. Then, those features are fed into 10 DGFM-SAB blocks to obtain the deep visual features.

Especially, the DGFM-SAB block is composed of DGFM concatenated with SAB. Considering the indicative function of ROI, we calculate the correlation matrix between LDCT and its mask and then acquire the response matrix between the correlation matrix and the average CT image by multi-heads attention mechanism:

$$F_i = Softmax[Prj(F_i^{SAB})^T \times Prj(F_{mask})] \times Prj(F_{AVG}) \tag{1}$$

where, F_i^{SAB} are the output of i-th SAB. F_{mask} and F_{AVG} represent the shallow features of the input ROI mask and the average CT image respectively. Meanwhile, $Prj(\cdot)$ is the projection function, $Softmax[\cdot]$ means the softmax function and F_i are the output features of the i-th DGFM. The DGFM helps the backbone to focus on the ROI and tiny structural information by continuously introducing additional guidance information.

Furthermore, to take advantage of the multi-scale information which is essential for obtaining the response matrix containing the connections between different levels of features, as shown in Fig. 1(b), we design the sampling attention block (SAB) which introduces the resampling features into middle connection to fuse the multi-scale information. In the SAB, the input features are up-sampled and down-sampled simultaneously and then down-sampled and up-sampled to recover the spatial resolution, which can effectively extract multi-scale features. In addition, as shown in Fig. 1(c), we introduce the channel attention module (CAM) to focus on those channels with high response values, leading to detailed features with high differentiation to different regions.

Shared Heads Mechanism. Singly using the SR head that consists of Pixel Shuffle layer and convolution layer fails to suppress the residual artifacts because of its poor noise removal ability. To deal with this problem, we develop a dual-path architecture by introducing the shared denoising head into SR task where the parameters of SR heads and denoising heads in different paths are shared respectively. Two paths are designed to process the deep features extracted from our backbone: (1) The SR path transfers the deep features to those with high-frequency information and reconstructs the SR result, and (2) the denoising path migrates the deep features to those without noise and recovers the clean result secondly. Especially, the parameters of those two paths are shared and optimized by multiple supervised strategy simultaneously. This process could be formulated as:

$$
\begin{aligned}
I_{final} &= \frac{I_{f1} + I_{f2}}{2} = \frac{DN[H_{3\times3}(I_{SR})] + SR[H_{3\times3}(I_{DN})]}{2} \\
&= \frac{H_{1\times1}\langle H_{3\times3}\{PS[H_{3\times3}(F_n)]\}\rangle + PS\langle H_{3\times3}\{H_{3\times3}[H_{1\times1}(F_n)]\}\rangle}{2} \quad (2)
\end{aligned}
$$

where, F_n is the output of our backbone, $H_{k\times k}$ means $k \times k$ convolutional layer, $SR(\cdot)$ represents SR head, $DN(\cdot)$ represents denoising head, $PS(\cdot)$ expresses Pixel Shuffle layer, I_{SR} is the result of SR head, I_{DN} is the result of denoising head and I_{final} is the final reconstructed result.

2.2   Target Function

Following the multiple supervision strategy, the target function L_{total} is calculated as:

$$
\begin{aligned}
L_{total} &= \lambda_1 L_{SR} + \lambda_2 L_{DN} + L_{final} \\
&= \lambda_1 \|I_{gt} - I_{SR}\|_1 + \lambda_2 \|BI(I_{gt}) - I_{DN}\|_1 + \|I_{gt} - I_{final}\|_1 \quad (3)
\end{aligned}
$$

where, I_{gt} is the ground truth, $BI(\cdot)$ means bicubic interpolation, $\|\cdot\|_1$ represents the L1 norm and λ_1, λ_2 are the weight parameters for adjusting the losses.

3   Experiments

3.1   Datasets and Experiment Setup

Datasets. Two widely-used public CT image datasets, 3D-IRCADB [5] and PANCREAS [5], are used for both training and testing. The 3D-IRCADB dataset is used for liver and its lesion detection which consists of 2823 512×512 CT files from 20 patients. We choose 1663 CT images from 16 patients for training, 226 CT images from 2 patients for validation and 185 CT images from 2 patients for testing. Similarly, the PANCREAS dataset is used for pancreas segmentation which consists of 19328 512×512 CT files from 82 patients. We choose 5638 CT images from 65 patients for training, 668 CT images from 8 patients for validation and 753 CT images from 9 patients for testing. All HU values are set as $[-135, 215]$. Following Zeng et al. [30], we set the blank flux as 0.5×10^5 to simulate the effect of low dose noise. And we use bicubic interpolation to degrade the HR images to 256×256 LR images and 128×128 LR images.

Table 1. Ablation experiments on PANCREAS dataset with the scale factor of 2 and 4

(a) Ablation experiments for dual-guidance on the PANCREAS dataset with the scale factor of 2 and 4

AVG CT	Mask	×2		×4	
		PSNR	SSIM	PSNR	SSIM
×	×	30.0282 ± 2.9426	0.8948 ± 0.0431	28.5120 ± 2.2875	0.8643 ± 0.0508
✓	×	29.9600 ± 3.2378	0.8950 ± 0.0419	28.1490 ± 2.3284	0.8592 ± 0.0543
×	✓	30.2991 ± 3.1391	0.8960 ± 0.0413	28.6589 ± 2.2497	0.8639 ± 0.0522
✓	✓	**30.4047** ± 3.1558	**0.8974** ± 0.0383	**28.7542** ± 2.2728	**0.8672** ± 0.0412

The best quantitative performance is shown in **bold** and the second-best in underlined.

(b) Ablation experiments for shared heads mechanism on the PANCREAS dataset with the scale factor of 2 and 4

Heads	Param (M) ×2/×4	×2		×4	
		PSNR	SSIM	PSNR	SSIM
SR Only	5.748/5.896	30.2904 ± 3.0620	0.8948 ± 0.0431	28.4422 ± 2.3707	0.8628 ± 0.0523
Unshared	6.009/6.304	30.3257 ± 3.2504	0.8940 ± 0.0442	28.5675 ± 2.2540	0.8645 ± 0.0529
Shared	5.795/5.934	**30.4047** ± 3.1558	**0.8974** ± 0.0383	**28.7542** ± 2.2728	**0.8672** ± 0.0412

Experiment Setup. All experiments are implemented on Ubuntu 16.04.12 with an NVIDIA RTX 3090 24G GPU using Pytorch 1.8.0 and CUDA 11.1.74. We augment the data by rotation and flipping first and then randomly crop them to 128×128 patches. Adam optimizer with $\beta_1 = 0.9$ and $\beta_2 = 0.99$ is used to minimize the target function. λ_1 and λ_2 of our target function are set as 0.2. The batch size is set to 16 and the learning rate is set to 10^{-4} which decreases to 5×10^{-5} at 200K iterations. Peak signal-to-noise (PSNR) and structural similarity (SSIM) are used as the quantitative indexes to evaluate the performance.

3.2 Ablation Study

Table 1a shows the experimental result of the dual-guidance ablation study. Introducing the average CT image guidance alone degrades performance compared with the model without guidance for both the scale factor of 2 and 4. And introducing mask guidance alone could improve the reconstruction effect. When the average CT image guidance and the mask guidance are both embedded, the performance will be promoted further. Table 1b presents the result of the shared heads mechanism ablation study. The experimental result proves that introducing the proposed dual-path architecture could promote the reconstruction performance and the model with shared heads is superior than that without them in both reconstruction ability and parameter amount.

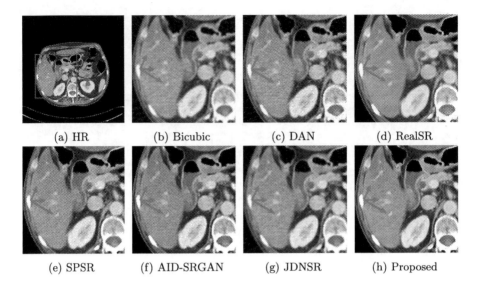

(a) HR   (b) Bicubic   (c) DAN   (d) RealSR

(e) SPSR   (f) AID-SRGAN   (g) JDNSR   (h) Proposed

Fig. 2. Qualitative results on the 3D-IRCADB dataset with the scale factor of 2. (a) is the HR image and its red rectangle region displays the liver and its lateral issues. (b) to (h) are the reconstruction results by different methods. (Color figure online)

3.3   Comparison with State-of-the-Art Methods

We compare the performance of our proposed method with other state-of-the-art methods, including Bicubic interpolation [16], DAN [11], RealSR [15], SPSR [21], AID-SRGAN [12] and JDNSR [10].

Figure 2 shows the qualitative comparison results on the 3D-IRCADB dataset with the scale factor of 2. All methods enhance the image quality to different extents compared with bicubic interpolation. However, for the calcifications within the liver which are indicated by the blue arrows, our method recovers the clearest edges. The results of DAN, SPSR and AID-SRGAN suffers from the artifacts. JDNSR blurs the issue structural information, e.g. the edges of liver and bone. For the inferior vena cava, portal vein, and gallbladder within the kidney, RealSR restores blurred details and textures though it could recover clear edges of calcifications. Figure 3 shows the qualitative comparison results on the PANCREAS dataset with the scale factor of 4. Figure 3 has similar observation as Fig. 2, that is, our method could suppress more artifacts than other methods, especially at the edges of the pancreas and the texture and structure of the issues with in the kidney. Therefore, our method reconstructs more detailed results than other methods.

Table 2 shows the quantitative comparison results of different state-of-the-art methods with two scale factors on two datasets. For the 3D-IRCADB and PANCREAS datasets, our method outperforms the second-best methods 1.6896/0.0157 and 1.7325/0.0187 on PSNR/SSIM with the scale factor of 2 respectively. Similarly, our method outperforms the second-best methods

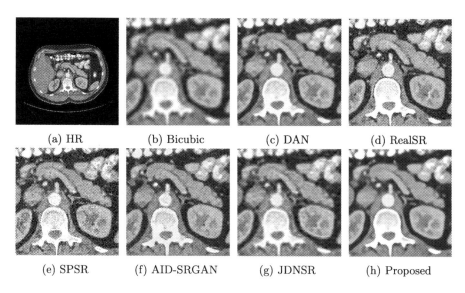

(a) HR	(b) Bicubic	(c) DAN	(d) RealSR
(e) SPSR	(f) AID-SRGAN	(g) JDNSR	(h) Proposed

Fig. 3. Qualitative results on PANCREAS dataset with the scale factor of 4. (a) is the HR image and its red rectangle region shows the pancreas and kidney. (b) to (h) are the reconstruction results by different methods. (Color figure online)

Table 2. Quantitative comparison on the 3D-IRCADB and PANCREAS datasets with other state-of-the-art methods

Method	Scale	3D-IRCADB		PANCREAS	
		PSNR↑	SSIM↑	PSNR↑	SSIM↑
Bicubic	2	28.2978 ± 1.0071	0.8613 ± 0.0400	26.9997 ± 1.7192	0.8362 ± 0.0452
DAN	2	29.4768 ± 1.7852	0.8636 ± 0.0465	28.1050 ± 2.8794	0.8602 ± 0.0510
RealSR	2	<u>30.0746</u> ± 1.6303	<u>0.8981</u> ± 0.0390	<u>28.6712</u> ± 2.8363	<u>0.8784</u> ± 0.4810
SPSR	2	29.8686 ± 1.8197	0.8809 ± 0.0417	28.3321 ± 2.9872	0.8755 ± 0.0508
AID-SRGAN	2	29.6212 ± 1.5741	0.8784 ± 0.0423	27.5311 ± 2.7042	0.8579 ± 0.0521
JDNSR	2	27.7927 ± 0.9000	0.8579 ± 0.0402	25.7842 ± 1.7106	0.8322 ± 0.0424
Proposed	2	**31.7642** ± 2.5292	**0.9138** ± 0.0362	**30.4047** ± 3.1558	**0.8974** ± 0.0383
Bicubic	4	25.9209 ± 0.5327	0.8071 ± 0.0452	23.7091 ± 1.4903	0.7700 ± 0.5860
DAN	4	<u>29.4389</u> ± 1.6800	<u>0.8703</u> ± 0.0471	<u>27.5829</u> ± 2.4628	<u>0.8491</u> ± 0.0534
RealSR	4	27.5951 ± 0.9666	0.8272 ± 0.0508	25.9374 ± 2.0109	0.8214 ± 0.0621
SPSR	4	25.8575 ± 0.5969	0.7948 ± 0.0544	25.5834 ± 2.3017	0.8105 ± 0.0675
AID-SRGAN	4	27.5824 ± 1.1192	0.8312 ± 0.0506	26.0488 ± 1.9642	0.8193 ± 0.0585
JDNSR	4	26.6793 ± 0.6690	0.8194 ± 0.0438	24.7504 ± 1.4858	0.7857 ± 0.0515
Proposed	4	**30.0436** ± 1.7803	**0.8811** ± 0.462	**28.7542** ± 2.2728	**0.8672** ± 0.0412

0.6047/0.0157 and 1.1813/0.0281 on PSNR/SSIM with the scale factor of 4 respectively. Those quantitative superiorities confirm our qualitative observations.

4 Conclusion

In this paper, we propose an LDCT image SR network with dual-guidance feature distillation and dual-path content communication. Facing the existing problem that reconstructed results suffer from residual artifacts, we design a dual-guidance feature distillation backbone which consists of DGFM and SAB to extract deep visual information. Especially, the DGFM could fuse the average CT image to take the advantage of the 3D spatial information of CT volume and the segmentation mask to focus on the ROI, which provides pixel-wise shallow information and deep semantic features for our backbone. The SAB leverages the essential multi-scale features to enhance the ability for feature extraction. Then, our shared heads mechanism reconstructs the deep features obtained by our backbone to satisfactory results. The experiments compared with 6 state-of-the-art methods on 2 public datasets demonstrate the superiority of our method.

Acknowledgements. This work was supported by National Natural Science Foundation of China under Grant 61901098, 61971118, Science and Technology Plan of Liaoning Province 2021JH1/10400051.

References

1. Bruno, D.M., Samit, B.: Distance-driven projection and backprojection in three dimensions. Phys. Med. Biol. **49**(11), 2463–2475 (2004)
2. Chen, H., et al.: Low-dose CT with a residual encoder-decoder convolutional neural network. IEEE Trans. Med. Imaging **36**(12), 2524–2535 (2017)
3. Chen, Y., Zheng, Q., Chen, J.: Double paths network with residual information distillation for improving lung CT image super resolution. Biomed. Sig. Process. Control **73**, 103412 (2022)
4. Chi, J., Sun, Z., Wang, H., Lyu, P., Yu, X., Wu, C.: CT image super-resolution reconstruction based on global hybrid attention. Comput. Biol. Med. **150**, 106112 (2022)
5. Clark, K., et al.: The cancer imaging archive (TCIA): maintaining and operating a public information repository. J. Digit. Imaging **26**(6), 1045–1057 (2013)
6. Dai, T., Cai, J., Zhang, Y., Xia, S.T., Zhang, L.: Second-order attention network for single image super-resolution. In: Proceedings of the IEEE/CVF Conference on Computer Vision and Pattern Recognition, pp. 11065–11074 (2019)
7. Dong, C., Loy, C.C., He, K., Tang, X.: Learning a deep convolutional network for image super-resolution. In: Fleet, D., Pajdla, T., Schiele, B., Tuytelaars, T. (eds.) ECCV 2014. LNCS, vol. 8692, pp. 184–199. Springer, Cham (2014). https://doi.org/10.1007/978-3-319-10593-2_13
8. Dong, C., Loy, C.C., He, K., Tang, X.: Image super-resolution using deep convolutional networks. IEEE Trans. Pattern Anal. Mach. Intell. **38**(2), 295–307 (2015)
9. Dong, C., Loy, C.C., Tang, X.: Accelerating the super-resolution convolutional neural network. In: Leibe, B., Matas, J., Sebe, N., Welling, M. (eds.) ECCV 2016. LNCS, vol. 9906, pp. 391–407. Springer, Cham (2016). https://doi.org/10.1007/978-3-319-46475-6_25
10. Hou, H., Jin, Q., Zhang, G., Li, Z.: Ct image quality enhancement via a dual-channel neural network with jointing denoising and super-resolution. Neurocomputing **492**, 343–352 (2022)

11. Huang, Y., Li, S., Wang, L., Tan, T., et al.: Unfolding the alternating optimization for blind super resolution. Adv. Neural. Inf. Process. Syst. **33**, 5632–5643 (2020)

12. Huang, Y., Wang, Q., Omachi, S.: Rethinking degradation: radiograph super-resolution via AID-SRGAN. In: Lian, C., Cao, X., Rekik, I., Xu, X., Cui, Z. (eds.) Machine Learning in Medical Imaging, MLMI 2022. LNCS, vol. 13583, pp. 43–52. Springer, Cham (2022). https://doi.org/10.1007/978-3-031-21014-3_5

13. Irani, M., Peleg, S.: Super resolution from image sequences. In: 1990 Proceedings of the 10th International Conference on Pattern Recognition, vol. 2, pp. 115–120. IEEE (1990)

14. Irani, M., Peleg, S.: Improving resolution by image registration. Graph. Models Image Process. (CVGIP) **53**(3), 231–239 (1991)

15. Ji, X., Cao, Y., Tai, Y., Wang, C., Li, J., Huang, F.: Real-world super-resolution via kernel estimation and noise injection. In: Proceedings of the IEEE/CVF Conference on Computer Vision and Pattern Recognition Workshops, pp. 466–467 (2020)

16. Keys, R.: Cubic convolution interpolation for digital image processing. IEEE Trans. Acoust. Speech Sig. Process. **29**(6), 1153–1160 (1981)

17. Kim, J., Lee, J.K., Lee, K.M.: Accurate image super-resolution using very deep convolutional networks. In: Proceedings of the IEEE Conference on Computer Vision and Pattern Recognition, pp. 1646–1654 (2016)

18. Li, B., et al.: Diagnostic value and key features of computed tomography in coronavirus disease 2019. Emerg. Microbes Infect. **9**(1), 787–793 (2020)

19. Liang, J., Cao, J., Sun, G., Zhang, K., Van Gool, L., Timofte, R.: SwinIR: image restoration using Swin transformer. In: Proceedings of the IEEE/CVF International Conference on Computer Vision, pp. 1833–1844 (2021)

20. Lim, B., Son, S., Kim, H., Nah, S., Mu Lee, K.: Enhanced deep residual networks for single image super-resolution. In: Proceedings of the IEEE Conference on Computer Vision and Pattern Recognition Workshops, pp. 136–144 (2017)

21. Ma, C., Rao, Y., Lu, J., Zhou, J.: Structure-preserving image super-resolution. IEEE Trans. Pattern Anal. Mach. Intell. **44**, 7898–7911 (2021)

22. Prasad, K., Cole, W., Haase, G.: Radiation protection in humans: extending the concept of as low as reasonably achievable (ALARA) from dose to biological damage. Br. J. Radiol. **77**(914), 97–99 (2004)

23. Ramani, S., Fessler, J.A.: A splitting-based iterative algorithm for accelerated statistical X-ray CT reconstruction. IEEE Trans. Med. Imaging **31**(3), 677–688 (2012)

24. Schultz, R.R., Stevenson, R.L.: Extraction of high-resolution frames from video sequences. IEEE Trans. Image Process. **5**(6), 996–1011 (1996)

25. Smith, P.: Bilinear interpolation of digital images. Ultramicroscopy **6**(2), 201–204 (1981)

26. Stark, H., Oskoui, P.: High-resolution image recovery from image-plane arrays, using convex projections. JOSA A **6**(11), 1715–1726 (1989)

27. Sun, K., Xiao, B., Liu, D., Wang, J.: Deep high-resolution representation learning for human pose estimation. In: Proceedings of the IEEE/CVF Conference on Computer Vision and Pattern Recognition, pp. 5693–5703 (2019)

28. Veronesi, G., et al.: Recommendations for implementing lung cancer screening with low-dose computed tomography in Europe. Cancers **12**(6), 1672 (2020)

29. Yin, X., et al.: Domain progressive 3D residual convolution network to improve low-dose CT imaging. IEEE Trans. Med. Imaging **38**(12), 2903–2913 (2019)

30. Zeng, D., et al.: A simple low-dose X-ray CT simulation from high-dose scan. IEEE Trans. Nucl. Sci. **62**(5), 2226–2233 (2015)

31. Zhang, Y., Li, K., Li, K., Wang, L., Zhong, B., Fu, Y.: Image super-resolution using very deep residual channel attention networks. In: Ferrari, V., Hebert, M., Sminchisescu, C., Weiss, Y. (eds.) ECCV 2018. LNCS, vol. 11211, pp. 294–310. Springer, Cham (2018). https://doi.org/10.1007/978-3-030-01234-2_18
32. Zhang, Y., Tian, Y., Kong, Y., Zhong, B., Fu, Y.: Residual dense network for image restoration. IEEE Trans. Pattern Anal. Mach. Intell. **43**(7), 2480–2495 (2020)

MEPNet: A Model-Driven Equivariant Proximal Network for Joint Sparse-View Reconstruction and Metal Artifact Reduction in CT Images

Hong Wang[1(✉)], Minghao Zhou[1,2], Dong Wei[1], Yuexiang Li[1], and Yefeng Zheng[1]

[1] Tencent Jarvis Lab, Shenzhen, People's Republic of China
{hazelhwang,hippomhzhou,donwei,vicyxli,yefengzheng}@tencent.com
[2] Xi'an Jiaotong University, Xi'an, Shaan'xi, People's Republic of China

Abstract. Sparse-view computed tomography (CT) has been adopted as an important technique for speeding up data acquisition and decreasing radiation dose. However, due to the lack of sufficient projection data, the reconstructed CT images often present severe artifacts, which will be further amplified when patients carry metallic implants. For this joint sparse-view reconstruction and metal artifact reduction task, most of the existing methods are generally confronted with two main limitations: 1) They are almost built based on common network modules without fully embedding the physical imaging geometry constraint of this specific task into the dual-domain learning; 2) Some important prior knowledge is not deeply explored and sufficiently utilized. Against these issues, we specifically construct a dual-domain reconstruction model and propose a model-driven equivariant proximal network, called MEPNet. The main characteristics of MEPNet are: 1) It is optimization-inspired and has a clear working mechanism; 2) The involved proximal operator is modeled via a rotation equivariant convolutional neural network, which finely represents the inherent rotational prior underlying the CT scanning that the same organ can be imaged at different angles. Extensive experiments conducted on several datasets comprehensively substantiate that compared with the conventional convolution-based proximal network, such a rotation equivariance mechanism enables our proposed method to achieve better reconstruction performance with fewer network parameters. We will release the code at https://github.com/hongwang01/MEPNet.

Keywords: Sparse-view reconstruction · Metal artifact reduction · Rotation equivariance · Proximal network · Generalization capability

---

Supplementary Information The online version contains supplementary material available at https://doi.org/10.1007/978-3-031-43999-5_11.

H. Greenspan et al. (Eds.): MICCAI 2023, LNCS 14229, pp. 109–120, 2023.
https://doi.org/10.1007/978-3-031-43999-5_11

1   Introduction

Computed tomography (CT) has been widely adopted in clinical applications. To reduce the radiation dose and shorten scanning time, sparse-view CT has drawn much attention in the community [10,34]. However, sparse data sampling inevitably degenerates the quality of CT images and leads to adverse artifacts. In addition, when patients carry metallic implants, such as hip prostheses and spinal implants [11,13,24], the artifacts will be further aggravated due to beam hardening and photon starvation. For the joint sparse-view reconstruction and metal artifact reduction task (SVMAR), how to design an effective method for artifact removal and detail recovery is worthy of in-depth exploration.

For the sparse-view (SV) reconstruction, the existing deep-learning (DL)-based methods can be roughly divided into three categories based on the information domain exploited, *e.g.*, sinogram domain, image domain, and dual domains. Specifically, for the sinogram-domain methods, sparse-view sinograms are firstly repaired based on deep networks, such as U-Net [10] and dense spatial-channel attention network [37], and then artifact-reduced CT images are reconstructed via the filtered-back-projection (FBP) process. For the image-domain methods, researchers have proposed to learn the clean CT images from degraded ones via various structures [18,34,35]. Alternatively, both sinogram and CT images are jointly exploited for the dual reconstruction [4,21,32,36].

For the metal artifact reduction (MAR) task, similarly, the current DL-based approaches can also be categorized into three types. To be specific, sinogram-domain methods aim to correct the sinogram for the subsequent CT image reconstruction [6,33]. Image-domain-based works have proposed different frameworks, such as simple residual network [8] and an interpretable structure [22,23,26], to learn artifact-reduced images from metal-affected ones. The dual-domain methods [12,24,25,36] focus on the mutual learning between sinogram and CT image.

Albeit achieving promising performance, these aforementioned methods are sub-optimal for the SVMAR task. The main reasons are: 1) Most of them do not consider the joint influence of sparse data sampling and MAR, and do not fully embed the physical imaging constraint between the sinogram domain and CT image domain under the SVMAR scenario; 2) Although a few works focus on the joint SVMAR task, such as [36], the network structure is empirically built based on off-the-shelf modules, *e.g.*, U-Net and gated recurrent units, and it does not fully investigate and embed some important prior information underlying the CT imaging procedure. However, for such a highly ill-posed restoration problem, the introduction of the proper prior is important and valuable for constraining the network learning and helping it evolve in a right direction [24].

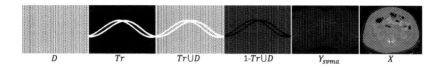

D $\quad$ Tr $\quad$ $Tr\cup D$ $\quad$ $1\text{-}Tr\cup D$ $\quad$ Y_{svma} $\quad$ X

Fig. 1. Illustration of the elements in the model Eq. (2) for easy understanding.

To alleviate these issues, in this paper, we propose a model-driven equivariant proximal network, called MEPNet, which is naturally constructed based on the CT imaging geometry constraint for this specific SVMAR task, and takes into account the inherent prior structure underlying the CT scanning procedure. Concretely, we first propose a dual-domain reconstruction model and then correspondingly construct an unrolling network framework based on a derived optimization algorithm. Furthermore, motivated by the fact that the same organ can be imaged at different angles making the reconstruction task equivariant to rotation [2], we carefully formulate the proximal operator of the built unrolling neural network as a rotation-equivariant convolutional neural network (CNN). Compared with the standard-CNN-based proximal network with only translation-equivariance property [3], our proposed method effectively encodes more prior knowledge, *e.g.*, rotation equivariance, possessed by this specific task. With such more accurate regularization, our proposed MEPNet can achieve higher fidelity of anatomical structures and has better generalization capability with fewer network parameters. This is finely verified by comprehensive experiments on several datasets of different body sites. To the best of our knowledge, we should be the first to study rotation equivariance in the context of SVMAR and validate its utility, which is expected to make insightful impacts on the community.

2   Preliminary Knowledge About Equivariance

Equivariance of a mapping w.r.t. a certain transformation indicates that executing the transformation on the input produces a corresponding transformation on the output [3,28]. Mathematically, given a group of transformations G, a mapping Φ from the input feature space to the output feature space is said to be group equivariant about G if

$$\Phi\big(T_g(f)\big) = T_g'\big(\Phi(f)\big), \forall g \in G, \tag{1}$$

where f is any input feature map in the input feature space; T_g and T_g' represent the actions of g on the input and output, respectively.

The prior work [3] has shown that adopting group equivariant CNNs to encode symmetries into networks would bring data efficiency and it can constrain the network learning for better generalization. For example, compared with the fully-connected layer, the translational equivariance property enforces weight sharing for the conventional CNN, which makes CNN use fewer parameters to preserve the representation capacity and then obtain better generalization ability. Recently, different types of equivariant CNNs have been designed to preserve more symmetries beyond current CNNs, such as rotation symmetry [1, 2, 27] and scale symmetry [7, 20]. However, most of these methods do not consider specific designs for the SVMAR reconstruction. In this paper, we aim to build a physics-driven network for the SVMAR task where rotation equivariance is encoded.

3   Dual-Domain Reconstruction Model for SVMAR

In this section, for the SVMAR task, we derive the corresponding dual domain reconstruction model and give an iterative algorithm for solving it.

Dual-Domain Reconstruction Model. Given the captured sparse-view metal-affected sinogram $Y_{svma} \in \mathbb{R}^{N_b \times N_p}$, where N_b and N_p are the number of detector bins and projection views, respectively, to guarantee the data consistency between the reconstructed clean CT image $X \in \mathbb{R}^{H \times W}$ and the observed sinogram Y_{svma}, we can formulate the corresponding optimization model as [36]:

$$\min_{X} \|(1 - Tr \cup D) \odot (\mathcal{P}X - Y_{svma})\|_F^2 + \mu R(X), \tag{2}$$

where $D \in \mathbb{R}^{N_b \times N_p}$ is the binary sparse downsampling matrix with 1 indicating the missing region; $Tr \in \mathbb{R}^{N_b \times N_p}$ is the binary metal trace with 1 indicating the metal-affected region; $\mathcal{P}$ is forward projection; $R(\cdot)$ is a regularization function for capturing the prior of X; $\cup$ is the union set; $\odot$ is the point-wise multiplication; H and W are the height and width of CT images, respectively; μ is a trade-off parameter. One can refer to Fig. 1 for easy understanding.

To jointly reconstruct sinogram and CT image, we introduce the dual regularizers $R_1(\cdot)$ and $R_2(\cdot)$, and further derive Eq. (2) as:

$$\min_{S,X} \|\mathcal{P}X - S\|_F^2 + \lambda \|(1 - Tr \cup D) \odot (S - Y_{svma})\|_F^2 + \mu_1 R_1(S) + \mu_2 R_2(X), \tag{3}$$

where S is the to-be-estimated clean sinogram; λ is a weight factor. Following [24], we rewrite S as $\bar{Y} \odot \bar{S}$ for stable learning, where $\bar{Y}$ and $\bar{S}$ are the normalization coefficient implemented via the forward projection of a prior image, and the normalized sinogram, respectively. Then we can get the final dual-domain reconstruction model for this specific SVMAR task as:

$$\min_{\bar{S},X} \|\mathcal{P}X - \bar{Y} \odot \bar{S}\|_F^2 + \lambda \|(1 - Tr \cup D) \odot (\bar{Y} \odot \bar{S} - Y_{svma})\|_F^2 + \mu_1 R_1(\bar{S}) + \mu_2 R_2(X). \tag{4}$$

As observed, given Y_{svma}, we need to jointly estimate $\bar{S}$ and X. For $R_1(\cdot)$ and $R_2(\cdot)$, the design details are presented below.

Iterative Optimization Algorithm. To solve the model (4), we utilize the classical proximal gradient technique [15] to alternatively update the variables $\bar{S}$ and X. At iterative stage k, we can get the corresponding iterative rules:

$$\begin{aligned}
\bar{S}_k &= \text{prox}_{\mu_1 \eta_1}\big(\bar{S}_{k-1} - \eta_1\big(\bar{Y} \odot (\bar{Y} \odot \bar{S}_{k-1} - \mathcal{P}X_{k-1}) + \lambda(1 - Tr \cup D) \odot \bar{Y} \odot (\bar{Y} \odot \bar{S}_{k-1} - Y_{svma})\big)\big), \\
X_k &= \text{prox}_{\mu_2 \eta_2}\big(X_{k-1} - \eta_2 \mathcal{P}^T\big(\mathcal{P}X_{k-1} - \bar{Y} \odot \bar{S}_k\big)\big),
\end{aligned} \tag{5}$$

where η_i is stepsize; $\text{prox}_{\mu_i\eta_i}(\cdot)$ is proximal operator, which relies on the regularization term $R_i(\cdot)$. For any variable, its iterative rule in Eq. (5) consists of two steps: an explicit gradient step to ensure data consistency and an implicit proximal computation $\text{prox}_{\mu_i\eta_i}(\cdot)$ which enforces the prior $R_i(\cdot)$ on the to-be-estimated variable. Traditionally, the prior form $R_i(\cdot)$ is empirically designed, e.g., l_1 penalty, which may not always hold in real complicated scenarios. Due to the high representation capability, CNN has been adopted to adaptively learn the proximal step in a data-driven manner for various tasks [5,14,30]. Motivated by their successes, in the next section, we will deeply explore the prior of this specific SVMAR task and carefully construct the network for $\text{prox}_{\mu_i\eta_i}(\cdot)$.

4 Equivariant Proximal Network for SVMAR

By unfolding the iterative rules (5) for K iterations, we can easily build the unrolling neural network. Specifically, at iteration k, the network structure is sequentially composed of:

$$\bar{S}_k = \text{proxNet}_{\theta_{\bar{s}}^{(k)}}\big(\bar{S}_{k-1} - \eta_1\big(\bar{Y}\odot\big(\bar{Y}\odot\bar{S}_{k-1}-\mathcal{P}X_{k-1}\big)+\lambda\big(1-Tr\cup D\big)\odot\bar{Y}\odot\big(\bar{Y}\odot\bar{S}_{k-1}-Y_{svma}\big)\big)\big),$$

$$X_k = \text{proxNet}_{\theta_x^{(k)}}\big(X_{k-1} - \eta_2\mathcal{P}^T\big(\mathcal{P}X_{k-1} - \bar{Y}\odot\bar{S}_k\big)\big),$$

$$(6)$$

where $\text{proxNet}_{\theta_{\bar{s}}^{(k)}}$ and $\text{proxNet}_{\theta_x^{(k)}}$ are proximal networks with parameters $\theta_{\bar{s}}^{(k)}$ and $\theta_x^{(k)}$ to execute the proximal operators $\text{prox}_{\mu_1\eta_1}$ and $\text{prox}_{\mu_2\eta_2}$, respectively.

To build $\text{proxNet}_{\theta_{\bar{s}}^{(k)}}$, we follow [24] and choose a standard-CNN-based structure with four $[Conv+BN+ReLU+Conv+BN+Skip\ Connection]$ residual blocks, which do not change image sizes. While for $\text{proxNet}_{\theta_x^{(k)}}$, we carefully investigate that during the CT scanning, the same body organ can be imaged at different rotation angles. However, the conventional CNN for modeling $\text{proxNet}_{\theta_x^{(k)}}$ in [24] has only the translation equivariance property and it cannot preserve such an intrinsic rotation equivariance structure [3]. Against this issue, we propose to replace the standard CNN in [24] with a rotation equivariant CNN. Then we can embed more useful prior, such as rotation equivariance, to constrain the network, which would further boost the quality of reconstructed CT images (refer to Sect. 5.2).

Specifically, from Eq. (1), for a rotation group G and any input feature map f, we expect to find a properly parameterized convolutional filter ψ which is group equivariant about G, satisfying

$$[T_\theta[f]] \star \psi = T_\theta[f \star \psi] = f \star \pi_\theta[\psi], \forall\theta \in G, \tag{7}$$

where π_θ is a rotation operator. Due to its solid theoretical foundation, the Fourier-series-expansion-based method [28] is adopted to parameterize ψ as:

$$\psi(x) = \sum_{m=0}^{p-1}\sum_{n=0}^{p-1} a_{mn}\varphi_{mn}^c(x) + b_{mn}\varphi_{mn}^s(x), \tag{8}$$

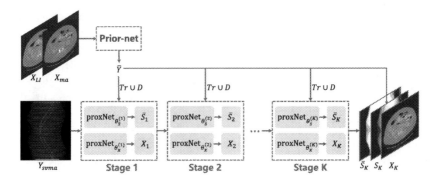

Fig. 2. The framework of the proposed MEPNet where "prior-net" is designed in [24].

where $x = [x_i, x_j]^T$ is 2D spatial coordinates; a_{mn} and b_{mn} are learnable expansion coefficients; $\varphi_{mn}^c(x)$ and $\varphi_{mn}^s(x)$ are 2D fixed basis functions as designed in [28]; p is chosen to be 5 in experiments. The action π_θ on ψ in Eq. (7) can be achieved by coordinate transformation as:

$$\pi_\theta[\psi](x) = \psi(U_\theta^{-1}x), \text{where } U_\theta = \begin{bmatrix} \cos\theta & \sin\theta \\ -\sin\theta & \cos\theta \end{bmatrix}, \forall \theta \in G. \tag{9}$$

Based on the parameterized filter in Eq. (8), we follow [28] to implement the rotation-equivariant convolution for the discrete domain. Compared with other types, $e.g.$, harmonics and partial-differential-operator-like bases [19,27], the basis in Eq. (8) has higher representation accuracy, especially when being rotated.

By implementing proxNet$_{\theta_{\tilde{s}}^{(k)}}$ and proxNet$_{\theta_x^{(k)}}$ in Eq. (6) with the standard CNN and the rotation-equivariant CNN with the $p8$ group,[1] respectively, we can then construct the model-driven equivariant proximal network, called MEP-Net, as shown in Fig. 2. The expansion coefficients, $\{\theta_{\tilde{s}}^{(k)}\}_{k=1}^K$, θ_{prior} for learning $\bar{Y}$ [24], η_1, η_2, and λ, are all flexibly learned from training data end-to-end.

Remark: Our MEPNet is indeed inspired by InDuDoNet [24]. However, MEP-Net contains novel and challenging designs: 1) It is specifically constructed based on the physical imaging procedure for the SVMAR task, leading to a clear working mechanism; 2) It embeds more prior knowledge, $e.g.$, rotation equivariance, via advanced filter parametrization method, which promotes better reconstruction; 3) It is desirable that the usage of more transformation symmetries would further decrease the number of model parameters and improve the generalization. These advantages are validated in the experiments below.

---

[1] Considering performance and efficiency, we follow [28] and chose p8 group for discretized equivariance convolution on CT images. The parameterized filters for eight different rotation orientations share a set of expansion coefficients, largely reducing the network parameters (validated in Sect. 5.2).

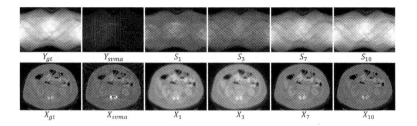

Fig. 3. The sinogram S_k and CT image X_k reconstructed by our MEPNet ($K = 10$).

5   Experiments

5.1   Details Description

Datasets and Metrics. Consistent with [24], we synthesize the training set by randomly selecting 1000 clean CT images from the public DeepLesion [29] and collecting 90 metals with various sizes and shapes from [33]. Specifically, following the CT imaging procedure with fan-beam geometry in [24,31,36], all the CT images are resized as 416×416 pixels where pixel spacing is used for normalization, and 640 fully-sampled projection views are uniformly spaced in $360°$. To synthesize sparse-view metal-affected sinogram Y_{svma}, similar to [36], we uniformly sample 80, 160, and 320 projection views to mimic 8, 4, and 2-fold radiation dose reduction. By executing the FBP process on Y_{svma}, we can obtain the degraded CT image X_{svma}.

The proposed method is tested on three datasets including DeepLesion-test (2000 pairs), Pancreas-test (50 pairs), and CLINIC-test (3397 pairs). Specifically, DeepLesion-test is generated by pairing another 200 clean CT images from DeepLesion [29] with 10 extra testing metals from [33]. Pancreas-test is formed by randomly choosing 5 patients with 50 slices from Pancreas CT [17] and pairing each slice with one randomly-selected testing metal. CLINIC-test is synthesized by pairing 10 volumes with 3397 slices randomly chosen from CLINIC [13] with one testing metal slice-by-slice. The 10 testing metals have different sizes as [35] in pixels. For evaluation on different sizes of metals as listed in Table 2 below, we merge the adjacent two sizes into one group. Following [12,24], we adopt peak signal-to-noise ratio (PSNR) and structured similarity index (SSIM) for quantitative analysis.

Implementation Details. Our MEPNet is trained end-to-end with a batch size of 1 for 100 epochs based on PyTorch [16] on an NVIDIA Tesla V100-SMX2 GPU card. An Adam optimizer with parameters $(\beta_1, \beta_2) = (0.5, 0.999)$ is exploited. The initial learning rate is 2×10^{-4} and it is decayed by 0.5 every 40 epochs. For a fair comparison, we adopt the same loss function as [24] and also select the total number of iterations K as 10.

116   H. Wang et al.

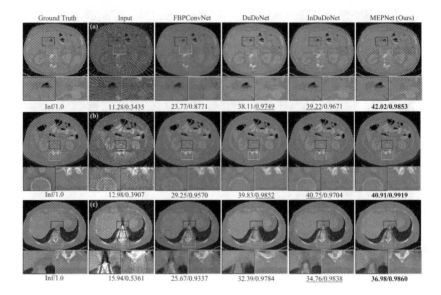

Fig. 4. DeepLesion-test: Artifact-reduced images (the corresponding PSNRs/SSIMs are shown below) of the comparing methods under different sparse-view under-sampling rates (a) ×8, (b) ×4, (c) ×2, and various sizes of metals marked by red pixels. (Color figure online)

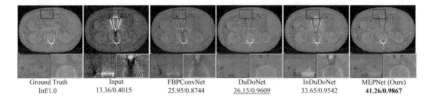

Fig. 5. Cross-domain: Reconstruction results with the corresponding PSNR/SSIM of different methods on Pancreas-test under ×4 under-sampling rate.

5.2 Performance Evaluation

Working Mechanism. Figure 3 presents the sinogram S_k and CT image X_k reconstructed by MEPNet at different stages. We can easily observe that S_k and X_k are indeed alternatively optimized in information restoration and artifact reduction, approaching the ground truth Y_{gt} and X_{gt}, respectively. This finely shows a clear working mechanism of our proposed MEPNet, which evolves in the right direction specified by Eq. (5).

Visual Comparison. Figure 4 shows the reconstructed results of different methods, including FBPConvNet [9], DuDoNet [12], InDuDoNet [24], and the proposed MEPNet, on three degraded images from DeepLesion-test with different sparse-view under-sampling rates and various sizes of metallic implants.[2] As seen, compared with these baselines, our proposed MEPNet can consistently produce cleaner outputs with stronger artifact removal and higher structural fidelity, especially around the metals, thus leading to higher PSNR/SSIM values.

Figure 5 presents the cross-domain results on Pancreas-test with ×4 under-sampling rate where DL-based methods are trained on synthesized DeepLesion. As seen, DuDoNet produces over-smoothed output due to the lack of physical geometry constraint on the final result. In contrast, MEPNet achieves more efficient artifact suppression and sharper detail preservation. Such favorable generalization ability is mainly brought by the dual-domain joint regularization and the fine utilization of rotation symmetry via the equivariant network, which can reduce the model parameters from 5,095,703 (InDuDoNet) to 4,723,309 (MEP-Net). Besides, as observed from the bone marked by the green box, MEPNet alleviates the rotational-structure distortion generally existing in other baselines. This finely validates the effectiveness of embedding rotation equivariance.

Quantitative Evaluation. Table 1 lists the average PSNR/SSIM on three testing sets. It is easily concluded that with the increase of under-sampling rates, all these comparison methods present an obvious performance drop. Nevertheless, our MEPNet still maintains higher PSNR/SSIM scores on different testing sets, showing good superiority in generalization capability. Table 2 reports the results on the DeepLesion-test with different sizes of metals under the ×4 sparse-view under-sampling rate. We can observe that MEPNet almost outperforms others, especially for the large metal setting, showing good generality.[3]

Table 1. Average PSNR (dB) and SSIM of different methods on three testing sets.

Methods	DeepLesion-test			Pancreas-test			CLINIC-test		
	×8	×4	×2	×8	×4	×2	×8	×4	×2
Input	12.65	13.63	16.55	12.37	13.29	16.04	13.95	14.99	18.04
	0.3249	0.3953	0.5767	0.3298	0.3978	0.5645	0.3990	0.4604	0.6085
FBPConvNet [9]	25.91	27.38	29.11	24.24	25.44	26.85	27.92	29.62	31.56
	0.8467	0.8851	0.9418	0.8261	0.8731	0.9317	0.8381	0.8766	0.9362
DuDoNet [12]	34.33	36.83	38.18	30.54	35.14	36.97	34.47	37.34	38.81
	0.9479	0.9634	0.9685	0.9050	0.9527	0.9653	0.9157	0.9493	0.9598
InDuDoNet [24]	37.50	40.24	40.71	36.86	38.17	38.22	38.39	39.67	40.86
	0.9664	0.9793	0.9890	0.9664	0.9734	0.9857	0.9572	0.9621	0.9811
MEPNet (Ours)	38.48	41.43	42.66	36.76	40.69	41.17	39.04	41.58	42.30
	0.9767	0.9889	0.9910	0.9726	0.9872	0.9896	0.9654	0.9820	0.9857

Table 2. Average PSNR (dB)/SSIM of the comparing methods on DeepLesion-test with the ×4 under-sampling rate and different sizes of metallic implants.

Methods	Large Metal	⟶	Small Metal			Average
Input	13.68/0.3438	13.63/0.3736	13.61/0.4046	13.61/0.4304	13.60/0.4240	13.63/0.3953
FBPConvNet [9]	26.15/0.7865	26.96/0.8689	27.77/0.9154	27.98/0.9216	28.03/0.9331	27.38/0.8851
DuDoNet [12]	31.73/0.9519	33.89/0.9599	37.81/0.9667	40.19/0.9688	40.54/0.9696	36.83/0.9634
InDuDoNet [24]	33.78/0.9540	38.15/0.9746	41.96/0.9873	43.48/0.9898	43.83/0.9910	40.24/0.9793
MEPNet (Ours)	37.51/0.9797	39.45/0.9879	42.78/0.9920	43.92/0.9924	43.51/0.9924	41.31/0.9889

---

[2] Here InDuDoNet is a particularly strong baseline and it is exactly an ablation study, which is the degenerated form of MEPNet with removing group equivariance.

[3] More experimental results are included in *supplementary material*.

6   Conclusion and Future Work

In this paper, for the SVMAR task, we have constructed a dual-domain reconstruction model and built an unrolling model-driven equivariant network, called MEPNet, with a clear working mechanism and strong generalization ability. These merits have been substantiated by extensive experiments. Our proposed method can be easily extended to more applications, including limited-angle and low-dose reconstruction tasks. A potential limitation is that consistent with [24,36], the data pairs are generated based on the commonly-adopted protocol, which would lead to a domain gap between simulation settings and clinical scenarios. In the future, we will try to collect clinical data captured in the sparse-view metal-inserted scanning configuration to evaluate our method.

Acknowledgements. This work was supported by the National Key R&D Program of China under Grant 2020AAA0109500/2020AAA0109501.

References

1. Celledoni, E., Ehrhardt, M.J., Etmann, C., Owren, B., Schönlieb, C.B., Sherry, F.: Equivariant neural networks for inverse problems. Inverse Prob. **37**(8), 085006 (2021)
2. Chen, D., Tachella, J., Davies, M.E.: Equivariant imaging: learning beyond the range space. In: Proceedings of the IEEE/CVF International Conference on Computer Vision, pp. 4379–4388 (2021)
3. Cohen, T., Welling, M.: Group equivariant convolutional networks. In: International Conference on Machine Learning, pp. 2990–2999 (2016)
4. Ding, Q., Ji, H., Gao, H., Zhang, X.: Learnable multi-scale Fourier interpolation for sparse view CT image reconstruction. In: Medical Image Computing and Computer Assisted Intervention, pp. 286–295 (2021)
5. Fu, J., Wang, H., Xie, Q., Zhao, Q., Meng, D., Xu, Z.: KXNet: a model-driven deep neural network for blind super-resolution. In: Avidan, S., Brostow, G., Cissé, M., Farinella, G.M., Hassner, T. (eds.) Computer Vision, ECCV 2022. LNCS, vol. 13679, pp. 235–253. Springer, Cham (2022). https://doi.org/10.1007/978-3-031-19800-7_14
6. Ghani, M.U., Karl, W.C.: Fast enhanced CT metal artifact reduction using data domain deep learning. IEEE Trans. Comput. Imag. **6**, 181–193 (2019)
7. Gunel, B., et al.: Scale-equivariant unrolled neural networks for data-efficient accelerated MRI reconstruction. In: Medical Image Computing and Computer Assisted Intervention, pp. 737–747 (2022)
8. Huang, X., Wang, J., Tang, F., Zhong, T., Zhang, Y.: Metal artifact reduction on cervical CT images by deep residual learning. Biomed. Eng. Online **17**(1), 1–15 (2018)
9. Jin, K.H., McCann, M.T., Froustey, E., Unser, M.: Deep convolutional neural network for inverse problems in imaging. IEEE Trans. Image Process. **26**(9), 4509–4522 (2017)
10. Lee, H., Lee, J., Kim, H., Cho, B., Cho, S.: Deep-neural-network-based sinogram synthesis for sparse-view CT image reconstruction. IEEE Trans. Radiat. Plasma Med. Sci. **3**(2), 109–119 (2018)

11. Liao, H., Lin, W.A., Zhou, S.K., Luo, J.: ADN: artifact disentanglement network for unsupervised metal artifact reduction. IEEE Trans. Med. Imaging **39**(3), 634–643 (2019)

12. Lin, W.A., et al.: DuDoNet: dual domain network for CT metal artifact reduction. In: Proceedings of the IEEE/CVF Conference on Computer Vision and Pattern Recognition, pp. 10512–10521 (2019)

13. Liu, P., et al.: Deep learning to segment pelvic bones: large-scale CT datasets and baseline models. arXiv preprint arXiv:2012.08721 (2020)

14. Liu, X., Xie, Q., Zhao, Q., Wang, H., Meng, D.: Low-light image enhancement by retinex-based algorithm unrolling and adjustment. IEEE Trans. Neural Netw. Learn. Syst. (2023)

15. Parikh, N., Boyd, S., et al.: Proximal algorithms. Found. Trends Optim. **1**(3), 127–239 (2014)

16. Paszke, A., et al.: Automatic differentiation in PyTorch (2017)

17. Roth, H.R., et al.: DeepOrgan: multi-level deep convolutional networks for automated pancreas segmentation. In: Medical Image Computing and Computer Assisted Intervention, pp. 556–564 (2015)

18. Shen, L., Pauly, J., Xing, L.: NeRP: implicit neural representation learning with prior embedding for sparsely sampled image reconstruction. IEEE Trans. Neural Netw. Learn. Syst. (2022)

19. Shen, Z., He, L., Lin, Z., Ma, J.: PDO-eConvs: partial differential operator based equivariant convolutions. In: International Conference on Machine Learning, pp. 8697–8706 (2020)

20. Sosnovik, I., Szmaja, M., Smeulders, A.: Scale-equivariant steerable networks. arXiv preprint arXiv:1910.11093 (2019)

21. Wang, C., Shang, K., Zhang, H., Li, Q., Hui, Y., Zhou, S.K.: DuDoTrans: dual-domain transformer provides more attention for sinogram restoration in sparse-view CT reconstruction. arXiv preprint arXiv:2111.10790 (2021)

22. Wang, H., Li, Y., He, N., Ma, K., Meng, D., Zheng, Y.: DICDNet: deep interpretable convolutional dictionary network for metal artifact reduction in CT images. IEEE Trans. Med. Imaging **41**(4), 869–880 (2021)

23. Wang, H., Li, Y., Meng, D., Zheng, Y.: Adaptive convolutional dictionary network for CT metal artifact reduction. arXiv preprint arXiv:2205.07471 (2022)

24. Wang, H., et al.: InDuDoNet: an interpretable dual domain network for CT metal artifact reduction. In: Medical Image Computing and Computer Assisted Intervention, pp. 107–118 (2021)

25. Wang, H., Li, Y., Zhang, H., Meng, D., Zheng, Y.: InDuDoNet+: a deep unfolding dual domain network for metal artifact reduction in CT images. Med. Image Anal. **85**, 102729 (2022)

26. Wang, H., Xie, Q., Li, Y., Huang, Y., Meng, D., Zheng, Y.: Orientation-shared convolution representation for CT metal artifact learning. In: International Conference on Medical Image Computing and Computer-Assisted Intervention, pp. 665–675 (2022)

27. Weiler, M., Hamprecht, F.A., Storath, M.: Learning steerable filters for rotation equivariant CNNs. In: Proceedings of the IEEE Conference on Computer Vision and Pattern Recognition, pp. 849–858 (2018)

28. Xie, Q., Zhao, Q., Xu, Z., Meng, D.: Fourier series expansion based filter parametrization for equivariant convolutions. IEEE Trans. Pattern Anal. Mach. Intell. **45**, 4537–4551 (2022)

29. Yan, K., et al.: Deep lesion graphs in the wild: relationship learning and organization of significant radiology image findings in a diverse large-scale lesion database. In: Proceedings of the IEEE Conference on Computer Vision and Pattern Recognition, pp. 9261–9270 (2018)
30. Yang, Y., Sun, J., Li, H., Xu, Z.: ADMM-Net: a deep learning approach for compressive sensing MRI. arXiv preprint arXiv:1705.06869 (2017)
31. Yu, L., Zhang, Z., Li, X., Xing, L.: Deep sinogram completion with image prior for metal artifact reduction in CT images. IEEE Trans. Med. Imaging **40**(1), 228–238 (2020)
32. Zhang, H., Liu, B., Yu, H., Dong, B.: MetaInv-Net: meta inversion network for sparse view CT image reconstruction. IEEE Trans. Med. Imaging **40**(2), 621–634 (2020)
33. Zhang, Y., Yu, H.: Convolutional neural network based metal artifact reduction in X-ray computed tomography. IEEE Trans. Med. Imaging **37**(6), 1370–1381 (2018)
34. Zhang, Z., Liang, X., Dong, X., Xie, Y., Cao, G.: A sparse-view CT reconstruction method based on combination of DenseNet and deconvolution. IEEE Trans. Med. Imaging **37**(6), 1407–1417 (2018)
35. Zhang, Z., Yu, L., Liang, X., Zhao, W., Xing, L.: TransCT: dual-path transformer for low dose computed tomography. In: Medical Image Computing and Computer Assisted Intervention, pp. 55–64 (2021)
36. Zhou, B., Chen, X., Zhou, S.K., Duncan, J.S., Liu, C.: DuDoDR-Net: dual-domain data consistent recurrent network for simultaneous sparse view and metal artifact reduction in computed tomography. Med. Image Anal. **75**, 102289 (2022)
37. Zhou, B., Zhou, S.K., Duncan, J.S., Liu, C.: Limited view tomographic reconstruction using a cascaded residual dense spatial-channel attention network with projection data fidelity layer. IEEE Trans. Med. Imaging **40**(7), 1792–1804 (2021)

MoCoSR: Respiratory Motion Correction and Super-Resolution for 3D Abdominal MRI

Weitong Zhang[1,2(✉)], Berke Basaran[1,2,3], Qingjie Meng[2], Matthew Baugh[2], Jonathan Stelter[7], Phillip Lung[6], Uday Patel[6], Wenjia Bai[2,3,4], Dimitrios Karampinos[7], and Bernhard Kainz[2,5]

[1] UKRI CDT in AI for Healthcare, Imperial College London, London, UK
weitong.zhang20@imperial.ac.uk
[2] Department of Computing, Imperial College London, London, UK
[3] Data Science Institute, Imperial College London, London, UK
[4] Department of Brain Sciences, Imperial College London, London, UK
[5] Friedrich-Alexander University Erlangen-Nürnberg, Erlangen, DE, Germany
[6] St Mark' Radiology, London North West University Healthcare NHS Trust, London, UK
[7] Department of Diagnostic and Interventional Radiology, Technical University of Munich, Munich, Germany

Abstract. Abdominal MRI is critical for diagnosing a wide variety of diseases. However, due to respiratory motion and other organ motions, it is challenging to obtain motion-free and isotropic MRI for clinical diagnosis. Imaging patients with inflammatory bowel disease (IBD) can be especially problematic, owing to involuntary bowel movements and difficulties with long breath-holds during acquisition. Therefore, this paper proposes a deep adversarial super-resolution (SR) reconstruction approach to address the problem of multi-task degradation by utilizing cycle consistency in a staged reconstruction model. We leverage a low-resolution (LR) latent space for motion correction, followed by super-resolution reconstruction, compensating for imaging artefacts caused by respiratory motion and spontaneous bowel movements. This alleviates the need for semantic knowledge about the intestines and paired data. Both are examined through variations of our proposed approach and we compare them to conventional, model-based, and learning-based MC and SR methods. Learned image reconstruction approaches are believed to occasionally hide disease signs. We investigate this hypothesis by evaluating a downstream task, automatically scoring IBD in the area of the terminal ileum on the reconstructed images and show evidence that our method does not suffer a synthetic domain bias.

Keywords: Abdominal MR · Motion Correction · Super-resolution · Deep Learning

Supplementary Information The online version contains supplementary material available at https://doi.org/10.1007/978-3-031-43999-5_12.

1 Introduction

Inflammatory bowel disease (IBD) is a relatively common, but easily overlooked disease. Its insidious clinical presentation [1] can lead to a long delay between the initial causative event and the diagnosis [2]. One of its manifestations, Crohn's disease, often exhibits symptoms such as abdominal pain, diarrhoea, fatigue, and cramping pain, which can be accompanied by severe complications [3]. Although Crohn's disease cannot be completely cured, early diagnosis can significantly reduce treatment costs and permanent physical damage [4]. MRI plays a crucial role in diagnosing and monitoring Crohn's disease. In clinical applications and research, high-resolution (HR) MRI is often preferred over endoscopy as it is non-invasive and visualises more details for the small bowel. MRI is preferred over computed tomography (CT) imaging as it does not use radiation, which is an important consideration in younger patients. Unfortunately, MR acquisition for patients with Crohn's disease can easily become compromised by respiratory motion.

As a result, many patients' images are degraded by respiration, involuntary movements and peristalsis. Furthermore, due to technical limitations, it is difficult to acquire HR images in all scan orientations. This limits the assessment of the complete volume in 3D. Given these problems, we aim to develop a novel method that can perform both motion correction (MC) and super-resolution (SR) to improve the quality of 3D IBD MRI and to support accurate interpretation and diagnosis.

Motion can cause multiple issues for MR acquisition. Abdominal MRI scans are usually 2D multi-slice acquisitions [5]. As a result, 3D bowel motion can lead to intra- and inter-plane corruptions [6], *e.g.*, slice misregistration, slice profile effects, and anisotropic spatial resolution. SR can be used to enhance these scans, but conventional methods often struggle with this type of anisotropic data or may unintentionally hide significant imaging findings.

Despite these challenges, MC and SR are crucial because corrupted MR images can lead to inaccurate interpretation and diagnosis [7]. Manual correction or enhancement of these volumes is not feasible.

Contribution: Our method (MoCoSR) alleviates the need for semantic knowledge and manual paired-annotation of individual structures and the requirement for acquiring multiple image stacks from different orientations, *e.g.*, [8]. There are several methodological contributions of our work: (1) First, to account for non-isotropic voxel sizes of abdominal images, we reconstruct spatial resolution from corrupted bowel MR images by enforcing cycle consistency. (2) Second, volumes are corrected by incorporating latent features in the LR domain. The complementary spatial information from unpaired quality images is exploited via cycle regularisation to provide an explicit constraint. Third, we conduct extensive evaluations on 200 subjects from a UK Crohn's disease study, and a public abdominal MRI dataset with realistic respiratory motion. (3) Experimental evaluation and analysis show that our MoCoSR is able to generate high-quality MR images and performs favourably against other, alternative methods. Furthermore, we explore confidence in the generated data and improvements to the diagnostic process. (4) Experiments with existing models for predicting the degree of small bowel inflammation in Crohn's disease patients show that MoCoSR can retain diagnostically relevant features and maintain the original HR feature distribution for downstream image analysis tasks.

Related Work: MRI SR. For learning-based MRI super-resolution, [9] discussed the feasibility of learning-based SR methods, where encoder-decoder methods [10–12] are commonly used to model a variety of complex structures while preserving details. Most single-forward [13–15] and adversarial methods [16,17] rely on paired data to learn the mapping and degradation processes, which is not acceptable in real-world scenarios where data are mismatched. [18] utilizes cyclic consistency structures to address unpaired degradation adaptation in brain SR, however abdominal data would be more complicated and suffer from motion corruption. Joint optimization of MC and SR remains challenging because of the high-dimensionality of HR image space, and LR latent space has been introduced in order to alleviate this issue. Recent studies on SR joint with other tasks (*e.g.*, reconstruction, denoising) have demonstrated improvements in the LR space [11,19,20]. For this purpose, we utilize a cycle consistency framework to handle unpaired data and joint tasks.

Automated Evaluation of IBD. In the field of machine learning and gastrointestinal disease, [21] used random forests to segment diseased intestines, which is the first time that image analysis support has been applied to bowel MRI. However, this technique requires radiologists to label and evaluate diseased bowel segments, and patients' scan times are long. In [22] residual networks focused on the terminal ileum to detect Crohn's disease. In this case, quality reconstruction data is extremely important for the detection of relevant structures.

2   Method

Problem Formulation and Preliminaries: In 3D SR, the degradation process is modeled with: $I_{LR} = \mathcal{D}(I_{HR}; k_I, \downarrow_s) + n$, $\mathcal{D}()$ represents the downsampling with the blur kernel k_I, scaling factor s, and noise n. In this work, we propose the motion corruption term M, which operates on LR latent space. And our MC-SR model can be refined to

$$I_{LR} = M(\mathcal{D}(I_{HR}; k_I, \downarrow_s); Z) + n \tag{1}$$

MoCoSR Concept: As shown in Fig. 1, our multi-task framework consists of three parts: a pair of corrupted LR (CLR) encoder and SR decoder on corrupted LR input, a pair of quality SR (QLR) encoder and learned LR (LLR) decoder on HR input and two task-specific discriminators. For the first two pairs, individual features are extracted and scaling is applied to provide our network with the ability to handle multiple tasks at different scales. The discriminators are then used to identify features for each scale.

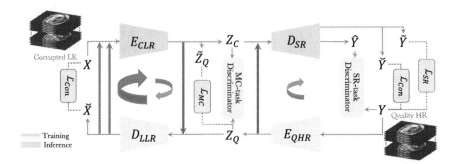

Fig. 1. During training, the input comprises corrupted MRI with motion artifacts and blurring effects. During inference, only the first pair of LR encoder E_{CLR} and SR decoder D_{SR} will be utilized to generate high-quality, motion-free, and super-resolved images.

A quality HR volume Y is fed first into the QHR encoder in order to obtain a quality latent feature map Z_Q upon which the corrupted LR can be trained. due to the unpaired scenario. For MC, Z_Q passes to LLR and CLR to gain $\tilde{Z}_Q$ with $\mathcal{L}_{MC}$, where LLR learns the corruption feature and degradation, and CLR encodes the multi-degradation from the corrupted LR input X. For SR, $\tilde{Y}$ is then generated after the SR decoder and ensure the SR purpose with $\mathcal{L}_{SR}$. To ensure training stability, two consistency loss functions $\mathcal{L}_{Con}$ are utilized for each resolution space. The arrows indicate the flow of the tasks. A pair of task-specific discriminators is used to improve the performance of each task-related encoder and decoder.

Loss Functions: Rather than aiming to reconstruct motion-free and HR images in high dimensional space with paired data, we propose to regularize in low dimensional latent space to obtain a high quality LR feature that can be used for upscaling. A $\mathcal{L}_{MC}$ between $\tilde{Z}_Q$ downsampled from QHR and Z_Q cycled after LLR and CLR, defines in an unpaired manner as follows:

$$\mathcal{L}_{MC} = E\left[\left|Z_Q - \tilde{Z}_Q\right|\right] \tag{2}$$

As the ultimate goal, SR focuses on the recovery and upscaling of detailed high-frequency feature, $\mathcal{L}_{SR}$ is proposed to optimize the reconstruction capability by passing through cycles at various spaces:

$$\mathcal{L}_{SR} = \mathcal{L}_1(Y, \tilde{Y}) \tag{3}$$

The dual adversarial $\mathcal{L}_{DAdv}$ is applied to improve the generation capability of the two sets of single generating processes in the cyclic network:

$$\mathcal{L}_{DAdv} = \mathcal{L}_{adv}(\check{X}) + \mathcal{L}_{adv}(\check{Y}) + \mathcal{L}_{adv}(\tilde{Y}) \tag{4}$$

The corresponding two task-specific discriminators $\mathcal{L}_{DMC}$ and $\mathcal{L}_{DSR}$ for discriminating between corrupted and quality images followed are used for the purpose of staged reconstruction Z_Q and $\hat{Y}$ of MC at latent space and SR at spatial space, respectively. Furthermore, a joint cycle consistency loss is used to improve the stability of training in both spaces:

$$\mathcal{L}_{Con} = \mathcal{L}_1(X, \check{X}) + \mathcal{L}_1(Y, \check{Y}) \tag{5}$$

For MoCoSR generators, the joint loss function is finally defined as follows:

$$\mathcal{L} = \mathcal{L}_{SR} + \lambda_1 \mathcal{L}_{MC} + \lambda_2 \mathcal{L}_{DAdv} + \lambda_3 \mathcal{L}_{Con} \tag{6}$$

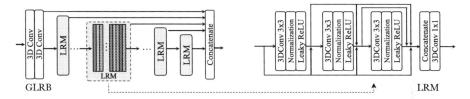

Fig. 2. GLRB is used to construct an effective feature extraction backbone based on encoders and decoders with different inputs. Additionally, the SR decoder and the QHR encoder include additional Pixel-shuffle and downsampling layers.

Network Architecture: In the paired encoder-decoder structure, we developed a 3D Global and Local Residual Blocks (GLRB) with Local Residual Modules (LRM) in Fig. 2 based on [23]. The GLRB is designed to extract local features and global structural information at 3D level, and then construct blocks for multi-scale features connected to the output of the first layer. The residual output is then added to the input using a residual connection to obtain a staged output. The model implements the extraction of local features while integrating all previous features through the connected blocks and compression layers. The decoders are employed with the upsampling prior to the convolution layers.

3   Experiments

Data Degradation: We use $64 \times 64 \times 64$ patches. For downsampling and the degradation associated with MRI scanning, (1) Gaussian noise with a standard deviation of 0.25 was added to the image. (2) Truncation at the k-space, retaining only the central region of the data. On top of this, we have developed a motion simulation process to represent the pseudo-periodic multi-factor respiratory motion (PMRM) that occurs during scanning, as shown in Fig. 3 (a). The simulated motion includes the influence of environmental factors that cause the respiratory intensity and frequency to fluctuate within a certain range. This lead to the presence of inter-slice misalignment in image domains.

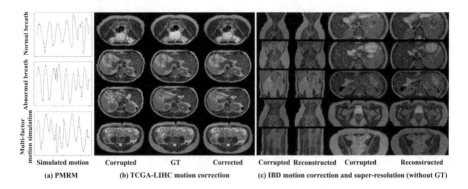

Fig. 3. (a) The proposed PMRM sequence of respiratory simulation for motion corruption for Crohn's diseases. (b) The staged reconstruction process involving application of PMRM, MC result, and ground truth (GT) on TCGA-LIHC dataset. The MC matrix is calculated and applied on HR for visual comparison with simulated motion perturbations directly. (c) The MoCoSR results on clinical IBD data without GT.

TCGA-LIHC Data Set, Abdominal MRI: A total of 237 MR exams are collected from The Cancer Imaging Archive Liver Hepatocellular Carcinoma (TCGA-LIHC). The data contains 908 MRI series from 97 patients. We applied simulated motion to TCGA MRI-Abdomen series to generate the motion-corrupted dataset with respiration-induced shift.

IBD Data Set, Inflammatory Bowel Disease: MRI sequences obtained include axial T2 images, coronal T2 images and axial postcontrast MRI data on a Philips Achieva 1.5 T MR scanner. Abdominal 2D-acquired images exhibit motion shifts between slices and fibrillation artefacts due to the difficulty of holding one's breath/body movement and suppressing random organ motion for extended periods. The dataset contains 200 available sample cases with four classes, healthy, mild, moderate, and severe small bowel Crohn's disease inflammation as shown in Table 1. The abnormal Crohn's disease sample cases, which could contain more than one segment of terminal ileum and small bowel Crohn's disease, were defined based on the review of clinical endoscopic, histological, and radiological images and by unanimous review by the same two radiologists (this criterion has been used in the recent METRIC trial investigating imaging in Crohn's disease [24]).

Setting: We compare with interpolation of bicubic and bilinear techniques, rapid and accurate MR image SR (RAISR-MR) with hashing-based learning [25], which we use as the representative of the model-based methods, and MRI SR with various generative adversarial networks (MRESR [17], CMRSR [16]) as the learning-based representatives. For training, with train/val/test with 70%/10%/20%. Various methods were included in the quantitative evaluation, including single forward WGAN (S-WGAN) and cycle RDB (C-RDB) for ablation experiments on the cycle consistency framework and the GLRB setting.

Table 1. IBD data set: the upper shows MRI data acquisition parameters. The lower shows the number of patients in each severity level of inflammation.

Planes	Sequence	TR [ms]/TE [ms]	Matrix	Slice [mm]	FOV	Time [s]
Axial	T1 FFE (e-THRIVE)	5.9/3.4	$512 \times 512 \times 96$	3.00	375	20.7×2
Axial Postcon	single-shot T2 TSE	587/120	$528 \times 528 \times 72$	3.50	375	22.3×2
Coronal	single-shot T2 TSE	554/120	$512 \times 512 \times 34$	3.00	375	21.1
Inflammation class		Healthy	Mild	Moderate	Severe	
Number of patients		100	49	42	9	

Table 2. Quantitative evaluation of extensive methods on TCGA-LIHC and IBD data sets in terms of SSIM and PSNR for 3D abdominal MRI SR

Method	TCGA-LIHC upper abdominal		IBD lower abdominal	
	SSIM ↑	PSNR ↑	SSIM ↑	PSNR ↑
Bicubic	0.8166	28.53	0.6582	25.30
Bilinear	0.8219	29.15	0.6749	26.27
RAISR-MR [25]	0.8940	32.24	0.8512	31.53
MRESR [17]	0.9042	33.74	0.8529	30.60
CMRSR [16]	0.9185	35.92	0.8662	32.90
S-WGAN	0.8521	30.91	0.7313	28.19
C-RDB	0.9102	35.15	0.8624	32.78
MoCoSR (C-GLRB)	0.9257	35.67	0.8719	33.56

Results: The quantitative results are presented in Table 2 on TCGA-LIHC and IBD data sets. There is a performance gap between interpolation and GAN-based methods, and the SR method based on learning has a significant mapping advantage for complex gastrointestinal images. MoCoSR achieves the best performance among all evaluation metrics. CMRSR and MRESR cannot guarantee the output quality of mapping from HR back to LR, resulting in poor performance on complex 3D bowel data. Representatives for the qualitative evaluation are shown in Fig. 4.

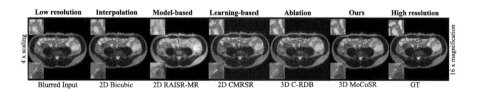

Fig. 4. Qualitative evaluation for SR on IBD dataset using representative methods of interpolation, model-based, learning-based, ablation, and MoCoSR.

Sensitivity Analysis: We evaluate if our method influences downstream analysis using the example of automatic scoring of Crohn's disease with existing deep networks. If features defining Crohn's disease are hidden by the method, this would affect disease scoring. We use a classifier with an attention mechanism similar to [22], trained on HR raw data. Our evaluation is based on the average possibility of normal and abnormal small bowel inflammation on MRI. The degree of small bowel inflammation on abnormal MRIs was classified by Radiologists as mild, moderate or severe. This outcome was compared against the results of the data constructed from different SR methods.

Complete results including LR degraded image, SR image reconstructed by MRESR, CMRSR, and MoCoSR, are shown in Table 3. We tracked and quantified the changes by performing a significance evaluation (t-test) based on p-values < 0.05. The ideal SR data can achieve classification results as close as possible to HR data with lower requirements. Our method obtains similar small-scale attenuation results on both healthy and abnormal samples. The p-value is larger than 0.05 for MoCoSR, *i.e.*, there is no statistically significant difference between the original and reconstructed data for the prediction results. The results of MRESR are volatile but present an unexpected improvement on healthy samples. CMRSR makes the predicted probability much lower than that of HR.

Discussion: According to the sensitivity analysis and comparison results, our MoCoSR method shows superior results compared to the forward adversarial reconstruction algorithms and encoder-decoder structures. Combining multi-scale image information in the feature space of different resolution image domains yields better results than inter-domain integration. The cycle consistency network splits the different resolution spaces and latent space, which facilitates the flexibility of the neural network to customize the MC according to the specific purpose and ensures consistency of the corrected data with the unpaired data. Furthermore, although these methods can obtain acceptable SSIM and PSNR, the key features used by the classifier for downstream tasks are potentially lost during the reconstructions. Conversely, the reconstruction result will implicitly cause domain shift. This leads to a distribution shift in the samples, which makes the disease prediction biased as shown in Fig. 5. The data generated by ours can

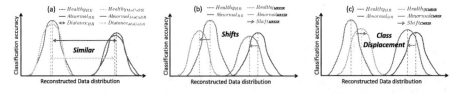

Fig. 5. Distribution relationships between reconstructed 3D bowel data and original HR data in a downstream Crohn's disease diagnosis task. (a) MoCoSR preserves diagnostic features and reconstructs a close representation of original data distribution. (b) The comparison method results in the loss and concealment of discriminative features. (c) Incorrectly reconstructed data misleads shifts in the distribution of SR data, which affects downstream task results.

Table 3. Crohn's disease classification performance on different 3D data. MoCoSR has a negligible effect on the downstream classification task as shown by high p-values in contrast to LR, MRESR, and CMRSR which produce significantly lower performance.

Data	Classification	Changes in/ Prediction	Significance p-value	CI 95% (Lower, Upper)
HR	Healthy	0.784	N/A	N/A
	Abnormal	0.730		
LR	Healthy	$-0.19^*\Downarrow$	<0.001	(0.13, 0.24)
	Abnormal	$-0.21^*\Downarrow$	<0.001	(0.14, 0.25)
MRESR [17]	Healthy	$-0.11^*\Downarrow$	<0.001	(0.05, 0.16)
	Abnormal	$-0.10^*\Downarrow$	0.002	(0.04, 0.15)
CMRSR [16]	Healthy	$+0.03\uparrow$	0.370	(-0.08, 0.03)
	Abnormal	$-0.08^*\downarrow$	0.012	(0.02, 0.13)
MoCoSR (ours)	Healthy	$-0.01\downarrow$	0.630	(-0.04, 0.06)
	Abnormal	$-0.02\downarrow$	0.561	(-0.04, 0.07)

reconstruct the results, retain likely all of the diagnostically valuable features, and maintain the original data distribution. The present sensitivity study is limited to the automatic classification from single domain and down-stream task framework, and future extensions will explore model-based and learning segmentation tasks across data domains and acquisitions.

4   Conclusion

MoCoSR is a DL-based approach to reconstruct high-quality SR MRI. MoCoSR is evaluated extensively and compared to the various image SR reconstruction algorithms on a public abdominal dataset, simulating different degrees of respiratory motion, and an IBD dataset with inherent motion. MoCoSR demonstrated superior performance. To test if our learned reconstruction preserves clinically relevant features, we tested on a downstream disease scoring method and found no decrease in disease prediction performance with MoCoSR.

Acknowledgements. This work was supported by the JADS programme at the UKRI Centre for Doctoral Training in Artificial Intelligence for Healthcare (EP/S023283/1) and HPC resources provided by the Erlangen National High Performance Computing Center (NHR@FAU) of the Friedrich-Alexander-Universität Erlangen-Nürnberg (FAU) under the NHR project b143dc. NHR funding is provided by federal and Bavarian state authorities. NHR@FAU hardware is partially funded by the German Research Foundation (DFG) - 440719683. Support was also received by the ERC - project MIA-NORMAL 101083647 and DFG KA 5801/2-1, INST 90/1351-1.

References

1. Gastrointestinal Unit Medical Services MGH, Andres, P.G., Friedman, L.S., et al.: Epidemiology and the natural course of inflammatory bowel disease. Gastroenterol. Clin. North Am. **28**(2), 255–281 (1999)
2. Sandler, R., Eisen, G.: Epidemiology of inflammatory bowel disease. In: Kirsner (ed.) Inflammatory Bowel Disease, p. 96 5th ed. WB Saunders, Philadelphia (2000)
3. Rosen, M.J., Dhawan, A., Saeed, S.A.: Inflammatory bowel disease in children and adolescents. JAMA Pediatrics. **169**(11), 1053–60 (2015)
4. Tielbeek, J.A., et al.: Grading Crohn disease activity with MRI: interobserver variability of MRI features, MRI scoring of severity, and correlation with Crohn disease endoscopic index of severity. AJR **201**(6), 1220–8 (2013)
5. Ebner, M., et al.: Point-spread-function-aware slice-to-volume registration: application to upper abdominal MRI super-resolution. In: Zuluaga, M.A., Bhatia, K., Kainz, B., Moghari, M.H., Pace, D.F. (eds.) RAMBO/HVSMR -2016. LNCS, vol. 10129, pp. 3–13. Springer, Cham (2017). https://doi.org/10.1007/978-3-319-52280-7_1
6. Zaitsev, M., Maclaren, J., Herbst, M.: Motion artifacts in MRI: a complex problem with many partial solutions. Magn. Reson. Imaging. **42**(4), 887–901 (2015)
7. Afaq, A., et al.: Pitfalls on PET/MRI. In: Seminars in Nuclear Medicine, vol. 51, pp. 529–39. Elsevier (2021)
8. Alansary, A., et al.: PVR: patch-to-volume reconstruction for large area motion correction of fetal MRI. IEEE Trans. Med. Imaging. **36**(10), 2031–44 (2017)
9. Wang, Z., Chen, J., Hoi, S.C.H.: Deep learning for image super-resolution: a survey. IEEE Trans. Pattern Anal. Mach. Intell. **43**(10), 3365–87 (2021)
10. Lim, B., Son, S., Kim, H., Nah, S., Mu Lee K.: Enhanced deep residual networks for single image super-resolution. In: CVPR, pp. 136–44 (2017)
11. Feng, C.-M., Yan, Y., Fu, H., Chen, L., Xu, Y.: Task transformer network for joint MRI reconstruction and super-resolution. In: de Bruijne, M., et al. (eds.) MICCAI 2021, Part VI. LNCS, vol. 12906, pp. 307–317. Springer, Cham (2021). https://doi.org/10.1007/978-3-030-87231-1_30
12. Feng, C.-M., Fu, H., Yuan, S., Xu, Y.: Multi-contrast MRI super-resolution via a multi-stage integration network. In: de Bruijne, M., et al. (eds.) MICCAI 2021, Part VI. LNCS, vol. 12906, pp. 140–149. Springer, Cham (2021). https://doi.org/10.1007/978-3-030-87231-1_14
13. Chen, Y., Shi, F., Christodoulou, A.G., Xie, Y., Zhou, Z., Li, D.: Efficient and accurate MRI super-resolution using a generative adversarial network and 3D multi-level densely connected network. In: Frangi, A.F., Schnabel, J.A., Davatzikos, C., Alberola-López, C., Fichtinger, G. (eds.) MICCAI 2018, Part I. LNCS, vol. 11070, pp. 91–99. Springer, Cham (2018). https://doi.org/10.1007/978-3-030-00928-1_11
14. Sánchez, I., Vilaplana, V.: Brain MRI super-resolution using 3D generative adversarial networks. arXiv preprint arXiv:1812.11440 (2018)
15. Georgescu, M.I., et al.: Multimodal multi-head convolutional attention with various kernel sizes for medical image super-resolution. In: Proceedings of the IEEE/CVF Winter Conference on Applications of Computer Vision, pp. 2195–205 (2023)
16. Zhao, M., Wei, Y., Wong, K.K.: A Generative Adversarial Network technique for high-quality super-resolution reconstruction of cardiac magnetic resonance images. Magn. Reson. Imaging. **85**, 153–60 (2022)
17. Do, H., Bourdon, P., Helbert, D., Naudin, M., Guillevin, R.: 7T MRI super-resolution with Generative Adversarial Network. Electronic Imaging. **2021**(18), 106–1 (2021)

18. Liu, J., Li, H., Huang, T., Ahn, E., Razi, A., Xiang, W.: Unsupervised representation learning for 3D MRI super resolution with degradation adaptation. arXiv preprint arXiv:2205.06891 (2022)

19. Wang, S., et al.: Joint motion correction and super resolution for cardiac segmentation via latent optimisation. In: de Bruijne, M., et al. (eds.) MICCAI 2021. LNCS, vol. 12903, pp. 14–24. Springer, Cham (2021). https://doi.org/10.1007/978-3-030-87199-4_2

20. Luo, Z., Huang, H., Yu, L., Li, Y., Fan, H., Liu, S.: Deep constrained least squares for blind image super-resolution. In: Proceedings of the IEEE/CVF Conference on Computer Vision and Pattern Recognition, pp. 17642–17652 (2022)

21. Mahapatra, D., Schüffler, P.J., Tielbeek, J.A., Makanyanga, J.C., Stoker, J., Taylor, S.A., et al.: Automatic detection and segmentation of Crohn's disease tissues from abdominal MRI. IEEE Trans. Med. Imaging. **32**(12), 2332–47 (2013)

22. Holland, R., Patel, U., Lung, P., Chotzoglou, E., Kainz, B.: Automatic detection of bowel disease with residual networks. In: Rekik, I., Adeli, E., Park, S.H. (eds.) PRIME 2019. LNCS, vol. 11843, pp. 151–159. Springer, Cham (2019). https://doi.org/10.1007/978-3-030-32281-6_16

23. He, K., Zhang, X., Ren, S., Sun, J.: Deep residual learning for image recognition. In: CVPR, pp. 770–778 (2016)

24. Taylor, S.A., et al.: Diagnostic accuracy of magnetic resonance enterography and small bowel ultrasound for the extent and activity of newly diagnosed and relapsed Crohn's disease (METRIC): a multicentre trial. Lancet Gastroenterol Hepatol. **3**(8), 548–58 (2018)

25. Romano, Y., Isidoro, J., Milanfar, P.: RAISR: rapid and accurate image super resolution. IEEE Trans. Comput. Imaging. **3**(1), 110–25 (2016)

Estimation of 3T MR Images from 1.5T Images Regularized with Physics Based Constraint

Prabhjot Kaur[1]($\boxtimes$) (ID), Atul Singh Minhas[2] (ID), Chirag Kamal Ahuja[3] (ID), and Anil Kumar Sao[4] (ID)

[1] Indian Institute of Technology Mandi, Mandi, India
kaurprabhjotinresearch@gmail.com
[2] Magnetica, Northgate, Australia
[3] Postgraduate Institute of Medical Education and Research, Chandigarh, India
[4] Indian Institute of Technology Bhilai, Bhilai, India

Abstract. Limited accessibility to high field MRI scanners (such as 7T, 11T) has motivated the development of post-processing methods to improve low field images. Several existing post-processing methods have shown the feasibility to improve 3T images to produce 7T-*like* images [3,18]. It has been observed that improving lower field (LF, ≤ 1.5T) images comes with additional challenges due to poor image quality such as the function mapping 1.5T and higher field (HF, 3T) images is more complex than the function relating 3T and 7T images [10]. Except for [10], no method has been addressed to improve ≤ 1.5T MRI images. Further, most of the existing methods [3,18] including [10] require example images, and also often rely on pixel to pixel correspondences between LF and HF images which are usually inaccurate for ≤ 1.5T images. The focus of this paper is to address the unsupervised framework for quality improvement of 1.5T images and avoid the expensive requirements of example images and associated image registration. The LF and HF images are assumed to be related by a linear transformation (LT). The unknown HF image and unknown LT are estimated in alternate minimization framework. Further, a physics based constraint is proposed that provides an additional non-linear function relating LF and HF images in order to achieve the desired high contrast in estimated HF image. This constraint exploits the fact that the T1 relaxation time of tissues increases with increase in field strength, and if it is incorporated in the LF acquisition the HF contrast can be simulated. The experimental results demonstrate that the proposed approach provides processed 1.5T images, i.e., estimated 3T-*like* images with improved image quality, and is comparably better than the existing methods addressing similar problems.

This work is financially supported by Ministry of Electronics and Information Technology, India.

Supplementary Information The online version contains supplementary material available at https://doi.org/10.1007/978-3-031-43999-5_13.

The improvement in image quality is also shown to provide better tissue segmentation and volume quantification as compared to scanner acquired 1.5T images. The same set of experiments have also been conducted for 0.25T images to estimate 1.5T images, and demonstrate the advantages of proposed work.

Keywords: Low field MRI · 0.25T · 1.5T · 3T · T1 relaxation times · Optimization

1   Introduction

Magnetic resonance imaging (MRI) has seen a tremendous growth over the past three decades as a preferred diagnostic imaging modality. Starting with the 0.25T MRI scanners in 19801980ss [11], clinical scanners have emerged from 1.5T and 3T to the recently approved 7T scanners [9]. Primary reason for the preference of higher magnetic field strength (FS) is the better image quality but it comes with high cost. Generally, the cost of clinical MRI scanners increases at USD 1million/Tesla [14]. Therefore, there is a clear divide in the distribution of MRI scanners across different countries in the world. With the FDA approval of clinical use of 7T MRI scanners- 3T and 7T MRI scanners have become preferred scanners in developed countries but inexpensive low field MRI scanners (<1T) and mid field MRI scanners (1.5T) are still a popular choice in developing countries [10,11].

Since the past decade, there is a growing interest towards improving the quality of images acquired with low FS MRI scanners by learning the features responsible for image quality from high FS MRI scanners [8,9,19]. Earlier methods in such image translational problems synthesized target image contrast using paired example images [13]. The first work to address estimation of 7T-*like* images was reported in 2016 [1]. It exploits canonical correlation analysis (CCA) [5] to relate paired 3T-7T images in a more correlated space. With advent of deep learning, 3D convolutional neural network (CNN) was proposed to learn non-linear mapping between 3T and 7T images [3]. Different architectures of deep learning were addressed with better performances than previous methods [2,7,8,18]. However, except Lin et al [10], none of the existing approaches address the problem of improving the image quality of ≤1.5T images to estimate high quality 3T-like MR images. Notably, to the best of our knowledge no approach is reported in literature to estimate the 3T (or 1.5T-like) images from 1.5T (or 0.25T) images. This problem is particularly challenging because of the severity of degradation present in ≤1T MR images [10]. In [10], the mapping between example 3T and simulated 0.36T images using CNNs to estimate 3T from scanner-acquired 0.36T images is learned, and it requires the a-priori knowledge of distribution of tissue-specific SNR for given FS. Since the databases for example images are scarce and SNR of tissues is not known a-priori [12], the unsupervised methods needs to be explored to improve ≤1.5T images.

In this work, we address the problem of estimating 3T-like images ($\mathbf{x}$) from 1.5T images ($\mathbf{y}$) in an unsupervised manner. To our best knowledge, this is the first work to develop a method for improving ≤1.5T images to estimate HF images without requiring example images. The proposed method formulates the estimation of 3T HF images as an inverse problem: $\mathbf{y} = f(\mathbf{x})$, where $f(.)$ is the unknown degradation model. The novel contributions of our work are as follows: (i) the alternate minimization (AM) framework is formulated to estimate $\mathbf{x}$ as well as the mapping kernel $f(.)$ relating $\mathbf{x}$ and $\mathbf{y}$, (ii) Acquisition physics based signal scaling is proposed to synthesize the desired image contrast similar to HF image, (iii) the simulated contrast image is used as a regularizer while estimating $\mathbf{x}$ from $\mathbf{y}$. The experimental results demonstrate that the proposed approach provides improved quality of MRI images in terms of image contrast and sharp tissue boundaries that is similar to HF images. The experiments further demonstrate the successful application of the improved quality images provided by proposed method in improved tissue segmentation and volume quantification.

2   Proposed Method

The proposed work formulates the estimation of HF image ($\mathbf{x}$) from LF image ($\mathbf{y}$) in an alternate minimization framework:

$$\min_{\mathbf{h},\mathbf{x}} \underbrace{||\mathbf{y} - \mathbf{h} * \mathbf{x}||_F^2,}_{\text{Data Fidelity}} +\lambda_1 \underbrace{||\mathbf{c} \odot \mathbf{y} - \mathbf{x}||_F^2}_{\text{Physics based Regularizer}} +\lambda_2 \underbrace{||\mathbf{h}||_2^2}_{\text{Regularizing } \mathbf{h}} \tag{1}$$

Here $\mathbf{x} \in \mathbb{R}^{m \times n}$ and $\mathbf{y} \in \mathbb{R}^{m \times n}$ represent the HF and LF MRI images, respectively. The matrix $\mathbf{h} \in \mathbb{R}^{p \times p}$ represents transformation kernel convolved using $*$ operator with each patch of $\mathbf{y}$ of size $p \times p$ and p is empirically chosen as 5. The matrix $\mathbf{c} \in \mathbb{R}^{m \times n}$ represent the pixel wise scale when multiplied with $\mathbf{y}$ generates the image with contrast similar to HF image. Here, $\mathbf{c} \odot \mathbf{y}$ represents the Hadamard product of pixel wise scale $\mathbf{c}$ and $\mathbf{y}$ the 1.5T (or 0.25T) image.

2.1   Physics Based Regularizer

The physics based regularizer exploits the fact that the T1 relaxation time increases with FS. The differences among T1 relaxation times of gray matter (GM), white matter (WM) and cereberospinal fluid (CSF) also increase with FS, leading to increased contrast in the T1-weighted images in the order 3T ≥1.5T ≥ 0.25T. This factor is used to simulate an image from $\mathbf{y}$ which convey similar information as $\mathbf{x}$, and is obtained by scaling the signal intensities of $\mathbf{y}$ based on changes in signal due to changes in T1 relaxation time with respect to FS- denoted by $\mathbf{r}$. The acquired signal for $\mathbf{y}$ denoted by $\mathbf{s}_l$ can be corrected with relaxation time to simulate the signal $\mathbf{s}_h$ that would have been acquired for $\mathbf{x}$ as:

$$\hat{\mathbf{s}}_h = \mathbf{r} \times \mathbf{s}_l. \tag{2}$$

For example, in the spin echo (SE) pulse sequence, acquired signal can be represented as, $\mathbf{s} = \mathbf{A}B_0^2 \frac{\left(1-e^{(-TR/T_1)}\sin\theta\right)}{\left(1-e^{(-TR/T_1)}\cos\theta\right)} e^{(-TE/T_2)}$ [6,16]. Here, $\mathbf{A}$ represents the proportionality constant. The field strengths for 3T (or 1.5T) and 1.5T (or 0.25T) scanners are denoted by B_0. The factor $\mathbf{r}$ for the given voxel for SE sequence by assuming long TE, and T2 to be same across different FS (assumption derived from literature [12]) be computed as

$$\frac{\left(1 - e^{(-TR_h/T_{1,h})}sin\theta_h\right)}{\left(1 - e^{(-TR_h/T_{1,h})}cos\theta_h\right)} \bigg/ \frac{\left(1 - e^{(-TR_l/T_{1,l})}sin\theta_l\right)}{\left(1 - e^{(-TR_l/T_{1,l})}cos\theta_l\right)}. \tag{3}$$

Here, T_1 and T_2 represent the T1/T2 relaxation times, TR-repetition time, TE-echo time and θ-flip angle (FA). The parameters with subscript h and l represent the parameters used for HF and LF MRI acquisitions, respectively. Similarly, respective mathematical formulations can be used for other pulse sequences such as fast SE (FSE) and gradient echo (GRE). After computing $\mathbf{r}$ using Eq. (3), it can be used to simulate HF acquisition $\hat{s}_h$ using Eq. (2)

The values of T_1/T_2 relaxation times differ with tissues, hence different values of $\mathbf{r}$ should be computed for every voxel considering the tissues present in that voxel. However, if we assign single tissue per voxel using any tissue segmentation technique such as FAST [15], compute $\mathbf{r}$ for each voxel to estimate $\hat{s}_h$, this could lead to discontinuity among tissue boundaries in simulated HF image, and is shown in Fig. S1 in supplementary material (SM).

Compute Relaxation Times for Voxels with More Than One Tissue: In real practice there exist many voxels with more than one tissue kind present in them, and is the reason for discontinuities present in Fig. S1. The challenge is that T1 relaxation time for such voxels is not known that is required to estimate $\mathbf{r}$. We address this issue by estimating the T1 relaxation time of such voxel as linear combination of T1 relaxation times of tissues present in the given voxel. The linear weights are directly proportional to the probability of tissues present in the voxel. Though this work performs well with probability maps but we use pixel intensity to denote the probability of tissue to avoid the additional and expensive tissue segmentation step as follows: The top 5 percentile of pixel intensities are assumed to belong to WM and bottom 20 percentile as GM. Consider w and q as T1 relaxation time and corresponding pixel intensity, respectively. We here approximate relation between w and q using linear equation as $\frac{w_{WM} - w_{GM}}{q_{WM} - q_{GM}} = \frac{w - w_{GM}}{q - q_{GM}}$. Here, subscript indicates the tissue type. Here, w_{WM} and w_{GM} are T1 relaxation times which are currently assumed to be constant, and taken from literature [12] whereas q_{WM} and q_{GM} are the pixel intensity values computed by averaging pixel intensities falling in mentioned percentiles for WM and GM, respectively. In a simplest case, say for an image q_{WM} and q_{GM} are 1 and 0, then the linear relationship is reduced to $w = (w_{WM} - w_{GM})q + w_{GM}$. Hence, for the given pixel intensity we can estimate the corresponding T1 relaxation time as a linear combination of T1 relaxation times of WM and GM. The range of pixel intensities in $\mathbf{y}$ is partitioned into several bins. This is followed by approximation of T1 relaxation times $(T_{1,l})$ for different bins for $\mathbf{y}$ by assuming w_{WM} and w_{GM} as T1 relaxation times at LF FS. In the similar way, $T1_{1,h}$ can be approximated.

After approximating $T_{1,l}$ and $T_{1,h}$ for all voxels (and empirically choosing the TR_h and θ_h) the $\mathbf{r}$ is computed using Eq. (3) and is used to estimate the HF acquisition using Eq. (2). The estimated $\hat{\mathbf{s}}_h$ is used to compute the pixel wise scale $\mathbf{c}$ as $\mathbf{c} = \hat{\mathbf{s}}_h/\mathbf{y}$, i.e., an element wise division operator. The computed $\mathbf{c}$ is used in Eq. (1) to constrain the solution space of estimated HF image $\mathbf{x}$.

3 Experimental Results

The proposed work has been demonstrated for (i) estimating 3T-*like* from 1.5T images, (ii) estimating 1.5T-*like* from 0.25T images, (iii) evaluating accuracy of tissue segmentation using improved images in (i) and (ii), and (iv) comparing (i), (ii) and (iii) with existing methods. The results related to (ii) are summarized in SM due to space constraints. The values for λ_1 and λ_2 are chosen as 1.2 and 0.4, respectively.

3.1 Data

The MRI images used to demonstrate the efficacy of proposed work were acquired from five healthy subjects of age 25 ± 10 years. Three different MRI scanners were used in this study: 0.25T (G-scan Brio, Esaote), 1.5T (Aera Siemens) and 3T (Verio, Siemens). The first three subjects were scanned using each of the three scanners while the other two subjects were scanned with only 1.5T and 3T scanners. The three scanners were located in Post Graduate Institute of Medical Education & Research (PGIMER) Chandigarh, India. Scanning was performed using the standard clinical protocols-optimized for both clinical requirement and work-load of clinical site. All the scans were performed according to the guidelines of the Declaration of Helsinki. The details of pulse sequence and scan parameters used to acquire data in each of the three scanners is mentioned in SM Table S1. The acquired T1 MR image volumes for each scanner and each human subject were pre-processed similar to the human connectome project (HCP) pre-processing pipelines for structural MR images [4]. Please note that proposed approach does not require the LF and HF images to be skull stripped or to have pixel to pixel correspondence. It is only done to provide reference based similarity scores of estimated image with respect to HF image.

3.2 Analysis/Ablation Study of Proposed Approach

The HF image is estimated at different stages of proposed approach to demonstrate the significance of each term in Eq. (1), and is shown in Fig. 1(a). In Fig. 1(b), the impact of proposed regularizer on estimation of HF image is demonstrated by changing the values of λ_1. It can be observed that the image Fig. 1(a)-(ii) obtained just from the data fidelity term leads to sharp image details but without any contrast improvement. However, HF image simulated by the physics based regularizer from Eq. (2) improves the contrast but details remain blurred, as is evident from image Fig. 1(a)-(iii). Once the data fidelity and the physics

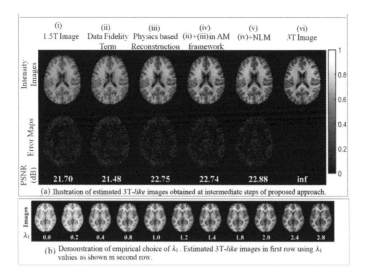

(a) Illustration of estimated 3T-*like* images obtained at intermediate steps of proposed approach.

(b) Demonstration of empirical choice of λ_1. Estimated 3T-*like* images in first row using λ_1 values as shown in second row.

Fig. 1. (a) Ablation study of proposed approach. (b) Demonstration of significance of proposed physics based constraint and associated parameter λ_1.

based regularizer is combined as in Eq. (1) in AM framework the corresponding image is shown in Fig. 1(a)-(iv) that is sharper as well as improved in contrast, and with image sharpness = 3.19, PIQE = 4.7, and SSIM = 0.0762. The image in Fig. 1(a)-(iv) is further smoothed in Fig. 1(a)-(v) using Non local means (NLM) approach to avoid any grainy effect if present due to division of pixel intensities into bins. The improvement in image quality is also evident from the PSNR values which increase from Fig. 1(a)-(ii) to (v).

3.3   Comparison with Existing Approaches

The performance of proposed approach is compared with existing methods that either address the contrast synthesis or estimation of HF images ScSR [17],CCA [5], MIMECS [13], ED [8]. The comparison is done in four ways: (i) *Objective* analysis using *reference based* metrics that requires the ground truth HF image, and pixel to pixel to correspondence between query image and ground truth image- Table 1, (ii) *Objective* analysis using *no-reference based* metrics which describe the quality of images solely based on edge sharpness and image contrast- Table 2, (i ii) *Subjective analysis* that includes rating of images by clinical experts in range 0 to 5, 5 and 0 being the highest and worst quality images, respectively - Table 3 and (iv) *Qualitative analysis* that includes the analysis of visual appearance of image details - Fig. 2. It can be observed that the existing methods provide better performance in terms of reference based metrics but perform inferior to proposed method in case of no-reference based metrics and subjective scores. This is due to the way the existing methods are designed, i.e., these methods are trained to minimize the mean square error

Table 1. OBJECTIVE ANALYSIS- NO-REFERENCE BASED METRICS

Metric		1.5T	HM	Supervised Approaches					Unsupervised Approaches		3T
				SCSR	CCA	MIMECS	EDp	ED	Proposed	Proposed (NLM)	
Signal Difference	WM-GM	0.1652	0.3414	0.1605	0.2079	0.2084	0.1291	0.1715	0.2166	0.2165	0.2449
	GM-CSF	0.3381	0.4536	0.3263	0.3154	0.3112	0.3345	0.3208	0.3207	0.3209	0.3144
Image Sharpness	Sharpness	55.62	95.16	58.09	54.81	54.25	60.39	52.06	63.01	64.13	64.6
	Edge width	0.1634	0.1559	0.1527	0.1837	0.1855	0.1545	0.1716	0.1812	0.1735	0.1776
	Edge height	9.08	14.81	8.87	10.07	10.06	9.32	8.93	11.42	11.12	11.46
PIQE	mean	67.57	60.63	51.52	74.32	75.86	58.76	82.81	57.78	72.53	70.25

Table 2. OBJECTIVE ANALYSIS- REFERENCE BASED METRICS

Metric		1.5T	HM	Supervised Approaches					Unsupervised Approaches		3T
				SCSR	CCA	MIMECS	EDp	ED	Proposed	Proposed (NLM)	
PSNR	mean	23.6	17.71	22.57	25.01	25.02	19.96	25.41	24.24	24.27	inf
	std.	0.1873	0.2244	0.1639	0.1441	0.1639	0.0426	0.1178	0.1849	0.1864	-
SSIM	mean	0.8314	0.8885	0.7639	0.8531	0.8465	0.6917	0.8511	0.8387	0.8465	1
	std.	0.0069	0.002	0.0047	0.0066	0.0072	0.0036	0.0058	0.0064	0.0064	-
UQI	mean	0.4854	0.8581	0.4300	0.6647	0.6607	0.4362	0.7053	0.5196	0.5263	1
	std.	0.101	0.0044	0.084	0.0275	0.0265	0.0065	0.0183	0.0755	0.095	-
VIF	mean	0.2697	0.5022	0.188	0.281	0.282	0.2784	0.3072	0.2451	0.251	1
	std.	0.0052	0.0074	0.002	0.0063	0.0063	0.0011	0.0067	0.0039	0.004	-

between estimated and HF images, thus they provide higher peak signal to noise ratio (PSNR) and structural similarity index metric (SSIM) but the image details are still blurred which lead to drop in edge sharpness. The validation of argument can also be derived by visually inspecting the images in Fig. 2. It has been observed that encoder-decoder based approach (ED [8]), MIMECS [13], CCA [5] and ScSR [17] provides blurred image details. The possible reasons are (i) minimizing MSE can provide perceptually blurred results, (ii) the weighted averaging involved in [5,13,17] induces blur, (iii) inaccurate pixel to pixel correspondences makes it difficult for supervised methods to learn the actual mapping relating input and target images. The drop in performances of existing approaches due to inaccuracies in image registration is more prominently observed when improving 0.25T images in Fig. S3. The robustness to such inaccuracies by proposed approach due to its unsupervised nature shows its clear advantages over existing methods.

Table 3. SUBJECTIVE ANALYSIS

Expert	1.5T	HM	Supervised Approaches					Unsupervised Approaches		3T
			SCSR	CCA	MIMECS	EDp	ED	Proposed	Proposed (NLM)	
(i)	3.25	4	2.5	4	4	2.25	3	4	4.25	5
(ii)	1	3	3	4	3	3	2	4	4	5
(iii)	1.75	3.75	1.5	0.75	1.5	0	2.25	5	4	4
(iv)	3.5	2.75	2.5	1.5	2	0	2.25	4.25	3	4.25
(v)	0.25	5	0.25	2	2.25	3	1.25	5	4.25	4.25

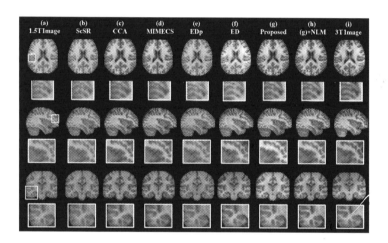

Fig. 2. Comparison of quality of HF images estimated by various methods.

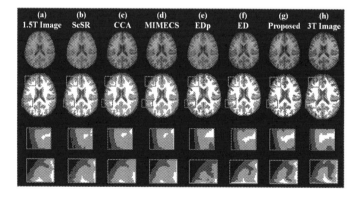

Fig. 3. Demonstration of improved segmentation from images estimated by proposed approach and its comparison with existing approaches

3.4   Application to Tissue Segmentation and Volume Quantification

The segmentation labels for WM, GM and CSF were computed using FAST tool-box in FSL software [15] for 3T-*like* images estimated by different approaches, and shown in Fig. 3. The improved tissue segmentation for 0.25T images is shown in Fig. S5 in SM. The zoomed windows in both figures indicate that the segmentation label of WM is improved for the estimated 3T reconstructed image by the proposed approach for estimated 3T image from 1.5T image. The quantitative measure used to evaluate performance of different methods is dice ratio, and is reported in Table S2, and its comparison is mentioned in Table S3. The ability to accurately segment tissues from image estimated by proposed approach is shown to be comparable both qualitatively and quantitatively to existing methods. Further, proposed method is shown to provide statistically significant with $p < 0.01$ improvement in accuracy of WM and GM tissue volume quantification for estimated 3T (and 1.5T images), and is shown in Fig. S6.

4   Summary

We propose a method to estimate HF images from ≤ 1.5T images in an unsupervised manner. Here, the knowledge of acquisition physics to simulate HF image is exploited, and used it in a novel way to regularize the estimation of HF image. The proposed method demonstrates the benefits over state of the art supervised methods that are severely effected by the inaccuracies if present in image registration process. Lower the FS image is, harder is to get accurate image registration, and thus proposed method proves to be a better choice. Further, it is also demonstrated that the proposed approach provides statistically significant accurate tissue segmentation. The code for this work is publicly shared on https://drive.google.com/drive/folders/1WbzkBJS1BWAje8aF0ty2SWYTQ9i0 B7Yr?usp=sharing.

References

1. Bahrami, K., Rekik, I., Shi, F., Gao, Y., Shen, D.: 7T-guided learning framework for improving the segmentation of 3T MR images. In: Ourselin, S., Joskowicz, L., Sabuncu, M.R., Unal, G., Wells, W. (eds.) MICCAI 2016. LNCS, vol. 9901, pp. 572–580. Springer, Cham (2016). https://doi.org/10.1007/978-3-319-46723-8_66
2. Bahrami, K., Rekik, I., Shi, F., Shen, D.: Joint reconstruction and segmentation of 7T-like MR images from 3T MRI based on cascaded convolutional neural networks. In: Descoteaux, M., Maier-Hein, L., Franz, A., Jannin, P., Collins, D.L., Duchesne, S. (eds.) MICCAI 2017. LNCS, vol. 10433, pp. 764–772. Springer, Cham (2017). https://doi.org/10.1007/978-3-319-66182-7_87
3. Bahrami, K., Shi, F., Rekik, I., Shen, D.: Convolutional neural network for reconstruction of 7T-like images from 3T MRI using appearance and anatomical features. In: Carneiro, G., et al. (eds.) LABELS/DLMIA-2016. LNCS, vol. 10008, pp. 39–47. Springer, Cham (2016). https://doi.org/10.1007/978-3-319-46976-8_5
4. Glasser, M.F., et al.: The minimal preprocessing pipelines for the human connectome project. Neuroimage **80**, 105–124 (2013). Mapping the Connectome

5. Huang, H., He, H., Fan, X., Zhang, J.: Super-resolution of human face image using canonical correlation analysis. Pattern Recogn. **43**(7), 2532–2543 (2010)
6. Jung, B.A., Weigel, M.: Spin echo magnetic resonance imaging. J. Magn. Reson. Imaging **37**(4), 805–817 (2013)
7. Kaur, P., Sao, A.K.: Single Image based reconstruction of high field-like MR images. In: Shen, D., et al. (eds.) MICCAI 2019. LNCS, vol. 11766, pp. 74–82. Springer, Cham (2019). https://doi.org/10.1007/978-3-030-32248-9_9
8. Kaur, P., Sharma, A., Nigam, A., Bhavsar, A.: MR-SRNET: transformation of low field MR images to high field MR images. In: 25th IEEE International Conference on Image Processing (ICIP), pp. 2057–2061, October 2018
9. der Kolk, A.G.V., Hendrikse, J., Zwanenburg, J.J., Visser, F., Luijten, P.R.: Clinical applications of 7T MRI in the brain. Eur. J. Radiol. **82**(5), 708–718 (2013)
10. Lin, H., et al.: Deep learning for low-field to high-field MR: image quality transfer with probabilistic decimation simulator. In: Knoll, F., Maier, A., Rueckert, D., Ye, J.C. (eds.) MLMIR 2019. LNCS, vol. 11905, pp. 58–70. Springer, Cham (2019). https://doi.org/10.1007/978-3-030-33843-5_6
11. Marques, J.P., Simonis, F.F., Webb, A.G.: Low-field MRI: an MR physics perspective. J. Magn. Reson. Imaging **49**(6), 1528–1542 (2019)
12. Nishimura, D.: Principles of Magnetic Resonance Imaging. Stanford (2010)
13. Roy, S., Carass, A., Prince, J.L.: Magnetic resonance image example-based contrast synthesis. IEEE Trans. Med. Imaging **32**(12), 2348–2363 (2013)
14. Sarracanie, M., Salameh, N.: Low-field MRI: how low can we go? A fresh view on an old debate. Front. Phys. **8**, 1–14 (2020)
15. Shi, F., Wang, L., Dai, Y., Gilmore, J.H., Lin, W., Shen, D.: Label: Pediatric brain extraction using learning-based meta-algorithm. Neuroimage **62**(3), 1975–1986 (2012)
16. Wu, Z., Chen, W., Nayak, K.S.: Minimum field strength simulator for proton density weighted MRI. PLoS ONE **11**(5), 1–15 (2016)
17. Yang, J., Wright, J., Huang, T.S., Ma, Y.: Image super-resolution via sparse representation. IEEE Trans. Image Process. **19**(11), 2861–2873 (2010)
18. Zhang, Y., Cheng, J.-Z., Xiang, L., Yap, P.-T., Shen, D.: Dual-domain cascaded regression for synthesizing 7T from 3T MRI. In: Frangi, A.F., Schnabel, J.A., Davatzikos, C., Alberola-López, C., Fichtinger, G. (eds.) MICCAI 2018. LNCS, vol. 11070, pp. 410–417. Springer, Cham (2018). https://doi.org/10.1007/978-3-030-00928-1_47
19. van der Zwaag, W., Schäfer, A., Marques, J.P., Turner, R., Trampel, R.: Recent applications of UHF-MRI in the study of human brain function and structure: a review. NMR Biomed. **29**(9), 1274–1288 (2016)

Feature-Conditioned Cascaded Video Diffusion Models for Precise Echocardiogram Synthesis

Hadrien Reynaud[1,2(✉)], Mengyun Qiao[2,3], Mischa Dombrowski[4],
Thomas Day[5,6], Reza Razavi[5,6], Alberto Gomez[5,7], Paul Leeson[7,8],
and Bernhard Kainz[2,4]

[1] UKRI CDT in AI for Healthcare, Imperial College London, London, UK
hadrien.reynaud19@imperial.ac.uk
[2] Department of Computing, Imperial College London, London, UK
[3] Department of Brain Sciences and DSI, Imperial College London, London, UK
[4] Friedrich–Alexander University Erlangen–Nürnberg, Erlangen, Germany
[5] School of BMEIS, King's College London, London, UK
[6] Guy's and St Thomas' NHS Foundation Trust, London, UK
[7] Ultromics Ltd., Oxford, UK
[8] John Radcliffe Hospital, Cardiovascular Clinical Research Facility, Oxford, UK

Abstract. Image synthesis is expected to provide value for the translation of machine learning methods into clinical practice. Fundamental problems like model robustness, domain transfer, causal modelling, and operator training become approachable through synthetic data. Especially, heavily operator-dependant modalities like Ultrasound imaging require robust frameworks for image and video generation. So far, video generation has only been possible by providing input data that is as rich as the output data, *e.g.*, image sequence plus conditioning in → video out. However, clinical documentation is usually scarce and only single images are reported and stored, thus retrospective patient-specific analysis or the generation of rich training data becomes impossible with current approaches. In this paper, we extend elucidated diffusion models for video modelling to generate plausible video sequences from single images and arbitrary conditioning with clinical parameters. We explore this idea within the context of echocardiograms by looking into the variation of the Left Ventricle Ejection Fraction, the most essential clinical metric gained from these examinations. We use the publicly available EchoNet-Dynamic dataset for all our experiments. Our image to sequence approach achieves an R^2 score of 93%, which is 38 points higher than recently proposed sequence to sequence generation methods. Code and weights are available at https://github.com/HReynaud/EchoDiffusion.

Keywords: Generative · Diffusion · Video · Cardiac · Ultrasound

Supplementary Information The online version contains supplementary material available at https://doi.org/10.1007/978-3-031-43999-5_14.

1 Introduction

Ultrasound (US) is widely used in clinical practice because of its availability, real-time imaging capabilities, lack of side effects for the patient and flexibility. US is a dynamic modality that heavily relies on operator experience and on-the-fly interpretation, which requires many years of training and/or Machine Learning (ML) support that can handle image sequences. However, clinical reporting is conventionally done via single, selected images that rarely suffice for clinical audit or as training data for ML. Simulating US from anatomical information, *e.g.* Computed Tomography (CT) [28], Magnetic Resonance Imaging (MRI) [25] or computational phantoms [11, 27], has been considered as a possible avenue to provide more US data for both operator and ML training. However, simulations are usually very computationally expensive due to complex scattering, reflection and refraction of sound waves at tissue boundaries during image generation. Therefore, the image quality of US simulations has not yet met the necessary quality to support tasks such as cross-modality registration, multi-modal learning, and robust decision support for image analysis during US examinations. More recently, generative deep learning methods have been proposed to address this issue. While early approaches show promising results, they either focus on generating individual images [16] or require video input data and further conditioning to provide useful results [17, 21]. Research in the field of image-conditioned video generation is very scarce [33] and, to the best of our knowledge, we are the first to apply it to medical imaging.

Contribution: In this paper, we propose a new method for video diffusion [7, 30] based on the Elucidated Diffusion Model (EDM) [13] that allows to synthesise plausible video data from single frames together with precise conditioning on interpretable clinical parameters, *e.g.*, Left Ventricular Ejection Fraction (LVEF) in echocardiography. This is the first time diffusion models have been extended for US image and video synthesis. Our contributions are three-fold: (1) We show that discarding the conventional text-embeddings [7, 20, 23, 24, 30] to control the reverse diffusion process is desirable for medical use cases where very specific elements must be precisely controlled; (2) We quantitatively improve upon existing methods [21] for counterfactual modelling, *e.g.*, when doctors try to answer questions like "how would the scan of this patient look like if we would change a given clinical parameter?"; (3) We show that fine-grained control of the conditioning leads to precise data generation with specific properties and outperforms the state-of-the-art when using such data, for example, for the estimation of LVEF in patients that are not commonly represented in training databases.

Related Work: Video Generation has been a research area within computer vision for many years now. Prior works can be organized in three categories: (1) pixel-level autoregressive models [2, 4, 12], (2) latent-level autoregressive model coupled with generators or up-samplers [1, 14] and (3) latent-variable transformer-based models with up-samplers [5,37]. Diffusion models have shown reasonable performance on low temporal and spatial resolutions [10] as well as on longer samples with high definition image quality [7,30] conditioned on text inputs. Recently, [38] combined an autoregressive pixel-level model with a diffusion-based pipeline that predicts a correction of the frame, while [3] presents an autoregressive latent diffusion model.

Ultrasound simulation has been attempted with three major approaches: (1) physics-based simulators [11,28], (2) cross-modality registration-based methods [15] and (3) deep-learning based methods, usually conditioned on US, MRI or CT image priors [25,34,35] to condition the anatomy of the generated US images. Cine-ultrasound has also attracted some interest. [17] presents a motion-transfer-based method for pelvic US video generation, while [21] proposes a causal model for generating echocardiograms conditioned on arbitrary LVEF.

LVEF is a major metric in the assessment of cardiac function and diagnosis of cardiomyopathy. The EchoNet-dynamic dataset [19] is used as the go-to benchmark for LVEF-regression methods. Various works [18,22] have attempted to improve on [19] but the most reproducible method remains the use of an R2+1D model trained over fixed-length videos. The `R2+1D_18` trained for this work achieves an R^2 score of 0.81 on samples of 64 frames spanning 2 s.

2 Method

Diffusion probabilistic models [8,31,32] are the most recent family of generative models. In this work, we follow the definition of the EDM from [13]. Let $q(\boldsymbol{x})$ represent the real distribution of our data, with a standard deviation of σ_q. A family of distributions $p(\boldsymbol{x}; \sigma)$ can be obtained by adding i.i.d Gaussian noise with a standard deviation of σ to the data. When $\sigma_{\max} \gg \sigma_q$, the distribution $p(\boldsymbol{x}; \sigma_{\max})$ is essentially the same as pure Gaussian noise. The core idea of diffusion models is to sample a pure noise data point $\boldsymbol{x}_0 \sim \mathcal{N}(\boldsymbol{0}, \sigma_{\max}^2 \mathbf{I})$ and then progressively remove the noise, generating images $\boldsymbol{x}_i$ with standard deviation σ_i such that $\sigma_{\max} = \sigma_0 > \sigma_1 > ... > \sigma_N = 0$, and $\boldsymbol{x}_i \sim p(\boldsymbol{x}; \sigma_i)$. The final image $\boldsymbol{x}_N$ produced by this process is thus distributed according to $q(\boldsymbol{x})$, the true distribution of the data. To perform the reverse diffusion process, we define a denoising function $D(\boldsymbol{x}, \sigma)$ trained to minimize the L_2 denoising error for all samples drawn from q for every σ such that:

$$\mathcal{L} = \mathbb{E}_{\boldsymbol{y} \sim q} \mathbb{E}_{\boldsymbol{n} \sim \mathcal{N}(\boldsymbol{0}, \sigma^2 \mathbf{I})} ||D(\boldsymbol{y} + \boldsymbol{n}; \sigma) - \boldsymbol{y}||_2^2 \tag{1}$$

where $\boldsymbol{y}$ is a training data point and $\boldsymbol{n}$ is noise. By following the definition of ordinary differential equations (ODE) we can continuously increase or decrease the noise level of our data point by moving it forward or backward in the diffusion process, respectively. To define the ODE we need a schedule $\sigma(t)$ that sets the noise level given the time step t, which we set to $\sigma(t) = t$. The probability flow ODE's characteristic property is that moving a sample $\boldsymbol{x}_a \sim p(\boldsymbol{x}_a; \sigma(t_a))$ from the diffusion step t_a to t_b with $t_a > t_b$ or $t_a < t_b$ should result in a sample $\boldsymbol{x}_b \sim p(\boldsymbol{x}_b; \sigma(t_b))$ and this requirement is satisfied by setting $d\boldsymbol{x} = -\dot{\sigma}(t)\sigma(t)\nabla_x \log p(\boldsymbol{x}; \sigma(t))dt$ where $\dot{\sigma}$ denotes the time derivative and $\nabla_x \log p(\boldsymbol{x}; \sigma)$ is the score function. From the score function, we can thus write $\nabla_x \log p(\boldsymbol{x}; \sigma) = (D(\boldsymbol{x}; \sigma) - \boldsymbol{x})/\sigma^2$ in the case of our denoising function, such that the score function isolates the noise from the signal $\boldsymbol{x}$ and can either amplify it or diminish it depending on the direction we take in the diffusion process. We define $D(\boldsymbol{x}; \sigma)$ to transform a neural network F, which can be trained inside D by following the loss described in Eq. (1). The EDM also defines a list of four important preconditionings which are defined as $c_{\text{skip}}(\sigma) = (\sigma_q^2)/(\sigma^2 + \sigma_q^2)$, $c_{\text{out}}(\sigma) = \sigma * \sigma_q * 1/(\sigma_q^2 * \sigma^2)^{0.5}$, $c_{\text{in}}(\sigma) = 1/(\sigma_q^2 * \sigma^2)^{0.5}$ and $c_{\text{noise}}(\sigma) = \log(\sigma_t)/4$ where σ_q is the standard deviation of the real data distribution. In this paper, we focus on generating temporally coherent and realistic-looking echocardiograms. We start by generating a low resolution, low-frame rate video $\boldsymbol{v}_0$ from noise and condition on arbitrary clinical parameters and an anatomy instead of the commonly used text-prompt embeddings [7,30]. Then, the video is used as conditioning for the following diffusion model, which generates a temporally and/or spatially upsampled video $\boldsymbol{v}_1$ resembling $\boldsymbol{v}_0$, following the Cascaded Diffusion Model (CDM) [9] idea. Compared to image diffusion models, the major change to the Unet-based architecture is to add time-aware layers, through attention, at various levels as well as 3D convolutions (see Fig. 1 and Appendix Fig. 1). For the purpose of this research, we extend [7] to handle our own set of conditioning inputs, which are a single image $\boldsymbol{I}_c$ and a scalar value λ_c, while following the EDM setup, which we apply to video generation. We formally define the denoising models in the cascade as D_{θ_s} where s defines the rank (stage) of the model in the cascade, and where D_{θ_0} is the base model. The base model is defined as:

$$D_{\theta_0}(\boldsymbol{x}; \sigma, \boldsymbol{I}_c, \lambda_c) = c_{\text{skip}}(\sigma)\boldsymbol{x} + c_{\text{out}}(\sigma)F_{\theta_0}(c_{\text{in}}(\sigma)\boldsymbol{x}; c_{\text{noise}}(\sigma), \boldsymbol{I}_c, \lambda_c)),$$

where F_{θ_0} is the neural network transformed by D_{θ_0} and D_{θ_0} outputs $\boldsymbol{v}_0$. For all subsequent models in the cascade, the conditioning remains similar, but the models also receive the output from the preceding model, such that:

$$D_{\theta_s}(\boldsymbol{x}; \sigma, \boldsymbol{I}_c, \lambda_c, \boldsymbol{v}_{s-1}) = c_{\text{skip}}(\sigma)\boldsymbol{x} + c_{\text{out}}(\sigma)F_{\theta_s}(c_{\text{in}}(\sigma)\boldsymbol{x}; c_{\text{noise}}(\sigma), \boldsymbol{I}_c, \lambda_c, \boldsymbol{v}_{s-1})).$$

This holds $\forall s > 0$ and inputs $\boldsymbol{I}_c, \boldsymbol{v}_{s-1}$ are rescaled to the spatial and temporal resolutions expected by the neural network F_{θ_s} as a pre-processing. We apply the robustness trick from [9], i.e, we add a small amount of noise to real videos $\boldsymbol{v}_{s-1}$

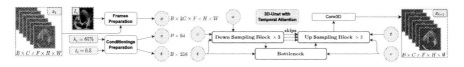

Fig. 1. Summarized view of our Model. Inputs (blue): a noised sample x_i, a diffusion step t_i, one anatomy image I_c, and one LVEF λ_c. Output (red): a slightly denoised version of x_i named x_{i+1}. See Appendix Fig. 1 for more details. (Color figure online)

during training, when using them as conditioning, in order to mitigate domain gaps with the generated samples $\boldsymbol{v}_{s-1}$ during inference.

Sampling from the EDM is done through a stochastic sampling method. We start by sampling a noise sample $\boldsymbol{x}_0 \sim \mathcal{N}(\boldsymbol{0}, t_0^2 \mathbf{I})$, where t comes from our previously defined $\sigma(t_i) = t_i$ and sets the noise level. We follow [13] and set constants $S_{\text{noise}} = 1.003, S_{t_{\min}} = 0.05, S_{t_{\max}} = 50$ and one constant S_{churn} dependent on the model. These are used to compute $\gamma_i(t_i) = \min(S_{\text{churn}}/N, \sqrt{2} - 1)$ $\forall t_i \in [S_{t_{\min}}, S_{t_{\max}}]$ and 0 otherwise, where N is the number of sampling steps. Then $\forall i \in \{0, ..., N-1\}$, we sample $\boldsymbol{\epsilon}_i \sim \mathcal{N}(\boldsymbol{0}, S_{\text{noise}}\mathbf{I})$ and compute a slightly increased noise level $\hat{t}_i = (\gamma_i(t_i) + 1)t_i$, which is added to the previous sample $\hat{\boldsymbol{x}}_i = \boldsymbol{x}_i + (\hat{t}_i^2 - t_i^2)^{0.5}\boldsymbol{\epsilon}_i$. We then execute the denoising model D_θ on that sample and compute the local slope $\boldsymbol{d}_i = (\hat{\boldsymbol{x}}_i - D_\theta(\hat{\boldsymbol{x}}_i; \hat{t}_i))/\hat{t}_i$ which is used to predict the next sample $\boldsymbol{x}_{i+1} = \hat{\boldsymbol{x}}_i + (\hat{t}_{i+1} - \hat{t}_i)\boldsymbol{d}_i$. At every step but the last (*i.e:* $\forall i \neq N-1$), we apply a correction to $\boldsymbol{x}_{i+1}$ such that: $\boldsymbol{d}_i' = (\boldsymbol{x}_{i+1} - D_\theta(\boldsymbol{x}_{i+1}; t_{i+1}))/t_{i+1}$ and $\boldsymbol{x}_{i+1} = \hat{\boldsymbol{x}}_i + (t_{i+1} - \hat{t}_i)(\boldsymbol{d}_i + \boldsymbol{d}_i')/2$. The correction step doubles the number of executions of the model, and thus the sampling time per step, compared to DDPM [8] or DDIM [32]. The whole sampling process is repeated sequentially for all models in the cascaded EDM. Models are conditioned on the previous output video $\boldsymbol{v}_{s-1}$ inputted at each step of the sampling process, with the frame conditioning $\boldsymbol{I}_c$ as well as the scalar value λ_c.

Conditioning: Our diffusion models are conditioned on two components. First, an *anatomy*, which is represented by a randomly sampled frame $\boldsymbol{I}_c$. It defines the patient's anatomy, but also all the information regarding the visual style and quality of the target video. These parameters cannot be explicitly disentangled, and we therefore limit ourselves to this approach. Second, we condition the model on clinical parameters λ_c. This is done by discarding the text-encoders that are used in [10,30] and directly inputting normalized clinical parameters into the conditional inputs of the Unets. By doing so, we give the model fine-grained control over the generated videos, which we evaluate using task-specific metrics.

Parameters: As video diffusion models are still in their early stage, there is no consensus on which are the best methods to train them. In our case, we define, depending on our experiment, 1-, 2- or 4-stages CDMs. We also experiment with various schedulers and parametrizations of the model. [26,32] show relatively fast sampling techniques which work fine for image sampling. However, in the case of video, we reach larger sampling times as we sample 64 frames at once. We therefore settled for the EDM [13], which presents a method to sample from the model in much fewer steps, largely reducing sampling times. We do not observe any particular speed-up in training and would argue, from our experience, that the v-parametrization [32] converges faster. We experimentally find our models to behave well with parameters close to those suggested in [13].

3   Experiments

Data: We use the EchoNet-Dynamic [19] dataset, a publicly available dataset that consists of 10,030 4-chamber cardiac ultrasound sequences, with a spatial resolution of 112×112 pixels. Videos range from 0.7 to 20.0 s long, with frame rates between 18 and 138 frames per second (fps). Each video has 3 channels, although most of them are greyscale. We keep the original data split of EchoNet-Dynamic which has 7465 training, 1288 validation and 1277 testing videos. We only train on the training data, and validate on the validation data. In terms of labels, each video comes with an LVEF score $\lambda \in [0, 100]$, estimated by a trained clinician. At every step of our training process, we pull a batch of videos, which are resampled to 32 fps. For each video, we retrieve its corresponding ground truth LVEF as well as a random frame. After that, the video is truncated or padded to 64 frames, in order to last 2 s, which is enough to cover any human heartbeat. The randomly sampled frame is sampled from the same original video as the 64-frames sample, but may not be contained in those 64 frames, as it may come from before or after that sub-sample.

Architectural Variants: We define three sets of models, and present them in details in Table 1 of the Appendix. We call the models X-Stage Cascaded Models (XSCM) and present the models' parameters at every stage. Every CDM starts with a *Base* diffusion model that is conditioned on the LVEF and one conditional frame. The subsequent models perform either temporal super resolution (TSR), spatial super resolution (SSR) or temporal and spatial super resolution (TSSR). TSR, SSR and TSSR models receive the same conditioning inputs as the Base model, along with the output of the previous-stage model. Note that [7] does not mention TSSR models and [30] states that extending an SSR model to perform simultaneous temporal and spatial up-sampling is too challenging.

Training: All our models are trained from scratch on individual cluster nodes, each with $8 \times$ NVIDIA A100. We use a per-GPU batch size of 4 to 8, resulting in batches of 32 to 64 elements after gradient accumulation. The distributed training is handled by the `accelerate` library from HuggingFace. We did not see any speed-up or memory usage reduction when enabling mixed precision and thus used full precision. As pointed out by [9] all models in a CDM can be trained in parallel which significantly speeds up experimentation. We empirically find that training with a learning rate up to $5*10^{-4}$ is stable and reaches good image quality. We use an Exponential Moving Average (EMA) copy of our model to smooth out the training. We train all our models' stages for 48h, *i.e.*, the 2SCM and 4SCM CDMs are proportionally more costly to train than the 1SCM. As noted by [7,30] training on images and videos improves the overall image quality. As our dataset only consists of videos, we simply deactivate the time attention layers in the Unet with a 25% chance during training, for all models.

Results: We evaluate our models' video synthesis capabilities on two objectives: LVEF accuracy (R^2, MAE, RMSE) and image quality (SSIM, LPIPS, FID, FVD). We formulate the task as counterfactual modelling, where we set

(1) a random conditioning frame as confounder, (2) the ground-truth LVEF as a factual conditioning, and (3) a random LVEF in the physiologically plausible range from 15% to 85% as counterfactual conditioning. For each ground truth video, we sample three random starting noise samples and conditioning frames. We use the LVEF regression model to create a feedback loop, following what [21] did, even though their model was run 100× per sample instead of 3×. For each ground truth video, we keep the sample with the best LVEF accuracy to compute all our scores over 1288 videos for each model.

The results in Table 1 show that increasing the frame rate improves model fidelity to the given LVEF, while adding more models to the cascade decreases image quality. This is due to a distribution gap between true low-resolution samples and sequentially generated samples during inference. This issue is partially addressed by adding noise to real low-resolution samples during training, but the 1SCM model with only one stage still achieves better image quality metrics. However, the 2SCM and 4SCM models perform equally well on LVEF metrics and outperform the 1SCM model thanks to their higher temporal resolution that precisely captures key frames of the heartbeat. The TSSR model, used in the 2SCM, yields the best compromise between image quality, LVEF accuracy, and sampling times, and is compared to previous literature.

We outperform previous work for LVEF regression: counterfactual video generation improves with our method by a large margin of 38 points for the R^2 score as shown in Table 2. The similarity between our factual and counterfactual results show that our time-agnostic confounding factor (*i.e.* an image instead of a video) prevents entanglement, as opposed to the approach taken in [21]. Our method does not score as high for SSIM as global image similarity metric, which is expected because of the stochasticity of the speckle noise. In [21] this was mitigated by their data-rich confounder. Our results also match other video diffusion models [3,7,10,30] as structure is excellent, while texture tends to be more noisy as shown in Fig. 2.

Table 1. Metrics for all CDMs. The *Gen.* task is the counterfactual generation comparable to [21], the *Rec.* task is the factual reconstruction task. *Frames* is the number of frames generated by the model, always spanning 2 s. [‡]Videos are temporally upsampled to 64 frames for metric computation. *S. time* is the sampling time for one video on an RTX A5000. R^2, MAE and RMSE are computed between the conditional LVEF λ_c and the regressed LVEF using the model described in Sect. 1. SSIM, LPIPS, FID and FVD are used to quantify the image quality. LPIPS is computed with VGG [29], FID [6] and FVD [36] with I3D. FID and FVD are computed over padded frames of 128×128 pixels.

Model	Task	Res.	Frames	S. time	$R^2 \uparrow$	MAE↓	RMSE↓	SSIM↑	LPIPS↓	FID↓	FVD↓
1SCM	Gen.	112	16[‡]	62 s[‡]	0.64	9.65	12.2	0.53	0.21	12.3	60.5
2SCM	Gen.	112	64	146 s	0.89	4.81	6.69	0.53	0.24	31.7	141
4SCM	Gen.	112	64	279 s	0.93	3.77	5.26	0.48	0.25	24.6	230
1SCM	Rec.	112	16[‡]	62 s[‡]	0.76	4.51	6.07	0.53	0.21	13.6	89.7
2SCM	Rec.	112	64	146 s	0.93	2.22	3.35	0.54	0.24	31.4	147
4SCM	Rec.	112	64	279 s	0.90	2.42	3.87	0.48	0.25	24.0	228

Table 2. Comparison of our 2SCM model to previous work. We try to reconstruct a ground truth video or to generate a new one. Our model is conditioned on a single frame and an LVEF, while [21] conditions on the entire video and an LVEF. In both cases the LVEF is either the ground truth LVEF (*Rec.*) or a randomly sampled LVEF (*Gen.*).

Method	Conditioning	Task	S. time	R^2 ↑	MAE ↓	RMSE ↓	SSIM ↑
Dartagnan [21]	Video+EF	Gen.	~1 s	0.51	15.7	18.4	**0.79**
2SCM	Image+EF	Gen.	146 s	**0.89**	**4.81**	**6.69**	0.53
Dartagnan [21]	Video+EF	Rec.	~1 s	0.87	2.79	4.45	**0.82**
2SCM	Image+EF	Rec.	146 s	**0.93**	**2.22**	**3.35**	0.54

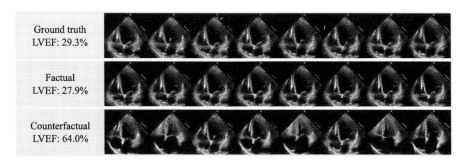

Fig. 2. Top: Ground truth frames with 29.3% LVEF. Middle: Generated factual frames, with estimated 27.9% LVEF. Bottom: Generated counterfactual frames, with estimated 64.0% LVEF. (Counter-)Factual frames are generated with the 1SCM, conditioned on the ground-truth anatomy.

Qualitative Study: We asked three trained clinicians (Consultant cardiologist > 10 years experience, Senior trainee in cardiology > 5 years experience, Chief cardiac physiologist > 15 years experience) to classify 100 samples, each, as *real* or *fake*. Experts were not given feedback on their performance during the evaluation process and were not shown fake samples beforehand. All samples were generated with the 1SCM model or were true samples from the EchoNet-Dynamic dataset, resampled to 32fps and 2 s. The samples were picked by alphabetical order from the validation set. Among the 300 samples evaluated, 130 (43.33%) were real videos, 89 (29.67%) were factual generated videos, and 81 (27.0%) were counterfactual generated videos. The average expert accuracy was 54.33%, with an inter-expert agreement of 50.0%. More precisely, experts detected real samples with an accuracy of 63.85%, 50.56% for factual samples and 43.21% for the counterfactual samples. The average time taken to evaluate each sample was 16.2 s. We believe that these numbers show the video quality that our model reaches, and can put in perspective the SSIM scores from Table 1.

Downstream Task: We train our LVEF regression model on rebalanced datasets and resampled datasets. We rebalance the datasets by using our 4SCM model to generate samples for LVEF values that have insufficient data. The resampled datasets are smaller datasets randomly sampled from the real training set. We show that, in small data regimes, using generated data to rebalance the dataset improves the overall performance. Training on 790 real data samples yields an R^2 score of 56% while the rebalanced datasets with 790 samples, ~50% of which are real, reaches a better 59% on a balanced validation set. This observation is mitigated when more data is available. See Appendix Table 2 for all our results.

Discussion: Generating echocardiograms is a challenging task that differs from traditional computer vision due to the noisy nature of US images and videos. However, restricting the training domain simplifies certain aspects, such as not requiring a long series of CDMs to reach the target resolution of 112×112 pixels and limiting samples to 2 s, which covers any human heartbeat. The limited pixel-intensity space of the data also allows for models with fewer parameters. In the future, we plan to explore other organs and views within the US domain, with different clinical conditionings and segmentation maps.

4   Conclusion

Our application of EDMs to US video generation achieves state-of-the-art performance on a counterfactual generation task, a data augmentation task, and a qualitative study by experts. This significant advancement provides a valuable solution for downstream tasks that could benefit from representative foundation models for medical imaging and precise medical video generation.

Acknowledgements. This work was supported by Ultromics Ltd., the UKRI Centre for Doctoral Training in Artificial Intelligence for Healthcare (EP/S023283/1) and HPC resources provided by the Erlangen National High Performance Computing Center (NHR@FAU) of the Friedrich-Alexander-Universität Erlangen-Nürnberg (FAU) under the NHR project b143dc. NHR funding is provided by federal and Bavarian state authorities. NHR@FAU hardware is partially funded by the German Research Foundation (DFG) - 440719683. We thank Phil Wang (https://github.com/lucidrains) for his open source implementation of [7]. Support was also received from the ERC - project MIA-NORMAL 101083647 and DFG KA 5801/2-1, INST 90/1351-1.

References

1. Babaeizadeh, M., Finn, C., Erhan, D., Campbell, R.H., Levine, S.: Stochastic variational video prediction. arXiv:1710.11252 (2018)
2. Babaeizadeh, M., Saffar, M.T., Nair, S., Levine, S., Finn, C., Erhan, D.: FitVid: overfitting in pixel-level video prediction. arXiv:2106.13195 (2021)
3. Esser, P., Chiu, J., Atighehchian, P., Granskog, J., Germanidis, A.: Structure and content-guided video synthesis with diffusion models. arXiv:2302.03011 (2023)
4. Finn, C., Goodfellow, I., Levine, S.: Unsupervised learning for physical interaction through video prediction. In: Advances in Neural Information Processing Systems, vol. 29 (2016)

5. Gupta, A., Tian, S., Zhang, Y., Wu, J., Martín-Martín, R., Fei-Fei, L.: MaskViT: masked visual pre-training for video prediction. arXiv:2206.11894 (2022)
6. Heusel, M., Ramsauer, H., Unterthiner, T., Nessler, B., Hochreiter, S.: GANs trained by a two time-scale update rule converge to a local Nash equilibrium. arXiv:1706.08500 (2018)
7. Ho, J., et al.: Imagen video: high definition video generation with diffusion models (2022). arXiv:2210.02303
8. Ho, J., Jain, A., Abbeel, P.: Denoising diffusion probabilistic models. In: Advances in Neural Information Processing Systems, vol. 33, pp. 6840–6851 (2020)
9. Ho, J., Saharia, C., Chan, W., Fleet, D.J., Norouzi, M., Salimans, T.: Cascaded diffusion models for high fidelity image generation. J. Mach. Learn. Res. **23**, 1–33 (2022)
10. Ho, J., Salimans, T., Gritsenko, A., Chan, W., Norouzi, M., Fleet, D.J.: Video diffusion models (2022). arXiv:2204.03458
11. Jensen, J.: Simulation of advanced ultrasound systems using Field II. In: 2004 2nd IEEE International Symposium on Biomedical Imaging: Nano to Macro (IEEE Cat No. 04EX821), pp. 636–639, vol. 1 (2004)
12. Kalchbrenner, N., et al.: Video pixel networks. In: ICML, pp. 1771–1779 (2017)
13. Karras, T., Aittala, M., Aila, T., Laine, S.: Elucidating the design space of diffusion-based generative models. arXiv:2206.00364 (2022)
14. Kumar, M., et al.: VideoFlow: a conditional flow-based model for stochastic video generation. arXiv:1903.01434 (2020)
15. Ledesma-Carbayo, M., et al.: Spatio-temporal nonrigid registration for ultrasound cardiac motion estimation. IEEE TMI **24**, 1113–1126 (2005)
16. Liang, J., et al.: Sketch guided and progressive growing GAN for realistic and editable ultrasound image synthesis. Med. Image Anal. **79**, 102461 (2022)
17. Liang, J., et al.: Weakly-supervised high-fidelity ultrasound video synthesis with feature decoupling. In: Wang, L., Dou, Q., Fletcher, P.T., Speidel, S., Li, S. (eds.) MICCAI 2022. LNCS, vol. 13434, pp. 310–319. Springer, Cham (2022). https://doi.org/10.1007/978-3-031-16440-8_30
18. Mokhtari, M., Tsang, T., Abolmaesumi, P., Liao, R.: EchoGNN: explainable ejection fraction estimation with graph neural networks. In: Wang, L., Dou, Q., Fletcher, P.T., Speidel, S., Li, S. (eds.) MICCAI 2022. LNCS, vol. 13434, pp. 360–369. Springer, Cham (2022). https://doi.org/10.1007/978-3-031-16440-8_35
19. Ouyang, D., et al.: Video-based AI for beat-to-beat assessment of cardiac function. Nature **580**, 252–256 (2020)
20. Ramesh, A., et al.: Zero-shot text-to-image generation. arXiv:2102.12092 (2021)
21. Reynaud, H., et al.: D'ARTAGNAN: counterfactual video generation. In: Wang, L., Dou, Q., Fletcher, P.T., Speidel, S., Li, S. (eds.) MICCAI 2022. LNCS, vol. 13438, pp. 599–609. Springer, Cham (2022). https://doi.org/10.1007/978-3-031-16452-1_57
22. Reynaud, H., Vlontzos, A., Hou, B., Beqiri, A., Leeson, P., Kainz, B.: Ultrasound video transformers for cardiac ejection fraction estimation. In: de Bruijne, M., et al. (eds.) MICCAI 2021. LNCS, vol. 12906, pp. 495–505. Springer, Cham (2021). https://doi.org/10.1007/978-3-030-87231-1_48
23. Rombach, R., Blattmann, A., Lorenz, D., Esser, P., Ommer, B.: High-resolution image synthesis with latent diffusion models. arXiv:2112.10752 (2022)
24. Saharia, C., et al.: Photorealistic text-to-image diffusion models with deep language understanding. arXiv:2205.11487 (2022)

25. Salehi, M., Ahmadi, S.-A., Prevost, R., Navab, N., Wein, W.: Patient-specific 3D ultrasound simulation based on convolutional ray-tracing and appearance optimization. In: Navab, N., Hornegger, J., Wells, W.M., Frangi, A.F. (eds.) MICCAI 2015. LNCS, vol. 9350, pp. 510–518. Springer, Cham (2015). https://doi.org/10.1007/978-3-319-24571-3_61
26. Salimans, T., Ho, J.: Progressive distillation for fast sampling of diffusion models. arXiv:2202.00512 (2022)
27. Segars, W.P., Sturgeon, G., Mendonca, S., Grimes, J., Tsui, B.M.W.: 4D XCAT phantom for multimodality imaging research. Med. Phys. **37**, 4902–4915 (2010)
28. Shams, R., Hartley, R., Navab, N.: Real-time simulation of medical ultrasound from CT images. In: Metaxas, D., Axel, L., Fichtinger, G., Székely, G. (eds.) MICCAI 2008. LNCS, vol. 5242, pp. 734–741. Springer, Heidelberg (2008). https://doi.org/10.1007/978-3-540-85990-1_88
29. Simonyan, K., Zisserman, A.: Very deep convolutional networks for large-scale image recognition. arXiv:1409.1556 (2015)
30. Singer, U., et al.: Make-a-video: text-to-video generation without text-video data. arXiv:2209.14792 (2022)
31. Sohl-Dickstein, J., Weiss, E.A., Maheswaranathan, N., Ganguli, S.: Deep unsupervised learning using nonequilibrium thermodynamics. arXiv:1503.03585 (2015)
32. Song, J., Meng, C., Ermon, S.: Denoising diffusion implicit models. arXiv:2010.02502 (2022)
33. Song, Y., Zhu, J., Li, D., Wang, X., Qi, H.: Talking face generation by conditional recurrent adversarial network. arXiv:1804.04786 (2019)
34. Teng, L., Fu, Z., Yao, Y.: Interactive translation in echocardiography training system with enhanced cycle-GAN. IEEE Access **8**, 106147–106156 (2020)
35. Tomar, D., Zhang, L., Portenier, T., Goksel, O.: Content-preserving unpaired translation from simulated to realistic ultrasound images. In: de Bruijne, M., et al. (eds.) MICCAI 2021. LNCS, vol. 12908, pp. 659–669. Springer, Cham (2021). https://doi.org/10.1007/978-3-030-87237-3_63
36. Unterthiner, T., Steenkiste, S.V., Kurach, K., Marinier, R., Michalski, M., Gelly, S.: FVD: a new metric for video generation. In: ICLR 2022 Workshop: Deep Generative Models for Highly Structured Data (2019)
37. Villegas, R., et al.: Phenaki: variable length video generation from open domain textual description. arXiv:2210.02399 (2022)
38. Yang, R., Srivastava, P., Mandt, S.: Diffusion probabilistic modeling for video generation. arXiv:2203.09481 (2022)

DULDA: Dual-Domain Unsupervised Learned Descent Algorithm for PET Image Reconstruction

Rui Hu[1,3], Yunmei Chen[2], Kyungsang Kim[3],
Marcio Aloisio Bezerra Cavalcanti Rockenbach[3,4], Quanzheng Li[3,4(✉)],
and Huafeng Liu[1(✉)]

[1] State Key Laboratory of Modern Optical Instrumentation, Department of Optical Engineering, Zhejiang University, Hangzhou 310027, China
liuhf@zju.edu.cn
[2] Department of Mathematics, University of Florida, Gainesville, FL 32611, USA
[3] The Center for Advanced Medical Computing and Analysis, Massachusetts General Hospital/Harvard Medical School, Boston, MA 02114, USA
li.quanzheng@mgh.harvard.edu
[4] Data Science Office, Massachusetts General Brigham, Boston, MA 02116, USA

Abstract. Deep learning based PET image reconstruction methods have achieved promising results recently. However, most of these methods follow a supervised learning paradigm, which rely heavily on the availability of high-quality training labels. In particular, the long scanning time required and high radiation exposure associated with PET scans make obtaining these labels impractical. In this paper, we propose a dual-domain unsupervised PET image reconstruction method based on learned descent algorithm, which reconstructs high-quality PET images from sinograms without the need for image labels. Specifically, we unroll the proximal gradient method with a learnable $l_{2,1}$ norm for PET image reconstruction problem. The training is unsupervised, using measurement domain loss based on deep image prior as well as image domain loss based on rotation equivariance property. The experimental results demonstrate the superior performance of proposed method compared with maximum-likelihood expectation-maximization (MLEM), total-variation regularized EM (EM-TV) and deep image prior based method (DIP).

Keywords: Image reconstruction · Positron emission tomography (PET) · Unsupervised learning · Model based deep learning · Dual-domain

1 Introduction

Positron Emission Tomography (PET) is a widely used modality in functional imaging for oncology, cardiology, neurology, and medical research [1]. However, PET images often suffer from a high level of noise due to several physical degradation factors as well as the ill-conditioning of the PET reconstruction problem.

© The Author(s), under exclusive license to Springer Nature Switzerland AG 2023
H. Greenspan et al. (Eds.): MICCAI 2023, LNCS 14229, pp. 153–162, 2023.
https://doi.org/10.1007/978-3-031-43999-5_15

As a result, the quality of PET images can be compromised, leading to difficulties in accurate diagnosis.

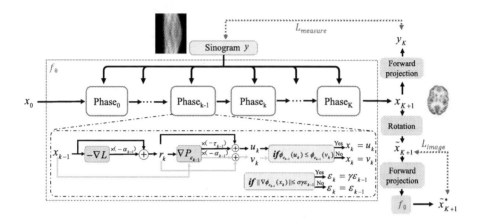

Fig. 1. Diagram of the proposed DULDA for PET image reconstruction. The LDA was unrolled into several phases with the learnable $l_{2,1}$ norm, where each phase includes the gradient calculation of both likelihood and regularization.

Deep learning (DL) techniques, especially supervised learning, have recently garnered considerable attention and show great promise in PET image reconstruction compared with traditional analytical methods and iterative methods. Among them, four primary approaches have emerged: DL-based post-denoising [2,3], end-to-end direct learning [4–6], deep learning regularized iterative reconstruction [7–10] and deep unrolled methods [11–13].

DL-based post denoising methods are relatively straightforward to implement but can not reduce the lengthy reconstruction time and its results are significantly affected by the pre-reconstruction algorithm. End-to-end direct learning methods utilize deep neural networks to learn the directing mapping from measurement sinogram to PET image. Without any physical constraints, these methods can be unstable and extremely data-hungry. Deep learning regularized iterative reconstruction methods utilize a deep neural network as a regularization term within the iterative reconstruction process to regularize the image estimate and guide the reconstruction process towards a more accurate and stable solution. Despite the incorporation of deep learning, the underlying mathematical framework and assumptions of deep learning regularized iterative methods still rely on the conventional iterative reconstruction methods. Deep unrolled methods utilize a DNN to unroll the iterative reconstruction process and to learn the mapping from sinogram to the reconstructed PET images, which potentially result in more accurate and explainable image reconstruction. Deep unrolled methods have demonstrated improved interpretabillity and yielded inspiring outcomes.

However, the aforementioned approaches for PET image reconstruction depend on high quality ground truths as training labels, which can be difficult and expensive to obtain. This challenge is further compounded by the high dose exposure associated with PET imaging. Unsupervised/self supervised learning has gained considerable interest in medical imaging, owing to its ability to mitigate the need for high-quality training labels. Gong et al. proposed a PET image reconstruction approach using the deep image prior (DIP) framework [15], which employed a randomly initialized Unet as a prior. In another study, Fumio et al. proposed a simplified DIP reconstruction framework with a forward projection model, which reduced the network parameters [16]. Shen et al. proposed a DeepRED framework with an approximate Bayesian framework for unsupervised PET image reconstruction [17]. These methods all utilize generative models to generate PET images from random noise or MRI prior images and use sinogram to design loss functions. However, these generative models tend to favor low frequencies and sometimes lack of mathematical interpretability. In the absence of anatomic priors, the network convergence can take a considerable amount of time, resulting in prolonged reconstruction times. Recently, equivariant property [18] of medical imaging system is proposed to train the network without labels, which shows the potential for the designing of PET reconstruction algorithms.

In this paper, we propose a dual-domain unsupervised learned descent algorithm for PET image reconstruction, which is the first attempt to combine unsupervised learning and deep unrolled method for PET image reconstruction. The main contributions of this work are summarized as follows: 1) a novel model based deep learning method for PET image reconstruction is proposed with a learnable $l_{2,1}$ norm for more general and robust feature sparsity extraction of PET images; 2) a dual domain unsupervised training strategy is proposed, which is plug-and-play and does not need paired training samples; 3) without any anatomic priors, the proposed method shows superior performance both quantitatively and visually.

2   Methods and Materials

2.1   Problem Formulation

As a typical inverse problem, PET image reconstruction can be modeled in a variational form and cast as an optimization task, as follows:

$$\min \phi(\boldsymbol{x}; \boldsymbol{y}, \boldsymbol{\theta}) = -L(\boldsymbol{y}|\boldsymbol{x}) + P(\boldsymbol{x}; \boldsymbol{\theta}) \tag{1}$$

$$L(\boldsymbol{y}|\boldsymbol{x}) = \sum_i y_i \log \overline{y}_i - \sum_i \overline{y}_i \tag{2}$$

$$\overline{\boldsymbol{y}} = \boldsymbol{A}\boldsymbol{x} + \boldsymbol{b} \tag{3}$$

where $\boldsymbol{y}$ is the measured sinogram data, $\overline{\boldsymbol{y}}$ is the mean of the measured sinogram. $\boldsymbol{x}$ is the PET activity image to be reconstructed, $L(\boldsymbol{y}|\boldsymbol{x})$ is the Poisson log-likelihood of measured sinogram data. $P(\boldsymbol{x}; \boldsymbol{\theta})$ is the penalty term with learnable parameter $\boldsymbol{\theta}$. $\boldsymbol{A} \in \mathbb{R}^{I \times J}$ is the system response matrix, with A_{ij} representing the probabilities of detecting an emission from voxel j at detector i.

We expect that the parameter $\boldsymbol{\theta}$ in penalty term P can be learned from the training data like many other deep unrolling methods. However, most of these methods directly replace the penalty term [14] or its gradient [11,13] with a network, which loses some mathematical rigor and interpretablities.

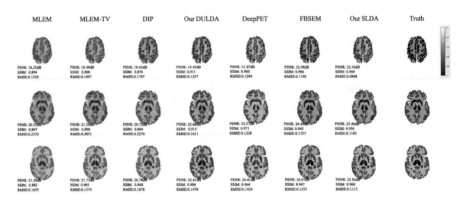

Fig. 2. Reconstruction results of MLEM, EMTV, DIP, proposed DULDA, DeepPET, FBSEM and proposed SLDA on different slices of the test set.

2.2 Parametric Form of Learnable Regularization

We choose to parameterize P as the $l_{2,1}$ norm with a feature extraction operator $g(x)$ to be learned in the training data. The smooth nonlinear mapping g is used to extract sparse features and the $l_{2,1}$ norm is used as a robust and effective sparse feature regularization. Specifically, we formulate P as follows [19]:

$$P(\boldsymbol{x}; \boldsymbol{\theta}) = ||\boldsymbol{g}_{\theta}(\boldsymbol{x})||_{2,1} = \sum_{i=1}^{m} ||\boldsymbol{g}_{i,\theta}(\boldsymbol{x})|| \tag{4}$$

where $\boldsymbol{g}_{i,\theta}(\boldsymbol{x})$ is i-th feature vector. We choose $\boldsymbol{g}$ as a multi-layered CNN with nonlinear activation function σ, and σ is a smoothed ReLU:

$$\sigma(x) = \begin{cases} 0, & \text{if } x \leqslant -\delta, \\ \dfrac{1}{4\delta}x^2 + \dfrac{1}{2}x + \dfrac{\sigma}{4}, & \text{if } -\delta < x < \delta, \\ x, & \text{if } x \geqslant \delta, \end{cases} \tag{5}$$

In this case, the gradient $\nabla \boldsymbol{g}$ can be computed directly. The Nesterov's smoothing technique is used in P for the derivative calculation of the $l_{2,1}$ norm through smooth approximation:

$$P_{\varepsilon}(\boldsymbol{x}) = \sum \frac{1}{2\varepsilon}||\boldsymbol{g}_i(\boldsymbol{x})||^2 + \sum (||\boldsymbol{g}_i(\boldsymbol{x}) - \frac{\varepsilon}{2}||) \tag{6}$$

$$\nabla P_\varepsilon(\boldsymbol{x}) = \sum \nabla g_i(\boldsymbol{x})^T \frac{g_i(\boldsymbol{x})}{\varepsilon} + \sum \nabla g_i(\boldsymbol{x})^T \frac{g_i(\boldsymbol{x})}{||g_i(\boldsymbol{x})||} \tag{7}$$

where parameter ε controls how close the approximation P_ε to the original P.

Algorithm 1. Learned Descent Algorithm for PET image reconstruction

---

Input: Image initialization $\boldsymbol{x_0}$, $\rho, \gamma \in (0,1)$, $\varepsilon_0, \sigma, \tau > 0$, maximum number of iteration I, total phase numbers K and measured Sinogram $\boldsymbol{y}$
1: **for** $i \in [1, I]$ **do**
2:    $\boldsymbol{r}_k = \boldsymbol{x}_{k-1} + \alpha_{k-1}(\sum \frac{\boldsymbol{A}^T \boldsymbol{y}}{\boldsymbol{A}\,\boldsymbol{x}_{k-1}+\boldsymbol{b}} - \sum \boldsymbol{A}^T \boldsymbol{1})$
3:    $\boldsymbol{u}_k = \boldsymbol{r}_k - \tau_{k-1} \nabla P_{\varepsilon_{k-1}}(\boldsymbol{r}_k)$
4:    **repeat**
5:       $\boldsymbol{v}_k = \boldsymbol{x}_{k-1} - \alpha_{k-1} \nabla(-L(\boldsymbol{y}|\boldsymbol{x}_{k-1})) - \alpha_{k-1} \nabla P_{\varepsilon_{k-1}}(\boldsymbol{x}_{k-1})$
6:    **until** $\phi_{\varepsilon_{k-1}}(\boldsymbol{v}_k) \leqslant \phi_{\varepsilon_{k-1}}(\boldsymbol{x}_{k-1})$
7:    If $\phi(\boldsymbol{u}_k) \leqslant \phi(\boldsymbol{v}_k)$, $\boldsymbol{x}_k = \boldsymbol{u}_k$; otherwise, $\boldsymbol{x}_k = \boldsymbol{v}_k$
8:    If $||\nabla\phi_{\varepsilon_{k-1}}(\boldsymbol{x}_k)|| < \sigma\gamma\varepsilon_{k-1}$, $\varepsilon_k = \gamma\varepsilon_{k-1}$; otherwise, $\varepsilon_k = \varepsilon_{k-1}$
9: **end for**
10: **return** x$_\text{K}$;

2.3   Learned Descent Algorithm for PET

With the parametric form of learnable regularization given above, we rewrite
Eq. 1 as the objective function:

$$\min \phi(\boldsymbol{x}; \boldsymbol{y}, \boldsymbol{\theta}) = -L(\boldsymbol{y}|\boldsymbol{x}) + P_\varepsilon(\boldsymbol{x}; \boldsymbol{\theta}) \tag{8}$$

We unrolled the learned descent algorithm in several phases as shown in Fig. 1.
In each phase $k - 1$, we apply the proximal gradient step in Eq. 8:

$$\boldsymbol{r}_k = \boldsymbol{x}_{k-1} - \alpha_{k-1}\nabla(-L(\boldsymbol{y}|\boldsymbol{x})) = \boldsymbol{x}_{k-1} + \alpha_{k-1}(\sum \frac{\boldsymbol{A}^T \boldsymbol{y}}{\boldsymbol{A}\boldsymbol{x}_{k-1} + \boldsymbol{b}} - \sum \boldsymbol{A}^T \boldsymbol{1}) \tag{9}$$

$$\boldsymbol{x}_k = \text{prox}_{\alpha_{k-1}P_{\varepsilon_{k-1}}}(\boldsymbol{r}_k) \tag{10}$$

where the proximal operator is defined as:

$$\text{prox}_{\alpha P}(\boldsymbol{r}) = \arg\min_{x}\{\frac{1}{2\alpha}||\boldsymbol{x} - \boldsymbol{r}||^2 + P(\boldsymbol{x})\} \tag{11}$$

In order to have a close form solution of the proximal operator, we perform a
Taylor approximation of $P_{\varepsilon_{k-1}}$:

$$\tilde{P}_{\varepsilon_{k-1}}(\boldsymbol{x}) = P_{\varepsilon_{k-1}}(\boldsymbol{r}_k) + (\boldsymbol{x} - \boldsymbol{r}_k) \cdot \nabla P_{\varepsilon_{k-1}}(\boldsymbol{r}_k) + \frac{1}{2\beta_{k-1}}||\boldsymbol{x} - \boldsymbol{r}_k||^2 \tag{12}$$

After discarding higher-order constant terms, we can simplify the Eq. 10 as:

$$\boldsymbol{u}_k = \text{prox}_{\alpha_{k-1}\tilde{P}_{\varepsilon_{k-1}}}(\boldsymbol{r}_k) = \boldsymbol{r}_k - \tau_{k-1}\nabla P_{\varepsilon_{k-1}}(\boldsymbol{r}_k) \tag{13}$$

where α_{k-1} and β_{k-1} are two parameters greater than 0 and $\tau_{k-1} = \frac{\alpha_{k-1}\beta_{k-1}}{\alpha_{k-1}+\beta_{k-1}}$.
We also calculate a close-form safeguard $\boldsymbol{v}_k$ as:

$$\boldsymbol{v}_k = \boldsymbol{x}_{k-1} - \alpha_{k-1}\nabla(-L(\boldsymbol{y}|\boldsymbol{x}_{k-1})) - \alpha_{k-1}\nabla P_{\varepsilon_{k-1}}(\boldsymbol{x}_{k-1}) \qquad (14)$$

The line search strategy is used by shrinking α_{k-1} to ensure objective function
decay. We choose the $\boldsymbol{u}_k$ or $\boldsymbol{v}_k$ with smaller objection function value $\phi_{\varepsilon_{k-1}}$ to
be the next $\boldsymbol{x}_k$. The smoothing parameter ε_{k-1} is shrinkage by $\gamma \in (0,1)$ if the
$||\nabla\phi_{\varepsilon_{k-1}}(\boldsymbol{x}_k)|| < \sigma\gamma\varepsilon_{k-1}$ is satisfied. The whole flow is shown in Algorithm 1.

1/4 dose	1/2 dose	Full dose
MLEM / Proposed	MLEM / Proposed	MLEM / Proposed

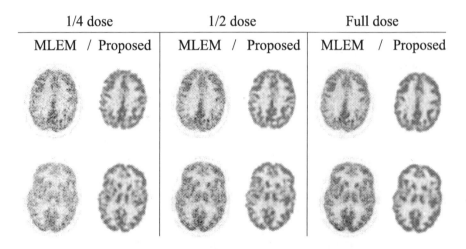

Fig. 3. The robust analysis on proposed DULDA with one clinical patient brain sample
with different dose level. From left to right: MLEM results and DULDA results with
quarter dose sinogram, half dose sinogram and full dose sinogram.

2.4 Dual-Domain Unsupervised Training

The whole reconstruction network is indicated by f_θ with learned parameter θ.
Inspired by Deep image prior [20] and equivariance [18] of PET imaging system,
the proposed dual-domain unsupervised training loss function is formulated as:

$$L_{dual} = L_{image} + \lambda L_{measure} \qquad (15)$$

where λ is the parameter that controls the ratio of different domain loss function,
which was set to 0.1 in the experiments. For image domain loss L_{image}, the
equivariance constraint is used. For example, if the test sample x_t first undergoes
an equivariant transformation, such as rotation, we obtain x_{tr}. Subsequently, we
perform a PET scan to obtain the sinogram data of x_{tr} and x_t. The image
reconstructed by the f_θ of these two sinogram should also keep this rotation
properties. The L_{image} is formulate as:

$$L_{image} = ||T_r \underbrace{f_\theta(\boldsymbol{y})}_{x_t} - f_\theta(A(\underbrace{T_r(f_\theta(\boldsymbol{y}))}_{x_{tr}}))||^2 \qquad (16)$$

Table 1. Quantitative analysis and bias-variance analysis for the reconstruction results of MLEM, EM-TV, DIP, Proposed DULDA, DeepPET, FBSEM and proposed SLDA.

Methods	PSNR(dB)↑	SSIM↑	RMSE↓	CRC↑	Bias↓	Variance↓
MLEM	20.02 ± 1.91	0.889 ± 0.015	0.160 ± 0.045	0.6517	0.5350	0.2311
EM-TV	20.28 ± 2.21	0.904 ± 0.014	0.154 ± 0.044	0.8027	0.5389	0.2340
DIP	19.96 ± 1.50	0.873 ± 0.012	0.187 ± 0.047	0.8402	0.2540	**0.2047**
Our DULDA	**20.80 ± 1.77**	**0.910 ± 0.011**	**0.148 ± 0.011**	**0.8768**	0.2278	0.2449
DeepPET	23.40 ± 2.87	0.962 ± 0.011	0.135 ± 0.021	0.8812	0.1470	**0.1465**
FBSEM	23.59 ± 1.50	0.954 ± 0.008	0.122 ± 0.017	0.8825	0.1593	0.2034
Our SLDA	**24.21 ± 1.83**	**0.963 ± 0.007**	**0.104 ± 0.013**	**0.9278**	**0.1284**	0.1820

where T_r denotes the rotation operator, A is the forward projection which also can be seen as a measurement operator. For sinogram domain loss $L_{measure}$, the data argumentation with random noise ξ is performed on y:

$$L_{measure} = ||(y + \xi) - Af_\theta(y + \xi)||^2 \tag{17}$$

2.5   Implementation Details and Reference Methods

We implemented DULDA using Pytorch 1.7 on a NVIDIA GeForce GTX Titan X. The Adam optimizer with a learning rate of 10^{-4} was used and trained for 100 epochs with batch size of 8. The total unrolled phase was 4. The image x_0 was initialized with the values of one. The smoothing parameter ε_0 and δ were initialized to be 0.001 and 0.002. The step-size α_0 and β_0 were initialized to be 0.01 and 0.02. The system matrix was computed by using Michigan Image Reconstruction Toolbox (MIRT) with a strip-integral model [21]. The proposed DULDA was compared with MLEM [22], total variation regularized EM (EM-TV) [23] and deep image prior method (DIP) [16]. For both MLEM and EM-TV, 25 iterations were adopted. The penalty parameter for EM-TV was $2e^{-5}$. For DIP, we used random noise as input and trained 14000 epochs with the same training settings as DULDA to get the best results before over-fitting. The proposed method can also be trained in a fully supervised manner (we call it SLDA). The loss is the mean square error between the output and the label image. To further demonstrate the effectiveness, we compared SLDA with DeepPET [5] and FBSEM [11], the training settings remained the same.

Table 2. Ablation study for different phase numbers and loss function type of DULDA on the test datasets.

Settings		PSNR↑	SSIM↑	MSE↓
phase numbers	2	14.53 ± 1.45	0.769 ± 0.024	0.314 ± 0.047
	4	20.80 ± 1.77	0.910 ± 0.011	0.148 ± 0.011
	6	20.29 ± 1.16	0.903 ± 0.014	0.156 ± 0.016
	8	19.94 ± 1.31	0.884 ± 0.012	0.180 ± 0.013
	10	15.33 ± 0.65	0.730 ± 0.020	0.313 ± 0.050
only L_{image}		15.41 ± 0.69	0.729 ± 0.008	0.324 ± 0.048
only $L_{measure}$		19.61 ± 1.49	0.881 ± 0.012	0.181 ± 0.011
$L_{image} + L_{measure}$		20.80 ± 1.77	0.910 ± 0.011	0.148 ± 0.011

3 Experiment and Results

3.1 Experimental Evaluations

Forty $128 \times 128 \times 40$ 3D Zubal brain phantoms [24] were used in the simulation study as ground truth, and one clinical patient brain images with different dose level were used for the robust analysis. Two tumors with different size were added in each Zubal brain phantom. The ground truth images were firstly forward-projected to generate the noise-free sinogram with count of 10^6 for each transverse slice and then Poisson noise were introduced. 20 percent of uniform random events were simulated. In total, 1600 (40×40) 2D sinograms were generated. Among them, 1320 (33 samples) were used in training, 200 (5 samples) for testing, and 80 (2 samples) for validation. A total of 5 realizations were simulated and each was trained/tested independently for bias and variance calculation [15]. We used peak signal to noise ratio (PSNR), structural similarity index (SSIM) and root mean square error (RMSE) for overall quantitative analysis. The contrast recovery coefficient (CRC) [25] was used for the comparison of reconstruction results in the tumor region of interest (ROI) area.

3.2 Results

Figure 2 shows three different slices of the reconstructed brain PET images using different methods. The DIP method and proposed DULDA have lower noise compared with MLEM and EM-TV visually. However, the DIP method shows unstable results cross different slices and fails in the recovery of the small cortex region. The proposed DULDA can recover more structural details and the white matter appears to be more sharpen. The quantitative and bias-variance results are shown in Table 1. We noticed that DIP method performs even worse than MLEM without anatomic priors. The DIP method demonstrates a certain ability to reduce noise by smoothing the image, but this leads to losses in important structural information, which explains the lower PSNR and SSIM. Both DIP method and DULDA have a better CRC and Bias performance compared with MLEM and EM-TV. In terms of supervised training, SLDA also performs best.

4   Discussion

To test the robustness of proposed DULDA, we forward-project one patient brain image data with different dose level and reconstructed it with the trained DULDA model. The results compared with MLEM are shown in Fig. 3. The patient is scanned with a GE Discovery MI 5-ring PET/CT system. The real image has very different cortex structure and some deflection compared with the training data. It can be observed that DULDA achieves excellent reconstruction results in both details and edges across different dose level and different slices.Table 2 shows the ablation study on phase numbers and loss function for DULDA. It can be observed that the dual domain loss helps improve the performance and when the phase number is 4, DULDA achieves the best performance.

5   Conclusions

In this work, we proposed a dual-domain unsupervised model-based deep learning method (DULDA) for PET image reconstruction by unrolling the learned descent algorithm. Both quantitative and visual results show the superior performance of DULDA when compared to MLEM, EM-TV and DIP based method. Future work will focus more on clinical aspects.

Acknowledgements. This work is supported in part by the National Key Technology Research and Development Program of China (No: 2021YFF0501503), the Talent Program of Zhejiang Province (No: 2021R51004), the Key Research and Development Program of Zhejiang Province (No: 2021C03029) and by NSF grants: DMS2152961.

References

1. Nordberg, A., Rinne, J., Kadir, A., Langström, B.: The use of PET in Alzheimer disease. Nat. Rev. Neurol. **6**, 78–87 (2010)
2. Cui, J., et al.: PET image denoising using unsupervised deep learning. Eur. J. Nucl. Med. Mol. Imaging **46**, 2780–2789 (2019)
3. Onishi, Y., et al.: Anatomical-guided attention enhances unsupervised PET image denoising performance. Med. Image Anal. **74**, 102226 (2021)
4. Zhu, B., Liu, J., Cauley, S., Rosen, B., Rosen, M.: Image reconstruction by domain-transform manifold learning. Nature **555**, 487–492 (2018)
5. Häggström, I., Schmidtlein, C., Campanella, G., Fuchs, T.: DeepPET: a deep encoder-decoder network for directly solving the PET image reconstruction inverse problem. Med. Image Anal. **54**, 253–262 (2019)
6. Li, Y., et al.: A deep neural network for parametric image reconstruction on a large axial field-of-view PET. Eur. J. Nucl. Med. Mol. Imaging **50**, 701–714 (2023)
7. Gong, K., et al.: Iterative PET image reconstruction using convolutional neural network representation. IEEE Trans. Med. Imaging **38**, 675–685 (2018)

8. Kim, K., et al.: Penalized PET reconstruction using deep learning prior and local linear fitting. IEEE Trans. Med. Imaging **37**, 1478–1487 (2018)

9. Li, S., Wang, G.: Deep kernel representation for image reconstruction in PET. IEEE Trans. Med. Imaging **41**, 3029–3038 (2022)

10. Li, S., Gong, K., Badawi, R., Kim, E., Qi, J., Wang, G.: Neural KEM: a kernel method with deep coefficient prior for PET image reconstruction. IEEE Trans. Med. Imaging **42**, 785–796 (2022)

11. Mehranian, A., Reader, A.: Model-based deep learning PET image reconstruction using forward-backward splitting expectation-maximization. IEEE Trans. Radiat. Plasma Med. Sci. **5**, 54–64 (2020)

12. Lim, H., Chun, I., Dewaraja, Y., Fessler, J.: Improved low-count quantitative PET reconstruction with an iterative neural network. IEEE Trans. Med. Imaging **39**, 3512–3522 (2020)

13. Hu, R., Liu, H.: TransEM: residual swin-transformer based regularized PET image reconstruction. In: Wang, L., Dou, Q., Fletcher, P.T., Speidel, S., Li, S. (eds.) MICCAI 2022, Part IV. LNCS, vol. 13434, pp. 184–193. Springer, Cham (2022)

14. Gong, K., et al.: MAPEM-Net: an unrolled neural network for Fully 3D PET image reconstruction. In: 15th International Meeting on Fully Three-dimensional Image Reconstruction in Radiology And Nuclear Medicine, vol. 11072, pp. 109–113 (2019)

15. Gong, K., Catana, C., Qi, J., Li, Q.: PET image reconstruction using deep image prior. IEEE Trans. Med. Imaging **38**, 1655–1665 (2018)

16. Hashimoto, F., Ote, K., Onishi, Y.: PET image reconstruction incorporating deep image prior and a forward projection model. IEEE Trans. Radiat. Plasma Med. Sci. **6**, 841–846 (2022)

17. Shen, C., et al.: Unsupervised Bayesian PET reconstruction. IEEE Trans. Radiat. Plasma Med. Sci. **7**, 75–190 (2022)

18. Chen, D., Tachella, J., Davies, M.: Robust equivariant imaging: a fully unsupervised framework for learning to image from noisy and partial measurements. In: Proceedings of the IEEE/CVF Conference on Computer Vision and Pattern Recognition, pp. 5647–5656 (2022)

19. Chen, Y., Liu, H., Ye, X., Zhang, Q.: Learnable descent algorithm for nonsmooth nonconvex image reconstruction. SIAM J. Imag. Sci. **14**, 1532–1564 (2021)

20. Ulyanov, D., Vedaldi, A., Lempitsky, V.: Deep image prior. In: Proceedings Of The IEEE Conference On Computer Vision And Pattern Recognition, pp. 9446–9454 (2018)

21. Noh, J., Fessler, J., Kinahan, P.: Statistical sinogram restoration in dual-energy CT for PET attenuation correction. IEEE Trans. Med. Imaging **28**, 1688–1702 (2009)

22. Shepp, L., Vardi, Y.: Maximum likelihood reconstruction for emission tomography. IEEE Trans. Med. Imaging **1**, 113–122 (1982)

23. Jonsson, E., Huang, S., Chan, T.: Total variation regularization in positron emission tomography. CAM Report. 9848 (1998)

24. Zubal, I., Harrell, C., Smith, E., Rattner, Z., Gindi, G., Hoffer, P.: Computerized three-dimensional segmented human anatomy. Med. Phys. **21**, 299–302 (1994)

25. Qi, J., Leahy, R.: A theoretical study of the contrast recovery and variance of MAP reconstructions from PET data. IEEE Trans. Med. Imaging **18**, 293–305 (1999)

Transformer-Based Dual-Domain Network for Few-View Dedicated Cardiac SPECT Image Reconstructions

Huidong Xie[1], Bo Zhou[1], Xiongchao Chen[1], Xueqi Guo[1], Stephanie Thorn[1], Yi-Hwa Liu[1], Ge Wang[2], Albert Sinusas[1], and Chi Liu[1(✉)]

[1] Yale University, New Haven, CT 06511, USA
{huidong.xie,chi.liu}@yale.edu
[2] Rensselaer Polytechnic Institute, Troy, NY 12180, USA

Abstract. Cardiovascular disease (CVD) is the leading cause of death worldwide, and myocardial perfusion imaging using SPECT has been widely used in the diagnosis of CVDs. The GE 530/570c dedicated cardiac SPECT scanners adopt a stationary geometry to simultaneously acquire 19 projections to increase sensitivity and achieve dynamic imaging. However, the limited amount of angular sampling negatively affects image quality. Deep learning methods can be implemented to produce higher-quality images from stationary data. This is essentially a few-view imaging problem. In this work, we propose a novel 3D transformer-based dual-domain network, called TIP-Net, for high-quality 3D cardiac SPECT image reconstructions. Our method aims to first reconstruct 3D cardiac SPECT images directly from projection data without the iterative reconstruction process by proposing a customized projection-to-image domain transformer. Then, given its reconstruction output and the original few-view reconstruction, we further refine the reconstruction using an image-domain reconstruction network. Validated by cardiac catheterization images, diagnostic interpretations from nuclear cardiologists, and defect size quantified by an FDA 510(k)-cleared clinical software, our method produced images with higher cardiac defect contrast on human studies compared with previous baseline methods, potentially enabling high-quality defect visualization using stationary few-view dedicated cardiac SPECT scanners.

Keywords: Cardiac SPECT · Few-view Imaging · Transformer

1 Introduction

The GE Discovery NM Alcyone 530c/570c [1] are dedicated cardiac SPECT systems with 19 cadmium zinc telluride (CZT) detector modules designed for

Supplementary Information The online version contains supplementary material available at https://doi.org/10.1007/978-3-031-43999-5_16.

stationary imaging. Limited amount of angular sampling on scanners of this type could affect image quality. Due to the unique geometry of Alcyone scanners, the centers of FOV vary at different angular positions. Hence, unlike CT scanners, there is no straightforward method to combine projections at different positions on Alcyone scanners. Xie *et al.* [12] proposed to incorporate the displacements between centers of FOV and the rotation angles into the iterative method for multi-angle reconstructions with registration steps within each iteration. Despite its promising results, acquiring multi-angle projections on this scanner is time-consuming and inconvenient in reality. Rotating the detectors also limits the capability of dynamic imaging. Thus, it is desirable to obtain high-quality rotation-based reconstruction directly from the stationary SPECT projection data. This is essentially a few-view reconstruction problem.

Previous works have attempted to address this problem by using deep-learning-based image-to-image networks. Xie *et al.* proposed a 3D U-net-like network to directly synthesize dense-view images from few-view counterparts [12]. Since convolutional networks have limited receptive fields due to small kernel size, Xie *et al.* further improved their method with a transformer-based image-to-image network for SPECT reconstruction [13]. Despite their promising reconstruction results, these methods use MLEM (maximum likelihood expectation maximization) reconstruction from one-angle few-view acquisition data as network input. The initial MLEM reconstruction may contain severe image artifacts with important image features lost during the iterative reconstruction process, thus would be challenging to be recovered with image-based methods. Learning to reconstruct images directly from the projection data could lead to improved quality.

There are a few previous studies proposed to learn the mapping between raw data and images. AUTOMAP [14] utilized fully-connected layers to learn the inverse Fourier transform between k-space data and the corresponding MRI images. While such a technique could be theoretically adapted to other imaging modalities, using a similar approach would require a significant amount of trainable parameters, and thus is infeasible for 3D data. Würfl *et al.* [10] proposed a back-projection operator to link projections and images to reduce memory burden for CT. However, their back-projection process is not learnable. There are also recent works [5,6,11] that tried to incorporate the embedded imaging physics and geometry of CT scanners into the neural networks to reduce redundant trainable parameters for domain mapping. However, these networks are unable to be generalized to other imaging modalities due to different physical properties. Moreover, these methods are hard to be extended to 3D cases because of the geometric complexity and additional memory/computational burden.

Here, we propose a novel 3D Transformer-based Dual-domain (projection & image) Network (TIP-Net) to address these challenges. The proposed method reconstructs high-quality few-view cardiac SPECT using a two-stage process. First, we develop a 3D projection-to-image transformer reconstruction network that directly reconstructs 3D images from the projection data. In the second stage, this intermediate reconstruction is combined with the original few-view reconstruction for further refinement, using an image-domain reconstruction network. Validated on physical phantoms, porcine, and human studies acquired on GE Alcyone 570c SPECT/CT scanners, TIP-Net demonstrated superior performance than previous baseline methods. Validated by cardiac catheterization (Cath) images, diagnostic results from nuclear cardiologists, and cardiac defect quantified by an FDA-510(k)-cleared software, we also show that TIP-Net produced images with higher resolution and cardiac defect contrast on human studies, as compared to previous baselines. Our method could be a clinically useful tool to improve cardiac SPECT imaging.

2 Methodology

Following the acquisition protocol described by Xie *et al.* [12], we acquired a total of eight porcine and two physical phantom studies prospectively with four angles of projections. The training target in this work is four-angle data with 76 (19×4) projections and the corresponding input is one-angle data with 19 projections. Twenty clinical anonymized ^{99m}Tc-tetrofosmin SPECT human studies were retrospectively included for evaluation. Since only one-angle data was available for the human studies, Cath images and clinical interpretations from nuclear cardiologists were used to determine the presence/absence of true cardiac defects and to assess the images reconstructed by different methods. The use of animal and human studies was approved by the Institutional Animal Care & Use Committee (IACUC) of Yale University.

2.1 Network Structure

The overall network structure is presented in Fig. 1. TIP-Net is divided into two parts. The transformer-based [2] projection-net (P-net) aims at reconstructing 3D SPECT volumes directly from 3D projections obtained from the scanner. Information from the system matrix (IMG_{bp}) is also incorporated in P-net as prior knowledge for image reconstruction. IMG_{bp} is obtained by multiplying the system matrix $S \in \mathbb{R}^{19,456 \times 245,000}$ with the projection data. The outputs from P-net (IMG_p) serve as an input for image-net (I-net).

   To reconstruct IMG_p in P-net, we may simply project the output from the Transformer block to the size of 3D reconstructed volume (i.e., $70 \times 70 \times 50$). However, such implementation requires a significant amount of GPU memory. To alleviate the memory burden, we proposed to learn the 3D reconstruction in a slice-by-slice manner. The output from the Transformer block is projected to a single slice (70×70), and the whole Transformer network (red rectangular in Fig. 1) is looped 50 times to produce a whole 3D volume ($70 \times 70 \times 50$). Since different slice has different detector sensitivity, all the 50 loops use different trainable parameters and the ith loop aims to reconstruct the ith slice in the 3D volume. Within each loop, the Transformer block takes the entire 3D projections as input to reconstruct the ith slice in the SPECT volume. With the self-attention mechanism, all 50 loops can observe the entire 3D projections for reconstruction.

The 70×70 slice is then combined with IMG_{bp} (only at i^{th} slice), and the resized 3D projection data. The resulting feature maps ($70 \times 70 \times 21$) are fed into a shallow 2D CNN to produce a reconstructed slice at the ith loop ($70 \times 70 \times 1$), which is expected to be the ith slice in the SPECT volume.

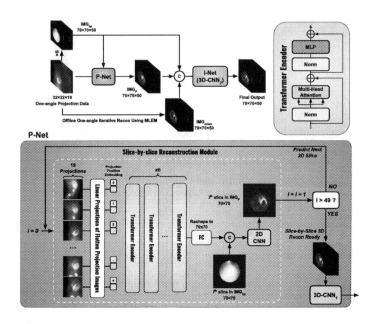

Fig. 1. Overall TIP-Net structure. The TIP-Net is divided into P-net and I-net. The back-projected image volume was generated by multiplying the 3D projections with the system matrix. FC layer: fully-connected layer. Both 3D-CNN$_1$ and 3D-CNN$_2$ share the same structure but with different trainable parameters.

I-net takes images reconstructed by P-net (IMG_p), prior knowledge from the system matrix (IMG_{bp}), and images reconstructed by MLEM (IMG_{mlem}) using one-angle data for final reconstructions. With such a design, the network can combine information from both domains for potentially better reconstructions compared with methods that only consider single-domain data.

Both 3D-CNN$_1$ and 3D-CNN$_2$ use the same network structure as the network proposed in [12], but with different trainable parameters.

2.2   Optimization, Training, and Testing

In this work, the TIP-Net was trained using a Wasserstein Generative Adversarial Network (WGAN) with gradient penalty [4]. A typical WGAN contains 2 separate networks, one generator (G) and a discriminator (D). In this work, the network G is the TIP-Net depicted in Fig. 1. Formulated below, the overall objective function for network G includes SSIM, MAE, and Adversarial loss.

$$\min_{\theta_G} L = \left\{ \ell(G(I_{\mathrm{one}}), I_{\mathrm{four}}) + \lambda_a \ell(\mathrm{P}_{\mathrm{net}}(I_{\mathrm{one}}), I_{\mathrm{four}}) - \underbrace{\lambda_b \, \mathbb{E}_{I_{\mathrm{one}}} [D(G(I_{\mathrm{one}}))]}_{\text{Adversarial Loss}} \right\},$$

$$(1)$$

where I_{one} and I_{four} denote images reconstructed by one-angle data and four-angle data respectively using MLEM. θ_G represents trainable parameters of network G. $\mathrm{P}_{\mathrm{net}}(I_{\mathrm{one}})$ represents the output of the P-net ($\mathrm{IMG_p}$). The function ℓ is formulated as below:

$$\ell(X, Y) = \ell_{\mathrm{MAE}}(X, Y) + \lambda_c \, \ell_{\mathrm{SSIM}}(X, Y) + \lambda_d \, \ell_{\mathrm{MAE}}(\mathrm{SO}(X), \mathrm{SO}(Y)), \qquad (2)$$

where X and Y represent two image volumes used for calculations. ℓ_{MAE} and ℓ_{SSIM} represent MAE and SSIM loss functions, respectively. The Sobel operator (SO) was used to obtain edge images, and the MAE between them was included as the loss function. $\lambda_a = 0.1$, $\lambda_b = 0.005$, $\lambda_c = 0.8$, and $\lambda_d = 0.1$ were fine-tuned experimentally. Network D shares the same structure as that proposed in [12].

The Adam method [8] was used for optimizations. The parameters were initialized using the Xavier method [3]. 250 volumes of simulated 4D extended cardiac-torso (XCAT) phantoms [9] were used for network pre-training. Leave-one-out testing process was used to obtain testing results for all the studies.

2.3   Evaluations

Reconstructed images were quantitatively evaluated using SSIM, RMSE, and PSNR. Myocardium-to-blood pool (MBP) ratios were also included for evaluation. For myocardial perfusion images, higher ratios are favorable and typically represent higher image resolution. Two ablated networks were trained and used for comparison. One network, denoted as 3D-CNN, shared the same structure as either 3D-CNN$_1$ or 3D-CNN$_2$ but only with $\mathrm{IMG_{mlem}}$ as input. The other network, denoted as Dual-3D-CNN, used the same structure as the TIP-Net outlined in Fig. 1 but without any projection-related inputs. Compared with these two ablated networks, we could demonstrate the effectiveness of the proposed projection-to-image module in the TIP-Net. We also compared the TIP-Net with another transformer-based network (SSTrans-3D) [13]. Since SSTrans-3D only considers image-domain data, comparing TIP-Net with SSTrans-3D could demonstrate the effectiveness of the transformer-based network for processing projection-domain information.

To compare the performance of cardiac defect quantifications, we used the FDA 510(k)-cleared Wackers-Liu Circumferential Quantification (WLCQ) software [7] to calculate the myocardial perfusion defect size (DS). For studies without cardiac defect, we should expect lower measured defect size as the uniformity of myocardium improves. For studies with cardiac defects, we should expect higher measured defect size as defect contrast improves.

Cath images and cardiac polar maps are also presented. Cath is an invasive imaging technique used to determine the presence/absence of obstructive lesions that results in cardiac defects. We consider Cath as the gold standard for the defect information in human studies. The polar map is a 2D representation of the 3D volume of the left ventricle. All the metrics were calculated based on the entire 3D volumes. All the clinical descriptions of cardiac defects in this paper were confirmed by nuclear cardiologists based on SPECT images, polar maps, WLCQ quantification, and Cath images (if available).

3 Results

3.1 Porcine Results

Results from one sample porcine study are presented in Fig. S1. This pig had a large post-occlusion defect created by inflating an angioplasty balloon in the left anterior descending artery. As pointed out by the green arrows in Fig. S1, all neural networks improved the defect contrast with higher MBP ratios compared with the one-angle images. The TIP-Net results were better than other network results in terms of defect contrast based on defect size measured by WLCQ. TIP-Net also produced images with clearer right ventricles, as demonstrated by the Full-width at Half Maximum (FHMW) values presented in Fig. S1.

Quantitative results for 8 porcine and 2 physical phantom studies are included in Table 1. Based on paired t-tests, all the network results are statistically better than one-angle results ($p < 0.001$), and the TIP-Net results are statistically better than all the other three network results ($p < 0.05$). The average MBP ratios shown in Table 1 indicate that the proposed TIP-Net produced images with higher resolution compared with image-based networks.

Average defect size measurements for porcine and physical phantom studies with cardiac defects are 35.9%, 42.5%, 42.3%, 43.6%, 47.0%, 46.2% for one-angle, 3D-CNN, Dual-3D CNN, TIP-Net, and four-angle image volumes, respectively. Images reconstructed by TIP-Net present overall higher defect contrast and the measured defect size is closest to the four-angle images.

For normal porcine and physical phantom studies without cardiac defect, these numbers are 16.0%, 12.0%, 14.7%, 11.2%, 11.5%, and 10.1%. All neural networks showed improved uniformity in the myocardium with lower measured defect size. Both transformer-based methods, TIP-Net and SSTrans-3D, showed better defect quantification than other methods on these normal subjects.

Table 1. Quantitative values for porcine and phantom results obtained using different methods (MEAN ± STD). Images reconstructed using four-angle data were used as the reference. Best values are marked in bold. $p < 0.05$ observed between all groups.

Evaluation	One-angle	3D-CNN	Dual-3D-CNN	SSTrans-3D	TIP-Net	Four-angle
SSIM↑	0.945 ± 0.009	0.951 ± 0.008	0.950 ± 0.008	0.947 ± 0.009	**0.953 ± 0.008**	\
PSNR↑	35.764 ± 3.546	36.763 ± 3.300	36.787 ± 3.208	36.613 ± 3.509	**38.114 ± 2.876**	\
RMSE↓	0.020 ± 0.005	0.0181 ± 0.005	0.0185 ± 0.005	0.0194 ± 0.005	**0.0179 ± 0.005**	\
MBP↑	5.16 ± 1.83	10.73 ± 2.98	10.21 ± 3.17	11.10 ± 2.78	**12.73 ± 3.73**	13.63 ± 3.91

3.2   Human Results

Chosen by a nuclear cardiologist, results from one representative human study with multiple cardiac defects are presented in Fig. 2. Note that there was no four-angle data for human studies. Based on clinical interpretations from the nuclear cardiologist and Cath results, this patient had multiple perfusion defects in the apical (blue arrows in Fig. 2) and basal (green and yellow arrows in Fig. 2) regions. As presented in Fig. 2, all the networks produced images with better defect contrasts. The proposed TIP-Net further improved the defect contrast, which is favorable and could make the clinical interpretations more confident.

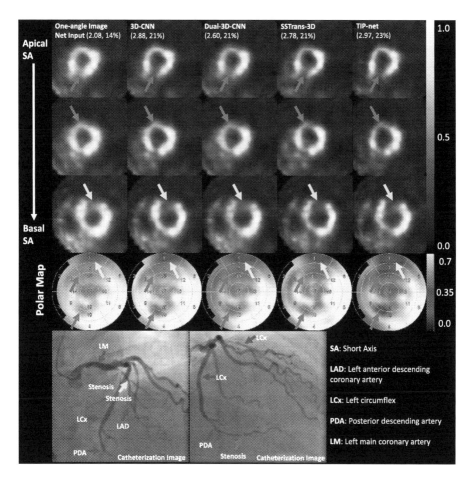

Fig. 2. Results from a sample stress human study. Red arrows point to different coronary vessels. Non-red arrows with the same color point to the same defect/stenosis in the heart. Numbers in the parentheses are the MBP ratios and the defect size (%LV). (Color figure online)

Another human study was selected and presented in Fig. 3. This patient had stenosis in the distal left anterior descending coronary artery, leading to a medium-sized defect in the entire apical region (light-green arrows). As validated by the Cath images, polar map, and interpretation from a nuclear cardiologist, TIP-Net produced images with higher apical defect contrast compared with other methods, potentially leading to a more confident diagnostic decision.

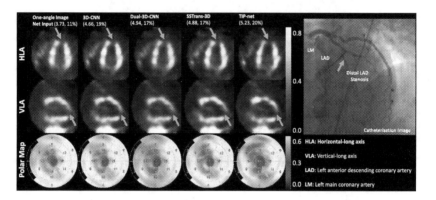

Fig. 3. Results from a sample stress human study. Red arrows point to different coronary vessels. Light-green arrows point to the same defect/stenosis in the heart. Numbers in the parentheses are the MBP ratios and the defect size (%LV). (Color figure online)

The average MBP ratios on human studies for one-angle, 3D-CNN, Dual-3D-CNN, SSTrans-3D, and TIP-Net images are 3.13 ± 0.62, 4.08 ± 0.83, 4.31 ± 1.07, 4.17 ± 0.99, and 4.62 ± 1.29, respectively (MEAN $\pm$ STD). The higher ratios in images produced by TIP-Net typically indicate higher image resolution.

14 patients in the testing data have cardiac defects, according to diagnostic results. The average defect size measurements for these patients are 16.8%, 22.6%, 21.0%, 22.4%, and 23.6% for one-angle, 3D-CNN, Dual-3D-CNN, SSTrans-3D, and TIP-Net results. The higher measured defect size of the TIP-Net indicates that the proposed TIP-Net produced images with higher defect contrast.

For the other 6 patients without cardiac defects, these numbers are 11.5%, 12.8%, 13.8%, 12.4%, and 11.8%. These numbers show that TIP-Net did not introduce undesirable noise in the myocardium, maintaining the overall uniformity for normal patients. However, other deep learning methods tended to introduce non-uniformity in these normal patients and increased the defect size.

3.3 Intermediate Network Output

To further show the effectiveness of P-net in the overall TIP-Net design, outputs from P-net (images reconstructed by the network directly from projections) are presented in Fig. S2. The presented results demonstrate that the network can learn the mapping between 3D projections to 3D image volumes directly without the iterative reconstruction process. In the porcine study, limited angular sampling introduced streak artifacts in the MLEM-reconstructed one-angle images (yellow arrows in Fig. S2). P-net produced images with fewer few-view artifacts and higher image resolution.

The human study presented in Fig. S2 has an apical defect according to the diagnostic results (blue arrows in Fig. S2). However, this apical defect is barely visible in the one-angle image. P-net produced images with higher resolution and improved defect contrast. Combining both outputs (one-angle images and IMG_p), TIP-Net further enhanced the defect contrast.

4   Discussion and Conclusion

We proposed a novel TIP-Net for 3D cardiac SPECT reconstruction. To the best of our knowledge, this work is the first attempt to learn the mapping from 3D realistic projections to 3D image volumes. Previous works in this direction [5,6,14] were 2D and only considered simulated projections with ideal conditions. 3D realistic projection data have more complex geometry are also affected by other physical factors that may not exist in simulated projections.

The proposed method was tested for myocardial perfusion SPECT imaging. Validated by nuclear cardiologists, diagnostic results, Cath images, and defect size measured by WLCQ, the proposed TIP-Net produced images with higher resolution and higher defect contrast for patients with perfusion defects. For normal patients without perfusion defects, TIP-Net maintained overall uniformity in the myocardium with higher image resolution. Similar performance was observed in porcine and physical phantom studies.

References

1. Bocher, M., Blevis, I., Tsukerman, L., Shrem, Y., Kovalski, G., Volokh, L.: A fast cardiac gamma camera with dynamic SPECT capabilities: design, system validation and future potential. Eur. J. Nucl. Med. Mol. Imaging **37**(10), 1887–1902 (2010). https://doi.org/10.1007/s00259-010-1488-z
2. Dosovitskiy, A., et al.: An image is worth 16×16 words: transformers for image recognition at scale. arXiv:2010.11929 [cs], June 2021. http://arxiv.org/abs/2010.11929
3. Glorot, X., Bengio, Y.: Understanding the difficulty of training deep feedforward neural networks. In: Teh, Y.W., Titterington, M. (eds.) Proceedings of the Thirteenth International Conference on Artificial Intelligence and Statistics. Proceedings of Machine Learning Research, vol. 9, pp. 249–256. PMLR, Chia Laguna Resort, Sardinia, Italy, 13–15 May 2010 (2010). http://proceedings.mlr.press/v9/glorot10a.html
4. Gulrajani, I., Ahmed, F., Arjovsky, M., Dumoulin, V., Courville, A.: Improved training of Wasserstein GANs. In: Advances in Neural Information Processing Systems, vol. 30. Curran Associates, Inc. (2017)

5. He, J., Wang, Y., Ma, J.: Radon inversion via deep learning. IEEE Trans. Med. Imaging **39**(6), 2076–2087 (2020). https://doi.org/10.1109/TMI.2020.2964266
6. Li, Y., Li, K., Zhang, C., Montoya, J., Chen, G.: Learning to reconstruct computed tomography (CT) images directly from sinogram data under a variety of data acquisition conditions. IEEE Trans. Med. Imaging **39**, 2469–2481 (2019). https://doi.org/10.1109/TMI.2019.2910760
7. Liu, Y., Sinusas, A., DeMan, P., Zaret, B., Wackers, F.: Quantification of SPECT myocardial perfusion images: methodology and validation of the Yale-CQ method. J. Nucl. Cardiol. **6**(2), 190–204 (1999). https://doi.org/10.1016/s1071-3581(99)90080-6
8. PK, D., B, J.: Adam: a method for stochastic optimization. In: 3rd International Conference on Learning Representations, ICLR 2015, Conference Track Proceedings, San Diego, CA, USA, 7–9 May 2015 (2015). http://arxiv.org/abs/1412.6980
9. Segars, W., Sturgeon, G., Mendonca, S., Grimes, J., Tsui, B.: 4D XCAT phantom for multimodality imaging research. Med. Phys. **37**(9), 4902–4915 (2010). https://doi.org/10.1118/1.3480985. https://aapm.onlinelibrary.wiley.com/doi/abs/10.1118/1.3480985
10. Würfl, T., et al.: Deep learning computed tomography: learning projection-domain weights from image domain in limited angle problems. IEEE Trans. Med. Imaging **37**(6), 1454–1463 (2018). https://doi.org/10.1109/TMI.2018.2833499
11. Xie, H., et al.: Deep efficient end-to-end reconstruction (DEER) network for few-view breast CT image reconstruction. IEEE Access **8**, 196633–196646 (2020). https://doi.org/10.1109/ACCESS.2020.3033795
12. Xie, H., et al.: Increasing angular sampling through deep learning for stationary cardiac SPECT image reconstruction. J. Nucl. Cardiol. **30**, 86–100 (2022). https://doi.org/10.1007/s12350-022-02972-z
13. Xie, H., et al.: Deep learning based few-angle cardiac SPECT reconstruction using transformer. IEEE Trans. Radiat. Plasma Med. Sci. **7**, 33–40. https://doi.org/10.1109/TRPMS.2022.3187595
14. Zhu, B., Liu, J.Z., Cauley, S.F., Rosen, B.R., Rosen, M.S.: Image reconstruction by domain-transform manifold learning. Nature **555**(7697), 487–492 (2018). https://doi.org/10.1038/nature25988

Learned Alternating Minimization Algorithm for Dual-Domain Sparse-View CT Reconstruction

Chi Ding[1(✉)], Qingchao Zhang[1], Ge Wang[2], Xiaojing Ye[3], and Yunmei Chen[1]

[1] University of Florida, Gainesville, FL 32611, USA
{ding.chi,qingchaozhang,yun}@ufl.edu
[2] Rensselaer Polytechnic Institute, Troy, NY 12180, USA
wangg6@rpi.edu
[3] Georgia State University, Atlanta, GA 30302, USA
xye@gsu.edu

Abstract. We propose a novel Learned Alternating Minimization Algorithm (LAMA) for dual-domain sparse-view CT image reconstruction. LAMA is naturally induced by a variational model for CT reconstruction with learnable nonsmooth nonconvex regularizers, which are parameterized as composite functions of deep networks in both image and sinogram domains. To minimize the objective of the model, we incorporate the smoothing technique and residual learning architecture into the design of LAMA. We show that LAMA substantially reduces network complexity, improves memory efficiency and reconstruction accuracy, and is provably convergent for reliable reconstructions. Extensive numerical experiments demonstrate that LAMA outperforms existing methods by a wide margin on multiple benchmark CT datasets.

Keywords: Learned alternating minimization algorithm ·
Convergence · Deep networks · Sparse-view CT reconstruction

1 Introduction

Sparse-view Computed Tomography (CT) is an important class of low-dose CT techniques for fast imaging with reduced X-ray radiation dose. Due to the significant undersampling of sinogram data, the sparse-view CT reconstruction problem is severely ill-posed. As such, applying the standard filtered-backprojection (FBP) algorithm, [1] to sparse-view CT data results in significant

This work was supported in part by National Science Foundation under grants DMS-1925263, DMS-2152960 and DMS-2152961 and US National Institutes of Health grants R01HL151561, R01CA237267, R01EB032716 and R01EB031885.

Supplementary Information The online version contains supplementary material available at https://doi.org/10.1007/978-3-031-43999-5_17.

severe artifacts in the reconstructed images, which are unreliable for clinical use. In recent decades, variational methods have become a major class of mathematical approaches that model reconstruction as a minimization problem. The objective function of the minimization problem consists of a penalty term that measures the discrepancy between the reconstructed image and the given data and a regularization term that enforces prior knowledge or regularity of the image. Then an optimization method is applied to solve for the minimizer, which is the reconstructed image of the problem. The regularization in existing variational methods is often chosen as relatively simple functions, such as total variation (TV) [2–4], which is proven useful in many instances but still far from satisfaction in most real-world image reconstruction applications due to their limitations in capturing fine structures of images. Hence, it remains a very active research area in developing more accurate and effective methods for high-quality sparse-view CT reconstruction in medical imaging.

Deep learning (DL) has emerged in recent years as a powerful tool for image reconstruction. Deep learning parameterizes the functions of interests, such as the mapping from incomplete and/or noisy data to reconstructed images, as deep neural networks. The parameters of the networks are learned by minimizing some loss functional that measures the mapping quality based on a sufficient amount of data samples. The use of training samples enables DL to learn more enriched features, and therefore, DL has shown tremendous success in various tasks in image reconstruction. In particular, DL has been used for medical image reconstruction applications [5–12], and experiments show that these methods often significantly outperform traditional variational methods.

DL-based methods for CT reconstruction have also evolved fast in the past few years. One of the most successful DL-based approaches is known as unrolling [10,13–16]. Unrolling methods mimic some traditional optimization schemes (such as proximal gradient descent) designed for variational methods to build the network structure but replace the term corresponding to the handcrafted regularization in the original variational model by deep networks. Most existing DL-based CT reconstruction methods use deep networks to extract features of the image or the sinogram [5,7,9–12,17–19]. More recently, dual-domain methods [6,8,15,18] emerged and can further improve reconstruction quality by leveraging complementary information from both the image and sinogram domains. Despite the substantial improvements in reconstruction quality over traditional variational methods, there are concerns with these DL-based methods due to their lack of theoretical interpretation and practical robustness. In particular, these methods tend to be memory inefficient and prone to overfitting. One major reason is that these methods only superficially mimic some known optimization schemes but lose all convergence and stability guarantees.

Recently, a new class of DL-based methods known as learnable descent algorithm (LDA) [16,19,20] have been developed for image reconstruction. These methods start from a variational model where the regularization can be parameterized as a deep network whose parameters can be learned. The objective function is potentially nonconvex and nonsmooth due to such parameterization. Then

LDA aims to design an efficient and convergent scheme to minimize the objective function. This optimization scheme induces a highly structured deep network whose parameters are completely inherited from the learnable regularization and trained adaptively using data while retaining all convergence properties. The present work follows this approach to develop a dual-domain sparse-view CT reconstruction method. Specifically, we consider learnable regularizations for image and sinogram as composite objectives, where they unroll parallel subnetworks and extract complementary information from both domains. Unlike the existing LDA, we will design a novel adaptive scheme by modifying the alternating minimization methods [21–25] and incorporating the residual learning architecture to improve image quality and training efficiency.

2 Learnable Variational Model

We formulate the dual-domain reconstruction model as the following two-block minimization problem:

$$\arg\min_{\mathbf{x},\mathbf{z}} \; \Phi(\mathbf{x},\mathbf{z};\mathbf{s},\Theta) := \frac{1}{2}\|\mathbf{A}\mathbf{x}-\mathbf{z}\|^2 + \frac{\lambda}{2}\|\mathbf{P}_0\mathbf{z}-\mathbf{s}\|^2 + R(\mathbf{x};\theta_1) + Q(\mathbf{z};\theta_2), \quad (1)$$

where $(\mathbf{x},\mathbf{z})$ are the image and sinogram to be reconstructed and $\mathbf{s}$ is the sparse-view sinogram. The first two terms in (1) are the data fidelity and consistency, where $\mathbf{A}$ and $\mathbf{P}_0\mathbf{z}$ represent the Radon transform and the sparse-view sinogram, respectively, and $\|\cdot\| \equiv \|\cdot\|_2$. The last two terms represent the regularizations, which are defined as the $l_{2,1}$ norm of the learnable convolutional feature extraction mappings in (2). If this mapping is the gradient operator, then the regularization reduces to total variation that has been widely used as a hand-crafted regularizer in image reconstruction. On the other hand, the proposed regularizers are generalizations and capable to learn in more adapted domains where the reconstructed image and sinogram become sparse:

$$R(\mathbf{x};\theta_1) = \left\|\mathbf{g}^R(\mathbf{x},\theta_1)\right\|_{2,1} := \sum_{i=1}^{m_R}\left\|\mathbf{g}_i^R(\mathbf{x},\theta_1)\right\|, \quad (2a)$$

$$Q(\mathbf{z};\theta_2) = \left\|\mathbf{g}^Q(\mathbf{z},\theta_2)\right\|_{2,1} := \sum_{j=1}^{m_Q}\left\|\mathbf{g}_j^Q(\mathbf{z},\theta_2)\right\|, \quad (2b)$$

where θ_1, θ_2 are learnable parameters. We use $\mathbf{g}^r(\cdot) \in \mathbb{R}^{m_r \times d_r}$ to present $\mathbf{g}^R(\mathbf{x},\theta_1)$ and $\mathbf{g}^Q(\mathbf{z},\theta_2)$, i.e. r can be R or Q. The d_r is the depth and $\sqrt{m_r} \times \sqrt{m_r}$ is the spacial dimension. Note $\mathbf{g}_i^r(\cdot) \in \mathbb{R}^{d_r}$ is the vector at position i across all channels. The feature extractor $\mathbf{g}^r(\cdot)$ is a CNN consisting of several convolutional operators separated by the smoothed ReLU activation function as follows:

$$\mathbf{g}^r(\mathbf{y}) = \mathbf{w}_l^r * a \cdots a(\mathbf{w}_2^r * a(\mathbf{w}_1^r * \mathbf{y})), \quad (3)$$

where $\{\mathbf{w}_i^r\}_{i=1}^l$ denote convolution parameters with d_r kernels. Kernel sizes are $(3,3)$ and $(3,15)$ for the image and sinogram networks, respectively. $a(\cdot)$ denotes smoothed ReLU activation function, which can be found in [16].

3   A Learned Alternating Minimization Algorithm

This section formally introduces the Learned Alternating Minimization Algorithm (LAMA) to solve the nonconvex and nonsmooth minimization model (1). LAMA incorporates the residue learning structure [26] to improve the practical learning performance by avoiding gradient vanishing in the training process with convergence guarantees. The algorithm consists of three stages, as follows:

The first stage of LAMA aims to reduce the nonconvex and nonsmooth problem in (1) to a nonconvex smooth optimization problem by using an appropriate smoothing procedure

$$r_\varepsilon(\mathbf{y}) = \sum_{i \in I_0^r} \frac{1}{2\varepsilon} \|\mathbf{g}_i^r(\mathbf{y})\|^2 + \sum_{i \in I_1^r} \left(\|\mathbf{g}_i^r(\mathbf{y})\| - \frac{\varepsilon}{2} \right), \quad \mathbf{y} \in Y =: m_r \times d_r, \quad (4)$$

where $(r, \mathbf{y})$ represents either $(R, \mathbf{x})$ or $(Q, \mathbf{z})$ and

$$I_0^r = \{i \in [m_r] \mid \|\mathbf{g}_i^r(\mathbf{y})\| \leq \varepsilon\}, \quad I_1^r = [m_r] \setminus I_0^r. \quad (5)$$

Note that the non-smoothness of the objective function (1) originates from the non-differentiability of the $l_{2,1}$ norm at the origin. To handle the non-smoothness, we utilize Nesterov's smoothing technique [27] as previously applied in [16]. The smoothed regularizations take the form of the Huber function, effectively removing the non-smoothness aspects of the problem.

The second stage solves the smoothed nonconvex problem with the fixed smoothing factor $\varepsilon = \varepsilon_k$, i.e.

$$\min_{\mathbf{x},\mathbf{z}} \{\Phi_\varepsilon(\mathbf{x}, \mathbf{z}) := f(\mathbf{x}, \mathbf{z}) + R_\varepsilon(\mathbf{x}) + Q_\varepsilon(\mathbf{z})\}. \quad (6)$$

where $f(\mathbf{x}, \mathbf{z})$ denotes the first two data fitting terms from (1). In light of the substantial improvement in practical performance by ResNet [26], we propose an inexact proximal alternating linearized minimization algorithm (PALM) [22] for solving (6). With $\varepsilon = \varepsilon_k > 0$, the scheme of PALM [22] is

$$\mathbf{b}_{k+1} = \mathbf{z}_k - \alpha_k \nabla_{\mathbf{z}} f(\mathbf{x}_k, \mathbf{z}_k), \quad \mathbf{u}_{k+1}^{\mathbf{z}} = \arg\min_{\mathbf{u}} \frac{1}{2\alpha_k} \|\mathbf{u} - \mathbf{b}_{k+1}\|^2 + Q_{\varepsilon_k}(\mathbf{u}), \quad (7)$$

$$\mathbf{c}_{k+1} = \mathbf{x}_k - \beta_k \nabla_{\mathbf{x}} f(\mathbf{x}_k, \mathbf{u}_{k+1}^{\mathbf{z}}), \quad \mathbf{u}_{k+1}^{\mathbf{x}} = \arg\min_{\mathbf{u}} \frac{1}{2\beta_k} \|\mathbf{u} - \mathbf{c}_{k+1}\|^2 + R_{\varepsilon_k}(\mathbf{u}), \quad (8)$$

where α_k and β_k are step sizes. Since the proximal point $\mathbf{u}_{k+1}^{\mathbf{x}}$ and $\mathbf{u}_{k+1}^{\mathbf{z}}$ are are difficult to compute, we approximate $Q_{\varepsilon_k}(\mathbf{u})$ and $R_{\varepsilon_k}(\mathbf{u})$ by their linear approximations at $\mathbf{b}_{k+1}$ and $\mathbf{c}_{k+1}$, i.e. $Q_{\varepsilon_k}(\mathbf{b}_{k+1}) + \langle \nabla Q_{\varepsilon_k}(\mathbf{b}_{k+1}), \mathbf{y} - \mathbf{b}_{k+1} \rangle$ and $R_{\varepsilon_k}(\mathbf{c}_{k+1}) + \langle \nabla R_{\varepsilon_k}(\mathbf{c}_{k+1}), \mathbf{u} - \mathbf{c}_{k+1} \rangle$, together with the proximal terms $\frac{1}{2p_k} \|\mathbf{u} - \mathbf{b}_{k+1}\|^2$ and $\frac{1}{2q_k} \|\mathbf{u} - \mathbf{c}_{k+1}\|^2$. Then by a simple computation, $\mathbf{u}_{k+1}^{\mathbf{x}}$ and $\mathbf{u}_{k+1}^{\mathbf{z}}$ are now determined by the following formulas

$$\mathbf{u}_{k+1}^{\mathbf{z}} = \mathbf{b}_{k+1} - \hat{\alpha}_k \nabla Q_{\varepsilon_k}(\mathbf{b}_{k+1}), \quad \mathbf{u}_{k+1}^{\mathbf{x}} = \mathbf{c}_{k+1} - \hat{\beta}_k \nabla R_{\varepsilon_k}(\mathbf{c}_{k+1}), \quad (9)$$

where $\hat{\alpha}_k = \frac{\alpha_k p_k}{\alpha_k + p_k}$, $\hat{\beta}_k = \frac{\beta_k q_k}{\beta_k + q_k}$. In deep learning approach, the step sizes α_k, $\hat{\alpha}_k$, β_k and $\hat{\beta}_k$ can also be learned. Note that the convergence of the sequence $\{(\mathbf{u}_{k+1}^{\mathbf{z}}, \mathbf{u}_{k+1}^{\mathbf{x}})\}$ is not guaranteed. We proposed that if $(\mathbf{u}_{k+1}^{\mathbf{z}}, \mathbf{u}_{k+1}^{\mathbf{x}})$ satisfy the following **Sufficient Descent Conditions** (SDC):

$$\Phi_{\varepsilon_k}(\mathbf{u}_{k+1}^{\mathbf{x}}, \mathbf{u}_{k+1}^{\mathbf{z}}) - \Phi_{\varepsilon_k}(\mathbf{x}_k, \mathbf{z}_k) \leq -\eta \left(\left\| \mathbf{u}_{k+1}^{\mathbf{x}} - \mathbf{x}_k \right\|^2 + \left\| \mathbf{u}_{k+1}^{\mathbf{z}} - \mathbf{z}_k \right\|^2 \right), \quad (10a)$$

$$\left\| \nabla \Phi_{\varepsilon_k}(\mathbf{x}_k, \mathbf{z}_k) \right\| \leq \frac{1}{\eta} \left(\left\| \mathbf{u}_{k+1}^{\mathbf{x}} - \mathbf{x}_k \right\| + \left\| \mathbf{u}_{k+1}^{\mathbf{z}} - \mathbf{z}_k \right\| \right), \quad (10b)$$

for some $\eta > 0$, we accept $\mathbf{x}_{k+1} = \mathbf{u}_{k+1}^{\mathbf{x}}$, $\mathbf{z}_{k+1} = \mathbf{u}_{k+1}^{\mathbf{z}}$. If one of (10a) and (10b) is violated, we compute $(\mathbf{v}_{k+1}^{\mathbf{z}}, \mathbf{v}_{k+1}^{\mathbf{x}})$ by the standard Block Coordinate Descent (BCD) with a simple line-search strategy to safeguard convergence: Let $\bar{\alpha}, \bar{\beta}$ be positive numbers in $(0, 1)$ compute

$$\mathbf{v}_{k+1}^{\mathbf{z}} = \mathbf{z}_k - \bar{\alpha} \left(\nabla_{\mathbf{z}} f(\mathbf{x}_k, \mathbf{z}_k) + \nabla Q_{\varepsilon_k}(\mathbf{z}_k) \right), \quad (11)$$

$$\mathbf{v}_{k+1}^{\mathbf{x}} = \mathbf{x}_k - \bar{\beta} \left(\nabla_{\mathbf{x}} f(\mathbf{x}_k, \mathbf{v}_{k+1}^{\mathbf{z}}) + \nabla R_{\varepsilon_k}(\mathbf{x}_k) \right). \quad (12)$$

Set $\mathbf{x}_{k+1} = \mathbf{v}_{k+1}^{\mathbf{x}}$, $\mathbf{z}_{k+1} = \mathbf{v}_{k+1}^{\mathbf{z}}$, if for some $\delta \in (0, 1)$, the following holds:

$$\Phi_{\varepsilon}(\mathbf{v}_{k+1}^{\mathbf{x}}, \mathbf{v}_{k+1}^{\mathbf{z}}) - \Phi_{\varepsilon}(\mathbf{x}_k, \mathbf{z}_k) \leq -\delta(\left\| \mathbf{v}_{k+1}^{\mathbf{x}} - \mathbf{x}_k \right\|^2 + \left\| \mathbf{v}_{k+1}^{\mathbf{z}} - \mathbf{z}_k \right\|^2). \quad (13)$$

Otherwise we reduce $(\bar{\alpha}, \bar{\beta}) \leftarrow \rho(\bar{\alpha}, \bar{\beta})$ where $0 < \rho < 1$, and recompute $\mathbf{v}_{k+1}^{\mathbf{x}}, \mathbf{v}_{k+1}^{\mathbf{z}}$ until the condition (13) holds.

The third stage checks if $\|\nabla \Phi_{\varepsilon}\|$ has been reduced enough to perform the second stage with a reduced smoothing factor ε. By gradually decreasing ε, we obtain a subsequence of the iterates that converges to a Clarke stationary point of the original nonconvex and nonsmooth problem. The algorithm is given below.

Algorithm 1. The Linearized Alternating Minimization Algorithm (LAMA)

Input: Initializations: $\mathbf{x}_0$, $\mathbf{z}_0$, δ, η, ρ, γ, ε_0, σ, λ

1: **for** $k = 0, 1, 2, \ldots$ **do**
2:     $\mathbf{b}_{k+1} = \mathbf{z}_k - \alpha_k \nabla_{\mathbf{z}} f(\mathbf{x}_k, \mathbf{z}_k)$, $\mathbf{u}_{k+1}^{\mathbf{z}} = \mathbf{b}_{k+1} - \hat{\alpha}_k \nabla Q_{\varepsilon_k}(\mathbf{b}_{k+1})$
3:     $\mathbf{c}_{k+1} = \mathbf{x}_k - \beta_k \nabla_{\mathbf{x}} f(\mathbf{x}_k, \mathbf{u}_{k+1}^{\mathbf{z}})$, $\mathbf{u}_{k+1}^{\mathbf{x}} = \mathbf{c}_{k+1} - \hat{\beta}_k \nabla R_{\varepsilon_k}(\mathbf{c}_{k+1})$
4:     **if** (10) holds **then**
5:         $(\mathbf{x}_{k+1}, \mathbf{z}_{k+1}) \leftarrow (\mathbf{u}_{k+1}^{\mathbf{x}}, \mathbf{u}_{k+1}^{\mathbf{z}})$
6:     **else**
7:         $\mathbf{v}_{k+1}^{\mathbf{z}} = \mathbf{z}_k - \bar{\alpha} \left[\nabla_{\mathbf{z}} f(\mathbf{x}_k, \mathbf{z}_k) + \nabla Q_{\varepsilon_k}(\mathbf{z}_k) \right]$
8:         $\mathbf{v}_{k+1}^{\mathbf{x}} = \mathbf{x}_k - \bar{\beta} \left[\nabla_{\mathbf{x}} f(\mathbf{x}_k, \mathbf{v}_{k+1}^{\mathbf{z}}) + \nabla R_{\varepsilon_k}(\mathbf{x}_k) \right]$
9:         **if** (13) **then** $(\mathbf{x}_{k+1}, \mathbf{z}_{k+1}) \leftarrow (\mathbf{v}_{k+1}^{\mathbf{x}}, \mathbf{v}_{k+1}^{\mathbf{z}})$ **else** $(\bar{\beta}, \bar{\alpha}) \leftarrow \rho(\bar{\beta}, \bar{\alpha})$ and **go to** 7
10:    **end if**
11:    **if** $\|\nabla \Phi_{\varepsilon_k}(\mathbf{x}_{k+1}, \mathbf{z}_{k+1})\| < \sigma \gamma \varepsilon_k$ **then** $\varepsilon_{k+1} = \gamma \varepsilon_k$ **else** $\varepsilon_{k+1} = \varepsilon_k$
12: **end for**
13: **return** $\mathbf{x}_{k+1}$

---

4   Network Architecture

The architecture of the proposed multi-phase neural networks follows LAMA exactly. Hence we also use LAMA to denote the networks as each phase corresponds to each iteration in Algorithm 1. The networks inherit all the convergence properties of LAMA such that the solution is stabilized. Moreover, the algorithm effectively leverages complementary information through the inter-domain connections shown in Fig. 1 to accurately estimate the missing data. The network is also memory efficient due to parameter sharing across all phases.

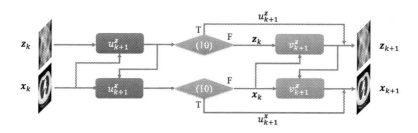

Fig. 1. Schematic illustration of one phase in LAMA, where (10) stands for the SDC.

5   Convergence Analysis

Since we deal with a nonconvex and nonsmooth optimization problem, we first need to introduce the following definitions based on the generalized derivatives.

Definition 1. *(Clarke subdifferential). Suppose that $f : \mathbb{R}^n \times \mathbb{R}^m \to (-\infty, \infty]$ is locally Lipschitz. The Clarke subdifferential of f at $(\mathbf{x}, \mathbf{z})$ is defined as*

$$\partial^c f(\mathbf{x}, \mathbf{z}) := \{(w_1, w_2) \in \mathbb{R}^n \times \mathbb{R}^m | \langle w_1, v_1 \rangle + \langle w_2, v_2 \rangle$$

$$\leq \limsup_{(z_1, z_2) \to (\mathbf{x}, \mathbf{z}), \, t \to 0_+} \frac{f(z_1 + tv_1, z_2 + tv_2) - f(z_1, z_2)}{t}, \quad \forall (v_1, v_2) \in \mathbb{R}^n \times \mathbb{R}^m\}.$$

where $\langle w_1, v_1 \rangle$ stands for the inner product in $\mathbb{R}^n$ and similarly for $\langle w_2, v_2 \rangle$.

Definition 2. *(Clarke stationary point) For a locally Lipschitz function f defined as in Definition 1, a point $X = (\mathbf{x}, \mathbf{z}) \in \mathbb{R}^n \times \mathbb{R}^m$ is called a Clarke stationary point of f, if $0 \in \partial f(X)$.*

We can have the following convergence result. All proofs are given in the supplementary material.

Theorem 1. *Let $\{Y_k = (\mathbf{x}_k, \mathbf{z}_k)\}$ be the sequence generated by the algorithm with arbitrary initial condition $Y_0 = (\mathbf{x}_0, \mathbf{z}_0)$, arbitrary $\varepsilon_0 > 0$ and $\varepsilon_{tol} = 0$. Let $\{\tilde{Y}_l\} =: (\mathbf{x}_{k_l+1}, \mathbf{z}_{k_l+1})\}$ be the subsequence, where the reduction criterion in the algorithm is met for $k = k_l$ and $l = 1, 2, \ldots$. Then $\{\tilde{Y}_l\}$ has at least one accumulation point, and each accumulation point is a Clarke stationary point.*

6   Experiments and Results

6.1   Initialization Network

The initialization $(\mathbf{x}_0, \mathbf{z}_0)$ is obtained by passing the sparse-view sinogram $\mathbf{s}$ defined in (1) through a CNN consisting of five residual blocks. Each block has four convolutions with 48 channels and kernel size $(3, 3)$, which are separated by ReLU. We train the CNN for 200 epochs using MSE, then use it to synthesize full-view sinograms $\mathbf{z}_0$ from $\mathbf{s}$. The initial image $\mathbf{x}_0$ is generated by applying FBP to $\mathbf{z}_0$. The resulting image-sinogram pairs are then provided as inputs to LAMA for the final reconstruction procedure. Note that the memory size of our method in Table 1 includes the parameters of the initialization network.

Table 1. Comparison of LAMA and existing methods on CT data with 64 and 128 views.

Data	Metric	Views	FBP [1]	DDNet [5]	LDA [16]	DuDoTrans [6]	Learn++ [15]	LAMA (Ours)
Mayo	PSNR	64	27.17 ± 1.11	35.70 ± 1.50	37.16 ± 1.33	37.90 ± 1.44	43.02 ± 2.08	$\mathbf{44.58 \pm 1.15}$
		128	33.28 ± 0.85	42.73 ± 1.08	43.00 ± 0.91	43.48 ± 1.04	49.77 ± 0.96	$\mathbf{50.01 \pm 0.69}$
	SSIM	64	$0.596 \pm 9\mathrm{e}{-4}$	$0.923 \pm 4\mathrm{e}{-5}$	$0.932 \pm 1\mathrm{e}{-4}$	$0.952 \pm 1.0\mathrm{e}{-4}$	$0.980 \pm 3\mathrm{e}{-5}$	$\mathbf{0.986 \pm 7\mathrm{e}{-6}}$
		128	$0.759 \pm 1\mathrm{e}{-3}$	$0.974 \pm 4\mathrm{e}{-5}$	$0.976 \pm 2\mathrm{e}{-5}$	$0.985 \pm 1\mathrm{e}{-5}$	$0.995 \pm 1\mathrm{e}{-6}$	$\mathbf{0.995 \pm 6\mathrm{e}{-7}}$
NBIA	PSNR	64	25.72 ± 1.93	35.59 ± 2.76	34.31 ± 2.20	35.53 ± 2.63	38.53 ± 3.41	$\mathbf{41.40 \pm 3.54}$
		128	31.86 ± 1.27	40.23 ± 1.98	40.26 ± 2.57	40.67 ± 2.84	43.35 ± 4.02	$\mathbf{45.20 \pm 4.23}$
	SSIM	64	$0.592 \pm 2\mathrm{e}{-3}$	$0.920 \pm 3\mathrm{e}{-4}$	$0.896 \pm 4\mathrm{e}{-4}$	$0.938 \pm 2\mathrm{e}{-4}$	$0.956 \pm 2\mathrm{e}{-4}$	$\mathbf{0.976 \pm 8\mathrm{e}{-5}}$
		128	$0.743 \pm 2\mathrm{e}{-3}$	$0.961 \pm 1\mathrm{e}{-4}$	$0.963 \pm 1\mathrm{e}{-4}$	$0.976 \pm 6\mathrm{e}{-5}$	$0.983 \pm 5\mathrm{e}{-5}$	$\mathbf{0.988 \pm 3\mathrm{e}{-5}}$
N/A	param	N/A	N/A	6e5	**6e4**	8e6	6e6	3e5

6.2   Experiment Setup

Our algorithm is evaluated on the *"2016 NIH-AAPM-Mayo Clinic Low-Dose CT Grand Challenge"* and the National Biomedical Imaging Archive (NBIA) datasets. We randomly select 500 and 200 image-sinogram pairs from AAPM-Mayo and NBIA, respectively, with 80% for training and 20% for testing. We evaluate algorithms using the peak signal-to-noise ratio (PSNR), structural similarity (SSIM), and the number of network parameters. The sinograms have 512 detector elements, each with 1024 evenly distributed projection views. The sinograms are downsampled into 64 or 128 views while the image size is 256×256, and we simulate projections and back-projections in fan-beam geometry using distance-driven algorithms [28,29] implemented in a PyTorch-based library CTLIB [30]. Given N training data pairs $\{(\mathbf{s}^{(i)}, \hat{\mathbf{x}}^{(i)})\}_{i=1}^{N}$, the loss function for training the regularization networks is defined as:

$$\mathcal{L}(\Theta) = \frac{1}{N} \sum_{i=1}^{N} \left\| \mathbf{x}_{k+1}^{(i)} - \hat{\mathbf{x}}^{(i)} \right\|^2 + \left\| \mathbf{z}_{k+1}^{(i)} - \mathbf{A}\hat{\mathbf{x}}^{(i)} \right\|^2 + \mu \left(1 - \mathrm{SSIM}\big(\mathbf{x}_{k+1}^{(i)}, \hat{\mathbf{x}}^{(i)}\big) \right),$$

(14)

where μ is the weight for SSIM loss set as 0.01 for all experiments, $\hat{\mathbf{x}}^{(i)}$ is ground truth image, and final reconstructions are $(\mathbf{x}_{k+1}^{(i)}, \mathbf{z}_{k+1}^{(i)}) := \mathrm{LAMA}(\mathbf{x}_0^{(i)}, \mathbf{z}_0^{(i)})$.

We use the Adam optimizer with learning rates of $1e-4$ and $6e-5$ for the image and sinogram networks, respectively, and train them with a warm-up approach. The training starts with three phases for 300 epochs, then adding two phases for 200 epochs each time until the number of phases reaches 15. The algorithm is implemented in Python using the PyTorch framework. Our experiments were run on a Linux server with an NVIDIA A100 Tensor Core GPU.

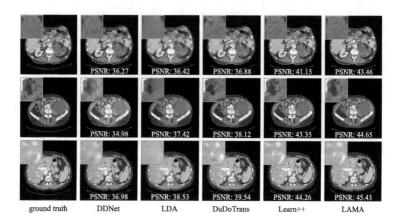

Fig. 2. Visual comparison for AAPM-Mayo dataset using 64-view sinograms.

6.3   Numerical and Visual Results

We perform an ablation study to compare the reconstruction quality of LAMA and BCD defined in (11), (12) versus the number of views and phases. Figure 3 illustrates that 15 phases strike a favorable balance between accuracy and computation. The residual architecture (9) introduced in LAMA is also proven to be more effective than solely applying BCD for both datasets. As illustrated in Sect. 5, the algorithm is also equipped with the added advantage of retaining convergence guarantees.

We evaluate LAMA by applying the pipeline described in Sect. 6.2 to sparse-view sinograms from the test set and compare with state-of-the-art methods

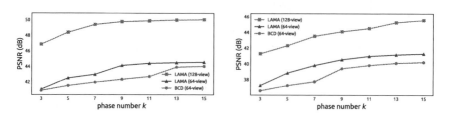

Fig. 3. PSNR of reconstructions obtained by LAMA or BCD over phase number k using 64-view or 128-view sinograms. *Left*: AAPM-Mayo. *Right*: NBIA.

where the numerical results are presented in Table 1. Our method achieves superior results regarding PSNR and SSIM scores while having the second-lowest number of network parameters. The numerical results indicate the robustness and generalization ability of our approach. Additionally, we demonstrate the effectiveness of our method in preserving structural details while removing noise and artifacts through Fig. 2. More visual results are provided in the supplementary materials. Overall, our approach significantly outperforms state-of-the-art methods, as demonstrated by both numerical and visual evaluations.

7   Conclusion

We propose a novel, interpretable dual-domain sparse-view CT image reconstruction algorithm LAMA. It is a variational model with composite objectives and solves the nonsmooth and nonconvex optimization problem with convergence guarantees. By introducing learnable regularizations, our method effectively suppresses noise and artifacts while preserving structural details in the reconstructed images. The LAMA algorithm leverages complementary information from both domains to estimate missing information and improve reconstruction quality in each iteration. Our experiments demonstrate that LAMA outperforms existing methods while maintaining favorable memory efficiency.

References

1. Kak, A.C, Slaney, M.: Principles of Computerized Tomographic Imaging. Society For Industrial And Applied Mathematics (2001)
2. Rudin, L.I., Osher, S., Fatemi, E.: Nonlinear total variation based noise removal algorithms. Physica D **60**(1), 259–268 (1992)
3. LaRoque, S.J., Sidky, E.Y., Pan, X.: Accurate image reconstruction from few-view and limited-angle data in diffraction tomography. J. Opt. Soc. Am. A **25**(7), 1772 (2008)
4. Kim, H., Chen, J., Wang, A., Chuang, C., Held, M., Pouliot, J.: Non-local total-variation (NLTV) minimization combined with reweighted l1-norm for compressed sensing CT reconstruction. Phys. Med. Biol. **61**(18), 6878 (2016)
5. Zhang, Z., Xiaokun Liang, X., Dong, Y.X., Cao, G.: A sparse-view CT reconstruction method based on combination of DenseNet and deconvolution. IEEE Trans. Med. Imaging **37**(6), 1407–1417 (2018)
6. Wang, C., Shang, K., Zhang, H., Li, Q., Hui, Y., Kevin Zhou, S.: DuDoTrans: dual-domain transformer provides more attention for sinogram restoration in sparse-view CT reconstruction (2021)
7. Lee, H., Lee, J., Kim, H., Cho, B., Cho, S.: Deep-neural-network-based sinogram synthesis for sparse-view CT image reconstruction. IEEE Trans. Radiat. Plasma Med. Sci. **3**(2), 109–119 (2019)
8. Weiwen, W., Dianlin, H., Niu, C., Hengyong, Yu., Vardhanabhuti, V., Wang, G.: Drone: dual-domain residual-based optimization network for sparse-view CT reconstruction. IEEE Trans. Med. Imaging **40**(11), 3002–3014 (2021)
9. Jin, K.H., McCann, M.T., Froustey, E., Unser, M.: Deep convolutional neural network for inverse problems in imaging. IEEE Trans. Image Process. **26**(9), 4509–4522 (2017)

10. Chen, H., et al.: Learn: learned experts' assessment-based reconstruction network for sparse-data CT. IEEE Trans. Med. Imaging **37**(6), 1333–1347 (2018)
11. Zhang, J., Yining, H., Yang, J., Chen, Y., Coatrieux, J.-L., Luo, L.: Sparse-view X-ray CT reconstruction with gamma regularization. Neurocomputing **230**, 251–269 (2017)
12. Chen, H., et al.: Low-dose CT with a residual encoder-decoder convolutional neural network. IEEE Trans. Med. Imaging **36**(12), 2524–2535 (2017)
13. Zhang, J., Ghanem, B.: Ista-Net: interpretable optimization-inspired deep network for image compressive sensing. In: 2018 IEEE/CVF Conference on Computer Vision and Pattern Recognition, pp. 1828–1837 (2018)
14. Monga, V., Li, Y., Eldar, Y.C.: Algorithm unrolling: interpretable, efficient deep learning for signal and image processing. IEEE Signal Process. Mag. **38**(2), 18–44 (2021)
15. Zhang, Y., et al.: Learn++: recurrent dual-domain reconstruction network for compressed sensing CT. IEEE Trans. Radiat. Plasma Med. Sci. **7**, 132–142 (2020)
16. Chen, Y., Liu, H., Ye, X., Zhang, Q.: Learnable descent algorithm for nonsmooth nonconvex image reconstruction. SIAM J. Imag. Sci. **14**(4), 1532–1564 (2021)
17. Xia, W., Yang, Z., Zhou, Q., Lu, Z., Wang, Z., Zhang, Y.: A Transformer-Based Iterative Reconstruction Model for Sparse-View CT Reconstruction. In: Wang, L., Dou, Q., Fletcher, P.T., Speidel, S., Li, S. (eds.) Medical Image Computing and Computer Assisted Intervention – MICCAI 2022. MICCAI 2022. Lecture Notes in Computer Science, vol. 13436, pp. 790–800. Springer, Cham (2022). https://doi.org/10.1007/978-3-031-16446-0_75
18. Ge, R., et al.: DDPNet: a novel dual-domain parallel network for low-dose CT reconstruction. In: Wang, L., Dou, Q., Fletcher, P.T., Speidel, S., Li, S. (eds.) Medical Image Computing and Computer Assisted Intervention – MICCAI 2022. MICCAI 2022. Lecture Notes in Computer Science, vol. 13436, pp. 748–757. Springer, Cham (2022). https://doi.org/10.1007/978-3-031-16446-0_71
19. Zhang, Q., Alvandipour, M., Xia, W., Zhang, Y., Ye, X., Chen, Y.: Provably convergent learned inexact descent algorithm for low-dose CT reconstruction (2021)
20. Bian, W., Zhang, Q., Ye, X., Chen, Y.: A learnable variational model for joint multimodal MRI reconstruction and synthesis. Lect. Notes Comput. Sci. **13436**, 354–364 (2022)
21. Pham, N.H., Nguyen, L.M., Phan, D.T., Tran-Dinh, Q.: ProxSARAH: an efficient algorithmic framework for stochastic composite nonconvex optimization. J. Mach. Learn. Res. **21**(110), 1–48 (2020)
22. Bolte, J., Sabach, S., Teboulle, M.: Proximal alternating linearized minimization for nonconvex and nonsmooth problems. Math. Program. **146** (2013)
23. Pock, T., Sabach, S.: Inertial proximal alternating linearized minimization (iPALM) for nonconvex and nonsmooth problems. SIAM J. Imag. Sci. **9**(4), 1756–1787 (2016)
24. Driggs, D., Tang, J., Liang, J., Davies, M., Schönlieb, C.-B.: A stochastic proximal alternating minimization for nonsmooth and nonconvex optimization. SIAM J. Imag. Sci. **14**(4), 1932–1970 (2021)
25. Yang, Y., Pesavento, M., Luo, Z.-Q., Ottersten, B.: Inexact block coordinate descent algorithms for nonsmooth nonconvex optimization. IEEE Trans. Signal Process. **68**, 947–961 (2020)
26. He, K., Zhang, X., Ren, S., Sun, J.: Deep residual learning for image recognition (2015)
27. Nesterov, Yu.: Smooth minimization of non-smooth functions. Math. Program. **103**(1), 127–152 (2004)

28. De Man, B., Basu, S.: Distance-driven projection and backprojection. In: 2002 IEEE Nuclear Science Symposium Conference Record, vol. 3, pp. 1477–1480 (2002)
29. De Man, B., Basu, S.: Distance-driven projection and backprojection in three dimensions. Phys. Med. Biol. **49**(11), 2463–2475 (2004)
30. Xia, W., et al.: Magic: manifold and graph integrative convolutional network for low-dose CT reconstruction. IEEE Trans. Med. Imaging **40**, 3459–3472 (2021)

TriDo-Former: A Triple-Domain Transformer for Direct PET Reconstruction from Low-Dose Sinograms

Jiaqi Cui[1], Pinxian Zeng[1], Xinyi Zeng[1], Peng Wang[1], Xi Wu[2], Jiliu Zhou[1,2], Yan Wang[1(✉)], and Dinggang Shen[3,4(✉)]

[1] School of Computer Science, Sichuan University, Chengdu, China
wangyanscu@hotmail.com
[2] School of Computer Science, Chengdu University of Information Technology, Chengdu, China
[3] School of Biomedical Engineering, ShanghaiTech University, Shanghai, China
dinggang.shen@gmail.com
[4] Department of Research and Development, Shanghai United Imaging Intelligence Co., Ltd., Shanghai, China

Abstract. To obtain high-quality positron emission tomography (PET) images while minimizing radiation exposure, various methods have been proposed for reconstructing standard-dose PET (SPET) images from low-dose PET (LPET) sinograms directly. However, current methods often neglect boundaries during sinogram-to-image reconstruction, resulting in high-frequency distortion in the frequency domain and diminished or fuzzy edges in the reconstructed images. Furthermore, the convolutional architectures, which are commonly used, lack the ability to model long-range non-local interactions, potentially leading to inaccurate representations of global structures. To alleviate these problems, in this paper, we propose a transformer-based model that unites triple domains of sinogram, image, and frequency for direct PET reconstruction, namely TriDo-Former. Specifically, the TriDo-Former consists of two cascaded networks, i.e., a sinogram enhancement transformer (SE-Former) for denoising the input LPET sinograms and a spatial-spectral reconstruction transformer (SSR-Former) for reconstructing SPET images from the denoised sinograms. Different from the vanilla transformer that splits an image into 2D patches, based specifically on the PET imaging mechanism, our SE-Former divides the sinogram into 1D projection view angles to maintain its inner-structure while denoising, preventing the noise in the sinogram from prorogating into the image domain. Moreover, to mitigate high-frequency distortion and improve reconstruction details, we integrate global frequency parsers (GFPs) into SSR-Former. The GFP serves as a learnable frequency filter that globally adjusts the frequency components in the frequency domain, enforcing the network to restore high-frequency details resembling real SPET images. Validations on a clinical dataset demonstrate that our TriDo-Former outperforms the state-of-the-art methods qualitatively and quantitatively.

Supplementary Information The online version contains supplementary material available at https://doi.org/10.1007/978-3-031-43999-5_18.

Keywords: Positron Emission Tomography (PET) · Triple-Domain · Vision Transformer · Global Frequency Parser · Direct Reconstruction

1 Introduction

As an in vivo nuclear medical imaging technique, positron emission tomography (PET) enables the visualization and quantification of molecular-level activity and has been extensively applied in hospitals for disease diagnosis and intervention [1, 2]. In clinic, to ensure that more diagnostic information can be retrieved from PET images, physicians prefer standard-dose PET scanning which is obtained by injecting standard-dose radioactive tracers into human bodies. However, the use of radioactive tracers inevitably induces potential radiation hazards. On the other hand, reducing the tracer dose during the PET scanning will introduce unintended noise, thus leading to degraded image quality with limited diagnostic information. To tackle this clinical dilemma, it is of high interest to reconstruct standard-dose PET (SPET) images from the corresponding low-dose PET (LPET) data (i.e., sinograms or images).

In the past decade, deep learning has demonstrated its promising potential in the field of medical images [3–6]. Along the research direction of PET reconstruction, most efforts have been devoted to indirect reconstruction methods [7–16] which leverage the LPET images pre-reconstructed from the original projection data (i.e., LPET sinograms) as the starting point to estimate SPET images. For example, inspired by the preeminent performance of generative adversarial network (GAN) in computer vision [17, 18], Wang *et al.* [9] proposed a 3D conditional generative adversarial network (3D-cGAN) to convert LPET images to SPET images. However, beginning from the pre-reconstructed LPET images rather than the original LPET sinograms, these indirect methods may lose or blur details such as edges and small-size organs in the pre-reconstruction process, leading to unstable and compromised performance.

To remedy the above limitation, several studies focus on the more challenging direct reconstruction methods [19–27] which complete the reconstruction from the original sinogram domain (i.e., LPET sinograms) to the image domain (i.e., SPET images). Particularly, Haggstrom *et al.* [19] proposed DeepPET, employing a convolutional neural network (CNN)-based encoder-decoder network to reconstruct SPET images from LPET sinograms. Although these direct methods achieve excellent performance, they still have the following limitations. First, due to the lack of consideration for the boundaries, the reconstruction from the sinogram domain to the image domain often leads to distortion of the reconstructed image in the high-frequency part of the frequency domain, which is manifested as blurred edges. Second, current networks ubiquitously employ CNN-based architecture which is limited in modeling long-range semantic dependencies in data. Lacking such non-local contextual information, the reconstructed images may suffer from missing or inaccurate global structure.

In this paper, to resolve the first limitation above, we propose to represent the reconstructed SPET images in the frequency domain, then encourage them to resemble the corresponding real SPET images in the high-frequency part. As for the second limitation, we draw inspiration from the remarkable progress of vision transformer [28] in medical image analysis [29, 30]. Owing to the intrinsic self-attention mechanism, the transformer can easily correlate distant regions within the data and capture non-local information. Hence, the transformer architecture is considered in our work.

Overall, we propose an end-to-end transformer model dubbed TriDo-Former that unites triple domains of sinogram, image, and frequency to directly reconstruct the clinically acceptable SPET images from LPET sinograms. Specifically, our TriDo-Former is comprised of two cascaded transformers, i.e., a sinogram enhancement transformer (SE-Former) and a spatial-spectral reconstruction transformer (SSR-Former). The SE-Former aims to predict denoised SPET-like sinograms from LPET sinograms, so as to prevent the noise in sinograms from propagating into the image domain. Given that each row of the sinogram is essentially the projection at a certain imaging views angle, dividing it into 2D patches and feeding them directly into the transformer will inevitably break the continuity of each projection view. Therefore, to retain the inner-structure of sinograms and filter the noise, we split a sinogram by rows and obtain a set of 1D sequences of different imaging view angles. Then, the relations between view angles are modeled via the self-attention mechanism in the SE-Former. Note that the SE-Former is designed specifically for the sinogram domain of LPET to effectively reduce noise based on the imaging mechanisms of PET. The denoised sinograms can serve as a better basis for the subsequent sinogram-to-image reconstruction. The SSR-Former is designed to reconstruct SPET images from the denoised sinograms. In pursuit of better image quality, we construct the SSR-Former by adopting the powerful swin transformer [31] as the backbone. To compensate for the easily lost high-frequency details, we propose a global frequency parser (GFP) and inject it into the SSR-Former. The GFP acts as a learnable frequency filter to globally modify the components of specific frequencies of the frequency domain, forcing the network to learn accurate high-frequency details and produce construction results with shaper boundaries. Through the above triple-domain supervision, our TriDo-Former exhausts the model representation capability, thereby achieving better reconstructions.

The contributions of our proposed method can be described as follows. (1) To fully exploit the triple domains of sinogram, image, and frequency while capturing global context, we propose a novel triple-domain transformer to directly reconstruct SPET images from LPET sinograms. *To our knowledge, we are the first to leverage both triple-domain knowledge and transformer for PET reconstruction.* (2) We develop a sinogram enhancement transformer (SE-Former) that is tailored for the sinogram domain of LPET to suppress the noise while maintaining the inner-structure, thereby preventing the noise in sinograms from propagating into the image domain during the sinogram-to-image reconstruction. (3) To reconstruct high-quality PET images with clear-cut details, we design a spatial-spectral transformer (SS-Former) incorporated with the global frequency parser (GFP) which globally calibrates the frequency components in the frequency domain for recovering high-frequency details. (4) Experimental results demonstrate the superiority of our method both qualitatively and quantitatively, compared with other state-of-the-art methods.

2 Methodology

The overall architecture of our proposed TriDo-Former is depicted in Fig. 1, which consists of two cascaded sub-networks, i.e., a sinogram enhancement transformer (SE-Former) and a spatial-spectral reconstruction transformer (SSR-Former). Overall, taking the LPET sinograms as input, the SE-Former first predicts the denoised SPET-like sinograms which are then sent to SSR-Former to reconstruct the estimated PET (denoted as EPET) images. A detailed description is given in the following sub-sections.

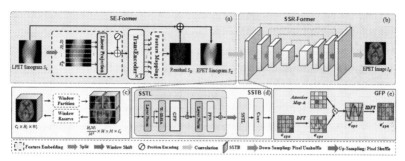

Fig. 1. Overview of the proposed TriDo-Former.

2.1 Sinogram Enhancement Transformer (SE-Former)

As illustrated in Fig. 1(a), the SE-Former which is responsible for denoising in the input LPET sinograms consists of three parts, i.e., a feature embedding module, transformer encoder (TransEncoder) blocks, and a feature mapping module. Given that each row of sinogram is the 1D projection at an imaging view angle, we first divide the LPET sinograms by rows and perform linear projection in the feature embedding module to obtain a set of 1D sequences, each contains consistent information of a certain view angle. Then, we perform self-attention in the TransEncoder blocks to model the interrelations between projection view angles, enabling the network to better model the general characteristics under different imaging views which is crucial for sinogram denoising. After that, the feature mapping module predicts the residual between the LPET and SPET sinograms which is finally added to the input LPET sinograms to generate the EPET sinograms as the output of SE-Former. We argue that the introduction of residual learning allows the SE-Former to focus only on learning the difference between LPET and SPET sinograms, facilitating faster convergence.

Feature Embedding: We denote the input LPET sinogram as $S_L \in \mathbb{R}^{C_s \times H_s \times W_s}$, where H_s, W_s are the height, width and C_s is the channel dimension. As each row of sinogram is a projection view angle, the projection at the i-th ($i = 1, 2, \ldots H_s$) row can be defined as $s_L^i \in \mathbb{R}^{C_s \times W_s}$. Therefore, by splitting the sinogram by rows, we obtain a set of 1D sequence data $S_L^* = \{s_L^i\}_{i=1}^{H_s} \in \mathbb{R}^{H_s \times D}$, where H_s is the number of projection view angles and $D = C_s \times W_s$ equals to the pixel number in each sequence data. Then, S_L^* is linearly projected to sequence $\tilde{S}_L^* \in \mathbb{R}^{H_s \times d}$, where d is the output dimension of the projection. To maintain the position information of different view angles, we introduce a learnable position embedding $S_{pos} = \{s_{pos}^i\}_{i=1}^{H_s} \in \mathbb{R}^{H_s \times d}$ and fuse it with $\tilde{S}_L^*$ by element-wise addition, thus creating the input feature embedding $F_0 = S_{pos} + \tilde{S}_L^*$ which is further sent to T stacked TransEncoder blocks to model global characteristics between view angles.

TransEncoder: Following the standard transformer architecture [28], each TransEncoder block contains a multi-head self-attention (MSA) module and a feed forward network (FFN) respectively accompanied by layer normalization (LN). For j-th ($j = 1, 2, \ldots, T$) TransEncoder block, the calculation process can be formulated as:

$$F_j = F_{j-1} + MSA\left(LN\left(F_{j-1}\right)\right) + FFN(LN(F_{j-1} + MSA(LN(F_{j-1})))), \qquad (1)$$

where F_j denotes the output of j-th TransEncoder block. After applying T identical TransEncoder blocks, the non-local relationship between projections at different view angles is accurately preserved in the output sequence $F_T \in \mathbb{R}^{H_s \times d}$.

Feature Mapping: The feature mapping module is designed for projecting the sequence data back to the sinogram. Concretely, F_T is first reshaped to $\mathbb{R}^{C\prime \times H_s \times W_s}$ ($C\prime = \frac{d}{W_s}$) and then fed into a linear projection layer to reduce the channel dimension from $C\prime$ to C_s. Through these operations, the residual sinogram $S_R \in \mathbb{R}^{C_s \times H_s \times W_s}$ of the same dimension as S_L, is obtained. Finally, following the spirit of residual learning, S_R is directly added to the input S_L to produce the output of SE-Former, i.e., the predicted denoised sinogram $S_E \in \mathbb{R}^{C_s \times H_s \times W_s}$.

2.2 Spatial-Spectral Reconstruction Transformer (SSR-Former)

The SSR-Former is designed to reconstruct the denoised sinogram obtained from the SE-Former to the corresponding SPET images. As depicted in Fig. 1 (b), SSR-Former adopts a 4-level U-shaped structure, where each level is formed by a spatial-spectral transformer block (SSTB). Furthermore, each SSTB contains two spatial-spectral transformer layers (SSTLs) and a convolution layer for both global and local feature extraction. Meanwhile, a 3×3 convolution is placed as a projection layer at the beginning and the end of the network. For detailed reconstruction and invertibility of sampling, we employ the pixel-unshuffle and pixel-shuffle operators for down-sampling and up-sampling. In addition, skip connections are applied for multi-level feature aggregation.

Spatial-Spectral Transformer Layer (SSTL): As shown in Fig. 1(d), an SSTL consists of a window-based spatial multi-head self-attention (W-SMSA) followed by FFN and LN. Following swin transformer [31], a window shift operation is conducted between the two SSTLs in each SSTB for cross-window information interactions. Moreover, to capture the high-frequency details which can be easily lost, we devise global frequency parsers (GFPs) that encourage the model to recover the high-frequency component of the frequency domain through the global adjustment of specific frequencies. Generally, the W-SMSA is leveraged to guarantee the essential global context in the reconstructed PET images, while GFP is added to enrich the high-frequency boundary details. The calculations of the core W-SMSA and GFP are described as follows.

Window-based Spatial Multi-Head Self-Attention (W-SMSA): Denoting the input feature embedding of certain W-SMSA as $e_{in} \in \mathbb{R}^{C_I \times H_I \times W_I}$, where H_I, W_I and C_I represent the height, width and channel dimension, respectively. As depicted in Fig. 1(c), a window partition operation is first conducted in spatial dimension with a window size of M. Thus, the whole input features are divided into N ($N = \frac{H_I \times W_I}{M^2}$) non-overlapping patches $e_{in}^{*} = \{e_{in}^{m}\}_{m=1}^{N}$. Then, a regular spatial self-attention is performed separately for each window after partition. After that, the output patches are gathered through the window reverse operation to obtain the spatial representative feature $e_{spa} \in \mathbb{R}^{C_I \times H_I \times W_I}$.

Global Frequency Parser (GFP): After passing the W-SMSA, the feature e_{spa} are already spatially representative, but still lack accurate spectral representations in the frequency domain. Hence, we propose a GFP module to rectify the high-frequency component in the frequency domain. As illustrated in Fig. 1(e), the GFP module is comprised of a 2D discrete Fourier transform (DFT), an element-wise multiplication between the frequency feature and the learnable global filter, and a 2D inverse discrete Fourier transform (IDFT). Our GFP can be regarded as a learnable version of frequency filters. The main idea is to learn a parameterized attentive map applying on the frequency domain features. Specifically, we first convert the spatial feature e_{spa} to the frequency domain via 2D DFT, obtaining the spectral feature $e_{spe} = DFT(e_{spa})$. Then, we modulate the frequency components of e_{spe} by multiplying a learnable parameterized attentive map $A \in \mathbb{R}^{C_I \times H_I \times W_I}$ to e_{spe}, which can be formulated as:

$$e_{spe\prime} = A \cdot e_{spe}, \tag{2}$$

The parameterized attentive map A can adaptively adjust the frequency components of the frequency domain and compel the network to restore the high-frequency part to resemble that of the supervised signal, i.e., the corresponding real SPET images (ground truth), in the training process. Finally, we reverse $e_{spe\prime}$ back to the image domain by adopting 2D IDFT, thus obtaining the optimized feature $e_{spa\prime} = DFT(e_{spe\prime})$. In this manner, more high-frequency details are preserved for generating shaper constructions.

2.3 Objective Function

The objective function for our TriDo-Former is comprised of two aspects: 1) a sinogram domain loss L_{sino} and 2) an image domain loss L_{img}.

The sinogram domain loss aims to narrow the gap between the real SPET sinograms S_S and the EPET sinograms S_E that are denoised from the input LPET sinograms. Considering the critical influence of sinogram quality, we apply the L2 loss to increase the error punishment, thus forcing a more accurate prediction. It can be expressed as:

$$L_{sino} = E_{S_S, S_E \sim p_{data}(S_S, S_E)} ||S_S - S_E||_2, \tag{3}$$

For the image domain loss, the L1 loss is leveraged to minimize the error between the SPET images I_S and the EPET images I_E while encouraging less blurring, which can be defined as:

$$L_{img} = E_{I_S, I_E \sim p_{data}(I_S, I_E)} ||I_S - I_E||_1, \tag{4}$$

Overall, the final objective function is formulated by the weighted sum of the above losses, which is defined as:

$$L_{total} = L_{sino} + \lambda L_{img}. \tag{5}$$

where λ is the hyper-parameters to balance these two terms.

2.4   Details of Implementation

Our network is implemented by Pytorch framework and trained on an NVIDIA GeForce GTX 3090 with 24 GB memory. The whole network is trained end-to-end for 150 epochs in total using Adam optimizer with the batch size of 4. The learning rate is initialized to 4e-4 for the first 50 epochs and decays linearly to 0 for the remaining 100 epochs. The number T of the TransEncoder in SE-Former is set to 2 and the window size M is set to 4 in the W-SMSA of the SSR-Former. The weighting coefficient λ in Eq. (6) is empirically set as 10.

3   Experiments and Results

Datasets: We train and validate our proposed TriDo-Former on a real human brain dataset including 8 normal control (NC) subjects and 8 mild cognitive impairment (MCI) subjects. All PET scans are acquired by a Siemens Biograph mMR system housed in Biomedical Research Imaging Center. A standard dose of 18F-Flurodeoxyglucose ($[^{18}F]$ FDG) was administered. According to standard protocol, SPET sinograms were acquired in a 12-minute period within 60-minute of radioactive tracer injection, while LPET sinograms were obtained consecutively in a 3-min shortened acquisition time to simulate the acquisition at a quarter of the standard dose. The SPET images which are utilized as the ground truth in this study were reconstructed from the corresponding SPET sinograms using the traditional OSEM algorithm [32].

Experimental Settings: Due to the limited computational resources, we slice each 3D scan of size $128 \times 128 \times 128$ into 128 2D slices with a size of 128×128. The Leave-One-Out Cross-Validation (LOOCV) strategy is applied to enhance the stability of the model with limited samples. To evaluate the performance, we adopt three typical quantitative evaluation metrics including peak signal-to-noise (PSNR), structural similarity index (SSIM), and normalized mean squared error (NMSE). Note that, we restack the 2D slices into complete 3D PET scans for evaluation.

Comparative Experiments: We compare our TriDo-Former with four direct reconstruction methods, including (1) OSEM [32] (applied on the input LPET sinograms, serving as the lower bound), (2) DeepPET [19], (3) Sino-cGAN [23], and (4) LCPR-Net [24] as well as one indirect reconstruction methods, i.e., (5) 3D-cGAN [9]. The comparison results are given in Table 1, from which we can see that our TriDo-Former achieves the best results among all the evaluation criteria. Compared with the current state-of-the-art LCPR-Net, our proposed method still enhances the PSNR and SSIM by 0.599 dB and 0.002 for NC subjects, and 0.681 dB and 0.002 for MCI subjects, respectively. Moreover, our model also has minimal parameters and GLOPs of 38 M and 16.05, respectively, demonstrating its speed and feasibility in clinical applications. We also visualize the results of our method and the compared approaches in Fig. 2, where the differences in global structure are highlighted with circles and boxes while the differences in edge details are marked by arrows. As can be seen, compared with other methods which have inaccurate structure and diminished edges, our TriDo-Former yields the best visual effect with minimal error in both global structure and edge details.

Table 1. Quantitative comparison with five PET reconstruction methods in terms of PSNR, SSIM, and NMSE. The best performance is marked as bold.

Method	NC subject			MCI subject			Params	GFLOPs
	PSNR	SSIM	NMSE	PSNR	SSIM	NMSE		
OSEM [32]	20.684	0.979	0.0530	21.541	0.977	0.0580	-	-
DeepPET [19]	23.991	0.982	0.0248	24.125	0.982	0.0272	60M	49.20
Sino-cGAN [23]	24.191	0.985	0.0254	24.224	0.985	0.0269	39M	19.32
LCPR-Net [24]	24.313	0.985	0.0227	24.607	0.985	0.0257	77M	77.26
3D-cGAN [9]	24.024	0.983	0.0231	24.617	0.981	0.0256	127M	70.38
Proposed	**24.912**	**0.987**	**0.0203**	**25.288**	**0.987**	**0.0228**	**38M**	**16.05**

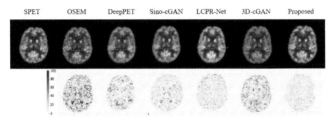

Fig. 2. Visual comparison of the reconstruction methods.

Evaluation on Clinical Diagnosis: To further prove the clinical value of our method, we further conduct an Alzheimer's disease diagnosis experiment as the downstream task. Specifically, a multi-layer CNN is firstly trained by real SPET images to distinguish between NC and MCI subjects with 90% accuracy. Then, we evaluate the PET images reconstructed by different methods on the trained classification model. Our insight is that, if the model can discriminate between NC and MCI subjects from the reconstructed images more accurately, the quality of the reconstructed images and SPET images (whose quality is preferred in clinical diagnosis) are closer. As shown in Fig. 3, the classification accuracy of our proposed method (i.e., 88.6%) is the closest to that of SPET images (i.e., 90.0%), indicating the huge clinical potential of our method in facilitating disease diagnosis.

Ablation Study: To verify the effectiveness of the key components of our TriDo-Former, we conduct the ablation studies with the following variants: (1) replacing SE-Former and SSR-Former with DnCNN [33] (the famous CNN-based denoising network) and vanilla U-Net (denoted as DnCNN + UNet), (2) replacing DnCNN with SE-Former

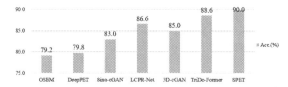

Fig. 3. Results of the clinical diagnosis of Alzheimer's disease (NC/MCI).

(denoted as SE-Former + UNet), (3) replacing the U-Net with our SSR-Former but removing GFP (denoted as Proposed w/o GFP), and (4) using the proposed TriDo-Former model (denoted as Proposed). According to the results in Table 2, the performance of our model progressively improves with the introduction of SE-Former and SSR-Former. Particularly, when we remove the GFP in SSR-Former, the performance largely decreases as the model fails to recover high-frequency details. Moreover, we conduct the clinical diagnosis experiment and the spectrum analysis to further prove the effectiveness of the GFP, and the results are included in supplementary material.

Table 2. Quantitative comparison with models constructed in the ablation study in terms of PSNR, SSIM, and NMSE.

Method	NC subjects			MCI subjects		
	PSNR	SSIM	NMSE	PSNR	SSIM	NMSE
DnCNN + UNet	23.872	0.981	0.0253	24.153	0.982	0.0266
SE-Former + UNet	24.177	0.982	0.0249	24.506	0.982	0.0257
Proposed w/o GFP	24.583	0.984	0.0235	24.892	0.984	0.0250
Proposed	**24.912**	**0.987**	**0.0203**	**25.288**	**0.987**	**0.0228**

4 Conclusion

In this paper, we innovatively propose a triple-domain transformer, named TriDo-Former, for directly reconstructing the high-quality PET images from LPET sinograms. Our model exploits the triple domains of sinogram, image, and frequency as well as the ability of the transformer in modeling long-range interactions, thus being able to reconstruct PET images with accurate global context and sufficient high-frequency details. Experimental results on the real human brain dataset have demonstrated the feasibility and superiority of our method, compared with the state-of-the-art PET reconstruction approaches.

Acknowledgement. This work is supported by the National Natural Science Foundation of China (NSFC 62071314), Sichuan Science and Technology Program 2023YFG0263, 2023NSFSC0497, 22YYJCYJ0086, and Opening Foundation of Agile and Intelligent Computing Key Laboratory of Sichuan Province.

References

1. Chen, W.: Clinical applications of PET in brain tumors. J. Nucl. Med. **48**(9), 1468–1481 (2007)
2. Wang, Y., Ma, G., An, L., et al.: Semi-supervised tripled dictionary learning for standard-dose PET image prediction using low-dose PET and multimodal MRI. IEEE Trans. Biomed. Eng. **64**(3), 569–579 (2016)
3. Zhou, T., Fu, H., Chen, G., et al.: Hi-net: hybrid-fusion network for multi-modal MR image synthesis. IEEE Trans. Med. Imaging **39**(9), 2772–2781 (2020)
4. Li, Y., Zhou, T., He, K., et al.: Multi-scale transformer network with edge-aware pre-training for cross-modality MR image synthesis. IEEE Trans. Med. Imaging (2023)
5. Wang, K., et al.: Tripled-uncertainty guided mean teacher model for semi-supervised medical image segmentation. In: de Bruijne, M., et al. (eds.) MICCAI 2021. LNCS, vol. 12902, pp. 450–460. Springer, Cham (2021). https://doi.org/10.1007/978-3-030-87196-3_42
6. Zhan, B., Xiao, J., Cao, C., et al.: Multi-constraint generative adversarial network for dose prediction in radiotherapy. Med. Image Anal. **77**, 102339 (2022)
7. Wang, Y., Zhang, P., Ma, g., et al: Predicting standard-dose PET image from low- dose PET and multimodal MR images using mapping-based sparse representation. Phys. Med. Biol. **61**(2), 791–812 (2016)
8. Spuhler, K., Serrano-Sosa, M., Cattell, R., et al.: Full-count PET recovery from low-count image using a dilated convolutional neural network. Med. Phys. **47**(10), 4928–4938 (2020)
9. Wang, Y., Yu, B., Wang, L., et al.: 3D conditional generative adversarial networks for high-quality PET image estimation at low dose. Neuroimage **174**, 550–562 (2018)
10. Wang, Y., Zhou, L., Yu, B., et al.: 3D auto-context-based locality adaptive multi-modality GANs for PET synthesis. IEEE Trans. Med. Imaging **38**(6), 1328–1339 (2018)
11. Wang, Y., Zhou, L., Wang, L., et al.: Locality adaptive multi-modality GANs for high-quality PET image synthesis. In: Frangi, A., et al. (eds.) MICCAI 2018, vol. 11070, pp. 329–337. Springer, Cham (2018)
12. Luo, Y., Wang, Y., Zu, C., et al.: 3D Transformer-GAN for high-quality PET reconstruction. In: de Bruijne, M., et al. (eds.) MICCAI 2021, vol. 12906, pp. 276–285. Springer, Cham (2021)
13. Luo, Y., Zhou, L., Zhan, B., et al.: Adaptive rectification based adversarial network with spectrum constraint for high-quality PET image synthesis. Med. Image Anal. **77**, 102335 (2022)
14. Fei, Y., Zu, C., Jiao, Z., et al.: Classification-aided high-quality PET image synthesis via bidirectional contrastive GAN with shared information maximization. In: Wang, L., et al. (eds.) MICCAI 2022, vol. 13436, pp. 527–537. Springer, Cham (2022)
15. Zeng, P., Zhou, L., Zu, C., et al.: 3D CVT-GAN: a 3D convolutional vision transformer-GAN for PET reconstruction. In: Wang, L., et al. (eds.) MICCAI 2022, vol. 13436, pp. 516–526. Springer, Cham (2022)
16. Jiang, C., Pan, Y., Cui, Z., et al: Reconstruction of standard-dose PET from low-dose PET via dual-frequency supervision and global aggregation module. In: Proceedings of the19th International Symposium on Biomedical Imaging Conference, pp. 1–5 (2022)
17. Cui, J., Jiao, Z., Wei, Z., et al.: CT-only radiotherapy: an exploratory study for automatic dose prediction on rectal cancer patients via deep adversarial network. Front. Oncol. **12**, 875661 (2022)
18. Li, H., Peng, X., Zeng, J., et al.: Explainable attention guided adversarial deep network for 3D radiotherapy dose distribution prediction. Knowl. Based Syst. **241**, 108324 (2022)
19. Häggström, I., Schmidtlein, C.R., et al.: DeepPET: A deep encoder-decoder network for directly solving the PET image reconstruction inverse problem. Med. Image Anal. **54**, 253–262 (2019)

20. Wang, B., Liu, H.: FBP-Net for direct reconstruction of dynamic PET images. Phys. Med. Biol. **65**(23), 235008 (2020)
21. Ma, R., Hu, J., Sari, H., et al.: An encoder-decoder network for direct image reconstruction on sinograms of a long axial field of view PET. Eur. J. Nucl. Med. Mol. Imaging **49**(13), 4464–4477 (2022)
22. Whiteley, W., Luk, W.K., et al.: DirectPET: full-size neural network PET reconstruction from sinogram data. J. Med. Imaging **7**(3), 32503 (2020)
23. Liu, Z., Ye, H., and Liu, H: Deep-learning-based framework for PET image reconstruction from sinogram domain. Appl. Sci. **12**(16), 8118 (2022)
24. Xue, H., Zhang, Q., Zou, S., et al.: LCPR-Net: low-count PET image reconstruction using the domain transform and cycle-consistent generative adversarial networks. Quant. Imaging Med. Surg. **11**(2), 749 (2021)
25. Feng, Q., Liu, H.: Rethinking PET image reconstruction: ultra-low-dose, sinogram and deep learning. In: Martel, A.L., et al. (eds.) MICCAI 2020, vol. 12267, pp. 783–792. Springer, Cham (2020)
26. Liu, Z., Chen, H., Liu, H.: Deep learning based framework for direct reconstruction of PET images. In: Shen, D., et al. (eds.) MICCAI 2019. LNCS, vol. 11766, pp. 48–56. Springer, Cham (2019). https://doi.org/10.1007/978-3-030-32248-9_6
27. Hu, R., Liu, H: TransEM: Residual swin-transformer based regularized PET image reconstruction. In: Wang, L., et al (eds.) MICCAI 2022, vol. 13434, pp. 184–193. Springer, Cham (2022)
28. Dosovitskiy, A., Beyer, L., Kolesnikov, A., et al.: An image is worth 16×16 words: transformers for image recognition at scale. In: Proceedings of the IEEE/CVF International Con-ference on Computer Vision. IEEE, Venice (2020)
29. Zhang, Z., Yu, L., Liang, X., et al.: TransCT: dual-path transformer for low dose computed tomography. In: de Bruijne, M., et al. (eds.) MICCAI 2021, vol. 12906, pp. 55–64. Springer, Cham (2021)
30. Zheng, H., Lin, Z., Zhou, Q., et al.: Multi-transSP: Multimodal transformer for survival prediction of nasopharyngeal carcinoma patients. In: Wang, L., et al. (eds.) MICCAI 2022, vol. 13437, pp. 234–243. Springer, Cham (2022)
31. Liu, Z., Lin, Y., Cao, Y., et al: Swin transformer: hierarchical vision transformer using shifted windows. In Proceedings of the IEEE/CVF International Conference on Computer Vision, pp. 10012–10022. IEEE, Montreal (2021)
32. Hudson, H., Larkin, R.: Accelerated image reconstruction using ordered subsets of projection data. IEEE Trans. Med. Imaging **13**, 601–609 (1994)
33. Zhang, K., Zuo, W., Chen, Y., Meng, D., Zhang, L.: Beyond a gaussian denoiser: residual learning of deep CNN for image denoising. IEEE Trans. Image Process. **26**(7), 3142-3155. (2017)

Computationally Efficient 3D MRI Reconstruction with Adaptive MLP

Eric Z. Chen[1], Chi Zhang[2], Xiao Chen[1], Yikang Liu[1], Terrence Chen[1], and Shanhui Sun[1(✉)]

[1] United Imaging Intelligence, Cambridge, MA, USA
shanhui.sun@uii-ai.com
[2] Department of Electrical and Computer Engineering, Center for Magnetic Resonance Research, University of Minnesota, Minneapolis, MN, USA

Abstract. Compared with 2D MRI, 3D MRI provides superior volumetric spatial resolution and signal-to-noise ratio. However, it is more challenging to reconstruct 3D MRI images. Current methods are mainly based on convolutional neural networks (CNN) with small kernels, which are difficult to scale up to have sufficient fitting power for 3D MRI reconstruction due to the large image size and GPU memory constraint. Furthermore, MRI reconstruction is a deconvolution problem, which demands long-distance information that is difficult to capture by CNNs with small convolution kernels. The multi-layer perceptron (MLP) can model such long-distance information, but it requires a fixed input size. In this paper, we proposed Recon3DMLP, a hybrid of CNN modules with small kernels for low-frequency reconstruction and adaptive MLP (dMLP) modules with large kernels to boost the high-frequency reconstruction, for 3D MRI reconstruction. We further utilized the circular shift operation based on MRI physics such that dMLP accepts arbitrary image size and can extract global information from the entire FOV. We also propose a GPU memory efficient data fidelity module that can reduce >50% memory. We compared Recon3DMLP with other CNN-based models on a high-resolution (HR) 3D MRI dataset. Recon3DMLP improves HR 3D reconstruction and outperforms several existing CNN-based models under similar GPU memory consumption, which demonstrates that Recon3DMLP is a practical solution for HR 3D MRI reconstruction.

Keywords: 3D MRI reconstruction · Deep learning · MLP

Contribution from Chi Zhang was carried out during his internship at United Imaging Intelligence, Cambridge, MA.

Supplementary Information The online version contains supplementary material available at https://doi.org/10.1007/978-3-031-43999-5_19.

1   Introduction

Compared with 2D MRI, 3D MRI has superior volumetric spatial resolution and signal-to-noise ratio. However, 3D MRI, especially high resolution (HR) 3D MRI (e.g., at least $1\,mm^3$ voxel size), often takes much longer acquisition time than 2D scans. Therefore, it is necessary to accelerate 3D MRI by acquiring sub-sampled k-space. However, it is more challenging to reconstruct HR 3D MRI images than 2D images. For example, HR 3D MRI data can be as large as $380{\times}294{\times}138{\times}64$, which is more than 100X larger than common 2D MRI data [13] (e.g., $320{\times}320{\times}1{\times}15$, hereafter data dimensions are defined as RO×PE×SPE×Coil, where RO stands for read-out, PE for phase-encoding, and SPE for slice-phase-encoding). Although deep learning (DL) based methods have shown superior reconstruction speed and image quality, they are constrained by GPU memory for 3D MRI reconstruction in the clinical setting.

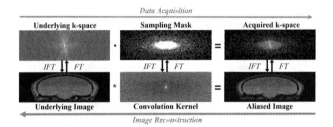

Fig. 1. Demonstration of k-space acquisition, which is equivalent to a convolution in the image domain, and reconstruction, which is a deconvolution process to recover the underlying image. The convolution kernel has the most energy at the center but spans the entire FOV, suggesting that global information is necessary for reconstruction. (Color figure online)

Due to the large 3D image size and computation constraint, the state-of-the-art methods for 2D MRI reconstruction [12,20] are not directly transferable to 3D MRI reconstruction. Instead of using 3D convolutions, [1] proposed a 2D CNN on the PE-SPE plane for 3D MRI reconstruction. [31] proposed to downsample the 3D volume and reconstruct the smaller 3D image, which is then restored to the original resolution by a super-resolution network. [3,23] used 3D CNN models to reconstruct each coil of 3D MRI data independently. [11] applied the gradient checkpointing technique to save the GPU memory during training. GLEAM [21] splits the network into modules and updates the gradient on each module independently, which reduces memory usage during training.

The previous works on 3D MRI reconstruction have several limitations. First, all these methods are based on CNN. In the context of 3D reconstruction, deep CNN networks require significant GPU memory and are difficult to scale. As a result, many models are designed to be relatively small to fit within available resources [1,3,23]. Given that a high-resolution 3D volume can contain over 100 million voxels, the model's fitting power is critical. Small models may lack the necessary fitting power, resulting in suboptimal performance in 3D MRI reconstruction. Second, due to network inductive bias, CNN prioritizes low-frequency information reconstruction and tends to generate smooth images [2,22]. Third, CNN has a limited receptive field due to highly localized convolutions using small kernels. The k-space sub-sampling is equivalent to convolving the underlying aliasing-free image using a kernel that covers the entire field of view (FOV) (orange arrow in Fig. 1). Therefore, the contribution of aliasing artifacts for a voxel comes from all other voxels globally in the sub-sampling directions. Then reconstruction is deconvolution and it is desirable to utilize the global information along the sub-sampled directions (green arrow in Fig. 1). Although convolution-based methods such as large kernels [17,29], dilation, deformable convolution [5] as well as attention-based methods such as Transformers [7,18] can enlarge the receptive field, it either only utilizes limited voxels within the FOV or may lead to massive computation [22]. Recently, multi-layer perceptron (MLP) based models have been proposed for various computer vision tasks [4,14,15,25–28,30]. MLP models have better fitting power and less inductive bias than CNN models [16]. MLP performs matrix multiplication instead of convolution, leading to enlarged receptive fields with lower memory and time cost than CNN and attention-based methods. However, MLP requires a fixed input image resolution and several solutions have been proposed [4,15,16,18]. Nevertheless, these methods were proposed for natural image processing and failed to exploit global information from the entire FOV. Img2ImgMixer [19] adapted MLP-Mixer [25] to 2D MRI reconstruction but on fixed-size images. AUTOMAP [32] employs MLP on whole k-space to learn the Fourier transform, which requires massive GPU memory and a fixed input size and thus is impractical even for 2D MRI reconstruction. Fourth, the methods to reduce GPU memory are designed to optimize gradient calculation for training, which is not beneficial for inference when deployed in clinical practice.

To tackle these problems, we proposed Recon3DMLP for 3D MRI reconstruction, a hybrid of CNN modules with small kernels for low-frequency reconstruction and adaptive MLP (dMLP) modules with large kernels to boost the high-frequency reconstruction. The dMLP improves the model fitting ability with almost the same GPU memory usage and a minor increase in computation time. We utilized the circular shift operation [18] based on MRI physics such that the proposed dMLP accepts arbitrary image size and can extract global information from the entire FOV. Furthermore, we propose a memory-efficient data fidelity (eDF) module that can reduce >50% memory. We also applied gradient checkpointing, RO cropping, and half-precision (FP16) to save GPU memory. We compared Recon3DMLP with other CNN-based models on an HR 3D multi-coil MRI dataset. The proposed dMLP improves HR 3D reconstruction and outperforms several existing CNN-based strategies under similar GPU memory consumption, which demonstrate that Recon3DMLP is a practical solution for HR 3D MRI reconstruction.

2   Method

2.1   Recon3DMLP for 3D MRI Reconstruction

The MRI reconstruction problem can be solved as

$$x = \arg\min_x ||y - MFSx||_2^2 + \lambda||x - g_\theta(x_u)||_2^2, \tag{1}$$

where y is the acquired measurements, x_u is the under-sampled image, M and S are the sampling mask and coil sensitivities, F denotes FFT and λ is a weighting scalar. g_θ is a neural network with the data fidelity (DF) module [24].

The proposed Recon3DMLP adapts the memory-friendly cascaded structure. Previous work has shown that convolutions with small kernels are essential for low-level tasks [27]. Therefore, we added the dMLP module with large kernels after each 3D CNN with small kernels (k = 3) to increase the fitting capacity and utilize the global information.

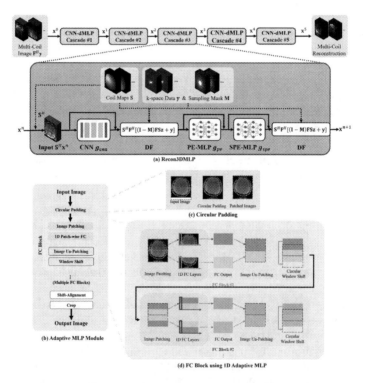

Fig. 2. (a) The proposed Recon3DMLP for 3D MRI reconstruction, which is a cascaded network and each cascade consists of a hybrid of CNN and dMLP modules. (b) The overall structure of dMLP module. (c) Circular padding is applied to ensure image can be patched. (d) Shared 1D FC layers is then applied to the patch dimension, followed by un-patch and shift operations. The FC blocks are stacked multiple times. The shift-alignment and crop operations are then applied to recover the original image shape.

2.2   Adaptive MLP for Flexible Image Resolution

The dMLP module includes the following operations (Fig. 2): 1) circular padding, 2) image patching, 3) FC layers, 4) circular shift, 5) shift alignment and 6) cropping. The input is circular-padded in order to be cropped into patches, and the shared 1D FC layers are applied over the patch dimension. The output is then un-patched into the original image shape. Next, the circular shift is applied along the patched dimension by a step size. The circular padding and shift are based on the DFT periodicity property of images. Then operations 2-4 (FC block) are stacked several times. Due to the shift operation in each FC block, the current patch contains a portion of information from two adjacent patches in the previous FC block, which allows information exchange between patches and thus dMLP can cover the entire FOV. In the end, the shift alignment is applied to roll back the previous shifts in the image domain. The padded region is then cropped out to generate the final output. Since the sub-sampling in k-space is a linear process that can be decomposed as 1D convolutions in the image domain along each sub-sampled direction, we use 1D dMLP for 3D reconstruction.

2.3   Memory Efficient Data Fidelity Module

In the naive implementation of the DF module

$$d_{DF} = S^H F^H [(I - M)FSz + y], \tag{2}$$

the coil combined image z is broadcasted to multi-coil data $(I - M)FSz$ and it increases memory consumption. Instead, we can process the data coil-by-coil

$$d_{eDF} = \sum_c S_c^H F^H [(I - M_c)FS_c z + y_c], \tag{3}$$

where c is the coil index. Together with eDF, we also employed RO cropping and gradient checkpointing for training and half-precision for inference.

2.4   Experiments

We collected a multi-contrast HR 3D brain MRI dataset with IRB approval, ranging from $224\times220\times96\times12$ to $336\times336\times192\times32$ [3]. There are 751 3D multi-coil images for training, 32 for validation, and 29 for testing.

We started with a small 3D CNN model (Recon3DCNN) with an expansion factor $e = 6$, where the channels increase from 2 to 12 in the first convolution layer and reduce to 2 in the last layer in each cascade. We then enlarged Recon3DCNN with increased width (e = 6,12,16,24) and depth (double convolution layers in each cascade). We also replaced the 3D convolution in Recon3DCNN with depth separable convolution [10] or separate 1D convolution for each 3D dimension. We also adapted the reparameterization technique [6] for Recon3DCNN such that the residual connection can be removed during inference to reduce the GPU memory. For comparison, we also adapted a 3D version of cascaded UNet, where each UNet

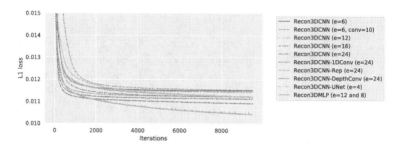

Fig. 3. The fitting power of various models on HR 3D MRI reconstruction. Models with lower loss indicate better fitting capacity.

Table 1. Evaluation of different models on HR 3D MRI reconstruction. The inference GPU memory and forward time were measured on a 3D image in $380\times294\times138\times24$.

Model	Memory Saving	Parameters (K)	GPU (G)	Time (S)	SSIM	PSNR
Recon3DCNN (e = 6)	None	65	>40	Fail	NA	NA
Recon3DCNN (e = 6)	FP16	65	35.5	1.17	0.9581	40.2790
Recon3DCNN (e = 6)	eDF	65	18.8	3.49	0.9581	40.2795
Recon3DCNN (e = 6)	FP16+eDF	65	11.5	3.04	0.9581	40.2785
Recon3DCNN (e = 6, conv = 10)	FP16+eDF	130	11.5	4.26	0.9597	40.5042
Recon3DCNN (e = 12)	FP16+eDF	247	11.6	3.14	0.9623	40.8118
Recon3DCNN (e = 16)	FP16+eDF	433	13.3	3.20	0.9636	40.9880
Recon3DCNN (e = 24)	FP16+eDF	960	15.2	3.67	0.9649	41.1503
Recon3DCNN-1DConv (e = 24)	FP16+eDF	386	16.6	5.51	0.9639	41.0473
Recon3DCNN-Rep (e = 24)	FP16+eDF	995	12.5	3.68	0.9613	40.4970
Recon3DCNN-DepthConv (e = 24)	FP16+eDF	111	17.2	4.11	0.9594	40.4367
Recon3DCNN-UNet (e = 4)	FP16+eDF	7,056	10.6	4.16	0.9565	40.4229
Recon3DMLP (e = 6/8, SKconv)	FP16+eDF	72	10.5	4.38	0.9617	41.0456
Recon3DMLP (e = 6/8, LKconv)	FP16+eDF	157	11.5	4.55	0.9620	41.0741
Recon3DMLP (e = 6/8, k = 3)	FP16+eDF	115	11.5	3.37	0.9622	41.0627
Recon3DMLP (e = 6/8, share)	FP16+eDF	1,465	11.5	3.36	0.9627	41.1455
Recon3DMLP (e = 6/8, no shift)	FP16+eDF	11,264	11.5	3.36	0.9619	41.0853
Recon3DMLP (e = 6/8, proposed)	FP16+eDF	11,264	11.5	3.38	0.9637	41.1953

has five levels with e = 4 at the initial layer and the channels were doubled at each level. To demonstrate the effectiveness of dMLP, we built Recon3DMLP by adding two 1D dMLP on PE (k = 64) and SPE (k = 16) to the smallest Recon3DCNN (e = 6). Since GELU [9,25] has larger memory overhead, we used leaky ReLU for all models. We performed ablation studies on Recon3DMLP by sharing the FC blocks among shifts, removing shifts, reducing patch size to 3 as well as replacing the dMLP with large kernel convolutions (LKconv) using k = 65 for PE and k = 17 for SPE, as well as small kernel convolutions (SKconv) using k = 3. We attempted to adapt ReconFormer[1], a transformer-based model, and Img2ImgMixer[2], an MLP based model. Both models require to specify a fixed input image size when constructing the model and failed to run on datasets with various sizes, indicating the limitation of these methods. Note that the two models were originally demonstrated on the 2D datasets with the same size [8,19]. All models were trained with loss = L1+SSIM and lr = 0.001 for 50 epochs using an NVIDIA A100 GPU with Pytorch 1.10 and CUDA 11.3. The pvalues were calculated by the Wilcoxon signed-ranks test.

---

[1] https://github.com/guopengf/ReconFormer.
[2] https://github.com/MLI-lab/imaging_MLPs.

3   Results

We first demonstrate the benefit of eDF and FP16 inference with a small CNN model Recon3DCNN (e = 6) (first and second panels in Table 1). Without eDF and FP16, the model takes >40G inference GPU memory and fails to reconstruct the test data, which indicates the challenge of HR 3D MRI reconstruction. FP16 and eDF reduce at least 11% and 53% inference memory. However, the model with only eDF is slower than the model with only FP16. By combining eDF and FP16, the inference GPU memory is reduced by 71% to 11.5G, which makes the model feasible to be deployed with a mid-range GPU in practice. Hereafter, we applied eDF and FP16 to all models.

Next, we aim to improve Recon3DCNN's performance by increasing the width and depth (third panel in Table 1 and Fig. 4). By making the model wider (increase e = 6 to e = 24), the PSNR/SSIM improves significantly ($p < 10^{-7}$). However, the inference GPU memory also increases by 33%. On the other hand, doubling the depth also improves the performance ($p < 10^{-5}$), but not as significantly as increasing the model width. Also, the former increases inference time (40%) more than the latter (21%). Also increasing the model depth does not affect the inference GPU memory. Next, we experimented with those commonly used techniques for efficient computation to modify the best CNN model Recon3DCNN (e = 24) (fourth panel in Table 1 and Fig. 4). All those variants lead to a performance drop compared to the original model ($p < 10^{-7}$), because such methods reduce the model's fitting capacity. Those variants also result in memory increase except Recon3DCNN with reparameterization technique. These results indicate such methods proposed for natural image processing are not suitable for HR 3D MRI reconstruction.

The performance of Recon3DCNN improves when becoming larger (i.e., more parameters), which indicates CNN models lack fitting power for HR 3D MR reconstruction. Therefore, we performed an overfitting experiment where models were trained and tested on one data. Figure 3 confirms that Recon3DCNN can not overfit one test data in 10K iterations and models with better fitting ability tend to have better PSNR/SSIM (Table 1). The variants of Recon3DCNN indeed have lower fitting power than the original model. This motivates us to build Recon3DMLP by adding dMLP to Recon3DCNN (e = 6) to increase its capacity while maintaining low memory usage. Recon3DMLP has better fitting ability and less reconstruction error than all models (Figs. 3 and 4). Compared to the smaller Recon3DCNN (e = 6), Recon3DMLP has similar GPU memory usage but better PSNR/SSIM ($p < 10^{-7}$). Compared to the larger Recon3DCNN (e = 24), Recon3DMLP has 24% less GPU memory usage and better PSNR ($p < 10^{-7}$) and only marginally worse SSIM ($p = 0.05$). The cascaded 3D UNet has less GPU memory consumption but lower fitting power, worse performance ($p < 10^{-7}$) and longer inference time than Recon3DCNN (e = 24) and Recon3DMLP.

To investigate the source of the gain, we perform ablation studies on Recon3DMLP (last panel in Table 1). By removing the shift operations, the dMLP module can only utilize the global information within the large patch, which leads to a drop in PSNR/SSIM ($p < 10^{-7}$). When reducing the patch

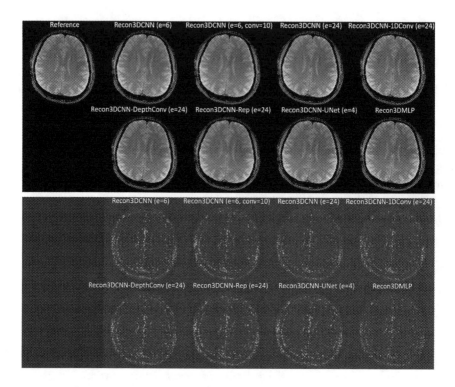

Fig. 4. Reconstruction results and corresponding error maps.

Fig. 5. The k-space difference between Recon3DMLP with and without dMLP across training iterations. Red areas in the outer k-space indicate Recon3DMLP with dMLP has recovered more high-frequency information and faster than that without dMLP. (Color figure online)

size to 3 but keeping the shift operations such that the model can only utilize the global information through the shift operations, the performance also drops ($p < 10^{-7}$) but less than the previous one. This indicates the shift operations can help the model to learn the global information and thus improve the reconstruction results. Also, models with and without shift operations do not significantly differ in GPU memory and time, suggesting the shift operations are computationally efficient. By sharing the FC parameters among shifts, the model has much fewer parameters and performance drops slightly ($p < 10^{-7}$) while GPU memory and time are similar to the original Recon3DMLP. We also replaced the dMLP modules in Recon3DMLP with convolutions using larger kernels and small kernels, respectively. Recon3DMLP (LKconv) and Recon3DMLP (SKconv)[3] have worse performance ($p < 10^{-3}$) as well as longer time than their counterpart Recon3DMLP and Recon3DMLP (small patch), indicating the dMLP is better than the convolutions for HR 3D MRI reconstruction. We compared the Recon3DMLP with and without dMLP modules and Fig. 5 shows that dMLP modules help to learn the high-frequency information faster.

4   Discussion and Conclusion

Although MLP has been proposed for vision tasks on natural images as well as 2D MRI reconstruction with fixed input size, we are the first to present a practical solution utilizing the proposed dMLP and eDF to overcome the computational constraint for HR 3D MRI reconstruction with various sizes. Compared with CNN based models, Recon3DMLP improves image quality with a little increase in computation time and similar GPU memory usage.

One limitation of our work is using the same shift and patch size without utilizing the multi-scale information. dMLP module that utilizes various patch and shift sizes will be investigated in future work. MLP-based models such as Recon3DMLP may fail if the training data is small.

References

1. Ahn, S., et al.: Deep learning-based reconstruction of highly accelerated 3D MRI. arXiv preprint arXiv:2203.04674 (2022)
2. Basri, R., Galun, M., Geifman, A., Jacobs, D., Kasten, Y., Kritchman, S.: Frequency bias in neural networks for input of non-uniform density. In: International Conference on Machine Learning, pp. 685–694. PMLR (2020)
3. Chen, E.Z., et al.: Accelerating 3D multiplex MRI reconstruction with deep learning. arXiv preprint arXiv:2105.08163 (2021)
4. Chen, S., Xie, E., Ge, C., Liang, D., Luo, P.: CycleMLP: A MLP-like architecture for dense prediction. arXiv preprint arXiv:2107.10224 (2021)
5. Dai, J., et al.: Deformable convolutional networks. In: Proceedings of the IEEE International Conference on Computer Vision, pp. 764–773 (2017)
6. Ding, X., Zhang, X., Ma, N., Han, J., Ding, G., Sun, J.: RepVGG: Making VGG-style convnets great again. In: Proceedings of the IEEE/CVF Conference on Computer Vision and Pattern Recognition, pp. 13733–13742 (2021)
7. Dosovitskiy, A., et al.: An image is worth 16×16 words: transformers for image recognition at scale. arXiv preprint arXiv:2010.11929 (2020)
8. Guo, P., Mei, Y., Zhou, J., Jiang, S., Patel, V.M.: ReconFormer: accelerated MRI reconstruction using recurrent transformer. arXiv preprint arXiv:2201.09376 (2022)
9. Hendrycks, D., Gimpel, K.: Gaussian error linear units (GELUs). arXiv preprint arXiv:1606.08415 (2016)
10. Howard, A.G., et al.: MobileNets: efficient convolutional neural networks for mobile vision applications. arXiv preprint arXiv:1704.04861 (2017)

---

[3] These are CNN models but we consider them as ablated models of Recon3DMLP and slightly abuse the notation.

11. Kellman, M., et al.: Memory-efficient learning for large-scale computational imaging. IEEE Trans. Comp. Imag. **6**, 1403–1414 (2020)

12. Knoll, F., et al.: Advancing machine learning for MR image reconstruction with an open competition: overview of the 2019 fastmri challenge. Magn. Reson. Med. **84**(6), 3054–3070 (2020)

13. Knoll, F., et al.: fastMRI: a publicly available raw k-space and DICOM dataset of knee images for accelerated MR image reconstruction using machine learning. Radiol. Artif. Intell. **2**(1), e190007 (2020)

14. Li, J., Hassani, A., Walton, S., Shi, H.: ConvMLP: hierarchical convolutional MLPs for vision. arXiv preprint arXiv:2109.04454 (2021)

15. Lian, D., Yu, Z., Sun, X., Gao, S.: As-MLP: an axial shifted MLP architecture for vision. arXiv preprint arXiv:2107.08391 (2021)

16. Liu, R., Li, Y., Tao, L., Liang, D., Zheng, H.T.: Are we ready for a new paradigm shift? a survey on visual deep MLP. Patterns **3**(7), 100520 (2022)

17. Liu, S., et al.: More convnets in the 2020s: scaling up kernels beyond 51×51 using sparsity. arXiv preprint arXiv:2207.03620 (2022)

18. Liu, Z., et al.: Swin transformer: hierarchical vision transformer using shifted windows. In: Proceedings of the IEEE/CVF International Conference on Computer Vision, pp. 10012–10022 (2021)

19. Mansour, Y., Lin, K., Heckel, R.: Image-to-image MLP-mixer for image reconstruction. arXiv preprint arXiv:2202.02018 (2022)

20. Muckley, M.J., et al.: Results of the 2020 fastMRI challenge for machine learning MR image reconstruction. IEEE Trans. Med. Imaging **40**(9), 2306–2317 (2021)

21. Ozturkler, B., et al.: Gleam: greedy learning for large-scale accelerated MRI reconstruction. arXiv preprint arXiv:2207.08393 (2022)

22. Rahaman, N., et al.: On the spectral bias of neural networks. In: International Conference on Machine Learning, pp. 5301–5310. PMLR (2019)

23. Ramzi, Z., Chaithya, G., Starck, J.L., Ciuciu, P.: NC-PDNet: a density-compensated unrolled network for 2D and 3D non-cartesian MRI reconstruction. IEEE Trans. Med. Imaging **41**(7), 1625–1638 (2022)

24. Schlemper, J., Caballero, J., Hajnal, J.V., Price, A., Rueckert, D.: A deep cascade of convolutional neural networks for MR image reconstruction. In: Niethammer, M., et al. (eds.) IPMI 2017. LNCS, vol. 10265, pp. 647–658. Springer, Cham (2017). https://doi.org/10.1007/978-3-319-59050-9_51

25. Tolstikhin, I.O., et al.: MLP-mixer: an all-MLP architecture for vision. Adv. Neural. Inf. Process. Syst. **34**, 24261–24272 (2021)

26. Touvron, H., et al.: ResMLP: feedforward networks for image classification with data-efficient training. IEEE Transactions on Pattern Analysis and Machine Intelligence (2022)

27. Tu, Z., et al.: Maxim: multi-axis MLP for image processing. In: Proceedings of the IEEE/CVF Conference on Computer Vision and Pattern Recognition, pp. 5769–5780 (2022)

28. Valanarasu, J.M.J., Patel, V.M.: UNeXt: MLP-based rapid medical image segmentation network. In: Wang, L., Dou, Q., Fletcher, P.T., Speidel, S., Li, S. (eds.) Medical Image Computing and Computer Assisted Intervention –MICCAI 2022. MICCAI 2022. Lecture Notes in Computer Science, vol.13435, pp. 23–33. Springer, Cham (2022). https://doi.org/10.1007/978-3-031-16443-9_3

29. Xu, L., Ren, J.S., Liu, C., Jia, J.: Deep convolutional neural network for image deconvolution. In: Advances in neural information processing systems, vol. 27 (2014)

30. Yang, H., et al.: Denoising of 3D MR images using a voxel-wise hybrid residual MLP-CNN model to improve small lesion diagnostic confidence. In: Wang, L., Dou, Q., Fletcher, P.T., Speidel, S., Li, S. (eds.) Medical Image Computing and Computer Assisted Intervention – MICCAI 2022. MICCAI 2022. Lecture Notes in Computer Science, vol. 13433, pp. 292–302. Springer, Cham (2022). https://doi.org/10.1007/978-3-031-16437-8_28

31. Zhang, H., Shinomiya, Y., Yoshida, S.: 3D MRI reconstruction based on 2D generative adversarial network super-resolution. Sensors **21**(9), 2978 (2021)

32. Zhu, B., Liu, J.Z., Cauley, S.F., Rosen, B.R., Rosen, M.S.: Image reconstruction by domain-transform manifold learning. Nature **555**(7697), 487–492 (2018)

Building a Bridge: Close the Domain Gap in CT Metal Artifact Reduction

Tao Wang[1], Hui Yu[1], Yan Liu[2], Huaiqiang Sun[3], and Yi Zhang[4(✉)]

[1] College of Computer Science, Sichuan University, Chengdu, China
[2] College of Electrical Engineering, Sichuan University, Chengdu, China
[3] Department of Radiology, West China Hospital of Sichuan University,
Chengdu, China
[4] School of Cyber Science and Engineering, Sichuan University, Chengdu, China
yzhang@scu.edu.cn

Abstract. Metal artifacts in computed tomography (CT) degrade the imaging quality, leading to a negative impact on the clinical diagnosis. Empowered by medical big data, many DL-based approaches have been proposed for metal artifact reduction (MAR). In supervised MAR methods, models are usually trained on simulated data and then applied to the clinical data. However, inferior MAR performance on clinical data is usually observed due to the domain gap between simulated and clinical data. Existing unsupervised MAR methods usually use clinical unpaired data for training, which often distort the anatomical structure due to the absence of supervision information. To address these problems, we propose a novel semi-supervised MAR framework. The clean image is employed as the bridge between the synthetic and clinical metal-affected image domains to close the domain gap. We also break the cycle-consistency loss, which is often utilized for domain transformation, since the bijective assumption is too harsh to accurately respond to the facts of real situations. To further improve the MAR performance, we propose a new Artifact Filtering Module (AFM) to eliminate features helpless in recovering clean images. Experiments demonstrate that the performance of the proposed method is competitive with several state-of-the-art unsupervised and semi-supervised MAR methods in both qualitative and quantitative aspects.

Keywords: Computed Tomography · Metal Artifact Reduction · Deep Learning · Domain Gap

This work was supported in part by the National Natural Science Foundation of China under Grant 62271335; in part by the Sichuan Science and Technology Program under Grant 2021JDJQ0024; and in part by the Sichuan University "From 0 to 1" Innovative Research Program under Grant 2022SCUH0016.

1   Introduction

Metal implants can heavily attenuate X-rays in computed tomography (CT) scans, leading to severe artifacts in reconstructed images. It is essential to remove metal artifacts in CT images for subsequent diagnosis.

Recently, with the emergence of deep learning (DL) [21,22], many DL-based approaches have been proposed for metal artifact reduction (MAR) and achieved encouraging results. These methods can be roughly classified into three groups: supervised, unsupervised, and semi-supervised MAR methods. Supervised MAR methods [2,8,10,12,19,20,23,24] directly learn the mapping from synthetic metal-affected data to metal-free one under the guidance of the desired data. Then the learned models are applied to the clinical data. Unfortunately, due to the domain gap between synthetic and clinical data, poor generalization performance usually occurs, leading to unexpected results. Unsupervised MAR methods [5,9,27] can avoid the problem since their training and application are both on the clinical data. Nonetheless, the absence of supervision information makes it easy to distort the anatomical structure in the corrected results. Recently, several semi-supervised MAR methods have been proposed. SFL-CNN [17] and a variant of ADN [9] denoted as SemiADN [14] are two representative works, which utilize the same network to deal with synthetic and clinical data simultaneously. These methods inherit the advantages of both supervised and unsupervised MAR methods, but the mentioned-above domain gap problem remains. In these works, the network attempts to find a balance between the synthetic and clinical data but ultimately results in an unsatisfactory outcome in both domains, leaving room for further improvement.

In this work, our goal is to explicitly reduce the domain gap between synthetic and clinical metal-corrupted CT images for improved clinical MAR performance. Some domain adaptation-based networks [7,16,18,25] are designed to close the domain gap and they usually assume that there is a one-to-one correspondence between two domains, i.e. bijection, which is implemented via the constraint of cycle consistency loss. However, this assumption is too harsh to accurately respond to the facts of real situations. Furthermore, when the model learns an identical transformation, this assumption is still met. Hence, maintaining the diversity of image generation is another challenge.

To close the domain gap, we propose a novel semi-supervised MAR framework. In this work, the clean image domain acts as the bridge, where the bijection is substituted with two simple mappings and the strict assumption introduced by the cycle-consistency loss is relaxed. Our goal is to convert simulated and clinical metal-corrupted data back and forth. As an intermediate product, clean images are our target. To improve the transformations of two metal-corrupted images into metal-free ones, we propose a feature selection mechanism, denoted as Artifact Filtering Module (AFM), where AFM acts as a filter to eliminate features helpless in recovering clean images.

2   Method

2.1   Problem Formulation

The metal corruption process can be formulated as a linear superposition model as [10,11]:

$$X_{ma} = X_{free} + A, \tag{1}$$

where X_{ma}, X_{free}, and A represent metal-affected CT images, metal-free CT images, and metal artifacts, respectively. X_{ma} is the observation signal and X_{free} is the target signal to be reconstructed.

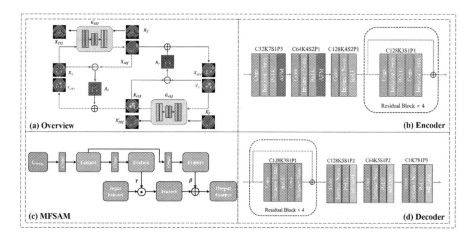

Fig. 1. Overview of the proposed method. C: channel number, K: kernel size, S: stride, and P: padding size. More details on AFM are in Sect. 2.3.

2.2   Overview

Figure 1 presents the overall architecture of our proposed method. Let I_s be the domain of all synthetic metal-corrupted CT images and I_c be the domain of all clinical metal-corrupted CT images. The generators aim to convert them to clean CT image domain I_f, where different generators take metal-corrupted CT images from different domains. I_f takes the role of a bridge to close the domain gap. The following subsections present the details of these image translation branches.

2.3   Image Translation

According to Eq. 1, if two metal artifact reduction translations are completed, the subsequent two transformations can be obtained by subtracting the output of the network from the original input. Therefore, only two translators are needed. The first translator is used to convert I_s into I_f, denoted as G_{s2f} and the second translator is used to convert I_c into I_f, denoted as G_{c2f}. In this work, two translators, G_{s2f} and G_{c2f}, share the same network architecture, consisting of one encoder and one decoder, as shown in Fig. 1 (b) and (d).

Encoding Stage. In most DL-based MAR methods, the networks take a metal-corrupted CT image as input and map it to a metal-free CT image domain. Noise-related features are involved in the entire process, which negatively affects the restoration of clean images. In the principal component analysis (PCA)-based image denoising method [1], by retaining only the most important features, noise and irrelevant information are eliminated. In this work, we propose an artifact filtering module (AFM) for feature selection. At the encoding step, feature maps that contribute little to the reconstruction are also considered to be related to noise, where the encoder acts as a filter and only allows useful information to pass through. Specifically, there are two criteria for feature selection: 1) the selected feature maps contain as much information as possible, which is assessed by Variance (Var), and 2) the correlation between the selected feature maps should be as small as possible, which is measured by covariance (Cov). Finally, each feature map gets a score as:

$$score = \frac{Var}{Cov + \lambda}, \tag{2}$$

where λ is a small constant to prevent division by zero and we set $\lambda = 1e - 7$ in this paper. Feature maps with high scores will be selected. Therefore, we can dynamically select different feature maps according to the inputs.

Decoding Stage. At the encoding stage, features that are helpless to reconstruct the clean image are filtered out. Decoder then maps the remaining features, which contain useful information, back into the image domain. To push the generated image to fall into the clean image domain, we employ conditional normalization layers [4,15] and propose a metal-free spatially aware module (MFSAM). The mean and variance of the features are modulated to match those of the metal-free image style by the MFSAM. The details of MFSAM are illustrated in Fig. 1 (c).

Metal Artifacts Reduction and Generation Stage. The framework consists of four image translation branches: two metal artifacts reduction branches $I_s \leftarrow I_f$ and $I_c \leftarrow I_f$, and two metal artifact generation branches $I_f \leftarrow I_s$ and $I_f \leftarrow I_c$.

(1) $I_s \leftarrow I_f$: In this transformation, we employ G_{s2f} to learn the mapping from the synthetic metal-affected image domain to the metal-free image domain, which is denoted as:

$$X_{s2f} = G_{s2f}(X_s), \tag{3}$$

where X_s is synthetic metal-affected CT image in I_s and X_{s2f} is the corrected result of X_s. According to Eq. 1, the metal artifacts A_s can be obtained as follows:

$$A_s = X_s - X_{s2f}. \tag{4}$$

(2) $I_c \leftarrow I_f$: In this transformation, we use G_{c2f} to learn the mapping from the clinical metal-affected image domain to the metal-free image domain, resulting

in metal-corrected CT image X_{c2f} and the metal artifacts A_c. This process is the same as the transformation of $I_s \leftarrow I_f$ and can be formulated as follows:

$$X_{c2f} = G_{c2f}(X_c), \tag{5}$$

$$A_c = X_c - X_{c2f}, \tag{6}$$

where X_c is clinical metal-affected CT image in I_c.

(3): $I_f \leftarrow I_s$: We use the artifacts of X_s to obtain a synthetic domain metal-corrupted image X_{c2s} by adding A_s to the learned metal-free CT image X_{c2f}:

$$X_{c2s} = X_{c2f} + A_s. \tag{7}$$

(4): $I_f \leftarrow I_c$: Synthesizing clinical domain metal-corrupted image X_{s2c} can be achieved by adding A_c to the learned metal-free CT image X_{s2f}:

$$X_{s2c} = X_{s2f} + A_c. \tag{8}$$

2.4   Loss Function

In our framework, the loss function contains two parts: adversarial loss and reconstruction loss.

Adversarial Loss. Due to the lack of paired clinical metal-corrupted and metal-free CT images, as well as paired clinical and synthetic metal-corrupted CT images, we use PatchGAN-based discriminators, D_f, D_s, and D_c, and introduce an adversarial loss for weak supervision. D_f learns to distinguish whether an image is a metal-free image, D_s learns to determine whether an image is a synthetic metal-affected CT image, and D_c learns to determine whether an image is a clinical metal-affected CT image. The total adversarial loss $\mathcal{L}_{adv}$ is written as:

$$\begin{aligned} \mathcal{L}_{adv} &= \mathbb{E}[log D_f(X_f)] + \mathbb{E}[1 - log D_f(X_{c2f})] \\ &+ \mathbb{E}[log D_s(X_s)] + \mathbb{E}[1 - log D_s(X_{c2s})] \\ &+ \mathbb{E}[log D_c(X_c)] + \mathbb{E}[1 - log D_c(X_{s2c})]. \end{aligned} \tag{9}$$

Reconstruction Loss. The label X_{gt} of X_s is employed to guide the G_{s2f} to reduce the metal artifacts. The reconstruction loss $\mathcal{L}_{s2f}$ on X_{x2f} can be formulated as:

$$\mathcal{L}_{s2f} = ||(X_{s2f} - X_{gt})||_1. \tag{10}$$

When X_{syn} is transformed into the clinical domain, G_{c2f} can also reduce the metal artifacts with the help of X_{gt}. The reconstruction loss $\mathcal{L}_{sc2f}$ on X_{sc2f} can be formulated as:

$$\mathcal{L}_{sc2f} = ||(X_{sc2f} - X_{gt})||_1, \tag{11}$$

where X_{sc2f} is the MAR results of X_{s2c}.

To obtain optimal MAR results, it is necessary to remove any noise-related features while preserving as much of the content information as possible. When the input image is already metal-free, the input image has no noise-related features, and the reconstructed image should not suffer from any information loss. Here, we employed the model error loss to realize this constrain:

$$\mathcal{L}_{model\_error} = ||(X_{f2f} - X_f)||_1 + ||(X'_{f2f} - X'_f)||_1, \tag{12}$$

where X_{f2f} is a reconstructed image from X_f using G_{s2f} and X'_{f2f} is a reconstructed image from X'_f using G_{c2f}.

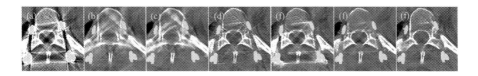

Fig. 2. Visual comparisons with the SOTA MAR methods on simulated data. (a): Uncorrected, (b): LI,(c): NMAR,(d): ADN, (e): β-cycleGAN, (f): SemiADN, (e): Ours. Display window: [−375,560] HU.

Table 1. Quantitative results of SOTA MAR methods on the simulated dataset.

PSNR(dB)/SSIM	Small	→	Middle	→	Large	Average	Params(M)
Uncorrected	16.23/0.7092	16.16/0.6513	16.19 /0.6594	16.14/0.6443	16.12/0.6288	16.18/0.6684	–
LI	28.68/0.8095	26.84/0.7549	25.37/0.7347	26.07/0.7309	26.06/0.7296	27.19/0.7664	–
NMAR	29.41/0.8294	28.19/0.7913	27.24/0.7752	27.80/0.7743	27.62/0.7711	28.43/0.7987	–
ADN	31.32/ 0.8425	30.28/0.8043	30.50/0.8089	29.98/0.7908	29.69/0.7806	30.54/0.8127	32.43
β−cycleGAN	30.34/0.8365	28.17/0.7811	27.99/0.7961	27.18/0.7710	26.45/0.7502	28.53/0.7958	**5.48**
SemiADN	31.07/0.8636	30.99/0.8393	**31.68**/0.8500	30.23/0.8322	29.37/0.8144	30.66/0.8435	32.43
Ours	**32.57/0.8789**	**31.28/0.8541**	31.29/**0.8562**	**30.83/0.8472**	**30.38/0.8402**	**31.54/0.8606**	6.59

Overall Loss. The overall loss function is defined as follows:

$$\mathcal{L} = \mathcal{L}_{adv} + \lambda_{recon}(\mathcal{L}_{s2f} + \mathcal{L}_{sc2f} + \mathcal{L}_{model\_error}), \tag{13}$$

where λ_{recon} is the weighting parameter and as suggested by [9,14], it was set as 20.0 in our work.

3   Experiments

3.1   Dataset and Implementation Details

In this work, we used one synthesized dataset and two clinical datasets, denoted as SY, CL1 and CL2, respectively. The proposed method was trained on SY and CL1. For data simulation, we followed the procedure of [26] and used the metal-free CT images of the Spineweb dataset [3]. We set the threshold to 2,000 HU [20,24] to obtain 116 metal masks from metal-affected CT images of the

Spineweb dataset, where 100 masks are for training and the remaining 16 masks are for testing. We used 6,000 synthesized pairs for training and 2,000 pairs for evaluation. For CL1, we randomly chose 6,000 metal-corrupted CT images and 6,000 metal-free CT images from Spineweb for training and another 224 metal-corrupted CT images for testing. For CL2, clinical metal-corrupted CT images were collected from our local hospital to investigate the generalization performance. The model was implemented with the *PyTorch* framework and optimized by the Adam optimizer with the parameters $(\beta_1, \beta_2) = (0.5, 0.999)$. The learning rate was initialized to 0.0001 and halved every 20 epochs. The network was trained with 60 epochs on one NVIDIA 1080Ti GPU with 11 GB memory, and the batch size was 2.

3.2   Comparison with State-of-the-Art Methods

The proposed method was compared with several classic and state-of-the-art (SOTA) MAR methods: LI [6], NMAR [13], ADN [9], β-cycleGAN [5] and SemiADN [14]. LI and NMAR are traditional MAR methods. ADN and β-cycleGAN are SOTA unsupervised MAR methods. SemiADN is a SOTA semi-supervised MAR method. Structural similarity (SSIM) and peak signal-to-noise ratio (PSNR) were adopted as quantitative metrics. Our intuition for determining the number of feature maps selected in AFM is based on the observation that the majority of information in an image is typically related to its content, while noise-related features are few. Therefore, in our work, during the encoding stage, we discarded 2 out of 32 and 4 out of 64 feature maps at the first and second down-sampling stages, respectively.

MAR Performance on SY: The quantitative scores are presented in Table 1. The sizes of the 16 metal implants in the testing dataset are: [254, 274, 270, 262, 267, 363, 414, 441, 438, 445, 527, 732, 845, 889, 837, 735] in pixels. They are divided into five groups according to their sizes. It is observed that all methods significantly improve both SSIM and PSNR scores compared with uncorrected CT images. Aided by supervision, SemiADN obtains higher quantitative scores than ADN. Compared with these SOTA unsupervised and semi-supervised MAR methods, our method achieves the best quantitative performance. Figure 2 shows the visual comparisons on SY. The proposed method outperforms all other methods in artifact suppression and effectively preserves anatomical structures around metallic implants, thereby demonstrating its effectiveness.

MAR Performance on CL1: Figure 3 presents three representative clinical metal-affected CT images with different metallic implant sizes from small to large. When the metal is small, all methods can achieve good MAR performance but there are differences in tissue detail preservation. ADN and β-cycleGAN are more prone to lose details around the metal. In SemiADN, the missing details are recovered with the help of the supervision signal. However, in the second case, more artifacts are retained in SemiADN than ADN and β-cycleGAN. Compared with these MAR methods, our method is well-balanced between detail preservation and artifact reduction. When the metallic implant gets larger, as shown in

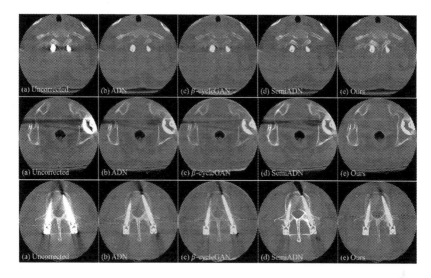

Fig. 3. Visual comparisons on CL1 dataset. Display window: [-1000,1000] HU.

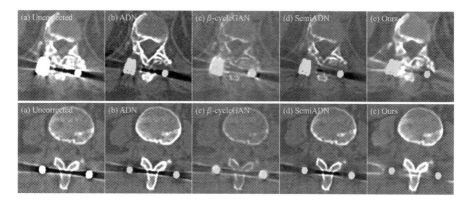

Fig. 4. Visual comparisons on CL2 dataset. Display window: $[-375,560]$ HU.

the third case, other methods are limited to reduce artifacts, and SemiADN even aggravates the impact of artifacts. Fortunately, our method is able to effectively suppress metal artifacts, thus demonstrating its potential for practical clinical use.

MAR Performance on CL2: To assess the generalization capability of our method, we further evaluated MAR performance on CL2 with the model trained on SYN and CL1. Two practical cases are presented in Fig. 4. ADN and Semi-ADN fail to deal with severe artifacts and even introduce a deviation of HU value, while β-cycleGAN shows a certain ability to suppress these artifacts. Nonetheless, our proposed method outperforms β-cycleGAN in terms of artifact suppression and detail preservation. It can be seen that our method exhibits good generalization ability, which means it can effectively address metal artifacts even in scenarios where the simulated data and clinical metal-corrected data have different clean image domains. It shows the robustness of our proposed method across different imaging configurations.

3.3   Ablation Study

In this section, we investigate the effectiveness of the proposed AFM. Table 2 shows the results of our ablation models, where M1 refers to the model without AFM, and M2 replaces the AFM with channel attention. Table 2 shows that AFM can improve the scores of M1. Although M2 integrates a channel attention mechanism to dynamically adjust the weight of different feature maps, our proposed AFM method achieves higher quantitative scores, which indicates its superior performance.

Table 2. Quantitative comparison of different variants of our method.

	M1	M2	Ours
PSNR(dB)	28.99	29.37	31.54
SSIM	0.8025	0.8353	0.8606

4   Conclusion

In this paper, we explicitly bridge the domain gap between synthetic and clinical metal-corrupted CT images. We employ the clean image domain as the bridge and break the cycle-consistency loss, thereby eliminating the necessity for strict bijection assumption. At the encoding step, feature maps with limited influence will be eliminated, where the encoder acts as a bottleneck only allowing useful information to pass through. Experiments demonstrate that the performance of the proposed method is competitive with several SOTA MAR methods in both qualitative and quantitative aspects. In particular, our method exhibits good generalization ability on clinical data.

References

1. Babu, Y.M.M., Subramanyam, M.V., Prasad, M.G.: PCA based image denoising. Signal Image Process. **3**(2), 236 (2012)
2. Ghani, M.U., Karl, W.C.: Fast enhanced CT metal artifact reduction using data domain deep learning. IEEE Tran. Comput. Imaging **6**, 181–193 (2019)
3. Glocker, B., Zikic, D., Konukoglu, E., Haynor, D.R., Criminisi, A.: Vertebrae localization in pathological spine CT via dense classification from sparse annotations. In: Proceedings of the Medical Image Computing and Computer-Assisted Intervention, pp. 262–270 (2013)

4. Huang, X., Belongie, S.: Arbitrary style transfer in real-time with adaptive instance normalization. In: Proceedings of the IEEE International Conference on Computer Vision (ICCV) (2017)
5. Lee, J., Gu, J., Ye, J.C.: Unsupervised CT metal artifact learning using attention-guided β-CycleGAN. IEEE Trans. Med. Imaging **40**(12), 3932–3944 (2021)
6. Lewitt, R.M., Bates, R.: Image reconstruction from projections III: projection completion methods. Optik **50**, 189–204 (1978)
7. Li, Y., Chang, Y., Gao, Y., Yu, C., Yan, L.: Physically disentangled intra- and inter-domain adaptation for varicolored haze removal. In: Proceedings of the IEEE/CVF Conference on Computer Vision and Pattern Recognition (CVPR), pp. 5841–5850 (2022)
8. Liao, H., et al.: Generative mask pyramid network for CT/CBCT metal artifact reduction with joint projection-sinogram correction. In: Shen, D., et al. (eds.) MICCAI 2019. LNCS, vol. 11769, pp. 77–85. Springer, Cham (2019). https://doi.org/10.1007/978-3-030-32226-7_9
9. Liao, H., Lin, W.A., Zhou, S.K., Luo, J.: ADN: artifact disentanglement network for unsupervised metal artifact reduction. IEEE Trans. Med. Imaging **39**(3), 634–643 (2019)
10. Lin, W.A., et al.: DuDoNet: dual domain network for CT metal artifact reduction. In: Proceedings of the IEEE/CVF Conference on Computer Vision and Pattern Recognition, pp. 10512–10521 (2019)
11. Lyu, Y., Fu, J., Peng, C., Zhou, S.K.: U-DuDoNet: unpaired dual-domain network for CT metal artifact reduction. In: de Bruijne, M., et al. (eds.) MICCAI 2021. LNCS, vol. 12906, pp. 296–306. Springer, Cham (2021). https://doi.org/10.1007/978-3-030-87231-1_29
12. Lyu, Y., Lin, W.-A., Liao, H., Lu, J., Zhou, S.K.: Encoding metal mask projection for metal artifact reduction in computed tomography. In: Martel, A.L., et al. (eds.) MICCAI 2020. LNCS, vol. 12262, pp. 147–157. Springer, Cham (2020). https://doi.org/10.1007/978-3-030-59713-9_15
13. Meyer, E., Raupach, R., Lell, M., Schmidt, B., Kachelriess, M.: Normalized metal artifact reduction (NMAR) in computed tomography. Med. Phys. **37**(10), 5482–5493 (2010)
14. Niu, C., et al.: Low-dimensional manifold constrained disentanglement network for metal artifact reduction. IEEE Trans. Radiat. Plasma Med. Sci. 1–1 (2021). https://doi.org/10.1109/TRPMS.2021.3122071
15. Park, T., Liu, M.Y., Wang, T.C., Zhu, J.Y.: Semantic image synthesis with spatially-adaptive normalization. In: Proceedings of the IEEE/CVF Conference on Computer Vision and Pattern Recognition, pp. 2337–2346 (2019)
16. Shao, Y., Li, L., Ren, W., Gao, C., Sang, N.: Domain adaptation for image dehazing. In: Proceedings of the IEEE/CVF Conference on Computer Vision and Pattern Recognition (CVPR) (2020)
17. Shi, Z., Wang, N., Kong, F., Cao, H., Cao, Q.: A semi-supervised learning method of latent features based on convolutional neural networks for CT metal artifact reduction. Med. Phys. **49**(6), 3845–3859 (2022). https://doi.org/10.1002/mp.15633
18. Wang, M., Lang, C., Liang, L., Lyu, G., Feng, S., Wang, T.: Attentive generative adversarial network to bridge multi-domain gap for image synthesis. In: 2020 IEEE International Conference on Multimedia and Expo (ICME), pp. 1–6 (2020). https://doi.org/10.1109/ICME46284.2020.9102761
19. Wang, T., et al.: IDOL-net: an interactive dual-domain parallel network for CT metal artifact reduction. IEEE Trans. Radiat. Plasma Med. Sci. 1–1 (2022). https://doi.org/10.1109/TRPMS.2022.3171440

20. Wang, T., et al.: DAN-net: dual-domain adaptive-scaling non-local network for CT metal artifact reduction. Phys. Med. Biol. **66**(15), 155009 (2021). https://doi.org/10.1088/1361-6560/ac1156
21. Wang, T., Yu, H., Lu, Z., Zhang, Z., Zhou, J., Zhang, Y.: Stay in the middle: a semi-supervised model for CT metal artifact reduction. In: ICASSP 2023–2023 IEEE International Conference on Acoustics, Speech and Signal Processing (ICASSP), pp. 1–5 (2023). https://doi.org/10.1109/ICASSP49357.2023.10095681
22. Yu, H., et al.: DESEG: auto detector-based segmentation for brain metastases. Phys. Med. Biol. **68**(2), 025002 (2023)
23. Yu, L., Zhang, Z., Li, X., Ren, H., Zhao, W., Xing, L.: Metal artifact reduction in 2D CT images with self-supervised cross-domain learning. Phys. Med. Biol. **66**(17), 175003 (2021)
24. Yu, L., Zhang, Z., Li, X., Xing, L.: Deep sinogram completion with image prior for metal artifact reduction in CT images. IEEE Trans. Med. Imaging **40**(1), 228–238 (2020)
25. Zhang, K., Li, Y.: Single image dehazing via semi-supervised domain translation and architecture search. IEEE Signal Process. Lett. **28**, 2127–2131 (2021). https://doi.org/10.1109/LSP.2021.3120322
26. Zhang, Y., Yu, H.: Convolutional neural network based metal artifact reduction in x-ray computed tomography. IEEE Trans. Med. Imaging **37**(6), 1370–1381 (2018)
27. Zhao, B., Li, J., Ren, Q., Zhong, Y.: Unsupervised reused convolutional network for metal artifact reduction. In: Yang, H., Pasupa, K., Leung, A.C.-S., Kwok, J.T., Chan, J.H., King, I. (eds.) ICONIP 2020. CCIS, vol. 1332, pp. 589–596. Springer, Cham (2020). https://doi.org/10.1007/978-3-030-63820-7_67

Geometric Ultrasound Localization Microscopy

Christopher Hahne[(✉)] and Raphael Sznitman

ARTORG Center, University of Bern, Bern, Switzerland
`christopher.hahne@unibe.ch`

Abstract. Contrast-Enhanced Ultra-Sound (CEUS) has become a viable method for non-invasive, dynamic visualization in medical diagnostics, yet Ultrasound Localization Microscopy (ULM) has enabled a revolutionary breakthrough by offering ten times higher resolution. To date, Delay-And-Sum (DAS) beamformers are used to render ULM frames, ultimately determining the image resolution capability. To take full advantage of ULM, this study questions whether beamforming is the most effective processing step for ULM, suggesting an alternative approach that relies solely on Time-Difference-of-Arrival (TDoA) information. To this end, a novel geometric framework for microbubble localization via ellipse intersections is proposed to overcome existing beamforming limitations. We present a benchmark comparison based on a public dataset for which our geometric ULM outperforms existing baseline methods in terms of accuracy and robustness while only utilizing a portion of the available transducer data.

Keywords: Ultrasound · Microbubble · Localization · Microscopy · Geometry · Parallax · Triangulation · Trilateration · Multilateration · Time-of-Arrival

1 Introduction

Ultrasound Localization Microscopy (ULM) has revolutionized medical imaging by enabling sub-wavelength resolution from images acquired by piezo-electric transducers and computational beamforming. However, the necessity of beamforming for ULM remains questionable. Our work challenges the conventional assumption that beamforming is the ideal processing step for ULM and presents an alternative approach based on geometric reconstruction from Time-of-Arrival (ToA) information.

The discovery of ULM has recently surpassed the diffraction-limited spatial resolution and enabled highly detailed visualization of the vascularity [8]. ULM borrows concepts from super-resolution fluorescence microscopy techniques to

---

Supplementary Information The online version contains supplementary material available at https://doi.org/10.1007/978-3-031-43999-5_21.

precisely locate individual particles with sub-pixel accuracy over multiple frames. By the accumulation of all localizations over time, ULM can produce a super-resolved image, providing researchers and clinicians with highly detailed representation of the vascular structure.

While Contrast-Enhanced Ultra-Sound (CEUS) is used in the identification of musculoskeletal soft tissue tumours [5], the far higher resolution capability offered by ULM has great potential for clinical translation to improve the reliability of cancer diagnosis (*i.e.*, enable differentiation of tumour types in kidney cancer [7] or detect breast cancer tissue [1]). Moreover, ULM has shown promise in imaging neurovascular activity after visual stimulation (functional ULM) [14]. The pioneering study by Errico *et al.* [8] initially demonstrated the potential of ULM by successfully localizing contrast agent particles (microbubbles) using a 2D point-spread-function model. In general, the accuracy in MicroBubble (MB) localization is the key to achieving sub-wavelength resolution [4], for which classical imaging methods [11,17], as well as deep neural networks [1,16], have recently been reported.

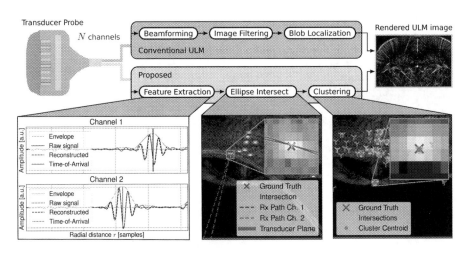

Fig. 1. Comparison of ULM processing pipelines: Classical ULM (top) employs computational beamforming from N channels and image filters to localize microbubbles. Our geometric ULM (bottom) consists of a cross-channel phase-consistent Time-of-Arrival detection (left) to form ellipses that intersect at a microbubble position (middle). As a refinement step, ellipse intersections are fused via clustering (right).

However, the conventional approach for ULM involves using computational beamformers, which may not be ideal for MB localization. For example, a recent study has shown that ultrasound image segmentation can be learned from radio-frequency data and thus without beamforming [13]. Beamforming techniques have been developed to render irregular topologies, whereas MBs exhibit a uniform geometric structure, for which ULM only requires information about its spatial position. Although the impact of adaptive beamforming has been studied for ULM to investigate its potential to refine MB localization [3], optimization of the Point-Spread Function (PSF) poses high demands on the transducer array, data storage, and algorithm complexity.

To this end, we propose an alternative approach for ULM, outlined in Fig. 1, that entirely relies on Time-Difference-of-Arrival (TDoA) information, omitting beamforming from the processing pipeline for the first time. We demonstrate a novel geometry framework for MB localization through ellipse intersections to overcome limitations inherent to beamforming. This approach provides a finer distinction between overlapping and clustered spots, improving localization precision, reliability, and computation efficiency. In conclusion, we challenge the conventional wisdom that beamforming is necessary for ULM and propose a novel approach that entirely relies on TDoA information for MB localization. Our proposed approach demonstrates promising results and indicates a considerable trade-off between precision, computation, and memory.

2   Method

Geometric modeling is a useful approach for locating landmarks in space. One common method involves using a Time-of-Flight (ToF) round-trip setup that includes a transmitter and multiple receivers [10]. This setup is analogous to the parallax concept in visual imaging, where a triangle is formed between the target, emitter, and receivers, as illustrated in Fig. 2. The target's location can be accurately estimated using trilateration by analyzing the time delay between the transmitted and received signals. However, the triangle's side lengths are unknown in the single receiver case, and all possible travel path candidates form triangles with equal circumferences fixed at the axis connecting the receiver and the source. These candidates reside on an elliptical shape. By adding a second receiver, its respective ellipse intersects with the first one resolving the target's 2-D position. Thus, the localization accuracy depends on the ellipse model, which is parameterized by the known transducer positions and the time delays we seek to estimate. This section describes a precise echo feature extraction, which is essential for building the subsequent ellipse intersection model. Finally, we demonstrate our localization refinement through clustering.

2.1   Feature Extraction

Feature extraction of acoustic signals has been thoroughly researched [9,18]. To leverage the geometric ULM localization, we wish to extract Time-of-Arrival (ToA) information (instead of beamforming) at sub-wavelength precision. Despite the popularity of deep neural networks, which have been studied for ToA detection [18], we employ an energy-based model [9] for echo feature extraction to demonstrate the feasibility of our geometric ULM at the initial stage. Ultimately, future studies can combine our proposed localization with a supervised network. Here, echoes $f(\mathbf{m}_k; t)$ are modeled as Multimodal Exponentially-Modified Gaussian Oscillators (MEMGO) [9],

$$f(\mathbf{m}_k; t) = \alpha_k \exp\left(-\frac{(t - \mu_k)^2}{2\sigma_k^2}\right)\left(1 + \mathrm{erf}\left(\eta_k \frac{t - \mu_k}{\sigma_k \sqrt{2}}\right)\right)\cos\left(\omega_k\left(t - \mu_k\right) + \phi_k\right),$$
$$(1)$$

where $t \in \mathbb{R}^T$ denotes the time domain with a total number of T samples and $\mathbf{m}_k = [\alpha_k, \mu_k, \sigma_k, \eta_k, \omega_k, \phi_k]^{\mathsf{T}} \in \mathbb{R}^6$ contains the amplitude α_k, mean μ_k, spread σ_k, skew η_k, angular frequency ω_k and phase ϕ_k for each echo k. Note that $\mathrm{erf}(\cdot)$ is the error function. To estimate these parameters iteratively, the cost function is given by,

$$\mathcal{L}_{\mathrm{E}}\left(\hat{\mathbf{m}}_n\right) = \left\|y_n(t) - \sum_{k=1}^{K} f\left(\mathbf{m}_k; t\right)\right\|_2^2,$$
$$(2)$$

where $y_n(t)$ is the measured signal from waveform channel $n \in \{1, 2, \ldots, N\}$ and the sum over k accumulates all echo components $\hat{\mathbf{m}}_n = [\mathbf{m}_1^{\mathsf{T}}, \mathbf{m}_2^{\mathsf{T}}, \ldots, \mathbf{m}_K^{\mathsf{T}}]^{\mathsf{T}}$. We get the best echo feature set $\hat{\mathbf{m}}_n^{\star}$ over all iterations j via,

$$\hat{\mathbf{m}}_n^{\star} = \underset{\hat{\mathbf{m}}_n^{(j)}}{\arg\min}\left\{\mathcal{L}_{\mathrm{E}}\left(\hat{\mathbf{m}}_n^{(j)}\right)\right\},$$
$$(3)$$

for which we use the Levenberg-Marquardt solver. Model-based optimization requires initial estimates to be nearby the solution space. For this, we detect initial ToAs via gradient-based analysis of the Hilbert-transformed signal to set $\hat{\mathbf{m}}_n^{(1)}$ as in [9].

Before geometric localization, one must ensure that detected echo components correspond to the same MB. In this work, echo matching is accomplished in a heuristic brute-force fashion. Given an echo component $\mathbf{m}_{n,k}^{\star}$ from a reference channel index n, a matching echo component from an adjacent channel index $n \pm g$ with gap $g \in \mathbb{N}$ is found by $k + h$ in the neighborhood of $h \in \{-1, 0, 1\}$. A corresponding phase-precise ToA $t_{n,k}^{\star}$ is obtained by $t_{n\pm g,k}^{\star} = \mu_{n\pm g,k+h}^{\star} + \phi_{n,k}^{\star} - \Delta$, which takes $\mu_{n,k}^{\star}$ and $\phi_{n,k}^{\star}$ from $\hat{\mathbf{m}}_n^{\star}$ for phase-precise alignment across transducer channels after upsampling. Here, Δ is a fixed offset to accurately capture the onset of the MB locations [2]. We validate echo correspondence through a re-projection error in adjacent channels and reject those with weak alignment.

2.2   Ellipse Intersection

While ellipse intersections can be approximated iteratively, we employ Eberly's closed-form solution [6] owing to its fast computation property. Although one might expect that the intersection of arbitrarily placed ellipses is straightforward, it involves advanced mathematical modelling due to the degrees of freedom in the ellipse positioning. An ellipse is drawn by radii (r_a, r_b) of the major and minor axes with,

$$r_a = \frac{t_{n,k}^{\star}}{2}, \quad \text{and} \quad r_b = \frac{1}{2}\sqrt{\left(t_{n,k}^{\star}\right)^2 - \|\hat{\mathbf{u}}_s - \mathbf{u}_n\|_2^2},$$
$$(4)$$

where the virtual transmitter $\hat{\mathbf{u}}_s \in \mathbb{R}^2$ and each receiver $\mathbf{u}_n \in \mathbb{R}^2$ with channel index n represent the focal points of an ellipse, respectively. For the intersection, we begin with the ellipse standard equation. Let any point $\mathbf{s} \in \mathbb{R}^2$ located on an ellipse and displaced by its center $\mathbf{c}_n \in \mathbb{R}^2$ such that,

$$(\mathbf{s} - \mathbf{c}_n)^{\mathsf{T}} \left(\frac{\|\mathbf{v}_n\|_2^2}{r_a^2 |\mathbf{v}_n|^2} + \frac{\|\mathbf{v}_n^{\perp}\|_2^2}{r_b^2 |\mathbf{v}_n^{\perp}|^2} \right) (\mathbf{s} - \mathbf{c}_n) = (\mathbf{s} - \mathbf{c}_n)^{\mathsf{T}} \mathbf{M} (\mathbf{s} - \mathbf{c}_n) = 1, \quad (5)$$

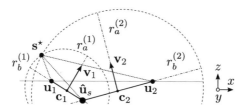

Fig. 2. Transducer geometry used for the ellipse intersection and localization of a MB position $\mathbf{s}^{\star}$ from virtual source $\hat{\mathbf{u}}_s$ and receiver positions $\mathbf{u}_n$, which span ellipses rotated by $\mathbf{v}_n$ around their centers $\mathbf{c}_n$.

where $\mathbf{M}$ contains the ellipse equation with $\mathbf{v}_n$ and $\mathbf{v}_n^{\perp}$ as a pair of orthogonal ellipse direction vectors, corresponding to their radial extents (r_0, r_1) as well as the squared norm $\|\cdot\|_2^2$ and vector norm $|\cdot|$. For subsequent root-finding, it is the goal to convert the standard Eq. (5) to a quadratic polynomial with coefficients b_j given by, $B(x, y) = b_0 + b_1 x + b_2 y + b_3 x^2 + b_4 xy = 0$, which, when written in vector-matrix form reads,

$$0 = \begin{bmatrix} x & y \end{bmatrix} \begin{bmatrix} b_3 & b_4/2 \\ b_4/2 & b_5 \end{bmatrix} \begin{bmatrix} x \\ y \end{bmatrix} + \begin{bmatrix} b_1 & b_2 \end{bmatrix} \begin{bmatrix} x \\ y \end{bmatrix} + b_0 = \mathbf{s}^{\mathsf{T}} \mathbf{B} \mathbf{s} + \mathbf{b}^{\mathsf{T}} \mathbf{s} + b_0, \quad (6)$$

where $\mathbf{B}$ and $\mathbf{b}$ carry high-order polynomial coefficients b_j found via matrix factorization [6]. An elaborated version of this is found in the supplementary material.

Let two intersecting ellipses be given as quadratic equations $A(x, y)$ and $B(x, y)$ with coefficients a_j and b_j, respectively. Their intersection is found via polynomial root-finding of the equation,

$$D(x, y) = d_0 + d_1 x + d_2 y + d_3 x^2 + d_4 xy = 0, \quad (7)$$

where $\forall j, d_j = a_j - b_j$. When defining $y = w - (a_2 + a_4 x)/2$ to substitute y, we get $A(x, w) = w^2 + (a_0 + a_1 x + a_3 x^2) - (a_2 + a_4 x)^2/4 = 0$ which after rearranging is plugged into (7) to yield an intersection point $\mathbf{s}_i^{\star} = [x_i, w_i]^{\mathsf{T}}$. We refer the interested reader to the insightful descriptions in [6] for further implementation details.

2.3 Clustering

Micro bubble reflections are dispersed across multiple waveform channels yielding groups of location candidates for the same target bubble. Localization deviations result from ToA variations, which can occur due to atmospheric conditions, receiver clock errors, and system noise. Due to the random distribution of corresponding ToA errors [8], we regard these candidates as clusters. Thus, we aim to find a centroid $\mathbf{p}^\star$ of each cluster using multiple bi-variate probability density functions of varying sample sizes by,

$$
m(\mathbf{p}^{(j)}) = \frac{\sum_{\mathbf{s}_i^\star \in \Omega^{(j)}} \exp\left(\|\mathbf{s}_i^\star - \mathbf{p}^{(j)}\|_2^2\right) \mathbf{s}_i^\star}{\sum_{\mathbf{s}_i^\star \in \Omega^{(j)}} \exp\left(\|\mathbf{s}_i^\star - \mathbf{p}^{(j)}\|_2^2\right)}
\tag{8}
$$

Here, the bandwidth of the kernel is set to $\lambda/4$. The Mean Shift algorithm updates the estimate $\mathbf{p}^{(j)}$ by setting it to the weighted mean density on each iteration j until convergence. In this way, we obtain the position of the target bubble.

3 Experiments

Dataset: We demonstrate the feasibility of our geometric ULM and present benchmark comparison outcomes based on the PALA dataset [11]. This dataset is chosen as it is publicly available, allowing easy access and reproducibility of our results. To date, it is the only public ULM dataset featuring Radio Frequency (RF) data as required by our method. Its third-party simulation data makes it possible to perform a numerical quantification and direct comparison of different baseline benchmarks for the first time, which is necessary to validate the effectiveness of our proposed approach.

Metrics: For MB localization assessment, the minimum Root Mean Squared Error (RMSE) between the estimated $\mathbf{p}^\star$ and the nearest ground truth position is computed. To align with the PALA study [11], only RMSEs less than $\lambda/4$ are considered true positives and contribute to the total RMSE of all frames. In cases where the RMSE distance is greater than $\lambda/4$, the estimated $\mathbf{p}^\star$ is a false positive. Consequently, ground truth locations without an estimate within the $\lambda/4$ neighbourhood are false negatives. We use the Jaccard Index to measure the MB detection capability, which considers both true positives and false negatives and provides a robust measure of each algorithm's performance. The Structural Similarity Index Measure (SSIM) is used for image assessment.

For a realistic analysis, we employ the noise model used in [11], which is given by,

$$
n(t) \sim \mathcal{N}(0, \sigma_p^2) \times \max(y_n(t)) \times 10^{(L_A + L_C)/20} \pm \max(y_n(t)) \times 10^{L_C/20}, \tag{9}
$$

where $\sigma_p = \sqrt{B \times 10^{P/10}}$ and $\mathcal{N}(0, \sigma_p^2)$ are normal distributions with mean 0 and variance σ_p^2. Here, L_C and L_A are noise levels in dB, and $n(t)$ is the array of length T containing the random values drawn from this distribution. The additive noise model is then used to simulate a waveform channel $y'_n(t) = y_n(t) + n(t) \circledast g(t, \sigma_f)$ suffering from noise, where $\circledast$ represents the convolution operator, and $g(t, \sigma_f)$ is the one-dimensional Gaussian kernel with standard deviation $\sigma_f = 1.5$. To mimic the noise reduction achieved through the use of sub-aperture beamforming with 16 transducer channels [11], we multiplied the RF data noise by a factor of 4 for an equitable comparison.

Baselines: We compare our approach against state-of-the-art methods that utilize beamforming together with classical image filterings [8], Spline interpolation [17], Radial Symmetry (RS) [11] and a deep-learning-based U-Net [16] for MB localization. To only focus on the localization performance of each algorithm, we conduct the experimental analysis without temporal tracking. We obtain the results for classical image processing approaches directly from the open-source code provided by the authors of the PALA dataset [11]. As there is no publicly available implementation of [16] to date, we model and train the U-Net [15] according to the paper description, including loss design, layer architecture, and the incorporation of dropout. Since the U-Net-based localization is a supervised learning approach, we split the PALA dataset into sequences 1–15 for testing and 16–20 for training and validation, with a split ratio of 0.9, providing a sufficient number of 4500 training frames.

Results: Table 1 provides the benchmark comparison results with state-of-the-art methods. Our proposed geometric inference indicates the best localization performance represented by an average RMSE of around one-tenth of a wavelength. Also, the Jaccard Index reflects an outperforming balance of true positive

Table 1. Summary of localization results using 15k frames of the PALA dataset [11]. The RMSE is reported as mean±std, best scores are bold and units are given in brackets.

Method	Channels [#]	RMSE [$\lambda/10$]	Jaccard [%]	Time [s]	SSIM [%]
Weighted Avg. [11]	128	1.287 ± 0.162	44.253	0.080	69.49
Lanczos [11]	128	1.524 ± 0.175	38.688	0.382	75.87
RS [11]	128	1.179 ± 0.172	50.330	0.099	72.17
Spline [17]	128	1.504 ± 0.174	39.370	0.277	75.72
2-D Gauss Fit [17]	128	1.240 ± 0.162	51.342	3.782	73.93
U-Net [16]	128	1.561 ± 0.154	52.078	**0.004**	90.07
G-ULM (proposed)	8	1.116 ± 0.206	38.113	0.268	79.74
	16	1.077 ± 0.136	66.414	0.485	87.10
	32	1.042 ± 0.125	72.956	0.945	92.18
	64	1.036 ± 0.124	73.175	1.317	**93.70**
	128	$\mathbf{0.967 \pm 0.109}$	**78.618**	3.747	92.02

and false negative MB detections by our approach. These results support the hypothesis that our proposed geometric localization inference is a considerable alternative to existing beamforming-based methods. Upon closer examination of the channels column in Table 1, it becomes apparent that our geometric ULM achieves reasonable localization performance with only a fraction of the 128 channels available in the transducer probe. Using more than 32 channels improves the Jaccard Index but at the expense of computational resources. This finding confirms the assumption that transducers are redundant for MB tracking. The slight discrepancy in SSIM scores between our 128-channel results and the 64-channel example may be attributed to the higher number of false positives in the former, which decreases the overall SSIM value.

We provide rendered ULM image regions for visual inspection in Fig. 3 with full frames in the supplementary material. To enhance visibility, all images are processed with sRGB and additional gamma correction using an exponent of 0.9. The presence of noisy points in Figs. 3b to 3d is attributed to the exces-

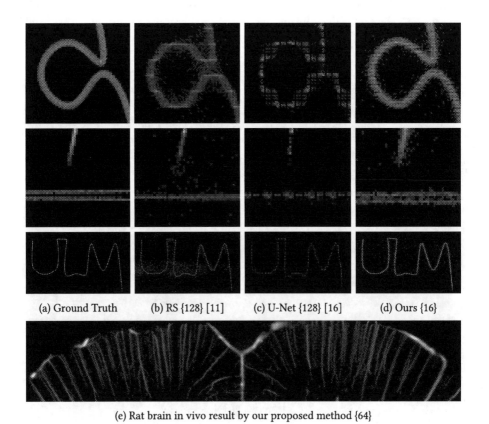

(a) Ground Truth   (b) RS {128} [11]   (c) U-Net {128} [16]   (d) Ours {16}

(e) Rat brain in vivo result by our proposed method {64}

Fig. 3. Rendered ULM regions from Table 1 in (a) to (d) and a rat brain result in (e) without temporal tracking. Numbers in curly brackets indicate the transducer number.

sive false positive localizations, resulting in poorer SSIM scores. Overall, these visual observations align with the numerical results presented in Table 1. An NVIDIA RTX2080 GPU was used for all computations and time measurements. To improve performance, signal processing chains are often pipelined, allowing for the simultaneous computation of subsequent processes. Table 1 lists the most time-consuming process for each method, which acts as the bottleneck. For our approach, the MEMGO feature extraction is the computationally most expensive process, followed by clustering. However, our method contributes to an overall efficient computation and acquisition time, as it skips beamforming and coherent compounding [12] with the latter reducing the capture interval by two-thirds.

Table 2 presents the results for the best-of-3 algorithms at various noise levels L_C. As the amount of noise from (9) increases, there is a decline in the Jaccard Index, which suggests that each method is more susceptible to false detections from noise clutter. Although our method is exposed to higher noise in the RF domain, it is seen that L_C has a comparable impact on our method. However, it is important to note that the U-Net yields the most steady and consistent results for different noise levels.

Table 2. Performance under noise variations from 128 (others) vs 16 (ours) transducers.

Noise	RMSE [$\lambda/10$]			Jaccard Index [%]		
L_C [dB]	RS [11]	U-Net [16]	Ours	RS [11]	U-Net [16]	Ours
-30	1.245 ± 0.171	1.564 ± 0.151	1.076 ± 0.136	54.036	51.032	65.811
-20	1.496 ± 0.223	1.459 ± 0.165	1.262 ± 0.249	27.037	45.647	27.962
-10	1.634 ± 0.542	1.517 ± 0.238	1.459 ± 0.564	2.510	18.162	3.045

4   Summary

This study explored whether a geometric reconstruction may serve as an alternative to beamforming in ULM. We employed an energy-based model for feature extraction in conjunction with ellipse intersections and clustering to pinpoint contrast agent positions from RF data available in the PALA dataset. We carried out a benchmark comparison with state-of-the-art methods, demonstrating that our geometric model provides enhanced resolution and detection reliability with fewer transducers. This capability will be a stepping stone for 3-D ULM reconstruction where matrix transducer probes typically consist of 32 transducers per row only. It is essential to conduct follow-up studies to evaluate the high potential of our approach in an extensive manner before entering a pre-clinical phase. The promising results from this study motivate us to expand our research to more RF data scenarios. We believe our findings will inspire further research in this exciting and rapidly evolving field.

Acknowledgment. This research is supported by the Hasler Foundation under project number 22027.

References

1. Bar-Shira, O., et al.: Learned super resolution ultrasound for improved breast lesion characterization. In: de Bruijne, M., et al. (eds.) MICCAI 2021. LNCS, vol. 12907, pp. 109–118. Springer, Cham (2021). https://doi.org/10.1007/978-3-030-87234-2_11

2. Christensen-Jeffries, K., et al.: Microbubble axial localization errors in ultrasound super-resolution imaging. IEEE Trans. Ultrason. Ferroelectr. Freq. Control **64**(11), 1644–1654 (2017)

3. Corazza, A., Muleki-Seya, P., Aissani, A.W., Couture, O., Basarab, A., Nicolas, B.: Microbubble detection with adaptive beamforming for ultrasound localization microscopy. In: 2022 IEEE International Ultrasonics Symposium (IUS), pp. 1–4. IEEE (2022)

4. Couture, O., Hingot, V., Heiles, B., Muleki-Seya, P., Tanter, M.: Ultrasound localization microscopy and super-resolution: a state of the art. IEEE Trans. Ultrason. Ferroelectr. Freq. Control **65**(8), 1304–1320 (2018)

5. De Marchi, A., et al.: Perfusion pattern and time of vascularisation with ceus increase accuracy in differentiating between benign and malignant tumours in 216 musculoskeletal soft tissue masses. Eur. J. Radiol. **84**(1), 142–150 (2015)

6. Eberly, D.: Intersection of ellipses. Technical report, Geometric Tools, Redmond WA 98052 (2020)

7. Elbanna, K.Y., Jang, H.-J., Kim, T.K., Khalili, K., Guimarães, L.S., Atri, M.: The added value of contrast-enhanced ultrasound in evaluation of indeterminate small solid renal masses and risk stratification of cystic renal lesions. Eur. Radiol. **31**(11), 8468–8477 (2021). https://doi.org/10.1007/s00330-021-07964-0

8. Errico, C., et al.: Ultrafast ultrasound localization microscopy for deep super-resolution vascular imaging. Nature **527**(7579), 499–502 (2015)

9. Hahne, C.: Multimodal exponentially modified gaussian oscillators. In: 2022 IEEE International Ultrasonics Symposium (IUS), pp. 1–4 (2022)

10. Hahne, C.: 3-dimensional sonic phase-invariant echo localization. In: 2023 IEEE International Conference on Robotics and Automation (ICRA), pp. 4121–4127 (2023)

11. Heiles, B., Chavignon, A., Hingot, V., Lopez, P., Teston, E., Couture, O.: Performance benchmarking of microbubble-localization algorithms for ultrasound localization microscopy. Nature Biomed. Eng. **6**(5), 605–616 (2022)

12. Montaldo, G., Tanter, M., Bercoff, J., Benech, N., Fink, M.: Coherent plane-wave compounding for very high frame rate ultrasonography and transient elastography. IEEE Trans. Ultrason. Ferroelectr. Freq. Control **56**(3), 489–506 (2009)

13. Nair, A.A., Tran, T.D., Reiter, A., Bell, M.A.L.: A deep learning based alternative to beamforming ultrasound images. In: 2018 IEEE International Conference on Acoustics, Speech and Signal Processing (ICASSP), pp. 3359–3363. IEEE (2018)

14. Renaudin, N., Demené, C., Dizeux, A., Ialy-Radio, N., Pezet, S., Tanter, M.: Functional ultrasound localization microscopy reveals brain-wide neurovascular activity on a microscopic scale. Nat. Methods **19**(8), 1004–1012 (2022)

15. Ronneberger, O., Fischer, P., Brox, T.: U-Net: convolutional networks for biomedical image segmentation. In: Navab, N., Hornegger, J., Wells, W.M., Frangi, A.F. (eds.) MICCAI 2015, Part III. LNCS, vol. 9351, pp. 234–241. Springer, Cham (2015). https://doi.org/10.1007/978-3-319-24574-4_28

16. van Sloun, R.J., et al.: Super-resolution ultrasound localization microscopy through deep learning. IEEE Trans. Med. Imaging **40**(3), 829–839 (2020)

17. Song, P., Manduca, A., Trzasko, J., Daigle, R., Chen, S.: On the effects of spatial sampling quantization in super-resolution ultrasound microvessel imaging. IEEE Trans. Ultrason. Ferroelectr. Freq. Control **65**(12), 2264–2276 (2018)

18. Zonzini, F., Bogomolov, D., Dhamija, T., Testoni, N., De Marchi, L., Marzani, A.: Deep learning approaches for robust time of arrival estimation in acoustic emission monitoring. Sensors **22**(3), 1091 (2022)

Global k-Space Interpolation for Dynamic MRI Reconstruction Using Masked Image Modeling

Jiazhen Pan[1]([✉]), Suprosanna Shit[1], Özgün Turgut[1], Wenqi Huang[1],
Hongwei Bran Li[1,2], Nil Stolt-Ansó[3], Thomas Küstner[4],
Kerstin Hammernik[3,5], and Daniel Rueckert[1,3,5]

[1] School of Medicine, Klinikum Rechts der Isar, Technical University of Munich,
Munich, Germany
jiazhen.pan@tum.de, suprosanna.shit@tum.de
[2] Department of Quantitative Biomedicine, University of Zurich, Zürich, Switzerland
[3] School of Computation, Information and Technology, Technical University of
Munich, Munich, Germany
[4] Medical Image and Data Analysis, University Hospital of Tübingen, Tübingen,
Germany
[5] Department of Computing, Imperial College London, London, UK

Abstract. In dynamic Magnetic Resonance Imaging (MRI), k-space is typically undersampled due to limited scan time, resulting in aliasing artifacts in the image domain. Hence, dynamic MR reconstruction requires not only modeling spatial frequency components in the x and y directions of k-space but also considering temporal redundancy. Most previous works rely on image-domain regularizers (priors) to conduct MR reconstruction. In contrast, we focus on interpolating the undersampled k-space before obtaining images with Fourier transform. In this work, we connect masked image modeling with k-space interpolation and propose a novel Transformer-based k-space Global Interpolation Network, termed k-GIN. Our k-GIN learns global dependencies among low- and high-frequency components of 2D+t k-space and uses it to interpolate unsampled data. Further, we propose a novel k-space Iterative Refinement Module (k-IRM) to enhance the high-frequency components learning. We evaluate our approach on 92 in-house 2D+t cardiac MR subjects and compare it to MR reconstruction methods with image-domain regularizers. Experiments show that our proposed k-space interpolation method quantitatively and qualitatively outperforms baseline methods. Importantly, the proposed approach achieves substantially higher robustness and generalizability in cases of highly-undersampled MR data.

Keywords: Cardiac MR Imaging Reconstruction · k-space
Interpolation · Masked Image Modeling · Masked Autoencoders ·
Transformers

Supplementary Information The online version contains supplementary material available at https://doi.org/10.1007/978-3-031-43999-5_22.

1   Introduction

CINE Cardiac Magnetic Resonance (CMR) imaging is widely recognized as the gold standard for evaluating cardiac morphology and function [19]. Raw data for CMR is acquired in the frequency domain (k-space). MR reconstruction from k-space data with high spatio-temporal resolutions throughout the cardiac cycle is an essential step for CMR. Short scan times, ideally within a single breath-hold, are preferable to minimize patient discomfort and reduce potential image artifacts caused by patient motion. Typically, due to the restricted scanning times, only a limited amount of k-space data can be obtained for each temporal frame. Note that while some k-space data are unsampled, the sampled ones are reliable sources of information. However, the Fourier transform of undersampled k-space corrupts a broad region of pixels in the image domain with aliasing artifacts because of violating the Nyquist-Shannon sampling theorem. Previous works [9,15,21,26,30] have attempted to remove the image artifacts primarily by regularizing on the image domain using conventional/learning-based image priors. However, after the Fourier transform the artifacts may completely distort and/or obscure tissues of interest before the image-domain regularizers kick in, making these methods challenging to recover true tissue structures.

On a different principle, k-space interpolation methods first attempt to estimate the full k-space leveraging redundancy in sampled frequency components before Fourier transform. Image domain methods rely on artifacts-specific image priors to denoise the corrupted pixels, making them susceptible to variability in artifact types arising from different undersampling factors. Unlike image domain methods, k-space-based methods have a consistent task of interpolating missing data from reliable sampled ones, even though the undersampling factor may vary. This makes k-space interpolation methods simple, robust and generic over multiple undersampling factors.

In this work, we are interested in learning an entirely k-space-based interpolation for the Cartesian undersampled dynamic MR data. An accurate learnable k-space interpolator can be achieved via (1) *a rich representation of the sampled k-space data*, which can facilitate the exploitation of the limited available samples, and (2) *global dependency modeling of k-space* to interpolate unsampled data from the learned representation. Modeling global dependencies are beneficial because a local structure in the image domain is represented by a wide range of frequency components in k-space. Furthermore, in the context of dynamic MR, the interpolator also has to exploit temporal redundancies.

In the recent past, masked image modeling [11,34] has emerged as a promising method for learning rich generalizable representation by reconstructing the whole image from a masked (undersampled) input. Masked Autoencoders (MAE) [11] are one such model that leverages the *global dependencies* of the undersampled input using Transformers and learns masked-based *rich feature representation*. Despite sharing the same reconstruction principle, MAE has not been explored in k-space interpolation of Cartesian undersampled data. In this work, we cast 2D+t k-space interpolation as a masked signal reconstruction problem and propose a novel Transformer-based method entirely in k-space. Further, we introduce a refinement module on k-space to boost the accuracy of high-frequency interpolation. **Our contributions can be summarized as follows**:

1. We propose a novel k-space Global Interpolation Network, termed k-GIN, leveraging masked image modeling for the first time in k-space. To the best of our knowledge, our work enables the first Transformer-based k-space interpolation for 2D+t MR reconstruction.
2. Next, we propose k-space Iterative Refinement Module, termed k-IRM that refines k-GIN interpolation by efficiently gathering spatio-temporal redundancy of the MR data. Crucially, k-IRM specializes in learning high-frequency details with the aid of customized High-Dynamic-Range (HDR) loss.
3. We evaluate our approach on 92 in-house CMR subjects and compare it to model-based reconstruction baselines using image priors. Our experiments show that the proposed k-space interpolator outperforms baseline methods with superior qualitative and quantitative results. Importantly, our method demonstrates improved robustness and generalizability regarding varying undersampling factors than the model-based counterparts.

2   Related Work

Reconstruction with image priors is broadly used with either image-only denoising [17,31,36] or with model-based approaches to incorporate the k-space consistency by solving an inverse problem. For the latter, the physics-based model can be formulated as a low-rank and a sparse matrix decomposition in CMR [14,26], or a motion-compensated MR reconstruction problem [3,27,28], or data consistency terms with convolution-based [9,30] or Transformers-based [12,18] image regularizers.

k-space-Domain Interpolation. Methods include works [8,22], which have introduced auto-calibration signals (ACS) in the multi-coil k-space center of Cartesian sampled data. RAKI [2,16] uses convolutional networks for optimizing imaging and scanner-specific protocols. Nevertheless, these methods have limited flexibility during the scanning process since they all require a fixed set of ACS data in k-space. [20,32] introduced k-space interpolation methods which do not require calibration signals. [10] proposed a k-space U-Net under residual learning setting. However, these methods heavily rely on local operators such as convolution and may overlook non-local redundancies. [7] uses Transformers applicable only on radial sampled k-space data. However, using masked image modeling with Transformers in k-space for dynamic MR imaging e.g. 2D+t CMR data has not been studied yet.

Hybrid Approaches. Combine information from both k-space and image-domain. KIKI-Net [6] employs an alternating optimization between the image domain and k-space. [25,33] use parallel architectures for k-space and image-domain simultaneously. However, their ablation shows limited contribution coming from the k-space compared to the image domain, implying an under-exploitation of the k-space. Concurrently, [37] use Transformers in k-space but their performance is heavily dependent on the image domain fine-tuning at the final stage.

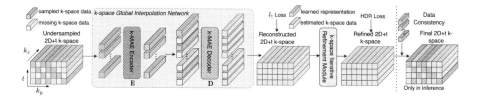

Fig. 1. The proposed k-space-based dynamic MR reconstruction framework consists of k-space Global Interpolation Network (k-GIN) (see 3.1) and k-space Iterative Refinement Module (k-IRM) for refining the k-GIN interpolation (see 3.2). In the final stage for the inference, we replace k-space estimation at the sampled position with ground-truth k-space values, ensuring the data consistency.

3 Method

Fully sampled complex 2D+t dynamic MR k-space data can be expressed as $\mathbf{y} \in \mathbb{C}^{XYT}$, X and Y are the height (k_x) and the width (k_y) of the k-space matrix, and T is the number of frames along time. In this work, we express k-space as 2 channels (real and imaginary) data $\mathbf{y} \in \mathbb{R}^{2XYT}$. For the MR acquisition, a binary Cartesian sampling mask $M \in \mathbb{Z}^{YT}|M_{ij} \in \{0,1\}$ is applied in the k_y-t plane, i.e. all k-space values along the k_x (readout direction) are sampled if the mask is 1, and remains unsampled if the mask is 0. Figure 1 shows a pictorial undersampled k-space data. Let us denote the collection of sampled k-space lines as $\mathbf{y}_s$ and unsampled lines as $\mathbf{y}_u$. The dynamic MR reconstruction task is to estimate $\mathbf{y}_u$ and reconstruct $\mathbf{y}$ using $\mathbf{y}_s$ only. In this work, we propose a novel Transformer-based reconstruction framework consisting of 1) k-GIN to learn global representation and 2) k-IRM to achieve refined k-space interpolation with a focus on high-frequency components.

3.1 k-space Global Interpolation Network (k-GIN)

In our proposed approach, we work on the k_y-t plane and consider k_x as the channel dimension. Further, we propose each point in the k_y-t plane to be an individual *token*. In total, we have YT number of tokens, out of which YT/R are sampled tokens for an undersampling factor of R. Our objective is to contextualize global dependencies among every sampled token. For that, we use a ViT/MAE [5,11] encoder $\mathbf{E}$ consisting of alternating blocks of multi-head self-attention and multi-layer-perceptrons. The encoder takes advantage of each token's position embedding to correctly attribute its location. Following ViT, we use LayerNorm and GELU activation. We obtain rich feature representation $\mathbf{f}_E = \mathbf{E}(\mathbf{y}_s)$ of the sampled k-space from the encoder.

Next, we want a preliminary estimate of the undersampled k-space data from the learned feature representation. To this end, we employ a decoder $\mathbf{D}$ of similar architecture as the encoder. We initialize all the unsampled tokens $\mathbf{y}_u$ with a single learnable token shared among them. Subsequently, we add their corresponding position embedding to these unsampled tokens. During the decoding

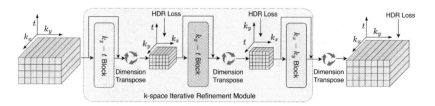

Fig. 2. k-space Iterative Refinement Module (k-IRM) refines high-frequency components of the k-space data (see 3.1). Its refinement Transformer blocks extract the spatio-temporal redundancy by operating the on k_y-t, k_x-t and k_x-k_y plane.

process, the unsampled tokens attend to the well-contextualized features $\mathbf{f}_E$ and produce an estimate of the whole k-space $\hat{\mathbf{y}}_r = \mathbf{D}\left([\mathbf{f}_E, \mathbf{y}_u]\right)$. Since our masking pattern includes more sampled data in low-frequency than high-frequency components, we observe k-GIN gradually learn from low-frequency to high-frequency. Note that the imbalance of magnitude in k-space results in more emphasis on low-frequency when ℓ_1 loss is applied between estimation and ground-truth. We leverage this property into the learning behavior of k-GIN and deliberately use ℓ_1 loss between $\hat{\mathbf{y}}_r$ and $\mathbf{y}$, read as $\mathcal{L}_{\ell_1} = \|\hat{\mathbf{y}}_r - \mathbf{y}\|_1$.

3.2 k-space Iterative Refinement Module (k-IRM)

Using ℓ_1 loss in k-GIN makes it focus more on the low-frequency components learning but the high-frequency estimation is still sub-optimal. Inspired by the iterative refinement strategy [38] which is widely used to improve estimation performance, we propose to augment k-GIN's expressive power, especially in high-frequency components, with k-space Iterative Refinement Module. This consists of three Transformer blocks that operate on three orthogonal planes. All three blocks are identical in architecture. The first block operates on k_y-t plane and treats k_x as channel dimension. The second block operates on k_x-t plane and considers k_y as channels, while the final block operates on k_x-k_y plane with t as channel dimension. Note that for the final refinement block uses 4×4 size token while the previous two blocks consider each point as a single token. These configurations enable scalable exploration of the spatio-temporal redundancy present in the output of the k-GIN. We denote $\hat{\mathbf{y}}_1$, $\hat{\mathbf{y}}_2$ and $\hat{\mathbf{y}}_3$ as the estimation after each block in the k-IRM. We apply the skip connection between each refinement block and iteratively minimize the residual error at each stage.

Inspired by [13,24], we applied an approximated logarithm loss function called High-Dynamic Range (HDR) loss for all the stages of k-IRM. HDR loss handles the large magnitude difference in the k-space data and makes the network pay more attention to high-frequency learning. The HDR loss function is defined as: $\mathcal{L}_{\mathrm{HDR}} = \sum_{i=1}^{3} \left\| \frac{\hat{\mathbf{y}}_i - \mathbf{y}}{s(\hat{\mathbf{y}}) + \epsilon} \right\|_2^2$ where $s(\cdot)$ is the stop-gradient operator preventing the network back-propagation of estimation in the denominator and ϵ controls the operational range of the logarithmic approximation.

3.3   Inference

The inference is identical to the training till obtaining the refined output from the k-IRM. Then we replace k-space estimation at the sampled position with ground-truth k-space values, ensuring the data-consistency. Note that this step is not done during the training as it deteriorates learning k-space representation. Once full k-space has been estimated, we use Fourier transform to obtain the image reconstruction during the inference. Note that image reconstruction is not needed during the training since our framework is entirely based in k-space.

4   Data and Experiments

Dataset. The training was performed on 81 subjects (a mix of patients and healthy subjects) of in-house acquired short-axis 2D CINE CMR, whereas testing was carried out on 11 subjects. Data were acquired with 30/34 multiple receiver coils and 2D balanced steady-state free precession sequence on a 1.5T MR (Siemens Aera with TE=1.06 ms, TR=2.12 ms, resolution=1.9×1.9mm^2 with 8mm slice thickness, 8 breath-holds of 15 s duration). The MR data were acquired with a matrix size of 192×156 with 25 temporal cardiac phases (40ms temporal resolution). Afterwards, these data were converted to single-coil MR imaging and k-space data using coil sensitivity maps, simulating a fully sampled single-coil acquisition. A stack of 12 slices along the long axis was collected, resulting in 415/86 image sequence (2D+t) for training/test.

Implementation Details. We use an NVIDIA A6000 GPU to train our framework. The batch size was set to 1 with a one-cycle learning-rate scheduler (max. learning rate 0.0001). We use 8 layers, 8 heads and 512 embedding dimensions for all of our Transformer blocks. We train our network with joint ℓ_1 and HDR-loss with ϵ tuned to 0.5. Training and inference were carried out on retrospectively undersampled images with masks randomly generated by VISTA [1]. We train the network with $R = 4$ undersampled data while we test our method on an undersampled factor $R = 4$, 6 and 8 during the inference. We can use this inference strategy to test our model's generalizability and robustness to different undersampling factors in comparison to the following baseline methods.

Baseline Methods and Metrics. We compare the proposed framework with three single-coil MR reconstruction methods that apply image priors: TV-norm Optimization (TV-Optim) which is widely used in reconstruction [21,23], L+S [26] and DcCNN [30]. TV-Optim reconstructs the image using TV-norm [29] as the image regularizer. L+S leverages compressed sensing techniques and addresses the reconstruction using low rankness and sparsity of the CMR as the image prior, whilst DcCNN employs 3D convolutional neural networks in the image domain together with data-consistency terms. We use the same training and inference strategy for DcCNN to test its model robustness.

Table 1. Quantitative analysis of reconstruction for accelerated CINE CMR (R=4, 6 and 8) using TV-Optim, L+S [26], DcCNN [30] and the proposed method. PSNR (sequence based), SSIM and NMSE are used to evaluate the reconstruction performance. The mean value with the standard deviations are shown. The best results are marked in bold.

Acc R	Methods	NMSE ↓	SSIM ↑	PSNR ↑
4	TV-Optim	0.120 ± 0.031	0.922 ± 0.026	36.743 ± 3.233
	L+S [26]	0.097 ± 0.026	0.949 ± 0.021	39.346 ± 2.911
	DcCNN [30]	$\mathbf{0.087 \pm 0.022}$	0.957 ± 0.019	40.293 ± 2.891
	Proposed	0.088 ± 0.022	$\mathbf{0.958 \pm 0.019}$	$\mathbf{40.368 \pm 3.030}$
6	TV-Optim	0.161 ± 0.047	0.887 ± 0.040	34.066 ± 3.605
	L+S	0.154 ± 0.042	0.901 ± 0.036	34.921 ± 3.174
	DcCNN	0.116 ± 0.029	0.932 ± 0.026	37.666 ± 2.768
	Proposed	$\mathbf{0.109 \pm 0.029}$	$\mathbf{0.940 \pm 0.026}$	$\mathbf{38.461 \pm 3.095}$
8	TV-Optim	0.289 ± 0.078	0.808 ± 0.067	29.052 ± 3.817
	L+S	0.245 ± 0.061	0.826 ± 0.047	30.413 ± 2.888
	DcCNN	0.276 ± 0.047	0.821 ± 0.040	29.778 ± 2.822
	Proposed	$\mathbf{0.151 \pm 0.041}$	$\mathbf{0.904 \pm 0.036}$	$\mathbf{35.674 \pm 3.293}$

We Fourier transform our interpolated full k-space to obtain the image reconstruction and utilize Peak Signal-to-Noise Ratio (PSNR), Structural Similarity Index (SSIM) and Normalized Mean Squared Error (NMSE) to evaluate the reconstruction performance with the baseline quantitatively.

5 Results and Discussion

The quantitative results in Table 1 show consistent superior performance of the proposed method across every single undersampling factor compared to all other baseline methods. Figure 3 shows a qualitative comparison for a typical test sample. It can be seen that the reconstruction methods with image priors can still provide comparable results at $R = 4$, however, suffer from a large performance drop when acceleration rates get higher, especially at $R = 8$. The non-trivial hyper-parameters tuning has to be carried out for L+S and TV-Optim to adapt to the specific image prior at different acceleration factors. It is also noteworthy that the proposed method and DcCNN are both trained only on $R = 4$ undersampled CMR. DcCNN demonstrates inferior reconstruction for $R = 8$ since there is a mismatch in artifact characteristics between $R = 4$ and $R = 8$. On the contrary, the task of interpolating k-space for $R = 4$ and $R = 8$ remains the same, i.e., to estimate missing data from sampled data. We efficiently leverage rich contextualized representation of k-GIN to interpolate full k-space even when a lesser number of sampled k-space data are given as input than seen during training. The observation confirms the superior robustness and generalizability of our proposed framework.

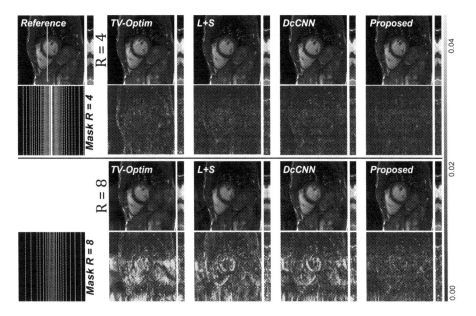

Fig. 3. Qualitative comparison of the proposed method with TV-Optim, L+S and DcCNN in the $R = 4$ and 8 undersampled data. Reference images, undersampling masks, reconstructed (x-y and y-t plane) images and their corresponding error maps are showcased. The selected y-axis is marked with a yellow line in the reference image.

Next, we conduct an ablation study to validate our architectural design. We carry out experiments to investigate the impact of applying k-IRM. We conduct the interpolation using 1) only k-GIN, 2) k-GIN + k_y-t plane refinement, 3) k-GIN + k_x-t plane refinement, 4) k-GIN + k_x-k_y refinement and 5) k-GIN with all three refinement blocks. Table 2 in supplementary presents quantitative comparisons amongst five configurations as above. We observe k-IRM offers the best performance when all 3 refinement blocks are used together. In the second ablation, Table 3 in supplementary shows the usefulness of applying ℓ_1 in k-GIN and HDR in k-IRM. HDR makes k-GIN's learning inefficient since HDR deviates from its learning principle of "first low-frequency then high-frequency". On the other hand, $\ell_1 + \ell_1$ combination hinders high-frequency learning.

Outlook. Previous works [6,25] have speculated limited usefulness coming from k-space in a hybrid setting. However, our work presents strong evidence of k-space representation power which can be leveraged in future work with hybrid reconstruction setup. Furthermore, one can utilize our work as a pre-training task since the image reconstruction itself is an "intermediate step" for downstream tasks e.g. cardiac segmentation and disease classification. In the future, one can reuse the learned encoder representation of k-GIN to directly solve downstream tasks without requiring image reconstruction.

Limitation. We also acknowledge some limitations of the work. First, we have not evaluated our method on prospectively collected data, which would be our focus in future work. Second, the current study only investigates the single coil setup due to hardware memory limitations. In the future, we will address the multi-coil scenario by applying more memory-efficient Transformers backbones e.g. [4,35].

6 Conclusion

In this work, we proposed a novel Transformer-based method with mask image modeling to solve the dynamic CMR reconstruction by only interpolating the k-space without any image-domain priors. Our framework leverages Transformers' global dependencies to exploit redundancies in all three k_x-, k_y- and t-domain. Additionally, we proposed a novel refinement module (k-IRM) to boost high-frequency learning in k-space. Together, k-GIN and k-IRM not only produce high-quality k-space interpolation and superior CMR reconstruction but also generalize significantly better than baselines for higher undersampling factors.

Acknowledgement. This work is partly supported by the European Research Council (ERC) with Grant Agreement no. 884622. Suprosanna Shit is supported by ERC with the Horizon 2020 research and innovation program (101045128-iBack-epic-ERC2021-COG). Hongwei Bran Li is supported by an Nvidia GPU research grant.

References

1. Ahmad, R., Xue, H., Giri, S., Ding, Y., Craft, J., Simonetti, O.P.: Variable density incoherent spatiotemporal acquisition (VISTA) for highly accelerated cardiac MRI. Magn. Reson. Med. **74**(5), 1266–1278 (2015)
2. Akçakaya, M., Moeller, S., Weingärtner, S., Uğurbil, K.: Scan-specific robust artificial-neural-networks for k-space interpolation (RAKI) reconstruction: database-free deep learning for fast imaging. Magn. Reson. Med. **81**(1), 439–453 (2019)
3. Batchelor, P., Atkinson, D., Irarrazaval, P., Hill, D., et al.: Matrix description of general motion correction applied to multishot images. Magn. Reson. Med. **54**, 1273–1280 (2005)
4. Dao, T., Fu, D., Ermon, S., Rudra, A., Ré, C.: FlashAttention: fast and memory-efficient exact attention with IO-awareness. In: Advances in Neural Information Processing Systems, vol. 35, pp. 16344–16359 (2022)
5. Dosovitskiy, A., et al.: An image is worth 16x16 words: transformers for image recognition at scale. arXiv preprint: arXiv:2010.11929 (2020)
6. Eo, T., Jun, Y., Kim, T., Jang, J., Lee, H.J., Hwang, D.: KIKI-Net: cross-domain convolutional neural networks for reconstructing undersampled magnetic resonance images. Magn. Reson. Med. **80**(5), 2188–2201 (2018)
7. Gao, C., Shih, S.F., Finn, J.P., Zhong, X.: A projection-based K-space transformer network for undersampled radial MRI reconstruction with limited training subjects. In: Wang, L., Dou, Q., Fletcher, P.T., Speidel, S., Li, S. (eds.) MICCAI 2022. Lecture Notes in Computer Science, vol. 13436. Springer, Cham (2022). https://doi.org/10.1007/978-3-031-16446-0_69

8. Griswold, M.A., et al.: Generalized autocalibrating partially parallel acquisitions (GRAPPA). Magn. Reson. Med. **47**(6), 1202–10 (2002)

9. Hammernik, K., Klatzer, T., Kobler, E., et al.: Learning a variational network for reconstruction of accelerated MRI data. Magn. Reson. Med. **79**(6), 3055–3071 (2018)

10. Han, Y., Sunwoo, L., Ye, J.C.: k-space deep learning for accelerated MRI. IEEE TMI **39**, 377–386 (2019)

11. He, K., Chen, X., Xie, S., Li, Y., Dollár, P., Girshick, R.: Masked autoencoders are scalable vision learners (2021). arXiv:2111.06377

12. Huang, J., Fang, Y., Wu, Y., Wu, H., et al.: Swin transformer for fast MRI. Neurocomputing **493**, 281–304 (2022)

13. Huang, W., Li, H.B., Pan, J., Cruz, G., Rueckert, D., Hammernik, K.: Neural implicit k-space for binning-free non-cartesian cardiac MR imaging. In: Frangi, A., de Bruijne, M., Wassermann, D., Navab, N. (eds.) IPMI 2023. Lecture Notes in Computer Science, vol. 13939. Springer, Cham (2023). https://doi.org/10.1007/978-3-031-34048-2_42

14. Huang, W., Ke, Z., Cui, Z.X., et al.: Deep low-rank plus sparse network for dynamic MR imaging. Med. Image Anal. **73**, 102190 (2021)

15. Jin, K.H., McCann, M.T., Froustey, E., Unser, M.: Deep convolutional neural network for inverse problems in imaging. IEEE TIP **26**(9), 4509–4522 (2017)

16. Kim, T., Garg, P., Haldar, J.: LORAKI: autocalibrated recurrent neural networks for autoregressive MRI reconstruction in k-space. arXiv preprint: arXiv:1904.09390 (2019)

17. Kofler, A., Dewey, M., Schaeffter, T., Wald, C., Kolbitsch, C.: Spatio-temporal deep learning-based undersampling artefact reduction for 2D radial cine MRI with limited training data. IEEE TMI **39**, 703–717 (2019)

18. Korkmaz, Y., et al.: Unsupervised MRI reconstruction via zero-shot learned adversarial transformers. IEEE TMI **41**(7), 1747–1763 (2022)

19. Lee, D., et al.: The growth and evolution of cardiovascular magnetic resonance: a 20-year history of the society for cardiovascular magnetic resonance (SCMR) annual scientific sessions. J. Cardiovasc. Magn. Reson. **20**(1), 1–11 (2018)

20. Lee, J., Jin, K., Ye, J.: Reference-free single-pass EPI Nyquist ghost correction using annihilating filter-based low rank Hankel matrix (ALOHA). Magn. Reson. Med. **76**(6), 1775–1789 (2016)

21. Lingala, S.G., Hu, Y., DiBella, E., Jacob, M.: Accelerated dynamic MRI exploiting sparsity and low-rank structure: k-t SLR. IEEE TMI **30**(5), 1042–1054 (2011)

22. Lustig, M., Pauly, J.M.: SPIRiT: iterative self-consistent parallel imaging reconstruction from arbitrary k-space. Magn. Reson. Med. **64**(2), 457–471 (2010)

23. Lustig, M., Donoho, D., Pauly, J.M.: Sparse MRI: the application of compressed sensing for rapid MR imaging. Magn. Reson. Med. **58**(6), 1182–1195 (2007)

24. Mildenhall, B., Hedman, P., Martin-Brualla, R., Srinivasan, P.P., Barron, J.T.: NeRF in the dark: high dynamic range view synthesis from noisy raw images. In: CVPR, pp. 16169–16178 (2022)

25. Nitski, O., Nag, S., McIntosh, C., Wang, B.: CDF-Net: cross-domain fusion network for accelerated MRI reconstruction. In: Martel, A.L., et al. (eds.) MICCAI 2020. LNCS, vol. 12262, pp. 421–430. Springer, Cham (2020). https://doi.org/10.1007/978-3-030-59713-9_41

26. Otazo, R., Candès, E., Sodickson, D.K.: Low-rank plus sparse matrix decomposition for accelerated dynamic MRI with separation of background and dynamic components. Magn. Reson. Med. **73**(3), 1125–1136 (2015)

27. Pan, J., Huang, W., Rueckert, D., Küstner, T., Hammernik, K.: Reconstruction-driven motion estimation for motion-compensated MR cine imaging. arXiv preprint: arXiv:2302.02504 (2023)
28. Pan, J., Rueckert, D., Kustner, T., Hammernik, K.: Learning-based and unrolled motion-compensated reconstruction for cardiac MR CINE imaging. In: Wang, L., Dou, Q., Fletcher, P.T., Speidel, S., Speidel, S. (eds.) MICCAI 2022. Lecture Notes in Computer Science, vol. 13436. Springer, Cham (2022). https://doi.org/10.1007/978-3-031-16446-0_65
29. Rudin, L.I., Osher, S., Fatemi, E.: Nonlinear total variation based noise removal algorithms. Physica D **60**(1), 259–268 (1992)
30. Schlemper, J., et al.: A deep cascade of convolutional neural networks for dynamic MR image reconstruction. IEEE TMI **37**(2), 491–503 (2018)
31. Shi, J., Liu, Q., Wang, C., Zhang, Q., Ying, S., Xu, H.: Super-resolution reconstruction of MR image with a novel residual learning network algorithm. Phys. Med. Biol. **63**(8), 085011 (2018)
32. Shin, P.J.: Calibrationless parallel imaging reconstruction based on structured low-rank matrix completion. Magn. Reson. Med. **72**(4), 959–970 (2014)
33. Singh, N.M., Iglesias, J.E., Adalsteinsson, E., Dalca, A.V., Golland, P.: Joint frequency- and image-space learning for fourier imaging. Mach. Learn. Biomed. Imaging (2022)
34. Xie, Z., et al.: SimMIM: a simple framework for masked image modeling. In: CVPR (2022)
35. Xiong, Y., et al.: Nyströmformer: a nyström-based algorithm for approximating self-attention. In: AAAI, vol. 35, pp. 14138–14148 (2021)
36. Yang, G., Yu, S., Dong, H., Slabaugh, G.G., et al.: DAGAN: deep de-aliasing generative adversarial networks for fast compressed sensing MRI reconstruction. IEEE TMI **37**(6), 1310–1321 (2018)
37. Zhao, Z., Zhang, T., Xie, T., Wang, Y., Zhang, Y.: K-space transformer for under-sampled MRI reconstruction. In: BMVC (2022)
38. Zhu, X., Su, W., Lu, L., Li, B., Wang, X., Dai, J.: Deformable DETR: deformable transformers for end-to-end object detection. arXiv preprint: arXiv:2010.04159 (2020)

Contrastive Diffusion Model with Auxiliary Guidance for Coarse-to-Fine PET Reconstruction

Zeyu Han[1], Yuhan Wang[1], Luping Zhou[2], Peng Wang[1], Binyu Yan[1], Jiliu Zhou[1,3], Yan Wang[1(✉)], and Dinggang Shen[4,5(✉)]

[1] School of Computer Science, Sichuan University, Chengdu, China
`wangyanscu@hotmail.com`
[2] School of Electrical and Information Engineering, University of Sydney, Sydney, Australia
[3] School of Computer Science, Chengdu University of Information Technology, Chengdu, China
[4] School of Biomedical Engineering, ShanghaiTech University, Shanghai, China
`dinggang.shen@gmail.com`
[5] Department of Research and Development, Shanghai United Imaging Intelligence Co., Ltd., Shanghai, China

Abstract. To obtain high-quality positron emission tomography (PET) scans while reducing radiation exposure to the human body, various approaches have been proposed to reconstruct standard-dose PET (SPET) images from low-dose PET (LPET) images. One widely adopted technique is the generative adversarial networks (GANs), yet recently, diffusion probabilistic models (DPMs) have emerged as a compelling alternative due to their improved sample quality and higher log-likelihood scores compared to GANs. Despite this, DPMs suffer from two major drawbacks in real clinical settings, i.e., the computationally expensive sampling process and the insufficient preservation of correspondence between the conditioning LPET image and the reconstructed PET (RPET) image. To address the above limitations, this paper presents a coarse-to-fine PET reconstruction framework that consists of a coarse prediction module (CPM) and an iterative refinement module (IRM). The CPM generates a coarse PET image via a deterministic process, and the IRM samples the residual iteratively. By delegating most of the computational overhead to the CPM, the overall sampling speed of our method can be significantly improved. Furthermore, two additional strategies, i.e., an auxiliary guidance strategy and a contrastive diffusion strategy, are proposed and integrated into the reconstruction process, which can enhance the correspondence between the LPET image and the RPET image, further improving clinical reliability. Extensive experiments on two human brain PET datasets demonstrate that our

---

Z. Han and Y. Wang—These authors contributed equally to this work.

Supplementary Information The online version contains supplementary material available at https://doi.org/10.1007/978-3-031-43999-5_23.

H. Greenspan et al. (Eds.): MICCAI 2023, LNCS 14229, pp. 239–249, 2023.
https://doi.org/10.1007/978-3-031-43999-5_23

method outperforms the state-of-the-art PET reconstruction methods. The source code is available at https://github.com/Show-han/PET-Reconstruction.

Keywords: Positron emission tomography (PET) · PET reconstruction · Diffusion probabilistic models · Contrastive learning

1   Introduction

Positron emission tomography (PET) is a widely-used molecular imaging technique that can help reveal the metabolic and biochemical functioning of body tissues. According to the dose level of injected radioactive tracer, PET images can be roughly classified as standard-(SPET) and low-dose PET (LPET) images. SPET images offer better image quality and more information in diagnosis compared to LPET images containing more noise and artifacts. However, the higher radiation exposure associated with SPET scanning poses potential health risks to the patient. Consequently, it is crucial to reconstruct SPET images from corresponding LPET images to produce clinically acceptable PET images.

In recent years, deep learning-based PET reconstruction approaches [7,9,13] have shown better performance than traditional methods. Particularly, generative adversarial networks (GANs) [8] have been widely adopted [12,14,15,18,26, 27] due to their capability to synthesize PET images with higher fidelity than regression-based models [29,30]. For example, Kand et al. [11] applied a Cycle-GAN model to transform amyloid PET images obtained with diverse radiotracers. Fei et al. [6] made use of GANs to present a bidirectional contrastive framework for obtaining high-quality SPET images. Despite the promising achievement of GAN, its adversarial training is notoriously unstable [22] and can lead to mode collapse [17], which may result in a low discriminability of the generated samples, reducing their confidence in clinical diagnosis.

Fortunately, likelihood-based generative models offer a new approach to address the limitations of GANs. These models learn the distribution's probability density function via maximum likelihood and could potentially cover broader data distributions of generated samples while being more stable to train. As an example, Cui et al. [3] proposed a model based on Nouveau variational autoencoder for PET image denoising. Among likelihood-based generative models, diffusion probabilistic models (DPMs) [10,23] are noteworthy for their capacity to outperform GANs in various tasks [5], such as medical imaging [24] and text-to-image generation [20]. DPMs consist of two stages: a forward process that gradually corrupts the given data and a reverse process that iteratively samples the original data from the noise. However, sampling from a diffusion model is computationally expensive and time-consuming [25], making it inconvenient for real clinical applications. Besides, existing conditional DPMs learn the input-output correspondence implicitly by adding a prior to the training objective, while this learned correspondence is prone to be lost in the reverse process [33], resulting in the RPET image missing crucial clinical information from the LPET image. Hence, the clinical reliability of the RPET image may be compromised.

Motivated to address the above limitations, in this paper, we propose a coarse-to-fine PET reconstruction framework, including a coarse prediction module (CPM) and an iterative refinement module (IRM). The CPM generates a coarse prediction by invoking a deterministic prediction network only once, while the IRM, which is the reverse process of the DPMs, iteratively samples the residual between this coarse prediction and the corresponding SPET image. By combining the coarse prediction and the predicted residual, we can obtain RPET images much closer to the SPET images. To accelerate the sampling speed of IRM, we manage to delegate most of the computational overhead to the CPM [2,28], hoping to narrow the gap between the coarse prediction and the SPET initially. Additionally, to enhance the correspondence between the LPET image and the generated RPET image, we propose an auxiliary guidance strategy at the input level based on the finding that auxiliary guidance can help to facilitate the reverse process of DPMs, and reinforce the consistency between the LPET image and RPET image by providing more LPET-relevant information to the model. Furthermore, at the output level, we suggest a contrastive diffusion strategy inspired by [33] to explicitly distinguish between positive and negative PET slices. To conclude, the contributions of our method can be described as follows:

- We introduce a novel PET reconstruction framework based on DPMs, which, to the best of our knowledge, is the first work that applies DPMs to PET reconstruction.
- To mitigate the computational overhead of DPMs, we employ a coarse-to-fine design that enhances the suitability of our framework for real-world clinical applications.
- We propose two novel strategies, i.e., an auxiliary guidance strategy and a contrastive diffusion strategy, to improve the correspondence between the LPET and RPET images and ensure that RPET images contain reliable clinical information.

2 Background: Diffusion Probabilistic Models

Diffusion Probabilistic Models (DPMs): DPMs [10,23] define a *forward process*, which corrupts a given image data $x_0 \sim q(x_0)$ step by step via a fixed Markov chain $q(x_t|x_{t-1})$ that gradually adds Gaussian noise to the data:

$$q(x_t|x_{t-1}) = \mathcal{N}(x_t; \sqrt{\alpha_t}x_{t-1}, (1-\alpha_t)I), t = 1, 2, \cdots, T, \quad (1)$$

where $\alpha_{1:T}$ is the constant variance schedule that controls the amount of noise added at each time step, and $q(x_T) \sim \mathcal{N}(x_T; 0, I)$ is the stationary distribution. Owing to the Markov property, a data x_t at an arbitrary time step t can be sampled in closed form:

$$q(x_t|x_0) = \mathcal{N}(x_t; \sqrt{\gamma_t}x_0, (1-\gamma_t)I); x_t = \sqrt{\gamma_t}x_0 + \sqrt{1-\gamma_t}\epsilon, \epsilon \sim \mathcal{N}(0, I), \quad (2)$$

where $\gamma_t = \prod_{i=1}^{t} \alpha_i$. Furthermore, we can derive the posterior distribution of x_{t-1} given (x_0, x_t) as $q(x_{t-1}|x_0, x_t) = \mathcal{N}(x_{t-1}; \hat{\mu}(x_0, x_t), \sigma_t^2 I)$, where $\hat{\mu}(x_0, x_t)$

and σ_t^2 are subject to x_0, x_t and $\alpha_{1:T}$. Based on this, we can leverage the *reverse process* from x_T to x_0 to gradually denoise the latent variables by sampling from the posterior distribution $q(x_{t-1}|x_0, x_t)$. However, since x_0 is unknown during inference, we use a transition distribution $p_\theta(x_{t-1}|x_t) := q(x_{t-1}|\mathcal{H}_\theta(x_t, t), x_t)$ to approximate $q(x_{t-1}|x_0, x_t)$, where $\mathcal{H}_\theta(x_t, t)$ manages to reconstruct x_0 from x_t and t, and it is trained by optimizing a variational lower bound of $log p_\theta(x)$.

Conditional DPMs: Given an image x_0 with its corresponding condition c, conditional DPMs try to estimate $p(x_0|c)$. To achieve that, condition c is concatenated with x_t [21] as the input of $\mathcal{H}_\theta$, denoted as $\mathcal{H}_\theta(c, x_t, t)$.

Simplified Training Objective: Instead of training $\mathcal{H}_\theta$ to reconstruct the x_0 directly, we use an alternative parametrization $\mathcal{D}_\theta$ named *denoising network* [10] trying to predict the noise vector $\epsilon \sim \mathcal{N}(0, I)$ added to x_0 in Eq. 2, and derive the following training objective:

$$\mathcal{L}_{DPM} = \mathbb{E}_{(c,x_0)\sim p_{train}} \mathbb{E}_{\epsilon\sim\mathcal{N}(0,I)} \mathbb{E}_{\gamma\sim p_\gamma} \|\mathcal{D}_\theta(c, \sqrt{\gamma}x_0 + \sqrt{1-\gamma}\epsilon, \gamma) - \epsilon\|_1, \quad (3)$$

where the distribution p_γ is the one used in WaveGrad [1]. Note that we also leverage techniques from WaveGrad to let the denoising network $\mathcal{D}_\theta$ conditioned directly on the noise schedule γ rather than time step t, and this gives us more flexibility to control the inference steps.

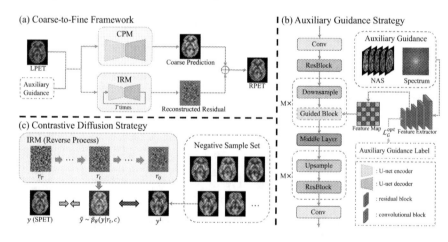

Fig. 1. Overall architecture of our proposed framework.

3   Methodology

Our proposed framework (Fig. 1(a)) has two modules, i.e., a coarse prediction module (CPM) and an iterative refinement module (IRM). The CPM predicts a coarse-denoised PET image from the LPET image, while the IRM models the residual between the coarse prediction and the SPET image iteratively. By combining the coarse prediction and residual, our framework can effectively generate high-quality RPET images. To improve the correspondence between the LPET image and the RPET image, we adopt an auxiliary guidance strategy (Fig. 1(b)) at the input level and a contrastive diffusion strategy (Fig. 1(c)) at the output level. The details of our method are described in the following subsections.

3.1   Coarse-to-Fine Framework

To simplify notation, we use a single conditioning variable c to represent the input required by both CPM and IRM, which includes the LPET image x_{lpet} and the auxiliary guidance x_{aux}. During inference, CPM first generates a coarse prediction $x_{cp} = \mathcal{P}_\theta(c)$, where $\mathcal{P}_\theta$ is the deterministic prediction network in CPM. The IRM, which is the reverse process of DPM, then tries to sample the residual r_0 (i.e., x_0 in Sect. 2) between the coarse prediction x_{cp} and the SPET image y via the following iterative process:

$$r'_{t-1} \sim p_\theta(r_{t-1}|r_t, c), t = T, T-1, \cdots, 1. \tag{4}$$

Herein, the prime symbol above the variable indicates that it is sampled from the reverse process instead of the forward process. When $t = 1$, we can obtain the final sampled residual r_0, and the RPET image y' can be derived by $r'_0 + x_{cp}$.

In practice, both CPM and IRM use the same network architecture shown in Fig. 1(c). CPM generates the coarse prediction x_{cp} by using $\mathcal{P}_\theta$ only once, but the denoising network $\mathcal{D}_\theta$ in IRM will be invoked multiple times during inference. Therefore, it is rational to delegate more computation overhead to $\mathcal{P}_\theta$ to obtain better initial results while keeping $\mathcal{D}_\theta$ small, since the reduction in computation cost in $\mathcal{D}_\theta$ will be accumulated by multiple times. To this end, we set the channel number in $\mathcal{P}_\theta$ much larger than that in the denoising network $\mathcal{D}_\theta$. This leads to a larger network size for $\mathcal{P}_\theta$ compared to $\mathcal{D}_\theta$.

3.2   Auxiliary Guidance Strategy

In this section, we will describe our auxiliary guidance strategy in depth which is proposed to enhance the reconstruction process at the input level by incorporating two auxiliary guidance, i.e., neighboring axial slices (NAS) and the spectrum. Our findings indicate that incorporating NAS provides insight into the spatial relationship between the current slice and its adjacent slices, while incorporating the spectrum imposes consistency in the frequency domain.

To effectively incorporate these two auxiliary guidances, as illustrated in Fig. 1(c), we replace the ResBlock in the encoder with a Guided ResBlock as done in [19]. During inference, the auxiliary guidance x_{aux} is first downsampled by a factor of 2^k as x_{aux}^k, where $k = 1, \cdots, M$, and M is the number of downsampling operations in the U-net encoder. Then x_{aux}^k is fed into a feature extractor $\mathcal{F}_\theta$ to generate its corresponding feature map $f_{aux}^k = \mathcal{F}_\theta(x_{aux}^k)$, which is next injected into the Guided ResBlock matching its resolution through 1×1 convolution.

To empower the feature extractor to contain information of its high-quality counterpart y_{aux}, we constrain it with $\mathcal{L}_1$ loss through a convolution layer $\mathcal{C}_\theta(\cdot)$:

$$\mathcal{L}_G^{opt} = \sum_{k=1}^{M} \|\mathcal{C}_\theta(\mathcal{F}_\theta(x_{aux}^k)) - y_{aux}^k\|_1, \tag{5}$$

where $opt \in \{\text{NAS, spectrum}\}$ denotes the kind of auxiliary guidance.

3.3 Contrastive Diffusion Strategy

In addition to the auxiliary guidance at the input level, we also develop a contrastive diffusion strategy at the output level to amplify the correspondence between the condition LPET image and the corresponding RPET image. In detail, we introduce a set of negative samples $Neg = \{y^1, y^2, ..., y^N\}$, which consists of N SPET slices, each from a randomly selected subject that is not in the current batch for training. Then, for the noisy latent residual r_t at time step t, we obtain its corresponding intermediate RPET $\tilde{y}$, and draw it close to the corresponding SPET y while pushing it far from the negative sample $y^i \in Neg$. Before this, we need to estimate the intermediate residual corresponding to r_t firstly, denoted as $\tilde{r}_0$. According to Sect. 2, the denoising network $\mathcal{D}_\theta$ manages to predict the Gaussian noise added to r_0, enabling us to calculate $\tilde{r}_0$ directly from r_t:

$$\tilde{r}_0 = \frac{r_t - (\sqrt{1 - \gamma_t})\mathcal{D}_\theta(c, r_t, \gamma_t)}{\sqrt{\gamma_t}}. \tag{6}$$

Then $\tilde{r}_0$ is added to the coarse prediction x_{cp} to obtain the intermediate RPET $\tilde{y} = x_{cp} + \tilde{r}_0$. Note that $\tilde{y}$ is a one-step estimated result rather than the final RPET y'. Herein, we define a generator $\tilde{p}_\theta(y|r_t, c)$ to represent the above process. Subsequently, the contrastive learning loss $\mathcal{L}_{CL}$ is formulated as:

$$\mathcal{L}_{CL} = \mathbb{E}_{q(y)}[-log\tilde{p}_\theta(y|r_t, c)] - \sum_{y^i \in Neg} \mathbb{E}_{q(y^i)}[-log\tilde{p}_\theta(y^i|r_t, c)]. \tag{7}$$

Intuitively, as illustrated in Fig. 1(b), the $\mathcal{L}_{CL}$ aims to minimize the discrepancy between the training label y and the intermediate RPET $\tilde{y}$ at each time step (first term), while simultaneously ensuring that $\tilde{y}$ is distinguishable from the negative samples, i.e., the SPET images of other subjects (second term). The contrastive diffusion strategy extends contrastive learning to each time step, which allows LPET images to establish better associations with their corresponding RPET images at different denoising stages, thereby enhancing the mutual information between the LPET and RPET images as done in [33].

3.4 Training Loss

Following [28], we modify the objective $\mathcal{L}_{DPM}$ in Eq. 3, and train CPM and IRM jointly by minimizing the following loss function:

$$\mathcal{L}_{main} = \mathbb{E}_{(c,y)\sim p_{train}}\mathbb{E}_{\epsilon\sim\mathcal{N}(0,I)}\mathbb{E}_{\gamma\sim p_\gamma}\|\mathcal{D}_\theta(c, \sqrt{\gamma}(y - \mathcal{P}_\theta(c)) + \sqrt{1-\gamma}\epsilon, \gamma) - \epsilon\|_1. \tag{8}$$

In summary, the final loss function is:

$$\mathcal{L}_{total} = \mathcal{L}_{main} + m\mathcal{L}_G^{NAS} + n\mathcal{L}_G^{spectrum} + k\mathcal{L}_{CL}, \tag{9}$$

where m, n and k are the hyper-parameters controlling the weights of each loss.

3.5   Implementation Details

The proposed method is implemented by the Pytorch framework using an NVIDIA GeForce RTX 3090 GPU with 24GB memory. The IRM in our framework is built upon the architecture of SR3 [21], a standard conditional DPM. The number of downsampling operations M is 3, and the negative sample set number N is 10. 4 neighboring slices are used as the NAS guidance and the spectrums are obtained through discrete Fourier transform. As for the weights of each loss, we set $m = n = 1$, and $k = 5e-5$ following [33]. We train our model for 500,000 iterations with a batch size of 4, using an Adam optimizer with a learning rate of $1e-4$. The total diffusion steps T are 2,000 during training and 10 during inference.

4   Experiments and Results

Datasets and Evaluation: We conducted most of our low-dose brain PET image reconstruction experiments on a public brain dataset, which is obtained from the Ultra-low Dose PET Imaging Challenge 2022 [4]. Out of the 206 18F-FDG brain PET subjects acquired using a Siemens Biograph Vision Quadra, 170 were utilized for training and 36 for evaluation. Each subject has a resolution of $128 \times 128 \times 128$, and 2D slices along the z-coordinate were used for training and evaluation. To simulate LPET images, we applied a dose reduction factor of 100 to each SPET image. To quantify the effectiveness of our method, we utilized three common evaluation metrics: the peak signal-to-noise (PSNR), structural similarity index (SSIM), and normalized mean squared error (NMSE). Additionally, we also used an in-house dataset, which was acquired on a Siemens Biograph mMR PET-MR system. This dataset contains PET brain images collected from 16 subjects, where 8 subjects are normal control (NC) and 8 subjects are mild cognitive impairment (MCI). To evaluate the generalizability of our method, all the experiments on this in-house dataset are conducted in a cross-dataset manner, i.e., training exclusively on the public dataset and inferring on the in-house dataset. Furthermore, we perform NC/MCI classification on this dataset as the clinical diagnosis experiment. ***Please refer to the supplementary materials for the experimental results on the in-house dataset.***

Comparison with SOTA Methods: We compare the performance of our method with 6 SOTA methods, including DeepPET [9] (regression-based method), Stack-GAN [27], Ea-GAN [31], AR-GAN [16], 3D CVT-GAN [32] (GAN-based method) and NVAE [3] (likelihood-based method) on the public

Table 1. Quantitative comparison results on the public dataset. *: We implemented this method ourselves as no official implementation was provided.

		PSNR↑	SSIM↑	NMSE↓	MParam.
regression-based method	DeepPET [9]	23.078	0.937	0.087	**11.03**
GAN-based method	Stack-GAN [27]	23.856	0.959	0.071	83.65
	Ea-GAN [31]	24.096	0.962	0.064	41.83
	AR-GAN [16]	24.313	0.961	0.055	43.27
	3D CVT-GAN [32]	25.080	0.971	0.039	28.72
likelihood-based method	*NVAE [3]	23.629	0.956	0.064	58.24
	Ours	25.638	0.974	0.033	34.10
	Ours-AMS	**25.876**	**0.975**	**0.032**	34.10

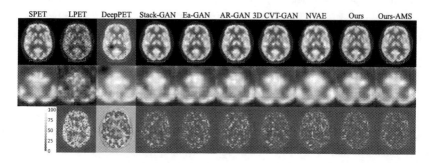

Fig. 2. Visual comparison with SOTA methods.

dataset. Since the IRM contains a stochastic process, we can also average multiple sampled (AMS) results to obtain a more stable reconstruction, which is denoted as Ours-AMS. Results are provided in Table 1. As can be seen, our method significantly outperforms all other methods in terms of PSNR, SSIM, and NMSE, and the performance can be further amplified by averaging multiple samples. Specifically, compared with the current SOTA method 3D CVT-GAN, our method (or ours-AMS) significantly boosts the performance by 0.558 dB (or 0.796 dB) in terms PSNR, 0.003 (or 0.004) in terms of SSIM, and 0.006 (or 0.007) in terms of NMSE. Moreover, 3D CVT-GAN uses 3D PET images as input. Since 3D PET images contain much more information than 2D PET images, our method has greater potential for improvement when using 3D PET images as input. Visualization results are illustrated in Fig. 2. Columns from left to right show the SPET, LPET, and RPET results output by different methods. Rows from top to bottom display the reconstructed results, zoom-in details, and error maps. As can be seen, our method generates the lowest error map while the details are well-preserved, consistent with the quantitative results.

Ablation Study: To thoroughly evaluate the impact of each component in our method, we perform an ablation study on the public dataset by breaking down

Table 2. Quantitative results of the ablation study on the public dataset.

	Single Sampling					Averaged Multiple Sampling			
	PSNR↑	SSIM↑	NMSE↓	MParam.	BFLOPs	PSNR↑	SSIM↑	NMSE↓	SD
(a) baseline	23.302	0.962	0.058	128.740	5973	23.850	0.968	0.052	6.16e−3
(b) CPM	24.354	0.963	0.049	**24.740**	**38**	−	−	−	−
(c) +IRM	24.015	0.966	0.044	31.020	132	24.339	0.967	0.041	3.78e−3
(d) +NAS	24.668	0.969	0.046	33.040	140	24.752	0.970	0.044	3.41e−3
(e) +spec	25.208	0.972	0.044	34.100	145	25.376	0.973	0.043	3.30e−3
(f)+$\mathcal{L}_{CL}$	**25.638**	**0.974**	**0.033**	34.100	145	**25.876**	**0.975**	**0.032**	**2.49e−3**

our model into several submodels. We begin by training the SR3 model as our baseline (a). Then, we train a single CPM with an L2 loss (b), followed by the incorporation of the IRM to calculate the residual (c), and the addition of the auxiliary NAS guidance (d), the spectrum guidance (e), and the $\mathcal{L}_{CL}$ loss term (f). Quantitative results are presented in Table 2. By comparing the results of (a) and (c), we observe that our coarse-to-fine design can significantly reduce the computational overhead of DPMs by decreasing MParam from 128.740 to 31.020 and BFLOPs from 5973 to 132, while achieving better results. The residual generated in (c) also helps to improve the result of the CPM in (b), leading to more accurate PET images. Moreover, our proposed auxiliary guidance strategy and contrastive learning strategy further improve the reconstruction quality, as seen by the increase in PSNR, SSIM, and NMSE scores from (d) to (f). Additionally, we calculate the standard deviation (SD) of the averaged multiple sampling results to measure the input-output correspondence. The standard deviation (SD) of (c) (6.16e−03) is smaller compared to (a) (3.78e−03). This is because a coarse RPET has been generated by the deterministic process. As such, the stochastic process IRM only needs to generate the residual, resulting in less output variability. Then, the SD continues to decrease (3.78e−03 to 2.49e−03) as we incorporate more components into the model, demonstrating the improved input-output correspondence.

5   Conclusion

In this paper, we propose a DPM-based PET reconstruction framework to reconstruct high-quality SPET images from LPET images. The coarse-to-fine design of our framework can significantly reduce the computational overhead of DPMs while achieving improved reconstruction results. Additionally, two strategies, i.e., the auxiliary guidance strategy and the contrastive diffusion strategy, are proposed to enhance the correspondence between the input and output, further improving clinical reliability. Extensive experiments on both public and private datasets demonstrate the effectiveness of our method.

Acknowledgement. This work is supported by the National Natural Science Foundation of China (NSFC 62371325, 62071314), Sichuan Science and Technology Program 2023YFG0263, 2023YFG0025, 2023NSFSC0497.

References

1. Chen, N., Zhang, Y., Zen, H., Weiss, R.J., Norouzi, M., Chan, W.: WaveGrad: estimating gradients for waveform generation. arXiv preprint arXiv:2009.00713 (2020)
2. Chung, H., Sim, B., Ye, J.C.: Come-closer-diffuse-faster: accelerating conditional diffusion models for inverse problems through stochastic contraction. In: Proceedings of the IEEE/CVF Conference on Computer Vision and Pattern Recognition, pp. 12413–12422 (2022)
3. Cui, J., et al.: Pet denoising and uncertainty estimation based on NVAE model using quantile regression loss. In: Wang, L., Dou, Q., Fletcher, P.T., Speidel, S., Li, S. (eds.) MICCAI 2022, Part IV. LNCS, vol. 13434, pp. 173–183. Springer, Cham (2022). https://doi.org/10.1007/978-3-031-16440-8_17
4. MICCAI challenges: Ultra-low dose pet imaging challenge 2022 (2022). https://doi.org/10.5281/zenodo.6361846
5. Dhariwal, P., Nichol, A.: Diffusion models beat GANs on image synthesis. Adv. Neural. Inf. Process. Syst. **34**, 8780–8794 (2021)
6. Fei, Y., et al.: Classification-aided high-quality pet image synthesis via bidirectional contrastive GAN with shared information maximization. In: Wang, L., Dou, Q., Fletcher, P.T., Speidel, S., Li, S. (eds.) MICCAI 2022, Part VI. LNCS, vol. 13436, pp. 527–537. Springer, Cham (2022). https://doi.org/10.1007/978-3-031-16446-0_50
7. Gong, K., Guan, J., Liu, C.C., Qi, J.: Pet image denoising using a deep neural network through fine tuning. IEEE Trans. Radiat. Plasma Med. Sci. **3**(2), 153–161 (2018)
8. Goodfellow, I., et al.: Generative adversarial networks. Commun. ACM **63**(11), 139–144 (2020)
9. Häggström, I., Schmidtlein, C.R., Campanella, G., Fuchs, T.J.: DeepPET: a deep encoder-decoder network for directly solving the pet image reconstruction inverse problem. Med. Image Anal. **54**, 253–262 (2019)
10. Ho, J., Jain, A., Abbeel, P.: Denoising diffusion probabilistic models. Adv. Neural. Inf. Process. Syst. **33**, 6840–6851 (2020)
11. Kang, S.K., Choi, H., Lee, J.S., Initiative, A.D.N., et al.: Translating amyloid pet of different radiotracers by a deep generative model for interchangeability. Neuroimage **232**, 117890 (2021)
12. Kaplan, S., Zhu, Y.M.: Full-dose pet image estimation from low-dose pet image using deep learning: a pilot study. J. Digit. Imaging **32**(5), 773–778 (2019)
13. Kim, K., et al.: Penalized pet reconstruction using deep learning prior and local linear fitting. IEEE Trans. Med. Imaging **37**(6), 1478–1487 (2018)
14. Lei, Y., et al.: Whole-body pet estimation from low count statistics using cycle-consistent generative adversarial networks. Phys. Med. Biol. **64**(21), 215017 (2019)
15. Luo, Y., et al.: 3D transformer-GAN for high-quality PET reconstruction. In: de Bruijne, M., et al. (eds.) MICCAI 2021, Part VI. LNCS, vol. 12906, pp. 276–285. Springer, Cham (2021). https://doi.org/10.1007/978-3-030-87231-1_27
16. Luo, Y., et al.: Adaptive rectification based adversarial network with spectrum constraint for high-quality pet image synthesis. Med. Image Anal. **77**, 102335 (2022)

17. Metz, L., Poole, B., Pfau, D., Sohl-Dickstein, J.: Unrolled generative adversarial networks. arXiv preprint arXiv:1611.02163 (2016)
18. Ouyang, J., Chen, K.T., Gong, E., Pauly, J., Zaharchuk, G.: Ultra-low-dose pet reconstruction using generative adversarial network with feature matching and task-specific perceptual loss. Med. Phys. **46**(8), 3555–3564 (2019)
19. Ren, M., Delbracio, M., Talebi, H., Gerig, G., Milanfar, P.: Image deblurring with domain generalizable diffusion models. arXiv preprint arXiv:2212.01789 (2022)
20. Rombach, R., Blattmann, A., Lorenz, D., Esser, P., Ommer, B.: High-resolution image synthesis with latent diffusion models. In: Proceedings of the IEEE/CVF Conference on Computer Vision and Pattern Recognition, pp. 10684–10695 (2022)
21. Saharia, C., Ho, J., Chan, W., Salimans, T., Fleet, D.J., Norouzi, M.: Image super-resolution via iterative refinement. IEEE Trans. Pattern Anal. Mach. Intell. **45**, 4713–4726 (2022)
22. Salimans, T., Goodfellow, I., Zaremba, W., Cheung, V., Radford, A., Chen, X.: Improved techniques for training GANs. Advances in Neural Inf. Process. Syst. **29** (2016)
23. Sohl-Dickstein, J., Weiss, E., Maheswaranathan, N., Ganguli, S.: Deep unsupervised learning using nonequilibrium thermodynamics. In: International Conference on Machine Learning, pp. 2256–2265. PMLR (2015)
24. Song, Y., Shen, L., Xing, L., Ermon, S.: Solving inverse problems in medical imaging with score-based generative models. arXiv preprint arXiv:2111.08005 (2021)
25. Ulhaq, A., Akhtar, N., Pogrebna, G.: Efficient diffusion models for vision: a survey. arXiv preprint arXiv:2210.09292 (2022)
26. Wang, Y., et al.: 3D conditional generative adversarial networks for high-quality pet image estimation at low dose. Neuroimage **174**, 550–562 (2018)
27. Wang, Y., et al.: 3D auto-context-based locality adaptive multi-modality GANs for pet synthesis. IEEE Trans. Med. Imaging **38**(6), 1328–1339 (2018)
28. Whang, J., Delbracio, M., Talebi, H., Saharia, C., Dimakis, A.G., Milanfar, P.: Deblurring via stochastic refinement. In: Proceedings of the IEEE/CVF Conference on Computer Vision and Pattern Recognition, pp. 16293–16303 (2022)
29. Xiang, L., et al.: Deep auto-context convolutional neural networks for standard-dose pet image estimation from low-dose PET/MRI. Neurocomputing **267**, 406–416 (2017)
30. Xu, J., Gong, E., Pauly, J., Zaharchuk, G.: 200x low-dose pet reconstruction using deep learning. arXiv preprint arXiv:1712.04119 (2017)
31. Yu, B., Zhou, L., Wang, L., Shi, Y., Fripp, J., Bourgeat, P.: EA-GANs: edge-aware generative adversarial networks for cross-modality mr image synthesis. IEEE Trans. Med. Imaging **38**(7), 1750–1762 (2019)
32. Zeng, P., et al.: 3D CVT-GAN: a 3D convolutional vision transformer-GAN for pet reconstruction. In: Wang, L., Dou, Q., Fletcher, P.T., Speidel, S., Li, S. (eds.) MICCAI 2022, Part VI. LNCS, vol. 13436, pp. 516–526. Springer, Cham (2022). https://doi.org/10.1007/978-3-031-16446-0_49
33. Zhu, Y., Wu, Y., Olszewski, K., Ren, J., Tulyakov, S., Yan, Y.: Discrete contrastive diffusion for cross-modal and conditional generation. arXiv preprint arXiv:2206.07771 (2022)

FreeSeed: Frequency-Band-Aware and Self-guided Network for Sparse-View CT Reconstruction

Chenglong Ma[1], Zilong Li[2], Junping Zhang[2], Yi Zhang[3],
and Hongming Shan[1,4,5(✉)] (iD)

[1] Institute of Science and Technology for Brain-Inspired Intelligence and MOE Frontiers Center for Brain Science, Fudan University, Shanghai, China
hmshan@fudan.edu.cn
[2] Shanghai Key Lab of Intelligent Information Processing and School of Computer Science, Fudan University, Shanghai, China
[3] School of Cyber Science and Engineering, Sichuan University, Chengdu, China
[4] Key Laboratory of Computational Neuroscience and Brain-Inspired Intelligence (Fudan University), Ministry of Education, Shanghai, China
[5] Shanghai Center for Brain Science and Brain-Inspired Technology, Shanghai, China

Abstract. Sparse-view computed tomography (CT) is a promising solution for expediting the scanning process and mitigating radiation exposure to patients, the reconstructed images, however, contain severe streak artifacts, compromising subsequent screening and diagnosis. Recently, deep learning-based image post-processing methods along with their dual-domain counterparts have shown promising results. However, existing methods usually produce over-smoothed images with loss of details due to i) the difficulty in accurately modeling the artifact patterns in the image domain, and ii) the equal treatment of each pixel in the loss function. To address these issues, we concentrate on the image post-processing and propose a simple yet effective FREquency-band-awarE and SElf-guidED network, termed FreeSeed, which can effectively remove artifacts and recover missing details from the contaminated sparse-view CT images. Specifically, we first propose a frequency-band-aware artifact modeling network (FreeNet), which learns artifact-related frequency-band attention in the Fourier domain for better modeling the globally distributed streak artifact on the sparse-view CT images. We then introduce a self-guided artifact refinement network (SeedNet), which leverages the predicted artifact to assist FreeNet in continuing to refine the severely corrupted details. Extensive experiments demonstrate the superior performance of FreeSeed and its dual-domain counterpart over the state-of-the-art sparse-view CT reconstruction methods. Source code is made available at https://github.com/Masaaki-75/freeseed.

Supplementary Information The online version contains supplementary material available at https://doi.org/10.1007/978-3-031-43999-5_24.

Keywords: Sparse-view CT · CT reconstruction · Fourier convolution

1 Introduction

X-ray computed tomography (CT) is an established diagnostic tool in clinical practice; however, there is growing concern regarding the increased risk of cancer induction associated with X-ray radiation exposure [14]. Lowering the dose of CT scans has been widely adopted in clinical practice to address this issue, following the "as low as reasonably achievable" (ALARA) principle in the medical community [9]. Sparse-view CT is one of the effective solutions, which reduces the radiation by only sampling part of the projection data for image reconstruction. Nevertheless, images reconstructed by the conventional filtered back-projection (FBP) present severe artifacts, thereby compromising their clinical value.

In recent years, the success of deep learning has attracted much attention in the field of sparse-view CT reconstruction. Existing learning-based approaches mainly include image-domain methods [2,4,18] and dual-domain ones [7,13,16], both involving image post-processing to restore a clean CT image from the low-quality one with streak artifacts. For the image post-processing, residual learning [3] is often employed to encourage learning the artifacts hidden in the residues, which has become a proven paradigm for enhancing the performance [2,4,6,16]. Unfortunately, existing image post-processing methods may fail to model the globally distributed artifacts within the image domain. They can also produce over-smoothed images due to the lack of differentiated supervision for each pixel. In this paper, we advance image post-processing to benefit both classical image-domain methods and the dominant dual-domain ones.

Motivation. We view the sparse-view CT image reconstruction as a two-step task: artifact removal and detail recovery. For the former, few work has investigated the fact that the artifacts exhibit similar pattern across different sparse-view scenarios, which is evident in *Fourier domain* as shown in Fig. 1: they are aggregated mainly in the mid-frequency band and gradually migrate from low to high frequencies as the number of views increases. Inspired by this, we propose a frequency-band-aware artifact modeling network (FreeNet) that learns the artifact-concentrated frequency components to remove the artifacts efficiently using learnable band-pass attention maps in the Fourier domain.

While Fourier domain band-pass maps help capture the pattern of the artifacts, restoring the image detail contaminated by strong artifacts may still be difficult due to the entanglement of artifacts and details in the residues. Consequently, we propose a self-guided artifact refinement network (SeedNet) that provides supervision signals to aid FreeNet in refining the image details contaminated by the artifacts. With these novel designs, we introduce a simple yet effective model termed FREquency-band-awarE and SElf-guidED network (FreeSeed), which enhances the reconstruction by modeling the pattern of artifacts from a frequency perspective and utilizing the artifact to restore the details. FreeSeed achieves promising results with only image data and can be further enhanced once the sinogram is available.

Our contributions can be summarized as follows: 1) a novel frequency-band-aware network is introduced to efficiently capture the pattern of global artifacts in the Fourier domain among different sparse-view scenarios; 2) to promote the restoration of heavily corrupted image detail, we propose a self-guided artifact refinement network that ensures targeted refinement of the reconstructed image and consistently improves the model performance across different scenarios; and 3) quantitative and qualitative results demonstrate the superiority of FreeSeed over the state-of-the-art sparse-view CT reconstruction methods.

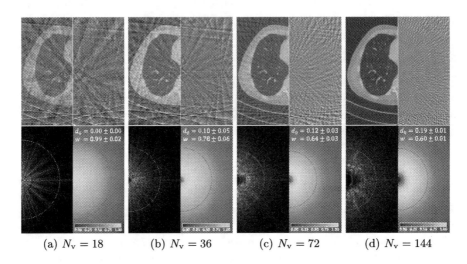

(a) $N_v = 18$ (b) $N_v = 36$ (c) $N_v = 72$ (d) $N_v = 144$

Fig. 1. First row: sparse-view CT images (left half) and the corresponding artifacts (right half); second row: real Fourier amplitude maps of artifacts (left half) and the learned band-pass attention maps (right half, with inner radius and bandwidth respectively denoted by d_0 and w. Values greater than 0.75 are bounded by red dotted line) given different number of views N_v. (Color figure online)

2 Methodology

2.1 Overview

Given a sparse-view sinogram with projection views N_v, let I_s and I_f denote the directly reconstructed sparse- and full- view images by FBP, respectively. In this paper, we aim to construct an image-domain model to effectively recover I_s with a level of quality close to I_f.

The proposed framework of FreeSeed is depicted in Fig. 2, which mainly consists of two designs: FreeNet that learns to remove the artifact and is built with band-pass Fourier convolution blocks that better capture the pattern of the artifact in Fourier domain; and SeedNet as a proxy module that enables FreeNet to refine the image detail under the guidance of the predicted artifact. Note that SeedNet is involved only in the training phase, additional computational cost will not be introduced in the application. The parameters of FreeNet and SeedNet in FreeSeed are updated in an iterative fashion.

2.2   Frequency-Band-Aware Artifact Modeling Network

To learn the globally distributed artifact, FreeNet uses band-pass Fourier convolution blocks as the basic unit to encode artifacts from both spatial and frequency aspects. Technically, Fourier domain knowledge is introduced by fast Fourier convolution (FFC) [1], which benefits from the non-local receptive field and has shown promising results in various computer vision tasks [12,17]. The features fed into FFC are split evenly along the channel into a spatial branch composed of vanilla convolutions and a spectral branch that applies convolution after real Fourier transform, as shown in Fig. 2.

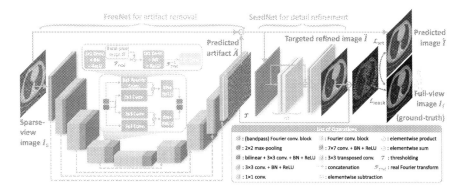

Fig. 2. Overview of the proposed FreeSeed.

Despite the effectiveness, a simple Fourier unit in FFC could still preserve some low-frequency information that may interfere with the learning of artifacts, which could fail to accurately capture the banded pattern of the features of sparse-view artifacts in the frequency domain. To this end, we propose to incorporate learnable band-pass attention maps into FFC. Given an input spatial-domain feature map $X_{in} \in \mathbb{R}^{C_{in} \times H \times W}$, the output $X_{out} \in \mathbb{R}^{C_{out} \times H \times W}$ through the Fourier unit with learnable band-pass attention maps is obtained as follows:

$$Z_{in} = \mathcal{F}_{real}\{X_{in}\}, \qquad Z_{out} = f\left(Z_{in} \odot H\right), \qquad X_{out} = \mathcal{F}_{real}^{-1}\{Z_{out}\}, \qquad (1)$$

where $\mathcal{F}_{real}$ and $\mathcal{F}_{real}^{-1}$ denote the real Fourier transform and its inverse version, respectively. f denotes vanilla convolution. "$\odot$" is the Hadamard product. Specifically, for c-th channel frequency domain feature $Z_{in}^{(c)} \in \mathbb{C}^{U \times V} (c = 1, ..., C_{in})$, the corresponding band-pass attention map $H^{(c)} \in \mathbb{R}^{U \times V}$ is defined by the following

Gaussian transfer function:

$$H^{(c)} = \exp\left[-\left(\frac{D^{(c)2} - d_0^{(c)2}}{w^{(c)} D^{(c)} + \epsilon}\right)^2\right], \tag{2}$$

$$D_{u,v}^{(c)} = \sqrt{\frac{(u - U/2)^2 + (v - V/2)^2}{\max_{u',v'}[(u' - U/2)^2 + (v' - V/2)^2]}}, \tag{3}$$

where $D^{(c)}$ is the c-th channel of the normalized distance map with entries denoting the distance from any point (u, v) to the origin. Two learnable parameters, $w^{(c)} > 0$ and $d_0^{(c)} \in [0, 1]$, represent the bandwidth and the normalized inner radius of the band-pass map, respectively, and are initialized as 1 and 0, respectively. ϵ is set to 1×10^{-12} to avoid division by zero. The right half part of the second row of Fig. 1 shows some samples of the band-pass maps.

The pixel-wise difference between the predicted artifact $\widehat{A}$ of FreeNet and the groundtruth artifact $A_f = I_s - I_f$ is measured by ℓ_2 loss:

$$\mathcal{L}_{\text{art}} = \|A_f - \widehat{A}\|_2. \tag{4}$$

2.3   Self-guided Artifact Refinement Network

Areas heavily obscured by the artifact should be given more attention, which is hard to achieve using only FreeNet. Therefore, we propose a proxy network SeedNet that provides supervision signals to focus FreeNet on refining the clinical detail contaminated by the artifact under the guidance of the artifact itself. SeedNet consists of residual Fourier convolution blocks. Concretely, given sparse-view CT images I_s, FreeNet predicts the artifact $\widehat{A}$ and restored image $\widehat{I} = I_s - \widehat{A}$; the latter is fed into SeedNet to produce targeted refined result $\widetilde{I}$. To guide the network on refining the image detail obscured by heavy artifacts, we design the transformation $\mathcal{T}$ that turns $\widehat{A}$ into a mask M using its mean value as threshold: $M = \mathcal{T}(\widehat{A})$, and define the following masked loss for SeedNet:

$$\mathcal{L}_{\text{mask}} = \|(I_f - \widetilde{I}) \odot M\|_2. \tag{5}$$

2.4   Loss Function for FreeSeed

FreeNet and SeedNet in our proposed FreeSeed are trained in an iterative fashion, where SeedNet is updated using $\mathcal{L}_{\text{mask}}$ defined in Eq. (5), and FreeNet is trained under the guidance of the total loss:

$$\mathcal{L}_{\text{total}} = \mathcal{L}_{\text{art}} + \alpha \mathcal{L}_{\text{mask}}, \tag{6}$$

where $\alpha > 0$ is empirically set as 1. The pseudo-code for the training process and the exploration on the selection of α can be found in our Supplementary Material. Once the training is complete, SeedNet can be dropped and the prediction is done by FreeNet.

2.5   Extending FreeSeed to Dual-Domain Framework

Dual-domain methods are effective in the task of sparse-view CT reconstruction when the sinogram data are available. To further enhance the image reconstruction quality, we extend FreeSeed to the dominant dual-domain framework by adding the sinogram-domain sub-network from DuDoNet [7], where the resulting dual-domain counterpart shown in Fig. 3 is called FreeSeed^{DUDO}. The sinogram-domain sub-network involves a mask U-Net that takes in the linearly interpolated sparse sinogram S_s, where a binary sinogram mask M_s that outlines the unseen part of the sparse-view sinogram is concatenated to each stage of the U-Net encoder. The mask U-Net is trained using sinogram loss $\mathcal{L}_{\text{sino}}$ and Radon consistency loss $\mathcal{L}_{\text{rc}}$. We refer the readers to Lin $et\ al.$ [7] for more information.

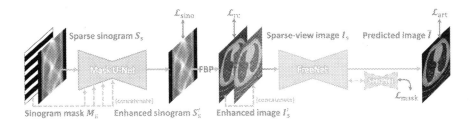

Fig. 3. Overview of dual-domain counterpart of FreeSeed.

3   Experiments

3.1   Experimental Settings

We conduct experiments on the dataset of "the 2016 NIH-AAPM Mayo Clinic Low Dose CT Grand Challenge" [8], which contains 5,936 CT slices in 1 mm image thickness from 10 anonymous patients, where a total of 5,410 slices from 9 patients, resized to 256 × 256 resolution, are randomly selected for training and the 526 slices from the remaining one patient for testing without patient overlap. Fan-beam CT projection under 120 kVp and 500 mA is simulated using TorchRadon toolbox [11]. Specifying the distance from the X-ray source to the rotation center as 59.5 cm and the number of detectors as 672, we generate sinograms from full-dose images with multiple sparse views $N_v \in \{18, 36, 72, 144\}$ uniformly sampled from full 720 views covering $[0, 2\pi]$.

The models are implemented in PyTorch [10] and are trained for 30 epochs with a mini-batch size of 2, using Adam optimizer [5] with $(\beta_1, \beta_2) = (0.5, 0.999)$ and a learning rate that starts from 10^{-4} and is halved every 10 epochs. Experiments are conducted on a single NVIDIA V100 GPU using the same setting. All sparse-view CT reconstruction methods are evaluated quantitatively in terms of root mean squared error (RMSE), peak signal-to-noise ratio (PSNR), and structural similarity (SSIM) [15].

3.2   Overall Performance

We compare our models (FreeSeed and FreeSeed^{DUDO}) with the following recon-struction methods: direct FBP, **DDNet** [18], **FBPConv** [4], **DuDoNet** [7], and **DuDoTrans** [13]. FBPConv and DDNet are image-domain methods, while DuDoNet and DuDoTrans are state-of-the-art dual-domain methods effective for CT image reconstruction. Table 1 shows the quantitative evaluation.

Not surprisingly, we find that the performance of conventional image-domain methods is inferior to the state-of-the-art dual-domain method, mainly due to the failure of removing the global artifacts. We notice that dual-domain methods underperform FBPConv when $N_v = 18$ because of the secondary artifact induced by the inaccurate sinogram restoration in the ultra-sparse scenario.

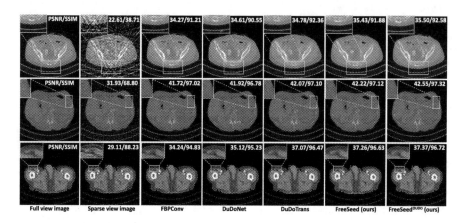

Fig. 4. Visual comparison of state-of-the-art methods. From top to bottom: N_v=36, 72 and 144; the display windows are [0,2000], [500,3000] and [50,500] HU, respectively.

Notably, FreeSeed outperforms the dual-domain methods in most scenar-ios. Figure 4 provides the visualization results for different methods. In general, FreeSeed successfully restores the tiny clinical structures (the spines in the first row, and the ribs in the second row) while achieving more comprehensive artifact removal (see the third row). Note that when the sinogram data are available, dual-domain counterpart FreeSeed^{DUDO} gains further improvements, showing the great flexibility of our model.

3.3   Ablation Study

Table 2 presents the effectiveness of each component in FreeSeed, where seven variants of FreeSeed are: (1) FBPConv upon which FreeNet is built (baseline); (2) FreeNet without band-pass attention maps nor SeedNet guidance $\mathcal{L}_{\text{mask}}$ (baseline + Fourier); (3) FBPConv trained with $\mathcal{L}_{\text{mask}}$ (baseline + SeedNet); (4) FreeNet trained without $\mathcal{L}_{mask}$ (FreeNet); (5) FreeNet trained with simple masked loss $\mathcal{L}_{1+mask} = \|(A_f - \widehat{A}) \odot (1 + M)\|_2$ (FreeNet$_{1+mask}$); (6) FreeNet trained with $\mathcal{L}_{mask}$ using ℓ_1 norm (FreeSeed$_{\ell_1}$); and (7) FreeNet trained with $\mathcal{L}_{mask}$ using ℓ_2 norm, $i.e.$, the full version of our model (FreeSeed).

By comparing the first two rows of Table 2, we find that simply applying FFC provides limited performance gains. Interestingly, we observe that the advantage of band-pass attention becomes more pronounced given more views, which can be seen in the last row of Fig. 1 where the attention maps are visualized by averaging all inner radii and bandwidths in different stages of FreeNet and calculating the map following Eq. (2). Figure 1 shows that these maps successfully capture the banded pattern of the artifact, especially in the cases of $N_v = 36, 72, 144$ where artifacts are less entangled with the image content and present a banded shape in the frequency domain. Thus, the band-pass attention maps lead to better convergence.

Table 1. Quantitative evaluation for state-of-the-art methods in terms of PSNR [dB], SSIM [%], and RMSE [$\times 10^{-2}$]. The best results are highlighted in **bold** and the second-best results are underlined.

Methods	$N_v = 18$			$N_v = 36$			$N_v = 72$			$N_v = 144$		
	PSNR	SSIM	RMSE	PSNR	SSIM	RMSE	PSNR	SSIM	RMSE	PSNR	SSIM	RMSE
FBP	22.88	36.59	7.21	26.44	49.12	4.78	31.63	66.23	2.63	38.51	86.23	1.19
DDNet	34.07	90.63	1.99	37.15	93.50	1.40	40.05	95.18	1.03	45.09	98.37	0.56
FBPConv	**35.04**	91.19	**1.78**	37.63	93.65	1.32	41.95	97.40	0.82	45.96	98.53	0.51
DuDoNet	34.42	91.07	1.91	38.18	93.45	1.24	42.80	97.21	0.73	47.79	98.96	0.41
DuDoTrans	34.89	91.08	1.81	38.55	**94.82**	1.19	43.13	97.67	0.70	48.42	99.15	0.38
FreeSeed	35.01	91.46	1.79	38.63	94.46	1.18	43.42	97.82	0.68	48.79	99.19	0.37
FreeSeed^{DUDO}	35.03	**91.81**	**1.78**	**38.80**	94.78	**1.16**	**43.78**	**97.90**	**0.65**	**49.06**	**99.23**	**0.35**

Table 2. PSNR value of variants of FreeSeed. The best results are highlighted in **bold** and the second-best results are underlined.

Variants	$N_v = 18$	$N_v = 36$	$N_v = 72$	$N_v = 144$
(1) baseline	**35.04**	37.63	41.95	45.96
(2) baseline + Fourier	34.78	38.23	42.33	47.32
(3) baseline + SeedNet	34.49	38.35	42.89	48.64
(4) FreeNet	34.77	38.42	43.06	48.63
(5) FreeNet$_{1+mask}$	34.54	38.17	42.94	48.73
(6) FreeSeed$_{\ell_1}$	34.79	38.45	43.06	**49.00**
(7) FreeSeed (ours)	35.01	**38.63**	**43.42**	48.79

The effectiveness of SeedNet can be seen by comparing Rows (1) and (3) and also Rows (4) and (7). Both the baseline and FreeNet can benefit from the

SeedNet supervision. Visually, clinical details in the image that are obscured by the heavy artifacts can be further refined by FreeNet; please refer to Fig. S1 in our Supplementary Material for more examples and ablation study. We also find that FreeNet$_{1+mask}$ does not provide stable performance gains, probably because directly applying a mask on the pixel-wise loss leads to the discontinuous gradient that brings about sub-optimal results, which, however, can be circumvented with the guidance of SeedNet. In addition, we trained FreeSeed with Eq. (6) using ℓ_1 norm. From the last two rows in Table 2 we find that ℓ_1 norm does not ensure stable performance gains when FFC is used.

4   Conclusion

In this paper, we proposed FreeSeed, a simple yet effective image-domain method for sparse-view CT reconstruction. FreeSeed incorporates Fourier knowledge into the reconstruction network with learnable band-pass attention for a better grasp of the globally distributed artifacts, and is trained using a self-guided artifact refinement network to further refine the heavily damaged image details. Extensive experiments show that both FreeSeed and its dual-domain counterpart outperformed the state-of-the-art methods. In future, we will explore FFC-based network for sinogram interpolation in sparse-view CT reconstruction.

Acknowledgement. This work was supported in part by National Natural Science Foundation of China (No. 62101136), Shanghai Sailing Program (No. 21YF1402800), Shanghai Municipal Science and Technology Major Project (No. 2018SHZDZX01) and ZJLab, Shanghai Municipal of Science and Technology Project (No. 20JC1419500), and Shanghai Center for Brain Science and Brain-inspired Technology.

References

1. Chi, L., Jiang, B., Mu, Y.: Fast fourier convolution. In: Advances in Neural Information Processing Systems, vol. 33, pp. 4479–4488 (2020)
2. Han, Y.S., Yoo, J., Ye, J.C.: Deep residual learning for compressed sensing CT reconstruction via persistent homology analysis. arXiv preprint: arXiv:1611.06391 (2016)
3. He, K., Zhang, X., Ren, S., Sun, J.: Deep residual learning for image recognition. In: Proceedings of the IEEE/CVF Conference on Computer Vision and Pattern Recognition, pp. 770–778 (2016)
4. Jin, K.H., McCann, M.T., Froustey, E., Unser, M.: Deep convolutional neural network for inverse problems in imaging. IEEE Trans. Image Process. **26**(9), 4509–4522 (2017)
5. Kingma, D.P., Ba, J.: Adam: a method for stochastic optimization. In: International Conference on Learning Representations (2015)
6. Lee, H., Lee, J., Kim, H., Cho, B., Cho, S.: Deep-neural-network-based sinogram synthesis for sparse-view CT image reconstruction. IEEE Trans. Radiat. Plasma Med. Sci. **3**(2), 109–119 (2018)

7. Lin, W.A., et al.: DuDoNet: dual domain network for CT metal artifact reduction. In: Proceedings of the IEEE/CVF Conference on Computer Vision and Pattern Recognition, pp. 10512–10521 (2019)

8. McCollough, C.: TU-FG-207A-04: overview of the low dose CT grand challenge. Med. Phys. **43**(6), 3759–3760 (2016)

9. Miller, D.L., Schauer, D.: The ALARA principle in medical imaging. Philosophy **44**, 595–600 (1983)

10. Paszke, A., et al.: PyTorch: an imperative style, high-performance deep learning library. In: Advances in Neural Information Processing Systems, vol. 32 (2019)

11. Ronchetti, M.: TorchRadon: fast differentiable routines for computed tomography. arXiv preprint: arXiv:2009.14788 (2020)

12. Suvorov, R., et al.: Resolution-robust large mask inpainting with Fourier convolutions. In: 2022 IEEE/CVF Winter Conference on Applications of Computer Vision, pp. 2149–2159 (2022)

13. Wang, C., Shang, K., Zhang, H., Li, Q., Zhou, S.K.: DuDoTrans: dual-domain transformer for sparse-view CT reconstruction. In: Machine Learning for Medical Image Reconstruction, pp. 84–94 (2022)

14. Wang, G., Yu, H., De Man, B.: An outlook on X-ray CT research and development. Med. Phys. **35**(3), 1051–1064 (2008)

15. Wang, Z., Bovik, A.C., Sheikh, H.R., Simoncelli, E.P.: Image quality assessment: from error visibility to structural similarity. IEEE Trans. Image Process. **13**(4), 600–612 (2004)

16. Wu, W., Hu, D., Niu, C., Yu, H., Vardhanabhuti, V., Wang, G.: DRONE: dual-domain residual-based optimization network for sparse-view CT reconstruction. IEEE Trans. Med. Imaging **40**(11), 3002–3014 (2021)

17. Zhang, D., Huang, F., Liu, S., Wang, X., Jin, Z.: SwinFIR: revisiting the SwinIR with fast fourier convolution and improved training for image super-resolution. arXiv preprint: arXiv:2208.11247 (2022)

18. Zhang, Z., Liang, X., Dong, X., Xie, Y., Cao, G.: A sparse-view CT reconstruction method based on combination of DenseNet and deconvolution. IEEE Trans. Med. Imaging **37**(6), 1407–1417 (2018)

Topology-Preserving Computed Tomography Super-Resolution Based on Dual-Stream Diffusion Model

Yuetan Chu[1], Longxi Zhou[1], Gongning Luo[1(✉)], Zhaowen Qiu[2(✉)], and Xin Gao[1(✉)]

[1] Computational Bioscience Research Center (CBRC), King Abdullah University of Science and Technology (KAUST), Thuwal, Saudi Arabia
{gongning.luo,xin.gao}@kaust.edu.sa
[2] Institute of Information and Computer Engineering, Northeast Forestry University, Harbin 150040, China
qiuzw@nefu.edu.cn

Abstract. X-ray computed tomography (CT) is indispensable for modern medical diagnosis, but the degradation of spatial resolution and image quality can adversely affect analysis and diagnosis. Although super-resolution (SR) techniques can help restore lost spatial information and improve imaging resolution for low-resolution CT (LRCT), they are always criticized for topology distortions and secondary artifacts. To address this challenge, we propose a dual-stream diffusion model for super-resolution with topology preservation and structure fidelity. The diffusion model employs a dual-stream structure-preserving network and an imaging enhancement operator in the denoising process for image information and structural feature recovery. The imaging enhancement operator can achieve simultaneous enhancement of vascular and blob structures in CT scans, providing the structure priors in the super-resolution process. The final super-resolved CT is optimized in both the convolutional imaging domain and the proposed vascular structure domain. Furthermore, for the first time, we constructed an ultra-high resolution CT scan dataset with a spatial resolution of 0.34×0.34 mm^2 and an image size of 1024×1024 as a super-resolution training set. Quantitative and qualitative evaluations show that our proposed model can achieve comparable information recovery and much better structure fidelity compared to the other state-of-the-art methods. The performance of high-level tasks, including vascular segmentation and lesion detection on super-resolved CT scans, is comparable to or even better than that of raw HRCT. The source code is publicly available at https://github.com/Arturia-Pendragon-Iris/UHRCT_SR.

Keywords: Computed tomography · Super resolution · Diffusion model · Image enhancement

Supplementary Information The online version contains supplementary material available at https://doi.org/10.1007/978-3-031-43999-5_25.

1   Introduction

Computed tomography (CT) is a prevalent imaging modality with applications in biology, disease diagnosis, interventional imaging, and other areas. High-resolution CT (HRCT) is beneficial for clinical diagnosis and surgical planning because it can provide detailed spatial information and specific features, usually employed in advanced clinical routines [1]. HRCT usually requires high-precision CT machines to scan for a long time with high radiation doses to capture the internal structures, which is expensive and can impose the risk of radiation exposure [2]. These factors make HRCT relatively less available, especially in towns and villages, compared to low-resolution CT (LRCT). However, degradation in spatial resolution and imaging quality brought by LRCT can interfere with the original physiological and pathological information, adversely affecting the diagnosis [3]. Consequently, how to produce high-resolution CT scans at a smaller radiation dose level with lower scanning costs is a holy grail of the medical imaging field (Fig. 1).

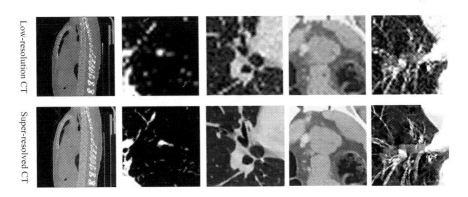

Fig. 1. Low-resolution CT scans can adversely affect the diagnosis. Restoring the original structural information while improving the spatial resolution is a great challenge.

With the advancement of artificial intelligence, super-resolution (SR) techniques based on neural networks indicate new approaches to this problem. By inferring detailed high-frequency features from LRCT, super-resolution can introduce additional knowledge and restore lost information due to low-resolution scanning. Deep-learning (DL) based methods, compared to traditional methods, can incorporate hierarchical features and representations from prior knowledge, resulting in improved results in SR tasks [4]. According to different neural-network frameworks, these SR methods can be broadly categorized into two classes: 1) convolutional neural network (CNN) based model [5–7], and 2) generative adversarial network (GAN) based model [2,8,9]. Very recently, the diffusion model is emerging as the most promising deep generative model [11], which usually consists of two stages: a forward stage to add noises and a reverse stage to separate noises and recover the original images. The diffusion model shows impressive generative capabilities for many tasks, including image generation, inpainting, translation, and super-resolution [10,12,13].

While DL-based methods can generate promising results, there can still be geometric distortions and artifacts along with structural edges in the super-resolved results [15,16]. These structural features always represent essential physiological structures, including vasculature, fibrosis, tumor, and other lesions. The distortion and infidelity of these features can lead to potential misjudgment for diagnosis, which is unacceptable for clinical application. Moreover, the target image size and spatial resolution of HRCT for most existing SR methods is about 512×512 and 0.8×0.8 mm^2. With the progress in hardware settings, ultra-high-resolution CT (UHRCT) with an image size of 1024×1024 and spatial resolution of 0.3×0.3 mm^2 can be available very recently [17]. Though UHRCT can provide much more detailed information, to our best knowledge, SR tasks targeting UHRCT have rarely been discussed and reported.

In this paper, we propose a novel dual-stream conditional diffusion model for CT scan super-resolution to generate UHRCT results with high image quality and structure fidelity. The conditional diffusion model takes the form $p(y|x)$, where x is the LRCT, and y is the targeted UHRCT [14]. The novel diffusion model incorporates a dual-stream structure-preserving network and a novel imaging enhancement operator in the denoising process. The imaging enhancement operator can simultaneously extract the vascular and blob structures in the CT scans and provide structure prior to the dual-stream network. The dual-stream network can fully exploit the prior information with two branches. One branch optimizes the SR results in the image domain, and the other branch optimizes the results in the structure domain. In practice, we use a convolution-based lightweight module to simulate the filtering operations, which enables faster and easier back-propagation in the training process. Furthermore, we constructed a new ultra-high resolution CT scan dataset obtained with the most advanced CT machines. The dataset contained 87 UHRCT scans with a spatial resolution of 0.34×0.34 mm^2 and an image size of 1024×1024. Extensive experiments, including qualitative and quantitative comparisons in both image consistency, structure fidelity, and high-level tasks, demonstrated the superiority of our method. Our contributions can be summarized as follows:

1) We proposed a novel dual-stream diffusion model framework for CT super-resolution. The framework incorporates a dual-stream structure-preserving network in the denoising process to realize better physiological structure restoration.

2) We designed a new image enhancement operator to model the vascular and blob structures in medical images. To avoid non-derivative operations in image enhancement, we proposed a novel enhancement module consisting of lightweight convolutional layers to replace the filtering operation for faster and easier back-propagation in structural domain optimization.

3) We established an ultra-high-resolution CT scan dataset with a spatial resolution of 0.34×0.34 mm^2 and an image size of 1024×1024 for training and testing the SR task.

4) We have conducted extensive experiments and demonstrated the excellent performance of the proposed SR methods in both the image and structure domains. In addition, we have evaluated our proposed method on high-level tasks, including vascular-system segmentation and lesion detection on the SRCT, indicating the reliability of our SR results.

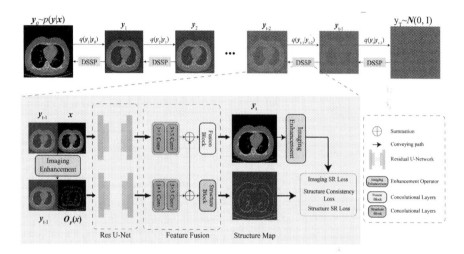

Fig. 2. Overview of the proposed Dual-stream Diffusion Model. The forward diffusion process q gradually adds Gaussian noises to the target HRCT. The reverse process p iteratively denoises the target image, conditioned on the source LRCT x, which is realized with the dual-stream structure-preserving network.

2   Method

2.1   Dual-Stream Diffusion Model for Structure-Preserving Super-Resolution

To preserve the structure and topology relationship in the denoising progress, we designed a novel Dual-Stream Diffusion Model (DSDM) for better super-resolution and topology restoration (Fig. 2). In the DSDM framework, given a HRCT slice y, we generate a noisy version $\tilde{y}$, and train the network G_{DSSP} to denoise $\tilde{y}$ with the corresponding LRCT slice x and a noise level indicator γ. The optimization is defined as

$$E_{(x,y)}E_{\epsilon \sim \mathcal{N}(0,I)}E_{\gamma} \left\| G_{\text{DSSP}}\left(x, \sqrt{\gamma}y + \sqrt{1-\gamma}\epsilon, \gamma\right) - \epsilon \right\|^{\mathrm{p}} \tag{1}$$

In the denoising process, we used a dual-stream structure-preserving (DSSP) network for supervised structure restoration. The DSSP network optimizes the denoised results in the image domain and the structure domain, respectively. The structural domain, obtained with the image enhancement operator, is concatenated with the LRCT slice as the input of the structure branch. The final SR results are obtained after the feature fusion model between the image map and the structure map.

2.2   Imaging Enhancement Operator for Vascular and Blob Structure

We introduced an enhancement operator in the DSSP network to model the vascular and blob structures, which can represent important physiological information according to clinical experience, and provide the prior structural information for the DSSP network. For one pixel $x = [x_1, x_2]^T$ in the CT slice, let $I(x)$ denote the imaging intensity at this point. The 2×2 Hessian matrix at the scale s is defined as [18]

$$H_{i,j}(x) = s^2 I(x) \frac{\partial^2 G(x, s)}{\partial x_i \partial x_j} \tag{2}$$

where $G(x, s)$ is the 2D Gaussian kernel. The two eigenvalues of the Hessian matrix are denoted as $\lambda = (\lambda_1, \lambda_2)$ and here we agree that $|\lambda_1| <= |\lambda_2|$. The eigenvalues of the Hessian matrix can reflect the geometric shape, curvature, and brightness of the local images. For the blob-like structures, the three eigenvalues are about the same, $\lambda_1 \approx \lambda_2$; for the vascular-like structures, λ_2 can be much larger than the absolute value of λ_1, $|\lambda_2| >> |\lambda_1|$ [19]. The eigenvalue relations at scale s can be indicated by several different functions. Here we proposed a novel structure kernel function, which is defined as

$$\mathcal{V}_C = \frac{\lambda_2 (\kappa_1 \lambda_1 + \lambda_\tau)(\kappa_2 \lambda_2 + \lambda_\tau)}{(\lambda_1 + \lambda_2 + \lambda_\tau)^3} \tag{3}$$

κ_1 and κ_2 are the parameters to control the sensitivity for the vascular-like structures and blob-like structures, respectively. λ_τ is the self-regulating factor.

$$\lambda_\tau = (\lambda_2 - \lambda_1) \frac{e^\gamma - e^{-\gamma}}{e^\gamma + e^{-\gamma}} + \lambda_1 \tag{4}$$

$$\gamma = \left| \frac{\lambda_2}{\lambda_1} \right| - 1 \tag{5}$$

When λ_1 is about the same with λ_2, λ_τ is closed to λ_1; and when λ_1 is much smaller to λ_2, λ_τ is closed to λ_2, which can achieve a balance between two conditions.

2.3   Objective Function

We designed a new loss function to ensure that the final SR result can be optimized in both the image domain and the structure domain. Denoting the reference image as y^t, the pixel-wise loss in the imaging domain is formulated as

$$\mathcal{L}_{SR}^{pixel} = \left\| G_{DSSP}^{image}\left(y^t\right) - y^t \right\|^2 + \lambda_1 \left\| G_{DSSP}^{image}\left(y^t\right) - y^t \right\|^1 \tag{6}$$

$G_{DSSP}^{image}\left(y^t\right)$ is the recovered image from the image-domain branch. L1 loss yields a significantly higher consistency and lower diversity, while L2 loss can better capture the outliers. Here we used a parameter λ_{L1} to balance these two losses.

In the meantime, a structure-constraint loss is also necessary to help the network achieve better performance in structure consistency. The loss function consists of two parts, which measure the consistency of the image-domain branch and the structure-domain branch. Denoting the structure-domain output as $G_{DSSP}^{struct}\left(y^t\right)$, the structure-constraint loss can be presented as

$$\mathcal{L}_{SR}^{struct} = \left\| F_c \circ G_{DSSP}^{image}\left(y^t\right) - F_c \circ y^t \right\|^1 + \left\| G_{DSSP}^{struct}\left(y^t\right) - F_c \circ y^t \right\|^1 \tag{7}$$

However, the image enhancement described above involves overly complex calculations, making back-propagation difficult in the training process. Here we utilized a convolution-based operator $O_{F_c}\left(\cdot\right)$ to simplify the calculation, which consists of several lightweight convolutional layers to simulate the operation of image enhancement. In this way, we transform the complex filtering operation into a simple convolution operation, thus back-propagation can be easily processed. The loss function is then modified as

$$\mathcal{L}_{SR}^{struct} = \left\| O_{F_c} \circ G_{DSSP}^{image}\left(y^t\right) - O_{F_c} \circ y^t \right\|^1 + \left\| G_{DSSP}^{struct}\left(y^t\right) - O_{F_c} \circ y^t \right\|^1 \tag{8}$$

The total objective function is the sum of two losses.

$$\mathcal{L}_{SR} = \mathcal{L}_{SR}^{pixel} + \lambda_2 \mathcal{L}_{SR}^{struct} \tag{9}$$

3   Experiments and Conclusion

3.1   Datasets and Evaluation

We constructed three datasets for framework training and evaluation; two of them were in-house data collected from two CT scanners(the ethics number is 20220359), and the other was the public Luna16 dataset [22]. More details about the two in-house datasets are described in the Supplementary Materials. We evaluated our SR model on three CT datasets:

- Dataset 1: 2D super-resolution from 256×256 to 1024×1024, with the spatial resolution from 1.36×1.36 mm^2 to 0.34×0.34 mm^2.

- Dataset 2: 3D super-resolution from $256 \times 256 \times 1X$ to $512 \times 512 \times 5X$, with the spatial resolution from $1.60 \times 1.60 \times 5.00$ mm^3 to $0.80 \times 0.80 \times 1.00$ mm^3.
- Dataset 3: 2D super-resolution from 256×256 to 512×512 on the Luna16 dataset.

We compare our model with other SOTA super-resolution methods, including bicubic interpolation, SRCNN [7], SRResNet [6], Cycle-GAN [2], and SR3 [12]. Performance is assessed qualitatively and quantitatively, using PSNR, SSIM [23], Visual Information Fidelity (VIF) [24], and Structure Mean Square Error (SMSE). VIF value is correlated well with the human perception of SR images, which can measure diagnostic acceptance and information maintenance. SMSE is proposed to evaluate the structure difference between the ground truth and SRCT. Specially, we obtained the structural features of the ground truth and SRCT with Frangi filtering and then calculated the pixel-wise difference [20].

$$SMSE = \|F_{\text{Frangi}}(HRCT) - F_{\text{Frangi}}(SRCT)\|^2 \tag{10}$$

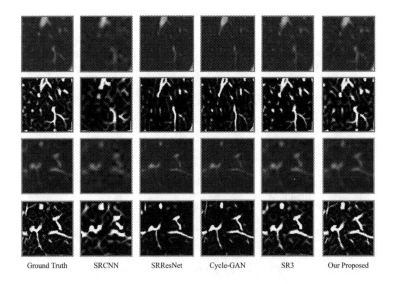

Ground Truth   SRCNN   SRResNet   Cycle-GAN   SR3   Our Proposed

Fig. 3. Visual comparison of super-resolved CT from the ultra-high-resolution dataset. The display window is $[-600, 400]$ HU. The restored images are shown in the first and third lines, and the structural features are shown in the second and fourth lines.

3.2   Qualitative and Quantitative Results

Qualitative comparisons are shown in Fig. 3 and the quantitative results are shown in Table 1. The super-resolution results with our proposed methods achieve the highest scores in both image restoration and structure consistency for most indices, and there are no obvious secondary artifacts introduced in the SR results. Although the GAN-based methods and SR3 can produce sharp details, they tend to generate artifacts for the vascular systems, which is more evident in the structure-enhanced figures. The problem of inconsistent structure is also reflected in the value of VIF and SMSE on both GAN-based methods and SR3.

Lesion Detection and Vessel Segmentation on Super-Resolved CT. To further evaluate the information maintenance of our SR methods, we conducted some high-level tasks, including lung nodules detection and pulmonary airway and blood vessel segmentation on the super-resolved CT scans. We compared the performance of different methods on SRCT and the ground truth. For nodule detection, these methods included U-Net, V-Net [25], ResNet [26], DCNN [27] and 3D-DCNN [28]. For the vessel segmentation, these methods included 3D U-Net, V-Net [25], nnUNet [31], Nardelli et al. [30] and Qin et al. [29]. Figure 4 shows the quantitative results of the performance comparison. The performance of these high-level tasks on the SR results is comparable to or even better than that on the ground-truth CT. Such results, to some extent, demonstrated that our SR method does not introduce artifacts or structural inconsistencies and cause misjudgment, while the improved spatial resolution and image quality generated by our proposed results shows great potential in improving the performance of high-level tasks.

Table 1. Quantitative evaluation of state-of-the-art super-resolution algorithms.

	Dataset 1				Dataset 2				Dataset 3			
	PSNR	SSIM	VIF	SMSE	PSNR	SSIM	VIF	SMSE	PSNR	SSIM	VIF	SMSE
Interpolation	28.27	0.926	0.583	1.46	22.73	0.817	0.516	1.76	23.84	0.776	0.679	1.51
SRCNN [7]	31.71	0.957	0.743	1.43	28.39	0.875	0.559	1.27	31.62	0.842	0.738	1.43
SRResNet [6]	32.69	0.96	0.762	0.992	29.6	0.854	0.685	1.03	32.84	0.897	0.796	0.892
Cycle-GAN [2]	37.32	**0.993**	0.901	0.462	32.31	0.918	0.881	0.822	37.82	0.921	0.915	0.282
SR3 [12]	37.18	0.974	0.812	0.474	36.85	0.957	0.916	0.859	**39.57**	0.968	0.902	0.274
Our Proposed	**40.75**	0.992	**0.977**	**0.162**	**38.76**	**0.979**	**0.967**	**0.274**	38.91	**0.977**	**0.974**	**0.162**

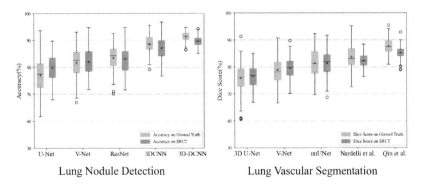

Lung Nodule Detection   Lung Vascular Segmentation

Fig. 4. Comparison of the detection and segmentation performances on HRCT and SRCT.

4   Conclusion

In this paper, we have established a dual-stream diffusion model framework to address the problem of topology distortion and artifact introduction that generally exists in the medical super-resolution results. We first propose a novel image enhancement operator to model the vessel and blob structures in the CT slice, which can provide a structure prior to the SR framework. Then, we design a dual-stream diffusion model that employs a dual-stream ream structure-preserving network in the denoising process. The final SR outputs are optimized not only by convolutional image-space losses but also by the proposed structure-space losses. Extensive experiments have shown that our SR methods can achieve high performance in both image restoration and structure fidelity, demonstrating the promising performance of information preservation and the potential of applying our SR results to downstream tasks.

Acknowledgment. This publication is based upon work supported by the King Abdullah University of Science and Technology (KAUST) Office of Research Administration (ORA) under Award No URF/1/4352-01-01, FCC/1/1976-44-01, FCC/1/1976-45-01, REI/1/5234-01-01, REI/1/5414-01-01.

References

1. Akagi, M., Nakamura, Y., Higaki, T., et al.: Deep learning reconstruction improves image quality of abdominal ultra-high-resolution CT. Eur. Radiol. **29**, 6163–6171 (2019)
2. You, C., Li, G., Zhang, Y., et al.: CT super-resolution GAN constrained by the identical, residual, and cycle learning ensemble (GAN-CIRCLE). IEEE Trans. Med. Imaging **39**(1), 188–203 (2019)
3. Wolterink, J.M., Leiner, T., Viergever, M.A., et al.: Generative adversarial networks for noise reduction in low-dose CT. IEEE Trans. Med. Imaging **36**(12), 2536–2545 (2017)
4. Wang, Z., Chen, J., Hoi, S.C.H.: Deep learning for image super-resolution: a survey. IEEE Trans. Pattern Anal. Mach. Intell. **43**(10), 3365–3387 (2020)
5. Ren, H., El-Khamy, M., Lee, J.: CT-SRCNN: cascade trained and trimmed deep convolutional neural networks for image super resolution. In: 2018 IEEE Winter Conference on Applications of Computer Vision (WACV), pp. 1423–1431. IEEE (2018)
6. Wu, Z., Shen, C., Van Den Hengel, A.: Wider or deeper: revisiting the resnet model for visual recognition. Pattern Recogn. **90**, 119–133 (2019)
7. Georgescu, M.I., Ionescu, R.T., Verga, N.: Convolutional neural networks with intermediate loss for 3D super-resolution of CT and MRI scans. IEEE Access **8**, 49112–49124 (2020)
8. Xie, Y., Franz, E., Chu, M., et al.: tempoGAN: a temporally coherent, volumetric GAN for super-resolution fluid flow. ACM Trans. Graph. (TOG) **37**(4), 1–15 (2018)
9. Lyu, Q., You, C., Shan, H., et al.: Super-resolution MRI and CT through GAN-circle. In: Developments in X-ray tomography XII, vol. 11113, pp. 202–208. SPIE (2019)

10. Su, X., Song, J., Meng, C., et al.: Dual diffusion implicit bridges for image-to-image translation. International Conference on Learning Representations (2022)
11. Nair, N.G., Mei, K., Patel, V.M.: At-ddpm: restoring faces degraded by atmospheric turbulence using denoising diffusion probabilistic models. In: Proceedings of the IEEE/CVF Winter Conference on Applications of Computer Vision, pp. 3434–3443 (2023)
12. Saharia, C., Ho, J., Chan, W., et al.: Image super-resolution via iterative refinement. IEEE Trans. Pattern Anal. Mach. Intell. (2022)
13. Saharia, C., Chan, W., Chang, H., et al.: Palette: Image-to-image diffusion models. In: ACM SIGGRAPH on Conference Proceedings 2022, pp. 1–10 (2022)
14. Lyu, Q., Wang, G.: Conversion Between CT and MRI Images Using Diffusion and Score-Matching Models. arXiv preprint arXiv:2209.12104 (2022)
15. Shi, Y., Wang, K., Chen, C., et al.: Structure-preserving image super-resolution via contextualized multitask learning. IEEE Trans. Multimedia **19**(12), 2804–2815 (2017)
16. Ma, C., Rao, Y., Lu, J., et al.: Structure-preserving image super-resolution. IEEE Trans. Pattern Anal. Mach. Intell. **44**(11), 7898–7911 (2021)
17. Oostveen, L.J., Boedeker, K.L., Brink, M., et al.: Physical evaluation of an ultra-high-resolution CT scanner. Eur. Radiol. **30**, 2552–2560 (2020)
18. Jerman, T., Pernuš, F., Likar, B., et al.: Blob enhancement and visualization for improved intracranial aneurysm detection. IEEE Trans. Visual Comput. Graph. **22**(6), 1705–1717 (2015)
19. Jerman, T., Pernuš, F., Likar, B., et al.: Enhancement of vascular structures in 3D and 2D angiographic images. IEEE Trans. Med. Imaging **35**(9), 2107–2118 (2016)
20. Frangi, A.F., Niessen, W.J., Vincken, K.L., Viergever, M.A.: Multiscale vessel enhancement filtering. In: Wells, W.M., Colchester, A., Delp, S. (eds.) MICCAI 1998. LNCS, vol. 1496, pp. 130–137. Springer, Heidelberg (1998). https://doi.org/10.1007/BFb0056195
21. Jiang, M., Wang, G., Skinner, M.W., Rubinstein, J.T., Vannier, M.W.: Blind deblurring of spiral CT images. IEEE Trans. Med. Imaging **22**(7), 837–845 (2003). https://doi.org/10.1109/TMI.2003.815075
22. Armato, S.G., et al.: The lung image database consortium (LIDC) and image database resource initiative (IDRI): a completed reference database of lung nodules on CT scans. Med. Phys. **38**, 915–931 (2011)
23. Horé, A., Ziou, D.: Image Quality Metrics: PSNR vs. SSIM. In: 2010 20th International Conference on Pattern Recognition, Istanbul, Turkey, pp. 2366–2369 (2010). https://doi.org/10.1109/ICPR.2010.579
24. Mahmoudpour, S., Kim, M.: A study on the relationship between depth map quality and stereoscopic image quality using upsampled depth maps. In: Emerging Trends in Image Processing, Computer Vision and Pattern Recognition, pp. 149–160. Morgan Kaufmann (2015)
25. Abdollahi, A., Pradhan, B., Alamri, A.: VNet: an end-to-end fully convolutional neural network for road extraction from high-resolution remote sensing data. IEEE Access **8**, 179424–179436 (2020)
26. He, K., Zhang, X., Ren, S., Sun, J.: Identity mappings in deep residual networks. In: Leibe, B., Matas, J., Sebe, N., Welling, M. (eds.) ECCV 2016. LNCS, vol. 9908, pp. 630–645. Springer, Cham (2016). https://doi.org/10.1007/978-3-319-46493-0_38
27. Jin, H., Li, Z., Tong, R., et al.: A deep 3D residual CNN for false-positive reduction in pulmonary nodule detection. Med. Phys. **45**(5), 2097–2107 (2018)
28. Naseer, I., Akram, S., Masood, T., et al.: Performance analysis of state-of-the-art CNN architectures for luna16. Sensors **22**(12), 4426 (2022)

29. Qin, Y., Zheng, H., Gu, Y., et al.: Learning tubule-sensitive cnns for pulmonary airway and artery-vein segmentation in ct. IEEE Trans. Med. Imaging **40**(6), 1603–1617 (2021)
30. Nardelli, P., Jimenez-Carretero, D., Bermejo-Pelaez, D., et al.: Pulmonary artery-vein classification in CT images using deep learning. IEEE Trans. Med. Imaging **37**(11), 2428–2440 (2018)
31. Isensee, F., Jaeger, P.F., Kohl, S.A.A., et al.: nnU-Net: a self-configuring method for deep learning-based biomedical image segmentation. Nat. Methods **18**(2), 203–211 (2021)
32. Trinh, D.H., Luong, M., Dibos, F., et al.: Novel example-based method for super-resolution and denoising of medical images. IEEE Trans. Image Process. **23**(4), 1882–1895 (2014)

MRIS: A Multi-modal Retrieval Approach for Image Synthesis on Diverse Modalities

Boqi Chen[✉] and Marc Niethammer

Department of Computer Science, University of North Carolina at Chapel Hill,
Chapel Hill, USA
bqchen@cs.unc.edu

Abstract. Multiple imaging modalities are often used for disease diagnosis, prediction, or population-based analyses. However, not all modalities might be available due to cost, different study designs, or changes in imaging technology. If the differences between the types of imaging are small, data harmonization approaches can be used; for larger changes, direct image synthesis approaches have been explored. In this paper, we develop an approach, MRIS, based on multi-modal metric learning to synthesize images of diverse modalities. We use metric learning via multi-modal image retrieval, resulting in embeddings that can relate images of different modalities. Given a large image database, the learned image embeddings allow us to use k-nearest neighbor (k-NN) regression for image synthesis. Our driving medical problem is knee osteoarthritis (KOA), but our developed method is general after proper image alignment. We test our approach by synthesizing cartilage thickness maps obtained from 3D magnetic resonance (MR) images using 2D radiographs. Our experiments show that the proposed method outperforms direct image synthesis and that the synthesized thickness maps retain information relevant to downstream tasks such as progression prediction and Kellgren-Lawrence grading (KLG). Our results suggest that retrieval approaches can be used to obtain high-quality and meaningful image synthesis results given large image databases. Our code is available at https://github.com/uncbiag/MRIS.

Keywords: Metric learning · k-nearest neighbor · osteoarthritis

1 Introduction

Recent successes of machine learning algorithms in computer vision and natural language processing suggest that training on large datasets is beneficial for model performance [2,5,18,21]. While several efforts to collect very large medical image datasets are underway [12,19], collecting large *homogeneous* medical

Supplementary Information The online version contains supplementary material available at https://doi.org/10.1007/978-3-031-43999-5_26.

image datasets is hampered by: a) cost, b) advancement of technology through-out long study periods, and c) general heterogeneity of acquired images across studies, making it difficult to utilize all data. Developing methods accounting for different imaging types would help make the best use of available data.

Although image harmonization and synthesis [3,14,15,22] methods have been explored to bridge the gap between different types of imaging, these methods are often applied to images of the same geometry. On the contrary, many stud-ies acquire significantly more diverse images; e.g., the OAI image dataset[1] [9] contains both 3D MR images of different sequences and 2D radiographs. Simi-larly, the UK Biobank [19] provides different 3D MR image acquisitions and 2D DXA images. Ideally, a machine learning system can make use of all data that is available. As a related first step in this direction, we explore the feasibility of predicting information gleaned from 3D geometry using 2D projection images. Being able to do so would allow a) pooling datasets that drastically differ in image types or b) relating information from a cheaper 2D screening to more readily interpretable 3D quantities that are difficult for a human observer.

We propose an image synthesis method for diverse modalities based on multi-modal metric learning and k-NN regression. To learn the metric, we use image retrieval as the target task, which aims at embedding images such that matching pairs of different modalities are close in the embedding space. We use a triplet loss [24] to contrastively optimize the gap between positive and negative pairs based on the cosine distance over the learned deep features. In contrast to the typical learning process, we carefully design the training scheme to avoid inter-ference when training with longitudinal image data. Given the learned embed-ding, we can synthesize images between diverse image types by k-NN regression through a weighted average based on their distances measured in the embed-ding space. Given a large database, this strategy allows for a quick and simple estimation of one image type from another.

We use knee osteoarthritis as the driving medical problem and evaluate our proposed approach using the OAI image data. Specifically, we predict cartilage thickness maps obtained from 3D MR images using 2D radiographs. This is a highly challenging task and therefore is a good test case for our approach for the following reasons: 1) cartilage is not explicitly visible on radiographs. Instead, the assessment is commonly based on joint space width (JSW), where decreases in JSW suggest decreases in cartilage thickness [1]; 2) the difficulty in predicting information obtained from a 3D image using only the 2D projection data; 3) the large appearance difference between MR images and thickness maps; 4) the need to capture fine-grained details within a small region of the input radiograph. While direct regression via deep neural networks is possible, such approaches lack interpretability and we show that they can be less accurate for diverse images.

The main contributions of our work are as follows.

1. We propose an image synthesis method for diverse modalities based on multi-modal metric learning using image retrieval and k-NN regression. We carefully construct the learning scheme to account for longitudinal data.

---

[1] https://nda.nih.gov/oai/.

2. We extensively test our approach for osteoarthritis, where we synthesize cartilage thickness maps derived from 3D MR using 2D radiographs.
3. Experimental results show the superiority of our approach over commonly used image synthesis methods, and the synthesized images retain sufficient information for downstream tasks of KL grading and progression prediction.

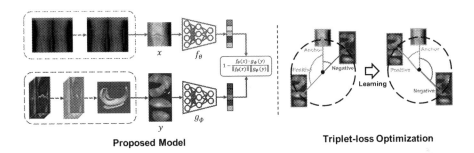

Proposed Model   **Triplet-loss Optimization**

Fig. 1. Proposed multi-modal metric learning model (left) trained using a triplet loss (right). Left top: encoding the region of interest from radiographs, extracted using the method from [26]. Left bottom: encoding thickness maps, extracted from MR images using the method from [11]. Features are compared using cosine similarity. Right: applying triplet loss on cosine similarity, where nonpaired data is moved away from paired data.

2   Method

In this work, we use multi-modal metric learning followed by k-NN regression to synthesize images of diverse modalities. Our method requires 1) a database containing matched image pairs; 2) target images aligned to an atlas space.

2.1   Multi-modal Longitudinally-Aware Metric Learning

Let $\{(x_a^i, y_a^i)\}$ be a database of multiple paired images with each pair containing two modalities x and y of the a-th subject and i-th timepoint if longitudinal data is available. We aim to learn a metric that allows us to reliably identify related image pairs, which in turn relate structures of different modalities. Specifically, we train our deep neural network via a triplet loss so that matching image pairs are encouraged to obtain embedding vectors closer to each other than mismatched pairs. Figure 1 illustrates the proposed multi-modal metric learning approach, which uses two convolutional neural networks (CNNs), each for extracting the features of one modality. The two networks may share the same architecture, but unlike Siamese networks [4], our CNNs have independent sets of weights. This is because the two modalities differ strongly in appearance.

Denoting the two CNNs as $f(\cdot;\theta)$ and $g(\cdot;\phi)$, where θ and ϕ are the CNN parameters, we measure the feature distance between two images x and y using cosine similarity

$$d(x,y) = 1 - \frac{f(x;\theta) \cdot g(y;\phi)}{\|f(x;\theta)\| \, \|g(y;\phi)\|} , \qquad (1)$$

where the output of f and g are vectors of the same dimension[2]. Given a mini-batch of N paired images, our goal is to learn a metric such that $f(x_a^i)$ and $g(y_a^i)$ are close (that is, for the truly matching image pair), while $f(x_a^i)$ and $g(y_b^j)$ are further apart, where $a \neq b$ and i, j are arbitrary timepoints of subjects a, b, respectively. *We explicitly avoid comparing across timepoints of the same subject to avoid biasing longitudinal trends.* This is because different patients have different disease progression speeds. For those with little to no progression, images may look very similar across timepoints and should therefore result in similar embeddings. It would be undesirable to view them as negative pairs. Therefore, our multi-modal longitudinally-aware triplet loss becomes

$$loss(\{(x_a^i, y_a^i)\}) = \sum_{(a,i)} \sum_{(b,j), b \neq a} \max[d(f_\theta(x_a^i), g_\phi(y_a^i)) - d(f_\theta(x_a^i), g_\phi(y_b^j)) + m, 0],$$

$$(2)$$

where m is the margin for controlling the minimum distance between positive and negative pairs. We sum over all subjects at all timepoints for each batch.

To avoid explicitly tracking the subjects in a batch, we can simplify the above equation by randomly picking one timepoint per subject during each training epoch. This then simplifies our multi-modal longitudinally aware triplet loss to a standard triplet loss of the form

$$loss(\{(x_a, y_b)\}) = \sum_{a=1}^{N} \sum_{b=1, b \neq a}^{N} \max[d(f_\theta(x_a), g_\phi(y_a)) - d(f_\theta(x_a), g_\phi(y_b)) + m, 0].$$

$$(3)$$

2.2   Image Synthesis

After learning the embedding space, it can be used to find the most relevant images with a new input, as shown in Fig. 2. Specifically, the features of a query image x are first extracted by the CNN model f_θ we described previously. Given a database of images of the target modality $\mathcal{S}^I = \{y_a^i\}$ and their respective embeddings $\mathcal{S}^F = \{g(y_a^i)\}$, we can then select the top k images with the smallest cosine distance, which will be the most similar images given this embedding. Denoting these k most similar images as $\mathcal{K} = \{\tilde{y}^k\}$ we can synthesize an image, $\hat{y}$ based on a query image, x as a weighted average of the form

$$\hat{y} = \sum_{i=1}^{K} w_i \tilde{y}^i \quad where \quad w_i = \frac{1 - d(x, \tilde{y}^i)}{\sum_{j=1}^{K}(1 - d(x, \tilde{y}^j))} , \qquad (4)$$

---

[2] For notational clarity we will suppress the dependency of f on θ and will write $f_\theta(\cdot)$ instead of $f(\cdot;\theta)$.

where the weights are normalized weights based on the cosine similarities. This requires us to work in an atlas space for the modality y, where all images in the database $\mathcal{S}^I$ are spatially aligned. However, images of the modality x do not need to be spatially aligned, as long as sensible embeddings can be captured by f_θ. As we will see, this is particularly convenient for our experimental setup, where the modality x is a 2D radiograph and the modality y is a cartilage thickness map derived from a 3D MR image, which can easily be brought into a common atlas space. As our synthesized image, $\hat{y}$, is a weighted average of multiple spatially aligned images, it will be smoother than a typical image of the target modality. However, we show in Sect. 3 that the synthesized images still retain the general disease patterns and retain predictive power.

Note also that our goal is not image retrieval or image reidentification, where one wants to find a known image in a database. Instead, we want to synthesize an image for a patient who is not included in our image database. Hence, we expect that no perfectly matched image exists in the database and therefore set $k > 1$. Based on theoretical analyses of k-NN regression [6], we expect the regression results to improve for larger image databases.

3   Experimental Results

This section focuses on investigating the following questions on the OAI dataset:

1. *How good is our retrieval performance?* We calculate recall values to determine the performance to retrieve the correct image;

2. *How accurate are our estimated images?* We compare the predicted cartilage thickness maps with those obtained from 3D MR images;

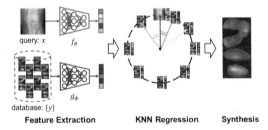

Fig. 2. Image synthesis by k-NN regression from the database. Given an unseen image x, we extract its features $f_\theta(x)$, find the k nearest neighbors in the database $\{y\}$ based on these features, and use them for a weighted k-NN regression.

3. *Does our prediction retain disease-relevant information for downstream tasks?* We test the performance of our predicted cartilage thickness maps in predicting KLG and osteoarthritis progressors;

4. *How does our approach compare to existing image synthesis models?* We show that our approach based on simple k-NN regression compares favorably to direct image synthesis approaches.

3.1   Dataset

We perform a large-scale validation of our method using the Osteoarthritis Initiative (OAI) dataset on almost 40,000 image pairs. This dataset includes $4,796$

patients between the ages of 45 to 79 years at the time of recruitment. Each patient is longitudinally followed for up to 96 months.

Images. The OAI acquired images of multiple modalities, including T2 and DESS MR images, as well as radiographs. We use the paired DESS MR images and radiographs in our experiments. After excluding all timepoints when patients do not have complete MR/radiograph pairs, we split the dataset into three sets by patient (i.e., data from the same patient are in the same sets): Set 1) to train the image retrieval model (2,000 patients; 13,616 pairs). This set also acts as a database during image synthesis; Set 2) to train the downstream task (1,750 patients; 16,802 pairs); Set 3) to test performance (897 patients; 8,418 pairs).

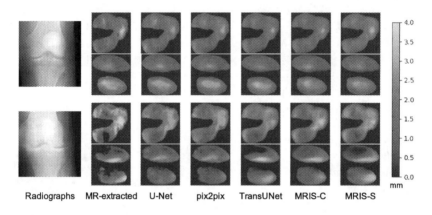

Radiographs   MR-extracted   U-Net   pix2pix   TransUNet   MRIS-C   MRIS-S

Fig. 3. Thickness map predictions for different methods and different severity. Our approach shows a better match of cartilage thickness with the MR-extracted thickness map than the other approaches. See more examples in the appendix.

Preprocessing. As can be seen from the purple dashed box in Fig. 1, we extract cartilage thickness maps from the DESS MR images using a deep segmentation network [27], register them to a common 3D atlas space [25], and then represent them in a common flattened 2D atlas space [11]. These 2D cartilage thickness maps are our target modality, which we want to predict from the 2D radiographs. Unlike MR images for which a separate scan is obtained for the left and right knees, OAI radiographs include both knees and large areas of the femur and tibia. To separate them, we apply the method proposed in [26], which automatically detects keypoints between the knee joint. As shown in the blue dashed box in Fig. 1, the region of interest for each side of the knee is being extracted using a region of 140 mm * 140 mm around the keypoints.

We normalize all input radiographs by linearly scaling the intensities so that the smallest 99% values are mapped to $[0, 0.99]$. We horizontally flip all right knees to the left as done in [11], randomly rotate images up to 15 degrees, add Gaussian noise, and adjust contrast. Unlike the radiographs, we normalize the cartilage thickness map by dividing all values by 3, which is approximately the 95-th percentile of cartilage thickness. All images are resized to $256 * 256$.

Table 1. Thickness map retrieval recall percentage on the testing set. R@k shows the percentage of queries for which the correct one is retrieved within the top k nearest neighbors.

Method	R@1 ↑	R@5 ↑	R@10 ↑	R@20 ↑
Femoral	28.26	58.19	71.13	82.11
Tibial	30.49	61.48	73.36	83.33
Combined	**45.21**	**75.53**	**84.73**	**90.64**

Table 2. Median ± MAD absolute error for both femoral and tibial cartilage between the predicted thickness maps and those extracted from MR images. We stratify the result by KLG. Larger KLG results in less accurate synthesis.

Median ± MAD		KLG01 ↓	KLG2 ↓	KLG3 ↓	KLG4 ↓	All ↓
Femoral Cartilage	U-Net	0.288 ± 0.173	0.324 ± 0.195	0.358 ± 0.214	0.410 ± 0.252	0.304 ± 0.183
	pix2pix	0.289 ± 0.173	0.326 ± 0.196	0.360 ± 0.216	0.411 ± 0.253	0.306 ± 0.183
	TransUNet	0.260 ± 0.157	0.300 ± 0.180	0.326 ± 0.195	0.384 ± 0.235	0.277 ± 0.167
	MRIS-C	0.265 ± 0.158	0.298 ± 0.178	0.319 ± 0.191	0.377 ± 0.226	0.279 ± 0.167
	MRIS-S	**0.259 ± 0.155**	**0.295 ± 0.176**	**0.319 ± 0.191**	**0.373 ± 0.223**	**0.275 ± 0.164**
Tibial Cartilage	U-Net	0.304 ± 0.181	0.324 ± 0.193	0.364 ± 0.216	0.428 ± 0.270	0.316 ± 0.188
	pix2pix	0.306 ± 0.182	0.325 ± 0.194	0.367 ± 0.219	0.433 ± 0.272	0.319 ± 0.190
	TransUNet	0.269 ± 0.160	0.288 ± 0.172	0.325 ± 0.192	**0.371 ± 0.254**	0.281 ± 0.167
	MRIS-C	0.271 ± 0.160	0.291 ± 0.171	0.319 ± 0.188	0.385 ± 0.225	0.282 ± 0.166
	MRIS-S	**0.265 ± 0.157**	**0.283 ± 0.168**	**0.313 ± 0.187**	0.379 ± 0.226	**0.276 ± 0.163**

3.2   Network Training

During multi-modal metric learning, our two branches use the ResNet-18 [10] model with initial parameters obtained by ImageNet pre-training [8]. We fine-tune the networks using AdamW [20] with initial learning rate 10^{-4} for radiographs and 10^{-5} for the thickness maps. The output embedding dimensions of both networks are 512. We train the networks with a batch size of 64 for a total of 450 epochs with a learning rate decay of 80% for every 150 epochs. We set the margin $m = 0.1$ in all our experiments.

For both downstream tasks, we fine-tune our model on a ResNet-18 pretrained network with the number of classes set to 4 for KLG prediction and 2 for progression prediction. Both tasks are trained with AdamW for 30 epochs, batch size 64, and learning rate decay by 80% for every 10 epochs. The initial learning rate is set to 10^{-5} for KLG prediction and 10^{-4} for progression prediction.

3.3   Results

This section shows our results for image retrieval, synthesis, and downstream tasks based on the questions posed above. All images synthesized from MRIS are based on the weighted average of the retrieved top $k = 20$ thickness maps.

Image Retrieval. To show the importance of the learned embedding space, we perform image retrieval on the test set, where our goal is to correctly find the corresponding matching pair. Since our training process does not compare images of the same patient at different timepoints, we test using only the baseline images for each patient (1,794 pairs). During training, we created two thickness map variants: 1) combining the femoral and tibial cartilage thickness maps (Combined); 2) separating the femoral and tibial thickness maps (Femoral/Tibial), which requires training two networks. Table 1 shows the image retrieval recall, where R@k represents the percentage of radiographs for which the correct thickness map is retrieved within the k-nearest neighbors in the embedding space. Combined achieves better results than retrieving femoral and tibial cartilage separately. This may be because more discriminative features can be extracted when both cartilages are provided, which simplifies the retrieval task. In addition, tibial cartilage appears to be easier to retrieve than femoral cartilage.

Table 3. Results on the downstream tasks of KLG and progression prediction. Our synthesis methods overall perform better than other synthesis methods and obtain a comparable result with the MR-extracted thickness maps.

Method	KLG Prediction (accuracy) ↑					Progression Prediction	
	KLG01	KLG2	KLG3	KLG4	overall	average precision ↑	roc auc ↑
U-Net	0.819	0.321	0.778	0.545	0.719	0.242	0.606
pix2pix	0.805	0.396	0.735	0.654	0.722	0.225	0.625
TransUNet	0.797	**0.528**	0.763	**0.865**	0.746	0.286	0.654
MRIS-C	0.865	0.469	0.757	0.673	0.781	0.299	0.713
MRIS-S	**0.869**	0.479	**0.786**	0.718	**0.789**	**0.307**	0.702
MR-extracted	0.842	0.523	0.727	0.795	0.775	0.286	**0.739**

Image Synthesis. To directly measure the performance of our synthesized images on the testing dataset, we show the median $\pm$ MAD (median absolute deviation) absolute error compared to the thickness map extracted by MR in Table 2. We created two variants by combining or separating the femoral and tibial cartilage, corresponding to MRIS-C(ombined) and MRIS-S(eparate). Unlike the image retrieval recall results, MRIS-S performs better than MRIS-C (last column of Table 2). This is likely because it should be beneficial to mix and match separate predictions for synthesizing femoral and tibial cartilage. Moreover, MRIS-S outperforms all baseline image synthesis methods [7,13,23].

Osteoarthritis is commonly assessed via Kellgren-Lawrence grade [16] on radiographs by assessing joint space width and the presence of osteophytes.

KLG=0 represents a healthy knee, while KLG=4 represents severe osteoarthritis. KLG=0 and 1 are often combined because knee OA is considered definitive only when KLG$\geq$ 2 [17]. To assess prediction errors by OA severity, we stratify our results in Table 2 by KLG. Both variants of our approach perform well, outperforming the simpler pix2pix and U-Net baselines for all KLG. The TransUNet approach shows competitive performance, but overall our MRIS-S achieves better results regardless of our much smaller model size. Figure 3 shows examples of images synthesized for the different methods for different severity of OA.

Downstream Tasks. The ultimate question is whether the synthesized images can still retain information for downstream tasks. Therefore, we test the ability to predict KLG and OA progression, where we define OA progression as whether or not the KLG will increase within the next 72 months. Table 3 shows that our synthesized thickness maps perform on par with the MR-extracted thickness maps for progression prediction and we even outperform on predicting KLG. MRIS overall performs better than U-Net [23], pix2pix [13] and TransUNet [7].

4   Conclusion

In this work, we proposed an image synthesis method using metric learning via multi-modal image retrieval and k-NN regression. We extensively validated our approach using the large OAI dataset and compared it with direct synthesis approaches. We showed that our method, while conceptually simple, can effectively synthesize alignable images of diverse modalities. More importantly, our results on the downstream tasks showed that our approach retains disease-relevant information and outperforms approaches based on direct image regression. Potential shortcomings of our approach are that the synthesized images tend to be smoothed due to the weight averaging and that spatially aligned images are required for the modality to be synthesized.

Acknowledgements. This work was supported by NIH 1R01AR072013; it expresses the views of the authors, not of NIH. Data and research tools used in this manuscript were obtained/analyzed from the controlled access datasets distributed from the Osteoarthritis Initiative (OAI), a data repository housed within the NIMH Data Archive. OAI is a collaborative informatics system created by NIMH and NIAMS to provide a worldwide resource for biomarker identification, scientific investigation and OA drug development. Dataset identifier: NIMH Data Archive Collection ID: 2343.

References

1. Altman, R.D., et al.: Radiographic assessment of progression in osteoarthritis. Arthritis Rheumatism: Official J. Am. College Rheumatol. **30**(11), 1214–1225 (1987)
2. Bao, H., Dong, L., Piao, S., Wei, F.: Beit: Bert pre-training of image transformers. arXiv preprint arXiv:2106.08254 (2021)
3. Boulanger, M., et al.: Deep learning methods to generate synthetic CT from MRI in radiotherapy: a literature review. Physica Med. **89**, 265–281 (2021)

4. Bromley, J., Guyon, I., LeCun, Y., Säckinger, E., Shah, R.: Signature verification using a Siamese time delay neural network. In: Advances in Neural Information Processing Systems 6 (1993)
5. Brown, T., et al.: Language models are few-shot learners. Adv. Neural. Inf. Process. Syst. **33**, 1877–1901 (2020)
6. Chen, G.H., Shah, D., et al.: Explaining the success of nearest neighbor methods in prediction. Foundat. Trends Mach. Learn. **10**(5–6), 337–588 (2018)
7. Chen, J., et al.: Transunet: transformers make strong encoders for medical image segmentation. arXiv preprint arXiv:2102.04306 (2021)
8. Deng, J., Dong, W., Socher, R., Li, L.J., Li, K., Fei-Fei, L.: Imagenet: a large-scale hierarchical image database. In: 2009 IEEE Conference on Computer Vision and Pattern Recognition, pp. 248–255. Ieee (2009)
9. Eckstein, F., Wirth, W., Nevitt, M.C.: Recent advances in osteoarthritis imaging-the osteoarthritis initiative. Nat. Rev. Rheumatol. **8**(10), 622–630 (2012)
10. He, K., Zhang, X., Ren, S., Sun, J.: Deep residual learning for image recognition. In: Proceedings of the IEEE Conference on Computer Vision and Pattern Recognition, pp. 770–778 (2016)
11. Huang, C., et al.: DADP: dynamic abnormality detection and progression for longitudinal knee magnetic resonance images from the osteoarthritis initiative. In: Medical Image Analysis, p. 102343 (2022)
12. Ikram, M.A., et al.: Objectives, design and main findings until 2020 from the Rotterdam study. Eur. J. Epidemiol. **35**(5), 483–517 (2020)
13. Isola, P., Zhu, J.Y., Zhou, T., Efros, A.A.: Image-to-image translation with conditional adversarial networks. In: Proceedings of the IEEE Conference on Computer Vision and Pattern Recognition, pp. 1125–1134 (2017)
14. Kasten, Y., Doktofsky, D., Kovler, I.: End-To-end convolutional neural network for 3D reconstruction of knee bones from Bi-planar X-Ray images. In: Deeba, F., Johnson, P., Würfl, T., Ye, J.C. (eds.) MLMIR 2020. LNCS, vol. 12450, pp. 123–133. Springer, Cham (2020). https://doi.org/10.1007/978-3-030-61598-7_12
15. Kawahara, D., Nagata, Y.: T1-weighted and T2-weighted MRI image synthesis with convolutional generative adversarial networks. Reports Practical Oncol. Radiotherapy **26**(1), 35–42 (2021)
16. Kellgren, J.H., Lawrence, J.: Radiological assessment of osteo-arthrosis. Ann. Rheum. Dis. **16**(4), 494 (1957)
17. Kohn, M.D., Sassoon, A.A., Fernando, N.D.: Classifications in brief: Kellgren-Lawrence classification of osteoarthritis. Clin. Orthop. Relat. Res. **474**(8), 1886–1893 (2016)
18. Li, J., Li, D., Savarese, S., Hoi, S.: BLIP-2: bootstrapping language-image pre-training with frozen image encoders and large language models. arXiv preprint arXiv:2301.12597 (2023)
19. Littlejohns, T.J., Sudlow, C., Allen, N.E., Collins, R.: UK Biobank: opportunities for cardiovascular research. Eur. Heart J. **40**(14), 1158–1166 (2019)
20. Loshchilov, I., Hutter, F.: Decoupled weight decay regularization. arXiv preprint arXiv:1711.05101 (2017)
21. Radford, A., et al.: Learning transferable visual models from natural language supervision. In: International Conference on Machine Learning, pp. 8748–8763. PMLR (2021)
22. Ren, M., Dey, N., Fishbaugh, J., Gerig, G.: Segmentation-renormalized deep feature modulation for unpaired image harmonization. IEEE Trans. Med. Imaging **40**(6), 1519–1530 (2021)

23. Ronneberger, O., Fischer, P., Brox, T.: U-Net: convolutional networks for biomedical image segmentation. In: Navab, N., Hornegger, J., Wells, W.M., Frangi, A.F. (eds.) MICCAI 2015. LNCS, vol. 9351, pp. 234–241. Springer, Cham (2015). https://doi.org/10.1007/978-3-319-24574-4_28

24. Schroff, F., Kalenichenko, D., Philbin, J.: Facenet: a unified embedding for face recognition and clustering. In: Proceedings of the IEEE Conference on Computer Vision and Pattern Recognition, pp. 815–823 (2015)

25. Shen, Z., Han, X., Xu, Z., Niethammer, M.: Networks for joint affine and non-parametric image registration. In: Proceedings of the IEEE/CVF Conference on Computer Vision and Pattern Recognition, pp. 4224–4233 (2019)

26. Tiulpin, A., Melekhov, I., Saarakkala, S.: KNEEL: knee anatomical landmark localization using hourglass networks. In: Proceedings of the IEEE/CVF International Conference on Computer Vision Workshops (2019)

27. Xu, Z., Shen, Z., Niethammer, M.: Contextual additive networks to efficiently boost 3d image segmentations. In: DLMIA/ML-CDS -2018. LNCS, vol. 11045, pp. 92–100. Springer, Cham (2018). https://doi.org/10.1007/978-3-030-00889-5_11

Dual Arbitrary Scale Super-Resolution for Multi-contrast MRI

Jiamiao Zhang[1], Yichen Chi[1], Jun Lyu[2], Wenming Yang[1(✉)], and Yapeng Tian[3]

[1] Shenzhen International Graduate School, Tsinghua University, Beijing, China
`yang.wenming@sz.tsinghua.edu.cn`
[2] School of Nursing, The Hong Kong Polytechnic University, Hung Hom, Hong Kong
[3] Department of Computer Science, The University of Texas at Dallas, Richardson, USA
`https://github.com/jmzhang79/Dual-ArbNet`

Abstract. Limited by imaging systems, the reconstruction of Magnetic Resonance Imaging (MRI) images from partial measurement is essential to medical imaging research. Benefiting from the diverse and complementary information of multi-contrast MR images in different imaging modalities, multi-contrast Super-Resolution (SR) reconstruction is promising to yield SR images with higher quality. In the medical scenario, to fully visualize the lesion, radiologists are accustomed to zooming the MR images at arbitrary scales rather than using a fixed scale, as used by most MRI SR methods. In addition, existing multi-contrast MRI SR methods often require a fixed resolution for the reference image, which makes acquiring reference images difficult and imposes limitations on arbitrary scale SR tasks. To address these issues, we proposed an implicit neural representations based dual-arbitrary multi-contrast MRI super-resolution method, called Dual-ArbNet. First, we decouple the resolution of the target and reference images by a feature encoder, enabling the network to input target and reference images at arbitrary scales. Then, an implicit fusion decoder fuses the multi-contrast features and uses an Implicit Decoding Function (IDF) to obtain the final MRI SR results. Furthermore, we introduce a curriculum learning strategy to train our network, which improves the generalization and performance of our Dual-ArbNet. Extensive experiments in two public MRI datasets demonstrate that our method outperforms state-of-the-art approaches under different scale factors and has great potential in clinical practice.

Keywords: MRI Super-resolution · Multi-contrast · Arbitrary scale · Implicit nerual representation

1 Introduction

Magnetic Resonance Imaging (MRI) is one of the most widely used medical imaging modalities, as it is non-invasive and capable of providing superior soft

Supplementary Information The online version contains supplementary material available at https://doi.org/10.1007/978-3-031-43999-5_27.

tissue contrast without causing ionizing radiation. However, it is challenging to acquire high-resolution MR images in practical applications [8] due to the inherent shortcomings of the systems [19,23] and the inevitable motion artifacts of the subjects during long acquisition sessions.

Super-resolution (SR) techniques are a promising way to improve the quality of MR images without upgrading hardware facilities. Clinically, multi-contrast MR images, e.g., T1, T2 and PD weighted images are obtained from different pulse sequences [14,21], which can provide complementary information to each other [3,7]. Although weighted images reflect the same anatomy, they excel at demonstrating different physiological and pathological features. Different time is required to acquire images with different contrast. In this regard, it is promising to leverage an HR reference image with a shorter acquisition time to reconstruct the modality with a longer scanning time. Recently, some efforts have been dedicated to multi-contrast MRI SR reconstruction. Zeng *et al.* proposed a deep convolution neural network to perform single- and multi-contrast SR reconstruction [27]. Dar *et al.* concatenated information from two modalities into the generator of a generative adversarial network (GAN) [6], and Lyu *et al.* introduced a GAN-based progressive network to reconstruct multi-contrast MR images [15]. Feng *et al.* used a multi-stage feature fusion mechanism for multi-contrast SR [7]. Li *et al.* adopted a multi-scale context matching and aggregation scheme to gradually and interactively aggregate multi-scale matched features [12]. Despite their effectiveness, these networks impose severe restrictions on the resolution of the reference image, largely limiting their applications. In addition, most existing multi-contrast SR methods only work with fixed integer scale factors and treat different scale factors as independent tasks. For example, they train a single model for a certain integer scale factor ($\times 2$, $\times 4$). In consequence, using these fixed models for arbitrary scale SR is inadequate. Furthermore, in practical medical applications, it is common for radiologists to zoom in on MR images at will to see localized details of the lesion. Thus, there is an urgent need for an efficient and novel method to achieve super-resolution of arbitrary scale factors in a single model.

In recent years, several methods have been explored for arbitrary scale super-resolution tasks on natural images, such as Meta-SR [9] and Arb-SR [24]. Although they can perform arbitrary up-sampling within the training scales, their generalization ability is limited when exceeding the training distribution, especially for large scale factors. Inspired by the success of implicit neural representation in modeling 3D shapes [5,10,16,18,20], several works perform implicit neural representations to the 2D image SR problem [4,17]. Since these methods can sample pixels at any position in the spatial domain, they can still perform well beyond the distribution of the training scale. Also, there is an MRI SR method that combines the meta-upscale module with GAN and performs arbitrary scale SR [22]. However, the GAN-based method generates unrealistic textures, which affects the diagnosis accuracy.

To address these issues, we propose an arbitrary-scale multi-contrast MRI SR framework. Specifically, we introduce the implicit neural representation to multi-contrast MRI SR and extend the concept of arbitrary scale SR to the reference image domain. Our contributions are summarized as follows:

284   J. Zhang et al.

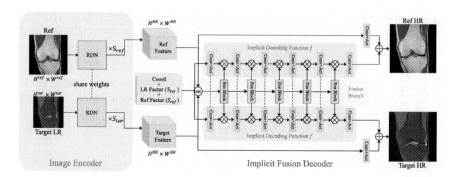

Fig. 1. Overall architecture of the proposed Dual-ArbNet. Our Dual-ArbNet includes a share-weighted image encoder and an implicit fusion decoder which contains a lightweight fusion branch and an implicit decoding function.

1) We propose a new paradigm for multi-contrast MRI SR with the implicit neural representation, called Dual-ArbNet. It allows arbitrary scale SR at any resolution of reference images.
2) We introduce a curriculum learning [2] strategy called Cur-Random to improve the stability, generalization, and multi-contrast fusion performance of the network.
3) Our extensive experiments can demonstrate the effectiveness of our method. Our Dual-ArbNet outperforms several state-of-the-art approaches on two benchmark datasets: fastMRI [26] and IXI [1].

2   Methodology

2.1   Background: Implicit Neural Representations

As we know, computers use 2D pixel arrays to store and display images discretely. In contrast to the traditional discrete representation, the Implicit Neural Representation (INR) can represent an image $I \in R^{H \times W}$ in the latent space $F \in R^{H \times W \times C}$, and use a local neural network (*e.g.*, convolution with kernel 1) to continuously represent the pixel value at each location. This local neural network fits the implicit function of the continuous image, called Implicit Decoding Function (IDF). In addition, each latent feature represents a local piece of continuous image [4], which can be used to decode the signal closest to itself through IDF. Thus, by an IDF $f(\cdot)$ and latent feature F, we can arbitrarily query pixel value at any location, and restore images of arbitrary resolution.

2.2   Network Architecture

The overall architecture of the proposed Dual-ArbNet is shown in Fig. 1. The network consists of an encoder and an implicit fusion decoder. The encoder performs feature extraction and alignment of the target LR and the reference image. The implicit fusion decoder predicts the pixel values at any coordinate by fusing the features and decoding through IDF, thus achieving reconstruction.

Encoder. In the image encoder, Residual Dense Network (RDN) [29] is used to extract image latent features for the network, and the reference image branch shares weights with the target LR image branch to achieve consistent feature extraction and reduce parameters. To aggregate the neighboring information in the reconstruction process, we further unfold the features of 3×3 neighborhoods around each pixel, expanding the feature channels nine times.

Since the resolution of target LR and reference image are different, we have to align them to target HR scale for further fusion. With the target image shaped $H^{tar} \times W^{tar}$ and reference image shaped $H^{ref} \times W^{ref}$, we use nearest interpolation to efficiently up-sample their feature maps to the target HR scale $H^{HR} \times W^{HR}$ by two different factors S_{ref} and S_{tar}:

$$F_{z\uparrow} = Upsample(RDN(I_z), S_z) \tag{1}$$

where $z \in \{ref, tar\}$ indicates the reference and target image, I_{tar} and I_{ref} are the input target LR and reference image. In this way, we obtain the latent feature nearest to each HR pixel for further decoding, and our method can handle Arbitrary scale SR for target images with Arbitrary resolution of reference images (Dual-Arb).

Decoder. As described in Sect. 2.1, the INR use a local neural network to fit the continuous image representation, and the fitting can be referred to as Implicit Decoding Function (IDF). In addition, we propose a fusion branch to efficiently fuse the target and reference latent features for IDF decoding. The overall decoder includes a fusion branch and a shared IDF, as shown in Fig. 1(see right).

Inspired by [25, 29], to better fuse the reference and target features in different dimensions, we use ResBlock with Channel Attention (CA) and Spatial Attention (SA) in our fusion branch. This 5 layers lightweight architecture can capture channel-wise and spatial-wise attention information and fuse them efficiently. The fusion process can be expressed as:

$$\begin{aligned} F_{fusion}^{(0)} &= cat(F_{tar\uparrow}, F_{ref\uparrow}) \\ F_{fusion}^{(i)} &= L_i(F_{fusion}^{(i-1)}) + F_{fusion}^{(i-1)}, \quad i = 1, 2, ..., 5 \end{aligned} \tag{2}$$

where L_i indicates the i-th fusion layer. Then, we equally divide the fused feature $F_{fusion}^{(i)}$ by channel into $F_{fusion,tar}^{(i)}$ and $F_{fusion,ref}^{(i)}$ for decoding respectively.

The IDF in our method is stacked by convolution layer with kernel size 1 ($conv_1$) and sin activation function $sin(\cdot)$. The $conv_1$ and $sin(\cdot)$ are used to transform these inputs to higher dimension space [17], thus achieving a better representation of the IDF. Since $conv_1(x)$ can be written as $W \cdot x + b$ without using any adjacent features, this decoding function can query SR value at any given coordinate. Akin to many previous works [4, 17], relative coordinate information

$P(x, y)$ and scale factors S_{ref}, S_{tar} are necessary for the IDF to decode results continuously. At each target pixel (x, y), we only use local fused feature F_{fusion}, which represents a local piece of continuous information, and coordinate $P(x, y)$ relative to the nearest fused feature, as well as scale factors $\{S_{ref}, S_{tar}\}$, to query in the IDF. Corresponding to the fusion layer, we stack 6 convolution with activation layers. i-th layer's decoding function $f^{(i)}$ can be express as:

$$f^{(0)}(x, y, z) = sin\left(W^{(0)} \cdot cat\left(S_{tar}, S_{ref}, P(x, y)\right) + b^{(0)}\right)$$
$$f^{(i)}(x, y, z) = sin\left(\left(W^{(i)} \cdot f^{(i-1)}(x, y, z) + b^{(i)}\right) \odot F^{(i)}_{fusion, z}(x, y)\right) \quad (3)$$

where (x, y) is the coordinate of each pixel, and $z \in \{ref, tar\}$ indicates the reference and target image. $\odot$ denotes element-wise multiplication, and cat is the concatenate operation. $W^{(i)}$ and $b^{(i)}$ are weight and bias of i-th convolution layer. Moreover, we use the last layer's output $f^{(5)}(\cdot)$ as the overall decoding function $f(\cdot)$. By introducing the IDF above, the pixel value at any coordinates $I_{z,SR}(x, y)$ can be reconstructed:

$$I_{z,SR}(x, y) = f(x, y, z) + Skip(F_{z\uparrow}) \quad (4)$$

where $Skip(\cdot)$ is skip connection branch with $conv_1$ and $sin(\cdot)$, $z \in \{ref, tar\}$.

Loss Function. An L1 loss between target SR results $I_{target, SR}$ and HR images I_{HR} is utilized as reconstruction loss to improve the overall detail of SR images, named as L_{rec}. The reconstructed SR images may lose some frequency information in the original HR images. K-Loss [30] is further introduced to alleviate the problem. Specifically, K_{SR} and K_{HR} denote the fast Fourier transform of $I_{target, SR}$ and I_{HR}. In k-space, the value of mask M is set to 0 in the high-frequency cut-off region mentioned in Sect. 3, otherwise set to 1. L2 loss is used to measure the error between K_{SR} and K_{HR}. K-Loss can be expressed as:

$$L_K = \|(K_{SR} - K_{HR}) \cdot M\|_2 \quad (5)$$

To this end, the full objective of the Dual-ArbNet is defined as:

$$L_{full} = L_{rec} + \lambda_K L_K \quad (6)$$

We set $\lambda_K = 0.05$ empirically to balance the two losses.

2.3   Curriculum Learning Strategy

Curriculum learning [2] has shown powerful capabilities in improving model generalization and convergence speed. It mimics the human learning process by allowing the model to start with easy samples and gradually progress to complex samples. To achieve this and stabilize the training process with different references, we introduce curriculum learning to train our model, named Cur-Random. This training strategy is divided into three phases, including warm-up, pre-learning, and full-training. Although our image encoder can be fed with reference images of arbitrary resolution, it is more common to use LR-ref (scale as target LR) or HR-ref (scale as target HR) in practice. Therefore, these two scales of reference images are used as our settings.

In the warm-up stage, we fix the integer SR scale to integer ($2\times$, $3\times$ and $4\times$) and use HR-Ref to stable the training process. Then, in the pre-learning stage, we use arbitrary scale target images and HR reference images to quickly improve the network's migration ability by learning texture-rich HR images. Finally, in the full-training stage, we train the model with a random scale for reference and target images, which further improves the generalization ability of the network.

3 Experiments

Datasets. Two public datasets are utilized to evaluate the proposed Dual-ArbNet network, including fastMRI [26] (PD as reference and FS-PD as target) and IXI dataset [1] (PD as reference and T2 as target). All the complex-valued images are cropped to integer multiples of 24 (as the smallest common multiple of the test scale). We adopt a commonly used down-sampling treatment to crop the k-space. Concretely, we first converted the original image into the k-space using Fourier transform. Then, only data in the central low-frequency region are kept, and all high-frequency information is cropped out. For the down-sampling factors k, only the central $\frac{1}{k^2}$ frequency information is kept. Finally, we used the inverse Fourier transform to convert the down-sampled data into the image domain to produce the LR image.

We compared our Dual-ArbNet with several recent state-of-the-art methods, including two multi-contrast SR methods: McMRSR [12], WavTrans [13], and three arbitrary scale image SR methods: Meta-SR [9], LIIF [4], Diinn [17].

Experimental Setup. Our proposed Dual-ArbNet is implemented in PyTorch with NVIDIA GeForce RTX 2080 Ti. The Adam optimizer is adopted for model training, and the learning rate is initialized to 10^{-4} at the full-training stage for all the layers and decreases by half for every 40 epochs. We randomly extract 6 LR patches with the size of 32×32 as a batch input. Following the setting in [9], we augment the patches by randomly flipping horizontally or vertically and rotating $90°$. The training scale factors of the Dual-ArbNet vary from 1 to 4 with stride 0.1, and the distribution of the scale factors is uniform. The performance of the SR reconstruction is evaluated by PSNR and SSIM.

Quantitative Results. Table 1 reports the average SSIM and PSNR with respect to different datasets under in-distribution and out-of-distribution large scales. Since the SR scale of McMRSR [12] and WavTrans [13] is fixed to $2\times$ and $4\times$, we use a $2\times$ model and down-sample the results when testing $1.5\times$. We use the $4\times$ model and up-sample the results to test $6\times$ and $8\times$, and down-sample the results to test $3\times$ results. Here, we provide the results with the reference image at HR resolution. As can be seen, our method yields the best results in all

Table 1. Quantitative comparison with other methods. Best and second best results are **highlighted** and underlined.

Dataset	Methods	In distribution				Out-of distribution		Average	
		×1.5	×2	×3	×4	×6	×8	PSNR	SSIM
fast	McMRSR [12]	37.773	34.546	31.087	30.141	27.859	26.200	31.268	0.889
	WavTrans [13]	36.390	32.841	31.153	30.197	28.360	**26.722**	30.944	0.890
	Meta-SR [9]	37.243	33.867	31.047	29.604	27.552	24.536	30.642	0.880
	LIIF [4]	37.868	34.320	31.717	30.301	28.485	26.273	31.494	0.892
	Diinn [17]	37.405	34.182	31.666	30.243	28.382	24.804	31.114	0.887
	Ours	**38.139**	**34.722**	**32.046**	**30.707**	**28.693**	26.419	**31.788**	**0.896**
IXI	McMRSR [12]	37.450	37.046	34.416	33.910	29.765	27.239	33.304	0.914
	WavTrans [13]	39.118	38.171	37.670	35.805	31.037	27.832	34.940	0.958
	Meta-SR [9]	42.740	36.115	32.280	29.219	25.129	23.003	31.414	0.916
	LIIF [4]	41.724	36.818	33.001	30.366	26.502	24.194	32.101	0.934
	Diinn [17]	43.277	37.231	33.285	30.575	26.585	24.458	32.569	0.936
	Ours	**43.964**	**40.768**	**38.241**	**36.816**	**33.186**	**29.537**	**37.085**	**0.979**

Table 2. Ablation study on different training strategies (top) and key components (bottom) under fastMRI dataset. Best results are **highlighted**.

TrainRef	TestRef	×1.5	×2	×3	×4	×6	×8	average
LR	LR	37.911	34.475	31.705	30.219	28.137	24.245	31.115
LR	HR	36.954	34.232	31.615	30.031	27.927	24.455	30.869
HR	LR	35.620	33.007	30.268	28.789	26.624	24.942	29.875
HR	HR	36.666	34.274	31.916	**30.766**	28.392	26.359	31.395
Random	LR	**38.143**	34.423	31.669	30.173	27.975	25.182	31.261
Random	HR	38.140	34.640	32.025	30.712	28.647	26.355	31.753
Cur-Random	LR	38.063	34.489	31.684	30.177	28.038	25.264	31.286
Cur-Random	HR	38.139	**34.722**	**32.046**	30.707	**28.693**	26.419	**31.788**

Setting	Ref	Scales	Coord	×1.5	×2	×3	×4	×6	×8	average
w/o ref	✗	✗	✓	37.967	34.477	31.697	30.214	28.154	24.996	31.251
w/o scale	✓	✗	✓	37.951	34.663	32.063	30.681	28.623	26.413	31.732
w/o coord	✓	✓	✗	38.039	34.706	32.036	30.702	28.592	26.288	31.727
Dual-ArbNet	✓	✓	✓	**38.139**	**34.722**	**32.046**	**30.707**	**28.693**	**26.419**	**31.788**

datasets. Notably, for out-of-distribution scales, our method performs even significantly better than existing methods. The results confirm that our framework outperforms the state-of-the-art in terms of performance and generalizability.

Qualitative Evaluation. Figure 2 provides the reconstruction results and the corresponding error maps of the in-distribution scale (4×) and out-of-distribution scale (6×). The more obvious the texture in the error map, the worse the reconstruction means. As can be observed, our reconstructed images

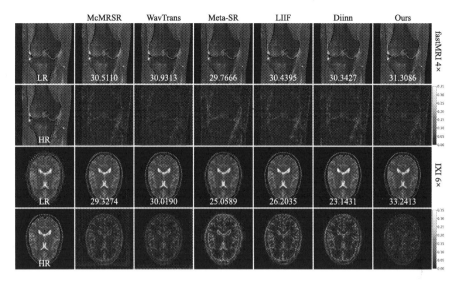

Fig. 2. Qualitative results and error maps of different SR methods on fastMRI and IXI dataset. The color bar on the right indicates the value of the error map. Our method can reconstruct fewer blocking artifacts and sharper texture details.

can eliminate blurred edges, exhibit fewer blocking artifacts and sharper texture details, especially in out-of-distribution scales.

Ablation Study on Different Training Strategies. We conduct experiments on different training strategies and reference types to demonstrate the performance of Dual-ArbNet and the gain of Cur-Random, as shown in Table 2(top). Regarding the type of reference image, we use HR, LR, Random, Cur-Random for training, and HR, LR for testing. As can be seen, the domain gap appears in inconsistent training-testing pairs, while Random training can narrow this gap and enhance the performance. In addition, the HR-Ref performs better than the LR-Ref due to its rich detail and sharp edges, especially in large scale factors. Based on the Random training, the Cur-Random strategy can further improve the performance and achieve balanced SR results.

Ablation Study on Key Components. In Table 2(bottom), to evaluate the validity of the key components of Dual-ArbNet, we conducted experiments without introducing coordinate information, thus verifying the contribution of coordinate in the IDF, named w/o coord. The setting without introducing scale factors in implicit decoding is designed to verify the effect of scale factors on model performance, named w/o scale. To verify whether the reference image can effectively provide auxiliary information for image reconstruction and better restore SR images, we further designed a single-contrast variant model without considering the reference image features in the model, named w/o ref. All the settings use Cur-Random training strategy.

As can be seen that the reconstruction results of w/o coord and w/o scale are not optimal because coordinates and scale can provide additional information for the implicit decoder. We observe that w/o ref has the worst results, indicating that the reference image can provide auxiliary information for super-resolving the target image.

4   Conclusion

In this paper, we proposed the Dual-ArbNet for MRI SR using implicit neural representations, which provided a new paradigm for multi-contrast MRI SR tasks. It can perform arbitrary scale SR on LR images at any resolution of reference images. In addition, we designed a new training strategy with reference to the idea of curriculum learning to further improve the performance of our model. Extensive experiments on multiple datasets show that our Dual-ArbNet achieves state-of-the-art results both within and outside the training distribution. We hope our work can provide a potential guide for further studies of arbitrary scale multi-contrast MRI SR.

Acknowledgements. This work was partly supported by the National Natural Science Foundation of China (Nos. 62171251 & 62311530100), the Special Foundations for the Development of Strategic Emerging Industries of Shenzhen (Nos. JCYJ20200109143010272 & CJGJZD20210408092804011) and Oversea Cooperation Foundation of Tsinghua.

References

1. Ixi dataset. http://brain-development.org/ixi-dataset/. Accessed 20 Feb 2023
2. Bengio, Y., Louradour, J., Collobert, R., Weston, J.: Curriculum learning. In: Proceedings of the 26th Annual International Conference on Machine Learning, pp. 41–48 (2009)
3. Chen, W., et al.: Accuracy of 3-t MRI using susceptibility-weighted imaging to detect meniscal tears of the knee. Knee Surg. Sports Traumatol. Arthrosc. **23**, 198–204 (2015)
4. Chen, Y., Liu, S., Wang, X.: Learning continuous image representation with local implicit image function. In: Proceedings of the IEEE/CVF Conference on Computer Vision and Pattern Recognition, pp. 8628–8638 (2021)
5. Chen, Z., Zhang, H.: Learning implicit fields for generative shape modeling. In: Proceedings of the IEEE/CVF Conference on Computer Vision and Pattern Recognition, pp. 5939–5948 (2019)
6. Dar, S.U., Yurt, M., Shahdloo, M., Ildız, M.E., Tınaz, B., Cukur, T.: Prior-guided image reconstruction for accelerated multi-contrast MRI via generative adversarial networks. IEEE J. Sel. Top. Signal Process. **14**(6), 1072–1087 (2020)
7. Feng, C.-M., Fu, H., Yuan, S., Xu, Y.: Multi-contrast MRI super-resolution via a multi-stage integration network. In: de Bruijne, M., et al. (eds.) MICCAI 2021. LNCS, vol. 12906, pp. 140–149. Springer, Cham (2021). https://doi.org/10.1007/978-3-030-87231-1_14

8. Feng, C.M., Wang, K., Lu, S., Xu, Y., Li, X.: Brain MRI super-resolution using coupled-projection residual network. Neurocomputing **456**, 190–199 (2021)
9. Hu, X., Mu, H., Zhang, X., Wang, Z., Tan, T., Sun, J.: Meta-SR: a magnification-arbitrary network for super-resolution. In: Proceedings of the IEEE/CVF Conference on Computer Vision and Pattern Recognition, pp. 1575–1584 (2019)
10. Jiang, C., et al.: Local implicit grid representations for 3D scenes. In: Proceedings of the IEEE/CVF Conference on Computer Vision and Pattern Recognition, pp. 6001–6010 (2020)
11. Lee, J., Jin, K.H.: Local texture estimator for implicit representation function. In: Proceedings of the IEEE/CVF Conference on Computer Vision and Pattern Recognition, pp. 1929–1938 (2022)
12. Li, G., et al.: Transformer-empowered multi-scale contextual matching and aggregation for multi-contrast MRI super-resolution. In: Proceedings of the IEEE/CVF Conference on Computer Vision and Pattern Recognition, pp. 20636–20645 (2022)
13. Li, G., Lyu, J., Wang, C., Dou, Q., Qin, J.: Wavtrans: synergizing wavelet and cross-attention transformer for multi-contrast mri super-resolution. In: Wang, L., Dou, Q., Fletcher, P.T., Speidel, S., Li, S. (eds.) MICCAI 2022, Part VI. LNCS, vol. 13436, pp. 463–473. Springer, Cham (2022). https://doi.org/10.1007/978-3-031-16446-0_44
14. Liu, X., Wang, J., Sun, H., Chandra, S.S., Crozier, S., Liu, F.: On the regularization of feature fusion and mapping for fast mr multi-contrast imaging via iterative networks. Magn. Reson. Imaging **77**, 159–168 (2021)
15. Lyu, Q., et al.: Multi-contrast super-resolution MRI through a progressive network. IEEE Trans. Med. Imaging **39**(9), 2738–2749 (2020)
16. Mescheder, L., Oechsle, M., Niemeyer, M., Nowozin, S., Geiger, A.: Occupancy networks: learning 3D reconstruction in function space. In: Proceedings of the IEEE/CVF Conference on Computer Vision and Pattern Recognition, pp. 4460–4470 (2019)
17. Nguyen, Q.H., Beksi, W.J.: Single image super-resolution via a dual interactive implicit neural network. In: Proceedings of the IEEE/CVF Winter Conference on Applications of Computer Vision, pp. 4936–4945 (2023)
18. Park, J.J., Florence, P., Straub, J., Newcombe, R., Lovegrove, S.: DeepsDF: learning continuous signed distance functions for shape representation. In: Proceedings of the IEEE/CVF Conference on Computer Vision and Pattern Recognition, pp. 165–174 (2019)
19. Plenge, E., et al.: Super-resolution methods in MRI: can they improve the trade-off between resolution, signal-to-noise ratio, and acquisition time? Magn. Reson. Med. **68**(6), 1983–1993 (2012)
20. Sitzmann, V., Martel, J., Bergman, A., Lindell, D., Wetzstein, G.: Implicit neural representations with periodic activation functions. Adv. Neural. Inf. Process. Syst. **33**, 7462–7473 (2020)
21. Sun, H., et al.: Extracting more for less: multi-echo mp2rage for simultaneous t1-weighted imaging, t1 mapping, mapping, SWI, and QSM from a single acquisition. Magn. Reson. Med. **83**(4), 1178–1191 (2020)
22. Tan, C., Zhu, J., Lio', P.: Arbitrary scale super-resolution for brain MRI images. In: Maglogiannis, I., Iliadis, L., Pimenidis, E. (eds.) AIAI 2020. IAICT, vol. 583, pp. 165–176. Springer, Cham (2020). https://doi.org/10.1007/978-3-030-49161-1_15
23. Van Reeth, E., Tham, I.W., Tan, C.H., Poh, C.L.: Super-resolution in magnetic resonance imaging: a review. Concepts Magn. Reson. Part A **40**(6), 306–325 (2012)

24. Wang, L., Wang, Y., Lin, Z., Yang, J., An, W., Guo, Y.: Learning a single network for scale-arbitrary super-resolution. In: Proceedings of the IEEE/CVF International Conference on Computer Vision, pp. 4801–4810 (2021)
25. Woo, S., Park, J., Lee, J.Y., Kweon, I.S.: CBAM: convolutional block attention module. In: Proceedings of the European Conference on Computer vision (ECCV), pp. 3–19 (2018)
26. Zbontar, J., et al.: fastMRI: an open dataset and benchmarks for accelerated mri. arXiv preprint arXiv:1811.08839 (2018)
27. Zeng, K., Zheng, H., Cai, C., Yang, Y., Zhang, K., Chen, Z.: Simultaneous single and multi-contrast super-resolution for brain MRI images based on a convolutional neural network. Comput. Biol. Med. **99**, 133–141 (2018)
28. Zhang, Y., Li, K., Li, K., Wang, L., Zhong, B., Fu, Y.: Image super-resolution using very deep residual channel attention networks. In: Proceedings of the European Conference on Computer Vision (ECCV) (2018)
29. Zhang, Y., Tian, Y., Kong, Y., Zhong, B., Fu, Y.: Residual dense network for image super-resolution. In: Proceedings of the IEEE Conference on Computer Vision and Pattern Recognition, pp. 2472–2481 (2018)
30. Zhou, B., Zhou, S.K.: DudorNet: learning a dual-domain recurrent network for fast MRI reconstruction with deep t1 prior. In: Proceedings of the IEEE/CVF Conference on Computer Vision and Pattern Recognition, pp. 4273–4282 (2020)

Dual Domain Motion Artifacts Correction for MR Imaging Under Guidance of K-space Uncertainty

Jiazhen Wang[1], Yizhe Yang[1], Yan Yang[1], and Jian Sun[1,2,3(✉)]

[1] Xi'an Jiaotong University, Xi'an, China
{jzwang,yyz0022}@stu.xjtu.edu.cn, {yangyan,jiansun}@xjtu.edu.cn
[2] Pazhou Laboratory (Huangpu), Guangzhou, China
[3] Peng Cheng Laboratory, Shenzhen, China

Abstract. Magnetic resonance imaging (MRI) may degrade with motion artifacts in the reconstructed MR images due to the long acquisition time. In this paper, we propose a dual domain motion correction network (D^2MC-Net) to correct the motion artifacts in 2D multi-slice MRI. Instead of explicitly estimating the motion parameters, we model the motion corruption by k-space uncertainty to guide the MRI reconstruction in an unfolded deep reconstruction network. Specifically, we model the motion correction task as a dual domain regularized model with an uncertainty-guided data consistency term. Inspired by its alternating iterative optimization algorithm, the D^2MC-Net is composed of multiple stages, and each stage consists of a k-space uncertainty module (KU-Module) and a dual domain reconstruction module (DDR-Module). The KU-Module quantifies the uncertainty of k-space corruption by motion. The DDR-Module reconstructs motion-free k-space data and MR image in both k-space and image domain, under the guidance of the k-space uncertainty. Extensive experiments on fastMRI dataset demonstrate that the proposed D^2MC-Net outperforms state-of-the-art methods under different motion trajectories and motion severities.

Keywords: Magnetic resonance imaging · Motion artifacts correction · Dual domain reconstruction · K-space uncertainty

1 Introduction

Magnetic resonance imaging (MRI) is a widely used non-invasive imaging technique. However, MRI is sensitive to subject motion due to the long time for k-space data acquisition [16]. Motion artifacts, appearing as ghosting or blurring artifacts in MR images, degrade the MR image quality [23] and affect the

---

J. Wang and Y. Yang—Both authors contributed equally to this work.

clinical diagnosis. During the scan, it is hard for subjects to remain still, especially for pediatrics or neuro-degenerative patients. Therefore, the correction of motion artifacts in MRI has a great clinical demand.

The typical methods for motion artifacts correction in MRI include the prospective and retrospective methods. The prospective methods measure the subject motion using external tracking devices or navigators during the scan for motion correction [11]. The retrospective motion correction methods either explicitly model and correct the motion in the image reconstruction algorithm, or learn the mapping from MR image with motion artifacts to the motion-free MR image using deep learning approach. Specifically, the methods in [2,4,6,9,15] are based on a forward model of subject motion, and jointly estimate the motion parameters and MR image using the optimization algorithm. The methods in [8,12,14] introduce convolutional neural networks (CNNs) into the joint optimization procedure to learn the MR image prior. The deep learning methods in [3,10,18,19,21] directly learn the mapping from motion-corrupted MR image to motion-free MR image by designing various deep networks. Some other methods correct the motion artifacts using additional prior information, such as the different contrasts of the same object [13], self-assisted adjacent slices priors [1].

In this paper, we propose a dual domain motion correction network (i.e., D^2MC-Net) to correct the motion artifacts in 2D multi-slice MRI. Instead of explicitly estimating motion parameters, we design a dual domain regularized model with an uncertainty-guided data consistency term, which models the motion corruption by k-space uncertainty to guide the MRI reconstruction. Then the alternating iterative algorithm of the model is unfolded to be a novel deep network, i.e., D^2MC-Net. As shown in Fig. 1, the D^2MC-Net contains multiple stages, and each stage consists of two key components, i.e., k-space uncertainty module (KU-Module) and dual domain reconstruction module (DDR-Module). The KU-Module measures the uncertainty of k-space data corrupted by the motion. The DDR-Module reconstructs motion-free k-space data and MR image in both k-space and image domain under the guidance of the k-space uncertainty. Extensive experiments on fastMRI dataset demonstrate that the proposed D^2MC-Net achieves the state-of-the-art results under different motion trajectories and motion severities. For example, under severe corruption with piecewise constant motion trajectory, our result in PSNR is at least 2.11 dB higher than the existing methods, e.g., Autofocusing+ [12].

Different from the optimization-based methods [2,4,6,8,9,12,14,15], our model is based on modeling the motion corruption by k-space uncertainty without explicitly estimating the motion parameters. Different from the deep learning methods [1,3,10,18,19,21], D^2MC-Net incorporates an uncertainty-guided data consistency term into the unfolded network to guide MRI reconstruction.

2   Methods

2.1   Problem Formulation

In our approach, we model the motion corruption by measuring the uncertainty of k-space data. Specifically, we assume that the distribution of motion-corrupted k-space data $\hat{y} \in \mathbb{C}^N$ at each position obeys a non-i.i.d. and pixel-wise Gaussian distribution, where N is the number of the k-space data. Specifically, considering the i-th position of the $\hat{y}$, we have

$$p(\hat{y}_{[i]}|\boldsymbol{x}, \sigma_{[i]}) \sim \mathcal{N}(\hat{y}_{[i]}|(\boldsymbol{\mathcal{F}}\boldsymbol{x})_{[i]}, \sigma_{[i]}^2), \tag{1}$$

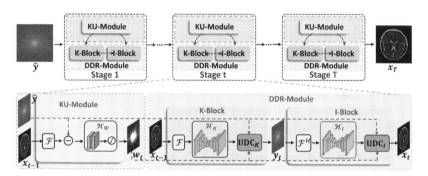

Fig. 1. The architecture of the proposed dual domain motion correction network, i.e. $D^2MC\text{-}Net$. Each stage consists of two components, i.e., k-space uncertainty module (KU-Module) and dual domain reconstruction module (DDR-Module).

where $\boldsymbol{x} \in \mathbb{C}^N$ denotes the motion-free image. $\boldsymbol{\mathcal{F}} \in \mathbb{C}^{N \times N}$ is the Fourier transform matrix. $\sigma_{[i]} \in [1, \infty)$ is the standard deviation at i-th position, which gives larger values to k-space data severely corrupted by the motion and smaller values to k-space data less affected by the motion.

Based on the above distribution model of the motion-corrupted k-space data, we propose the following maximum log-posterior estimation model:

$$\max_{\boldsymbol{x},\boldsymbol{y},\boldsymbol{w}} \log p(\boldsymbol{x}, \boldsymbol{y}, \boldsymbol{w}|\hat{\boldsymbol{y}}) = \max_{\boldsymbol{x},\boldsymbol{y},\boldsymbol{w}} \log p(\hat{\boldsymbol{y}}|\boldsymbol{x}, \boldsymbol{w}) + \log p(\boldsymbol{x}) + \log p(\boldsymbol{y}) + \text{const} \tag{2}$$

where $\boldsymbol{w} \in [0,1]^N$ represents the k-space uncertainty with the elements $w_{[i]} = 1/\sigma_{[i]}$. $p(\boldsymbol{x})$ and $p(\boldsymbol{y})$ are the prior distributions of the motion-free data in image domain and k-space domain. The likelihood distribution $\log p(\hat{\boldsymbol{y}}|\boldsymbol{x}, \boldsymbol{w}) = \prod_i p(\hat{y}_{[i]}|\boldsymbol{x}, \sigma_{[i]})$ has been modeled by Eq. (1). Then the solution of Eq. (2) can be converted to a dual domain regularized model with an uncertainty-guided data consistency term to correct the motion-related artifacts:

$$\boldsymbol{x}^*, \boldsymbol{y}^*, \boldsymbol{w}^* = \arg\min_{\boldsymbol{x},\boldsymbol{y},\boldsymbol{w}} \frac{1}{2} \|\boldsymbol{w} \odot \boldsymbol{\mathcal{F}}\boldsymbol{x} - \boldsymbol{w} \odot \hat{\boldsymbol{y}}\|_2^2 + \frac{\rho}{2} \|\boldsymbol{x} - \mathcal{H}_I(\boldsymbol{x}; \theta_I)\|_2^2$$

$$+ \frac{\lambda}{2} \|\boldsymbol{y} - \mathcal{H}_K(\boldsymbol{y}; \theta_K)\|_2^2 - \sum_i^N \log w_{[i]}, \qquad s.t. \quad \boldsymbol{y} = \boldsymbol{\mathcal{F}}\boldsymbol{x} \tag{3}$$

where $\mathcal{H}_I$ and $\mathcal{H}_K$ are learnable denoisers with parameters θ_I and θ_K, which adopt the U-Net [18] architecture in this paper. λ and ρ are trade-off parameters. The first term is the uncertainty-guided data consistency term corresponding to the log-likelihood $\log p(\hat{\boldsymbol{y}}|\boldsymbol{x}, \boldsymbol{w})$ which enforces consistency between the k-space data of reconstructed MR image and its motion-corrupted k-space data under the guidance of the uncertainty $\boldsymbol{w}$. The second and third terms are regularizations for imposing image-space prior $p(\boldsymbol{x})$ and k-space prior $p(\boldsymbol{y})$.

2.2   Dual Domain Motion Correction Network

Our proposed D^2MC-Net is designed based on the alternating optimization algorithm to solve Eq. (3). As shown in Fig. 1, taking the motion-corrupted k-space data as input, it reconstructs the motion-free MR images with T stages. Each stage consists of the k-space uncertainty module (KU-Module) and the dual domain reconstruction module (DDR-Module), respectively corresponding to the sub-problems for optimizing the k-space uncertainty $\boldsymbol{w}$, and the dual domain data including k-space data $\boldsymbol{y}$ and MR image $\boldsymbol{x}$. The KU-Module estimates the k-space uncertainty $\boldsymbol{w}$, quantifying the uncertainty of k-space data corrupted by motion. The DDR-Module is responsible for reconstructing the k-space data $\boldsymbol{y}$ and MR image $\boldsymbol{x}$, under the guidance of the k-space uncertainty $\boldsymbol{w}$. Details of these two modules at t-th stage are as follows.

K-space Uncertainty Module. This module is designed to update k-space uncertainty $\boldsymbol{w}$ in Eq. (3). If directly optimizing $\boldsymbol{w}$ in Eq. (3), $\boldsymbol{w}_t = 1/|\boldsymbol{\mathcal{F}}\boldsymbol{x}_{t-1}-\hat{\boldsymbol{y}}|$ at t-th stage, which depends on the difference between the k-space data of reconstructed image $\boldsymbol{\mathcal{F}}\boldsymbol{x}_{t-1}$ and the motion-corrupted k-space data $\hat{\boldsymbol{y}}$. We extend this estimate to be a learnable module defined as:

$$\boldsymbol{w}_t \triangleq \mathcal{H}_W(\boldsymbol{\mathcal{F}}\boldsymbol{x}_{t-1}, \hat{\boldsymbol{y}}; \theta_W), \tag{4}$$

where $\mathcal{H}_W$ is the sub-network with parameters θ_W. When $t=1$, we only send $\hat{\boldsymbol{y}}$ into the KU-Module because we do not have the estimate of the reconstructed MR images in such case.

Dual Domain Reconstruction Module. This module is designed to update k-space data $\boldsymbol{y}$ and MR image $\boldsymbol{x}$ in Eq. (3) under the guidance of the uncertainty $\boldsymbol{w}$. Specifically, given the reconstructed MR image $\boldsymbol{x}_{t-1}$ from $(t-1)$-th stage and the k-space uncertainty $\boldsymbol{w}_t$, the k-space data at t-th stage is updated by:

$$\begin{aligned}
\boldsymbol{y}_t &= \arg\min_{\boldsymbol{y}} \frac{1}{2}\|\boldsymbol{W}_t\boldsymbol{y} - \boldsymbol{W}_t\hat{\boldsymbol{y}}\|_2^2 + \frac{\lambda}{2}\|\boldsymbol{y} - \mathcal{H}_K(\boldsymbol{\mathcal{F}}\boldsymbol{x}_{t-1}; \theta_K)\|_2^2 \\
&= (\boldsymbol{W}_t^\top\boldsymbol{W}_t + \lambda\boldsymbol{I})^{-1}(\boldsymbol{W}_t^\top\boldsymbol{W}_t\hat{\boldsymbol{y}} + \lambda\mathcal{H}_K(\boldsymbol{\mathcal{F}}\boldsymbol{x}_{t-1}; \theta_K)) \\
&\triangleq \mathrm{UDC}_K \circ \mathcal{H}_K(\boldsymbol{\mathcal{F}}\boldsymbol{x}_{t-1}; \theta_K),
\end{aligned} \tag{5}$$

where $\boldsymbol{W}_t = \mathrm{diag}(\boldsymbol{w}_t) \in [0,1]^{N \times N}$ is a diagonal matrix, thus the matrix inversion in Eq. (5) can be computed efficiently. Equation (5) is defined as k-space reconstruction block (K-Block), solving the sub-problem for optimizing k-space data $\boldsymbol{y}$ in Eq. (3). Equation (5) can be implemented by firstly computing $\mathcal{H}_K$, followed by the k-space uncertainty-guided data consistency operator UDC_K in Eq. (5). Similarly, given the updated uncertainty $\boldsymbol{w}_t$ and k-space data $\boldsymbol{y}_t$, the MR image at t-th stage is updated by:

$$
\begin{aligned}
\boldsymbol{x}_t &= \arg\min_{\boldsymbol{x}} \frac{1}{2} \|\boldsymbol{W}_t\mathcal{F}\boldsymbol{x} - \boldsymbol{W}_t\hat{\boldsymbol{y}}\|_2^2 + \frac{\rho}{2}\|\boldsymbol{x} - \mathcal{H}_I(\mathcal{F}^H\boldsymbol{y}_t; \theta_I)\|_2^2 \\
&= \mathcal{F}^H(\boldsymbol{W}_t^\top\boldsymbol{W}_t + \rho\boldsymbol{I})^{-1}(\boldsymbol{W}_t^\top\boldsymbol{W}_t\hat{\boldsymbol{y}} + \rho\mathcal{F}\mathcal{H}_I(\mathcal{F}^H\boldsymbol{y}_t; \theta_I)) \\
&\triangleq \mathrm{UDC}_I \circ \mathcal{H}_I(\mathcal{F}^H\boldsymbol{y}_t; \theta_I).
\end{aligned}
\tag{6}
$$

Equation (6) is defined as image reconstruction block (I-Block), solving the sub-problem for optimizing MR image $\boldsymbol{x}$ in Eq. (3). Equation (6) can be implemented by firstly computing $\mathcal{H}_I$, followed by the image domian uncertainty-guided data consistency operator UDC_I in Eq. (6). The K-Block and I-Block are combined as the dual domain reconstruction module (DDR-Module) to sequentially reconstruct the k-space data $\boldsymbol{y}_t$ and MR image $\boldsymbol{x}_t$ at t-th stage.

In summary, by connecting the k-space uncertainty module and dual domain reconstruction module alternately, we construct a multi-stage deep network (i.e., $\mathrm{D}^2\mathrm{MC}$-Net) for motion artifacts correction as shown in Fig. 1.

2.3   Network Details and Training Loss

In the proposed $\mathrm{D}^2\mathrm{MC}$-Net, we use $T = 3$ stages for speed and accuracy trade-off. Each stage has three sub-networks (i.e., $\mathcal{H}_W$, $\mathcal{H}_K$ and $\mathcal{H}_I$) as shown in Fig. 1. $\mathcal{H}_K$ and $\mathcal{H}_I$ adopt U-Net [18] architecture which contains five encoder blocks and four decoder blocks followed by a 1×1 convolution layer for the final output. Each block consists of two 3×3 convolution layers, an instance normalization (IN) layer and a ReLU activation function. The average pooling and bilinear interpolation layers are respectively to reduce and increase the resolution of the feature maps. The number of output feature channels of the encoder and decoder blocks in U-Net are successively 32, 64, 128, 256, 512, 256 128, 64, 32. The structure of $\mathcal{H}_W$ is Conv→IN→ReLU→Conv→Sigmoid, where Conv denotes a 3×3 convolution layer. The number of output feature channels for these two convolution layers are 64 and 2, respectively.

The overall loss function in image space and k-space is defined as:

$$
\mathcal{L} = \sum_{t=1}^{T} \gamma\|\boldsymbol{y}_t - \boldsymbol{y}_{gt}\|_1 + \|\boldsymbol{x}_t - \boldsymbol{x}_{gt}\|_1 + (1 - \mathrm{SSIM}(\boldsymbol{x}_t, \boldsymbol{x}_{gt})),
\tag{7}
$$

where $\boldsymbol{x}_t$ and $\boldsymbol{y}_t$ are the reconstructed MR image and k-space data at t-th stage. $\boldsymbol{x}_{gt}$ and $\boldsymbol{y}_{gt}$ are the motion-free MR image and k-space data. SSIM [22] is the structural similarity loss. γ is a hyperparameter to balance the different losses in dual domain, and we set $\gamma = 0.001$. The Adam optimizer with mini-batch size of 4 is used to optimize the network parameters. The initial value of the learning rate is 1×10^{-4} and divided by 10 at 40-th epoch. We implement the proposed $\mathrm{D}^2\mathrm{MC}$-Net using PyTorch on one Nvidia Tesla V100 GPU for 50 epochs.

3   Experiments

Dataset. We evaluate our method on the T2-weighted brain images from the fastMRI dataset [7], and we randomly select 78 subjects for training and 39 subjects for testing. The in-plane matrix size of the subjects is resized to 384 × 384, and the number of slices varies from the subjects. Sensitivity maps are estimated using the ESPIRiT algorithm [20] for coil combination.

Motion Artifacts Simulation. We simulate in-plane and through-plane motion according to the forward model $\hat{y} = M\mathcal{F}T_\theta x$ [2], where $T_\theta \in \mathbb{R}^{N \times N}$ is the rigid-body motion matrix parameterized by a vector of translations and rotations $\theta \in \mathbb{R}^3 \times [-\pi, \pi]^3$. $M \in \{0,1\}^{N \times N}$ is the diagonal mask matrix in k-space. And we keep 7% of the k-space lines in the center for preventing excessive distortion of the images. The motion vectors are randomly selected from a Gaussian distribution $\mathcal{N}(0, 10)$. We follow the motion trajectories (i.e., piecewise constant, piecewise transient and Gaussian) used in the paper [5] to simulate motion. In addition, to generate various motion severities, each motion level has a series of motion-corrupted k-space lines: 0–30%, 0–50%, and 0–70% of the total of k-space lines for mild, moderate, and severe, respectively. Finally, the motion-corrupted volume k-space data is cut into slice data and sent to the proposed D^2MC-Net.

Table 1. Quantitative comparison of different methods on fastMRI under different motion trajectories and motion severities, in PSNR (dB), SSIM and NRMSE.

Motion Trajectories	Methods	Mild			Moderate			Severe		
		PSNR	SSIM	NRMSE	PSNR	SSIM	NRMSE	PSNR	SSIM	NRMSE
Piecewise Constant	Corrupted	33.76	0.9081	0.1344	30.71	0.8563	0.1895	28.52	0.8134	0.2366
	U-Net	35.84	0.9571	0.1023	32.65	0.9345	0.1494	32.14	0.9168	0.1527
	UPGAN	36.06	0.9537	0.0986	34.01	0.9287	0.1246	32.19	0.8781	0.1530
	SU-Net	35.92	0.9541	0.1012	34.00	0.9378	0.1254	32.97	0.9241	0.1389
	Alternating	37.08	0.9538	0.0879	34.51	0.9305	0.1186	32.33	0.9064	0.1506
	Autofocusing+	37.43	0.9559	0.0847	35.57	0.9356	0.1044	33.17	0.9115	0.1360
	Ours	**41.00**	**0.9761**	**0.0567**	**37.79**	**0.9594**	**0.0806**	**35.28**	**0.9399**	**0.1066**
Piecewise Transient	Corrupted	32.78	0.8317	0.1407	30.04	0.7750	0.1934	28.56	0.7469	0.2301
	U-Net	35.58	0.9511	0.1053	33.67	0.9338	0.1310	32.49	0.9217	0.1494
	UPGAN	37.57	0.9526	0.0809	35.50	0.9339	0.1026	34.20	0.9220	0.1191
	SU-Net	37.50	0.9540	0.0815	35.28	0.9363	0.1052	34.56	0.9335	0.1269
	Alternating	37.09	0.9447	0.0854	35.15	0.9264	0.1068	34.02	0.9170	0.1217
	Autofocusing+	37.21	0.9415	0.0850	35.58	0.9271	0.1021	34.37	0.9138	0.1169
	Ours	**38.94**	**0.9607**	**0.0691**	**37.37**	**0.9493**	**0.0828**	**35.96**	**0.9380**	**0.0973**
Gaussian	Corrupted	32.71	0.8293	0.1419	30.12	0.7749	0.1915	28.67	0.7444	0.2270
	U-Net	34.78	0.9484	0.1174	33.83	0.9357	0.1299	34.05	0.9268	0.1222
	UPGAN	36.25	0.9477	0.0938	35.94	0.9363	0.0974	34.42	0.9208	0.1160
	SU-Net	37.06	0.9523	0.0864	34.92	0.9402	0.1100	34.49	0.9290	0.1169
	Alternating	37.02	0.9432	0.0861	34.72	0.9194	0.1121	34.43	0.9196	0.1160
	Autofocusing+	37.75	0.9425	0.0792	35.50	0.9275	0.1033	34.67	0.9153	0.1129
	Ours	**39.40**	**0.9615**	**0.0654**	**37.58**	**0.9502**	**0.0807**	**36.21**	**0.9396**	**0.0945**

Performance Evaluation. We compare the proposed D^2MC-Net with four deep learning methods (i.e., U-Net [18], UPGAN [21], SU-Net [1], and Alternating [17]), and an optimization-based method (i.e., Autofocusing+ [12]). The motion-corrupted image without motion correction is denoted as "Corrupted". In Table 1, we show the quantitative results of different methods under different motion trajectories and motion severities. Compared with "Corrupted", these deep learning methods improve the reconstruction performance. By explicitly estimating motion parameters, Autofocusing+ produces better results than deep learning methods. Our method achieves the best results in all experiments, mainly because the uncertainty-guided data consistency term is introduced into the unfolded deep network to guide MRI reconstruction. The qualitative comparison results under the severe corruption with piecewise constant motion trajectory are shown in Fig. 2. In comparison, our method has the smallest reconstruction error and recovers finer image details while suppressing undesired artifacts. The PSNR and SSIM values in Fig. 2 also demonstrate the superiority of our method. For example, the PSNR value of our method is 3.06 dB higher than that of SU-Net [1].

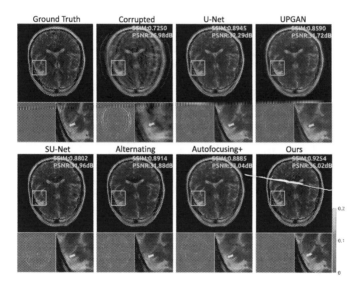

Fig. 2. Qualitative results of different methods under severe corruption with the piecewise constant motion trajectory.

Table 2. Ablation study of the key components of D^2MC-Net.

Methods	KU-Module	K-Block	I-Block	PSNR	SSIM	NRMSE
Baseline			✓	34.53	0.9416	0.1172
w/o K-Block	✓		✓	36.17	0.9503	0.0968
Ours ($w = 0$)		✓	✓	35.18	0.9438	0.1089
Ours ($w = 1$)		✓	✓	36.84	0.9521	0.0880
Ours	✓	✓	✓	**37.79**	**0.9594**	**0.0806**

Effectiveness of the Key Components. We evaluate the effectiveness of these key components, including KU-Module, K-Block, and I-Block in Fig. 1, under the moderate corruption with piecewise constant motion trajectory. In Table 2, (A) "Baseline" denotes the reconstruction model $x_t = \mathcal{H}_I(x_{t-1}; \theta_I)), t = 1 \cdots T$. (B) "$w/o$ K-Blcok" denotes our D^{2}MC-Net without K-Blocks. (C) "Ours ($w = \mathbf{0}$)" denotes our D^{2}MC-Net without KU-Modules, and the k-space uncertainty $w = \mathbf{0}$. (D) "Ours ($w = \mathbf{1}$)" denotes our D^{2}MC-Net without KU-Modules and the k-space uncertainty $w = \mathbf{1}$. (E) "Ours" is our full D^{2}MC-Net equipped with KU-Modules, K-Blocks and I-Blocks. As shown in Table 2, our results are better than all the compared variants, showing the effectiveness of the k-space uncertainty and dual-domain reconstruction. Compared with methods that do not use motion-corrupted k-space data (i.e., "Ours ($w = \mathbf{0}$)") and fully use motion-corrupted k-space data (i.e., "Ours ($w = \mathbf{1}$)") in reconstruction, our method selectively uses the motion-corrupted k-space data under the guidance of the learned k-space uncertainty w, and achieves higher performance.

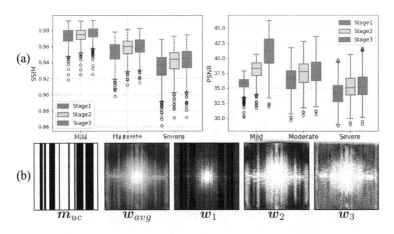

Fig. 3. (a) Qualitative results from different stages of the D^{2}MC-Net under different severities with the piecewise constant motion trajectory. (b) The visual results of w from different stages under moderate piecewise constant motion.

Table 3. Ablation study of the hyperparameters in the loss function.

Hyperparameters	$\gamma = 1$	$\gamma = 0.1$	$\gamma = 0.01$	$\gamma = 0.001$	$\gamma = 0.0001$	$\gamma = 0$
PSNR	37.19 dB	37.20 dB	37.41 dB	**37.79 dB**	37.44 dB	37.10 dB

Effect of Number of Stages. We evaluate the effect of the number of stage T in D^{2}MC-Net. Figure 3(a) reports the results of different T under different motion severities with piecewise constant motion trajectory. We observe that increasing the number of stages in D^{2}MC-Net achieves improvement in PSNR and SSIM metrics, but costs more memory and computational resources.

Visualization of the Uncertainty w. The estimated k-space uncertainties of all stages are visualized in Fig. 3(b). As we can see, the averaged k-space uncertainty w_{avg} over all stages approximates the real motion trajectory mask m_{uc} with ones indicating the un-corrupted k-space lines.

Effect of Different Loss Functions. We also investigate the effect of the k-space loss by adjusting the values of hyperparameters γ in Eq. (7). The PSNR results under the piecewise constant moderate motion are shown in Table 3. From these results, our method achieves the best performance at $\gamma = 0.001$.

4   Conclusion

In this paper, we proposed a novel dual domain motion correction network (D^2MC-Net) to correct the motion artifacts in MRI. The D^2MC-Net consists of KU-Modules and DDR-Modules. KU-Module measures the uncertainty of k-space data corrupted by motion. DDR-Module reconstructs the motion-free MR images in k-space and image domains under the guidance of the uncertainty estimated by KU-Module. Experiments on fastMRI dataset show the superiority of the proposed D^2MC-Net. In the future work, we will extend the D^2MC-Net to be a 3D motion correction method for 3D motion artifacts removal.

Acknowledgements. This work is supported by National Key R&D Program of China (2022YFA1004201), National Natural Science Foundation of China (12090021, 12125104, 61721002, U20B2075).

References

1. Al-masni, M.A., et al.: Stacked u-nets with self-assisted priors towards robust correction of rigid motion artifact in brain mri. Neuroimage **259**, 119411 (2022)
2. Alexander, L., Hannes, N., Pohmannand, R., Bernhard, S.: Blind retrospective motion correction of MR images. Magn. Reson. Med. **70**(6), 1608–1618 (2013)
3. Armanious, K., et al.: MedGAN: medical image translation using GANs. Comput. Med. Imaging Graph. **79**, 101684 (2020)
4. Atkinson, D., Hill, D.L.G., Stoyle, P.N.R., Summers, P.E., Keevil, S.F.: Automatic correction of motion artifacts in magnetic resonance images using an entropy focus criterion. IEEE Trans. Med. Imaging **16**(6), 903–910 (1997)
5. Ben, A.D., et al.: Retrospective motion artifact correction of structural MRI images using deep learning improves the quality of cortical surface reconstructions. Neuroimage **230**, 117756 (2021)
6. Daniel, P., et al.: Scout accelerated motion estimation and reduction (SAMER). Magn. Reson. Med. **87**(1), 163–178 (2022)
7. Florian, K., et al.: fastMRI: a publicly available raw k-space and DICOM dataset of knee images for accelerated MR image reconstruction using machine learning. Radiol. Artif. Intell. **2**(1) (2020)
8. Haskell, M.W., et al.: Network accelerated motion estimation and reduction (NAMER): convolutional neural network guided retrospective motion correction using a separable motion model. Magn. Reson. Med. **82**(4), 1452–1461 (2019)

9. Haskell, M.W., Cauley, S.F., Wald, L.L.: Targeted motion estimation and reduction (TAMER): data consistency based motion mitigation for MRI using a reduced model joint optimization. IEEE Trans. Med. Imaging **37**(5), 1253–1265 (2018)
10. Junchi, L., Mehmet, K., Mark, S., Jie, D.: Motion artifacts reduction in brain MRI by means of a deep residual network with densely connected multi-resolution blocks (DRN-DCMB). Magn. Reson. Imaging **71**, 69–79 (2020)
11. Kay, N., Peter, B.: Prospective correction of affine motion for arbitrary MR sequences on a clinical scanner. Magn. Reson. Med. **54**(5), 1130–1138 (2005)
12. Kuzmina, E., Razumov, A., Rogov, O.Y., Adalsteinsson, E., White, J., Dylov, D.V.: Autofocusing+: Noise-resilient motion correction in magnetic resonance imaging. In: Wang, L., Dou, Q., Fletcher, P.T., Speidel, S., Li, S. (eds.) MICCAI 2022. LNCS, vol. 13436, pp. 365–375. Springer, Cham (2022). https://doi.org/10.1007/978-3-031-16446-0_35
13. Lee, J., Kim, B., Park, H.: MC2-Net: motion correction network for multi-contrast brain MRI. Magn. Reson. Med. **86**(2), 1077–1092 (2021)
14. Levac, B., Jalal, A., Tamir, J.I.: Accelerated motion correction for MRI using score-based generative models. arXiv (2022). https://arxiv.org/abs/2211.00199
15. Lucilio, C.G., Teixeira, R.P.A.G., Hughes, E.J., Hutter, J., Price, A.N., Hajnal, J.V.: Sensitivity encoding for aligned multishot magnetic resonance reconstruction. IEEE Trans. Comput. Imaging **2**(3), 266–280 (2016)
16. Maxim, Z., Julian, M., Michael, H.: Motion artifacts in MRI: a complex problem with many partial solutions. J. Magn. Reson. Imaging **42**(4), 887–901 (2015)
17. Singh, N.M., Iglesias, J.E., Adalsteinsson, E., Dalca, A.V., Golland, P.: Joint frequency and image space learning for MRI reconstruction and analysis. J. Mach. Learn. Biomed. Imaging (2022)
18. Ronneberger, O., Fischer, P., Brox, T.: U-Net: convolutional networks for biomedical image segmentation. In: Navab, N., Hornegger, J., Wells, W.M., Frangi, A.F. (eds.) MICCAI 2015. LNCS, vol. 9351, pp. 234–241. Springer, Cham (2015). https://doi.org/10.1007/978-3-319-24574-4_28
19. Thomas, K., Karim, A., Jiahuan, Y., Bin, Y., Fritz, S., Sergios, G.: Retrospective correction of motion-affected MR images using deep learning frameworks. Magn. Reson. Med. **82**(4), 1527–1540 (2019)
20. Uecker, M., et al.: ESPIRiT-an eigenvalue approach to autocalibrating parallel MRI: where sense meets grappa. Magn. Reson. Med. **71**(3), 990–1001 (2014)
21. Upadhyay, U., Chen, Y., Hepp, T., Gatidis, S., Akata, Z.: Uncertainty-guided progressive GANs for medical image translation. In: de Bruijne, M., et al. (eds.) MICCAI 2021. LNCS, vol. 12903, pp. 614–624. Springer, Cham (2021). https://doi.org/10.1007/978-3-030-87199-4_58
22. Wang, Z., Bovik, A.C., Sheikh, H.R., Simoncelli, E.P.: Image quality assessment: from error visibility to structural similarity. IEEE Trans. Image Process. **13**(4), 600–612 (2004)
23. Wood, M.L., Henkelman, R.M.: MR image artifacts from periodic motion. Med. Phys. **12**(2), 143–151 (1985)

Trackerless Volume Reconstruction from Intraoperative Ultrasound Images

Sidaty El hadramy[1,2], Juan Verde[3], Karl-Philippe Beaudet[2], Nicolas Padoy[2,3], and Stéphane Cotin[1(✉)]

[1] Inria, Strasbourg, France
stephane.cotin@inria.fr
[2] ICube, University of Strasbourg, CNRS, Strasbourg, France
[3] IHU Strasbourg, Strasbourg, France

Abstract. This paper proposes a method for trackerless ultrasound volume reconstruction in the context of minimally invasive surgery. It is based on a Siamese architecture, including a recurrent neural network that leverages the ultrasound image features and the optical flow to estimate the relative position of frames. Our method does not use any additional sensor and was evaluated on *ex vivo* porcine data. It achieves translation and orientation errors of 0.449 ± 0.189 mm and $1.3 \pm 1.5°$ respectively for the relative pose estimation. In addition, despite the predominant non-linearity motion in our context, our method achieves a good reconstruction with final and average drift rates of 23.11% and 28.71% respectively. To the best of our knowledge, this is the first work to address volume reconstruction in the context of intravascular ultrasound. Source code of this work is publicly available at https://github.com/Sidaty1/IVUS_Trakerless_Volume_Reconstruction.

Keywords: Intraoperative Ultrasound · Liver Surgery · Volume Reconstruction · Recurrent Neural Networks

1 Introduction

Liver cancer is the most prevalent indication for liver surgery, and although there have been notable advancements in oncologic therapies, surgery remains as the only curative approach overall [20].

Liver laparoscopic resection has demonstrated fewer complications compared to open surgery [21], however, its adoption has been hindered by several reasons, such as the risk of unintentional vessel damage, as well as oncologic concerns such as tumor detection and margin assessment. Hence, the identification of intrahepatic landmarks, such as vessels, and target lesions is crucial for successful and safe surgery, and intraoperative ultrasound (IOUS) is the preferred technique to accomplish this task. Despite the increasing use of IOUS in surgery, its integration into laparoscopic workflows (i.e., laparoscopic intraoperative ultrasound) remains challenging due to combined problems.

H. Greenspan et al. (Eds.): MICCAI 2023, LNCS 14229, pp. 303–312, 2023.
https://doi.org/10.1007/978-3-031-43999-5_29

Performing IOUS during laparoscopic liver surgery poses significant challenges, as laparoscopy has poor ergonomics and narrow fields of view, and on the other hand, IOUS demands skills to manipulate the probe and analyze images. At the end, and **despite its real-time capabilities**, IOUS images are **intermittent and asynchronous** to the surgery, requiring multiple iterations and repetitive steps (probe-in → instruments-out → probe-out → instruments-in). Therefore, any method enabling a continuous and synchronous US assessment throughout the surgery, with minimal iterations required would significantly improve the surgical workflow, as well as its efficiency and safety.

To overcome these limitations, the use of intravascular ultrasound (IVUS) images has been proposed, enabling **continuous and synchronous inside-out imaging** during liver surgery [19]. With an intravascular approach, an overall view and full-thickness view of the liver can quickly and easily be obtained through mostly rotational movements of the catheter, while this is constrained to the lumen of the *inferior vena cava*, and with no interaction with the tissue (contactless, a.k.a. standoff technique) as illustrated in Fig. 1.

Fig. 1. *left*: IVUS catheter positioned in the lumen of the *inferior vena cava* in the posterior surface of the organ, and an example of the lateral firing and longitudinal beam-forming images; *middle*: anterior view of the liver and the rotational movements of the catheter providing full-thickness images; *right*: inferior view showing the rotational US acquisitions

However, to benefit from such a technology in a computer-guided solution, the different US images would need to be tracked and possibly integrated into a volume for further processing. External US probes are often equipped with an electromagnetic tracking system to track its position and orientation in real-time. This information is then used to register the 3D ultrasound image with the patient's anatomy. The use of such an electromagnetic tracking system in laparoscopic surgery is more limited due to size reduction. The tracking system may add additional complexity and cost to the surgical setup, and the tracking accuracy may be affected by metallic devices in the surgical field [22].

Several approaches have been proposed to address this limitation by proposing a trackerless ultrasound volume reconstruction. Physics-based methods have exploited speckle correlation models between different adjacent frames [6–8] to estimate their relative position. With the recent advances in deep learning, recent works have proposed to learn a higher order nonlinear mapping between adjacent frames and their relative spatial transformation. *Prevost et al.* [9] first demonstrated the effectiveness of a convolution neural network to learn the relative motion between a pair of US images. *Xie et al.* [10] proposed a pyramid warping layer that exploits the optical flow features in addition to the ultrasound features in order to reconstruct the volume. To enable a smooth 3D reconstruction, a case-wise correlation loss based on 3D CNN and Pearson correlation coefficient was proposed in [10,12]. *Qi et al.* [13] leverages past and future frames to estimate the relative transformation between each pair of the sequence; they used the consistency loss proposed in [14]. Despite the success of these approaches, they still suffer significant cumulative drift errors and mainly focus on linear probe motions. Recent work [15,16] proposed to exploit the acceleration and orientation of an inertial measurement unit (IMU) to improve the reconstruction performance and reduce the drift error. Motivated by the weakness of the state-of-the-art methods when it comes to large non-linear probe motions, and the difficulty of integrating IMU sensors in the case of minimally invasive procedures, we introduce a new method for pose estimation and volume reconstruction in the context of minimally invasive trackerless ultrasound imaging. We use a Siamese architecture based on a Sequence to Vector(Seq2Vec) neural network that leverages image and optical flow features to learn relative transformation between a pair of images.

Our method improves upon previous solutions in terms of robustness and accuracy, particularly in the presence of rotational motion. Such motion is predominant in the context highlighted above and is the source of additional non-linearity in the pose estimation problem. To the best of our knowledge, this is the first work that provides a clinically sound and efficient 3D US volume reconstruction during minimally invasive procedures. The paper is organized as follows: Sect. 2 details the method and its novelty, Sect. 3 presents our current results on *ex vivo* porcine data, and finally, we conclude in Sect. 4 and discuss future work.

2   Method

In this work, we make the assumption that the organ of interest does not undergo deformation during the volume acquisition. This assumption is realistic due to the small size of the probe. Let $I_0, I_1...I_{N-1}$ be a sequence of N frames. Our aim is to find the relative spatial transformation between each pair of frames I_i and I_j with $0 \leq i \leq j \leq N - 1$. This transformation is denoted $T_{(i,j)}$ and is a six degrees of freedom vector representing three translations and three Euler angles. To achieve this goal, we propose a Siamese architecture that leverages the optical flow in the sequences in addition to the frames of interest in order to provide a mapping with the relative frames spatial transformation. The overview of our method is presented in Fig. 2.

We consider a window of $2k + 3$ frames from the complete sequence of length N, where $0 \leq k \leq \lfloor \frac{N-3}{2} \rfloor$ is a hyper-parameter that denotes the number of frames

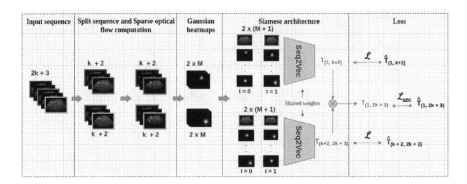

Fig. 2. Overview of the proposed method. The input sequence is split into two equal sequences with a common frame. Both are used to compute a sparse optical flow. Gaussian heatmaps tracking M points are then combined with the first and last frame of each sequence to form the network's input. We use a Siamese architecture based on Sequence to Vector (Seq2Vec) network. The learning is done by minimising the mean square error between the output and ground truth transformations.

between two frames of interest. Our method predicts two relative transformations between the pairs of frames (I_1, I_{k+2}) and (I_{k+2}, I_{2k+3}). The input window is divided into two equal sequences of length $k + 2$ sharing a common frame. Both deduced sequences are used to compute a sparse optical flow allowing to track the trajectory of M points. Then, Gaussian heatmaps are used to describe the motion of the M points in an image-like format(see Sect. 2.2). Finaly, a Siamese architecture based on two shared weights Sequence to Vector (Seq2Vec) network takes as input the Gaussian heatmaps in addition to the first and last frames and predicts the relative transformations. In the following we detail our pipeline.

2.1 Sparse Optical Flow

Given a sequence of frames I_i and I_{i+k+1}, we aim at finding the trajectory of a set of points throughout the sequence. We choose the M most prominent points from the first frame using the feature selection algorithm proposed in [3]. Points are then tracked throughout each pair of adjacent frames in the sequence by solving Eq. 1 which is known as the Optical flow equation. We use the pyramidal implementation of Lucas-Kanade method proposed in [4] to solve the equation. Thus, yielding a trajectory matrix $A \in \mathbb{R}^{M \times (k+2) \times 2}$ that contains the position of each point throughout the sequence. Figure 3 illustrates an example where we track two points in a sequence of frames.

$$I_i(x, y, t) = I_i(x + dx, y + dy, t + dt) \tag{1}$$

2.2 Gaussian Heatmaps

After obtaining the trajectory of M points in the sequence $\{I_i | 1 \leq i \leq k + 2\}$ we only keep the first and last position of each point, which corresponds to the

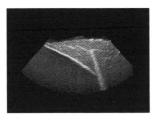
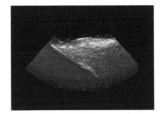

(a) First frame in the sequence   (b) Last frame in the sequence

Fig. 3. Sparse Optical tracking of two points in a sequence, red points represent the chosen points to track, while the blue lines describe the trajectory of the points throughout the sequence. (Color figure online)

positions in our frames of interest. We use Gaussian heatmaps $\mathcal{H} \in \mathbb{R}^{H \times W}$ with the same dimension as the ultrasound frames to encode these points, they are more suitable as input for the convolutional networks. For a point with a position (x_0, y_0), the corresponding heatmap is defined in the Eq. 2.

$$\mathcal{H}(x, y) = \frac{1}{\sigma^2 \sqrt{2\pi}} e^{-\frac{(x-x_0)^2 + (y-y_0)^2}{2\sigma^2}} \tag{2}$$

Thus, each of our M points are converted to a pair of heatmaps that represent the position in the first and last frames of the ultrasound sequence. These pairs concatenated with the ultrasound first and last frames form the recurrent neural network sequential input of size $(M + 1, H, W, 2)$, where $M + 1$ is the number of channels (M heatmaps and one ultrasound frame), H and W are the height and width of the frames and finally 2 represents the temporal dimension.

2.3 Network Architecture

The Siamese architecture is based on a sequence to vector network. Our network maps a sequence of two images having $M + 1$ channel each to a six degrees of freedrom vector (three translations and three rotation angles). The architecture of Seq2Vec is illustrated in the Fig. 4. It contains five times the same block composed of two Convolutional LSTMs (ConvLSTM) [5] followed by a Batch Normalisation. Their output is then flattened and mapped to a six degrees of freedom vector through linear layers; ReLU is the chosen activation function for the first linear layer. We use an architecture similar to the one proposed in [5] for the ConvLSTM layers. Seq2Vec networks share the same weights.

2.4 Loss Function

In the training phase, given a sequence of $2k + 3$ frames in addition to their ground truth transformations $\hat{T}_{(1,k+2)}$, $\hat{T}_{(k+2,2k+3)}$ and $\hat{T}_{(1,2k+3)}$, the Seq2Vec's weights are optimized by minimising the loss function given in the Eq. 3. The loss

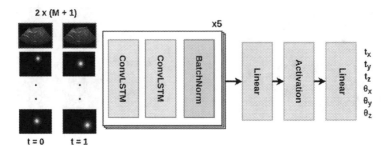

Fig. 4. Architecture of Seq2Vec network. We use five blocks that contains each two ConvLSTM followed by Batch Normalisation. The output is flattened and mapped to a six degree-of-freedom translation and rotation angles through linear layers. The network takes as input a sequence of two images with $M + 1$ channel each, M heatmaps and an ultrasound frame. The output corresponds to the relative transformation between the blue and red frames. (Color figure online)

contains two terms. The first represents the mean square error (MSE) between the estimated transformations $(T_{(1,k+2)}, T_{(k+2,2k+3)})$ at each corner point of the frames and their respective ground truth. The second term represents the accumulation loss that aims at reducing the error of the volume reconstruction, the effectiveness of the accumulation loss have been proven in the literature [13]. It is written as the MSE between the estimated $T_{(1,2k+3)} = T_{(k+2,2k+3)} \times T_{(1,k+2)}$ at the corner points of the frames and the ground truth $\hat{T}_{(1,2k+3)}$.

$$\mathcal{L} = ||T_{(1,k+2)} - \hat{T}_{(1,k+2)}||_2 + ||T_{(k+2,2k+3)} - \hat{T}_{(k+2,2k+3)}||_2 + ||T_{(1,2k+3)} - \hat{T}_{(1,2k+3)}||_2 \tag{3}$$

3   Results and Discussion

3.1   Dataset and Implementation Details

To validate our method, six tracked sequences were acquired from an *ex vivo* swine liver. A manually manipulated IVUS catheter was used (8 Fr lateral firing AcuNav^{TM} 4–10 MHz) connected to an ultrasound system (ACUSON S3000 HELX Touch, Siemens Healthineers, Germany), both commercially available. An electromagnetic tracking system (trakSTAR^{TM}, NDI, Canada) was used along with a 6 DoF sensor (Model 130) embedded close to the tip of the catheter, and the PLUS toolkit [17] along with 3D Slicer [18] were used to record the sequences. The frame size was initially 480 × 640. Frames were cropped to remove the patient and probe characteristics, then down-sampled to a size of 128 × 128 with an image spacing of 0.22 mm per pixel. First and end stages of the sequences were removed from the six acquired sequences, as they were considered to be largely stationary, and aiming to avoid training bias. Clips were created by sliding a window of 7 frames (corresponding to a value of $k = 2$) with a stride of 1 over each continuous sequence, yielding a data set that contains a total of 13734 clips. The tracking was provided for each frame as a 4×4 transformation matrix. We have converted each to a vector of six degrees of freedom that corresponds to three translations in mm and three Euler angles in $degrees$. For each clip, relative frame to frame transformations were computed for the frames number 0, 3 and 6. The distribution of the relative transformation between the frames in our clips is illustrated in the Fig. 5. It is clear that our data mostly contains rotations, in particular over the axis x. Heatmaps were calculated for two points ($M = 2$) and with a quality level of 0.1, a minimum distance of 7 and a block size of 7 for the optical flow algorithm (see [4] for more details). The number of heatmaps M and the frame jump k were experimentally chosen among 0, 2, 4, 6. The data was split into train, validation and test sets by a ratio of 7:1.5:1.5. Our method is implemented in $Pytorch$[1] $1.8.2$, trained and evaluated on a $GeForce$ RTX 3090. We use an Adam optimizer with a learning rate of 10^{-4}. The training process converges in 40 epochs with a batch size of 16. The model with the best performance on the validation data was selected and used for the testing.

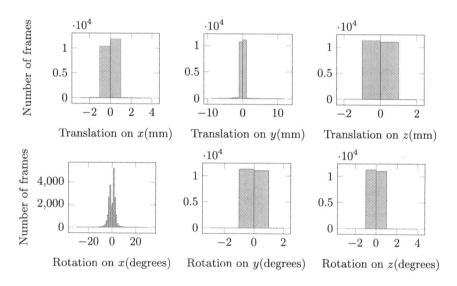

Fig. 5. The distribution of the relative rotations and translations over the dataset

3.2   Evaluation Metrics and Results

The test data was used to evaluate our method, it contains 2060 clips over which our method achieved a translation error of $\epsilon_{translation}$ of 0.449 ± 0.189 mm, and an orientation error of $\epsilon_{orientation}$ $1.3 \pm 1.5°$. We have evaluated our reconstruction with a commonly used in state-of-the-art metric called final drift error, which measures the distance between the center point of the final frame

---

[1] https://pytorch.org/docs/stable/index.html.

according to the real relative position and the estimated one in the sequence. On this basis, each of the following metrics was reported over the reconstructions of our method. **Final drift rate (FDR):** the final drift divided by the sequence length. **Average drift rate (ADR):** the average cumulative drift of all frames divided by the length from the frame to the starting point of the sequence. Table 1 shows the evaluation of our method over these metrics compared to the state-of-the-art methods MoNet [15] and CNN [9]. Both state-of-the-art methods use IMU sensor data as additional input to estimate the relative transformation between two relative frames. Due to the difficulty of including an IMU sensor in our IVUS catheter, the results of both methods were reported from the MoNet paper where the models have been trained on arm scans, see [15] for more details.

Table 1. The mean and standard deviation FDR and ADR of our method compared with state-of-the-art models MoNet [15] and CNN [9]

Models	FDR(%)	ADR(%)
CNN [9]	**31.88 (15.76)**	**39.71 (14.88)**
MoNet [15]	**15.67 (8.37)**	**25.08 (9.34)**
Ours	**23.11 (11.6)**	**28.71 (12.97)**

As the Table 1 shows, our method is comparable with state-of-the-art methods in terms of drift errors without using any IMU and with non-linear probe motion as one may notice in our data distribution in the Fig. 5. Figure 6 shows the volume reconstruction of two sequences of different sizes with our method in red against the ground truth slices. Despite the non-linearity of the probe motion, the relative pose estimation results obtained by our method remains very accurate. However, one may notice that the drift error increases with respect to the sequence length. This remains a challenge for the community even in the case of linear probe motions.

 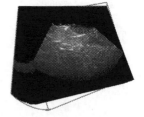

(a) Sequence of length 50 frames   (b) Sequence of length 300 frames

Fig. 6. The reconstruction of two sequences of lengths 50 and 300 respectively with our method in red compared with the ground truth sequences. (Color figure online)

4   Conclusion

In this paper, we proposed the first method for trackerless ultrasound volume reconstruction in the context of minimally invasive surgery. Our method does not use any additional sensor data and is based on a Siamese architecture that leverages the ultrasound image features and the optical flow to estimate relative transformations. Our method was evaluated on *ex vivo* porcine data and achieved translation and orientation errors of 0.449 ± 0.189 mm and $1.3 \pm 1.5°$ respectively with a fair drift error. In the future work, we will extend our work to further improve the volume reconstruction and use it to register a pre-operative CT image in order to provide guidance during interventions.

Aknowledgments. This work was partially supported by French state funds managed by the ANR under reference ANR-10-IAHU-02 (IHU Strasbourg).

References

1. De Gottardi, A., Keller, P.-F., Hadengue, A., Giostra, E., Spahr, L.: Transjugular intravascular ultrasound for the evaluation of hepatic vasculature and parenchyma in patients with chronic liver disease. BMC. Res. Notes **5**, 77 (2012)
2. Urade, T., Verde, J., Vázquez, A.G., Gunzert, K., Pessaux, P., et al.: Fluoroless intravascular ultrasound image-guided liver navigation in porcine models. BMC Gastroenterol. **21**, 24 (2021). https://doi.org/10.1186/s12876-021-01600-3
3. Shi, J., Tomasi, C.: Good features to track. In: 1994 Proceedings of the IEEE Computer Society Conference on Computer Vision and Pattern Recognition, CVPR 1994, pp. 593–600. IEEE (1994)
4. Bouguet, J.-Y.: Pyramidal implementation of the affine Lucas Kanade feature tracker description of the algorithm. Intel Corporation **5**, 4 (2001)
5. Shi, X., et al.: Convolutional LSTM network: a machine learning approach for precipitation nowcasting. In: Advances in Neural Information Processing Systems, vol. 28 (2015)
6. Mercier, L., Lang, T., Lindseth, F., Collins, D.L.: A review of calibration techniques for freehand 3-D ultrasound systems. Ultras. Med. Biol. **31**, 449–471 (2005)
7. Mohamed, F., Siang, C.V.: A survey on 3D ultrasound reconstruction techniques. Artif. Intell. Appl. Med. Biol. (2019)
8. Mozaffari, M.H., Lee, W.S.: Freehand 3-D ultrasound imaging: a systematic review. Ultras. Med. Biol. **43**(10), 2099–2124 (2017)
9. Prevost, R., Salehi, M., Sprung, J., Ladikos, A., Bauer, R., Wein, W.: Deep learning for sensorless 3D freehand ultrasound imaging. In: Descoteaux, M., Maier-Hein, L., Franz, A., Jannin, P., Collins, D.L., Duchesne, S. (eds.) MICCAI 2017. LNCS, vol. 10434, pp. 628–636. Springer, Cham (2017). https://doi.org/10.1007/978-3-319-66185-8_71
10. Xie, Y., Liao, H., Zhang, D., Zhou, L., Chen, F.: Image-based 3D ultrasound reconstruction with optical flow via pyramid warping network. In: Annual International Conference on IEEE Engineering in Medicine & Biology Society (2021)
11. Guo, H., Xu, S., Wood, B., Yan, P.: Sensorless freehand 3D ultrasound reconstruction via deep contextual learning. In: Martel, A.L., et al. (eds.) MICCAI 2020. LNCS, vol. 12263, pp. 463–472. Springer, Cham (2020). https://doi.org/10.1007/978-3-030-59716-0_44

12. Guo, H., Chao, H., Xu, S., Wood, B.J., Wang, J., Yan, P.: Ultrasound volume reconstruction from freehand scans without tracking. IEEE Trans. Biomed. Eng. **70**(3), 970–979 (2023)
13. Li, Q., et al.: Trackerless freehand ultrasound with sequence modelling and auxiliary transformation over past and future frames. arXiv:2211.04867v2 (2022)
14. Miura, K., Ito, K., Aoki, T., Ohmiya, J., Kondo, S.: Probe localization from ultrasound image sequences using deep learning for volume reconstruction. In: Proceedings of the SPIE 11792, International Forum on Medical Imaging in Asia 2021, p. 117920O (2021)
15. Luo, M., Yang, X., Wang, H., Du, L., Ni, D.: Deep motion network for freehand 3D ultrasound reconstruction. In: Wang, L., Dou, Q., Fletcher, P.T., Speidel, S., Li, S. (eds.) MICCAI 2022. LNCS, vol. 13434, pp. 290–299. Springer, Cham (2022). https://doi.org/10.1007/978-3-031-16440-8_28
16. Ning, G., Liang, H., Zhou, L., Zhang, X., Liao, H.: Spatial position estimation method for 3D ultrasound reconstruction based on hybrid transfomers. In: 2022 IEEE 19th International Symposium on Biomedical Imaging (ISBI), Kolkata, India (2022)
17. Lasso, A., Heffter, T., Rankin, A., Pinter, C., Ungi, T., Fichtinger, G.: PLUS: open-source toolkit for ultrasound-guided intervention systems. IEEE Trans. Biomed. Eng. **61**(10), 2527–2537 (2014)
18. Fedorov, A., et al.: 3D slicer as an image computing platform for the quantitative imaging network. Magn. Reson. Imaging **30**(9), 1323–1341 (2012). PMID: 22770690, PMCID: PMC3466397
19. Urade, T., Verde, J.M., García Vázquez, A., et al.: Fluoroless intravascular ultrasound image-guided liver navigation in porcine models. BMC Gastroenterol. **21**(1), 24 (2021)
20. Aghayan, D.L., Kazaryan, A.M., Dagenborg, V.J., et al.: Long-term oncologic outcomes after laparoscopic versus open resection for colorectal liver metastases: a randomized trial. Ann. Intern. Med. **174**(2), 175–182 (2021)
21. Fretland, Å.A., Dagenborg, V.J., Bjørnelv, G.M.W., et al.: Laparoscopic versus open resection for colorectal liver metastases: the OSLO-COMET randomized controlled trial. Ann. Surg. **267**(2), 199–207 (2018)
22. Franz, A.M., Haidegger, T., Birkfellner, W., Cleary, K., Peters, T.M., Maier-Hein, L.: Electromagnetic tracking in medicine a review of technology, validation, and applications. IEEE Trans. Med. Imaging **33**(8), 1702–1725 (2014)

Accurate Multi-contrast MRI Super-Resolution via a Dual Cross-Attention Transformer Network

Shoujin Huang[1], Jingyu Li[1], Lifeng Mei[1], Tan Zhang[1], Ziran Chen[1], Yu Dong[2], Linzheng Dong[2], Shaojun Liu[1(✉)], and Mengye Lyu[1(✉)]

[1] Shenzhen Technology University, Shenzhen, China
liusj14@tsinghua.org.cn, lvmengye@sztu.edu.cn
[2] Shenzhen Samii Medical Center, Shenzhen, China

Abstract. Magnetic Resonance Imaging (MRI) is a critical imaging tool in clinical diagnosis, but obtaining high-resolution MRI images can be challenging due to hardware and scan time limitations. Recent studies have shown that using reference images from multi-contrast MRI data could improve super-resolution quality. However, the commonly employed strategies, e.g., channel concatenation or hard-attention based texture transfer, may not be optimal given the visual differences between multi-contrast MRI images. To address these limitations, we propose a new Dual Cross-Attention Multi-contrast Super Resolution (DCAMSR) framework. This approach introduces a dual cross-attention transformer architecture, where the features of the reference image and the up-sampled input image are extracted and promoted with both spatial and channel attention in multiple resolutions. Unlike existing hard-attention based methods where only the most correlated features are sought via the highly down-sampled reference images, the proposed architecture is more powerful to capture and fuse the shareable information between the multi-contrast images. Extensive experiments are conducted on fastMRI knee data at high field and more challenging brain data at low field, demonstrating that DCAMSR can substantially outperform the state-of-the-art single-image and multi-contrast MRI super-resolution methods, and even remains robust in a self-referenced manner. The code for DCAMSR is avaliable at https://github.com/Solor-pikachu/DCAMSR.

Keywords: Magnetic resonance imaging · Super-resolution · Multi-contrast

1 Introduction

Magnetic Resonance Imaging (MRI) has revolutionized medical diagnosis by providing a non-invasive imaging tool with multiple contrast options [1,2]. However,

---

S. Huang and J. Li contribute equally to this work.

Supplementary Information The online version contains supplementary material available at https://doi.org/10.1007/978-3-031-43999-5_30.

generating high-resolution MRI images can pose difficulties due to hardware limitations and lengthy scanning times [3,4]. To tackle this challenge, super-resolution techniques have been developed to improve the spatial resolution of MRI images [5]. However, while several neural network-based super-resolution methods (e.g., EDSR [6], SwinIR [7], and ELAN [8]) have emerged from the computer vision field, they primarily utilize single-contrast data, ignoring the valuable complementary multi-contrast information that is easily accessible in MRI.

Recent studies have shown that multi-contrast data routinely acquired in MRI examinations can be used to develop more powerful super-resolution methods tailored for MRI by using fully sampled images of one contrast as a reference (Ref) to guide the recovery of high-resolution (HR) images of another contrast from low-resolution (LR) inputs [9]. In this direction, MINet [10] and SANet [11] have been proposed and demonstrated superior performance over previous single-image super-resolution approaches. However, these methods rely on relatively simple techniques, such as channel concatenation or spatial addition between LR and Ref images, or using channel concatenation followed by self-attention to identify similar textures between LR and Ref images. These approaches may overlook the complex relationship between LR and Ref images and lead to inaccurate super-resolution.

Recent advances in super-resolution techniques have led to the development of hard-attention-based texture transfer methods (such as TTSR [12], MASA [13], and McMRSR [14]) using the texture transformer architecture [12]. However, these methods may still underuse the rich information in multi-contrast MRI data. As illustrated in Fig. 1(a), these methods focus on spatial attention and only seek the most relevant patch for each query. They also repetitively use low-resolution attention maps from down-sampled Ref images ($\text{Ref}_{\downarrow\uparrow}$), which may not be sufficient to capture the complex relationship between LR and Ref images, potentially resulting in suboptimal feature transfer. These limitations can be especially problematic for noisy low-field MRI data, where down-sampling the Ref images (as the key in the transformer) can cause additional image blurring and information loss.

As shown in Fig. 1(b), our proposed approach is inspired by the transformer-based cross-attention approach [15], which provides a spatial cross-attention mechanism using full-powered transformer architecture without Ref image downsampling, as well as the UNETR++ architecture [16], which incorporates channel attention particularly suitable for multi-contrast MRI images that are anatomically aligned. Building upon these developments, the proposed Dual Cross-Attention Multi-contrast Super Resolution (DCAMSR) method can flexibly search the reference images for shareable information with multi-scale attention maps and well capture the information both locally and globally via spatial and channel attention. Our contributions are summarized as follows: 1) We present a novel MRI super-resolution framework different from existing hard-attention-based methods, leading to efficient learning of shareable multi-contrast information for more accurate MRI super-resolution. 2) We introduce a dual cross-attention transformer to jointly explore spatial and channel information, substantially improving the feature extraction and fusion processes. 3) Our proposed method robustly outperforms the current state-of-the-art single-image as well as multi-contrast MRI super-resolution methods, as demonstrated by extensive experiments on the high-field fastMRI [17] and more challenging low-field M4Raw [18] MRI datasets.

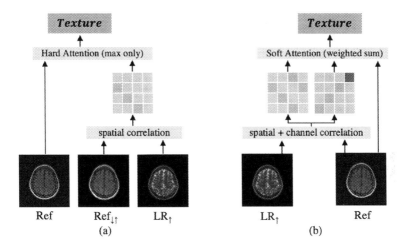

Fig. 1. (a) Illustration of Texture Transformer. (b) Illustration of the proposed Dual Cross-Attention Transformer.

2 Methodology

Overall Architecture. Our goal is to develop a neural network that can restore an HR image from an LR image and a Ref image. Our approach consists of several modules, including an encoder, a dual cross-attention transformer (DCAT) and a decoder, as shown in Fig. 2. Firstly, the LR is interpolated to match the resolution of HR. Secondly, we use the encoder to extract multi-scale features from both the up-sampled LR and Ref, resulting in features F_{LR} and F_{Ref}. Thirdly, the DCAT, which contains of dual cross-attention (DCA), Layer Normalization (LN) and feed-forward network (FFN), is used to search for texture features from F_{LR} and F_{Ref}. Fourthly, the texture features are aggregated with F_{LR} through the Fusion module at each scale. Finally, a simple convolution is employed to generate SR from the fused feature.

Encoder. To extract features from the up-sampled LR, we employ an encoder consisting of four stages. The first stage uses the combination of a depth-wise convolution and a residual block. In stages 2–4, we utilize a down-sampling layer and a residual block to extract multi-scale features. In this way, the multi-scale features for the $LR_\uparrow$ are extracted as $F_{LR}^{H \times W}$, $F_{LR}^{\frac{H}{2} \times \frac{W}{2}}$, $F_{LR}^{\frac{H}{4} \times \frac{W}{4}}$ and $F_{LR}^{\frac{H}{8} \times \frac{W}{8}}$,

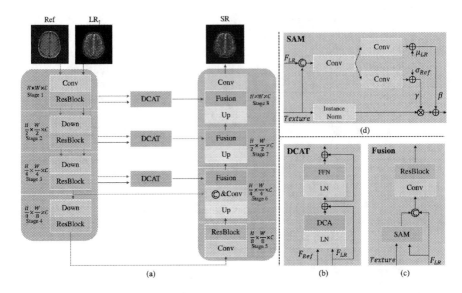

Fig. 2. (a) Network architecture of the proposed Dual Cross-attention Multi-contrast Super Resolution (DCAMSR). (b) Details of Dual Cross-Attention Transformer (DCAT). (c) Details of Fusion block. (d) Details of Spatial Adaptation Module (SAM).

respectively. Similarly, the multi-scale features for Ref are extracted via the same encoder in stages 1–3 and denoted as $F_{Ref}^{H \times W}$, $F_{Ref}^{\frac{H}{2} \times \frac{W}{2}}$ and $F_{Ref}^{\frac{H}{4} \times \frac{W}{4}}$, respectively.

Dual Cross-Attention Transformer (DCAT). The DCAT consists of a DCA module, 2 LNs, and a FFN comprising several 1×1 convolutions.

The core of DCAT is dual cross-attention mechanism, which is diagrammed in Fig. 3. Firstly, we project F_{LR} and F_{Ref} to q, k and v. For the two cross-attention branches, the linear layer weights for q and k are shared, while those for v are different:

$$q_{share} = W_{share}^q(F_{LR}), k_{share} = W_{share}^k(F_{Ref}), \tag{1}$$

$$v_{spatial} = W_{spatial}^v(F_{Ref}), v_{channel} = W_{channel}^v(F_{Ref}), \tag{2}$$

where $q_{share}, k_{share}, v_{spatial}$ and $v_{channel}$ are the parameter weights for shared queries, shared keys, spatial value layer, and channel value layer, respectively. In spatial cross-attention, we further project k_{share} and $v_{spatial}$ to $k_{project}$ and $v_{project}$ through linear layers, to reduce the computational complexity. The spatial and channel attentions are calculated as:

$$X_{spatial} = softmax(\frac{q_{share} \cdot k_{share}^T}{\sqrt{d}}) \cdot v_{project}, \tag{3}$$

$$X_{channel} = softmax(\frac{q_{share}^T \cdot k_{share}}{\sqrt{d}}) \cdot v_{channel}^T. \tag{4}$$

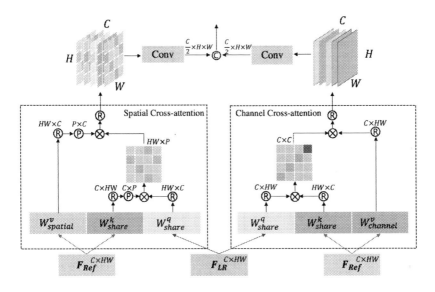

Fig. 3. Details of Dual Cross-Attention (DCA).

Finally, $X_{spatial}$ and $X_{channel}$ are reduced to half channel via 1×1 convolutions, and then concatenate to obtain the final feature:

$$X = Concat(Conv(X_{spatial}), Conv(X_{channel})). \tag{5}$$

For the whole DCAT, the normalized features $LN(F_{LR})$ and $LN(F_{Ref})$ are fed to the DCA and added back to F_{LR}. The obtained feature is then processed by the FFN in a residual manner to generate the texture feature. Specifically, the DCAT is summarized as:

$$X = F_{LR} + DCA(LN(F_{LR}), LN(F_{Ref})), \tag{6}$$

$$Texture = X + FFN(LN(X)). \tag{7}$$

Feeding the multi-scale features of $LR_\uparrow$ and Ref to DCAT, we can generate the texture features in multi-scales, denoted as $Texture^{H \times W}$, $Texture^{\frac{H}{2} \times \frac{W}{2}}$, and $Texture^{\frac{H}{4} \times \frac{W}{4}}$.

Decoder. In the decoder, we start from the feature $F_{LR}^{\frac{H}{8} \times \frac{W}{8}}$ and process it with a convolution and a residual block. Then it is up-sampled and concatenated with $F_{LR}^{\frac{H}{4} \times \frac{W}{4}}$, and then feed to a convolution to further incorporate the both information. Next, the incorporated feature is fed to the Fusion module along with $Texture^{\frac{H}{4} \times \frac{W}{4}}$, to produce the fused feature at $\frac{H}{4} \times \frac{W}{4}$ scale, denoted as $Fused^{\frac{H}{4} \times \frac{W}{4}}$. $Fused^{\frac{H}{4} \times \frac{W}{4}}$ is then up-sampled and feed to Fusion along with $Texture^{\frac{H}{2} \times \frac{W}{2}}$, generating $Fused^{\frac{H}{2} \times \frac{W}{2}}$. Similarly, $Fused^{\frac{H}{2} \times \frac{W}{2}}$ is up-sampled and feed to Fusion along with $Texture^{H \times W}$, generating $Fused^{H \times W}$. Finally, $Fused^{H \times W}$ is processed with a 1×1 convolution to generate SR.

In the Fusion module, following [13], the texture feature $Texture$ and input feature F_{LR} are first fed to Spatial Adaptation Module (SAM), a learnable structure ensuring the distributions of $Texture$ consistent with F_{LR}, as shown in Fig. 2(d). The corrected texture feature is then concatenated with the input feature F_{LR} and further incorporated via a convolution and a residual block, as shown in Fig. 2(c).

Loss Function. For simplicity and without loss of generality, L_1 loss between the restored SR and ground-truth is employed as the overall reconstruction loss.

3   Experiments

Datasets and Baselines. We evaluated our approach on two datasets: 1) fastMRI, one of the largest open-access MRI datasets. Following the settings of SANet [10,11], 227 and 24 pairs of PD and FS-PDWI volumes are selected for training and validation, respectively. 2) M4Raw, a publicly available dataset including multi-channel k-space and single-repetition images from 183 participants, where each individual haves multiple volumes for T1-weighted, T2-weighted and FLAIR contrasts [18]. 128 individuals/6912 slices are selected for training and 30 individuals/1620 slices are reserved for validation. Specifically, T1-weighted images are used as reference images to guide T2-weighted images. To generate the LR images, we first converted the original image to k-space and cropped the central low-frequency region. For down-sampling factors of 2× and 4×, we kept the central 25% and 6.25% values in k-space, respectively, and then transformed them back into the image domain using an inverse Fourier transform. The proposed method is compared with SwinIR [7], ELAN [8], SANet (the journal version of MINet) [11], TTSR [12], and MASA [13].

Implementation Details. All the experiments were conducted using Adam optimizer for 50 epochs with a batch size of 4 on 8 Nvidia P40 GPUs. The initial learning rate for SANet was set to 4×10^{-5} according to [11], and 2×10^{-4} for the other methods. The learning rate was decayed by a factor of 0.1 for the last 10 epochs. The performance was evaluated for enlargement factors of 2× and 4× in terms of PNSR and SSIM.

Quantitative Results. The quantitative results are summarized in Table 1. The proposed method achieves the best performance across all datasets for both single image super-resolution (SISR) and multi-contrast super-resultion (MCSR). Specifically, our LR-guided DCAMSR version surpasses state-of-the-art methods such as ELAN and SwinIR in SISR, and even outperforms SANet (a MCSR method). Among the MCSR methods, neither SANet, TTSR or MASA achieves better results than the proposed method. In particular, the PSNR for

Table 1. Quantitative results on two datasets with different enlargement scales, in terms of PSNR and SSIM. SISR means single image super resolution, MCSR means multi-contrast super resolution. The best results are marked in for multi-contrast super resolution, and in blue for single image super resolution. Note that TTSR and MASA are not applicable to 2× enlargement based on their official implementation.

Dataset		fastMRI				M4Raw			
Scale		2×		4×		2×		4×	
Metrics		PSNR	SSIM	PSNR	SSIM	PSNR	SSIM	PSNR	SSIM
SISR	ELAN	32.00	0.715	30.45	0.619	31.71	0.770	28.70	0.680
	SwinIR	32.04	0.717	30.58	0.624	32.08	0.775	29.42	0.701
	DCAMSR	32.07	0.717	30.71	0.627	32.19	0.777	29.74	0.709
MCSR	SANet	32.00	0.716	30.40	0.622	32.06	0.775	29.48	0.704
	TTSR	NA	NA	30.67	0.628	NA	NA	29.84	0.712
	MASA	NA	NA	30.78	0.628	NA	NA	29.52	0.704
	DCAMSR	32.20	0.721	30.97	0.637	32.31	0.779	30.48	0.728

MASA is even 0.18 dB lower than our SISR version of DCAMSR at 4× enlargement on M4Raw dataset. We attribute this performance margin to the difficulty of texture transformers in extracting similar texture features between Ref and $Ref_{\downarrow\uparrow}$. Despite the increased difficulty of super-resolution at 4× enlargement, our model still outperforms other methods, demonstrating the powerful texture transfer ability of the proposed DCA mechanism.

Qualitative Evaluation. Visual comparison is shown in Fig. 4, where the upsampled LR, the ground-truth HR, the restored SR and the error map for each method are visualized for 4× enlargement on both datasets. The error map depicts the degree of restoration error, where the more prominent texture indicating the poorer restoration quality. As can be seen, the proposed method produces the least errors compared with other methods.

Ablation Study. We conducted ablation experiments on the M4Raw dataset and the results are shown in Table 2. Three variations are tested: *w/o reference,*

Table 2. Ablation study on the M4Raw dataset with 4× enlargement.

Variant	Modules			Metrics		
	reference	multi-scale attention	channel attention	PSNR↑	SSIM↑	NMSE↓
w/o reference	✗	✓	✓	29.74	0.709	0.035
w/o multi-scale attention	✓	✗	✓	30.40	0.725	0.031
w/o channel attention	✓	✓	✗	29.79	0.712	0.035
DCAMSR	✓	✓	✓	30.48	0.728	0.029

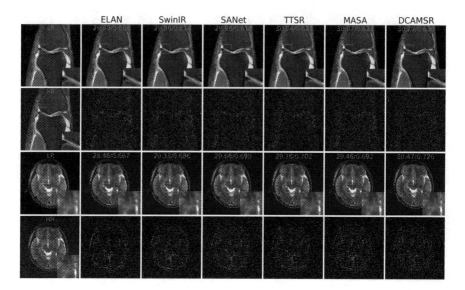

Fig. 4. Visual comparison of reconstruction results and error maps for 4× enlargement on both datasets. The upper two rows are fastMRI and the lower two rows are M4Raw.

where $LR_\uparrow$ is used as the reference instead of *Ref*; *w/o multi-scale attention*, where only the lowest-scale attention is employed and interpolated to other scales; and *w/o channel attention*, where only spatial attention is calculated. The improvement from *w/o reference* to DCAMSR demonstrates the effectiveness of MCSR compared with SISR. The performance degradation of *w/o multi-scale attention* demonstrates that the lowest-scale attention is not robust. The improvement from *w/o channel attention* to DCAMSR shows the effectiveness of the channel attention. Moreover, our encoder and decoder have comparable parameter size to MASA but we achieved higher scores, as shown in Table 1, demonstrating that the spatial search ability of DCAMSR is superior to the original texture transformer.

Discussion. Our reported results on M4Raw contain instances of slight inter-scan motion [18], demonstrating certain resilience of our approach to image misalignment, but more robust solutions deserve further studies. Future work may also extend our approach to 3D data.

4   Conclusion

In this study, we propose a Dual Cross-Attention Multi-contrast Super Resolution (DCAMSR) framework for improving the spatial resolution of MRI images. As demonstrated by extensive experiments, the proposed method outperforms existing state-of-the-art techniques under various conditions, proving a powerful and flexible solution that can benefit a wide range of medical applications.

Acknowledgments. This work was supported in part by the National Natural Science Foundation of China under Grant 62101348, the Shenzhen Higher Education Stable Support Program under Grant 20220716111838002, and the Natural Science Foundation of Top Talent of Shenzhen Technology University under Grants 20200208, GDRC202117, and GDRC202134.

References

1. Plenge, E., et al.: Super-resolution methods in MRI: can they improve the trade-off between resolution, signal-to-noise ratio, and acquisition time? Magn. Reson. Med. **68**(6), 1983–1993 (2012)
2. Van Reeth, E., Tham, I.W., Tan, C.H., Poh, C.L.: Super-resolution in magnetic resonance imaging: a review. Concepts Magn. Reson. Part A **40**(6), 306–325 (2012)
3. Feng, C.M., Wang, K., Lu, S., Xu, Y., Li, X.: Brain MRI super-resolution using coupled-projection residual network. Neurocomputing **456**, 190–199 (2021)
4. Li, G., Lv, J., Tong, X., Wang, C., Yang, G.: High-resolution pelvic MRI reconstruction using a generative adversarial network with attention and cyclic loss. IEEE Access **9**, 105951–105964 (2021)
5. Chen, Y., Xie, Y., Zhou, Z., Shi, F., Christodoulou, A.G., Li, D.: Brain MRI super resolution using 3D deep densely connected neural networks. In: 2018 IEEE 15th International Symposium on Biomedical Imaging (ISBI 2018), pp. 739–742. IEEE (2018)
6. Lim, B., Son, S., Kim, H., Nah, S., Mu Lee, K.: Enhanced deep residual networks for single image super-resolution. In: Proceedings of the IEEE Conference on Computer Vision and Pattern Recognition Workshops, pp. 136–144 (2017)
7. Liang, J., Cao, J., Sun, G., Zhang, K., Van Gool, L., Timofte, R.: SwinIR: image restoration using Swin transformer. In: Proceedings of the IEEE/CVF International Conference on Computer Vision, pp. 1833–1844 (2021)
8. Zhang, X., Zeng, H., Guo, S., Zhang, L.: Efficient long-range attention network for image super-resolution. In: Avidan, S., Brostow, G., Cissé, M., Farinella, G.M., Hassner, T. (eds.) ECCV 2022, Part XVII. LNCS, vol. 13677, pp. 649–667. Springer, Cham (2022)
9. Lyu, Q., et al.: Multi-contrast super-resolution MRI through a progressive network. IEEE Trans. Med. Imaging **39**(9), 2738–2749 (2020)
10. Feng, C.M., Fu, H., Yuan, S., Xu, Y.: Multi-contrast MRI super-resolution via a multi-stage integration network. In: Avidan, S., Brostow, G., Cissé, M., Farinella, G.M., Hassner, T. (eds.) MICCAI, Part VI. LNCS, vol. 13677, pp. 140–149. Springer, Cham (2021)
11. Feng, C.M., Yan, Y., Yu, K., Xu, Y., Shao, L., Fu, H.: Exploring separable attention for multi-contrast MR image super-resolution. arXiv preprint arXiv:2109.01664 (2021)
12. Yang, F., Yang, H., Fu, J., Lu, H., Guo, B.: Learning texture transformer network for image super-resolution. In: Proceedings of the IEEE/CVF Conference on Computer Vision and Pattern Recognition, pp. 5791–5800 (2020)

13. Lu, L., Li, W., Tao, X., Lu, J., Jia, J.: Masa-SR: matching acceleration and spatial adaptation for reference-based image super-resolution. In: Proceedings of the IEEE/CVF Conference on Computer Vision and Pattern Recognition, pp. 6368–6377 (2021)
14. Li, G., et al.: Transformer-empowered multi-scale contextual matching and aggregation for multi-contrast mri super-resolution. In: Proceedings of the IEEE/CVF Conference on Computer Vision and Pattern Recognition, pp. 20636–20645 (2022)
15. Jaegle, A., et al.: Perceiver IO: a general architecture for structured inputs & outputs. arXiv preprint arXiv:2107.14795 (2021)
16. Shaker, A., Maaz, M., Rasheed, H., Khan, S., Yang, M.H., Khan, F.S.: Unetr++: delving into efficient and accurate 3d medical image segmentation. arXiv preprint arXiv:2212.04497 (2022)
17. Zbontar, J., et al.: fastMRI: an open dataset and benchmarks for accelerated MRI. arXiv preprint arXiv:1811.08839 (2018)
18. Lyu, M., et al.: M4raw: a multi-contrast, multi-repetition, multi-channel MRI k-space dataset for low-field MRI research. Sci. Data **10**(1), 264 (2023)

DiffuseIR: Diffusion Models for Isotropic Reconstruction of 3D Microscopic Images

Mingjie Pan[1], Yulu Gan[2], Fangxu Zhou[1], Jiaming Liu[2], Ying Zhang[1],
Aimin Wang[3], Shanghang Zhang[2], and Dawei Li[1,4]([envelope])

[1] College of Future Technology, Peking University, Beijing, China
lidawei@pku.edu.cn
[2] School of Computer Science, Peking University, Beijing, China
[3] Department of Electronics, Peking University, Beijing, China
[4] Beijing Transcend Vivoscope Biotech, Beijing, China

Abstract. Three-dimensional microscopy is often limited by anisotropic spatial resolution, resulting in lower axial resolution than lateral resolution. Current State-of-The-Art (SoTA) isotropic reconstruction methods utilizing deep neural networks can achieve impressive super-resolution performance in fixed imaging settings. However, their generality in practical use is limited by degraded performance caused by artifacts and blurring when facing unseen anisotropic factors. To address these issues, we propose DiffuseIR, an unsupervised method for isotropic reconstruction based on diffusion models. First, we pre-train a diffusion model to learn the structural distribution of biological tissue from lateral microscopic images, resulting in generating naturally high-resolution images. Then we use low-axial-resolution microscopy images to condition the generation process of the diffusion model and generate high-axial-resolution reconstruction results. Since the diffusion model learns the universal structural distribution of biological tissues, which is independent of the axial resolution, DiffuseIR can reconstruct authentic images with unseen low-axial resolutions into a high-axial resolution without requiring re-training. The proposed DiffuseIR achieves SoTA performance in experiments on EM data and can even compete with supervised methods.

Keywords: Isotropic reconstruction · Unsupervised method · Diffusion model

1 Introduction

Three-dimensional (3D) microscopy imaging is crucial in revealing biological information from the nanoscale to the microscale. Isotropic high resolution across

---

M. Pan and Y. Gan—Equal contribution.

Supplementary Information The online version contains supplementary material available at https://doi.org/10.1007/978-3-031-43999-5_31.

H. Greenspan et al. (Eds.): MICCAI 2023, LNCS 14229, pp. 323–332, 2023.
https://doi.org/10.1007/978-3-031-43999-5_31

all dimensions is desirable for visualizing and analyzing biological structures. Most 3D imaging techniques have lower axial than lateral resolution due to physical slicing interval limitation or time-saving consideration [8,18,23,28,31]. Effective isotropic super-resolution algorithms are critical for high-quality 3D image reconstructions, such as electron microscopy and fluorescence microscopy.

Recently, deep learning methods have made significant progress in image analysis [9,13,14,25]. To address the isotropic reconstruction problem, [9] employs isotropic EM images to generate HR-LR pairs at axial and train a super-resolution model in a supervised manner, demonstrating the feasibility of inferring HR structures from LR images. [29,30] use 3D point spread function (PSF) as a prior for self-supervised super-resolution. However, isotropic high-resolution images or 3D point spread function (PSF) physical priors are difficult to obtain in practical settings, thus limiting these algorithms. Some methods like [3,21] have skillfully used cycleGAN [32] architecture to train axial super-resolution models without depending on isotropic data or physical priors. They learn from unpaired matching between high-resolution 2D slices in the lateral plane and low-resolution 2D slices in the axial plane, achieving impressive performance. However, these methods train models in fixed imaging settings and suffer from degraded performance caused by artifacts and blurring when facing unseen anisotropic factors. This limits their generality in practice [6]. In conclusion, a more robust paradigm needs to be proposed. Recently, with the success of the diffusion model in the image generation field [4,11,17,19,26], researchers applied the diffusion model to various medical image generation tasks and achieved impressive results [1,12,20,22,25]. Inspired by these works, we attempt to introduce diffusion models to address the isotropic reconstruction problem.

This paper proposes DiffuseIR, an unsupervised method based on diffusion models, to address the isotropic reconstruction problem. Unlike existing methods, DiffuseIR does not train a specific super-resolution model from low-axial-resolution to high-axial-resolution. Instead, we pre-train a diffusion model ϵ_θ to learn the structural distribution $p_\theta(X_{lat})$ of biological tissue from lateral microscopic images X_{lat}, which resolution is naturally high. Then, as shown in Fig. 1, we propose a Sparse Spatial Condition Sampling (SSCS) to condition the reverse-diffusion process of ϵ_θ. SSCS extracts sparse structure context from low-axial-resolution slice x_{axi} and generate reconstruction result $x_0 \sim p_\theta(X_{lat}|x_{axi})$. Since ϵ_θ learns the universal structural distribution p_θ, which is independent of the axial resolution, DiffuseIR can leverage the flexibility of SSCS to reconstruct authentic images with unseen anisotropic factors without requiring re-training. To further improve the quality of reconstruction, we propose a Refine-in-loop strategy to enhance the authenticity of image details with fewer sampling steps.

To sum up, our contributions are as follows:

(1) We are the first to introduce diffusion models to isotropic reconstruction and propose DiffuseIR. Benefiting from the flexibility of SSCS, DiffuseIR is naturally robust to unseen anisotropic spatial resolutions. (2) We propose a Refine-in-loop strategy, which maintains performance with fewer sampling steps and better preserves the authenticity of the reconstructed image details. (3) We perform extensive experiments on EM data with different imaging settings

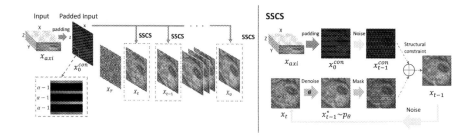

Fig. 1. Method Pipeline. DiffuseIR progressively conditions the denoising process with SSCS. For SSCS, we perform intra-row padding on input X_{lat} using the anisotropy factor α to obtain spatially aligned structural context, which is then merged with the diffusion model's output. Iterative SSCS refines reconstruction.

and achieve SOTA performance. Our unsupervised method is competitive with supervised methods and has much stronger robustness.

2   Methodology

As shown in Fig. 1, DiffuseIR address isotropic reconstruction by progressively conditions the denoising process of a pre-trained diffusion model ϵ_θ. Our method consists of three parts: DDPM pre-train, Sparse Spatial Condition Sampling and Refine-in-loop strategy.

DDPM Pretrain on Lateral. Our method differs from existing approaches that directly train super-resolution models. Instead, we pre-train a diffusion model to learn the distribution of high-resolution images at lateral, avoiding being limited to a specific axial resolution. Diffusion models [10,19] employ a Markov Chain diffusion process to transform a clean image x_0 into a series of progressively noisier images during the forward process. This process can be simplified as:

$$q(x_t|x_0) = N(x_t; \sqrt{\overline{\alpha}_t}x_0, (1 - \overline{\alpha}_t)I), \tag{1}$$

where $\overline{\alpha}_t$ controls the scale of noises. During inference, the model ϵ_θ predicts x_{t-1} from x_t. A U-Net ϵ_θ is trained for denoising process p_θ, which gradually reverses the diffusion process. This denoising process can be represented as:

$$p_\theta(x_{t-1}|x_t) = N(x_{t-1}; \epsilon_\theta(x_t, t), \sigma_t^2 I), \tag{2}$$

During training, we use 2D lateral slices, which is natural high-resolution to optimize ϵ_θ by mean-matching the noisy image obtained in Eq. 1 using the MSE loss [10]. Only HR slices at lateral plane X_{lat} were used for training, so the training process is unsupervised and independent of the specific axial resolution. So that ϵ_θ learns the universal structural distribution of biological tissues and can generate realistic HR images following $p_\theta(X_{lat})$.

Sparse Spatial Condition Sampling on Axial. We propose Sparse Spatial Condition Sampling (SSCS) to condition the generation process of ϵ_θ and generate high-axial-resolution reconstruction results. SSCS substitutes every reverse-diffusion step Eq. 2. We first transform the input axial LR slice x_{axi} to match the lateral resolution by intra-row padding: $(\alpha - 1)$ rows of zero pixels are inserted between every two rows of original pixels, where α is the anisotropic spatial factor. We denote M as the mask for original pixels in x_0^{con}, while $(1 - M)$ represents those empty pixels inserted. In this way, we obtain x_0^{con}, which reflects the sparse spatial content at axial, and further apply Eq. 1 to transform noise level:

Algorithm 1: Isotropic reconstruction using basic DiffuseIR

Input: axial slice x_{axi}, anisotropic factor α, refine-in-loop counts K

1 $x_0^{con}, M \leftarrow padding(x_{axi}, \alpha)$
2 **for** $t = T, ..., 1$ **do**
3      $x_{t-1}^{con} \sim N(\sqrt{\overline{\alpha}_t} x_0^{con}, (1 - \overline{\alpha}_t)I)$
4   **for** $i = 1, ..., K$ **do**
5          $x_{t-1}^* \sim N(x_{t-1}; \epsilon_\theta(x_t, t), \sigma_t^2 I)$
6          $x_{t-1} = M * x_{t-1}^{con} + (1 - M) * x_{t-1}^*$
7   **if** $t > 1$ *and* $i < K$ **then**
8              $x_t \sim N(\sqrt{1 - \beta_t} x_{t-1}, \beta_t I)$
9   **end**
10   **end**
11 **end**
12 return x_0

$$x_{t-1}^{con} \sim N(\sqrt{\overline{\alpha}_t} x_0^{con}, (1 - \overline{\alpha}_t)I) \tag{3}$$

Then, SSCS sample x_{t-1} at any time step t, conditioned on x_{t-1}^{con}. The process can be described as follows:

$$x_{t-1} = M \odot x_{t-1}^{con} + (1 - M) \odot x_{t-1}^*) \tag{4}$$

where x_{t-1}^* is obtained by sampling from the model ϵ_θ using Eq. 2, with x_t of the previous iteration. x_{t-1}^* and x_{t-1}^{con} are combined with M. By iterative denoising, we obtain the reconstruction result x_0. It conforms to the distribution $p_\theta(X_{lat})$ learned by the pre-trained diffusion model and maintains semantic consistency with the input LR axial slice. Since SSCS is parameter-free and decoupled from the model training process, DiffuseIR can adapt to various anisotropic spatial resolutions by modifying the padding factor according to α while other methods require re-training. This makes DiffuseIR a more practical and versatile solution for isotropic reconstruction.

Refine-in-Loop Strategy. We can directly use SSCS to generate isotropic results, but the reconstruction quality is average. The diffusion model is capable of extracting context from the sparse spatial condition. Still, we have discovered a phenomenon of texture discoordination at the mask boundaries, which reduces the reconstruction quality. For a certain time step t, the content of x_{t-1}^* may be unrelated to x_{t-1}^{con}, resulting in disharmony in x_{t-1} generated by SSCS. During the denoising of the next time step $t-1$, the model tries to repair the disharmony of x_{t-1} to conform to p_θ distribution. Meanwhile, this process will introduce new inconsistency and cannot converge on its own. To overcome this problem, we propose the Refine-in-loop strategy: For x_{t-1} generated by SSCS at time step t, we apply noise to it again and obtain a new x_t and then repeat SSCS at time step t. Our discovery showed that this uncomplicated iterative refinement method addresses texture discoordination significantly and enhances semantic precision.

The total number of inference steps in DiffuseIR is given by $T_{total} = T \cdot K$. As T_{total} increases, it leads to a proportional increase in the computation time of our method. However, larger T_{total} means more computational cost. Recent works such as [15,16,24] have accelerated the sampling process of diffusion models by reducing T while maintaining quality. For DiffuseIR, adjusting the sampling strategy is straightforward. Lowering T and raising refinement iterations K improves outcomes with a fixed T_{total}. We introduce and follow the approach presented in DDIM [24] as an example and conducted detailed ablation experiments in Sec. 3 to verify this. Our experiments show that DiffuseIR can benefit from advances in the community and further reduce computational overhead in future work.

3 Experiments and Discussion

Dataset and Implement Details. To evaluate the effectiveness of our method, we conducted experiments on two widely used public EM datasets, FIB-25 [27] and Cremi [5]. FIB-25 contains isotropic drosophila medulla connectome data obtained with FIB-SEM. We partitioned it into subvolumes of $256 \times 256 \times 256$ as ground truth and followed [9] to perform average-pooling by factor α(x2,x4,x8) along the axis to obtain downsampled anisotropic data. Cremi consists of drosophila brain data with anisotropic spatial resolution. We followed [3] to generate LR images with a degradation network and conduct experiments on lateral slices. All resulting images were randomly divided into the training (70%), validation (15%) and test (15%) set. For the pre-training of the diffusion model, we follow [19] by using U-Net with multi-head attention and the same training hyper-parameters. We use 256×256 resolution images with a batch size of 4 and train the model on 8×V100 GPUs. For our sampling setting, we set $T, K = 25, 40$, which is a choice selected from the ablation experiments in Sec. 3 that balances performance and speed.

Quantitative and Visual Evaluation. To evaluate the effectiveness of our method, we compared DiffuseIR with SoTA methods and presented the quantitative results in Table 1. We use PSNR and SSIM to evaluate results. PSNR is calculated using the entire 3D stack. SSIM is evaluated slice by slice in XZ and YZ viewpoints, with scores averaged for the final 3D stack score, measuring quality and 3D consistency. We use cubic interpolation as a basic comparison. 3DSRUNet [9] is a seminal isotropic reconstruction method, which requires HR and LR pairs as ground truth for supervised training. CycleGAN-IR [3] proposed an unsupervised approach using a CycleGAN [32] architecture, learning from unpaired axial and lateral slices. These methods train specialized models for a fixed anisotropic spatial setting and must be retrained for unseen anisotropic factors α, shown in Table 1. Despite being trained solely for denoising task and not exposed to axial slices, DiffuseIR outperforms unsupervised baselines and is competitive with supervised methods [9]. As shown in Fig. 2, our refine-in-loop strategy produces results with greater visual similarity to the Ground Truth, avoiding distortion and blurriness of details. Notably, the versatility afforded by

Table 1. Quantitative evaluation of DiffuseIR against baselines. PSNR↑ and SSIM↑ are used as evaluation metrics. We evaluated the FIB25 and Cremi datasets, considering three anisotropic spatial resolutions, $\alpha = 2, 4, 8$. Unlike other baselines which train a dedicated model for each α, our method only trains a single, generalizable model.

Methzod		FIB25			Cremi		
		×2	×4	×8	×2	×4	×8
Interplation	PSNR	33.21	30.29	29.19	31.44	29.34	28.27
	SSIM	0.854	0.722	0.538	0.782	0.574	0.451
†3DSRUNet	PSNR	**33.84**	32.31	30.97	**32.04**	31.12	**30.28**
	SSIM	0.877	0.824	0.741	**0.820**	0.761	0.719
CycleGAN-IR	PSNR	33.54	31.77	29.94	31.71	30.47	29.04
	SSIM	0.869	0.798	0.640	0.794	0.721	0.560
DiffuseIR (ours)	PSNR	33.81	**32.37**	**31.09**	31.97	**31.24**	30.24
	SSIM	**0.881**	**0.832**	**0.774**	0.819	**0.783**	**0.726**

† Supervised method.

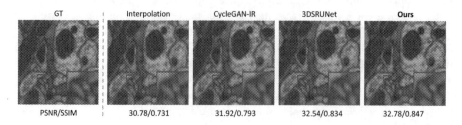

GT	Interpolation	CycleGAN-IR	3DSRUNet	Ours
PSNR/SSIM	30.78/0.731	31.92/0.793	32.54/0.834	32.78/0.847

Fig. 2. Visual comparisons on FIB-25 dataset ($\alpha = 4$). DiffuseIR can generate competitive results compared to supervised methods, and the results appear more visually realistic.

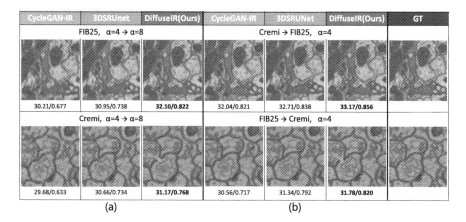

Fig. 3. Analysis on robustness. (a) Test on unseen anisotropic factor α. (b) Test on different datasets with domain shifts (e.g., train on FIB25, test on Cremi). Our method is robust against various anisotropic factors and domain shifts between two datasets.

SSCS allows DiffuseIR to achieve excellent results using only one model, even under different isotropic resolution settings. This indicates that DiffuseIR overcomes the issue of generalization to some extent in practical scenarios, as users no longer need to retrain the model after modifying imaging settings.

Further Analysis on Robustness. We examined the robustness of our model to variations in both Z-axis resolutions and domain shifts. Specifically, we investigated the following: **(a) Robustness to unseen anisotropic spatial factors.** The algorithm may encounter unseen anisotropic resolution due to the need for different imaging settings in practical applications. To assess the model's robustness to unseen anisotropic factors, we evaluated the model trained with the anisotropic factor $\alpha = 4$. Then we do inference under the scenario of anisotropic factor $\alpha = 8$. For those methods with a fixed super-resolution factor, we use cubic interpolation to upsample the reconstructed result by 2x along the axis. **(b) Robustness to the domain shifts.** When encountering unseen data in the real world, domain shifts often exist, such as differences in biological structure features and physical resolution, which can impact the model's performance [2,7]. To evaluate the model's ability to handle those domain shifts, we trained our model on one dataset and tested it on another dataset. **Analysis**: As shown in Fig. 3, DiffuseIR shows greater robustness than other methods. In scenario **(a)**, other methods are trained on specific anisotropic factors for super-resolution of axial LR to lateral HR. This can result in model fragility during testing with unseen anisotropic resolutions. In contrast, DiffuseIR directly learns the universal structural distribution at lateral through generation task, applicable to various axial resolutions. All methods exhibit decreased performance in scenario **(b)**. However, DiffuseIR shows a small performance degradation with the help of the multi-step generation of the diffusion model and sparse spatial constraints imposed by SSCS at each reverse-diffusion step. Further, compared to the previous methods predicting the result by one step, DiffuseIR makes the generating process more robust and controllable by adding constraints at each step to prevent the model from being off-limit.

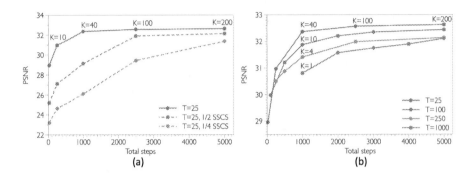

Fig. 4. Ablation Study: (a)ablation on SSCS frequency. Experimental results demonstrates the importance of SSCS. When reducing the frequency of SSCS usage, performance will severely decline. **(b)ablation on different refine-in-loop settings.** The results show that when the number of total steps is fixed, increase K will lead to higher PSNR.

Ablation Study. We conducted extensive ablation experiments Fig. 4. First, to demonstrate the effectiveness of SSCS, we use it only in partially alternate reverse-diffusion steps, such as $1/4$ or $1/2$ steps. As shown in Fig. 4 (a), increasing the frequency of SSCS significantly improves PSNR while bringing negligible additional computational costs. This indicates that SSCS have a vital effect on the model's performance. Second, for the Refine-in-loop strategy, results show that keeping the total number of steps unchanged (reducing the number of time steps T while increasing the refine iterations K) can markedly improve performance. Figure 4 (b) have the following settings: $T = \{25, 100, 250, 1000\}$ with $K\{40, 10, 4, 1\}$ to achieve a total of 5000 steps. The results show that the model performs best when $T = 25$ and PSNR gradually increases with the increase of K. A balanced choice is $\{T = 25, K = 40\}$, which improves PSNR by 1.56dB compared to $\{T = 1000, K = 1\}$ without using the Refine-in-loop strategy.

4   Conclusion

We introduce DiffuseIR, an unsupervised method for isotropic reconstruction based on diffusion models. To the best of our knowledge, We are the first to introduce diffusion models to solve this problem. Our approach employs Sparse Spatial Condition Sampling (SSCS) and a Refine-in-loop strategy to generate results robustly and efficiently that can handle unseen anisotropic resolutions. We evaluate DiffuseIR on EM data. Experiments results show our methods achieve SoTA methods and yield comparable performance to supervised methods. Additionally, our approach offers a novel perspective for addressing Isotropic Reconstruction problems and has impressive robustness and generalization abilities.

5 Limitation

DiffuseIR leverages the powerful generative capabilities of pre-trained Diffusion Models to perform high-quality isotropic reconstruction. However, this inevitably results in higher computational costs. Fortunately, isotropic reconstruction is typically used in offline scenarios, making DiffuseIR's high computational time tolerable. Additionally, the community is continuously advancing research on accelerating Diffusion Model sampling, from which DiffuseIR can benefit.

Acknowledgments. The authors acknowledge supports from Beijing Nova Program and Beijing Transcend Vivoscope Biotech Co., Ltd.

References

1. Chung, H., Chul Ye, J.: Score-based diffusion models for accelerated MRI. Med. Image Anal. **80**, 102479 (2023)
2. Csurka, G.: Domain adaptation for visual applications: a comprehensive survey. arXiv: Computer Vision and Pattern Recognition (2017)
3. Deng, S., et al.: Isotropic reconstruction of 3D EM images with unsupervised degradation learning. In: Martel, A.L., et al. (eds.) MICCAI 2020. LNCS, vol. 12265, pp. 163–173. Springer, Cham (2020). https://doi.org/10.1007/978-3-030-59722-1_16
4. Dhariwal, P., Nichol, A.: Diffusion models beat GANs on image synthesis. In: Neural Information Processing Systems (2021)
5. Funke, J., S.: cremi.org. http://cremi.org/
6. González-Ruiz, V., García-Ortiz, J., Fernández-Fernández, M., Fernández, J.J.: Optical flow driven interpolation for isotropic FIB-SEM reconstructions. Comput. Meth. Programs Biomed. **221**, 106856 (2022)
7. Guan, H., Liu, M.: Domain adaptation for medical image analysis: a survey (2021)
8. Hayworth, K.J., et al.: Ultrastructurally smooth thick partitioning and volume stitching for large-scale connectomics. Nat. Methods **12**, 319–322 (2015)
9. Heinrich, L., Bogovic, J.A., Saalfeld, S.: Deep learning for isotropic super-resolution from non-isotropic 3d electron microscopy. medical image computing and computer assisted intervention (2017)
10. Ho, J., Jain, A., Abbeel, P.: Denoising diffusion probabilistic models. In: Neural Information Processing Systems (2020)
11. Kawar, B., Elad, M., Ermon, S., Song, J.: Denoising diffusion restoration models (2023)
12. Kim, B., Chul, J.: Diffusion deformable model for 4d temporal medical image generation. In: Wang, L., Dou, Q., Fletcher, P.T., Speidel, S., Li, S. (eds.) MICCAI 2022. LNCS, vol. 13431, pp. 539–548. Springer, Cham (2023). https://doi.org/10.1007/978-3-031-16431-6_51
13. Li, X., et al.: Efficient meta-tuning for content-aware neural video delivery. In: Avidan, S., Brostow, G., Cissé, M., Farinella, G.M., Hassner, T. (eds.) ECCV 2022. LNCS, vol. 13678, pp. 308–324. Springer, Cham (2022). https://doi.org/10.1007/978-3-031-19797-0_18
14. Liu, J., et al.: Overfitting the data: compact neural video delivery via content-aware feature modulation. In: 2021 IEEE/CVF International Conference on Computer Vision (ICCV) (2021). https://doi.org/10.1109/iccv48922.2021.00459

15. Lu, C., Zhou, Y., Bao, F., Chen, J., Li, C., Zhu, J.: DPM-solver: a fast ode solver for diffusion probabilistic model sampling in around 10 steps (2022)
16. Lu, C., Zhou, Y., Bao, F., Chen, J., Li, C., Zhu, J.: Dpm-solver++: fast solver for guided sampling of diffusion probabilistic models (2022)
17. Lugmayr, A., Danelljan, M., Romero, A., Yu, F., Timofte, R., Gool, L.V.: Repaint: inpainting using denoising diffusion probabilistic models (2023)
18. Mikula, S.: Progress towards mammalian whole-brain cellular connectomics. Front. Neuroanat. **10**, 62 (2016)
19. Nichol, A., Dhariwal, P.: Improved denoising diffusion probabilistic models. arXiv: Learning (2021)
20. Özbey, M., et al.: Unsupervised medical image translation with adversarial diffusion models (2022)
21. Park, H., et al.: Deep learning enables reference-free isotropic super-resolution for volumetric fluorescence microscopy. Nature Commun. **13**, 3297 (2021)
22. Peng, C., Guo, P., Zhou, S.K., Patel, V., Chellappa, R.: Towards performant and reliable undersampled MR reconstruction via diffusion model sampling (2023)
23. Schrödel, T., Prevedel, R., Aumayr, K., Zimmer, M., Vaziri, A.: Brain-wide 3d imaging of neuronal activity in caenorhabditis elegans with sculpted light. Nat. Meth. **10**, 1013–1020 (2013)
24. Song, J., Meng, C., Ermon, S.: Denoising diffusion implicit models. arXiv: Learning (2020)
25. Song, Y., Shen, L., Xing, L., Ermon, S.: Solving inverse problems in medical imaging with score-based generative models. Cornell University - arXiv (2021)
26. Su, X., Song, J., Meng, C., Ermon, S.: Dual diffusion implicit bridges for image-to-image translation (2023)
27. ya Takemura, S., et al.: Synaptic circuits and their variations within different columns in the visual system of drosophila. In: Proceedings of the National Academy of Sciences of the United States of America (2015)
28. Verveer, P.J., Swoger, J., Pampaloni, F., Greger, K., Marcello, M., Stelzer, E.H.K.: High-resolution three-dimensional imaging of large specimens with light sheet-based microscopy. Nat. Methods **4**, 311–313 (2007)
29. Weigert, M., Royer, L., Jug, F., Myers, G.: Isotropic reconstruction of 3D fluorescence microscopy images using convolutional neural networks. arXiv: Computer Vision and Pattern Recognition (2017)
30. Weigert, M., et al.: Content-aware image restoration: pushing the limits of fluorescence microscopy. bioRxiv (2018)
31. Wu, Y., et al.: Three-dimensional virtual refocusing of fluorescence microscopy images using deep learning. Nature Methods **16**, 1323–1331 (2019)
32. Zhu, J.Y., Park, T., Isola, P., Efros, A.A.: Unpaired image-to-image translation using cycle-consistent adversarial networks. In: International Conference on Computer Vision (2017)

Physics-Informed Neural Networks for Tissue Elasticity Reconstruction in Magnetic Resonance Elastography

Matthew Ragoza[1]([✉]) [iD] and Kayhan Batmanghelich[2] [iD]

[1] University of Pittsburgh, Pittsburgh, PA 15213, USA
mtr22@pitt.edu
[2] Boston University, Boston, MA 02215, USA

Abstract. Magnetic resonance elastography (MRE) is a medical imaging modality that non-invasively quantifies tissue stiffness (elasticity) and is commonly used for diagnosing liver fibrosis. Constructing an elasticity map of tissue requires solving an inverse problem involving a partial differential equation (PDE). Current numerical techniques to solve the inverse problem are noise-sensitive and require explicit specification of physical relationships. In this work, we apply physics-informed neural networks to solve the inverse problem of tissue elasticity reconstruction. Our method does not rely on numerical differentiation and can be extended to learn relevant correlations from anatomical images while respecting physical constraints. We evaluate our approach on simulated data and *in vivo* data from a cohort of patients with non-alcoholic fatty liver disease (NAFLD). Compared to numerical baselines, our method is more robust to noise and more accurate on realistic data, and its performance is further enhanced by incorporating anatomical information.

Keywords: Physics-informed learning · Magnetic resonance elastography · Elasticity reconstruction · Deep learning · Medical imaging

1 Introduction

Tissue elasticity holds enormous diagnostic value for detecting pathological conditions such as liver fibrosis [1,2] and can be mapped by an imaging procedure called magnetic resonance elastography (MRE). During MRE, a mechanical stress is applied to the region of interest and an image is captured of the resulting tissue deformation, then the elasticity is inferred by solving the inverse problem of a partial differential equation (PDE). However, conventional methods for elasticity reconstruction are sensitive to noise, do not incorporate anatomical information, and are often only evaluated on artificial data sets [3–8].

Supplementary Information The online version contains supplementary material available at https://doi.org/10.1007/978-3-031-43999-5_32.

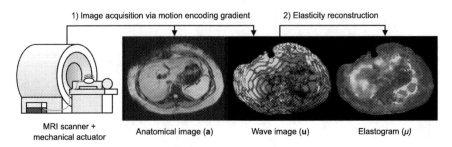

Fig. 1. During MRE, a mechanical actuator induces shear waves while a motion-encoding gradient captures the tissue deformation as a wave image. Then, an inverse problem is solved to reconstruct a map of tissue elasticity called an elastogram.

Elasticity reconstruction methods utilize a variety of numerical techniques and physical models [9]. Algebraic Helmholtz inversion (AHI) makes simplifying assumptions that enable an algebraic solution to the governing equations [3,5]. However, AHI relies on finite differences, which amplify noise. The finite element method (FEM) requires fewer physical assumptions and solves a variational formulation of the PDE [4,6–8,10], making it more flexible and typically more robust than AHI. However, neither method uses anatomical images, which have been successfully used to predict elasticity with deep learning [11–13].

Physics-informed neural networks (PINNs) are a recent deep learning framework that uses neural networks to solve PDEs [14]. PINNs represent unknown function(s) in a boundary value problem as neural networks. The boundary conditions and PDE are treated as loss functions and the problem is solved using gradient-based optimization. PINNs have been applied to elasticity reconstruction in other contexts [15–18], but evaluations have been limited to artificial data sets and prior work has not combined physics-informed learning with automated learning from anatomical MRI.

In this work, we develop a method for enhanced tissue elasticity reconstruction in MR elastography using physics-informed learning. We use PINNs to solve the equations of linear elasticity as an optimization problem for a given wave image. Our model simultaneously learns continuous representations of the measured displacement field and the latent elasticity field. We evaluate the method on a numerical simulation and on patient liver MRE data, where we demonstrate improved noise robustness and overall accuracy than AHI or FEM-based inversion. In addition, we show that augmenting our method with an anatomically-informed loss function further improves reconstruction quality.

2   Background

Magnetic Resonance Elastography (MRE). An MRE procedure involves placing the patient in an MRI scanner and using a mechanical actuator to induce shear waves in the region of interest (Fig. 1). A motion-encoding gradient (MEG) synchronizes with the mechanical vibration, causing phase shifts in the captured

Table 1. Physical equations relating the displacement field $\mathbf{u}$ to the shear modulus of elasticity μ during a steady-state harmonic motion from the theory of linear elasticity.

	Assumptions		Equation
Name	$\nabla \cdot \mathbf{u} = 0$	$\nabla \mu = 0$	$\mathcal{D}(\mu, \mathbf{u}, \mathbf{x}) = 0$
general form			$\nabla \cdot \left[\mu \left(\nabla \mathbf{u} + \nabla \mathbf{u}^\top \right) + \lambda (\nabla \cdot \mathbf{u}) \mathbf{I} \right] + \rho \omega^2 \mathbf{u} = 0$
heterogeneous	✓		$\mu \nabla^2 \mathbf{u} + \left(\nabla \mathbf{u} + \nabla \mathbf{u}^\top \right) \nabla \mu + \rho \omega^2 \mathbf{u} = 0$
Helmholtz	✓	✓	$\mu \nabla^2 \mathbf{u} + \rho \omega^2 \mathbf{u} = 0$

signal based on the tissue displacement [19]. A *wave image* that encodes the full 3D displacement field can be acquired using MEGs in each dimension [1]. Next, a map of tissue stiffness called an *elastogram* is recovered from the wave image by solving an inverse problem. This requires 1) choosing a physical model that relates the motion of an elastic body to its material properties, and 2) solving the governing equation(s) for the unknown material parameters.

Linear Elasticity. Physical models of MRE typically assume there is harmonic motion and a linear, isotropic stress-strain relation. Then tissue displacement is a complex vector field $\mathbf{u} : \Omega \to \mathbb{C}^3$ defined on spatial domain $\Omega \subset \mathbb{R}^3$, and shear elasticity is characterized by the *shear modulus*, a complex scalar field $\mu : \Omega \to \mathbb{C}$. The first equation of motion translates into the *general form* PDE shown in Table 1. The mass density ρ and actuator frequency ω are prescribed based on prior knowledge. The Lamé parameter λ can be ignored if we assume tissue is incompressible ($\nabla \cdot \mathbf{u} = 0$), reducing the PDE to the *heterogeneous form*. Finally, assuming that the shear modulus is locally homogeneous ($\nabla \mu = 0$) simplifies the PDE into the *Helmholtz equation*. The empirical validity of the homogeneity assumption has been criticized [20,21] and is explored in our experiments.

3 Proposed Method

We use physics-informed neural networks (PINNs) to encode the solution space of the inverse problem. Our PINN framework (Fig. 2) learns continuous representations of the displacement field $\mathbf{u}(\mathbf{x})$ and elastic modulus $\mu(\mathbf{x})$ while respecting a PDE from Table 1. We incorporate conventional MRI images by including an additional anatomically-informed loss function. In the following sections, we explain the PINN framework and how we incorporate the anatomical images.

Physics-Informed Neural Networks. We use a dual-network approach to reconstruct tissue elasticity with PINNs. First, we train a neural network $\hat{\mathbf{u}}(\mathbf{x}; \theta^{\mathbf{u}})$ to learn a mapping from spatial coordinates $\mathbf{x}$ to displacement vectors $\mathbf{u}$ in the wave image by minimizing the mean squared error $\mathcal{L}_{wave}$. The continuous representation of the displacement field enables automatic spatial differentiation.

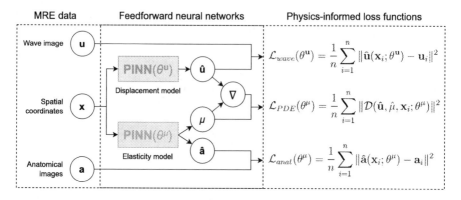

Fig. 2. In our framework, one PINN learns to map from spatial coordinates to displacement vectors by fitting to the wave image, while another learns to recover the shear elasticity at the corresponding position by minimizing a PDE residual. The elasticity model can also predict anatomical MRI features that are correlated with elasticity. Note that the PINNs are trained to fit a set of images for a single patient/phantom and that n represents the number of (spatial coordinate, image value) pairs per batch.

Then, we train a second neural network $\hat{\mu}(\mathbf{x}; \theta^{\mu})$ to map from spatial coordinates to the shear modulus μ by minimizing the residual of a PDE, defined by some differential operator $\mathcal{D}$. The PINNs and their spatial derivatives are used to evaluate the differential operator, which is minimized as a loss function $\mathcal{L}_{PDE}$ to recover the elasticity field. We combine the loss functions as follows:

$$\mathcal{L}(\theta^{\mathbf{u}}, \theta^{\mu}) = \lambda_{wave}\mathcal{L}_{wave}(\theta^{\mathbf{u}}) + \lambda_{PDE}\mathcal{L}_{PDE}(\theta^{\mu})$$

We train PINNs using either the Helmholtz equation (PINN-HH) or the heterogeneous PDE (PINN-het) as the differential operator $\mathcal{D}$ in our experiments. The loss weight hyperparameters λ_{wave} and λ_{PDE} control the contribution of each loss function to the overall objective. We initialized λ_{PDE} to a very low value and slowly stepped it up as the quality of $\hat{\mathbf{u}}$ improved during training.

Incorporating Anatomical Information. Prior work has demonstrated that tissue elasticity can be accurately predicted from anatomical MRI [22]. We include an additional output head $\hat{\mathbf{a}}(\mathbf{x}; \theta^{\mu})$ from the elasticity PINN that predicts anatomical features $\mathbf{a}(\mathbf{x})$ at the corresponding position, as encoded in standard MRI imaging sequences. Then we introduce an additional loss function $\mathcal{L}_{anat}$ to minimize the mean squared error between the predicted and true anatomical features. We explore how the relative weight of this loss function λ_{anat} affects elasticity reconstruction performance in our *in vivo* experiments.

We designed our models based on the SIREN architecture, which uses sine activation functions to better represent high spatial frequencies [23]. Both networks $\hat{\mathbf{u}}$ and $\hat{\mu}$ had five linear layers with 128 hidden units per layer and dense connections from all previous layer inputs. The first layer input was scaled by a hyperparameter ω_0 that biases the initial spatial frequency distribution. We employed the weight initialization scheme described in [23] to improve training convergence. We also extended the input vector with polar coordinates when training on patient data. We trained all models for 100,000 total iterations with the Adam optimizer using PyTorch v1.12.1 and DeepXDE v1.5.1 [24–26]. The code used for this work is available at https://github.com/batmanlab/MRE-PINN.

4   Related Work

Algebraic Helmholtz Inversion (AHI). One of the most common methods for elasticity reconstruction is algebraic inversion of the Helmholtz equation (AHI) [3]. This approach assumes incompressibility and local homogeneity, uses finite differences to compute the Laplacian of the wave image, and then solves the Helmholtz equation as a linear system to estimate the shear modulus. Despite its simplicity, AHI has an established track record in both research and clinical settings [9]. Filtering is often required to reduce the impact of noise on AHI [5].

Finite Element Method (FEM). Many techniques have been introduced for elasticity reconstruction based on the finite element method [4,6,7,10]. These use variational formulations to reduce the order of differentiation in the PDE. Then, they specify a mesh over the domain and represent unknown fields in terms of compact basis functions. This results in a linear system that can be solved for the elasticity coefficients either directly or iteratively [9]. Direct FEM inversion is more efficient and accurate [21], though it depends more on data quality [8]. We implemented direct FEM inversion of the Helmholtz equation (FEM-HH) and the heterogeneous PDE (FEM-het) with FEniCSx v0.5.1 [27].

5   Experiments and Results

We compare our methods (PINN-HH, PINN-het) to algebraic Helmholtz inversion (AHI) and direct FEM inversion (FEM-HH, FEM-het) on simulated data and liver data from a cohort of patients with NAFLD. We evaluate the overall reconstruction fidelity of each method and the impact of the homogeneity assumption. We assess their robustness to noise on the simulated data, and we study whether incorporating anatomical images enhances performance on patient liver data.

5.1   Robustness to Noise on Simulated Data

We obtained a numerical FEM simulation of an elastic wave in an incompressible rectangular domain containing four stiff targets of decreasing size from the BIOQIC research group (described in Barnhill et. al. 2018 [28]). Six wave fields were generated at frequencies ranging from 50–100 Hz in 10 Hz increments. We applied each reconstruction method to each single-frequency wave image after adding varying levels of Gaussian noise. Then we computed the contrast transfer efficiency (CTE) [29] in each target region as the ratio between the target-background contrast in the predicted elastogram and the true contrast, where a CTE of 100% is ideal. We include functions to download the simulation data set in the project codebase.

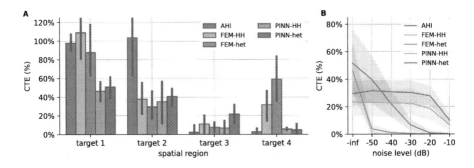

Fig. 3. Reconstruction performance on simulated data. Figure 3A shows the contrast transfer efficiency of each method in each target region of the simulation. Figure 3B shows how the contrast is affected by the noise level in the wave image.

Experiment Results. Figure 3A compares the CTE of each method in the different target regions of the simulation, which decrease in size from left to right. On target 1, AHI performs best with 98% CTE followed by FEM-HH and FEM-het with 109% and 88%. PINN-HH and PINN-het had 46% and 51% contrast on target 1. AHI also performed best on target 2, with 104% CTE. Next were PINN-het with 41% contrast and FEM-IIH with 38%. PINN-het outperformed the other methods on target 3 with 22% CTE followed by FEM-HH with 11%. Only the two FEM methods appear to have decent contrast on target 4, though this seems to be a false positive due to background variance seen in Fig. 4.

Figure 3B shows the effect of wave image noise on the contrast transfer efficiency. The reconstruction quality decays at –50 db of noise for both FEM-HH ($p = -1.7e{-}5$) and FEM-het ($p = 9.1e{-}6$), each retaining less than 5% contrast. AHI is also sensitive to –50 dB of noise ($p = 6.1e{-}4$) but its performance drops more gradually. The contrast from PINN-HH does not decrease significantly until –20 dB ($p = 8.8e{-}4$) and PINN-het is insensitive until –10 dB ($p = 4.8e{-}5$). This indicates that PINNs are more robust to noise than AHI or direct FEM.

Figure 4 displays reconstructed elastograms from each method using the 90 Hz simulated wave image, displayed at far left, next to the ground truth. AHI produces the clearest boundaries between the targets and background. The two FEM methods contain high variability within homogeneous regions, though the heterogeneous PDE appears to decrease the variance. PINN-HH displays textural artifacts that reduce the resolution. PINN-het has fewer artifacts and better resolution of the smaller targets due to the lack of homogeneity assumption.

5.2   Incorporating Anatomical Information on Patient Data

For our next experiment, we obtained abdominal MRE from a study at the University of Pittsburgh Medical Center that included patients at least 18 years old who were diagnosed with non-alcoholic fatty liver disease (NAFLD) and underwent MRE between January 2016–2019 (demographic and image acquisition details can be found in Pollack et al. 2021 [22]). 155 patients had high-quality

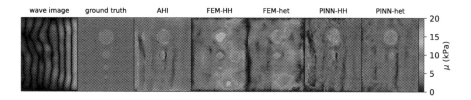

Fig. 4. Elasticity reconstruction examples on simulated data with no noise.

elastography, wave images, and anatomical MRI sequences. The wave images contained only one real displacement component and we did not have ground truth elasticity, so we used proprietary elastograms collected during the study as the "gold standard." We registered MRI sequences (T1 pre in-phase, T1 pre water, T1 pre out-phase, T1 pre fat, and T2) to the MRE using SimpleITK v2.0.0 [30] and incorporated them as the anatomical features **a** shown in Fig. 2. Liver regions were segmented using a previously reported deep learning model [22].

We performed elasticity reconstruction on each of the 155 patient wave images using each method. We investigated the influence of anatomical information when training PINNs by varying the anatomical loss weight λ_{anat}. Reconstruction fidelity was assessed using the Pearson correlation (R) between the predicted elasticity and the gold standard elasticity in the segmented liver regions.

Experiment Results. Figure 5A shows correlation distributions between predicted and gold standard elasticity across the different patients. PINN-het had the highest median correlation and least variation between patients (median = 0.84, IQR = 0.04). PINN-HH came in second (median = 0.76, IQR = 0.05) while FEM-het came in third, but had the most variability (median = 0.74, IQR = 0.19). FEM-HH performed slightly worse than FEM-het, but was less variable (median = 0.70, IQR = 0.11). AHI performed the worst on *in vivo* data (median = 0.63, IQR = 0.12) and had the greatest number of low outliers.

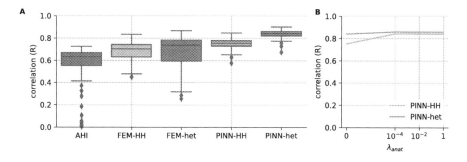

Fig. 5. Reconstruction performance on *in vivo* liver data. Figure 5A shows Pearson's correlations between predicted and gold standard elasticity across patients. Figure 5B shows the effect of the anatomic loss weight on PINN elasticity reconstruction performance.

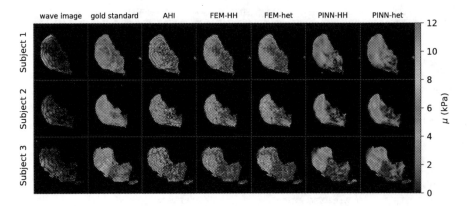

Fig. 6. Elasticity reconstruction examples for 3 patients from the NAFLD data set.

Figure 5B shows the effect of increasing the anatomic loss weight on PINN elasticity reconstruction quality. There was significant improvement in the correlation with the gold standard when the loss weight was increased from 0 to 1e–4 for both PINN-HH ($p = 4.9e–56$) and PINN-het ($p = 3.0e–14$), but no significant difference from increasing it further to 1e–2 for either method (PINN-HH: $p = 0.51$, PINN-het: $p = 0.23$). Raising the anatomic loss weight to 1e-4 increased the median correlation from 0.76 to 0.85 in PINN-HH and from 0.84 to 0.87 in PINN-het. This suggests that there is a synergistic effect from including physical constraints and anatomical imaging data in elasticity reconstruction.

Figure 6 displays reconstructed elastograms for three randomly selected patients. AHI, FEM-HH and FEM-het all tend to overestimate the stiffness and have artifacts around nulls in the wave image. In contrast, PINN-HH and PINN-het more closely resemble the gold standard elastography, especially in regions close to the clinical threshold for fibrosis [31]. Furthermore, neither PINN reconstruction method shows signs of instabilities around wave amplitude nulls.

6   Conclusion

PINNs have several clinically significant advantages over conventional methods for tissue elasticity reconstruction in MRE. They are more robust to noise, which is pervasive in real MRE data. Furthermore, they can leverage anatomical information from other MRI sequences that are standard practice to collect during an MRE exam, and doing so significantly improves reconstruction fidelity. Limitations of this work include the use of the incompressibility assumption to simplify the training framework, and the relatively poor contrast on simulated data. This underscores how accurate reconstruction on simulated data does not always translate to real data, and vice versa. In future work, we will evaluate PINNs for solving the general form of the PDE to investigate the effect of the incompressibility assumption. We will also extend to an operator learning framework in which the model learns to solve the PDE in a generalizable fashion without the need to retrain on each wave image. This would reduce the computation cost and enable further integration of physics-informed and data-driven learning.

Acknowledgements & Data Use. This work was supported by the Pennsylvania Department of Health (grant number 41000873310), National Institutes of Health (grant number R01HL141813), the National Science Foundation (grant number 1839332) and Tripod+X. This work used the Bridges-2 system, which is supported by NSF award number OAC-1928147 at the Pittsburgh Supercomputing Center (PSC).

The patient MRE data was acquired by Amir A. Borhani, MD while he was at University of Pittsburgh. We thank him for his collaboration and guidance during this project.

References

1. Manduca, A., et al.: Magnetic resonance elastography: non-invasive mapping of tissue elasticity. Med. Image Anal. **5**(4), 237–254 (2001). https://doi.org/10.1016/s1361-8415(00)00039-6

2. Petitclerc, L., Sebastiani, G., Gilbert, G., Cloutier, G., Tang, A.: Liver fibrosis: review of current imaging and MRI quantification techniques. J. Magn. Reson. Imaging **45**(5), 1276–1295 (2016)

3. Oliphant, T.E., Manduca, A., Ehman, R.L., Greenleaf, J.F.: Complex-valued stiffness reconstruction for magnetic resonance elastography by algebraic inversion of the differential equation. Magn. Reson. Med. **45**(2), 299–310 (2001). https://doi.org/10.1002/1522-2594(200102)45:2⟨299::aid-mrm1039⟩3.0.co;2-o

4. Park, E., Maniatty, A.M.: Shear modulus reconstruction in dynamic elastography: time harmonic case. Phys. Med. Biol. **51**, 3697 (2006). https://doi.org/10.1088/0031-9155/51/15/007

5. Papazoglou, S., Hamhaber, U., Braun, J., Sack, I.: Algebraic Helmholtz inversion in planar magnetic resonance elastography. Phys. Med. Biol. **53**(12), 3147–3158 (2008). https://doi.org/10.1088/0031-9155/53/12/005

6. Eskandari, H., Salcudean, S.E., Rohling, R., Bell, I.: Real-time solution of the finite element inverse problem of viscoelasticity. Inverse Prob. **27**(8), 085002 (2011). https://doi.org/10.1088/0266-5611/27/8/085002

7. Honarvar, M., Sahebjavaher, R., Sinkus, R., Rohling, R., Salcudean, S.E.: Curl-based finite element reconstruction of the shear modulus without assuming local homogeneity: Time harmonic case. IEEE Tran. Med. Imaging **32**(12), 2189–99 (2013). https://doi.org/10.1109/TMI.2013.2276060

8. Honarvar, M., Rohling, R., Salcudean, S.E.: A comparison of direct and iterative finite element inversion techniques in dynamic elastography. Phys. Med. Biol. **61**(8), 3026–48 (2016). https://doi.org/10.1088/0031-9155/61/8/3026

9. Fovargue, D., Nordsletten, D., Sinkus, R.: Stiffness reconstruction methods for MR elastography. NMR Biomed. **31**(10), e3935 (2018). https://doi.org/10.1002/nbm.3935

10. Fovargue, D., Kozerke, S., Sinkus, R., Nordsletten, D.: Robust MR elastography stiffness quantification using a localized divergence free finite element reconstruction. Med. Image Anal. **44**, 126–142 (2018)

11. Murphy, M.C., Manduca, A., Trzasko, J.D., Glaser, K.J., Huston III, J., Ehman, R.L.: Artificial neural networks for stiffness estimation in magnetic resonance elastography. Magn. Reson. Med. **80**(1), 351–360 (2017)

12. Solamen, L., Shi, Y., Amoh, J.: Dual objective approach using a convolutional neural network for magnetic resonance elastography. arXiv preprint: 1812.00441 [physics.med-ph] (2018)

13. Ni, B., Gao, H.: A deep learning approach to the inverse problem of modulus identification in elasticity. MRS Bull. **46**(1), 19–25 (2021). https://doi.org/10.1557/s43577-020-00006-y

14. Raissi, M., Perdikaris, P., Karniadakis, G.E.: Physics-informed neural networks: a deep learning framework for solving forward and inverse problems involving nonlinear partial differential equations. J. Comput. Phys. **378**, 686–707 (2019). https://doi.org/10.1016/j.jcp.2018.10.045

15. Haghighat, E., Raissi, M., Moure, A., Gomez, H., Juanes, R.: A physics-informed deep learning framework for inversion and surrogate modeling in solid mechanics. Comput. Methods Appl. Mech. Eng., 113741 (2021). https://doi.org/10.1016/j.cma.2021.113741

16. Zhang, E., Yin, M., Karniadakis, G.E.: Physics-informed neural networks for nonhomogeneous material identification in elasticity imaging. arXiv preprint: 2009.04525 [cs.LG] (2020). https://doi.org/10.48550/arXiv.2009.04525

17. Mallampati, A., Almekkawy, M.: Measuring tissue elastic properties using physics based neural networks. In: 2021 IEEE UFFC Latin America Ultrasonics Symposium (LAUS), pp. 1–4. IEEE, Gainesville (2021). https://doi.org/10.1109/LAUS53676.2021.9639231

18. Kamali, A., Sarabian, M., Laksari, K.: Elasticity imaging using physics-informed neural networks: spatial discovery of elastic modulus and Poisson's ratio. Acta Biomater. **155**, 400–409 (2023). https://doi.org/10.1016/j.actbio.2022.11.024

19. Wymer, D.T., Patel, K.P., Burke, W.F., III., Bhatia, V.K.: Phase-contrast MRI: physics, techniques, and clinical applications. RadioGraphics **40**(1), 122–140 (2020)

20. Sinkus, R., Daire, J.L., Beers, B.E.V., Vilgrain, V.: Elasticity reconstruction: beyond the assumption of local homogeneity. Comptes Rendus Mécanique **338**(7), 474–479 (2010). https://doi.org/10.1016/j.crme.2010.07.014

21. Honarvar, M.: Dynamic elastography with finite element-based inversion. Ph.D. thesis, University of British Columbia (2015). https://doi.org/10.14288/1.0167683

22. Pollack, B.L., et al.: Deep learning prediction of voxel-level liver stiffness in patients with nonalcoholic fatty liver disease. Radiology: AI **3**(6) (2021). https://doi.org/10.1148/ryai.2021200274

23. Sitzmann, V., Martel, J.N.P., Bergman, A.W., Lindell, D.B., Wetzstein, G.: Implicit neural representations with periodic activation functions (2020)

24. Kingma, D.P., Ba, J.L.: Adam: a method for stochastic optimization. In: Proceedings of 3rd International Conference Learning Representations (2015)

25. Paszke, A., et al.: PyTorch: an imperative style, high-performance deep learning library. In: Advance Neural Information Processing System, vol. 32, pp. 8024–8035. Curran Associates, Inc. (2019)

26. Lu, L., Meng, X., Mao, Z., Karniadakis, G.E.: DeepXDE: a deep learning library for solving differential equations. SIAM Rev. **63**(1), 208–228 (2021). https://doi.org/10.1137/19M1274067

27. Scroggs, M.W., Dokken, J.S., Richardson, C.N., Wells, G.N.: Construction of arbitrary order finite element degree-of-freedom maps on polygonal and polyhedral cell meshes. ACM Trans. Math. Softw. **48**, 1–23 (2022). https://doi.org/10.1145/3524456

28. Barnhill, E., Davies, P.J., Ariyurek, C., Fehlner, A., Braun, J., Sack, I.: Heterogeneous multifrequency direct inversion (HMDI) for magnetic resonance elastography with application to a clinical brain exam. Med. Image Anal. **46**, 180–188 (2018). https://doi.org/10.1016/j.media.2018.03.003

29. Kallel, F., Bertrand, M., Ophir, J.: Fundamental limitations on the contrast-transfer efficiency in elastography: an analytic study. Ultrasound Med. Biol. **22**(4), 463–470 (1996). https://doi.org/10.1016/0301-5629(95)02079-9
30. Lowekamp, B.C., Chen, D.T., Ibáñez, L., Blezek, D.: The design of SimpleITK. Front. Neuroinf. **7**(45) (2013). https://doi.org/10.3389/fninf.2013.00045
31. Mueller, S., Sandrin, L.: Liver stiffness: a novel parameter for the diagnosis of liver disease. Hepat. Med. **2**, 49–67 (2010). https://doi.org/10.2147/hmer.s7394

CT Kernel Conversion Using Multi-domain Image-to-Image Translation with Generator-Guided Contrastive Learning

Changyong Choi[1,2], Jiheon Jeong[1,2], Sangyoon Lee[2], Sang Min Lee[3], and Namkug Kim[2,3]([✉])

[1] Department of Biomedical Engineering, AMIST, Asan Medical Center, University of Ulsan College of Medicine, Seoul, Republic of Korea
[2] Department of Convergence Medicine, Asan Medical Center, University of Ulsan College of Medicine, Seoul, Republic of Korea
namkugkim@gmail.com
[3] Department of Radiology, Asan Medical Center, University of Ulsan College of Medicine, Seoul, Republic of Korea

Abstract. Computed tomography (CT) image can be reconstructed by various types of kernels depending on what anatomical structure is evaluated. Also, even if the same anatomical structure is analyzed, the kernel being used differs depending on whether it is qualitative or quantitative evaluation. Thus, CT images reconstructed with different kernels would be necessary for accurate diagnosis. However, once CT image is reconstructed with a specific kernel, the CT raw data, sinogram is usually removed because of its large capacity and limited storage. To solve this problem, many methods have been proposed by using deep learning approach using generative adversarial networks in image-to-image translation for kernel conversion. Nevertheless, it is still challenging task that translated image should maintain the anatomical structure of source image in medical domain. In this study, we propose CT kernel conversion method using multi-domain image-to-image translation with generator-guided contrastive learning. Our proposed method maintains the anatomical structure of the source image accurately and can be easily utilized into other multi-domain image-to-image translation methods with only changing the discriminator architecture and without adding any additional networks. Experimental results show that our proposed method can translate CT images from sharp into soft kernels and from soft into sharp kernels compared to other image-to-image translation methods. Our code is available at https://github.com/cychoi97/GGCL.

Keywords: CT · Kernel conversion · Image-to-image translation · Contrastive learning · Style transfer

---

C. Choi and J. Jeong—Contributed equally.

Supplementary Information The online version contains supplementary material available at https://doi.org/10.1007/978-3-031-43999-5_33.

H. Greenspan et al. (Eds.): MICCAI 2023, LNCS 14229, pp. 344–354, 2023.
https://doi.org/10.1007/978-3-031-43999-5_33

1   Introduction

Computed tomography (CT) image is reconstructed from sinogram, which is tomographic raw data collected from detectors. According to kernels being used for CT image reconstruction, there is a trade-off between spatial resolution and noise, and it affects intensity and texture quantitative values [1]. When CT image is reconstructed with sharp kernel, spatial resolution and noise increase, and abnormality can be easily detected in bones or lung. In contrast, with soft kernel, spatial resolution and noise reduce, and abnormality can be easily detected in soft tissues or mediastinum. In other words, CT image is reconstructed depending on what anatomical structure is evaluated. Also, even if the same anatomical structure is analyzed, the kernel being used differs depending on whether it is qualitative or quantitative evaluation. For example, CT images reconstructed with soft kernel is required to evaluate quantitative results for lung instead of sharp kernel. Thus, CT images reconstructed with different kernels would be necessary for accurate diagnosis.

However, once CT image is reconstructed with a specific kernel, sinogram is usually removed because of its large capacity and limited storage. Therefore, clinicians have difficulty to analyze qualitative or quantitative results without CT image reconstructed with different kernels, and this limitation reveals on retrospective or longitudinal studies that cannot control technical parameters, particularly [2]. Besides, there is another problem that patients should be scanned again and exposed to radiation.

Recently, many studies have achieved improvement in kernel conversion [2–5] using image-to-image translation methods [6–13] based on deep learning, especially generative adversarial networks (GANs) [14]. Nevertheless, it remains challenging that translated image should maintain its anatomical structure of source image in medical domain [9]. It is important for quantitative evaluation as well as qualitative evaluation. To solve this problem, we focus on improving maintenance of structure when the source image is translated.

Our contributions are as follows: (1) we propose multi-domain image-to-image translation with generator-guided contrastive learning (GGCL) for CT kernel conversion, which maintains the anatomical structure of the source image accurately; (2) Our proposed GGCL can be easily utilized into other multi-domain image-to-image translation with only changing the discriminator architecture and without adding any additional networks; (3) Experimental results showed that our method can translate CT images from sharp into soft kernels and from soft into sharp kernels compared to other image-to-image translation methods.

2   Method

2.1   Related Work

In deep learning methods for CT kernel conversion, there were proposed methods using convolutional neural networks [2, 3], but they were trained in a supervised manner. Recently, Yang et al. [5] proposed a new method using the adaptive instance normalization (AdaIN) [15] in an unsupervised manner and it showed significant performance, however, this method still has limitations that the target image for the test phase and additional architecture for AdaIN are needed.

Generator-guided discriminator regularization (GGDR) [16] is discriminator regularization method that intermediate feature map in the generator supervises semantic representations by matching with semantic label map in the discriminator for unconditional image generation. It has advantages that we don't need any ground-truth semantic segmentation masks and can improve fidelity as much as conditional GANs [17–19].

Recently, it has been shown that dense contrastive learning can have a positive effect on learning dense semantic labels. In dense prediction tasks such as object detection and semantic segmentation [20, 21], both global and local contrastive learning have been proposed to embed semantic information. Furthermore, it has been demonstrated that patch-wise contrastive learning performs well in style transfer for unsupervised image-to-image translation [12]. This motivated our experiments as it demonstrates that intermediate features can be learned through contrastive learning when learning dense semantic labels.

2.2 Generator-Guided Contrastive Learning

GGDR [16] uses cosine distance loss between the feature map and the semantic label map for unconditional image generation. However, unlike image generation, the generator has a structure with an encoder and a decoder in image-to-image translation [11], and this is quite important to maintain the structure of source image while translating the style of target image. Thus, it might be helpful for discriminator to inform more fine detail semantic representations by comparing similarity using patch-based contrastive learning [12] (see Fig. 1).

Multi-Domain Image-To-Image Translation. We apply generator-guided contrastive learning (GGCL) to StarGAN [6] as base architecture which is one of the multi-domain image-to-image translation model to translate kernels into all directions at once and show stability of GGCL. Basically, StarGAN uses adversarial loss, domain classification loss and cycle consistency loss [13] as follows:

$$\mathcal{L}_D = -\mathcal{L}_{adv} + \lambda_{cls}\mathcal{L}_{cls}^r, \tag{1}$$

$$\mathcal{L}_G = \mathcal{L}_{adv} + \lambda_{cls}\mathcal{L}_{cls}^f + \lambda_{cyc}\mathcal{L}_{cyc}, \tag{2}$$

where $\mathcal{L}_D$ and $\mathcal{L}_G$ are the discriminator and generator losses, respectively. They both have $\mathcal{L}_{adv}$, which is the adversarial loss. $\mathcal{L}_{cls}^r$ and $\mathcal{L}_{cls}^f$ are the domain classification losses for a real and fake image, respectively. $\mathcal{L}_{cyc}$, which is the cycle consistency loss, has an importance for the translated image to maintain the structure of source image.

Patch-Based Contrastive Learning. Our method is to add PatchNCE loss [12] between "positive" and "negative" patches from the feature map of the decoder in the generator and "query" patch from the semantic label map in the discriminator. The query patch is the same location with positive patch and different locations with N negative patches. So, the positive patch is learned to associate to the query patch more than the N negative patches. GGCL loss is the same as PatchNCE loss, which is the cross-entropy loss calculated for an $(N + 1)$-way classification, and it follows as:

$$\mathcal{L}_{ggcl} = \mathbb{E}_v\left[-\log\frac{\exp(v \cdot v^+/\tau)}{\exp(v \cdot v^+/\tau) + \sum_{n=1}^{N}\exp(v \cdot v_n^-/\tau)}\right], \tag{3}$$

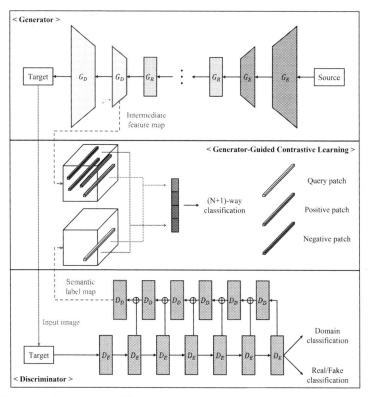

Fig. 1. Overview of generator-guided contrastive learning (GGCL) framework. The proposed method is to add patch-based contrastive learning between the intermediate feature map from the generator and the semantic label map from the discriminator to solve $(N + 1)$-way classification. G_E, G_R and G_D are the encoder, residual and decoder blocks of the generator, respectively. D_E and D_D are the encoder and decoder blocks of the discriminator, respectively. This method can be applied to any multi-domain image-to-image translation methods.

where v, v^+ and v_n^- are the vectors which are mapped from query, positive and n-th negative patches, respectively. $\tau = 0.07$ is the same configuration as CUT [12]. Since we use the features from the generator and the discriminator themselves, this requires no additional auxiliary networks and no feature encoding process.

Total Objective. GGCL follows the concept of GGDR, which the generator supervises the semantic representations to the discriminator, so it is a kind of the discriminator regularization. Discriminator conducts real/fake classification, domain classification and semantic label map segmentation, so it can be also a kind of the multi-task learning [22]. Our total objective functions for the discriminator and generator are written, respectively, as:

$$\mathcal{L}_D = -\mathcal{L}_{adv} + \lambda_{cls}\mathcal{L}_{cls}^r + \lambda_{ggcl}\mathcal{L}_{ggcl}, \tag{4}$$

$$\mathcal{L}_G = \mathcal{L}_{adv} + \lambda_{cls}\mathcal{L}_{cls}^f + \lambda_{cyc}\mathcal{L}_{cyc}, \tag{5}$$

where λ_{cls}, λ_{cyc} and λ_{ggcl} are hyper-parameters that weight the importance of domain classification loss, cycle consistency loss and GGCL loss, respectively. We used $\lambda_{cls} = 1$, $\lambda_{cyc} = 10$ and $\lambda_{ggcl} = 2$ in our experiments.

3   Experiments and Results

3.1   Datasets and Implementation

Datasets. For train dataset, chest CT images were reconstructed with B30f, B50f and B70f kernels, from soft to sharp, in Siemens Healthineers. We collected chest CT images from 102 (63 men and 39 women; mean age, 62.1 ± 12.8 years), 100 (47 men and 53 women; mean age, 64.9 ± 13.7 years) and 104 (64 men and 40 women; mean age, 62.2 ± 12.9 years) patients for B30f, B50f and B70f kernels from Siemens (see Table 1).

For test dataset, we collected chest CT images from paired 20 (15 men and 5 women; mean age, 67.1 ± 7.4 years) patients for each kernel in Siemens for quantitative and qualitative evaluation.

Table 1. CT acquisition parameters of dataset according to kernels.

	Kernel	Patients	Slices	Age (year)	Sex (M:F)	Slice Thickness	kVp
Test	B30f	102	36199	62.1 ± 12.8	63:39	1.0	120
	B50f	100	32795	64.9 ± 13.7	47:53	1.0	120
	B70f	104	36818	62.2 ± 12.9	64:40	1.0	120
	Kernel	Patients	Slices	Age (year)	Sex (M:F)	Slice Thickness	kVp
Test	B30f	20	6897	67.1 ± 7.4	15:5	1.0	120
	B50f	20	6897	67.1 ± 7.4	15:5	1.0	120
	B70f	20	6897	67.1 ± 7.4	15:5	1.0	120

Implementation Details. We maintained the original resolution 512×512 of CT images and normalized their Hounsfield unit (HU) range from [-1024HU ~ 3071HU] to [-1 ~ 1] for pre-processing. For training, the generator and the discriminator were optimized by Adam [23] with $\beta_1 = 0.5$, $\beta_2 = 0.999$, learning rate 1e-4 and the batch size is 2. We used WGAN-GP [24] and set $n_{critic} = 5$, where n_{critic} is the number of discriminator updates per each generator update. The feature map and the semantic label map were extracted in 256×256 size and resized 128×128 using averaging pooling. The number of patches for contrastive learning is 64. All experiments were conducted using single NVIDIA GeForce RTX 3090 24GB GPU for 400,000 iterations. We used peak signal-to-noise ratio (PSNR) [25] and structural similarity index measure (SSIM) [26] for quantitative assessment.

Architecture Improvements. Instead of using original StarGAN [6] architecture, we implemented architecture ablations to sample the best quality results, empirically. In generator, original StarGAN runs 4×4 transposed convolutional layers for upsampling. However, it causes degradation of visual quality of the translated image because of checkerboard artifact [27]. By using 3×3 convolutional layers and 2×2 pixelshuffle [28], we could prevent the artifact. In discriminator, we changed the discriminator to U-Net architecture [29] with skip connection, which consists of seven encoder layers for playing the role of patchGAN [8] and six decoder layers for extracting semantic label map, to utilize GGCL. For each decoder layer, we concatenated the feature from the encoder and the decoder layer with the same size, and ran 1×1 convolutional layer, then ran 2×2 pixelshuffle for upsampling. At the end of the decoder, it extracts semantic label map to compare with the feature map from the decoder layer of the generator. Lastly, we added spectral normalization [30] and leakyReLU activation function in all layers of the discriminator.

3.2 Comparison with Other Image-to-Image Translation Methods

We compared GGCL with two-domain image-to-image translation methods such as CycleGAN [13], CUT [12], UNIT [10] and multi-domain image-to-image translation methods such as AttGAN [7], StarGAN and StarGAN with GGDR [16] to show the effectiveness of GGCL. In this section, qualitative and quantitative results were evaluated for the translation into B30f, B50f and B70f kernels, respectively.

Qualitative Results. We showed the qualitative results of image-to-image translation methods including GGDR and GGCL applied to StarGAN for kernel conversion from Siemens into Siemens (see Fig. 2). For visualization, window width and window level were set 1500 and -700, respectively. While UNIT could not maintain the global structure of the source image and translate the kernel style of the target image, the other methods showed plausible results. However, they could not maintain the fine details like airway wall and vessel in specific kernel conversion, e.g., B50f to B30f, B30f to B50f and B50f to B70f. It could be observed through difference map between the target image and the translated image (see **Supplementary** Fig. 1). GGCL showed stability of translation for kernel conversion with any directions and maintained the fine details including airway wall, vessel and even noise pattern as well as the global structure of the source image.

Quantitative Results. We showed the quantitative results of image-to-image translation methods including GGDR and GGCL applied to StarGAN for kernel conversion from Siemens into the Siemens (see Table 2). In case of two-domain image-to-image translation methods, they showed high PSNR and SSIM performance in translation from B70f into B30f and from B30f into B70f, and UNIT showed the best performance in translation from B30f into B70f. However, they showed low performance in translation into the other kernels, especially soft into sharp, and it indicates that two-domain methods are unstable and cannot maintain the structure of the source image well. In case of multi-domain image-to-image translation methods, their performance still seemed unstable, however, when applying GGDR to StarGAN, it showed quite stable and improved the performance in translation into sharp kernels. Furthermore, when applying GGCL, it outperformed GGDR in translation into many kernels, especially from B30f into B70f and from B50f into B70f.

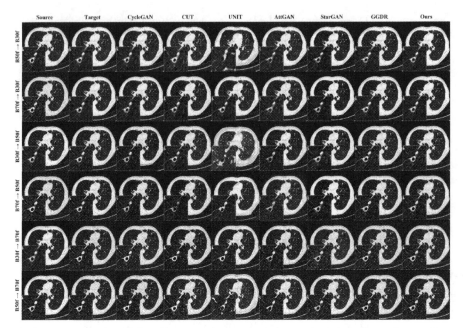

Fig. 2. The qualitative results of image-to-image translation methods including GGDR and our method for kernel conversion from Siemens into Siemens.

3.3 Ablation Study

We implemented ablation studies about the number of patches, size of pooling and loss weight for GGCL to find out the best performance. We evaluated our method while preserving the network architecture. Ablation studies were also evaluated by PSNR and SSIM. All ablation studies were to change one factor and the rest of them were fixed with their best configurations. The results about the number of patches showed improvement when the number of patches was 64 (see Table 3). The size of pooling also affected the performance improvement, and 2 was appropriate (see **Supplementary** Table 1). Lastly, the results of the loss weight for GGCL showed that 2 was the best performance (see **Supplementary** Table 2).

4 Discussion and Conclusion

In this paper, we proposed CT kernel conversion method using multi-domain image-to-image translation with generator-guided contrastive learning (GGCL). In medical domain image-to-image translation, it is important to maintain anatomical structure of the source image while translating style of the target image. However, GAN based generation has limitation that the training process may be unstable, and the results may be inaccurate so that some fake details may be generated. Especially in unsupervised manner, the anatomical structure of the translated image relies on cycle consistency mainly. If trained unstably, as the translated image to the target domain would be inaccurate, the reversed translated image to the original domain would be inaccurate as well. Then,

Table 2. The quantitative results of image-to-image translation methods including GGDR and our method from Siemens to Siemens.

| Method | Sharp to Soft | | | | | |
| | B50f → B30f | | B70f → B30f | | B70f → B50f | |
	PSNR	SSIM	PSNR	SSIM	PSNR	SSIM
CycleGAN	35.672	0.950	44.741	0.974	32.553	0.905
CUT	37.261	0.956	40.394	0.961	34.595	0.910
UNIT	24.545	0.672	44.964	0.979	22.868	0.540
AttGAN	38.685	0.927	37.435	0.900	32.596	0.733
StarGAN	37.262	0.930	36.024	0.903	31.660	0.799
w/ GGDR	47.659	0.987	**45.213**	0.979	**41.391**	**0.950**
w/ GGCL (ours)	**47.831**	**0.989**	44.943	**0.981**	40.332	0.944
Method	Soft to Sharp					
	B30f → B50f		B30f → B70f		B50f → B70f	
	PSNR	SSIM	PSNR	SSIM	PSNR	SSIM
CycleGAN	28.536	0.830	31.544	0.754	30.719	0.758
CUT	35.320	0.902	29.659	0.660	31.402	0.834
UNIT	22.994	0.511	**34.733**	**0.869**	23.288	0.563
AttGAN	32.604	0.753	28.293	0.556	28.662	0.564
StarGAN	31.738	0.836	28.531	0.601	28.527	0.601
w/ GGDR	**41.606**	**0.961**	31.062	0.757	34.547	0.869
w/ GGCL (ours)	41.279	0.958	32.584	0.818	**34.857**	**0.872**

the cycle consistency would fail to lead the images to maintain the anatomical structure. CycleGAN [13], CUT [12] and UNIT [10] showed this limitation (see Fig. 2 and Table 2), but GGCL solved this problem without any additional networks.

The benefit of GGCL was revealed at the translation from soft into sharp kernels. It is a more difficult task than the translation from sharp into soft kernels because spatial resolution should be increased and noise patterns should be clear, so this benefit can be meaningful. Nevertheless, the improvements from GGCL were quite slight compared to GGDR [16] (see Table 2) and inconsistent according to the number of patches (see Table 3). Besides, we did not show results about the different kernels from the external manufacturer. In future work, we will collect different types of kernels from the external manufacturer, and conduct experiments to show better improvements and stability of GGCL.

Table 3. Ablation studies about the number of patches.

Patch Num	Sharp to Soft					
	B50f → B30f		B70f → B30f		B70f → B50f	
	PSNR	SSIM	PSNR	SSIM	PSNR	SSIM
64	**47.831**	**0.989**	44.943	0.981	**40.332**	**0.944**
128	47.788	0.989	**45.318**	0.981	38.484	0.913
256	47.513	0.989	45.282	**0.983**	40.190	0.937
Patch Num	Soft to Sharp					
	B30f → B50f		B30f → B70f		B50f → B70f	
	PSNR	SSIM	PSNR	SSIM	PSNR	SSIM
64	41.279	0.958	32.584	0.818	**34.857**	**0.872**
128	40.294	0.948	31.812	0.781	31.676	0.764
256	**41.375**	**0.958**	**33.208**	**0.830**	34.710	0.867

Acknowledgement. This work was supported by a grant of the Korea Health Technology R&D Project through the Korea Health Industry Development Institute (KHIDI), funded by the Ministry of Health & Welfare, Republic of Korea (HI18C0022) and by Institute of Information & communications Technology Planning & Evaluation (IITP) grant funded by the Korea government (1711134538, 20210003930012002).

References

1. Mackin, D., et al.: Matching and homogenizing convolution kernels for quantitative studies in computed tomography. Invest. Radiol. **54**(5), 288 (2019)
2. Lee, S.M., et al.: CT image conversion among different reconstruction kernels without a sinogram by using a convolutional neural network. Korean J. Radiol. **20**(2), 295–303 (2019)
3. Eun, D.-I., et al.: CT kernel conversions using convolutional neural net for super-resolution with simplified squeeze-and-excitation blocks and progressive learning among smooth and sharp kernels. Comput. Meth. Programs Biomed. **196**, 105615 (2020)
4. Gravina, M., et al.: Leveraging CycleGAN in Lung CT Sinogram-free Kernel Conversion. In: Sclaroff, S., Distante, C., Leo, M., Farinella, G.M., Tombari, F. (eds.) Image Analysis and Processing – ICIAP 2022: 21st International Conference, Lecce, Italy, May 23–27, 2022, Proceedings, Part I, pp. 100–110. Springer International Publishing, Cham (2022). https://doi.org/10.1007/978-3-031-06427-2_9
5. Yang, S., Kim, E.Y., Ye, J.C.: Continuous conversion of CT kernel using switchable CycleGAN with AdaIN. IEEE Trans. Med. Imaging **40**(11), 3015–3029 (2021)
6. Choi, Y., et al.: Stargan: Unified generative adversarial networks for multi-domain image-to-image translation. In: Proceedings of the IEEE Conference on Computer Vision and Pattern Recognition (2018)
7. He, Z., et al.: Attgan: facial attribute editing by only changing what you want. IEEE Trans. Image Process. **28**(11), 5464–5478 (2019)

8. Isola, P., et al.: Image-to-image translation with conditional adversarial networks. In: Proceedings of the IEEE Conference on Computer Vision and Pattern Recognition (2017)

9. Kong, L., et al.: Breaking the dilemma of medical image-to-image translation. Adv. Neural. Inf. Process. Syst. **34**, 1964–1978 (2021)

10. Liu, M.-Y., Breuel, T., Kautz, J.: Unsupervised image-to-image translation networks. Adv. Neural Inform. Process. Syst. **30** (2017)

11. Pang, Y., et al.: Image-to-image translation: methods and applications. IEEE Trans. Multimedia **24**, 3859–3881 (2021)

12. Park, T., et al. (eds.): Computer Vision – ECCV 2020: 16th European Conference, Glasgow, UK, August 23–28, 2020, Proceedings, Part IX, pp. 319–345. Springer International Publishing, Cham (2020). https://doi.org/10.1007/978-3-030-58545-7_19

13. Zhu, J.-Y., et al.: Unpaired image-to-image translation using cycle-consistent adversarial networks. In: Proceedings of the IEEE International Conference on Computer Vision (2017)

14. Goodfellow, I., et al.: Generative adversarial networks. Commun. ACM **63**(11), 139–144 (2020)

15. Huang, X., Belongie, S.: Arbitrary style transfer in real-time with adaptive instance normalization. In: Proceedings of the IEEE International Conference on Computer Vision (2017)

16. Lee, G., et al.: Generator knows what discriminator should learn in unconditional GANs. In: Avidan, Shai, Brostow, G., Cissé, M., Farinella, G.M., Hassner, T. (eds.) Computer Vision – ECCV 2022: 17th European Conference, Tel Aviv, Israel, October 23–27, 2022, Proceedings, Part XVII, pp. 406–422. Springer Nature Switzerland, Cham (2022). https://doi.org/10.1007/978-3-031-19790-1_25

17. Mirza, M., Osindero, S.: Conditional generative adversarial nets. arXiv preprint arXiv:1411.1784 (2014)

18. Park, T., et al.: Semantic image synthesis with spatially-adaptive normalization. In: Proceedings of the IEEE/CVF Conference on Computer Vision and Pattern Recognition (2019)

19. Sushko, V., et al.: You only need adversarial supervision for semantic image synthesis. arXiv preprint arXiv:2012.04781 (2020)

20. Wang, X., et al.: Dense contrastive learning for self-supervised visual pre-training. In: Proceedings of the IEEE/CVF Conference on Computer Vision and Pattern Recognition (2021)

21. Xie, Z., et al.: Propagate yourself: exploring pixel-level consistency for unsupervised visual representation learning. In: Proceedings of the IEEE/CVF Conference on Computer Vision and Pattern Recognition (2021)

22. Zhang, Y., Yang, Q.: A survey on multi-task learning. IEEE Trans. Knowl. Data Eng. **34**(12), 5586–5609 (2021)

23. Kingma, D.P, Ba, J.: Adam: a method for stochastic optimization. arXiv preprint arXiv:1412.6980 (2014)

24. Gulrajani, I., et al.: Improved training of wasserstein gans. Adv. Neural Inform. Process. Syst. **30** (2017)

25. Fardo, F.A., et al.: A formal evaluation of PSNR as quality measurement parameter for image segmentation algorithms. arXiv preprint arXiv:1605.07116 (2016)

26. Wang, Z., et al.: Image quality assessment: from error visibility to structural similarity. IEEE Trans. Image Process. **13**(4), 600–612 (2004)

27. Odena, A., Dumoulin, V., Olah, C.: Deconvolution and checkerboard artifacts. Distill. **1**(10), e3 (2016)

28. Shi, W., et al.: Real-time single image and video super-resolution using an efficient sub-pixel convolutional neural network. In: Proceedings of the IEEE Conference on Computer Vision and Pattern Recognition (2016)

29. Ronneberger, O., Fischer, P., Brox, T.: U-net: convolutional networks for biomedical image segmentation. In: Medical Image Computing and Computer-Assisted Intervention–MICCAI 2015: 18th International Conference, Munich, Germany, October 5–9, 2015, Proceedings, Part III 18. Springer (2015)
30. Miyato, T., et al.: Spectral normalization for generative adversarial networks. arXiv preprint arXiv:1802.05957 (2018)

ASCON: Anatomy-Aware Supervised Contrastive Learning Framework for Low-Dose CT Denoising

Zhihao Chen[1], Qi Gao[1], Yi Zhang[2], and Hongming Shan[1,3,4(✉)] 🔟

[1] Institute of Science and Technology for Brain-Inspired Intelligence and MOE Frontiers Center for Brain Science, Fudan University, Shanghai, China
hmshan@fudan.edu.cn

[2] School of Cyber Science and Engineering, Sichuan University, Chengdu, China

[3] Key Laboratory of Computational Neuroscience and Brain-Inspired Intelligence (Fudan University), Ministry of Education, Shanghai, China

[4] Shanghai Center for Brain Science and Brain-Inspired Technology, Shanghai, China

Abstract. While various deep learning methods have been proposed for low-dose computed tomography (CT) denoising, most of them leverage the normal-dose CT images as the ground-truth to supervise the denoising process. These methods typically ignore the inherent correlation within a single CT image, especially the anatomical semantics of human tissues, and lack the interpretability on the denoising process. In this paper, we propose a novel **A**natomy-aware **S**upervised **CON**trastive learning framework, termed ASCON, which can explore the anatomical semantics for low-dose CT denoising while providing anatomical interpretability. The proposed ASCON consists of two novel designs: an efficient self-attention-based U-Net (ESAU-Net) and a multi-scale anatomical contrastive network (MAC-Net). First, to better capture global-local interactions and adapt to the high-resolution input, an efficient ESAU-Net is introduced by using a channel-wise self-attention mechanism. Second, MAC-Net incorporates a patch-wise non-contrastive module to capture inherent anatomical information and a pixel-wise contrastive module to maintain intrinsic anatomical consistency. Extensive experimental results on two public low-dose CT denoising datasets demonstrate superior performance of ASCON over state-of-the-art models. Remarkably, our ASCON provides anatomical interpretability for low-dose CT denoising for the first time. Source code is available at https://github. com/hao1635/ASCON.

Keywords: CT denoising · Deep learning · Self-attention · Contrastive learning · Anatomical semantics

---

Supplementary Information The online version contains supplementary material available at https://doi.org/10.1007/978-3-031-43999-5_34.

H. Greenspan et al. (Eds.): MICCAI 2023, LNCS 14229, pp. 355–365, 2023.
https://doi.org/10.1007/978-3-031-43999-5_34

1 Introduction

With the success of deep learning in the field of computer vision and image processing, many deep learning-based methods have been proposed and achieved promising results in low-dose CT (LDCT) denoising [1, 4–6, 9, 12, 23, 24, 26]. Typically, they employ a supervised learning setting, which involves a set of image pairs, LDCT images and their normal-dose CT (NDCT) counterparts. These methods typically use a pixel-level loss (e.g. mean squared error or MSE), which can cause over-smoothing problems.

To address this issue, a few studies [23, 26] used a structural similarity (SSIM) loss or a perceptual loss [11]. However, they all perform in a sample-to-sample manner and ignore the inherent anatomical semantics, which could blur details in areas with low noise levels. Previous studies have shown that the level of noise in CT images varies depending on the type of tissues [17]; see an example in Fig. S1 in Supplementary Materials. Therefore, it is crucial to characterize the anatomical semantics for effectively denoising diverse tissues.

In this paper, we focus on taking advantage of the inherent anatomical semantics in LDCT denoising from a contrastive learning perspective [7, 25, 27]. To this end, we propose a novel **A**natomy-aware **S**upervised **CON**trastive learning framework (ASCON), which consists of two novel designs: an efficient self-attention-based U-Net (ESAU-Net) and a multi-scale anatomical contrastive network (MAC-Net). First, to ensure that MAC-Net can effectively extract anatomical information, diverse global-local contexts and a larger input size are necessary. However, operations on full-size CT images with self-attention are computationally unachievable due to potential GPU memory limitations [20]. To address this limitation, we propose an ESAU-Net that utilizes a channel-wise self-attention mechanism [2, 22, 28] which can efficiently capture both local and global contexts by computing cross-covariance across feature channels.

Second, to exploit inherent anatomical semantics, we present the MAC-Net that employs a disentangled U-shaped architecture [25] to produce global and local representations. Globally, a patch-wise non-contrastive module is designed to select neighboring patches with similar semantic context as positive samples and align the same patches selected in denoised CT and NDCT which share the same anatomical information, using an optimization method similar to the BYOL method [7]. This is motivated by the prior knowledge that adjacent patches often share common semantic contexts [27]. Locally, to further improve the anatomical consistency between denoised CT and NDCT, we introduce a pixel-wise contrastive module with a hard negative sampling strategy [21], which randomly selects negative samples from the pixels with high similarity around the positive sample within a certain distance. Then we use a local InfoNCE loss [18] to pull the positive pairs and push the negative pairs.

Our contributions are summarized as follows. 1) We propose a novel ASCON framework to explore inherent anatomical information in LDCT denoising, which is important to provide interpretability for LDCT denoising. 2) To better explore anatomical semantics in MAC-Net, we design an ESAU-Net, which utilizes a channel-wise self-attention mechanism to capture both local and global contexts.

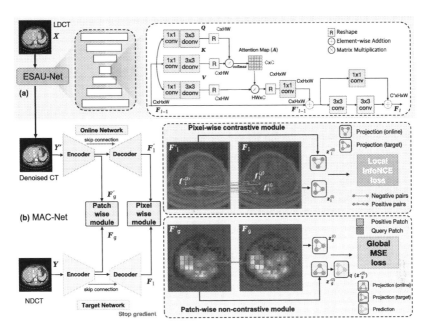

Fig. 1. Overview of our proposed ASCON. (a) Efficient self-attention-based U-Net (ESAU-Net); and (b) multi-scale anatomical contrastive network (MAC-Net).

3) We propose a MAC-Net that employs a disentangled U-shaped architecture and incorporates both global non-contrastive and local contrastive modules. This enables the exploitation of inherent anatomical semantics at the patch level, as well as improving anatomical consistency at the pixel level. 4) Extensive experimental results demonstrate that our ASCON outperforms other state-of-the-art methods, and provides anatomical interpretability for LDCT denoising.

2   Methodology

2.1   Overview of the Proposed ASCON

Figure 1 presents the overview of the proposed ASCON, which consists of two novel components: ESAU-Net and MAC-Net. First, given an LDCT image, $X \in \mathbb{R}^{1 \times H \times W}$, where $H \times W$ denotes the image size. X is passed through the ESAU-Net to capture both global and local contexts using a channel-wise self-attention mechanism and obtain a denoised CT image $Y' \in \mathbb{R}^{1 \times H \times W}$.

Then, to explore inherent anatomical semantics and remain inherent anatomical consistency, the denoised CT Y' and NDCT Y are passed to the MAC-Net to compute a global MSE loss $\mathcal{L}_{\text{global}}$ in a patch-wise non-contrastive module and a local infoNCE loss $\mathcal{L}_{\text{local}}$ in a pixel-wise contrastive module. During training, we use an alternate learning strategy to optimize ESAU-Net and MAC-Net separately, which is similar to GAN-based methods [10]. Please refer to Algorithm S1 in Supplementary Materials for a detailed optimization.

2.2 Efficient Self-attention-Based U-Net

To better leverage anatomical semantic information in MAC-Net and adapt to the high-resolution input, we design the ESAU-Net that can capture both local and global contexts during denoising. Different from previous works that only use self-attention in the coarsest level [20], we incorporate a channel-wise self-attention mechanism [2,28] at each up-sampling and down-sampling level in the U-Net [22] and add an identity mapping in each level, as shown in Fig. 1(a).

Specifically, in each level, given the feature map F_{l-1} as the input, we first apply a 1×1 convolution and a 3×3 depth-wise convolution to aggregate channel-wise contents and generate query (Q), key (K), and value (V) followed by a reshape operation, where $Q \in \mathbb{R}^{C\times HW}$, $K \in \mathbb{R}^{C\times HW}$, and $V \in \mathbb{R}^{C\times HW}$ (see Fig. 1(a)). Then, a channel-wise attention map $A \in \mathbb{R}^{C\times C}$ is generated through a dot-product operation by the reshaped query and key, which is more efficient than the regular attention map of size $HW \times HW$ [3], especially for high-resolution input. Overall, the process is defined as

$$\mathrm{Attention}(F) = w(V^T A) = w(V^T \cdot \mathrm{Softmax}(KQ^T/\alpha)), \tag{1}$$

where $w(\cdot)$ first reshapes the matrix back to the original size $C \times H \times W$ and then performs 1×1 convolution; α is a learnable parameter to scale the magnitude of the dot product of K and Q. We use multi-head attention similar to the standard multi-head self-attention mechanism [3]. The output of the channel-wise self-attention is represented as: $F'_{l-1} = \mathrm{Attention}(F_{l-1}) + F_{l-1}$. Finally, the output F_l of each level is defined as: $F_l = \mathrm{Conv}(F'_{l-1}) + \mathrm{Iden}(F_{l-1})$, where $\mathrm{Conv}(\cdot)$ is a two-layer convolution and $\mathrm{Iden}(\cdot)$ is an identity mapping using a 1×1 convolution; refer to Fig. S2(a) for the details of ESAU-Net.

2.3 Multi-scale Anatomical Contrastive Network

Overview of MAC-Net. The goal of our MAC-Net is to exploit anatomical semantics and maintain anatomical embedding consistency, First, a disentangled U-shaped architecture [22] is utilized to learn global representation $F_g \in \mathbb{R}^{512\times\frac{H}{16}\times\frac{W}{16}}$ after four down-sampling layers, and learn local representation $F_l \in \mathbb{R}^{64\times H\times W}$ by removing the last output layer. And we cut the connection between the coarsest feature and its upper level to make F_g and F_l more independent [25] (see Fig. S2(b)). The online network and the target network, both using the same architecture above, handle denoised CT Y' and NDCT Y, respectively, with F'_g and F'_l generated by the online network, and F_g and F_l generated by the target network (see Fig. 1(b)). The parameters of the target network are an exponential moving average of the parameters in the online network, following the previous works [7,8]. Next, a patch-wise non-contrastive module uses F_g and F'_g to compute a global MSE loss $\mathcal{L}_{\mathrm{global}}$, while a pixel-wise contrastive module uses F_l and F'_l to compute a local infoNCE loss $\mathcal{L}_{\mathrm{local}}$. Let us describe these two loss functions specifically.

Patch-Wise Non-contrastive Module. To better learn anatomical representations, we introduce a patch-wise non-contrastive module, also shown in Fig. 1(b). Specifically, for each pixel $\boldsymbol{f}_g^{(i)} \in \mathbb{R}^{512}$ in the $\boldsymbol{F}_g$ where $i \in \{1, 2, \ldots, \frac{HW}{256}\}$ is the index of the pixel location, it can be considered as a patch due to the expanded receptive field achieved through a sequence of convolutions and down-sampling operations [19]. To identify positive patch indices, we adopt a neighboring positive matching strategy [27], assuming that a semantically similar patch $\boldsymbol{f}_g^{(j)}$ exists in the vicinity of the query patch $\boldsymbol{f}_g^{(i)}$, as neighboring patches often share a semantic context with the query. We empirically consider a set of 8 neighboring patches. To sample patches with similar semantics around the query patch $\boldsymbol{f}_g^{(i)}$, we measure the semantic closeness between the query patch $\boldsymbol{f}_g^{(i)}$ and its neighboring patches $\boldsymbol{f}_g^{(j)}$ using the cosine similarity, which is formulated as

$$s(i, j) = (\boldsymbol{f}_g^{(i)})^\top (\boldsymbol{f}_g^{(j)}) / \|\boldsymbol{f}_g^{(i)}\|_2 \|\boldsymbol{f}_g^{(j)}\|_2. \tag{2}$$

We then select the top-4 positive patches $\{\boldsymbol{f}_g^{(j)}\}_{j \in \mathcal{P}^{(i)}}$ based on $s(i, j)$, where $\mathcal{P}^{(i)}$ is a set of selected patches (i.e., $|\mathcal{P}^{(i)}| = 4$). To obtain patch-level features $\boldsymbol{g}^{(i)} \in \mathbb{R}^{512}$ for each patch $\boldsymbol{f}_g^{(i)}$ and its positive neighbors, we aggregate their features using global average pooling (GAP) in the patch dimension. For the local representation of $\boldsymbol{f}_g'^{(i)}$, we select positive patches as same as $\mathcal{P}^{(i)}$, i.e., $\{\boldsymbol{f}_g'^{(j)}\}_{j \in \mathcal{P}^{(i)}}$. Formally,

$$\boldsymbol{g}^{(i)} := \mathrm{GAP}(\boldsymbol{f}_g^{(i)}, \{\boldsymbol{f}_g^{(j)}\}_{j \in \mathcal{P}^{(i)}}), \quad \boldsymbol{g}'^{(i)} := \mathrm{GAP}(\boldsymbol{f}_g'^{(i)}, \{\boldsymbol{f}_g'^{(j)}\}_{j \in \mathcal{P}^{(i)}}). \tag{3}$$

From the patch-level features, the online network outputs a projection $\boldsymbol{z}_g'^{(i)} = p_g'(\boldsymbol{g}'^{(i)})$ and a prediction $q(\boldsymbol{z}_g'^{(i)})$ while target network outputs the target projection $\boldsymbol{z}_g^{(i)} = p_g(\boldsymbol{g}^{(i)})$. The projection and prediction are both multi-layer perceptron (MLP). Finally, we compute the global MSE loss between the normalized prediction and target projection [7],

$$\mathcal{L}_{\mathrm{global}} = \sum_{i \in \mathcal{N}_{\mathrm{pos}}^g} \left\| q'(\boldsymbol{z}_g'^{(i)}) - \boldsymbol{z}_g^{(i)} \right\|_2^2 = \sum_{i \in \mathcal{N}_{\mathrm{pos}}^g} 2 - 2 \cdot \frac{\langle q'(\boldsymbol{z}_g'^{(i)}), \boldsymbol{z}_g^{(i)} \rangle}{\|q'(\boldsymbol{z}_g'^{(i)})\|_2 \cdot \|\boldsymbol{z}_g^{(i)}\|_2}, \tag{4}$$

where $\mathcal{N}_{\mathrm{pos}}^g$ is the indices set of positive samples in the patch-level embedding.

Pixel-Wise Contrastive Module. In this module, we aim to improve anatomical consistency between the denoised CT and NDCT using a local InfoNCE loss [18] (see Fig. 1(b)). First, for a query $\boldsymbol{f}_l'^{(i)} \in \mathbb{R}^{64}$ in the $\boldsymbol{F}_l'$ and its positive sample $\boldsymbol{f}_l^{(i)} \in \mathbb{R}^{64}$ in the $\boldsymbol{F}_l$ ($i \in \{1, 2, \ldots, HW\}$ is the location index), we use a hard negative sampling strategy [21] to select "difficult" negative samples with high probability, which enforces the model to learn more from the fine-grained details. Specifically, candidate negative samples are randomly sampled from $\boldsymbol{F}_l$ as long as their distance from $\boldsymbol{f}_l^{(i)}$ is less than m pixels ($m = 7$). We also use cosine similarity in Eq. (2) to select a set of semantically closest pixels, i.e. $\{\boldsymbol{f}_l^{(j)}\}_{j \in \mathcal{N}_{\mathrm{neg}}^{(i)}}$. Then we concatenate $\boldsymbol{f}_l'^{(i)}, \boldsymbol{f}_l^{(i)}$, and $\{\boldsymbol{f}_l^{(j)}\}_{j \in \mathcal{N}_{\mathrm{neg}}^{(i)}}$ and map them to a K-dimensional vector ($K=256$) through a two-layer MLP, obtaining $z_1^{(i)} \in \mathbb{R}^{(2+|\mathcal{N}_{\text{neg}}^{(i)}|)\times 256}$. The local InfoNCE loss in the pixel level is defined as

$$\mathcal{L}_{\text{local}} = -\sum_{i \in \mathcal{N}_{\text{pos}}^1} \log \frac{\exp\left(v'_1^{(i)} \cdot v_1^{(i)} / \tau\right)}{\exp\left(v'_1^{(i)} \cdot v_1^{(i)} / \tau\right) + \sum_{j=1}^{|\mathcal{N}_{\text{neg}}^{(i)}|} \exp\left(v'_1^{(i)} \cdot v_1^{(j)} / \tau\right)}, \tag{5}$$

where $\mathcal{N}_{\text{pos}}^1$ is the indices set of positive samples in the pixel level. $v'_1^{(i)}$, $v_1^{(i)}$, and $v_1^{(j)} \in \mathbb{R}^{256}$ are the query, positive, and negative sample in $z_1^{(i)}$, respectively.

2.4   Total Loss Function

The final loss is defined as $\mathcal{L} = \mathcal{L}_{\text{global}} + \mathcal{L}_{\text{local}} + \lambda \mathcal{L}_{\text{pixel}}$, where $\mathcal{L}_{\text{pixel}}$ consists of two common supervised losses: MSE and SSIM, defined as $\mathcal{L}_{\text{pixel}} = \mathcal{L}_{\text{MSE}} + \mathcal{L}_{\text{SSIM}}$. λ is empirically set to 10.

3   Experiments

3.1   Dataset and Implementation Details

We use two publicly available low-dose CT datasets released by the NIH AAPM-Mayo Clinic Low-Dose CT Grand Challenge in 2016 [15] and lately released in 2020 [16], denoted as Mayo-2016 and Mayo-2020, respectively. There is no overlap between the two datasets. Mayo-2016 includes normal-dose abdominal CT images of 10 anonymous patients and corresponding simulated quarter-dose CT images. Mayo-2020 provides the abdominal CT image data of 100 patients with 25% of the normal dose, and we randomly choose 20 patients for our experiments.

For the Mayo-2016, we choose 4800 pairs of 512×512 images from 8 patients for the training and 1136 pairs from the rest 2 patients as the test set. For the Mayo-2020, we employ 9600 image pairs with a size of 256×256 from randomly selected 16 patients for training and 580 pairs of 512×512 images from rest 4 patients for testing. The use of large-size training is to adapt our MAC-Net to exploit inherent semantic information. The default sampling hyper-parameters for Mayo-2016 are $|\mathcal{N}_{\text{pos}}^1| = 32$, $|\mathcal{N}_{\text{pos}}^g| = 512$, $|\mathcal{N}_{\text{neg}}^{(i)}| = 24$, while $|\mathcal{N}_{\text{pos}}^1| = 16$, $|\mathcal{N}_{\text{pos}}^g| = 256$, $|\mathcal{N}_{\text{neg}}^{(i)}| = 24$ for Mayo-2020. We use a binary function to filter the background while selecting queries in MAC-Net. For the training strategy, we employ a window of $[-1000, 2000]$ HU. We train our network for 100 epochs on 2 NVIDIA GeForce RTX 3090, and use the AdamW optimizer [14] with the momentum parameters $\beta_1 = 0.9$, $\beta_2 = 0.99$ and the weight decay of 1.0×10^{-9}. We initialize the learning rate as 1.0×10^{-4}, gradually reduced to 1.0×10^{-6} with the cosine annealing [13]. Since MAC-Net is only implemented during training, the testing time of ASCON is close to most of the compared methods.

3.2 Performance Comparisons

Quantitative Evaluations. We use three widely-used metrics including peak signal-to-noise ratio (PSNR), root-mean-square error (RMSE), and SSIM. Table 1 presents the testing results on Mayo-2016 and Mayo-2020 datasets. We compare our methods with 5 state-of-the-art methods, including RED-CNN [1], WGAN-VGG [26], EDCNN [12], DU-GAN [9], and CNCL [6]. Table 1 shows that our ESAU-Net with MAC-Net achieves the best performance on both the Mayo-2016 and the Mayo-2020 datasets. Compared to the ESAU-Net, ASCON further improves the PSNR by up to 0.54 dB on Mayo-2020, which demonstrates the effectiveness of the proposed MAC-Net and the importance of the inherent anatomical semantics during CT denoising. We also compute the contrast-to-noise ratio (CNR) to assess the detectability of a selected area of low-contrast lesion and our ASCON achieves the best CNR in Fig. S3.

Table 1. Performance comparison on the Mayo-2016 and Mayo-2020 datasets in terms of PSNR [dB], RMSE [$\times 10^{-2}$], and SSIM [%]

Methods	Mayo-2016			Mayo-2020		
	PSNR↑	RMSE↓	SSIM↑	PSNR↑	RMSE↓	SSIM↑
U-Net [22]	$44.13_{\pm1.19}$	$0.64_{\pm0.12}$	$97.38_{\pm1.09}$	$47.67_{\pm1.64}$	$0.43_{\pm0.09}$	$99.19_{\pm0.23}$
RED-CNN [1]	$44.23_{\pm1.26}$	$0.62_{\pm0.09}$	$97.34_{\pm0.86}$	$48.05_{\pm2.14}$	$0.41_{\pm0.11}$	$99.28_{\pm0.18}$
WGAN-VGG [26]	$42.49_{\pm1.28}$	$0.76_{\pm0.12}$	$96.16_{\pm1.30}$	$46.88_{\pm1.81}$	$0.46_{\pm0.10}$	$98.15_{\pm0.20}$
EDCNN [12]	$43.14_{\pm1.27}$	$0.70_{\pm0.11}$	$96.45_{\pm1.36}$	$47.90_{\pm1.27}$	$0.41_{\pm0.08}$	$99.14_{\pm0.17}$
DU-GAN [9]	$43.06_{\pm1.22}$	$0.71_{\pm0.10}$	$96.34_{\pm1.12}$	$47.21_{\pm1.52}$	$0.44_{\pm0.10}$	$99.00_{\pm0.21}$
CNCL [6]	$43.06_{\pm1.07}$	$0.71_{\pm0.10}$	$96.68_{\pm1.11}$	$45.63_{\pm1.34}$	$0.53_{\pm0.11}$	$98.92_{\pm0.59}$
ESAU-Net (ours)	$\underline{44.38}_{\pm1.26}$	$\underline{0.61}_{\pm0.09}$	$\underline{97.47}_{\pm0.87}$	$\underline{48.31}_{\pm1.87}$	$\underline{0.40}_{\pm0.12}$	$\underline{99.30}_{\pm0.18}$
ASCON (ours)	$\mathbf{44.48}_{\pm1.32}$	$\mathbf{0.60}_{\pm0.10}$	$\mathbf{97.49}_{\pm0.86}$	$\mathbf{48.84}_{\pm1.68}$	$\mathbf{0.37}_{\pm0.11}$	$\mathbf{99.32}_{\pm0.18}$

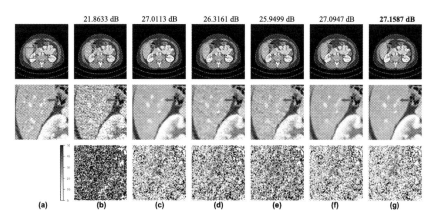

Fig. 2. Transverse CT images and corresponding difference images from the Mayo-2016 dataset: (a) NDCT; (b) LDCT; (c) RED-CNN [1]; (d) EDCNN [12]; (e) DU-GAN [9]; (f) ESAU-Net (ours); and (g) ASCON (ours). The display window is $[-160, 240]$ HU.

Qualitative Evaluations. Figure 2 presents qualitative results of three representative methods and our ESAU-Net with MAC-Net on Mayo-2016. Although ASCON and RED-CNN produce visually similar results in low-contrast areas after denoising. However, RED-CNN results in blurred edges between different tissues, such as the liver and blood vessels, while ASCON smoothed the noise and maintained the sharp edges. They are marked by arrows in the regions-of-interest images. We further visualize the corresponding difference images between NDCT and the generated images by our method as well as other methods as shown in the third row of Fig. 2. Note that our ASCON removes more noise components than other methods; refer to Fig. S4 for extra qualitative results on Mayo-2020.

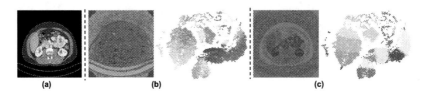

(a)   (b)   (c)

Fig. 3. Visualization of inherent semantics; (a) NDCT; (b) clustering and t-SNE results of ASCON w/o MAC-Net; and (c) clustering and t-SNE results of ASCON.

Table 2. Ablation results of Mayo-2020 on the different types of loss functions.

Loss	PSNR↑	RMSE↓	SSIM↑
$\mathcal{L}_{MSE}$	$48.34_{\pm 2.22}$	$0.40_{\pm 0.11}$	$99.27_{\pm 0.18}$
$\mathcal{L}_{MSE} + \mathcal{L}_{Perceptual}$	$47.83_{\pm 1.99}$	$0.42_{\pm 0.10}$	$99.13_{\pm 0.19}$
$\mathcal{L}_{MSE} + \mathcal{L}_{SSIM}$	$48.31_{\pm 1.87}$	$0.40_{\pm 0.12}$	$99.30_{\pm 0.18}$
$\mathcal{L}_{MSE} + \mathcal{L}_{SSIM} + \mathcal{L}_{global}$	$48.58_{\pm 2.12}$	$0.39_{\pm 0.10}$	$99.31_{\pm 0.17}$
$\mathcal{L}_{MSE} + \mathcal{L}_{SSIM} + \mathcal{L}_{local}$	$48.48_{\pm 2.37}$	$0.38_{\pm 0.11}$	$99.31_{\pm 0.18}$
$\mathcal{L}_{MSE} + \mathcal{L}_{SSIM} + \mathcal{L}_{local} + \mathcal{L}_{global}$	$\mathbf{48.84}_{\pm 1.68}$	$\mathbf{0.37}_{\pm 0.11}$	$\mathbf{99.32}_{\pm 0.18}$

Visualization of Inherent Semantics. To demonstrate that our MAC-Net can exploit inherent anatomical semantics of CT images during denoising, we select the features before the last layer in ASCON without MAC-Net and ASCON from Mayo-2016. Then we cluster these two feature maps respectively using a K-means algorithm and visualize them in the original dimension, and finally visualize the clustering representations using t-SNE, as shown in Fig. 3. Note that ASCON produces a result similar to organ semantic segmentation after clustering and the intra-class distribution is more compact, as well as the inter-class separation is more obvious. To the best of our knowledge, this is the first time that anatomical semantic information has been demonstrated in a CT denoising task, providing interpretability to the field of medical image reconstruction.

Ablation Studies. We start with a ESAU-Net using MSE loss and gradually insert some loss functions and our MAC-Net. Table 2 presents the results of different loss functions. It shows that both the global non-contrastive module and local contrastive module are helpful in obtaining better metrics due to the capacity of exploiting inherent anatomical information and maintaining anatomical consistency. Then, we add our MAC-Net to two supervised models: RED-CNN [1] and U-Net [22] but it is less effective, which demonstrates the importance of our ESAU-Net that captures both local and global contexts during denoising in Table S1. In addition, we evaluate the effectiveness of the training strategies including alternate learning, neighboring positive matching and hard negative sampling in Table S2.

4   Conclusion

In this paper, we explore the anatomical semantics in LDCT denoising and take advantage of it to improve the denoising performance. To this end, we propose an **A**natomy-aware **S**upervised **CON**trastive learning framework (ASCON), consisting of an efficient self-attention-based U-Net (ESAU-Net) and a multi-scale anatomical contrastive network (MAC-Net), which can capture both local and global contexts during denoising and exploit inherent anatomical information. Extensive experimental results on Mayo-2016 and Mayo-2020 datasets demonstrate the superior performance of our method, and the effectiveness of our designs. We also validated that our method introduces interpretability to LDCT denoising.

Acknowledgements. This work was supported in part by National Natural Science Foundation of China (No. 62101136), Shanghai Sailing Program (No. 21YF1402800), Shanghai Municipal Science and Technology Major Project (No. 2018SHZDZX01) and ZJLab, Shanghai Municipal of Science and Technology Project (No. 20JC1419500), and Shanghai Center for Brain Science and Brain-inspired Technology.

References

1. Chen, H., et al.: Low-dose CT with a residual encoder-decoder convolutional neural network. IEEE Trans. Med. Imaging **36**(12), 2524–2535 (2017)
2. Chen, Z., Niu, C., Wang, G., Shan, H.: LIT-Former: Linking in-plane and through-plane transformers for simultaneous CT image denoising and deblurring. arXiv preprint arXiv:2302.10630 (2023)
3. Dosovitskiy, A., et al.: An image is worth 16x16 words: Transformers for image recognition at scale. arXiv preprint arXiv:2010.11929 (2020)
4. Gao, Q., Li, Z., Zhang, J., Zhang, Y., Shan, H.: CoreDiff: Contextual error-modulated generalized diffusion model for low-dose CT denoising and generalization. arXiv preprint arXiv:2304.01814 (2023)
5. Gao, Q., Shan, H.: CoCoDiff: a contextual conditional diffusion model for low-dose CT image denoising. In: Developments in X-Ray Tomography XIV, vol. 12242. SPIE (2022)

6. Geng, M., et al.: Content-noise complementary learning for medical image denoising. IEEE Trans. Med. Imaging **41**(2), 407–419 (2021)
7. Grill, J.B., et al.: Bootstrap your own latent-a new approach to self-supervised learning. Proc. Adv. Neural Inf. Process. Syst. **33**, 21271–21284 (2020)
8. He, K., Fan, H., Wu, Y., Xie, S., Girshick, R.: Momentum contrast for unsupervised visual representation learning. In: Proceedings of the IEEE/CVF Conference on Computer Vision and Pattern Recognition, pp. 9729–9738 (2020)
9. Huang, Z., Zhang, J., Zhang, Y., Shan, H.: DU-GAN: generative adversarial networks with dual-domain U-Net-based discriminators for low-dose CT denoising. IEEE Trans. Instrum. Meas. **71**, 1–12 (2021)
10. Isola, P., Zhu, J.Y., Zhou, T., Efros, A.A.: Image-to-image translation with conditional adversarial networks. In: Proceedings of the IEEE Conference on Computer Vision and Pattern Recognition, pp. 1125–1134 (2017)
11. Johnson, J., Alahi, A., Fei-Fei, L.: Perceptual losses for real-time style transfer and super-resolution. In: Leibe, B., Matas, J., Sebe, N., Welling, M. (eds.) ECCV 2016. LNCS, vol. 9906, pp. 694–711. Springer, Cham (2016). https://doi.org/10.1007/978-3-319-46475-6_43
12. Liang, T., Jin, Y., Li, Y., Wang, T.: EDCNN: edge enhancement-based densely connected network with compound loss for low-dose CT denoising. In: 2020 15th IEEE International Conference on Signal Processing, vol. 1, pp. 193–198. IEEE (2020)
13. Loshchilov, I., Hutter, F.: SGDR: Stochastic gradient descent with warm restarts. arXiv preprint arXiv:1608.03983 (2016)
14. Loshchilov, I., Hutter, F.: Decoupled weight decay regularization. arXiv preprint arXiv:1711.05101 (2017)
15. McCollough, C.H., et al.: Low-dose CT for the detection and classification of metastatic liver lesions: results of the 2016 low dose CT grand challenge. Med. Phys. **44**(10), e339–e352 (2017)
16. Moen, T.R., et al.: Low-dose CT image and projection dataset. Med. Phys. **48**(2), 902–911 (2021)
17. Mussmann, B.R., et al.: Organ-based tube current modulation in chest CT. A comparison of three vendors. Radiography **27**(1), 1–7 (2021)
18. Oord, A.V.D., Li, Y., Vinyals, O.: Representation learning with contrastive predictive coding. arXiv preprint arXiv:1807.03748 (2018)
19. Park, T., Efros, A.A., Zhang, R., Zhu, J.-Y.: Contrastive learning for unpaired image-to-image translation. In: Vedaldi, A., Bischof, H., Brox, T., Frahm, J.-M. (eds.) ECCV 2020. LNCS, vol. 12354, pp. 319–345. Springer, Cham (2020). https://doi.org/10.1007/978-3-030-58545-7_19
20. Petit, O., Thome, N., Rambour, C., Themyr, L., Collins, T., Soler, L.: U-Net transformer: self and cross attention for medical image segmentation. In: Lian, C., Cao, X., Rekik, I., Xu, X., Yan, P. (eds.) MLMI 2021. LNCS, vol. 12966, pp. 267–276. Springer, Cham (2021). https://doi.org/10.1007/978-3-030-87589-3_28
21. Robinson, J., Chuang, C.Y., Sra, S., Jegelka, S.: Contrastive learning with hard negative samples. arXiv preprint arXiv:2010.04592 (2020)
22. Ronneberger, O., Fischer, P., Brox, T.: U-Net: convolutional networks for biomedical image segmentation. In: Navab, N., Hornegger, J., Wells, W.M., Frangi, A.F. (eds.) MICCAI 2015. LNCS, vol. 9351, pp. 234–241. Springer, Cham (2015). https://doi.org/10.1007/978-3-319-24574-4_28
23. Shan, H., et al.: Competitive performance of a modularized deep neural network compared to commercial algorithms for low-dose CT image reconstruction. Nat. Mach. Intell. **1**(6), 269–276 (2019)

24. Shan, H., et al.: 3-D convolutional encoder-decoder network for low-dose CT via transfer learning from a 2-D trained network. IEEE Trans. Med. Imaging **37**(6), 1522–1534 (2018)
25. Yan, K., et al.: SAM: self-supervised learning of pixel-wise anatomical embeddings in radiological images. IEEE Trans. Med. Imaging **41**(10), 2658–2669 (2022)
26. Yang, Q., et al.: Low-dose CT image denoising using a generative adversarial network with Wasserstein distance and perceptual loss. IEEE Trans. Med. Imaging **37**(6), 1348–1357 (2018)
27. Yun, S., Lee, H., Kim, J., Shin, J.: Patch-level representation learning for self-supervised vision transformers. In: Proceedings of the IEEE/CVF Conference on Computer Vision and Pattern Recognition, pp. 8354–8363 (2022)
28. Zamir, S.W., Arora, A., Khan, S., Hayat, M., Khan, F.S., Yang, M.H.: Restormer: Efficient transformer for high-resolution image restoration. In: Proceedings of the IEEE/CVF Conference on Computer Vision and Pattern Recognition, pp. 5728–5739 (2022)

Generating High-Resolution 3D CT with 12-Bit Depth Using a Diffusion Model with Adjacent Slice and Intensity Calibration Network

Jiheon Jeong[1], Ki Duk Kim[2], Yujin Nam[1], Kyungjin Cho[1], Jiseon Kang[2], Gil-Sun Hong[3], and Namkug Kim[2,3]([✉])

[1] Department of Biomedical Engineering, Asan Medical Institute of Convergence Science and Technology, University of Ulsan College of Medicine, Seoul 05505, South Korea
[2] Department of Convergence Medicine, Asan Medical Center, Seoul 05505, South Korea
namkugkim@gmail.com
[3] Department of Radiology, Asan Medical Center, Seoul 05505, South Korea

Abstract. Since the advent of generative models, deep learning-based methods for generating high-resolution, photorealistic 2D images have made significant successes. However, it is still difficult to create precise 3D image data with 12-bit depth used in clinical settings that capture the anatomy and pathology of CT and MRI scans. Using a score-based diffusion model, we propose a slice-based method that generates 3D images from previous 2D CT slices along the inferior direction. We call this method stochastic differential equations with adjacent slice-based conditional iterative inpainting (ASCII). We also propose an intensity calibration network (IC-Net) that adjusts the among slices intensity mismatch caused by 12-bit depth image generation. As a result, Frechet Inception Distance (FIDs) scores of FID-Ax, FID-Cor and FID-Sag of ASCII(2) with IC-Net were 14.993, 19.188 and 19.698, respectively. Anatomical continuity of the generated 3D image along the inferior direction was evaluated by an expert radiologist with more than 15 years of experience. In the analysis of eight anatomical structures, our method was evaluated to be continuous for seven of the structures.

Keywords: Score-based Diffusion Model · Adjacent Slice-based 3D Generation · Intensity Calibration Network · 12-bit Depth DICOM

1 Introduction

Unlike natural images that are typically processed in 8-bit depth, medical images, including X-ray, CT, and MR images, are processed in 12-bit or 16-bit depth to retain more detailed information. Among medical images, CT images are scaled using a quantitative measurement known as the Hounsfield unit (HU), which ranges from -1024 HU to 3071

J. Jeong and K. D. Kim—Contributed equally.

Supplementary Information The online version contains supplementary material available at https://doi.org/10.1007/978-3-031-43999-5_35.

HU in 12-bit depth. However, in both clinical practice and research, the dynamic range of HU is typically clipped to emphasize the region of interest (ROI). Such clipping of CT images, called windowing, can increase the signal-to-noise ratio (SNR) in the ROI. Therefore, most research on CT images performs windowing as a pre-processing method [1, 2].

Recent advancements in computational resources have enabled the development of 3D deep learning models such as 3D classification and 3D segmentation. 3D models have attracted much attention in the medical domain because they can utilize the 3D integrity of anatomy and pathology. However, the access to 3D medical imaging datasets is severely limited due to the patient privacy. The inaccessibility problem of 3D medical images can be addressed by generating high quality synthetic data. Some researches have shown that data insufficiency or data imbalance can be overcome using a well-trained generative model [3, 4]. However, generating images with intact 3D integrity is very difficult. Moreover, generating high-quality images [5] in the 12-bit depth, which is used in real clinical settings, is even more challenging.

The present study proposes a 2D-based 3D-volume generation method. To preserve the 3D integrity and transfer spatial information across adjacent axial slices, prior slices are utilized to generate each adjacent axial slice. We call this method Adjacent Slice-based Conditional Iterative Inpainting, ASCII. Experiments demonstrated that ASCII could generate 3D volumes with intact 3D integrity. Recently, score-based diffusion models have shown promising results in image generation [6–12], super resolution [13] and other tasks [14–17]. Therefore, ASCII employs a score-based diffusion model to generate images in 12-bit depth. However, since the images were generated in 12-bit depth, errors in the average intensity arose when the images were clipped to the brain parenchymal windowing range. To solve this issue, we propose a trainable intensity-calibration network (IC-Net) that matches the intensity of adjacent slices, which is trained in a self-supervised manner.

2 Related Works

Score-based generative models [7–9] and denoising diffusion probabilistic models (DDPMs) [6, 16] can generate high-fidelity data without an auxiliary network. In contrast, generative adversarial networks (GANs) [18] require a discriminator and variational auto-encoders (VAEs) [19] require a Gaussian encoder. Score-based generative models and diffusion models have two processing steps: a forward process that creates perturbed data with random noise taken from a pre-defined noise distribution in each step, and a backward process that denoises the perturbed data using a score network. The perturbation methods were defined as the stochastic differential equation (SDE) in [8]. A continuous process was defined as $\{x(t)\}_{t=0}^{T}$ with $x(0) \sim p_{data}$ and $x(T) \sim p_{T}$, where $t \in [0, 1]$ and p_{data}, p_{T} are the data distribution and prior noise distribution, respectively. The forward process was defined as the following SDE:

$$dx = f(x, t)dt + g(t)dw, \tag{1}$$

where f and g are the coefficients of the drift and diffusion terms in the SDE, respectively, and w induces the Wiener process (i.e., Brownian motion). The backward process was defined as the following reverse-SDE:

$$dx = \left[f(x,t) - g(t)^2 \nabla x log p_t(x)\right] dt + g(t) dw, \tag{2}$$

where w is the backward Wiener process. We define each variance σ_t as a monotonically increasing function. To solve the reverse-SDE given by above equation we train a score network $S_\theta(x, t)$ to estimate the score function of the perturbation kernel $\nabla x log p_t(x_t|x_0)$. Therefore, the objective of the score network is to minimize the following loss function:

$$\theta^* = \underset{\theta}{argmin} \int \lambda(t) \mathbb{E}_{x(0)} \mathbb{E}_{x(t)|x(0)} \|S_\theta(x(t), t) - \nabla x log p_t(x(t)|x(0))\|_2^2 dt, \tag{3}$$

where $\lambda(t)$ is a coefficient function depending on SDE. When the score network is trained using above equation, we approximate the score network $S_\theta(x, t)$ as $\nabla x log p_t(x_t|x_0)$. The model that generates using the predictor-corrector (PC) sampler, which alternately applies a numerical solver called predictor and Langevin MCMC called corrector, is called a *score-based diffusion model*. We also apply a perturbation kernel $p(x_t|x_0) = N(x_t; x_0, \sigma_t^2)$ and set the drift and diffusion coefficients of the SDE to $f = 0$ and $g = \sqrt{d\lceil \sigma_t^2 \rceil / dt}$, respectively. We call this variance the exploding SDE (VESDE).

3   Adjacent Slice-Based Conditional Iterative Inpainting, ASCII

3.1   Generating 12-Bit Whole Range CT Image

VESDEs experimented on four noise schedules and two GAN models [20–22] were compared for 12-bit whole range generation. As shown in Fig. 1, setting σ_{min} to 0.01 generated a noisy image (third and fifth column) whereas the generated image was well clarified when σ_{min} was reduced to 0.001 (forth and last column). However, in two GAN models, the important anatomical structures, such as white matter and grey matter, collapsed in the 12-bit generation. The coefficient of variation (CV), which is a measure used to compare variations while eliminating the influence of the mean, measured in the brain parenchyma of the generated images (excluding the bones and air from the windowing) at both noise levels [23]. It is observed that the anatomical structures tended to be more distinguishable when the CV was low. Furthermore, a board-certified radiologist also qualitatively assessed that the images generated with a lower σ_{min} showed a cleaner image. It can be interpreted that reducing σ_{min} lowers CV and therefore improves the image quality.

For quantitative comparison, we randomly generated 1,000 slices by each parameter and measured the CV. As shown Fig. 1, the variance of CV was lowest when σ_{min} and σ_{max} were set to 0.001 and 68, respectively. Also, setting σ_{max} to 1,348 is theoretically plausible according to previous study [9] because we preprocessed CT slices in the range of -1 to 1. Finally, we fixed σ_{min} and σ_{max} to 0.001 and 1,348, respectively in subsequent experiments.

In the case of GAN, as the convolution operator can be described as a high-pass filter [24], GANs trained through the discriminator's gradient are vulnerable to generating precise details in the low-frequency regions. Therefore, while anatomy such as bone (+1000HU) and air (−500HU) with strong contrast and high SNR are well-generated, anatomy such as parenchyma (20HU–30HU) with low contrast and SNR are difficult to generate accurately. As shown in the first and second column of Fig. 1, we can see that the GAN models seem to generate anatomy with strong contrast and high SNR well but failed to generate the others.

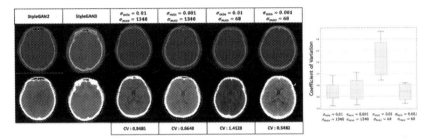

Fig. 1. Qualitative and quantitative generation results according to σ_{min} and σ_{max}

The SDE's scheduling can impact how the score-based diffusion model generates the fine-grained regions of images. To generate high-quality 12-bit images, the diffusion coefficient must be set to distinguish 1HU (0.00049). By setting σ_{max} to 1,348 and σ_{min} to 0.001, the final diffusion coefficient (0.00017) was set to a value lower than 1HU. In other words, setting the diffusion coefficient (0.00155) to a value greater than 1HU can generate noisy images. The detailed description and calculation of diffusion coefficient was in **Supplementary Material.**

3.2 Adjacent Slice-Based Conditional Iterative Inpainting

To generate a 3D volumetric image in a 2D slice-wise manner, a binary mask was used, which was moved along the channel axis. A slice $x^0 = \{-1024HU\}^D$ filled with intensity of air was padded before the first slice x^1 of CT and used as the initial seed. Then, the input of the model was given by $\left[x^t : x^{t+K-1}\right]$, where $t \in [0, N_s - K + 1]$ and N_s and K are the total slice number of CT and the number of contiguous slices, respectively. In addition, we omit augmentation because the model itself might generate augmented images. After training, the first slice was generated through a diffusion process using the initial seed. The generated first slice was then used as a seed to generate the next slice in an autoregressive manner. Subsequently, the next slices were generated through the same process. We call this method *adjacent slice-based conditional iterative inpainting, ASCII.*

Two experiments were conducted. The first experiment preprocessed the CT slices by clipping it with brain windowing [−10HU, 70HU] and normalizing it to [−1, 1] with σ_{min} set to 0.01 and σ_{max} set to 1,348, while the second experiment preprocessed the whole windowing [−1024HU, 3071HU] and normalized it to [−1, 1] with σ_{min} set to 0.001 and σ_{max} set to 1,348.

ASCII with windowing range   ASCII with whole range

Fig. 2. Results of ASCII with windowing range (red) and 12-bit whole range (blue)

As shown Fig. 2, the white matter and gray matter in the first experiment (brain windowing) could be clearly distinguished with maintained continuity of slices, whereas they were indistinguishable and remained uncalibrated among axial slices in the second experiment (whole range).

3.3 Intensity Calibration Network (IC-Net)

It was noted that the intensity mismatch problem only occurs in whole range generation. To address this issue, we first tried a conventional non-trainable post-processing, such as histogram matching. However, since each slice has different anatomical structure, the histogram of each slice image was fitted to their subtle anatomical variation. Therefore, anatomical regions were collapsed when the intensities of each slice of 3D CT were calibrated using histogram matching. Finally, we propose a solution for this intensity mismatching, a trainable intensity calibration network: IC-Net.

To calibrate the intensity mismatch, we trained the network with a self-supervised manner. First, adjacent two slices from real CT images, x^t, x^{t+1} were clipped using the window of which every brain anatomy HU value can be contained. Second, the intensity of x^{t+1} in ROI is randomly changed and the result is $\hat{x}_c^{t+1} = \left(x^{t+1} - \overline{x^{t+1}} \right) * \mu + \overline{x^{t+1}}$, where $\overline{x^{t+1}}$ and μ are the mean of x^{t+1} and shifting coefficient, respectively. And μ was configured to prevent the collapse of anatomical structures.

Finally, *intensity calibration network,* IC-Net was trained to calibrate the intensity of x^{t+1} to the intensity of x^t. The objective of IC-Net was only to calibrate the intensity of x^t and preserve both the subtle texture and the shape of a generated slice. The IC-Net uses the prior slice to calibrate the intensity of generated slice. The objective function of IC-Net is given by,

$$\mathcal{L}_{IC} = \mathbb{E}_t \mathbb{E}_{\mu \sim \mathcal{U}[-0.7, 1.3]} \left[\left| ICNet\left(\hat{x}_\mu^{t+1}, x^t \right) - x^{t+1} \right| \right] \tag{4}$$

As shown in Fig. 3, some important anatomical structures, such as midbrain, pons, medulla oblongata, and cerebellar areas, are blurred and collapsed when histogram matching was used. It can be risky as the outcomes vary depending on the matching seed image. On the other hand, the anatomical structure of the IC-Net matched images did not collapse. Also, the IC-Net does not require to set the matching seed image because it normalizes using the prior adjacent slice.

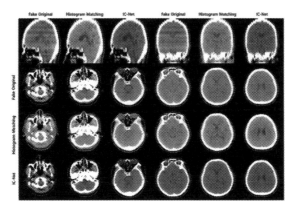

Fig. 3. Post-processing results of ASCII(2) in windowed range of whole range generation.

4 Experiments

4.1 ASCII with IC-Net in 12-Bit Whole Range CT Generation

The description of the dataset and model architecture is available in the **Supplementary Material**. We experimented ASCII on continuous K slices of K = 2 and 3 and called them ASCII(2) and ASCII(3), respectively. We generated a head & neck CT images via ASCII(2) and ASCII(3) with and without IC-Net, and slice-to-3D VAE [25]. Figure 4 demonstrate the example qualitative images. The 3D generated images were shown both in whole range and brain windowing range. The results showed that the both ASCII(2) and ASCII(3) were well calibrated using IC-Net. Also, anatomical continuity and the 3D integrity is preserved while the images were diverse enough. However, there was no significant visual difference between ASCII(2) and ASCII(3). Although the results in whole range appear to be correctly generated all models, the results in brain windowing range showed the differences. The same drawback of convolution operation addressed in the 12-bit generation of GAN based models, which was shown in Fig. 1, was also shown in slice-to-3D VAE.

The quantitative results in whole range are shown in Table 1. The mid-axial slice, mid-sagittal slice, and mid-coronal slice of the generated volumes were used to evaluate the Fréchet Inception Distance (FID) score, which we designated as FID-Ax, FID-Sag, and FID-Cor, respectively. And multi-scales structural similarity index measure (MS-SSIM) and batch-wise squared Maximum Mean Discrepancy (bMMD^2) were also evaluated for quantitative metrics. In general, quantitative results indicate that ASCII(2) performs better than ASCII(3). It's possible that ASCII(3) provides too much information from prior slices, preventing it from generating sufficiently diverse images. Additionally, IC-Net significantly improved generation performance, especially in the windowing range. The FID-Ax of ASCIIs was improved by IC-Net from 15.250 to 14.993 and 18.127 to 16.599 in the whole range, respectively. Also, the performance of FID-Cor and FID-Sag had significantly improved when IC-Net was used. The MS-SSIM showed that ASCIIs can generated it diverse enough.

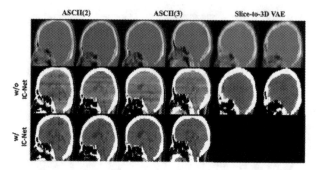

Fig. 4. Qualitative comparison among ASCII(2) and ASCII(3) with/without IC-Net calibration and Slice-to-3D VAE models.

The FID-Ax, FID-Cor, and FID-Sag scores of ASCIIs with IC-Net were improved in windowing range. The FID-Ax of ASCIIs was improved by IC-Net from 15.770 to 14.656 and 20.145 to 15.232 in the windowing range, respectively. On the other hand, ASCIIs without IC-Net had poor performance in the windowing range and this means that even when IC-Net is used, structures do not collapse.

Table 1. Quantitative comparison of ASCII(2) and ASCII(3) with/without IC-Net calibration and Slice-to-3D VAE in whole range and windowing range. Whole range and windowing range are set to [−1024HU, 3071HU] and [−10HU, 70HU], respectively.

	ASCII(2) w/ IC-Net	ASCII(2) w/o IC-Net	ASCII(3) w/ IC-Net	ASCII(3) w/o IC-Net	Slice-to-3D VAE [25]
Whole Range					
FID-Ax	**14.993**	15.250	16.599	18.127	29.137
FID-Cor	19.188	**19.158**	20.930	21.224	28.263
FID-Sag	19.698	**19.631**	21.991	22.311	29.024
MS-SSIM	**0.6271**	0.6275	0.6407	0.6406	0.9058
bMMD2	425704	429120	428045	432665	**311080**
Windowing Range					
FID-Ax	**14.656**	15.770	15.232	20.145	28.682
FID-Cor	**18.920**	19.830	19.996	24.230	28.828
FID-Sag	**18.569**	19.675	19.840	24.511	29.912
MS-SSIM	**0.5287**	0.5384	0.5480	0.5447	0.8609
bMMD^2	1975336	**1854921**	2044218	1858850	1894911

4.2   Calibration Robustness of IC-Net on Fixed Value Image Shift

To demonstrate the performance of IC-Net, we conducted experiments with the 7, 14, 21 and 28th slices, which sufficiently contain complex structures to show the calibration performances. The previous slice was used as an input to the IC-Net along with the target slice whose pixel values were to be shifted. And the absolute errors were measured between GT and predicted slice using IC-Net.

As shown in Fig. 5, it worked well for most shifting coefficients. The mean absolute error was measured from 1HU to 2HU when the shifting coefficient was set from 0.7 to 1.1. However, the errors were exploded when shifting coefficient was set to 1.2 or 1.3. It was because the images were collapsed when shifting coefficient increases than 1.2 since the intensity deviates from the ROI range $[-150HU, 150HU]$. Nevertheless, IC-Net can calibrate intensity to some extent even in the collapsed images as shown Fig. 5.

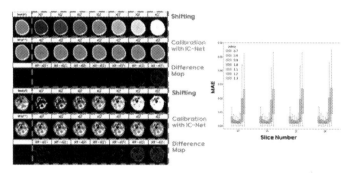

Fig. 5. (Left) IC-Net calibration results for fixed value image shift. Shifted images, IC-Net calibration result, and difference map are presented, respectively. Images were shifted with fixed value from 0.7 to 1.3. (Right) IC-Net calibration results for fixed value image shift. All slices were normalized from $[-150HU, 150HU]$ to $[-1, 1]$.

4.3   Visual Scoring Results of ASCII with IC-Net

Due to the limitations of slice-based methods in maintaining connectivity and 3D integrity, an experienced radiologist with more than 15 years of experience evaluated the images. Seeded with the 13th slices of real CT scans, in which the ventricle appears, ASCII(2) with IC-Net generated a total of 15 slices. Visual scoring shown in Table. 2 on a three-point scale was conducted blindly for 50 real and 50 fake CT scans, focusing on the continuity of eight anatomical structures. Although most of the fake regions were scored similarly to the real ones, the basilar arteries were evaluated with broken continuity. The basilar artery was frequently not generated, because it is a small region. As the model was trained on 5-mm thickness non-contrasted enhanced CT scans, preserving the continuity of the basilar artery is excessively demanding.

Table 2. Visual scoring results of integrity evaluation

Anatomy	Real	Fake
Skull (bone morphology, suture line)	3.00	2.98
Skull base (foramina and fissure)	3.00	2.84
Facial bone	3.00	2.98
Ventricles	3.00	2.92
Brain sulci and fissure	3.00	2.98
Basilar artery	2.92	1.38
Cerebra venous sinus	3.00	3.00
Ascending & descending nerve tract through internal capsule	3.00	3.00

Note: Scale 1 – discontinuity, Scale 2 – strained continuity, Scale 3 – well preserved continuity

5 Conclusion

We proposed a high-performance slice-based 3D generation method (ASCII) and combined it with IC-Net, which is trained in a self-supervised manner without any annotations. In our method, ASCII generates a 3D volume by iterative generation using previous slices and automatically calibrates the intensity mismatch between the previous and next slices using IC-Net. This pipeline is designed to generate high-quality medical image, while preserving 3D integrity and overcoming intensity mismatch caused in 12-bit generation.

ASCII had shown promising results in 12-bit depth whole range and windowing range, which are crucial in medical contexts. The integrity of the generated images was also confirmed in qualitative and quantitative assessment of 3D integrity evaluations by an expert radiologist. Therefore, ASCII can be used in clinical practice, such as anomaly detection in normal images generated from a seed image [26]. In addition, the realistic 3D images generated by ASCII can be used to train deep learning models [3, 4] in medical images, which frequently suffer from data scarcity.

References

1. Gerard, S.E., et al.: CT image segmentation for inflamed and fibrotic lungs using a multi-resolution convolutional neural network. Sci. Rep. **11**(1), 1–12 (2021)
2. Lassau, N., et al.: Integrating deep learning CT-scan model, biological and clinical variables to predict severity of COVID-19 patients. Nat. Commun. **12**(1), 1–11 (2021)
3. Frid-Adar, M., et al.: GAN-based synthetic medical image augmentation for increased CNN performance in liver lesion classification. Neurocomputing **321**, 321–331 (2018)
4. Bowles, C., et al.: Gan augmentation: augmenting training data using generative adversarial networks. arXiv preprint arXiv:1810.10863 (2018)
5. Hong, S., et al.: 3d-stylegan: a style-based generative adversarial network for generative modeling of three-dimensional medical images. In: Deep Generative Models, and Data Augmentation, Labelling, and Imperfections, pp. 24–34. Springer (2021)

6. Ho, J., Jain, A., Abbeel, P.: Denoising diffusion probabilistic models. Adv. Neural. Inf. Process. Syst. **33**, 6840–6851 (2020)
7. Song, Y., Ermon, S.: Generative modeling by estimating gradients of the data distribution. Adv. Neural Inform. Process. Syst. **32** (2019)
8. Song, Y., et al.: Score-based generative modeling through stochastic differential equations. arXiv preprint arXiv:2011.13456 (2020)
9. Song, Y., Ermon, S.: Improved techniques for training score-based generative models. Adv. Neural. Inf. Process. Syst. **33**, 12438–12448 (2020)
10. Meng, C., et al.: Sdedit: Image synthesis and editing with stochastic differential equations. arXiv preprint arXiv:2108.01073 (2021)
11. Nichol, A.Q., Dhariwal, P.: Improved denoising diffusion probabilistic models. In: International Conference on Machine Learning. PMLR (2021)
12. Hyvärinen, A., Dayan, P.: Estimation of non-normalized statistical models by score matching. J. Mach. Learn. Res. **6**(4) (2005)
13. Saharia, C., et al.: Image super-resolution via iterative refinement. arXiv preprint arXiv:2104.07636 (2021)
14. Kong, Z., et al.: Diffwave: a versatile diffusion model for audio synthesis. arXiv preprint arXiv:2009.09761 (2020)
15. Chen, N., et al.: WaveGrad: estimating gradients for waveform generation. arXiv preprint arXiv:2009.00713 (2020)
16. Luo, S., Hu, W.: Diffusion probabilistic models for 3d point cloud generation. In: Proceedings of the IEEE/CVF Conference on Computer Vision and Pattern Recognition (2021)
17. Mittal, G., et al.: Symbolic music generation with diffusion models. arXiv preprint arXiv:2103.16091 (2021)
18. Goodfellow, I., et al.: Generative adversarial nets. Adv. Neural Inform. Process. Syst. **27** (2014)
19. Kingma, D.P., Welling, M.: Auto-encoding variational bayes. arXiv preprint arXiv:1312.6114 (2013)
20. Karras, T., et al.: Analyzing and improving the image quality of stylegan. In: Proceedings of the IEEE/CVF Conference on Computer Vision and Pattern Recognition (2020)
21. Karras, T., et al.: Training generative adversarial networks with limited data. Adv. Neural. Inf. Process. Syst. **33**, 12104–12114 (2020)
22. Karras, T., et al.: Alias-free generative adversarial networks. Adv. Neural. Inf. Process. Syst. **34**, 852–863 (2021)
23. Brunel, N., Hansel, D.: How noise affects the synchronization properties of recurrent networks of inhibitory neurons. Neural Comput. **18**(5), 1066–1110 (2006)
24. Park, N., Kim, S.: How Do Vision Transformers Work? arXiv preprint arXiv:2202.06709 (2022)
25. Volokitin, A., Erdil, Ertunc, Karani, Neerav, Tezcan, Kerem Can, Chen, Xiaoran, Van Gool, Luc, Konukoglu, Ender: Modelling the distribution of 3D brain MRI using a 2D slice VAE. In: Martel, A.L., Abolmaesumi, Purang, Stoyanov, Danail, Mateus, Diana, Zuluaga, Maria A., Kevin Zhou, S., Racoceanu, Daniel, Joskowicz, Leo (eds.) Medical Image Computing and Computer Assisted Intervention – MICCAI 2020: 23rd International Conference, Lima, Peru, October 4–8, 2020, Proceedings, Part VII, pp. 657–666. Springer International Publishing, Cham (2020). https://doi.org/10.1007/978-3-030-59728-3_64
26. Schlegl, T., et al.: f-AnoGAN: Fast unsupervised anomaly detection with generative adversarial networks. Med. Image Anal. **54**, 30–44 (2019)

3D Teeth Reconstruction from Panoramic Radiographs Using Neural Implicit Functions

Sihwa Park[1], Seongjun Kim[1], In-Seok Song[2], and Seung Jun Baek[1(✉)]

[1] Korea University, Seoul, South Korea
{sihwapark,iamsjune,sjbaek,densis}@korea.ac.kr
[2] Korea University Anam Hospital, Seoul, South Korea

Abstract. Panoramic radiography is a widely used imaging modality in dental practice and research. However, it only provides flattened 2D images, which limits the detailed assessment of dental structures. In this paper, we propose Occudent, a framework for 3D teeth reconstruction from panoramic radiographs using neural implicit functions, which, to the best of our knowledge, is the first work to do so. For a given point in 3D space, the implicit function estimates whether the point is occupied by a tooth, and thus implicitly determines the boundaries of 3D tooth shapes. Firstly, Occudent applies multi-label segmentation to the input panoramic radiograph. Next, tooth shape embeddings as well as tooth class embeddings are generated from the segmentation outputs, which are fed to the reconstruction network. A novel module called Conditional eXcitation (CX) is proposed in order to effectively incorporate the combined shape and class embeddings into the implicit function. The performance of Occudent is evaluated using both quantitative and qualitative measures. Importantly, Occudent is trained and validated with actual panoramic radiographs as input, distinct from recent works which used synthesized images. Experiments demonstrate the superiority of Occudent over state-of-the-art methods.

Keywords: Panoramic radiographs · 3D reconstruction · Teeth segmentation · Neural implicit function

1 Introduction

Panoramic radiography (panoramic X-ray, or PX) is a commonly used technique for dental examination and diagnosis. While PX produces 2D images from panoramic scanning, Cone-Beam Computed Tomography (CBCT) is an alternative imaging modality which provides 3D information on dental, oral, and maxillofacial structures. Despite providing more comprehensive information than

Supplementary Information The online version contains supplementary material available at https://doi.org/10.1007/978-3-031-43999-5_36.

PX, CBCT is more expensive and exposes patients to a greater dose of radiation [3]. Thus, 3D teeth reconstruction from PX is of significant value, e.g., 3D visualization can aid clinicians with dental diagnosis and treatment planning. Other applications include treatment simulation and interactive virtual reality for dental education [12].

Previous 3D teeth reconstruction methods from 2D PX have relied on additional information such as tooth landmarks or tooth crown photographs. For example, [14] developed a model which uses landmarks on PX images to estimate 3D parametric models for tooth shapes, while [1] reconstructed a single tooth using a shape prior and reflectance model based on the corresponding crown photograph. Recent advances in deep neural networks have significantly impacted research on 3D teeth reconstruction. X2Teeth [13] performs 3D reconstruction of the entire set of teeth from PX based on 2D segmentation using convolutional neural networks. Oral-3D [22] generated 3D oral structures without supervised segmentation from PX using a GAN model [8]. Yet, those methods relied on synthesized images as input instead of real-world PX images, where the synthesized images are obtained from 2D projections of CBCT [27]. The 2D segmentation of teeth from PX is useful for 3D reconstruction in order to identify and isolate teeth individually. Prior studies on 2D teeth segmentation [11,28] focused on binary segmentation determining the presence of teeth. However, this information alone is insufficient for the construction of individual teeth. Instead, we leverage recent frameworks [17,21] on multi-label segmentation of PX into 32 classes including wisdom teeth.

In this paper, we propose *Occudent*, an end-to-end model to reconstruct 3D teeth from 2D PX images. Occudent consists of a multi-label 2D segmentation followed by 3D teeth reconstruction using *neural implicit functions* [15]. The function aims to learn the **occu**pancy of **dent**al structures, i.e., whether a point in space lies within the boundaries of 3D tooth shapes. Learning implicit functions is computationally advantageous over conventional encoder-decoder models outputting explicit 3D representations such as voxels, e.g., implicit models do not require large memory footprints to store and process voxels. Considering that 3D tooth shapes are characterized by tooth classes, we generate embeddings for tooth classes as well as segmented 2D tooth shapes. The combined class and shape embeddings are infused into the reconstruction network by a novel module called Conditional eXcitation (CX). CX performs learnable scaling of occupancy features conditioned on tooth class and shape embeddings. The performance of Occudent is evaluated with actual PX as input images, which differs from recent works using synthesized PX images [13,22]. Experiments show Occudent outperforms state-of-the-art baselines both quantitatively and qualitatively. The main contributions are summarized as follows: (1) the first use of a neural implicit function for 3D teeth reconstruction, (2) novel strategies to inject tooth class and shape information into implicit functions, (3) the superiority over existing baselines which is demonstrated with real-world PX images.

2   Methods

The proposed model, Occudent, consists of two main components: 2D teeth segmentation and 3D teeth reconstruction. The former performs the segmentation of 32 teeth from PX using UNet++ model [29]. The individually segmented tooth and the tooth class are subsequently passed to the latter for the reconstruction. The reconstruction process estimates the 3D representation of the tooth based on a neural implicit function. The overall architecture of Occudent is depicted in Fig. 1.

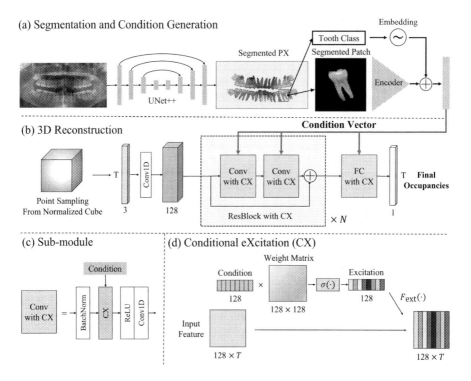

Fig. 1. (a) The PX image is segmented into 32 teeth classes using UNet++ model. For each tooth, a segmented patch is generated by cropping input PX with the predicted segmentation mask, which is subsequently encoded via an image encoder. The tooth class is encoded by an embedding layer. The patch and class embeddings are added together to produce the condition vector. (b) The reconstruction process consists of N ResBlocks to compute occupancy features of points sampled from 3D space. The condition vector from PX is processed by a Conditional eXcitation (CX) module incorporated in the ResBlocks. (c) The Conv with CX sub-module is composed of batch normalization, CX, ReLU, and a convolutional layer. The FC with CX is similar to the Conv with CX where Conv1D layer is replaced by fully connected layer. (d) CX injects condition information into the reconstruction network using excitation values. CX uses a trainable weight matrix to encode the condition vector into an excitation vector via a gating function. The input feature is scaled using the excitation vector through component-wise multiplication.

2.1   2D Teeth Segmentation

The teeth in input PX are segmented into 32 teeth classes. The 32 classes correspond to the traditional numbering of teeth, which includes incisors, canines, premolars and molars in both upper and lower jaws. We pose 2D teeth segmentation as a *multi-label segmentation* problem [17], since nearby teeth can overlap with each other in the PX image, i.e., a single pixel of the input image can be classified into two or more classes.

The input of the model is $H \times W$ size PX image. The segmentation output has dimension $C \times H \times W$, where channel dimension $C = 33$ represents the number of tooth classes: one class for the background and 32 classes for teeth similar to [17]. Hence, the $H \times W$ output at each channel is a segmentation output for each tooth class. The segmentation outputs are used to generate tooth patches for reconstruction, which is explained later in detail. For the segmentation, we adopt pre-trained UNet++ [29] as the base model. UNet++ is advantageous for medical image segmentation due to its modified skip pathways, which results in better performance compared to the vanilla UNet [20].

2.2   3D Teeth Reconstruction

Neural Implicit Representation. Typical representations of 3D shapes are point-based [7,19], voxel-based [4,26], or mesh-based methods [25]. These methods represent 3D shapes explicitly through a set of discrete points, vertices, and faces. Recently, implicit representation methods based on a continuous function which defines the boundary of 3D shapes have become increasingly popular [15,16,18]. Occupancy Networks [15] is a pioneering work which utilizes neural networks to approximate the implicit function of an object's occupancy. The term occupancy refers to whether a point in space lies in the interior or exterior of object boundaries. The occupancy function maps a 3D point to either 0 or 1, indicating the occupancy of the point. Let o_A denote the occupancy function for an object A as follows:

$$o_A : \mathbb{R}^3 \rightarrow \{0, 1\} \tag{1}$$

In practice, o_A can be estimated only by a set of observations of object A, denoted by $\mathcal{X}_A$. Examples of observations are projected images or point cloud data obtained from the object. Our objective is to estimate the occupancy function conditioned on $\mathcal{X}_A$. Specifically, we would like to find function f_θ which estimates the occupancy probability of a point in 3D space based on $\mathcal{X}_A$ [15]:

$$f_\theta : \mathbb{R}^3 \times \mathcal{X}_A \rightarrow [0, 1] \tag{2}$$

Inspired by the aforementioned framework, we leverage segmented tooth patch and tooth class as observations denoted by condition vector c. Specifically, the input to the function is a set of T randomly sampled locations within a unit cube, and the function outputs the occupancy probability of the input. Thus, the function is given by $f_\theta : (x, y, z, c) \rightarrow [0, 1]$.

The model for f_θ is depicted in Fig. 1 (b). The sampled locations are projected to 128 dimensional feature vectors using 1D convolution. Next, the features are processed by a sequence of ResNet blocks followed by FC (fully connected) layers. Conditional vector c is used for each block through Conditional eXcitation (CX) which we will explain later.

Class-Specific Conditional Features. A distinctive feature of the tooth reconstruction task is that teeth with the same number share properties such as surface and root shapes. Hence, we propose to use tooth class information in combination with a segmented tooth patch from PX. The tooth class is processed by a learnable embedding layer which outputs a class embedding vector.

Next, we create a square patch of the tooth using the segmentation output as follows. A binary mask of the segmented tooth is generated by applying thresholding to the segmentation output. A tooth patch is created by cropping out the tooth region from the input PX, i.e., the binary mask is applied (bitwise AND) to the input PX to obtain the patch. The segmented tooth patch is subsequently encoded using a pre-trained ResNet18 model [9], which outputs a patch embedding vector. The patch and class embeddings are added to yield the condition vector for the reconstruction model. This process is depicted in Fig. 1 (a).

Our approach differs from previous approaches, such as Occupancy Networks [15] which uses only single-view images for 3D reconstruction. X2Teeth [13] also addresses the task of 3D teeth reconstruction from 2D PX. However, X2Teeth only uses segmented image features for the reconstruction. By contrast, Occudent leverages a class-specific encoding method to boost the reconstruction performance, as demonstrated in ablation analysis in Supplementary Materials.

Conditional eXcitation. To effectively inject 2D observations into the reconstruction network, we propose Conditional eXcitation (CX) inspired by Squeeze-and-Excitation Network (SENet) [10]. In SENet, excitation refers to scaling input features according to their importance. In Occudent, the concept of excitation is extended to incorporating conditional features into the network. Firstly, the condition vector is encoded into excitation vector e. Next, the excitation is applied to input feature by scaling the feature components by e. The CX procedure can be expressed as:

$$e = \alpha \cdot \sigma(Wc), \tag{3}$$

$$y = F_{\text{ext}}(e, x) \tag{4}$$

where c is the condition vector, σ is a gating function, W is a learnable weight matrix, α is a hyperparameter for the excitation result, and F_{ext} is the excitation function. We use sigmoid function for σ, and component-wise multiplication for the excitation, $F_{\text{ext}}(e, x) = e \otimes x$. The CX module is depicted in Fig. 1 (d). CX differs from SENet in that CX derives the excitation from the condition vector, whereas SENet derives it from input features. Our approach also differs from Occupancy Networks which used Conditional Batch Normalization (CBN) [5,6] which combines conditioning with batch normalization. However, the conditioning process should be independent of input batches because those components serve different purposes in deep learning models. Thus, we propose to separate conditioning from batch normalization, as is done by CX.

3 Experiments

Dataset. The pre-training of the segmentation model was done with a dataset of 4000 PX images, sourced from 'The Open AI Dataset Project (AI-Hub, S. Korea)'. All data information can be accessed through 'AI-Hub (www.aihub.or.kr)'. The dataset consisted of two image sizes, 1976×976 and 2988×1468, which were resized to 256×768 to train the UNet++ model.

For the main experiments for reconstruction, we used a set of 39 PX images and matched CBCT images, obtained from Korea University Anam Hospital. This study was approved by the Institutional Review Board at Korea University (IRB number: 2020AN0410). The panoramic radiographs were of dimensions 1536×2860 and were resized to 600×1200 and randomly cropped of 592×1184 size for the segmentation training. The CBCT images were of size $768 \times 768 \times 576$, capturing cranial bones. The teeth labels for 2D PX and CBCT were manually annotated by two experienced annotators and subsequently verified by a board-certified dentist. To train and evaluate the model, the dataset was partitioned into training (30 cases), validation (2 cases), and testing (7 cases) subsets.

Implementation Details. For the pre-training of the segmentation model, we utilized a combination of cross-entropy and dice loss. For the main segmentation training, we used only dice loss. The segmentation and reconstruction models were trained separately. Following the completion of the segmentation model training, we fixed this model to predict its output for the reconstruction model.

Each 3D tooth label was fit in $144 \times 80 \times 80$ size tensor which was then regarded as $[-0.5, 0.5]^3$ normalized cube in 3D space. For the training of the neural implicit function, a set of $T = 100,000$ points was sampled from the unit cube. The preprocessing was consistent with that used in [23]. We trained all the other baseline models with these normalized cubes. For example, for 3D-R2N2 [4], we voxelized the cube to 128^3 size. For a fair comparison, the final meshes produced by each model were extracted and compared using four different metrics. The detailed configuration of our model is provided in Supplementary Materials.

Baselines. We considered several state-of-the-art models as baselines, including 3D-R2N2 [4], DeepRetrieval [13,24], Pix2Vox [26], PSGN [7], Occupancy Networks (OccNet) [15], and X2Teeth [13]. To adapt the 3D-R2N2 model to single-view reconstruction, we removed its LSTM component, following the approach in [15]. As for the DeepRetrieval method, we employed the same encoder architecture as 3D-R2N2, and utilized the encoded feature vector of the test image

Table 1. Comparison with baseline methods. The format of results is $mean \pm std$ obtained from 10 repetitions of experiments.

Method	IoU	Chamfer-L_1	NC	Precision
3D-R2N2	0.585 ± 0.005	0.382 ± 0.008	0.617 ± 0.009	0.634 ± 0.010
PSGN	0.606 ± 0.015	0.342 ± 0.016	0.829 ± 0.012	0.737 ± 0.018
Pix2Vox	0.562 ± 0.005	0.388 ± 0.008	0.599 ± 0.007	0.664 ± 0.009
DeepRetrieval	0.564 ± 0.005	0.394 ± 0.006	0.824 ± 0.003	0.696 ± 0.005
X2Teeth	0.592 ± 0.006	0.361 ± 0.017	0.618 ± 0.002	0.670 ± 0.009
OccNet	0.611 ± 0.006	0.353 ± 0.008	0.872 ± 0.003	0.691 ± 0.011
Occudent (Ours)	$\mathbf{0.651 \pm 0.004}$	$\mathbf{0.298 \pm 0.006}$	$\mathbf{0.890 \pm 0.001}$	$\mathbf{0.739 \pm 0.008}$

query. Subsequently, we compared each encoded vector from the test image to the encoded vectors from the training set, and retrieved the tooth with the minimum Euclidean distance of encoded vectors from the training set.

Evaluation Metrics. The evaluation of the proposed method was conducted using the following metrics: volumetric Intersection over Union (IoU), Chamfer-L_1 distance, and Normal Consistency (NC), as outlined in prior work [15]. In addition, we used volumetric precision [13] as a metric given by $|D \cap G|/|D|$ where G denotes the ground-truth set of points occupied by the object, and D denotes the set of points predicted as the object.

Quantitative Comparison. Table 1 presents a quantitative comparison of the proposed model with several baseline models. The results demonstrate that Occudent surpasses the other methods across all the metrics, and the methods based on neural implicit functions (Occudent and OccNet) perform better compared to conventional encoder-decoder approaches, such as Pix2Vox and 3D-R2N2. The performance gap between Occudent and X2Teeth is presumably because real PX images are used as input data. X2Teeth used *synthesized* images generated from the 2D projections of CBCT in [27]. Thus, both the input 2D shape and the target 3D shape come from the same modality (CBCT). However, the distribution of real PX images may differ significantly from that of 2D-projected CBCT. Explicit methods can be more sensitive to such differences than implicit methods, because typically in explicit methods, input features are directly encoded and subsequently decoded to predict the target shapes [4,13,26]. Overall, the differences in the IoU performances among the baselines are somewhat small. This is because all the baselines are moderately successful in generating coarse tooth shapes. However, Occudent is significantly better at generating details such as root shapes, which will be shown in the subsequent section.

Qualitative Comparison. Figure 2A illustrates the qualitative results of the proposed method in generating 3D teeth mesh outputs. From our model, each tooth is generated, and generated teeth are combined along with an arch curve based on a beta function [2]. Figure 2A demonstrates that our proposed method generates the most similar-looking outputs compared to the ground truth. For instance, our model can reconstruct a plausible shape for all tooth types including detailed shapes of molar roots. 3D-R2N2 produces larger and less detailed tooth shapes. PSGN and OccNet are better at generating rough shapes than 3D-R2N2, however, lack in detailed root shapes.

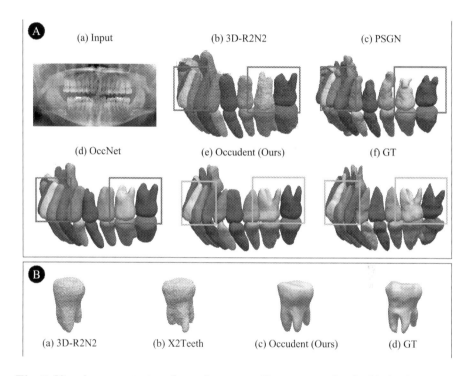

Fig. 2. Visual representation of sample outputs. Boxes are used to highlight the incisors and molars in the upper jaw.

As illustrated in Fig. 2B, Occudent produces a more refined mesh of tooth shape representation than voxel-based methods like 3D-R2N2 or X2Teeth. One of the limitations of voxel-based methods is that they heavily depend on the resolution of the output. For example, increasing the output size leads to an exponential increase in model size. By contrast, Occudent employs continuous neural implicit functions to represent shape boundaries, which enables us to generate smoother output and to be robust to the target size.

4   Conclusion

In this paper, we present a framework for 3D teeth reconstruction from a single PX. To the best of our knowledge, our method is the first to utilize a neural implicit function for 3D teeth reconstruction. The performance of our proposed framework is evaluated quantitatively and qualitatively, demonstrating its superiority over state-of-the-art techniques. Importantly, our framework is capable of accommodating two distinct modalities, PX, and CBCT. Our framework has the potential to be valuable in clinical practice and also can support virtual simulation or educational tools. In the future, further improvements can be made, such as incorporating additional imaging modalities or exploring neural architectures for more robust reconstruction.

Acknowledgements. This work was supported by the Korea Medical Device Development Fund grant funded by the Korea Government (the Ministry of Science and ICT, the Ministry of Trade, Industry and Energy, the Ministry of Health & Welfare, the Ministry of Food and Drug Safety) (Project Number: 1711195279, RS-2021-KD000009); the National Research Foundation of Korea (NRF) Grant through the Ministry of Science and ICT (MSIT), Korea Government, under Grant 2022R1A5A1027646; the National Research Foundation of Korea (NRF) grant funded by the Korea government (MSIT) (No. 2021R1A2C1007215); the MSIT, Korea, under the ICT Creative Consilience program (IITP-2023-2020-0-01819) supervised by the IITP (Institute for Information & communications Technology Planning & Evaluation)

References

1. Abdelrehim, A.S., Farag, A.A., Shalaby, A.M., El-Melegy, M.T.: 2D-PCA shape models: application to 3D reconstruction of the human teeth from a single image. In: Menze, B., Langs, G., Montillo, A., Kelm, M., Müller, H., Tu, Z. (eds.) MCV 2013. LNCS, vol. 8331, pp. 44–52. Springer, Cham (2014). https://doi.org/10.1007/978-3-319-05530-5_5
2. Braun, S., Hnat, W.P., Fender, D.E., Legan, H.L.: The form of the human dental arch. Angle Orthod. **68**(1), 29–36 (1998)
3. Brooks, S.L.: CBCT dosimetry: orthodontic considerations. In: Seminars in Orthodontics, vol. 15, pp. 14–18. Elsevier (2009)
4. Choy, C.B., Xu, D., Gwak, J.Y., Chen, K., Savarese, S.: 3D-R2N2: a unified approach for single and multi-view 3D object reconstruction. In: Leibe, B., Matas, J., Sebe, N., Welling, M. (eds.) ECCV 2016. LNCS, vol. 9912, pp. 628–644. Springer, Cham (2016). https://doi.org/10.1007/978-3-319-46484-8_38
5. De Vries, H., Strub, F., Mary, J., Larochelle, H., Pietquin, O., Courville, A.C.: Modulating early visual processing by language. In: Advances in Neural Information Processing Systems, vol. 30 (2017)
6. Dumoulin, V., et al.: Adversarially learned inference. arXiv preprint: arXiv:1606.00704 (2016)
7. Fan, H., Su, H., Guibas, L.J.: A point set generation network for 3D object reconstruction from a single image. In: Proceedings of the IEEE Conference on Computer Vision and Pattern Recognition, pp. 605–613 (2017)

8. Goodfellow, I., et al.: Generative adversarial networks. Commun. ACM **63**(11), 139–144 (2020)
9. He, K., Zhang, X., Ren, S., Sun, J.: Deep residual learning for image recognition. In: Proceedings of the IEEE Conference on Computer Vision and Pattern Recognition, pp. 770–778 (2016)
10. Hu, J., Shen, L., Sun, G.: Squeeze-and-excitation networks. In: Proceedings of the IEEE Conference on Computer Vision and Pattern Recognition, pp. 7132–7141 (2018)
11. Koch, T.L., Perslev, M., Igel, C., Brandt, S.S.: Accurate segmentation of dental panoramic radiographs with U-Nets. In: 2019 IEEE 16th International Symposium on Biomedical Imaging (ISBI 2019), pp. 15–19. IEEE (2019)
12. Li, Y., et al.: The current situation and future prospects of simulators in dental education. J. Med. Internet Res. **23**(4), e23635 (2021)
13. Liang, Y., Song, W., Yang, J., Qiu, L., Wang, K., He, L.: X2Teeth: 3D Teeth reconstruction from a single panoramic radiograph. In: Martel, A.L., et al. (eds.) MICCAI 2020. LNCS, vol. 12262, pp. 400–409. Springer, Cham (2020). https://doi.org/10.1007/978-3-030-59713-9_39
14. Mazzotta, L., Cozzani, M., Razionale, A., Mutinelli, S., Castaldo, A., Silvestrini-Biavati, A.: From 2D to 3D: construction of a 3D parametric model for detection of dental roots shape and position from a panoramic radiograph-a preliminary report. Int. J. Dent. **2013** (2013)
15. Mescheder, L., Oechsle, M., Niemeyer, M., Nowozin, S., Geiger, A.: Occupancy networks: learning 3D reconstruction in function space. In: Proceedings of the IEEE/CVF Conference on Computer Vision and Pattern Recognition, pp. 4460–4470 (2019)
16. Mildenhall, B., Srinivasan, P.P., Tancik, M., Barron, J.T., Ramamoorthi, R., Ng, R.: NeRF: representing scenes as neural radiance fields for view synthesis. In: Vedaldi, A., Bischof, H., Brox, T., Frahm, J.-M. (eds.) ECCV 2020. LNCS, vol. 12346, pp. 405–421. Springer, Cham (2020). https://doi.org/10.1007/978-3-030-58452-8_24
17. Nader, R., Smorodin, A., De La Fourniere, N., Amouriq, Y., Autrusseau, F.: Automatic teeth segmentation on panoramic X-rays using deep neural networks. In: 2022 26th International Conference on Pattern Recognition (ICPR), pp. 4299–4305. IEEE (2022)
18. Park, J.J., Florence, P., Straub, J., Newcombe, R., Lovegrove, S.: DeepSDF: learning continuous signed distance functions for shape representation. In: Proceedings of the IEEE/CVF Conference on Computer Vision and Pattern Recognition (CVPR) (2019)
19. Qi, C.R., Su, H., Mo, K., Guibas, L.J.: PointNet: deep learning on point sets for 3D classification and segmentation. In: Proceedings of the IEEE Conference on Computer Vision and Pattern Recognition, pp. 652–660 (2017)
20. Ronneberger, O., Fischer, P., Brox, T.: U-Net: convolutional networks for biomedical image segmentation. In: Navab, N., Hornegger, J., Wells, W.M., Frangi, A.F. (eds.) MICCAI 2015. LNCS, vol. 9351, pp. 234–241. Springer, Cham (2015). https://doi.org/10.1007/978-3-319-24574-4_28
21. Silva, B., Pinheiro, L., Oliveira, L., Pithon, M.: A study on tooth segmentation and numbering using end-to-end deep neural networks. In: 2020 33rd SIBGRAPI Conference on Graphics, Patterns and Images (SIBGRAPI), pp. 164–171. IEEE (2020)

22. Song, W., Liang, Y., Yang, J., Wang, K., He, L.: Oral-3D: reconstructing the 3D structure of oral cavity from panoramic x-ray. In: Thirty-Fifth AAAI Conference on Artificial Intelligence, AAAI 2021, Thirty-Third Conference on Innovative Applications of Artificial Intelligence, IAAI 2021, The Eleventh Symposium on Educational Advances in Artificial Intelligence, EAAI 2021, Virtual Event, 2-9 February 2021, pp. 566–573. AAAI Press (2021). https://ojs.aaai.org/index.php/AAAI/article/view/16135

23. Stutz, D., Geiger, A.: Learning 3D shape completion under weak supervision. CoRR **abs/1805.07290** (2018). http://arxiv.org/abs/1805.07290

24. Tatarchenko, M., Richter, S.R., Ranftl, R., Li, Z., Koltun, V., Brox, T.: What do single-view 3D reconstruction networks learn? In: Proceedings of the IEEE/CVF Conference on Computer Vision and Pattern Recognition, pp. 3405–3414 (2019)

25. Wang, N., Zhang, Y., Li, Z., Fu, Y., Liu, W., Jiang, Y.-G.: Pixel2Mesh: generating 3D mesh models from single RGB images. In: Ferrari, V., Hebert, M., Sminchisescu, C., Weiss, Y. (eds.) ECCV 2018. LNCS, vol. 11215, pp. 55–71. Springer, Cham (2018). https://doi.org/10.1007/978-3-030-01252-6_4

26. Xie, H., Yao, H., Sun, X., Zhou, S., Zhang, S.: Pix2Vox: context-aware 3D reconstruction from single and multi-view images. In: Proceedings of the IEEE/CVF International Conference on Computer Vision, pp. 2690–2698 (2019)

27. Yun, Z., Yang, S., Huang, E., Zhao, L., Yang, W., Feng, Q.: Automatic reconstruction method for high-contrast panoramic image from dental cone-beam CT data. Comput. Methods Programs Biomed. **175**, 205–214 (2019)

28. Zhao, Y., et al.: TSASNet: tooth segmentation on dental panoramic X-ray images by two-stage attention segmentation network. Knowl.-Based Syst. **206**, 106338 (2020)

29. Zhou, Z., Rahman Siddiquee, M.M., Tajbakhsh, N., Liang, J.: UNet++: a nested U-Net architecture for medical image segmentation. In: Stoyanov, D., et al. (eds.) DLMIA/ML-CDS -2018. LNCS, vol. 11045, pp. 3–11. Springer, Cham (2018). https://doi.org/10.1007/978-3-030-00889-5_1

DisC-Diff: Disentangled Conditional Diffusion Model for Multi-contrast MRI Super-Resolution

Ye Mao[1], Lan Jiang[2], Xi Chen[3], and Chao Li[1,2,4(✉)]

[1] Department of Clinical Neurosciences, University of Cambridge, Cambridge, UK
cl647@cam.ac.uk
[2] School of Science and Engineering, University of Dundee, Dundee, UK
[3] Department of Computer Science, University of Bath, Bath, UK
[4] School of Medicine, University of Dundee, Dundee, UK
https://github.com/Yebulabula/DisC-Diff

Abstract. Multi-contrast magnetic resonance imaging (MRI) is the most common management tool used to characterize neurological disorders based on brain tissue contrasts. However, acquiring high-resolution MRI scans is time-consuming and infeasible under specific conditions. Hence, multi-contrast super-resolution methods have been developed to improve the quality of low-resolution contrasts by leveraging complementary information from multi-contrast MRI. Current deep learning-based super-resolution methods have limitations in estimating restoration uncertainty and avoiding mode collapse. Although the diffusion model has emerged as a promising approach for image enhancement, capturing complex interactions between multiple conditions introduced by multi-contrast MRI super-resolution remains a challenge for clinical applications. In this paper, we propose a disentangled conditional diffusion model, DisC-Diff, for multi-contrast brain MRI super-resolution. It utilizes the sampling-based generation and simple objective function of diffusion models to estimate uncertainty in restorations effectively and ensure a stable optimization process. Moreover, DisC-Diff leverages a disentangled multi-stream network to fully exploit complementary information from multi-contrast MRI, improving model interpretation under multiple conditions of multi-contrast inputs. We validated the effectiveness of DisC-Diff on two datasets: the IXI dataset, which contains 578 normal brains, and a clinical dataset with 316 pathological brains. Our experimental results demonstrate that DisC-Diff outperforms other state-of-the-art methods both quantitatively and visually.

Keywords: Magnetic resonance imaging · Multi-contrast super-resolution · Conditional diffusion model

---

Y. Mao and L. Jiang—Contribute equally in this work.

1   Introduction

Magnetic Resonance Imaging (MRI) is the primary management tool for brain disorders [24–26]. However, high-resolution (HR) MRI with sufficient tissue contrast is not always available in practice due to long acquisition time [19], where low-resolution (LR) MRIs significantly challenge clinical practice.

Super-resolution (SR) techniques promise to enhance the spatial resolution of LR-MRI and restore tissue contrast. Traditional SR methods, e.g., bicubic interpolation [8], iterative deblurring algorithms [7] and dictionary learning-based methods [1] are proposed, which, however, are challenging to restore the high-frequency details of images and sharp edges due to the inability to establish the complex non-linear mapping between HR and LR images. In contrast, deep learning (DL) has outperformed traditional methods, owing to its ability to capture fine details and preserve anatomical structures accurately.

Earlier DL-based SR methods [3,6,13,14,20,23,28] focused on learning the one-to-one mapping between the single-contrast LR MRI and its HR counterpart. However, multi-contrast MRI is often required for diagnosing brain disorders due to the complexity of brain anatomy. Single-contrast methods are limited by their ability to leverage complementary information from multiple MRI contrasts, leading to inferior SR quality. As an improvement, multi-contrast SR methods [5,17,21,22,27] are proposed to improve the restoration of anatomical details by integrating additional contrast information. For instance, Zeng et al. designed a CNN consisting of two subnetworks to achieve multi-contrast SR [27]. Lyu et al. presented a progressive network to generate realistic HR images from multi-contrast MRIs by minimizing a composite loss of mean-squared-error, adversarial loss, perceptual loss etc. [17]. Feng et al. introduced a multi-stage integration network to extract complex interactions among multi-contrast features hierarchically, enhancing multi-contrast feature fusion [5]. Despite these advancements, most multi-contrast methods fail to 1) estimate restoration uncertainty for a robust model; 2) reduce the risk of mode collapse when applying adversarial loss to improve image fidelity.

Conditional diffusion models are a class of deep generative models that have achieved competitive performance in natural image SR [4,10,18]. The model incorporates a Markov chain-based diffusion process along with conditional variables, i.e., LR images, to restore HR images. The stochastic nature of the diffusion model enables the generation of multiple HR images through sampling, enabling inherent uncertainty estimation of super-resolved outputs. Additionally, the objective function of diffusion models is a variant of the variational lower bound that yields stable optimization processes. Given these advantages, conditional diffusion models promise to update MRI SR methods.

However, current diffusion-based SR methods are mainly single-contrast models. Several challenges remain for developing multi-contrast methods: 1) Integrating multi-contrast MRI into diffusion models increases the number of conditions. Traditional methods integrate multiple conditions via concatenation, which may not effectively leverage complementary information in multiple MRI contrasts, resulting in high-redundancy features for SR; 2) The noise and outliers in MRI can compromise the performance of standard diffusion models that use Mean Squared Error (MSE) loss to estimate the variational lower bound, leading to suboptimal results; 3) Diffusion models are often large-scale, and so are primarily intended for the generation of 2D images, i.e., treating MRI slices separately. Varied anatomical complexity across MRI slices can result in inconsistent diffusion processes, posing a challenge to efficient learning of SR-relevant features.

To address the challenges, we propose a novel conditional disentangled diffusion model (DisC-Diff). To the best of our knowledge, this is the first diffusion-based multi-contrast SR method. The main contribution of our work is fourfold:

- We propose a new backbone network disentangled U-Net for the conditional diffusion model, a U-shape multi-stream network composed of multiple encoders enhanced by disentangled representation learning.
- We present a disentanglement loss function along with a channel attention-based feature fusion module to learn effective and relevant shared and independent representations across MRI contrasts for reconstructing SR images.
- We tailor a Charbonnier loss [2] to overcome the drawbacks of the MSE loss in optimizing the variational lower bound, which could provide a smoother and more robust optimization process.
- For the first time, we introduce an entropy-inspired curriculum learning strategy for training diffusion models, which significantly reduces the impact of varied anatomical complexity on model convergence.

Our extensive experiments on the IXI and in-house clinical datasets demonstrate that our method outperforms other state-of-the-art methods.

2   Methodology

2.1   Overall Architecture

The proposed DisC-Diff is designed based on a conditional diffusion model implemented in [4]. As illustrated in Fig. 1, the method achieves multi-contrast

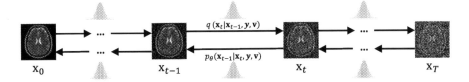

Fig. 1. Conceptual workflow of DisC-Diff on multi-contrast super-resolution. The forward diffusion process q (left-to-right) perturbs HR MRI $\mathbf{x}$ by gradually adding Gaussian noise. The backward diffusion process p (right-to-left) denoises the perturbed MRI, conditioning on its corresponding LR version $\mathbf{y}$ and other available MRI contrasts $\mathbf{v}$.

MRI SR through forward and reverse diffusion processes. Given an HR image $\mathbf{x}_0 \sim q(\mathbf{x}_0)$, the forward process gradually adds Gaussian noise to $\mathbf{x}_0$ over T diffusion steps according to a noise variance schedule $\beta_1, \ldots, \beta_T$. Specifically, each step of the forward diffusion process produces a noisier image $\mathbf{x}_t$ with distribution $q(\mathbf{x}_t \mid \mathbf{x}_{t-1})$, formulated as:

$$q\left(\mathbf{x}_{1:T} \mid \mathbf{x}_0\right) = \prod_{t=1}^{T} q\left(\boldsymbol{x}_t \mid \boldsymbol{x}_{t-1}\right), q\left(\mathbf{x}_t \mid \mathbf{x}_{t-1}\right) = \mathcal{N}\left(\mathbf{x}_t; \sqrt{1 - \beta_t}\mathbf{x}_{t-1}, \beta_t \mathbf{I}\right) \quad (1)$$

For sufficiently large T, the perturbed HR $\mathbf{x}_T$ can be considered a close approximation of isotropic Gaussian distribution. On the other hand, the reverse diffusion process p aims to generate a new HR image from $\mathbf{x}_T$. This is achieved by constructing the reverse distribution $p_\theta\left(\mathbf{x}_{t-1} \mid \mathbf{x}_t, \mathbf{y}, \mathbf{v}\right)$, conditioned on its associated LR image $\mathbf{y}$ and MRI contrast $\mathbf{v}$, expressed as follows:

$$p_\theta\left(\mathbf{x}_{0:T}\right) = p_\theta\left(\mathbf{x}_T\right) \prod_{t=1}^{T} p_\theta\left(\mathbf{x}_{t-1} \mid \mathbf{x}_t\right)$$
$$p_\theta\left(\mathbf{x}_{t-1} \mid \mathbf{x}_t, \mathbf{y}, \mathbf{v}\right) = \mathcal{N}\left(\mathbf{x}_{t-1}; \boldsymbol{\mu}_\theta\left(\mathbf{x}_t, \mathbf{y}, \mathbf{v}, t\right), \sigma_t^2 \mathbf{I}\right) \quad (2)$$

where p_θ denotes a parameterized model, θ is its trainable parameters and σ_t^2 can be either fixed to $\prod_{t=0}^{t} \beta_t$ or learned. It is challenging to obtain the reverse distribution via inference; thus, we introduce a disentangled U-Net parameterized model, shown in Fig. 2, which estimates the reverse distribution by learning disentangled multi-contrast MRI representations. Specifically, p_θ learns to conditionally generate HR image by jointly optimizing the proposed disentanglement loss $\mathcal{L}_{\text{disent}}$ and a Charbonnier loss $\mathcal{L}_{\text{charb}}$. Additionally, we leverage a curriculum learning strategy to aid model convergence of learning $\boldsymbol{\mu}_\theta\left(\mathbf{x}_t, \mathbf{y}, \mathbf{v}, t\right)$.

Disentangled U-Net. The proposed Disentangled U-Net is a multi-stream net composed of multiple encoders, separately extracting latent representations.

We first denote the representation captured from the HR-MRI $\mathbf{x}_t$ as $Z_{\mathbf{x}_t} \in \mathbb{R}^{H \times W \times 2C}$, which contains a shared representation $S_{\mathbf{x}_t}$ and an independent representation $I_{\mathbf{x}_t}$ (both with 3D shape $H \times W \times C$) extracted by two 3×3 convolutional filters. The same operations on $\mathbf{y}$ and $\mathbf{v}$ yield $S_{\mathbf{y}}, I_{\mathbf{y}}$ and $S_{\mathbf{v}}, I_{\mathbf{v}}$, respectively. Effective disentanglement minimizes disparity among shared representations while maximizing that among independent representations. Therefore, $S_{\mathbf{x}_t/\mathbf{y}/\mathbf{v}}$ are as close to each other as possible and can be safely reduced to a single representation S via a weighted sum, followed by the designed Squeeze-and-Excitation (SE) module (Fig. 2 B) that aims to emphasize the most relevant features by dynamically weighting the features in $I_{\mathbf{x}_t/\mathbf{y}/\mathbf{v}}$ or S, resulting in rebalanced disentangled representations $\hat{I}_{\mathbf{x}_t/\mathbf{y}/\mathbf{v}}$ and $\hat{S}$. Each SE module applies global average pooling to each disentangled representation, producing a length-C global descriptor. Two fully-connected layers activated by **SiLU** and **Sigmoid** are then applied to the descriptor to compute a set of weights $\mathbf{s} = [s_1, \ldots, s_i, s_{i+1}, \ldots, s_C]$,

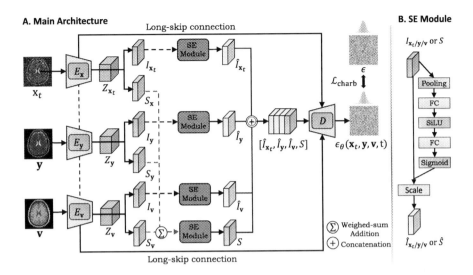

Fig. 2. (A) The disentangled U-Net consists of three encoders processing perturbed HR $\mathbf{x}_t$, LR $\mathbf{y}$, and additional contrast $\mathbf{v}$ respectively. The representations from each encoder are disentangled and concatenated as input to a single decoder D to predict the intermediate noise level $\epsilon_\theta(\mathbf{x}_t, \mathbf{y}, \mathbf{v}, t)$. The architecture includes SE Modules detailed in (B) for dynamically weighting disentangled representations.

where s_i represents the importance of the i-th feature map in the disentangled representation. Finally, the decoder block D shown in Fig. 2 performs upsampling on the concatenated representations $[\hat{I}_{\mathbf{x}_t}, \hat{I}_\mathbf{y}, \hat{I}_\mathbf{v}, \hat{S}]$ and outputs a noise prediction $\epsilon_\theta(\mathbf{x}_t, \mathbf{y}, \mathbf{v}, t)$ to compute the Gaussian mean in Eq. 3:

$$\boldsymbol{\mu}_\theta\left(\mathbf{x}_t, \mathbf{y}, \mathbf{v}, t\right) = \frac{1}{\sqrt{\alpha_t}}\left(\mathbf{x}_t - \frac{\beta_t}{\sqrt{1 - \bar{\alpha}_t}}\epsilon_\theta\left(\mathbf{x}_t, \mathbf{y}, \mathbf{v}, t\right)\right) \tag{3}$$

where $\alpha_t = 1 - \beta_t$ and $\bar{\alpha}_t = \prod_{s=0}^{t} \alpha_s$.

2.2 Design of Loss Functions

To effectively learn disentangled representations with more steady convergence in model training, a novel joint loss is designed in DisC-Diff as follows.

Disentanglement Loss. $\mathcal{L}_{\text{disent}}$ is defined as a ratio between $\mathcal{L}_{\text{shared}}$ and $\mathcal{L}_{\text{indep}}$, where $\mathcal{L}_{\text{shared}}$ measures the $\mathcal{L}_2$ distance between shared representations, and $\mathcal{L}_{\text{indep}}$ is the distance between independent representations:

$$\mathcal{L}_{\text{disent}} = \frac{\mathcal{L}_{\text{shared}}}{\mathcal{L}_{\text{indep}}} = \frac{\|S_{\mathbf{x}_t} - S_\mathbf{y}\|_2 + \|S_{\mathbf{x}_t} - S_\mathbf{v}\|_2 + \|S_\mathbf{y} - S_\mathbf{v}\|_2}{\|I_{\mathbf{x}_t} - I_\mathbf{y}\|_2 + \|I_{\mathbf{x}_t} - I_\mathbf{v}\|_2 + \|I_\mathbf{y} - I_\mathbf{v}\|_2} \tag{4}$$

Charbonnier Loss. $\mathcal{L}_{\text{charb}}$ is a smoother transition between $\mathcal{L}_1$ and $\mathcal{L}_2$ loss, facilitating more steady and accurate convergence during training [9]. It is less sensitive to the non-Gaussian noise in $\mathbf{x}_t$, $\mathbf{y}$ and $\mathbf{v}$, and encourages sparsity results, preserving sharp edges and details in MRIs. It is defined as:

$$\mathcal{L}_{\text{charb}} = \sqrt{(\epsilon_\theta(\mathbf{x}_t, \mathbf{y}, \mathbf{v}, t) - \epsilon)^2 + \gamma^2)} \tag{5}$$

where γ is a known constant. The total loss is the weighted sum of the above two losses:

$$\mathcal{L}_{\text{total}} = \lambda_1 \mathcal{L}_{\text{disent}} + \lambda_2 \mathcal{L}_{\text{charb}} \tag{6}$$

where λ_1, $\lambda_2 \in (0, 1]$ indicate the weights of the two losses.

2.3   Curriculum Learning

Our curriculum learning strategy improves the disentangled U-Net's performance on MRI data with varying anatomical complexity by gradually increasing the difficulty of training images, facilitating efficient learning of relevant features. All MRI slices are initially ranked based on the complexity estimated by Shannon-entropy values of their ground-truth HR-MRI, denoted as an ordered set $E = \{e_{\min}, \ldots, e_{\max}\}$. Each iteration samples N images whose entropies follow a normal distribution with $e_{\min} < \mu < e_{\max}$. As training progresses, μ gradually increases from $e_{\min}$ to $e_{\max}$, indicating increased complexity of the sampled images. The above strategy is used for the initial M iterations, followed by uniform sampling of all slices.

3   Experiments and Results

Datasets and Baselines. We evaluated our model on the public IXI dataset[1] and an in-house clinical brain MRI dataset. In both datasets, our setting is to utilize HR T1-weighted images HR_{T1} and LR T2-weighted image LR_{T2} created by k-space truncation [3] to restore 2× and 4× HR T2-weighted images, aligning with the setting in [5, 22, 27].

We split the 578 healthy brain MRIs in the IXI dataset into 500 for training, 6 for validation, and 70 for testing. We apply center cropping to convert each MRI into a new scan comprising 20 slices, each with a resolution 224 × 224. The processed IXI dataset is available for download at this link[2]. The clinical dataset is fully sampled using a 3T Siemens Magnetom Skyra scanner on 316 glioma patients. The imaging protocol is as follows: $\text{TR}_{T1} = 2300$ ms, $\text{TE}_{T1} = 2.98$ ms, $\text{FOV}_{T1} = 256 \times 240 \, \text{mm}^2$, $\text{TR}_{T2} = 4840$ ms, $\text{TE}_{T2} = 114$ ms, and $\text{FOV}_{T2} = 220 \times 165 \, \text{mm}^2$. The clinical dataset is split patient-wise into train/validation/test sets with a ratio of 7:1:2, and each set is cropped into $\mathbb{R}^{224 \times 224 \times 30}$. Both datasets are normalized using the min-max method without prior data augmentations for training.

---

[1] https://brain-development.org/ixi-dataset/.
[2] https://bit.ly/3yethO4.

We compare our method with three single-contrast SR methods (bicubic interpolation, EDSR [12], SwinIR [11]) and three multi-contrast SR methods (Guided Diffusion [4], MINet [5], MASA-SR [16]) (Table 1).

Table 1. Quantitative results on both datasets with 2× and 4× enlargement scales in terms of mean PSNR (dB) and SSIM. Bold numbers indicate the best results.

Dataset	IXI				Clinical Dataset			
Scale	2×		4×		2×		4×	
Metrics	PSNR	SSIM	PSNR	SSIM	PSNR	SSIM	PSNR	SSIM
Bicubic	32.84	0.9622	26.21	0.8500	34.72	0.9769	27.17	0.8853
EDSR [12]	36.59	0.9865	29.67	0.9350	36.89	0.9880	29.99	0.9373
SwinIR [11]	37.21	0.9856	29.99	0.9360	37.36	0.9868	30.23	0.9394
Guided Diffusion [4]	36.32	0.9815	30.21	0.9512	36.91	0.9802	29.35	0.9326
MINet [5]	36.56	0.9806	30.59	0.9403	37.73	0.9869	31.65	0.9536
MASA-SR [16]	–	–	30.61	0.9417	–	–	31.56	0.9532
DisC-Diff (Ours)	**37.64**	**0.9873**	**31.43**	**0.9551**	**37.77**	**0.9887**	**32.05**	**0.9562**

Implementation Details. DisC-Diff was implemented using PyTorch with the following hyperparameters: $\lambda_1 = \lambda_2 = 1.0$, diffusion steps $T = 1000$, 96 channels in the first layer, 2 BigGAN Residual blocks, and attention module at 28×28, 14×14, and 7×7 resolutions. The model was trained for 200,000 iterations ($M = 20,000$) on two NVIDIA RTX A5000 24 GB GPUs using the AdamW optimizer with a learning rate of 10^{-4} and a batch size of 8. Following the sampling strategy in [4], DisC-Diff learned the reverse diffusion process variances to generate HR-MRI in only 100 sampling steps. The baseline methods were retrained with their default hyperparameter settings. Guided Diffusion was modified to enable multi-contrast SR by concatenating multi-contrast MRI as input.

Quantitative Comparison. The results show that DisC-Diff outperforms other evaluated methods on both datasets at 2× and 4× enlargement scales. Specifically, on the IXI dataset with 4× scale, DisC-Diff achieves a PSNR increment of 1.44 dB and 0.82 dB and an SSIM increment of 0.0191 and 0.0134 compared to state-of-the-art single-contrast and multi-contrast SR methods [11,16]. The results show that without using disentangled U-Net as the backbone, Guided Diffusion performs much poorer than MINet and MASA-SR on the clinical dataset, indicating its limitation in recovering anatomical details of pathology-bearing brain. Our results suggest that disentangling multiple conditional contrasts could help DisC-Diff accurately control the HR image sampling process. Furthermore, the results indicate that integrating multi-contrast information inappropriately may damage the quality of super-resolved images, as evidenced by multi-contrast methods occasionally performs worse than single-contrast methods, e.g., EDSR showing higher SSIM than MINet on both datasets at 2× enlargement.

Visual Comparison and Uncertainty Estimation. Figure 3 shows the results and error maps for each method under IXI (2×) and Clinical (4×) settings, where less visible texture in the error map indicates better restoration. DisC-Diff outperforms all other methods, producing HR images with sharper edges and finer details, while exhibiting the least visible texture. Multi-contrast SR methods consistently generate higher-quality SR images than single-contrast SR methods, consistent with their higher PSNR and SSIM. Also, the lower variation between different restorations at the 2× scale compared to the 4× scale (Last column in Fig. 3) suggests higher confidence in the 2× restoration results.

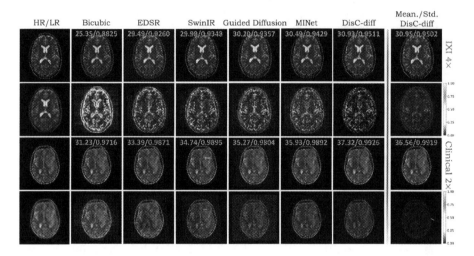

Fig. 3. Visual restoration results and error maps of different methods on IXI (4×) and glioma (2×) datasets, along with mean and standard deviation of our method's sampling results for indicating super-resolution uncertainty (Last column).

Ablation Study. We assess the contribution of three key components in DisC-Diff: 1) w/o $\mathcal{L}_{\text{disent}}$ - implement our model without disentanglement loss, 2) w/o $\mathcal{L}_{\text{charb}}$ -implement our model using MSE loss instead of charbonnier loss, and 3) w/o curriculum learning - training our model without curriculum learning. The 2× and 4× scale results on the IXI dataset are in Table 2. All three models perform worse than DisC-Diff, indicating that the components can enhance overall performance. The results of w/o $\mathcal{L}_{\text{disent}}$ demonstrate that disentangling representations are effective in integrating multi-contrast information. w/o $\mathcal{L}_{\text{charb}}$ performs the worst, consistent with our hypothesis that $\mathcal{L}_{\text{MSE}}$ is sensitive to noise in multi-contrast MRI and can cause the model to converge to local optima.

Table 2. Ablation Study on the IXI dataset with 2× and 4× enlargement scale.

Scale	2×		4×	
Metrics	PSNR	SSIM	PSNR	SSIM
$w/o\ \mathcal{L}_{\text{disent}}$	37.15	0.9834	31.08	0.9524
$w/o\ \mathcal{L}_{\text{charb}}$	36.70	0.9846	31.05	0.9532
w/o curriculum learning	37.58	0.9872	31.36	0.9533
DisC-Diff (Ours)	**37.64**	**0.9873**	**31.43**	**0.9551**

4 Conclusion

We present DisC-Diff, a novel disentangled conditional diffusion model for robust multi-contrast MRI super-resolution. While the sampling nature of the diffusion model has the advantage of enabling uncertainty estimation, proper condition sampling is crucial to ensure model accuracy. Therefore, our method leverages a multi-conditional fusion strategy based on representation disentanglement, facilitating a precise and high-quality HR image sampling process. Also, we experimentally incorporate a Charbonnier loss to mitigate the challenge of MRI noise and outliers on model performance. Future efforts will focus on embedding DisC-Diff's diffusion processes into a compact, low-dimensional latent space to optimize memory and training. We plan to integrate advanced strategies (e.g., DPM-Solver++ [15]) for faster image generation and develop a unified model that generalizes across various scales, eliminating iterative training.

References

1. Bhatia, K.K., Price, A.N., Shi, W., Hajnal, J.V., Rueckert, D.: Super-resolution reconstruction of cardiac MRI using coupled dictionary learning. In: 2014 IEEE 11th International Symposium on Biomedical Imaging (ISBI), pp. 947–950. IEEE (2014)
2. Charbonnier, P., Blanc-Feraud, L., Aubert, G., Barlaud, M.: Two deterministic half-quadratic regularization algorithms for computed imaging. In: Proceedings of 1st International Conference on Image Processing, vol. 2, pp. 168–172. IEEE (1994)
3. Chen, Y., Xie, Y., Zhou, Z., Shi, F., Christodoulou, A.G., Li, D.: Brain MRI super resolution using 3d deep densely connected neural networks. In: 2018 IEEE 15th International Symposium on Biomedical Imaging (ISBI 2018), pp. 739–742. IEEE (2018)
4. Dhariwal, P., Nichol, A.: Diffusion models beat GANs on image synthesis. In: Advances in Neural Information Processing Systems, vol. 34, pp. 8780–8794 (2021)
5. Feng, C.-M., Fu, H., Yuan, S., Xu, Y.: Multi-contrast MRI super-resolution via a multi-stage integration network. In: de Bruijne, M., et al. (eds.) MICCAI 2021. LNCS, vol. 12906, pp. 140–149. Springer, Cham (2021). https://doi.org/10.1007/978-3-030-87231-1_14

6. Feng, C.-M., Yan, Y., Fu, H., Chen, L., Xu, Y.: Task transformer network for joint MRI reconstruction and super-resolution. In: de Bruijne, M., et al. (eds.) MICCAI 2021. LNCS, vol. 12906, pp. 307–317. Springer, Cham (2021). https://doi.org/10.1007/978-3-030-87231-1_30

7. Hardie, R.: A fast image super-resolution algorithm using an adaptive wiener filter. IEEE Trans. Image Process. **16**(12), 2953–2964 (2007)

8. Khaledyan, D., Amirany, A., Jafari, K., Moaiyeri, M.H., Khuzani, A.Z., Mashhadi, N.: Low-cost implementation of bilinear and bicubic image interpolation for real-time image super-resolution. In: 2020 IEEE Global Humanitarian Technology Conference (GHTC), pp. 1–5. IEEE (2020)

9. Lai, W.S., Huang, J.B., Ahuja, N., Yang, M.H.: Fast and accurate image super-resolution with deep Laplacian pyramid networks. IEEE Trans. Pattern Anal. Mach. Intell. **41**(11), 2599–2613 (2018)

10. Li, H., et al.: SRDiff: single image super-resolution with diffusion probabilistic models. Neurocomputing **479**, 47–59 (2022)

11. Liang, J., Cao, J., Sun, G., Zhang, K., Van Gool, L., Timofte, R.: SwinIR: image restoration using Swin transformer. In: Proceedings of the IEEE/CVF International Conference on Computer Vision, pp. 1833–1844 (2021)

12. Lim, B., Son, S., Kim, H., Nah, S., Mu Lee, K.: Enhanced deep residual networks for single image super-resolution. In: Proceedings of the IEEE Conference on Computer Vision and Pattern Recognition Workshops, pp. 136–144 (2017)

13. Liu, C., Wu, X., Yu, X., Tang, Y., Zhang, J., Zhou, J.: Fusing multi-scale information in convolution network for MR image super-resolution reconstruction. Biomed. Eng. Online **17**(1), 1–23 (2018)

14. Liu, P., Li, C., Schönlieb, C.-B.: GANReDL: medical image enhancement using a generative adversarial network with real-order derivative induced loss functions. In: Shen, D., et al. (eds.) MICCAI 2019. LNCS, vol. 11766, pp. 110–117. Springer, Cham (2019). https://doi.org/10.1007/978-3-030-32248-9_13

15. Lu, C., Zhou, Y., Bao, F., Chen, J., Li, C., Zhu, J.: DPM-solver++: fast solver for guided sampling of diffusion probabilistic models. arXiv preprint: arXiv:2211.01095 (2022)

16. Lu, L., Li, W., Tao, X., Lu, J., Jia, J.: MASA-SR: matching acceleration and spatial adaptation for reference-based image super-resolution. In: Proceedings of the IEEE/CVF Conference on Computer Vision and Pattern Recognition, pp. 6368–6377 (2021)

17. Lyu, Q., et al.: Multi-contrast super-resolution MRI through a progressive network. IEEE Trans. Med. Imaging **39**(9), 2738–2749 (2020)

18. Rombach, R., Blattmann, A., Lorenz, D., Esser, P., Ommer, B.: High-resolution image synthesis with latent diffusion models. In: Proceedings of the IEEE/CVF Conference on Computer Vision and Pattern Recognition, pp. 10684–10695 (2022)

19. Shi, F., Cheng, J., Wang, L., Yap, P.T., Shen, D.: LRTV: MR image super-resolution with low-rank and total variation regularizations. IEEE Trans. Med. Imaging **34**(12), 2459–2466 (2015)

20. Shi, J., Liu, Q., Wang, C., Zhang, Q., Ying, S., Xu, H.: Super-resolution reconstruction of MR image with a novel residual learning network algorithm. Phys. Med. Biol. **63**(8), 085011 (2018)

21. Stimpel, B., Syben, C., Schirrmacher, F., Hoelter, P., Dörfler, A., Maier, A.: Multimodal super-resolution with deep guided filtering. In: Bildverarbeitung für die Medizin 2019. I, pp. 110–115. Springer, Wiesbaden (2019). https://doi.org/10.1007/978-3-658-25326-4_25

22. Tsiligianni, E., Zerva, M., Marivani, I., Deligiannis, N., Kondi, L.: Interpretable deep learning for multimodal super-resolution of medical images. In: de Bruijne, M., et al. (eds.) MICCAI 2021. LNCS, vol. 12906, pp. 421–429. Springer, Cham (2021). https://doi.org/10.1007/978-3-030-87231-1_41
23. Wang, J., Chen, Y., Wu, Y., Shi, J., Gee, J.: Enhanced generative adversarial network for 3D brain MRI super-resolution. In: Proceedings of the IEEE/CVF Winter Conference on Applications of Computer Vision, pp. 3627–3636 (2020)
24. Wei, Y., et al.: Multi-modal learning for predicting the genotype of glioma. IEEE Trans. Med. Imaging (2023)
25. Wei, Y., et al.: Structural connectome quantifies tumour invasion and predicts survival in glioblastoma patients. Brain **146**, 1714–1727 (2022)
26. Wei, Y., Li, C., Price, S.J.: Quantifying structural connectivity in brain tumor patients. In: de Bruijne, M., et al. (eds.) MICCAI 2021. LNCS, vol. 12907, pp. 519–529. Springer, Cham (2021). https://doi.org/10.1007/978-3-030-87234-2_49
27. Zeng, K., Zheng, H., Cai, C., Yang, Y., Zhang, K., Chen, Z.: Simultaneous single- and multi-contrast super-resolution for brain MRI images based on a convolutional neural network. Comput. Biol. Med. **99**, 133–141 (2018)
28. Zhang, Y., Li, K., Li, K., Fu, Y.: MR image super-resolution with squeeze and excitation reasoning attention network. In: Proceedings of the IEEE/CVF Conference on Computer Vision and Pattern Recognition, pp. 13425–13434 (2021)

CoLa-Diff: Conditional Latent Diffusion Model for Multi-modal MRI Synthesis

Lan Jiang[1], Ye Mao[2], Xiangfeng Wang[3], Xi Chen[4], and Chao Li[1,2,5(✉)]

[1] School of Science and Engineering, University of Dundee, Dundee, UK
[2] Department of Clinical Neurosciences, University of Cambridge, Cambridge, UK
[3] School of Computer Science and Technology,
East China Normal University, Shanghai, China
[4] Department of Computer Science, University of Bath, Bath, UK
[5] School of Medicine, University of Dundee, Dundee, UK
cl647@cam.ac.uk

Abstract. MRI synthesis promises to mitigate the challenge of missing MRI modality in clinical practice. Diffusion model has emerged as an effective technique for image synthesis by modelling complex and variable data distributions. However, most diffusion-based MRI synthesis models are using a single modality. As they operate in the original image domain, they are memory-intensive and less feasible for multi-modal synthesis. Moreover, they often fail to preserve the anatomical structure in MRI. Further, balancing the multiple conditions from multi-modal MRI inputs is crucial for multi-modal synthesis. Here, we propose the first diffusion-based multi-modality MRI synthesis model, namely Conditioned Latent Diffusion Model (CoLa-Diff). To reduce memory consumption, we perform the diffusion process in the latent space. We propose a novel network architecture, e.g., similar cooperative filtering, to solve the possible compression and noise in latent space. To better maintain the anatomical structure, brain region masks are introduced as the priors of density distributions to guide diffusion process. We further present auto-weight adaptation to employ multi-modal information effectively. Our experiments demonstrate that CoLa-Diff outperforms other state-of-the-art MRI synthesis methods, promising to serve as an effective tool for multi-modal MRI synthesis.

Keywords: Multi-modal MRI · Medical image synthesis · Latent space · Diffusion models · Structural guidance

1 Introduction

Magnetic resonance imaging (MRI) is critical to the diagnosis, treatment, and follow-up of brain tumour patients [26]. Multiple MRI modalities offer complementary information for characterizing brain tumours and enhancing patient

---

L. Jiang and Y. Mao—Contribute equally in this work.

© The Author(s), under exclusive license to Springer Nature Switzerland AG 2023
H. Greenspan et al. (Eds.): MICCAI 2023, LNCS 14229, pp. 398–408, 2023.
https://doi.org/10.1007/978-3-031-43999-5_38

management [4,27]. However, acquiring multi-modality MRI is time-consuming, expensive and sometimes infeasible in specific modalities, e.g., due to the hazard of contrast agent [15]. Trans-modal MRI synthesis can establish the mapping from the known domain of available MRI modalities to the target domain of missing modalities, promising to generate missing MRI modalities effectively. The synthetic methods leveraging multi-modal MRI, i.e., many-to-one translation, have outperformed single-modality models generating a missing modality from another available modality, i.e., one-to-one translation [23,33]. Traditional multi-modal methods [21,22], e.g., sparse encoding-based, patch-based and atlas-based methods, rely on the alignment accuracy of source and target domains and are poorly scalable. Recent generative adversarial networks (GANs) and variants, e.g., MM-GAN [23], DiamondGAN [13] and ProvoGAN [30], have been successful based on multi-modal MRI, further improved by introducing multi-modal coding [31], enhanced architecture [7], and novel learning strategies [29].

Despite the success, GAN-based models are challenged by the limited capability of adversarial learning in modelling complex multi-modal data distributions [25] Recent studies have demonstrated that GANs' performance can be limited to processing and generating data with less variability [1]. In addition, GANs' hyperparameters and regularization terms typically require fine-tuning, which otherwise often results in gradient vanish and mode collapse [2].

Diffusion model (DM) has achieved state-of-the-art performance in synthesizing natural images, promising to improve MRI synthesis models. It shows superiority in model training [16], producing complex and diverse images [9,17], while reducing risk of modality collapse [12].For instance, Lyu et al. [14] used diffusion and score-marching models to quantify model uncertainty from Monte-Carlo sampling and average the output using different sampling methods for CT-to-MRI generation; Özbey et al. [19] leveraged adversarial training to increase the step size of the inverse diffusion process and further designed a cycle-consistent architecture for unpaired MRI translation.

However, current DM-based methods focus on one-to-one MRI translation, promising to be improved by many-to-one methods, which requires dedicated design to balance the multiple conditions introduced by multi-modal MRI. Moreover, as most DMs operate in original image domain, all Markov states are kept in memory [9], resulting in excessive burden. Although latent diffusion model (LDM) [20] is proposed to reduce memory consumption, it is less feasible for many-to-one MRI translation with multi-condition introduced. Further, diffusion denoising processes tend to change the original distribution structure of the target image due to noise randomness [14], rendering DMs often ignore the consistency of anatomical structures embedded in medical images, leading to clinically less relevant results. Lastly, DMs are known for their slow speed of diffusion sampling [9,11,17], challenging its wide clinical application.

We propose a DM-based multi-modal MRI synthesis model, CoLa-Diff, which facilitates many-to-one MRI translation in latent space, and preserve anatomical structure with accelerated sampling. Our main contributions include:

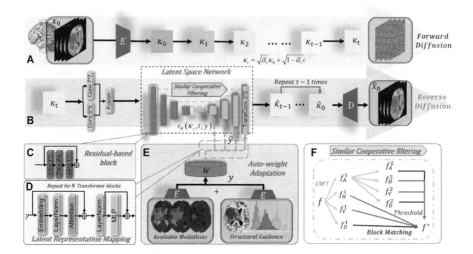

Fig. 1. Schematic diagram of CoLa-Diff. During the forward diffusion, Original images x_0 are compressed using encoder E to get κ_0, and after t steps of adding noise, the images turn into κ_t. During the reverse diffusion, the latent space network $\epsilon_\theta(\kappa_t, t, y)$ predicts the added noise, and other available modalities and anatomical masks as structural guidance are encoded to y, then processed by the auto-weight adaptation block W and embedded into the latent space network. Sampling from the distribution learned from the network gives $\hat{\kappa}_0$, then $\hat{\kappa}_0$ are decoded by D to obtain synthesized images.

– present a denoising diffusion probabilistic model based on multi-modal MRI. As far as we know, this is the first DM-based many-to-one MRI synthesis model.
– design a bespoke architecture, e.g., similar cooperative filtering, to better facilitate diffusion operations in the latent space, reducing the risks of excessive information compression and high-dimensional noise.
– introduce structural guidance of brain regions in each step of the diffusion process, preserving anatomical structure and enhancing synthesis quality.
– propose an auto-weight adaptation to balance multi-conditions and maximise the chance of leveraging relevant multi-modal information.

2   Multi-conditioned Latent Diffusion Model

Figure 1 illustrates the model design. As a latent diffusion model, CoLa-diff integrates multi-condition b from available MRI contrasts in a compact and low-dimensional latent space to guide the generation of missing modality $x \in \mathbb{R}^{H \times W \times 1}$. Precisely, b constitutes available contrasts and anatomical structure masks generated from the available contrasts. Similar to [9,20], CoLa-Diff involves a forward and a reverse diffusion process. During forward diffusion, x_0 is encoded by E to produce κ_0, then subjected to T diffusion steps to gradually add noise ϵ and generate a sequence of intermediate representations: $\{\kappa_0, \ldots, \kappa_T\}$. The t-th intermediate representation is denoted as κ_t, expressed as:

$$\kappa_t = \sqrt{\bar{\alpha}_t}\kappa_0 + \sqrt{1 - \bar{\alpha}_t}\epsilon, \quad \text{with } \epsilon \sim \mathcal{N}(0, \mathbf{I}) \tag{1}$$

where $\bar{\alpha}_t = \prod_{i=1}^{t} \alpha_i$, α_i denotes hyper-parameters related to variance.

The reverse diffusion is modelled by a latent space network with parameters θ, inputting intermediate perturbed feature maps κ_t and y (compressed b) to predict noise level $\epsilon_\theta(\kappa_t, t, y)$ for recovering feature maps $\hat{\kappa}_{t-1}$ from previous,

$$\hat{\kappa}_{t-1} = \sqrt{\bar{\alpha}_{t-1}}(\frac{\kappa_t - \sqrt{1 - \bar{\alpha}_t} \cdot \epsilon_\theta(\kappa_t, t, y)}{\sqrt{\bar{\alpha}_t}}) + \sqrt{1 - \bar{\alpha}_{t-1}} \cdot \epsilon_\theta(\kappa_t, t, y) \tag{2}$$

To enable effective learning of the underlying distribution of κ_0, the noise level needs to be accurately estimated. To achieve this, the network employs similar cooperative filtering and auto-weight adaptation strategies. $\hat{\kappa}_0$ is recovered by repeating Eq. 2 process for t times, and decoding the final feature map to generate synthesis images $\hat{x}_0$.

2.1   Latent Space Network

We map multi-condition to the latent space network for guiding noise prediction at each step t. The mapping is implemented by N transformer blocks (Fig. 1 (D)), including global self-attentive layers, layer-normalization and position-wise MLP. Following the latent diffusion model (LDM) [20], the network $\epsilon_\theta(\kappa_t, t, y)$ is trained to predict the noise added at each step using

$$\mathcal{L}_{\mathrm{E}} := \mathbb{E}_{E(x), y, \epsilon \sim \mathcal{N}(0,1), t} \left[\|\epsilon - \epsilon_\theta(\kappa_t, t, y)\|_2^2 \right] \tag{3}$$

To mitigate the excessive information losses that latent spaces are prone to, we replace the simple convolution operation with a residual-based block (three sequential convolutions with kernels $1 * 1$, $3 * 3$, $1 * 1$ and residual joins [8]), and enlarge the receptive field by fusion ($5 * 5$ and $7 * 7$ convolutions followed by AFF [6]) in the down-sampling section. Moreover, to reduce high-dimensional noise generated in the latent space, which can significantly corrupt the quality of multi-modal generation. We design a similar cooperative filtering detailed below.

Similar Cooperative Filtering. The approach has been devised to filter the downsampled features, with each filtered feature connected to its respective upsampling component (shown in Fig. 1 (F)). Given f, which is the downsampled feature of κ_t, suppose the 2D discrete wavelet transform ϕ [24] decomposes the features into low frequency component $f_A^{(i)}$ and high frequency components $f_H^{(i)}$, $f_V^{(i)}$, $f_D^{(i)}$, keep decompose $f_A^{(i)}$, where i is the number of wavelet transform layers. Previous work [5] has shown to effectively utilize global information by considering similar patches. However, due to its excessive compression, it is less suitable for LDM. Here, we group the components and further filter by similar block matching δ [18] or thresholding γ, use the inverse wavelet transform $\phi^{-1}(\cdot)$ to reconstruct the denoising results, given f^*.

$$f^* = \phi^{-1}(\delta(f_A^{(i)}), \delta(\sum_{j=1}^{i} f_D^{(i)}), \gamma(\sum_{j=1}^{i} f_H^{(i)}), \gamma(\sum_{j=1}^{i} f_V^{(i)})) \tag{4}$$

2.2   Structural Guidance

Unlike natural images, medical images encompass rich anatomical information. Therefore, preserving anatomical structure is crucial for MRI generation. However, DMs often corrupt anatomical structure, and this limitation could be due to the learning and sampling processes of DMs that highly rely on the probability density function [9], while brain structures by nature are overlapping in MRI density distribution and even more complicated by pathological changes.

Previous studies show that introducing geometric priors can significantly improve the robustness of medical image generation. [3,28]. Therefore, we hypothesize that incorporating structural prior could enhance the generation quality with preserved anatomy. Specifically, we exploit FSL-FAST [32] tool to segment four types of brain tissue: white matter, grey matter, cerebrospinal fluid, and tumour. The generated tissue masks and inherent density distributions (Fig. 1 (E)) are then used as a condition y_i to guide the reverse diffusion.

The combined loss function for our multi-conditioned latent diffusion is defined as

$$\mathcal{L}_{\text{MCL}} := \mathcal{L}_{\text{E}} + \mathcal{L}_{\text{KL}} \tag{5}$$

where KL is the KL divergence loss to measure similarity between real q and predicted p_θ distributions of encoded images.

$$\mathcal{L}_{\text{KL}} := \sum_{j=1}^{T-1} D_{KL}\left(q\left(\kappa_{j-1} \mid \kappa_j, \kappa_0\right) \| p_\theta\left(\kappa_{j-1} \mid \kappa_j\right)\right) \tag{6}$$

where D_{KL} is the KL divergence function.

2.3   Auto-Weight Adaptation

It is critical to balance multiple conditions, maximizing relevant information and minimising redundant information. For encoded conditions $y \in \mathbb{R}^{h \times w \times c}$, c is the number of condition channels. Set the value after auto-weight adaptation to $\tilde{y}$, the operation of this module is expressed as (shown in Fig. 1 (E))

$$\tilde{y} = F(y|\mu, \nu, o), \quad \text{with } \mu, \nu, o \in \mathbb{R}^c \tag{7}$$

The embedding outputs are adjusted by embedding weight μ. The auto-activation is governed by the learnable weight ν and bias o. y_c indicates each channel of y, where $y_c = [y_c^{m,n}]_{h \times w} \in R^{h \times w}$, $y_c^{m,n}$ is the eigenvalue at position (m, n) in channel c. We use large receptive fields and contextual embedding to avoid local ambiguities, providing embedding weight $\mu = [\mu_1, \mu_2..., \mu_c]$. The operation G_c is defined as:

$$G_c = \mu_c \|y_c\|_2 = \mu_c \left\{ \left[\sum_{m=1}^{h} \sum_{n=1}^{w} (y_c^{m,n})^2 \right] + \varpi \right\}^{\frac{1}{2}} \qquad (8)$$

where ϖ is a small constant added to the equation to avoid the issue of derivation at the zero point. The normalization method can establish stable competition between channels, $\mathbf{G} = \{G_c\}_{c=1}^{S}$. We use L_2 normalization for cross-channel operations:

$$\hat{G}_c = \frac{\sqrt{S}G_c}{\|\mathbf{G}\|_2} = \frac{\sqrt{S}G_c}{\left[\left(\sum_{c=1}^{S} G_c^2 \right) + \varpi \right]^{\frac{1}{2}}} \qquad (9)$$

where S denotes the scale. We use an activation mechanism for updating each channel to facilitate the maximum utilization of each condition during diffusion model training, and further enhance the synthesis performance. Given the learnable weight $\nu = [\nu_1, \nu_2, ..., \nu_c]$ and bias $\mathbf{o} = [o_1, o_2, ..., o_c]$ we compute

$$\tilde{y}_c = y_c[1 + S(\nu_c \hat{G}_c + o_c)] \qquad (10)$$

which gives new representations $\tilde{y}_c$ of each compressed conditions after the automatic weighting. $S(\cdot)$ denotes the Sigmoid activation function.

3   Experiments and Results

3.1   Comparisons with State-of-the-Art Methods

Datasets and Baselines. We evaluated CoLa-Diff on two multi-contrast brain MRI datasets: BRATS 2018 and IXI datasets. The BRATS 2018 contains MRI scans from 285 glioma patients. Each includes four modalities: T1, T2, T1ce, and FLAIR. We split them into (190:40:55) for training/validation/testing. For each subject, we automatically selected axial cross-sections based on the perceptible effective area of the slices, and then cropped the selected slices to a size of 224×224. The IXI[1] dataset consists of 200 multi-contrast MRIs from healthy brains, plit them into (140:25:35) for training/validation/testing. For preprocessing, we registered T2- and PD-weighted images to T1-weighted images using FSL-FLIRT [10], and other preprocessing are identical to the BRATS 2018.

We compared CoLa-Diff with four state-of-the-art multi-modal MRI synthesis methods: MM-GAN [23], Hi-Net [33], ProvoGan [30] and LDM [20].

Implementation Details. Our code is publicly available at https://github. com/SeeMeInCrown/CoLa_Diff_MultiModal_MRI_Synthesis. The hyperparameters of CoLa-Diff are defined as follows: diffusion steps to 1000; noise schedule to linear; attention resolutions to $32, 16, 8$; batch size to 8, learning rate to $9.6e-5$.

---

[1] https://brain-development.org/ixi-dataset/.

The noise variances were in the range of $\beta_1 = 10^{-4}$ and $\beta_T = 0.02$. An exponential moving average (EMA) over model parameters with a rate of 0.9999 was employed. The model is trained on 2 NVIDIA RTX A5000, 24 GB with Adam optimizer on PyTorch. An acceleration method [11] based on knowledge distillation was applied for fast sampling.

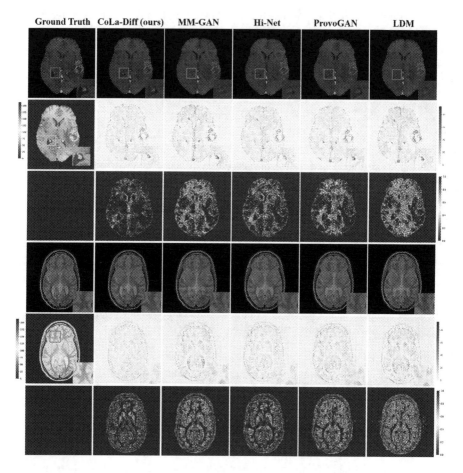

Fig. 2. Visualization of synthesized images, detail enlargements (row 1 and 4), corresponding error maps (row 2 and 5) and uncertainty maps (row 3 and 6).

Quantitative Results. We performed synthesis experiments for all modalities, with each modality selected as the target modality while remaining modalities and the generated region masks as conditions. Seven cases were tested in two datasets (Table 1). The results show that CoLa-Diff outperforms other models by up to 6.01 dB on PSNR and 5.74% on SSIM. Even when compared to the best of other models in each task, CoLa-Diff is a maximum of 0.81 dB higher in PSNR and 0.82% higher in SSIM.

Table 1. Performance in BRATS (top) and IXI (bottom). PSNR (dB) and SSIM (%) are listed as mean±std in the test set. **Boldface** marks the top models.

Model (BRATS 2018)	T2+T1ce+FLAIR →T1		T1+T1ce+FLAIR →T2		T2+T1+FLAIR →T1ce		T2+T1ce+T1 →FLAIR	
	PSNR	SSIM%	PSNR	SSIM%	PSNR	SSIM%	PSNR	SSIM%
MM-GAN	25.78±2.16	90.67±1.45	26.11±1.62	90.58±1.39	26.30±1.91	91.22±2.08	24.09±2.14	88.32±1.98
Hi-Net	27.42±2.58	93.46±1.75	25.64±2.01	92.59±1.42	27.02±1.26	93.35±1.34	25.87±2.82	91.22±2.13
ProvoGAN	27.79±4.42	93.51±3.16	26.72±2.87	92.98±3.91	29.26±2.50	93.96±2.34	25.64±2.77	90.42±3.13
LDM	24.55±2.62	88.34±2.51	24.79±2.67	88.47±2.60	25.61±2.48	89.18±2.55	23.12±3.16	86.90±3.24
CoLa-Diff (Ours)	**28.26±3.13**	**93.65±3.02**	**28.33±2.27**	**93.80±2.75**	**29.35±2.40**	**94.18±2.46**	**26.68±2.74**	**91.89±3.11**

Model (IXI)	T1+T2 →PD		T2+PD →T1		T1+PD →T2	
	PSNR	SSIM%	PSNR	SSIM%	PSNR	SSIM%
MM-GAN	30.61±1.64	95.42±1.90	27.32±1.70	92.35±1.58	30.87±1.75	94.68±1.42
Hi-Net	31.79±2.26	96.51±2.03	28.89±1.43	93.78±1.31	32.58±1.85	96.54±1.74
ProvoGAN	29.93±3.11	94.62±2.40	24.21±2.63	90.46±3.58	29.19±3.04	94.08±3.87
LDM	27.36±2.48	91.52±2.39	24.19±2.51	88.75±2.47	27.04±2.31	91.23±2.24
CoLa-Diff (Ours)	**32.24±2.95**	**96.95±2.26**	**30.20±2.38**	**94.49±2.15**	**32.86±2.83**	**96.57±2.27**

Qualitative Results. The first three and last three rows in Fig. 2 illustrate the synthesis results of T1ce from BRATS and PD from the IXI, respectively. From the generated images, we observe that CoLa-Diff is most comparable to the ground truth, with fewer errors shown in the heat maps. The synthesis uncertainty for each region is derived by performing 100 generations of the same slice and calculating the pixel-wise variance. From the uncertainty maps, CoLa-Diff is more confident in synthesizing the gray and white matter over other comparison models. Particularly, CoLa-Diff performs better in generating complex brain sulcus and tumour boundaries. Further, CoLa-Diff could better maintain the anatomical structure over comparison models.

3.2 Ablation Study and Multi-modal Exploitation Capabilities

We verified the effectiveness of each component in CoLa-Diff by removing them individually. We experimented on BRATS T1+T1ce+FLAIR→T2 task with four absence scenarios (Table 2 top). Our results show that each component contributes to the performance improvement, with Auto-weight adaptation bringing a PSNR increase of 1.9450dB and SSIM of 4.0808%.

To test the generalizability of CoLa-Diff under the condition of varied inputs, we performed the task of generating T2 on two datasets with progressively increasing input modalities (Table 2 bottom). Our results show that our model performance increases with more input modalities: SSIM has a maximum uplift value of 1.9603, PSNR rises from 26.6355 dB to 28.3126 dB in BRATS; from 32.164 dB to 32.8721 dB in IXI. The results could further illustrate the ability of CoLa-Diff to exploit multi-modal information.

Table 2. Ablation of four individual components (First four lines) and Multi-modal information utilisation (Last three lines). **Boldface** marks the best performing scenarios on each dataset.

	PSNR	SSIM%
w/o Modified latent diffusion network	27.1074	90.1268
w/o Structural guidance	27.7542	91.4865
w/o Auto-weight adaptation	26.3896	89.7129
w/o Similar cooperative filtering	27.9753	92.1584
T1 (BRATS)	26.6355	91.7438
T1+T1ce (BRATS)	27.3089	92.9772
T1+T1ce+Flair (BRATS)	**28.3126**	**93.7041**
T1 (IXI)	32.1640	96.0253
T1+PD (IXI)	**32.8721**	**96.5932**

4   Conclusion

This paper presents CoLa-Diff, a DM-based multi-modal MRI synthesis model with a bespoke design of network backbone, similar cooperative filtering, structural guidance and auto-weight adaptation. Our experiments support that CoLa-Diff achieves state-of-the-art performance in multi-modal MRI synthesis tasks. Therefore, CoLa-Diff could serve as a useful tool for generating MRI to reduce the burden of MRI scanning and benefit patients and healthcare providers.

References

1. Bau, D., et al.: Seeing what a GAN cannot generate. In: Proceedings of the IEEE/CVF International Conference on Computer Vision, pp. 4502–4511 (2019)
2. Berard, H., Gidel, G., Almahairi, A., Vincent, P., Lacoste-Julien, S.: A closer look at the optimization landscapes of generative adversarial networks. arXiv preprint: arXiv:1906.04848 (2019)
3. Brooksby, B.A., Dehghani, H., Pogue, B.W., Paulsen, K.D.: Near-infrared (NIR) tomography breast image reconstruction with a priori structural information from MRI: algorithm development for reconstructing heterogeneities. IEEE J. Sel. Top. Quantum Electron. **9**(2), 199–209 (2003)
4. Cherubini, A., Caligiuri, M.E., Péran, P., Sabatini, U., Cosentino, C., Amato, F.: Importance of multimodal MRI in characterizing brain tissue and its potential application for individual age prediction. IEEE J. Biomed. Health Inform. **20**(5), 1232–1239 (2016)
5. Dabov, K., Foi, A., Katkovnik, V., Egiazarian, K.: Image restoration by sparse 3D transform-domain collaborative filtering. In: Image Processing: Algorithms and Systems VI, vol. 6812, pp. 62–73. SPIE (2008)
6. Dai, Y., Gieseke, F., Oehmcke, S., Wu, Y., Barnard, K.: Attentional feature fusion. CoRR abs/2009.14082 (2020)
7. Dalmaz, O., Yurt, M., Çukur, T.: ResViT: residual vision transformers for multimodal medical image synthesis. IEEE Trans. Med. Imaging **41**(10), 2598–2614 (2022)
8. He, K., Zhang, X., Ren, S., Sun, J.: Deep residual learning for image recognition. In: Proceedings of the IEEE Conference on Computer Vision and Pattern Recognition (CVPR) (2016)

9. Ho, J., Jain, A., Abbeel, P.: Denoising diffusion probabilistic models. In: Advances in Neural Information Processing Systems, vol. 33, pp. 6840–6851 (2020)
10. Jenkinson, M., Smith, S.: A global optimisation method for robust affine registration of brain images. Med. Image Anal. **5**(2), 143–156 (2001)
11. Kong, Z., Ping, W.: On fast sampling of diffusion probabilistic models. arXiv preprint: arXiv:2106.00132 (2021)
12. Li, H., et al.: SRDiff: single image super-resolution with diffusion probabilistic models. Neurocomputing **479**, 47–59 (2022)
13. Li, H., et al.: DiamondGAN: unified multi-modal generative adversarial networks for MRI sequences synthesis. In: Shen, D., et al. (eds.) MICCAI 2019. LNCS, vol. 11767, pp. 795–803. Springer, Cham (2019). https://doi.org/10.1007/978-3-030-32251-9_87
14. Lyu, Q., Wang, G.: Conversion between CT and MRI images using diffusion and score-matching models. arXiv preprint: arXiv:2209.12104 (2022)
15. Merbach, A.S., Helm, L., Toth, E.: The Chemistry of Contrast Agents in Medical Magnetic Resonance Imaging. John Wiley & Sons, Hoboken (2013)
16. Müller-Franzes, G., et al.: Diffusion probabilistic models beat gans on medical images. arXiv preprint: arXiv:2212.07501 (2022)
17. Nichol, A.Q., Dhariwal, P.: Improved denoising diffusion probabilistic models. In: International Conference on Machine Learning, pp. 8162–8171. PMLR (2021)
18. Ourselin, S., Roche, A., Prima, S., Ayache, N.: Block matching: a general framework to improve robustness of rigid registration of medical images. In: Delp, S.L., DiGoia, A.M., Jaramaz, B. (eds.) MICCAI 2000. LNCS, vol. 1935, pp. 557–566. Springer, Heidelberg (2000). https://doi.org/10.1007/978-3-540-40899-4_57
19. Özbey, M., et al.: Unsupervised medical image translation with adversarial diffusion models. arXiv preprint: arXiv:2207.08208 (2022)
20. Rombach, R., Blattmann, A., Lorenz, D., Esser, P., Ommer, B.: High-resolution image synthesis with latent diffusion models. In: Proceedings of the IEEE/CVF Conference on Computer Vision and Pattern Recognition, pp. 10684–10695 (2022)
21. Roy, S., Carass, A., Prince, J.: A compressed sensing approach for MR tissue contrast synthesis. In: Székely, G., Hahn, H.K. (eds.) IPMI 2011. LNCS, vol. 6801, pp. 371–383. Springer, Heidelberg (2011). https://doi.org/10.1007/978-3-642-22092-0_31
22. Roy, S., Carass, A., Prince, J.L.: Magnetic resonance image example-based contrast synthesis. IEEE Trans. Med. Imaging **32**(12), 2348–2363 (2013)
23. Sharma, A., Hamarneh, G.: Missing MRI pulse sequence synthesis using multimodal generative adversarial network. IEEE Trans. Med. Imaging **39**(4), 1170–1183 (2019)
24. Shensa, M.J., et al.: The discrete wavelet transform: wedding the a trous and Mallat algorithms. IEEE Trans. Signal Process. **40**(10), 2464–2482 (1992)
25. Thanh-Tung, H., Tran, T.: Catastrophic forgetting and mode collapse in GANs. In: 2020 International Joint Conference on Neural Networks (IJCNN), pp. 1–10. IEEE (2020)
26. Vlaardingerbroek, M.T., Boer, J.A.: Magnetic Resonance Imaging: Theory and Practice. Springer Science & Business Media, Cham (2013)
27. Wei, Y., et al.: Multi-modal learning for predicting the genotype of glioma. IEEE Trans. Med. Imaging (2023)
28. Yu, B., Zhou, L., Wang, L., Shi, Y., Fripp, J., Bourgeat, P.: Ea-GANs: edge-aware generative adversarial networks for cross-modality MR image synthesis. IEEE Trans. Med. Imaging **38**(7), 1750–1762 (2019)

29. Yu, Z., Han, X., Zhang, S., Feng, J., Peng, T., Zhang, X.Y.: MouseGAN++: unsupervised disentanglement and contrastive representation for multiple MRI modalities synthesis and structural segmentation of mouse brain. IEEE Trans. Med. Imaging **42**, 1197–1209 (2022)
30. Yurt, M., Özbey, M., Dar, S.U., Tinaz, B., Oguz, K.K., Çukur, T.: Progressively volumetrized deep generative models for data-efficient contextual learning of MR image recovery. Med. Image Anal. **78**, 102429 (2022)
31. Zhan, B., Li, D., Wu, X., Zhou, J., Wang, Y.: Multi-modal MRI image synthesis via GAN with multi-scale gate mergence. IEEE J. Biomed. Health Inform. **26**(1), 17–26 (2022)
32. Zhang, Y., Brady, M., Smith, S.: Segmentation of brain MR images through a hidden Markov random field model and the expectation-maximization algorithm. IEEE Trans. Med. Imaging **20**(1), 45–57 (2001)
33. Zhou, T., Fu, H., Chen, G., Shen, J., Shao, L.: Hi-Net: hybrid-fusion network for multi-modal MR image synthesis. IEEE Trans. Med. Imaging **39**(9), 2772–2781 (2020)

JCCS-PFGM: A Novel Circle-Supervision Based Poisson Flow Generative Model for Multiphase CECT Progressive Low-Dose Reconstruction with Joint Condition

Rongjun Ge[1], Yuting He[2], Cong Xia[3], and Daoqiang Zhang[4]($\boxtimes$)

[1] School of Instrument Science and Engineering,
Southeast University, Nanjing, China
[2] School of Computer Science and Engineering, Southeast University, Nanjing, China
[3] Jiangsu Key Laboratory of Molecular and Functional Imaging, Department of Radiology, Zhongda Hospital, Medical School of Southeast University, Nanjing, China
[4] College of Computer Science and Technology,
Nanjing University of Aeronautics and Astronautics, Nanjing, China
dqzhang@nuaa.edu.cn

Abstract. Multiphase contrast-enhanced computed tomography (CECT) scan is clinically significant to demonstrate the anatomy at different phases. But such multiphase scans inherently lead to the accumulation of huge radiation dose for patients, and directly reducing the scanning dose dramatically decrease the readability of the imaging. Therefore, guided with **J**oint **C**ondition, a novel **C**ircle-**S**upervision based **P**oisson **F**low **G**enerative **M**odel (JCCS-PFGM) is proposed to promote the progressive low-dose reconstruction for multiphase CECT. JCCS-PFGM is constituted by three special designs: 1) a progressive low-dose reconstruction mechanism to leverages the imaging consistency and radiocontrast evolution along former-latter phases, so that enormously reduces the radiation dose needs and improve the reconstruction effect, even for the latter-phase scanning with extremely low dose; 2) a circle-supervision strategy embedded in PFGM to enhance the refactoring capabilities of normalized poisson field learned from the perturbed space to the specified CT image space, so that boosts the explicit reconstruction for noise reduction; 3) a joint condition to explore correlation between former phases and current phase, so that extracts the complementary information for current noisy CECT and guides the reverse process of diffusion jointly with multiphase condition for structure maintenance. The extensive experiments tested on the clinical dataset composed of 11436 images show that our JCCS-PFGM achieves promising PSNR up to 46.3dB, SSIM up to 98.5%, and MAE down to 9.67 HU averagely on phases I, II and III, in quantitative evaluations, as well as

Supplementary Information The online version contains supplementary material available at https://doi.org/10.1007/978-3-031-43999-5_39.

gains high-quality readable visualizations in qualitative assessments. All of these findings reveal our method a great potential in clinical multiphase CECT imaging.

1 Introduction

The substantial reduction of scanning radiation dose and its accurate reconstruction are of great clinical significance for multiphase contrast-enhanced computed tomography (CECT) imaging. 1) Multiphase CECT requires multiple scans at different phases, such as arterial phase, venous phase, delayed phase and etc., to demonstrate the anatomy and lesion with the contrast agent evolution intra human body over time [1]. But such multiphase scans inherently lead to the accumulation of huge radiation dose for patients [2,3]. As shown in Fig. 1(a), after triple-phase CECT scanning, the radiation damage suffered by the patient is three times that of the single phase. Combined with "as low as reasonably achievable" (ALARA) principle [4], it is thus extremely urgent to greatly reduce the radiation dose and risk for clinical multiphase CECT examination. 2) However, the low-dose acquired CT image also exists the problems of noise interference and unclear structure. As enlarged region shown in Fig. 1(b), the low-dose CECT behaves much lower signal-to-noise ratio than normal-dose CECT. It brings great difficulty to read the anatomical structure with high noise, especially for inexperienced radiologist. Therefore, the high-quality reconstruction with more readable pattern is clinically crucial for multi-phase low-dose CECT diagnosis.

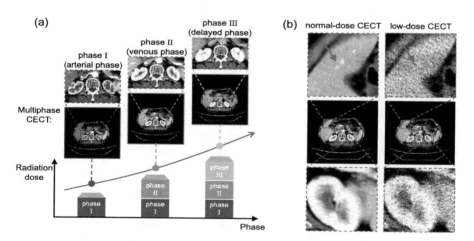

Fig. 1. (a) Multiphase CECT scans comprehensively enable the demonstration of anatomy and lesion with the contrast agent evolution at different phases, but inherently lead to the accumulation of huge radiation dose for patients. (b) Low-dose CECT effectively reduce the radiation risk, but it causes difficulty to read the anatomical structure with high noise, compared to normal-dose CECT.

As far as we know, most of the existing methods mainly focus on the single-phase low-dose CT (LDCT) reconstruction. Chen et al. [5] trained a deep CNN to transform LDCT images towards normal-dose CT images, patch by patch. In [6], a shortcut connections aided symmetrical CNN was adopt to predict noise distribution in LDCT. Shan et al. [7] attempted to transfer a trained 2D CNN to a 3D counterpart for low-dose CT image denoising. In [8], an attention residual dense network was developed for LDCT sinogram data denoising. In [9], low-dose sinogram- and image- domain networks were trained in a progressive way. Zhang et al. [10] further connected sinogram- and image- domain networks together for joint training. In [11] and [12], parallel network architectures were put forward for dual-domain information exchange and mutual optimization.

Multi-phase low-dose CT reconstruction is still ignored, though single-phase methods behave promising results on their issues [5–12]. Due to multiple scans in a short time, it has the inherent challenges: 1) The serious noise pollution is caused by the higher requirement of using much lower scanning dose to decrease multiphase radiation accumulation, compared to the single-phase imaging. Thus, how to elegantly learn such mapping relation from the lower-dose CECT with more serious noise to normal-dose CECT is extremely critical. 2) Complex multiphase correlation with redundancy and interference is induced by the evolution of contrast agent in the human body. Except redundancy and interference, strong causality also obviously exists among multiphase. But how to deeply explore such consistency and evolution along the multiphase for further reducing the dose of later phase and improving imaging quality is still an open challenge.

In this paper, guided with Joint Condition, a novel Circle-Supervision based Poisson Flow Generative Model (JCCS-PFGM) is proposed to make the progressive low-dose reconstruction for multiphase CECT. It deeply explores the correlation among multiphase and the mapping learning of PFGM, to progressively reduce the scanning radiation dose of multiphase CECT to the ultra low level of 5% dose, and achieve the high-quality reconstruction with noise reduction and structure maintenance. It thus significantly reduces the radiation risk of multiple CT scans in a short time, accompanied with clear multiphase CECT examination images. The main contributions of JCCS-PFGM can be summarized as: 1) an effectively progressive low-dose reconstruction mechanism is developed to leverages the imaging consistency and radiocontrast evolution along former-latter phases, so that enormously reduces the radiation dose needs and improve the reconstruction effect, even for the latter-phase scanning with extremely low dose; 2) a newly-designed circle-supervision strategy is proposed in PFGM to enhance the refactoring capabilities of normalized poisson field learned from the perturbed space to the specified CT image space, so that boosts the explicit reconstruction for noise reduction; 3) a novel joint condition is designed to explore correlation between former phases and current phase, so that extracts the complementary information for current noisy CECT and guides the reverse process of diffusion jointly with multiphase condition for structure maintenance.

2 Methodology

As shown in Fig. 2, the proposed JCCS-PFGM is progressively performed on multiphase low-dose CECT to reduce the radiation risk in multiple CT imaging and make the high-quality reconstruction with noise reduction and structure maintenance. It is conducted with three special designs: 1) the progressive low-dose reconstruction mechanism (detailed in Sect. 2.1) reasonably utilizes the consistency along the multiphase CECT imaging, via phase-by-phase reducing the radiation dose and introducing the priori knowledge from former-phase reconstruction; 2) the circle-supervision strategy (detailed in Sect. 2.2) embedded in PFGM makes further self-inspection on normal poisson field prediction, via penalizing the deviation between the same-perturbed secondary diffusion; and 3) the joint condition (detailed in Sect. 2.3) integrates the multi-phase consistency and evolution in guiding the reverse process of diffusion, via fusing the complementary information from former phases into current ultra low-dose CECT.

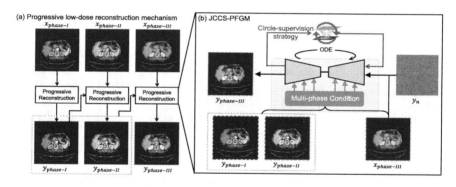

Fig. 2. JCCS-PFGM promotes the progressive low-dose reconstruction for multiphase CECT. It is composed by progressive low-dose reconstruction mechanism, circle-supervision strategy and joint condition.

2.1 Progressive Low-Dose Reconstruction Mechanism

The progressive low-dose reconstruction mechanism effectively promotes high-level base from former-phase to latter-phase for successively multiphase CECT reconstruction, instead of the casually equal dose reduction seriously breaking the structure in each phase. It further exploits the inherent consistency traceable along multiphase CECT to reduce the burden of multiphase reconstruction.

As show in Fig. 2(a), the reasonable-designed progressive low-dose reconstruction mechanism arranges the dose from relatively high to low along the causal multiphase of phases I, II and III. With such mechanism, the reconstruction of former phase acquire more scanning information, benefit from relatively high dose. And the latter phase is granted with much more reliable priori knowledge, benefit from the consistently traceable former-phase reconstruction. Denote the low-dose CECT at phases I, II and III as $x_{phase-I}$, $x_{phase-II}$ and $x_{phase-III}$, the procedure is formulated as:

$$\begin{cases} y_{phase-I} = \mathcal{R}_1(x_{phase-I}) \\ y_{phase-II} = \mathcal{R}_2(x_{phase-II}, y_{phase-I}) \\ y_{phase-III} = \mathcal{R}_3(x_{phase-III}, [y_{phase-II}, y_{phase-I}]) \end{cases} \quad (1)$$

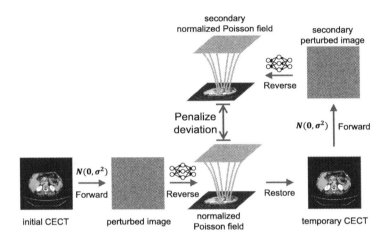

Fig. 3. The circle-supervision strategy robustly boosts the refactoring capabilities of normalized Poisson field learned by PFGM, via penalizing the deviation between the same-perturbed secondary diffusion.

where $y_{phase-I}$, $y_{phase-II}$ and $y_{phase-III}$ represent reconstruction result, as well as $\mathcal{R}_1(\cdot)$, $\mathcal{R}_2(\cdot)$ and $\mathcal{R}_3(\cdot)$ mean reconstruction model.

2.2   Circle-Supervision Strategy Embedded in PFGM

The circle-supervision strategy robustly boosts the refactoring capabilities of normalized Poisson field learned by PFGM. So that it further promotes the explicit reconstruction for noise reduction, instead of just CT-similar image generation.

PFGM is good at mapping a uniform distribution on a high-dimensional hemisphere into any data distribution [13]. It is inspired by electrostatics, and interpret initial image data points as electrical charges on the $z = 0$ hyperplane. Thus the initial image is able to be transformed into a uniform distribution on the hemisphere with radius $r \to \infty$. It estimates normalized Poisson field with deep neural network (DNN) and thus further uses backward ordinary differential equation (ODE) to accelerate sampling.

Here, we further propose the circle-supervision strategy on the normalized Poisson field which reflects the mapping direction from the perturbed space to the specified CT image space. It remedies the precise perception on the target initial CECT besides mapping direction learning, to enhance the crucial field components. As shown in Fig. 3, after randomly yield perturbed image through forward process, the DNN estimates the normalized Poisson field ϕ_1. Then according to the normalized field calculation in the forward process, the Poisson field is returned with denormalization operation, and further temporarily restore the perturbed image into the initial CECT space. The secondary diffusion is conducted with same perturbtion in forward process and DNN in reverse process. Finally, the normal Poisson field of the secondary diffusion ϕ_2 is estimated. The deviation between ϕ_1 and ϕ_2 is penalized to boost the refactoring capabilities. Besides the temporary CECT is also yield in the secondary diffusion to enhance the robustness.

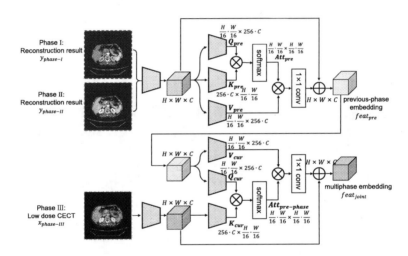

Fig. 4. The joint condition comprehensively fuses the consistency and evolution from the former phases to enhance the current ultra low-dose CECT (take the phase III low-dose CECT reconstruction for example). It is composed of self-fusion among former phases for consistency, and cross-fusion between former and current phases for evolution.

2.3   Joint Condition Fusing Multiphase Consistency and Evolution

The joint condition comprehensively fuses the consistency and evolution from the previous phases to enhance the current ultra low-dose CECT. With such multiphase fusion, the reverse process of diffusion is able to get the successive guide to perceive radiocontrast evolution for structure maintenance

As shown in Fig. 4, the joint condition consists of two parts: the self-fusion among previous phases for consistency and the cross-fusion between previous

Table 1. The quantitative analysis of the proposed method under different configurations. (PLDRM: Progressive low-dose reconstruction mechanism with direct concatenation, CS: Circle-supervision strategy)

Method	Phase I (30%dose)			Phase II (15%dose)			Phase III (5%dose)		
	MAE ($\downarrow$)	PSNR ($\uparrow$)	SSIM ($\uparrow$)	MAE ($\downarrow$)	PSNR ($\uparrow$)	SSIM ($\uparrow$)	MAE ($\downarrow$)	PSNR ($\uparrow$)	SSIM ($\uparrow$)
PFGM	7.80HU	47.4 db	97.3%	10.3HU	44.8 db	97.1%	16.0HU	42.5 db	96.8%
+PLDRM	7.80HU	47.4 db	97.3%	9.7HU	45.3 db	97.7%	15.3HU	43.1 db	97.2%
+CS	7.23HU	47.9 db	98.0%	9.2HU	45.6 db	98.0%	14.8HU	43.8 db	97.5%
Ours	**6.12HU**	**48.4 db**	**98.8%**	**8.49HU**	**46.0 db**	**98.7%**	**14.4HU**	**44.7 db**	**98.0%**

Table 2. The quantitative analysis of the proposed method compared with the existing methods.

Method	Phase I (30%dose)			Phase II (15%dose)			Phase III (5%dose)		
	MAE ($\downarrow$)	PSNR ($\uparrow$)	SSIM ($\uparrow$)	MAE ($\downarrow$)	PSNR ($\uparrow$)	SSIM ($\uparrow$)	MAE ($\downarrow$)	PSNR ($\uparrow$)	SSIM ($\uparrow$)
FBP	15.78HU	40.1 db	93.9%	21.5HU	37.6 db	89.2%	34.9HU	33.7 db	77.9%
RED-CNN	7.76HU	47.0 db	97.7%	11.8HU	44.2 db	96.2%	18.6HU	42.2 db	95.4%
CLEAR	8.61HU	45.9 db	96.2%	10.1HU	45.4 db	97.4%	19.7HU	41.5 db	95.9%
DDPNet	8.07HU	46.8 db	97.5%	10.4HU	45.8 db	98.0%	16.8HU	42.7 db	97.1%
Ours	**6.12HU**	**48.4 db**	**98.8%**	**8.49HU**	**46.0 db**	**98.7%**	**14.4HU**	**44.7 db**	**98.0%**

and current phases for evolution. 1) For the self-fusion among former phases, it firstly encodes the combination of the reconstruction results of the previous phase I $y_{phase-I}$ and phase II $y_{phase-II}$ into the feature domain. And key map K_{pre}, query map Q_{pre} and value map V_{pre} are then generated with further encoder and permutation. K_{pre} and Q_{pre} together establish the correlation weight, i.e., attention map Att_{pre}, among former phases combination to explore the inherent consistency. Att_{pre} further works on the value map V_{pre} to extract the consistent information which is finally added on the first feature representation to get previous-phases embedding $feat_{pre}$. The procedure is formulated as:

$$
\begin{cases}
K_{pre} = E_{K-pre}(E_{pre}([y_{phase-I}, y_{phase-II}])) \\
Q_{pre} = E_{Q-pre}(E_{pre}([y_{phase-I}, y_{phase-II}])) \\
V_{pre} = E_{V-pre}(E_{pre}([y_{phase-I}, y_{phase-II}])) \\
feat_{pre} = Conv(Softmax(Q_{pre}K_{pre}/\sqrt{d})V_{pre}) + E_{pre}([y_{phase-I}, y_{phase-II}])
\end{cases}
\tag{2}
$$

where $E_{pre}(\cdot), E_{K-pre}(\cdot), E_{Q-pre}(\cdot)$ and $E_{V-pre}(\cdot)$ are encoders, $Softmax(\cdot)$ means Softmax function, and $Conv((\cdot)$ represents 1×1 convolution.

2) For the cross-fusion between previous and current phases, it uses $feat_{pre}$ to generate query map Q_{cur} and value map V_{cur}. The current phase III low-dose CECT $x_{phase-III}$ is encoded for key map K_{cur}. Thus the evolution between the current phase and previous phases is explored between K_{cur} and Q_{cur} by attention map Att_{cur}. Then complimentary evolution from previous phases is extracted from value map V_{cur} with Att_{cur}, and then added into the current

phase. The procedure is formulated as:

$$
\begin{cases}
K_{cur} = E_{K-cur}(E_{cur}(x_{phase-III})) \\
Q_{cur} = E_{Q-cur}(feat_{pre}) \\
V_{cur} = E_{V-cur}(feat_{pre}) \\
feat_{cur} = Conv(Softmax(Q_{cur}K_{cur}/\sqrt{d})V_{cur}) + E_{cur}(x_{phase-III})
\end{cases}
\tag{3}
$$

3 Experiments

3.1 Materials and Configurations

A clinical dataset consists 38496 CECT images from 247 patient are used in the experiment. Each patient has triple-phase CECT. We randomly divide the dataset into 123 patients' CECT for training, 49 for validation, and 75 for testing. The basic DNN used in reverse process is same as PFGM. The joint condition is introduced in DNN by making cross attention with DNN feature. Mean Absolute Error (MAE), SSIM and Peak Signal-to-Noise- Ratio(PSNR) are used to evaluate the performance. The corresponding multiphase low-dose CECT is simulated by validated photon-counting model that incorporates the effect of the bowtie filter, automatic exposure control, and electronic noise [14].

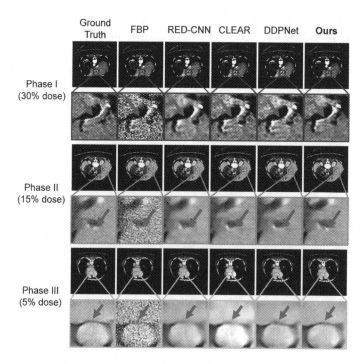

Fig. 5. Visual comparison with competing methods, the proposed JCCS-PFGM effectively keeps structural subtitles in all phases.

3.2 Results and Analysis

Overall Performance. As the last column shown in Table 1 and Table 2, the proposed JCCS-PFGM achieves high-quality multiphase low-dose CECT reconstruction with MAE down to 9.67 HU for information recovery, PSNR up to 46.3 dB for noise reduction, and SSIM up to 98.5% for structure maintaince, averagely on phases I, II and III.

Ablation Study. As shown in Table 1, the proposed JCCS-PFGM gains error decrease of 1.23HU, as well as increase of 1.05 dB PSNR and 1.06% SSIM, in average, compared to the basic PFGM with current-phase condition, and various configurations by successively adding progressive low-dose reconstruction mechanism and circle-supervision strategy. It reveals the effect of each special design. Especially for phase III with ultra low dose of 5%, it gets great improvement.

Comparison with Competing Methods. As shown in Table 2, the proposed JCCS-PFGM gains the best performance compared to FBP, RED-CNN [5], CLEAR [10] and DDPNet [12], with error decrease of 5.66 HU, PSNR increase of 3.62dB PSNR and SSIM improvement by 3.88% on average. Visually, Fig. 5 illustrates that the result from JCCS-PFGM preserved tiny structure in all the phases with the dose assignment scheme: phase I 30% of the total nominal dose, phase II 15% and phase III 5%. In the enlarged ROI where the interpretation is difficult with original LDCT images, our method revealed key details, such as the vessel indicated by the red arrow, much better than the compared methods.

4 Conclusion

In this paper, we propose JCCS-PFGM to make the progressive low-dose reconstruction for multiphase CECT. JCCS-PFGM t creatively consists of 1) the progressive low-dose reconstruction mechanism utilizes the consistency along the multiphase CECT imaging; 2) the circle-supervision strategy embedded in PFGM makes further self-inspection on normal poisson field prediction; 3) the joint condition integrates the multi-phase consistency and evolution in guiding the reverse process of diffusion. Extensive experiments with promising results from both quantitative evaluations and qualitative assessments reveal our method a great clinical potential in CT imaging.

Acknowledgements. This study was supported by Natural Science Foundation of Jiangsu Province (No. BK20210291), National Natural Science Foundation (No. 62101249), China Postdoctoral Science Foundation (No. 2021TQ0149, 2022M721611).

References

1. Meng, X.P., et al.: Radiomics analysis on multiphase contrast-enhanced CT: a survival prediction tool in patients with hepatocellular carcinoma undergoing transarterial chemoembolization. Front. Oncol. **10**, 1196 (2020)
2. Brenner, D.J., Hall, E.J.: Computed tomography an increasing source of radiation exposure. New England J. Med. **357**(22), 2277–2284 (2007)

3. Rastogi, S., et al.: Use of multiphase CT protocols in 18 countries: appropriateness and radiation doses. Can. Assoc. Radiol. J. **72**(3), 381–387 (2021)
4. Prasad, K.N., Cole, W.C., Haase, G.M.: Radiation protection in humans: extending the concept of as low as reasonably achievable (ALARA) from dose to biological damage. Br. J. Radiol. **77**(914), 97–99 (2004)
5. Chen, H., et al.: Low-dose CT denoising with convolutional neural network. In: 14th International Symposium on Biomedical Imaging (ISBI 2017), pp. 143–146. IEEE, Melbourne (2017)
6. Chen, H., et al.: Low-dose CT with a residual encoder-decoder convolutional neural network. IEEE Trans. Med. Imaging **36**(12), 2524–2535 (2017)
7. Shan, H., Zhang, Y., Yang, Q., Kruger, U., Kalra, M.K., Wang, G.: 3-D convolutional encoder-decoder network for low-dose CT via transfer learning from a 2-D trained network. IEEE Trans. Med. Imaging **37**(6), 1522–1534 (2018)
8. Ma, Y.J., Ren, Y., Feng, P., He, P., Guo, X.D., Wei, B.: Sinogram denoising via attention residual dense convolutional neural network for low-dose computed tomography. Nucl. Sci. Tech. **32**(4), 1–14 (2021)
9. Yin, X., et al.: Domain progressive 3D residual convolution network to improve low-dose CT imaging. IEEE Trans. Med. Imaging **38**(12), 2903–2913 (2019)
10. Zhang, Y., et al.: CLEAR: comprehensive learning enabled adversarial reconstruction for subtle structure enhanced low-dose CT imaging. IEEE Trans. Med. Imaging **40**(11), 3089–3101 (2021)
11. Ye, X., Sun, Z., Xu, R., Wang, Z., Li, H.: Low-dose CT reconstruction via dual-domain learning and controllable modulation. In: Wang, L., Dou, Q., Fletcher, P.T., Speidel, S., Li, S. (eds.) MICCAI 2022. LNCS, vol. 13436, pp. 549–559. Springer, Cham (2022). https://doi.org/10.1007/978-3-031-16446-0_52
12. Ge, R., et al.: DDPNet: a novel dual-domain parallel network for low-dose CT reconstruction. In: Wang, L., Dou, Q., Fletcher, P.T., Speidel, S., Li, S. (eds.) MICCAI 2022. LNCS, vol. 13436, pp. 748–757. Springer, Cham (2022). https://doi.org/10.1007/978-3-031-16446-0_71
13. Xu, Y., Liu, Z., Tegmark, M., Jaakkola, T. S.: Poisson flow generative models. In: Advances in Neural Information Processing Systems (2022)
14. Yu, L., Shiung, M., Jondal, D., McCollough, C.H.: Development and validation of a practical lower-dose-simulation tool for optimizing computed tomography scan protocols. J. Comput. Assist. Tomogr. **36**(4), 477–487 (2012)

Motion Compensated Unsupervised Deep Learning for 5D MRI

Joseph Kettelkamp[1]([✉]) [iD], Ludovica Romanin[2] [iD], Davide Piccini[2] [iD], Sarv Priya[1] [iD], and Mathews Jacob[1] [iD]

[1] University of Iowa, Iowa City, IA, USA
{joseph-kettelkamp,sarv-priya,mathews-jacob}@uiowa.edu
[2] Advanced Clinical Imaging Technology, Siemens Healthineers International AG, Lausanne, Switzerland
{ludovica.romanin,davide.piccini}@siemens-healthineers.com

Abstract. We propose an unsupervised deep learning algorithm for the motion-compensated reconstruction of 5D cardiac MRI data from 3D radial acquisitions. Ungated free-breathing 5D MRI simplifies the scan planning, improves patient comfort, and offers several clinical benefits over breath-held 2D exams, including isotropic spatial resolution and the ability to reslice the data to arbitrary views. However, the current reconstruction algorithms for 5D MRI take very long computational time, and their outcome is greatly dependent on the uniformity of the binning of the acquired data into different physiological phases. The proposed algorithm is a more data-efficient alternative to current motion-resolved reconstructions. This motion-compensated approach models the data in each cardiac/respiratory bin as Fourier samples of the deformed version of a 3D image template. The deformation maps are modeled by a convolutional neural network driven by the physiological phase information. The deformation maps and the template are then jointly estimated from the measured data. The cardiac and respiratory phases are estimated from 1D navigators using an auto-encoder. The proposed algorithm is validated on 5D bSSFP datasets acquired from two subjects.

Keywords: Free Running MRI · 5D MRI · Cardiac MRI

1 Introduction

Magnetic Resonance Imaging (MRI) is currently the gold standard for assessing cardiac function. It provides detailed images of the heart's anatomy and enables

This work is supported by grants NIH R01 AG067078 and R01 EB031169.

Supplementary Information The online version contains supplementary material available at https://doi.org/10.1007/978-3-031-43999-5_40.

accurate measurements of parameters such as ventricular volumes, ejection fraction, and myocardial mass. Current clinical protocols, which rely on serial breath-held imaging of the different cardiac slices with different views, often require long scan times and are associated with reduced patient comfort. Compressed sensing [1], deep learning [6,10], and motion-compensated approaches [13] were introduced to reduce the breath-hold duration in cardiac CINE MRI. Unfortunately, many subject groups, including pediatric and older subjects, cannot comply with even the shorter breath-hold durations.

5D free-breathing MRI approaches that rely on 3D radial readouts [3,11] have been introduced to overcome the above challenges. These methods resolve the respiratory and cardiac motion from either the center of k-space or Superior-Inferior (SI) k-space navigators. The k-space data is then binned into different cardiac/respiratory phases and jointly reconstructed using compressed sensing. The main benefit of this motion-resolved strategy is the ability to acquire the whole heart with isotropic spatial resolution as high as $1\,mm^3$. This approach allows the images to be reformatted into different views to visualize specific anatomical regions at different cardiac and/or respiratory phases. Despite the great potential of 5D MRI, current methods have some challenges that limit their use in routine clinical applications. Firstly, the motion-resolved compressed sensing reconstruction is very computationally intensive, and it can take several hours to have a dynamic 3D volume. And secondly, compressed sensing reconstructions require fine tuning of several regularization parameters, which greatly affect the final image quality, depending on the undersampling factor and the binning uniformity.

The main focus of this work is to introduce a motion-compensated reconstruction algorithm for 5D MRI. The proposed approach models the images at every time instance as a deformed version of a static image template. Such an image model may not be a good approximation in 2D schemes [13], where the organs may move in and out of the slice. However, the proposed model is more accurate for the 3D case. We introduce an auto-encoder to estimate the cardiac and respiratory phases from the superior-inferior (SI) k-t space navigators. We disentangle the latent variables to cardiac and respiratory phases by using the prior information of the cardiac and respiratory rates. The latent variables allow us to bin the data into different cardiac and respiratory phases. We use an unsupervised deep learning algorithm to recover the image volumes from the clustered data. The algorithm models the deformation maps as points on a smooth low-dimensional manifold in high dimensions, which is a non-linear function of the low-dimensional latent vectors. We model the non-linear mapping by a Convolutional Neural Network (CNN). When fed with the corresponding latent vectors, this CNN outputs the deformation maps corresponding to a specific cardiac or respiratory phase. We learn the parameters of the CNN and the image template from the measured k-t space data. We note that several manifold based approaches that model the images in the time series by a CNN were introduced in the recent years. [5,9,15]. All of these methods rely on motion resolved reconstruction, which is conceptually different from the proposed motion compensated reconstruction.

We validate the proposed scheme on cardiac MRI datasets acquired from two healthy volunteers. The results show that the approach is capable of resolving the cardiac motion, while offering similar image quality for all the different phases. In particular, the motion-compensated approach can combine the image information from all the motion states to obtain good quality images.

2   Methods

2.1   Acquisition Scheme

In vivo acquisitions were performed on a 1.5T clinical MRI scanner (MAGNE-TOM Sola, Siemens Healthcare, Erlangen, Germany). The free-running research sequence used in this work is a bSSFP sequence, in which all chemically shift-selective fat saturation pulses and ramp-up RF excitations were removed, in order to reduce the specific absorption rate (SAR) and to enable a completely uninterrupted acquisition [8]. K-space data were continuously sampled using a 3D golden angle kooshball phyllotayis trajectory [7], interleaved with the acquisition of a readout oriented along the superior-inferior (SI) direction for cardiac and respiratory self-gating [11]. The main sequence parameters were: radio frequency excitation angle of 55 with an axial slab-selective sinc pulse, resolution of $1.1\,mm^3$, FOV of $220\,mm^3$, TE/TR of $1.87/3.78\,ms$, and readout bandwidth of $898\,Hz/pixel$. The total fixed scan time was 7:58 min.

2.2   Forward Model

We model the measured k-space data at the time instant t as the multichannel Fourier measurements of $\boldsymbol{\rho}_t = \boldsymbol{\rho}(\mathbf{r}, t)$, which is the image volume at the time instance t:

$$\mathbf{b}_t = \underbrace{\mathbf{F}_{\mathbf{k}_t}\,\mathbf{C}\,\boldsymbol{\rho}_t}_{\mathcal{A}_t(\boldsymbol{\rho}_t)} \tag{1}$$

Here, $\mathbf{C}$ denotes the multiplication of the images by the multi-channel coil sensitivities, while $\mathbf{F}_{\mathbf{k}}$ denotes the multichannel Fourier operator. $\mathbf{k}_t$ denotes the k-space trajectory at the time instant t. In this work, we group 22 radial spokes corresponding to a temporal resolution of 88 ms.

An important challenge associated with the bSSFP acquisition without intermittent fat saturation pulses is the relatively high-fat signal compared to the myocardium and blood pool. Traditional parallel MRI and coil combination strategies often result in significant streaking artifacts from the fat onto the myocardial regions, especially in the undersampled setting considered in this work. We used the coil combination approach introduced in [4] to obtain virtual channels that are maximally sensitive to the cardiac region. A spherical region covering the heart was manually selected as the region of interest (ROI), while its complement multiplied by the distance function to the heart was chosen as the noise mask. We chose the number of virtual coils that preserve 75% of the energy within the ROI. This approach minimizes the strong fat signals, which are distant from the myocardium. We used the JSENSE algorithm [12,14] to compute the sensitivity maps of the virtual coils.

2.3 Image and Motion Models

The overview of the proposed scheme is shown in Fig. 1. The recovery of ρ_t from very few of their measurements $\mathbf{b}_t$ is ill-posed. To constrain the recovery, we model ρ_t as the deformed version of a static image template $\boldsymbol{\eta}(\mathbf{r})$:

$$\rho(\mathbf{r}, t) = \mathcal{I}\left[\boldsymbol{\eta}, \phi_t(\mathbf{r})\right] \tag{2}$$

Here, ϕ_t is the deformation map and the operator $\mathcal{I}$ denotes the deformation of $\boldsymbol{\eta}$. We implement (2) using cubic Bspline interpolation. This approach allows us to use the k-space data from all the time points to update the template, once the motion maps ϕ_t are estimated.

Classical MoCo approaches use image registration to estimate the motion maps ϕ_t from approximate (e.g. low-resolution) reconstructions of the images $\rho(\mathbf{r}, t)$. However, the quality of motion estimates depends on the quality of the reconstructed images, which are often low when we aim to recover the images at a fine temporal resolution (e.g. 88 ms).

We propose to estimate the motion maps directly from the measured $k - t$ space data. In particular, we estimate the motion maps ϕ_t such that the multi-channel measurements of $\rho(\mathbf{r}, t)$ specified by (2) match the measurements $\mathbf{b}_t$. We also estimate the template $\boldsymbol{\eta}$ from the k-t space data of all the time points. To constrain the recovery of the deformation maps, we model the deformation maps as the output of a convolutional neural network

$$\phi_t = \mathcal{G}_\theta[\mathbf{z}_t],$$

in response to low-dimensional latent vectors $\mathbf{z}_t$. Here, $\mathcal{G}_\theta$ is a convolutional neural network, parameterized by the weights θ. We note that this approach constrains the deformation maps as points on a low-dimensional manifold. They are obtained as non-linear mappings of the low-dimensional latent vectors $\mathbf{z}_t$, which capture the motion attributes. The non-linear mapping itself is modeled by the CNN.

2.4 Estimation of Latent Vectors from SI Navigators

We propose to estimate the latent vectors $\mathbf{z}_t$ from the SI navigators using an auto-encoder. In this work, we applied a low pass filter with cut-off frequency of 2.8 Hz to the SI navigators to remove high-frequency oscillations. Similarly, an eighth-degree Chebyshev polynomial is fit to each navigator voxel and is subtracted from the signal to remove drifts.

The auto-encoder involves an encoder that generates the latent vectors $\mathbf{z}_t = \mathcal{E}_\varphi(\mathbf{y}_t)$, are the navigator signals. The decoder reconstructs the navigator signals as $\mathbf{y}_t = \mathcal{D}_\psi(\mathbf{z}_t)$. In this work, we restrict the dimension of the latent space to three, two corresponding to respiratory motion and one corresponding to cardiac motion. To encourage the disentangling of the latent vectors to respiratory and cardiac signals, we use the prior information on the range of cardiac and respiratory frequencies as in [2]. We solve for the auto-encoder parameters from the navigator signals of each subject as

$$\{\varphi^*, \psi^*\} = \arg\min_{\varphi,\psi} \left\| \mathcal{F}\left\{ D_\psi\left(\underbrace{\mathcal{E}_\varphi(\mathbf{Y})}_{\mathbf{Z}} \right) - \mathbf{Y} \right\} \right\|_{l=1} + \lambda \left\| \mathbf{Z} \bigotimes \mathcal{B} \right\|_2^2 \quad (3)$$

Here, $\mathbf{Z} \in \mathbb{R}^{3 \times T}$ and $\mathbf{Y}$ are matrices whose columns are the latent vectors and the navigator signals at different time points. $\mathcal{F}$ is the Fourier transformation in the time domain. $\bigotimes$ denotes the convolution of the latent vectors with band-stop filters with appropriate stop bands. In particular, the stopband of the respiratory latent vectors was chosen to be 0.05–0.7 Hz, while the stopband was chosen as the complement of the respiratory bandstop filter Hz for the cardiac latent vectors. We observe that the median-seeking ℓ_1 loss in the Fourier domain is able to offer improved performance compared to the standard ℓ_2 loss used in conventional auto-encoders.

2.5   Motion Compensated Image Recovery

Once the auto-encoder parameters φ, ψ described in (3) are estimated from the navigator signals of the subject, we derive the latent vectors as $\mathbf{Z} = \mathcal{E}_{\varphi^*}(\mathbf{Y})$. Using the latent vectors, we pose the joint recovery of the static image template $\eta(\mathbf{r})$ and the deformation maps as

$$\{\boldsymbol{\eta}^*, \theta^*\} = \arg\min_{\eta,\theta} \sum_{t=1}^{T} \|\mathcal{A}_t\left(\rho(\mathbf{r}, t)\right) - \mathbf{b}_t\|^2 \quad \text{where} \quad \rho(\mathbf{r}, t) = \mathcal{I}\left(\boldsymbol{\eta}, \mathcal{G}_\theta[\mathbf{z}_t]\right) \quad (4)$$

The above optimization scheme can be solved using stochastic gradient optimization. Following optimization, one can generate real-time images shown in Fig. 3 and Fig. 3 as $\mathcal{I}\left(\boldsymbol{\eta}^*, \mathcal{G}_{\theta^*}[\mathbf{z}_t]\right)$.

The optimization scheme described in (4) requires T non-uniform fast Fourier transform steps per epoch. When the data is recovered with a high temporal resolution, this approach translates to a high computational complexity. To reduce computational complexity, we introduce a clustering scheme. In particular, we use k-means clustering to group the data to $N << T$ clusters. This approach allows us to pool the k-space data from multiple time points, all with similar latent codes.

$$\{\boldsymbol{\eta}^*, \theta^*\} = \arg\min_{\eta,\theta} \sum_{n=1}^{N} \|\mathcal{A}_n\left(\rho(\mathbf{r}, n)\right) - \mathbf{b}_n\|^2 \quad \text{where} \quad \rho(\mathbf{r}, n) = \mathcal{I}\left(\boldsymbol{\eta}, \mathcal{G}_\theta[\mathbf{z}_n]\right) \quad (5)$$

The above approach is very similar to (4). Here, $\mathbf{b}_n$, $\mathcal{A}_n$, and $\mathbf{z}_n$ are the grouped k-space data, the corresponding forward operator, and the centroid of the cluster, respectively. Note that once $N \to T$, both approaches become equivalent. In this work, we used $N = 30$. Once the training is done, one can still generate the real-time images as $\mathcal{I}\left(\boldsymbol{\eta}, \mathcal{G}_\theta[\mathbf{z}_t]\right)$.

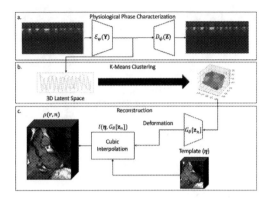

Fig. 1. Overview of the proposed reconstruction algorithm. In the first step shown in (a), we estimate the latent variables that capture the motion in the data using a constrained auto-encoder, as described in Fig. 3. The auto-encoder minimizes a cost function, which is the sum of an ℓ_1 data consistency term and a prior involving cardiac and frequency ranges. To reduce the computational complexity of the image reconstruction, we cluster the latent space using k-means algorithm as shown in (b). The cluster centers are fed in as inputs to the CNN denoted by $\mathcal{G}_\theta$, which outputs the deformation maps $\mathcal{G}_\theta[\mathbf{z}_n]$. We jointly optimize for both the template η and parameters θ of the generator.

2.6 Motion Resolved 5D Reconstruction for Comparison

We compare the proposed approach against a compressed sensing 5D reconstruction. In particular, we used the SI navigators to bin the data into 16 bins, consisting of four cardiac and four respiratory phases as described. We use a total variation regularization similar to [2] to constrain the reconstructions. We determined the regularization parameter manually to obtain the best reconstructions.

We note that the dataset with 6.15 min acquisition is a highly undersampled setting. In addition, because this dataset was not acquired with intermittent fat saturation pulses, it suffers from streaking artifacts that corrupt the reconstructions.

3   Results

We show the results from the two normal volunteers in Fig. 3 and 4, respectively. The images correspond to 2-D slices extracted from the 3D volume, corresponding to different cardiac and respiratory phases. We also show the time profile of the real-time reconstructions $\rho(\mathbf{r}, t) = \mathcal{I}(\eta, \mathcal{G}_\theta[\mathbf{z}_t])$ along the red line shown in the top row. We note that the approach can capture the cardiac and respiratory motion in the data. The different phase images shown in the figure were extracted manually from the real-time movies.

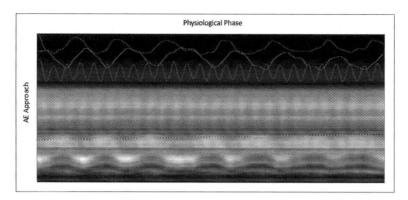

Fig. 2. Latent vectors estimated from the SI navigators (bottom curves) [11]. We note that the orange and the green curves estimated using the auto-encoder roughly follow the respiratory motion, while the blue curves capture the cardiac motion.

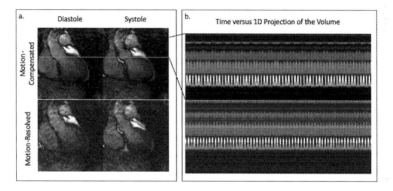

Fig. 3. Results from the first subject. (a) The top row shows a 2-D slice of the reconstructed 3D volume at diastole and systole, obtained using the proposed motion compensated approach. The bottom row shows the motion-resolved compressed sensing recovery of the same data. (b) shows the 1D projection versus time profile of the reconstructed datasets using the motion compensated (top) and motion resolved (bottom) approaches.

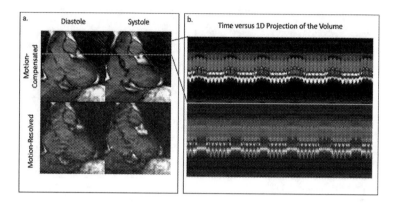

Fig. 4. Results from the second subject. (a) The top row shows a 2-D slice of the reconstructed 3D volume at diastole and systole, obtained using the proposed motion compensated approach. The bottom row shows the motion-resolved compressed sensing recovery of the same data. (b) shows the 1D projection versus time profile of the reconstructed datasets using the motion compensated (top) and motion resolved (bottom) approaches.

4   Discussion

The comparisons in Fig. 3 and 4 show that the proposed approach is able to offer improved reconstructions, where the cardiac phases are well-resolved. We note that the motion resolved reconstruction of the different phases have different image quality, depending on the number of spokes in the specific phases. By contrast, the proposed motion compensated reconstructions are able to combine the data from different motion states; the improved data efficiency translates to reconstructions with reduced streaking artifacts. Additionally, the auto-encoder accurately characterized the SI navigator and disentangled the cardiac and respiratory latent vectors Fig. 2.

We note that the comparison in this work is preliminary. The main focus of this work is to introduce the proposed motion-compensated reconstruction algorithm and the auto-encoder approach to estimate the latent vectors and to demonstrate its utility in 5D MRI. In our future work, we will focus on rigorous studies, including comparisons with 2D CINE acquisitions.

References

1. Bustin, A., Fuin, N., Botnar, R.M., Prieto, C.: From compressed-sensing to artificial intelligence-based cardiac MRI reconstruction. Front. Cardiovasc. Med. **7**, 17 (2020). https://doi.org/10.3389/FCVM.2020.00017/BIBTEX
2. Feng, L., Axel, L., Chandarana, H., Block, K.T., Sodickson, D.K., Otazo, R.: XD-grasp: golden-angle radial MRI with reconstruction of extra motion-state dimensions using compressed sensing. Magn. Reson. Med. **75**(2), 775–788 (2016)
3. Feng, L., et al.: 5D whole-heart sparse MRI. Magn. Reson. Med. **79**(2), 826–838 (2017). https://doi.org/10.1002/mrm.26745

4. Kim, D., Cauley, S.F., Nayak, K.S., Leahy, R.M., Haldar, J.P.: Region-optimized virtual (ROVir) coils: localization and/or suppression of spatial regions using sensor-domain beamforming. Magn. Reson. Med. **86**(1), 197–212 (2021). https://doi.org/10.1002/mrm.28706

5. Mohsin, Y.Q., Poddar, S., Jacob, M.: Free-breathing & ungated cardiac MRI using iterative SToRM (i-SToRM). IEEE Trans. Med. Imaging **38**(10), 2303–2313 (2019). https://doi.org/10.1109/tmi.2019.2908140

6. Oscanoa, J.A., et al.: Deep learning-based reconstruction for cardiac MRI: a review. Bioengineering **10**(3), 334 (2023). https://doi.org/10.3390/bioengineering10030334

7. Piccini, D., Littmann, A., Nielles-Vallespin, S., Zenge, M.O.: Spiral phyllotaxis: the natural way to construct a 3D radial trajectory in MRI. Magn. Reson. Med. **66**(4), 1049–1056 (2011). https://doi.org/10.1002/mrm.22898

8. Roy, C.W., et al.: Free-running cardiac and respiratory motion-resolved 5D whole-heart coronary cardiovascular magnetic resonance angiography in pediatric cardiac patients using ferumoxytol. J. Cardiovasc. Magn. Reson. **24**(1) (2022). https://doi.org/10.1186/s12968-022-00871-3

9. Rusho, R.Z., Zou, Q., Alam, W., Erattakulangara, S., Jacob, M., Lingala, S.G.: Accelerated pseudo 3D dynamic speech MR imaging at 3T using unsupervised deep variational manifold learning. In: Wang, L., Dou, Q., Fletcher, P.T., Speidel, S., Li, S. (eds.) MICCAI 2022. Lecture Notes in Computer Science, vol. 13436, pp. 697–706. Springer, Cham (2022). https://doi.org/10.1007/978-3-031-16446-0_66

10. Schlemper, J., Caballero, J., Hajnal, J.V., Price, A.N., Rueckert, D.: A deep cascade of convolutional neural networks for dynamic MR image reconstruction. IEEE Trans. Med. Imaging **37**, 491–503 (2018). https://doi.org/10.1109/TMI.2017.2760978, https://pubmed.ncbi.nlm.nih.gov/29035212/

11. Sopra, L.D., Piccini, D., Coppo, S., Stuber, M., Yerly, J.: An automated approach to fully self-gated free-running cardiac and respiratory motion-resolved 5d whole-heart MRI. Magn. Reson. Med. **82**(6), 2118–2132 (2019). https://doi.org/10.1002/mrm.27898

12. Uecker, M., Hohage, T., Block, K.T., Frahm, J.: Image reconstruction by regularized nonlinear inversion-joint estimation of coil sensitivities and image content. Magn. Reson. Med. **60**(3), 674–682 (2008). https://doi.org/10.1002/mrm.21691

13. Usman, M., et al.: Motion corrected compressed sensing for free-breathing dynamic cardiac MRI. Magn. Reson. Med. **70**, 504–516 (2013). https://doi.org/10.1002/MRM.24463

14. Ying, L., Sheng, J.: Joint image reconstruction and sensitivity estimation in SENSE (JSENSE). Magn. Reson. Med. **57**(6), 1196–1202 (2007). https://doi.org/10.1002/mrm.21245

15. Zou, Q., Torres, L.A., Fain, S.B., Higano, N.S., Bates, A.J., Jacob, M.: Dynamic imaging using motion-compensated smoothness regularization on manifolds (mocostorm). Phys. Med. Biol. **67** (2021). https://doi.org/10.1088/1361-6560/ac79fc, https://arxiv.org/abs/2112.03380

Differentiable Beamforming
for Ultrasound Autofocusing

Walter Simson[(✉)][iD], Louise Zhuang[iD], Sergio J. Sanabria[iD], Neha Antil[iD], Jeremy J. Dahl[iD], and Dongwoon Hyun[iD]

Stanford University, Stanford, CA 94305, USA
{waltersimson,dongwoon.hyun}@stanford.edu

Abstract. Ultrasound images are distorted by phase aberration arising from local sound speed variations in the tissue, which lead to inaccurate time delays in beamforming and loss of image focus. Whereas state-of-the-art correction approaches rely on simplified physical models (e.g. phase screens), we propose a novel physics-based framework called differentiable beamforming that can be used to rapidly solve a wide range of imaging problems. We demonstrate the generalizability of differentiable beamforming by optimizing the spatial sound speed distribution in a heterogeneous imaging domain to achieve ultrasound autofocusing using a variety of physical constraints based on phase shift minimization, speckle brightness, and coherence maximization. The proposed method corrects for the effects of phase aberration in both simulation and in-vivo cases by improving image focus while simultaneously providing quantitative speed-of-sound distributions for tissue diagnostics, with accuracy improvements with respect to previously published baselines. Finally, we provide a broader discussion of applications of differentiable beamforming in other ultrasound domains.

Keywords: Ultrasound · Image reconstruction · Optimization

1 Introduction

Ultrasound images are reconstructed by time sampling the reflected pressure signals measured by individual transducer elements in order to focus at specific spatial locations. The sample times are calculated so as to compensate for the time-of-flight from the elements to the desired spatial locations, often by assuming a constant speed of sound (SoS) in the medium, e.g., 1540 m m/s. However, the human body is highly heterogeneous, with slower SoS in adipose layers than in fibrous and muscular tissues. If unaccounted for, these differences lead to phase aberration, geometric distortions, and loss of focus and contrast

Supplementary Information The online version contains supplementary material available at https://doi.org/10.1007/978-3-031-43999-5_41.

[1]. This degradation is a fundamental limitation of current ultrasound image reconstruction and impacts downstream tasks such as diagnostics, volumetry, and registration.

Historically, phase aberration has been described using simplified phase-screen models [5,19], which assume that distortions generated from an unknown SoS can be modeled by a gross time delay offset at every element [1]. More recently, several methods have been proposed to estimate SoS distribution of the medium from aberration measurements as a step before actual image correction. A family of these methods still relies on simplified physical models of wave propagation to derive tractable inverse problems. These include assuming a horizontally layered medium [9] or coherent plane wavefront propagation at different angulations [17]. To reinforce specific assumptions about SoS heterogeneity, regularization is often introduced, including total variation for focal inclusion geometries [15] and Tikhonov regularization for smoothly varying layered SoS distributions [14,17]. While these methods perform well for one class of SoS inversion problems, it is challenging to generalize their applicability to arbitrary SoS distributions, which are generally found in clinical scenarios. Work has been carried out to find more generalizable estimation based on training neural network models to end-to-end learn SoS distributions or optimize the regularization function basis [4,16,18]. However, these methods require thousands of training instances, which can currently practically only be obtained from in-silico simulations and show challenges generalizing to real data.

Recent developments in artificial intelligence have been facilitated by the release of open-source tensor libraries, which can perform automatic differentiation of composable transformations on vector data. These libraries are the backbone of complex neural network architectures that use automatic reverse-mode differentiation (back-propagation) to iteratively optimize weights based on a set of training instances. These libraries also simplify and optimize portability to high-performance computing platforms. We hypothesize that such libraries can likewise be extended to model the pipeline of ultrasound image reconstruction as a composition of differentiable operations, allowing optimization based on a single data instance.

In this work, we propose an ultrasound imaging paradigm that jointly achieves sound speed estimation and image quality enhancement via differentiable beamforming. We formulate image reconstruction as a differentiable function of a spatially heterogeneous SoS map, and optimize it based on quality metrics extracted from the final reconstructed images (Fig. 1).

2   Methods

2.1   Beamforming Multistatic Synthetic Aperture Data

In ultrasound imaging, radiofrequency data (RF) represents the time series signal proportional to the pressure measured by each probe array sensor. A multistatic synthetic aperture dataset contains the RF pulse-echo responses of every pair of transmit and receive elements. We denote the signal due to the i-th transmit

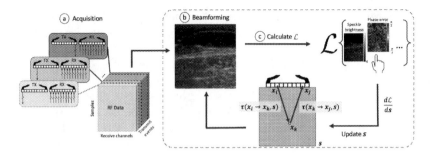

Fig. 1. Differentiable beamforming method for ultrasound autofocusing. Part (a) shows the initial full synthetic aperture data acquisition. The complete RF data is then used for beamforming in part (b) with an initial estimate of slowness, and afterwards, a desired loss is calculated in part (c). The loss is differentiated with respect to the slowness, which is then updated and used for the next iteration of beamforming. This process encapsulated in the box is then repeated until convergence is reached.

element, and j-th receive element as $u_{ij}(t)$. This signal can be focused to an arbitrary spatial location $\mathbf{x}_k$ by sampling $u_{ij}(t)$ at the time corresponding to the time-of-flight τ from the transmit element at $\mathbf{x}_i$ to $\mathbf{x}_k$ and back to the receive element at $\mathbf{x}_j$, achieved via 1D interpolation of the RF signal:

$$u_{ij}(\mathbf{x}_k) = u_{ij}\left(\tau(\mathbf{x}_i, \mathbf{x}_k) + \tau(\mathbf{x}_k, \mathbf{x}_j)\right). \tag{1}$$

(We describe our time-of-flight model in greater detail below in Sect. 2.3.) The interpolated signals are then summed across the transmit (N_t) and receive (N_r) apertures to obtain a focused ultrasound image:

$$u(\mathbf{x}_k) = \sum_{i=1}^{N_t} \sum_{j=1}^{N_r} u_{ij}(\mathbf{x}_k). \tag{2}$$

This process of interpolation and summation is called delay-and-sum (DAS) beamforming.

2.2 Differentiable Beamforming

DAS is composed of elementary differentiable operations and is consequently itself differentiable. Therefore, DAS can be incorporated into an automatic differentiation (AD) framework to allow for differentiation with respect to any desired input parameters $\boldsymbol{\theta}$. For a given loss function $\mathcal{L}(u(\mathbf{x}_k; \boldsymbol{\theta}))$ that measures the "quality" of the beamforming, $\boldsymbol{\theta}$ can be optimized using gradient descent to identify the optimal $\boldsymbol{\theta}^\star$ using update steps $\Delta\boldsymbol{\theta}$:

$$\boldsymbol{\theta}^\star = \arg\min_{\boldsymbol{\theta}} \mathcal{L}(u(\mathbf{x}_k; \boldsymbol{\theta})), \qquad \Delta\boldsymbol{\theta} = \boldsymbol{\theta} - \alpha\frac{\partial}{\partial\boldsymbol{\theta}}\mathcal{L}(u(\mathbf{x}_k; \boldsymbol{\theta})). \tag{3}$$

This differentiable framework is flexible, providing many ways to parameterize the beamforming. In this work, we will show the promise of differentiable beamforming on the task of sound speed estimation by optimizing for slowness $\mathbf{s}$ in a time of flight delay model (i.e. $\boldsymbol{\theta} = \mathbf{s}$).

2.3   Time of Flight Model

Here, we parameterize the slowness (i.c. the reciprocal of the sound speed) as a function of space. Specifically, we define the slowness at a set of control points as $\mathbf{s} = \{s(\mathbf{x}_k)\}_k$, which can be interpolated to obtain the slowness at arbitrary $\mathbf{x}$. The time-of-flight from $\mathbf{x}_1$ to $\mathbf{x}_2$ is the integral of the slowness along the path:

$$\tau(\mathbf{x}_1, \mathbf{x}_2; \mathbf{s}) = \int_{\mathbf{x}_1 \to \mathbf{x}_2} \mathbf{s} \, \mathrm{d}\mathbf{x}. \tag{4}$$

For simplicity and direct comparison with previous sound speed estimation models [17], a straight ray model of wave propagation is used.

2.4   Loss Functions for Sound Speed Optimization

Speckle Brightness Maximization. Diffuse ultrasound scattering produces an image texture called speckle. Speckle brightness can be used as a criterion of focus quality [13]. Written as a loss, this is the negative average pixel magnitude:

$$\mathrm{SB}(\mathbf{s}) = \frac{1}{N_k} \sum_k |u(\mathbf{x}_k; \mathbf{s})| = -\mathcal{L}_{\mathrm{SB}}(\mathbf{s}). \tag{5}$$

Coherence Factor Maximization. Coherence factor [6,11], also referred to as the F criterion or "focusing criterion", defined between 0 and 1, is the measure of the coherent signal sum over the incoherent signal sum of the receive aperture. When received signals are in focus (i.e. in equal phase), CF achieves the maximum value of 1. We use the negative CF as a loss:

$$\mathrm{CF}(\mathbf{s}) = \frac{1}{N_k} \sum_{k=1}^{N_k} \frac{\left| \sum_j \sum_i u_{ij}(\mathbf{x}_k; \mathbf{s}) \right|}{\sum_j |\sum_i u_{ij}(\mathbf{x}_k; \mathbf{s})|} = -\mathcal{L}_{\mathrm{CF}}(\mathbf{s}). \tag{6}$$

Phase-Error Minimization. The van Cittert Zernike theorem of optics [12] states that when imaging diffuse scatterers using a given transmit and receive sub-aperture $\mathcal{T}_a$ and $\mathcal{R}_a$ (i.e. subset of the available array elements), the resulting signal is almost perfectly correlated with the signal from a second set of apertures $\mathcal{T}_b$ and $\mathcal{R}_b$ when the two apertures share a common midpoint. The measured phase-shift between both signals should approach zero when aberration is corrected. Figure 2 illustrates this concept of phase error.

We estimate the phase shift as the complex angle between DAS signals u_a and u_b of the respective subapertures $(\mathcal{T}_a, \mathcal{R}_a)$ and $(\mathcal{T}_b, \mathcal{R}_b)$, calculated using (2):

$$\Delta\phi_{ab}(\mathbf{x_k}) = \angle\mathbb{E}[u_a(\mathbf{x_k}; \mathbf{s})u_b^*(\mathbf{x_k}; \mathbf{s})]. \tag{7}$$

The phase shift error (PE) is defined for a set of all aperture pairs (a, b) with common midpoint as

$$\text{PE}(\mathbf{s}) = \frac{1}{N_{(a,b)}} \sum_{(a,b)} |\Delta\phi_{ab}| = \mathcal{L}_{\text{PE}}(\mathbf{s}). \tag{8}$$

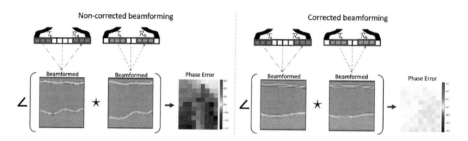

Fig. 2. Phase error minimization in correlated common mid-point sub-apertures. Phase error is computed as the angle of the cross correlation of complex beamformed signals from different sub-apertures sharing a common midpoint. When the correct slowness is used for the beamforming, the phase error is minimized.

3   Experimental Setup

3.1   Implementation of Differentiable Beamformer

A differentiable DAS beamformer was implemented in Python using JAX[1] [3], which provides out of the box GPU acceleration. DAS was parameterized by the slowness map, where the time-of-flights for beamforming were calculated via bilinear interpolation of the slowness along a discretized path from the transmitting element to a location of interest and from the location to a receiving element. The loss was computed on 5×5 pixel patches ($\lambda/2$ pixel spacing) on a regular 15×21 grid spanning the image. The sound speed map was then optimized via gradient descent. For the phase error loss, 17-element subapertures were used for beamforming. The beamformed data for every subaperture pair with a common midpoint were cross-correlated with a 5 × 5 path to compute the phase shift. We further leveraged acoustic reciprocity to combine the results for reciprocal transmit/receive subapertures. This phase-shift measurement was then used for the final phase error loss. The GPU-based implementation runs in ~300 s for 300 iterations on an NVIDIA RTX A6000. The code for this work can be found on GitHub[2].

---

[1] https://github.com/google/jax.
[2] https://github.com/waltsims/dbua.

3.2   Comparison with State-of-the-Art Methods

As a baseline for performance comparison, the Computed Ultrasound Tomography in Echo Mode (CUTE) method developed by Stähli et al. [17] was implemented in MATLAB; this method has been shown to achieve sound speed reconstruction of both layered and focal lesion geometries. The method shows some similarities in using phase error minimization from different apertures (albeit in the angular domain) and ray tracing paths. However, it relies on a coherent plane wavefront propagation model and Tikhonov regularization to build a tractable inverse problem.

3.3   Datasets

In-Silico. The CUDA-accelerated binaries of the k-Wave simulation suite [10] were used to generate multistatic RF data of 3D phantom model acquisitions. To compare with the baseline [17], simulations were first generated using plane-wave transmissions (115 transmits in steering range of -28.5°:0.5°:28.5°) and then converted to FSA format using REFoCUS [2] in the rtbf framework [7]. A linear 128 element linear probe was simulated, with a pitch of 0.3 mm and a center frequency of 4.8 MHz with a 100% bandwidth. The simulation domain was 60×51 x 7.4 mm^3. Iso-echoic phantoms were generated whereby the sound speed was modulated relative to the density of a region so the average brightness remained constant while the sound-speed variation introduced phase aberration.

In-Vivo. In-vivo data was collected on a Verasonics Vantage research system with a L12-3v linear transducer (192 elements, 0.2 mm pitch, 5 MHz center frequency). Three abdominal liver views, which contained subcutaneous adipose, musculoskeletal tissue and liver parenchyma, were collected from a healthy volunteer under a protocol approved by an institutional review board.

4   Results

Figure 3 shows SoS maps for in-silico phantom data. In the uncorrected (naive) B-modes, regions of darkening and smeared speckle can be seen as acoustic intensity diminishes due to aberration. In the quadrant phantom (a), a distinct spatial skewing can be observed from left to right. On the corrected images, image brightness is enhanced, iso-echogenic speckle distributions are revealed, aberrated regions are reduced, and the boundary between quadrants shows a congruent left-to-right and top-to-bottom transition. Similarly, in the inclusion phantom (b), characteristic triangles can be seen to the left and right of the inclusion in the naive B-mode. These triangular offshoots are artifacts produced by total wave reflection on the lateral lesion boundaries when the ultrasound wave encounters an SoS transition at grazing incidence. Moreover, diffraction of waves through the lesion lead to aberration errors behind the lesion. In the corrected B-mode, these dark regions are enhanced, and the image has an overall more homogeneous brightness pattern.

The sound speed distributions generated with differentiable beamforming are in general agreement with the ground truth sound speed distributions. Table 1 quantitatively compares the mean absolute error (MAE) and standard deviation (std) with respect to the ground truth. For all phantoms, differential beamforming achieved lower (better) error metrics than the baseline. Homogeneous phantoms were best reconstructed via CF loss function, while inhomogeneous phantoms were best reconstructed via PE loss function.

Figure 4 shows preliminary results with differential beamforming for the reconstructed in-vivo data. The SoS reconstruction successfully delineates abdominal layers including subcutaneous adipose fat (average 1494 m m/s), muscle (average 1551 m m/s) and liver parenchyma (average 1530 m m/s) in agreement with the literature values [8].

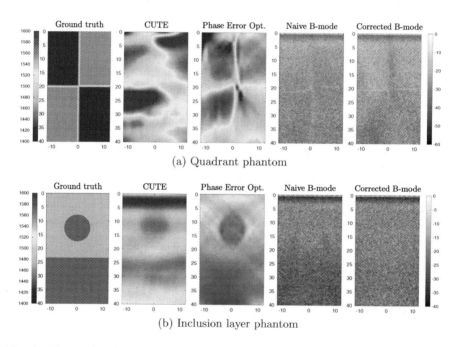

(a) Quadrant phantom

(b) Inclusion layer phantom

Fig. 3. The results of the imaging technique are shown. From left to right, each row shows: 1) the ground truth sound speed; 2) the CUTE method as a baseline; 3) our proposed phase error optimization; 4) a naive B-mode image formed assuming 1540 m m/s; and 5) the B-mode reconstructed according to our proposed sound speed estimates. (a) The geometric distortion at the tissue interfaces is corrected. (b) The B-mode image brightness becomes more homogeneous in the lower half of the image. Videos are provided in the supplementary material.

Table 1. Comparison of sound speed mean absolute error (MAE) ± standard error between state-of-the-art (CUTE) versus differential beamforming with speckle brightness, coherence factor, and phase error objective functions. Two layer, four layer, inclusion and inclusion layer definitions can be found in [17]. (figure_6a to 6e)

Phantom	Description	CUTE (baseline)	Speckle Brightness	Coherence Factor	Phase Error (proposed)
1420	homogenous	21.6 ± 21.4	3.9 ± 3.3	$\mathbf{3.2 \pm 2.6}$	4.8 ± 3.5
1465	homogenous	11.7 ± 18.8	4.5 ± 4.9	5.3 ± 4.6	$\mathbf{4.5 \pm 3.5}$
1480	homogenous	10.4 ± 18.5	6.1 ± 5.4	$\mathbf{4.1 \pm 4.2}$	4.7 ± 3.5
1510	homogenous	10.8 ± 17.0	6.1 ± 7.0	$\mathbf{4.4 \pm 4.5}$	4.8 ± 3.6
1540	homogenous	11.8 ± 15.8	7.8 ± 7.5	$\mathbf{5.1 \pm 4.4}$	6.1 ± 4.3
1555	homogenous	11.4 ± 15.3	$\mathbf{5.7 \pm 6.7}$	5.8 ± 4.7	5.9 ± 4.3
1570	homogenous	11.2 ± 14.8	7.5 ± 7.6	$\mathbf{4.9 \pm 4.7}$	6.5 ± 4.7
Quadrant	Fig. 3a	65.6 ± 36.3	63.2 ± 52.1	63.4 ± 47.7	$\mathbf{35.4 \pm 27.9}$
Two layer	[17]	40.2 ± 34.1	62.5 ± 54.2	33.2 ± 25.8	$\mathbf{13.4 \pm 14.7}$
Four layer	[17]	44.1 ± 27.5	50.5 ± 25.0	43.8 ± 23.2	$\mathbf{29.0 \pm 26.5}$
Inclusion	[17]	14.3 ± 16.4	8.3 ± 7.5	7.5 ± 5.9	$\mathbf{6.1 \pm 4.4}$
Inclusion layer	[17], Fig. 3b	19.8 ± 18.1	16.3 ± 14.8	15.0 ± 11.1	$\mathbf{7.5 \pm 5.0}$

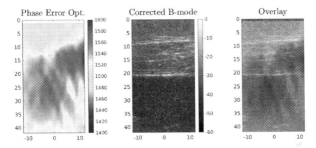

Fig. 4. A sample of in-vivo data reconstructed with the estimated sound speed via differentiable beamforming. Three layers consisting of subcutaneous adipose fat, muscle, and liver parenchyma are visible from top to bottom.

5   Discussion and Conclusion

Differentiable beamforming can be used to solve for unknown quantities with gradient descent. Here, we parameterized beamforming as a function of the slowness and optimized with respect to several candidate loss functions, showing that phase error was best for heterogeneous targets. The differentiable beamformer simultaneously provided B-mode image correction and quantitative sound speed characterization beyond the state-of-the-art across several challenging cases. Preliminary in-vivo quantitative SoS data for liver was shown, which has direct clinical applications such as in the noninvasive assessment of non-alcoholic fatty liver disease, as well as image enhancement in general.

Importantly, the differentiable beamformer allows us to incorporate fundamental physics principles like wave propagation, reducing the number of parameters to optimize. In the future, more complex wave propagation physics, such as refraction models, can be added to SoS optimization. In addition to sound speed, this work can be readily adapted to a broad set of applications such as beamforming with flexible arrays, where element positions are unknown, or passive cavitation mapping, where the origin of the signal is uncertain. Because the gradients flow through the entire imaging pipeline, the differentiable beamformer is also highly compatible with deep learning techniques. For instance, a model can be trained in a self-supervised fashion to identify optimal sound speed updates to accelerate convergence. Differentiable beamforming also enables the end-to-end optimization of imaging parameters for downstream tasks in computer-aided medical diagnostics.

Acknowledgements. This work was supported in part by the National Institute of Biomedical Imaging and Bioengineering under Grant K99-EB032230 and Grant R01-EB027100, as well as the National Science Foundation Graduate Research Fellowship under Grant No. DGE-1656518.

References

1. Ali, R., et al.: Aberration correction in diagnostic ultrasound: a review of the prior field and current directions. Z. Med. Phys. (2023). https://doi.org/10.1016/j.zemedi.2023.01.003
2. Ali, R., Herickhoff, C.D., Hyun, D., Dahl, J.J., Bottenus, N.: Extending retrospective encoding for robust recovery of the multistatic data set. IEEE Trans. Ultrason. Ferroelectr. Freq. Control **67**(5), 943–956 (2019)
3. Bradbury, J., et al.: JAX: composable transformations of Python+NumPy programs (2018). http://github.com/google/jax
4. Feigin, M., Freedman, D., Anthony, B.W.: A deep learning framework for single-sided sound speed inversion in medical ultrasound. IEEE Trans. Biomed. Eng. **67**(4), 1142–1151 (2020)
5. Hewish, A.: The diffraction of radio waves in passing through a phase-changing ionosphere. Proc. R. Soc. Lond. Ser. A. Math. Phys. Sci. **209**(1096), 81–96 (1951)
6. Hollman, K., Rigby, K., O'donnell, M.: Coherence factor of speckle from a multi-row probe. In: Proceedings 1999 IEEE Ultrasonics Symposium (Cat. No. 99CH37027), vol. 2, pp. 1257–1260. IEEE (1999)
7. Hyun, D., Li, Y.L., Steinberg, I., Jakovljevic, M., Klap, T., Dahl, J.J.: An open source GPU-based beamformer for real-time ultrasound imaging and applications. In: 2019 IEEE International Ultrasonics Symposium (IUS), pp. 20–23. IEEE (2019)
8. ITISFoundation: Tissue properties database v4-1 (2022). https://doi.org/10.13099/VIP21000-04-01
9. Jakovljevic, M., Hsieh, S., Ali, R., Chau Loo Kung, G., Hyun, D., Dahl, J.J.: Local speed of sound estimation in tissue using pulse-echo ultrasound: model-based approach. J. Acoust. Soc. Am. **144**, 254–266 (2018)
10. Jaros, J., Rendell, A.P., Treeby, B.E.: Full-wave nonlinear ultrasound simulation on distributed clusters with applications in high-intensity focused ultrasound. Int. J. High Perform. Comput. Appl. **30**(2), 137–155 (2016)
11. Mallart, R., Fink, M.: Adaptive focusing in scattering media through sound-speed inhomogeneities: the van Cittert Zernike approach and focusing criterion. J. Acoust. Soc. Am. **96**(6), 3721–3732 (1994)

12. Ng, G.C., Freiburger, P.D., Walker, W.F., Trahey, G.E.: A speckle target adaptive imaging technique in the presence of distributed aberrations. IEEE Trans. Ultrason. Ferroelectr. Freq. Control **44**(1), 140–151 (1997)
13. Nock, L., Trahey, G.E., Smith, S.W.: Phase aberration correction in medical ultrasound using speckle brightness as a quality factor. J. Acoust. Soc. Am. **85**(5), 1819–1833 (1989)
14. Sanabria, S.J., Brevett, T., Ali, R., Telichko, A., Dahl, J.: Direct speed of sound reconstruction from full-synthetic aperture data with dual regularization. In: 2022 IEEE International Ultrasonics Symposium (IUS), pp. 1–4 (2022)
15. Sanabria, S.J., Ozkan, E., Rominger, M., Goksel, O.: Spatial domain reconstruction for imaging speed-of-sound with pulse-echo ultrasound: simulation and in vivo study. Phys. Med. Biol. **63**(21), 215015 (2018)
16. Simson, W.A., Paschali, M., Sideri-Lampretsa, V., Navab, N., Dahl, J.J.: Investigating pulse-echo sound speed estimation in breast ultrasound with deep learning. arXiv preprint: arXiv:2302.03064 (2023)
17. Stähli, P., Kuriakose, M., Frenz, M., Jaeger, M.: Improved forward model for quantitative pulse-echo speed-of-sound imaging. Ultrasonics **108**, 106168 (2020)
18. Vishnevskiy, V., Sanabria, S.J., Goksel, O.: Image reconstruction via variational network for real-time hand-held sound-speed imaging. In: Knoll, F., Maier, A., Rueckert, D. (eds.) MLMIR 2018. LNCS, vol. 11074, pp. 120–128. Springer, Cham (2018). https://doi.org/10.1007/978-3-030-00129-2_14
19. Wild, A., Hobbs, R., Frenje, L.: Modelling complex media: an introduction to the phase-screen method. Phys. Earth Planet. Inter. **120**(3), 219–225 (2000)

InverseSR: 3D Brain MRI Super-Resolution Using a Latent Diffusion Model

Jueqi Wang[1(✉)], Jacob Levman[1,2,3], Walter Hugo Lopez Pinaya[4],
Petru-Daniel Tudosiu[4], M. Jorge Cardoso[4], and Razvan Marinescu[5]

[1] St. Francis Xavier University, Antigonish, Canada
{x2019cwn,jlevman}@stfx.ca
[2] Martinos Center for Biomedical Imaging, Department of Radiology,
Massachusetts General Hospital, Charlestown, USA
[3] Nova Scotia Health Authority, Halifax, Canada
[4] School of Biomedical Engineering and Imaging Sciences,
King's College London, London, UK
{walter.diaz_sanz,petru.tudosiu,m.jorge.cardoso}@kcl.ac.uk
[5] University of California, Santa Cruz, USA
ramarine@ucsc.edu

Abstract. High-resolution (HR) MRI scans obtained from research-grade medical centers provide precise information about imaged tissues. However, routine clinical MRI scans are typically in low-resolution (LR) and vary greatly in contrast and spatial resolution due to the adjustments of the scanning parameters to the local needs of the medical center. End-to-end deep learning methods for MRI super-resolution (SR) have been proposed, but they require re-training each time there is a shift in the input distribution. To address this issue, we propose a novel approach that leverages a state-of-the-art 3D brain generative model, the latent diffusion model (LDM) from [21] trained on UK BioBank, to increase the resolution of clinical MRI scans. The LDM acts as a generative prior, which has the ability to capture the prior distribution of 3D T1-weighted brain MRI. Based on the architecture of the brain LDM, we find that different methods are suitable for different settings of MRI SR, and thus propose two novel strategies: 1) for SR with more sparsity, we invert through both the decoder $\mathcal{D}$ of the LDM and also through a deterministic Denoising Diffusion Implicit Models (DDIM), an approach we will call InverseSR (LDM); 2) for SR with less sparsity, we invert only through the LDM decoder $\mathcal{D}$, an approach we will call InverseSR(Decoder). These two approaches search different latent spaces in the LDM model to find the optimal latent code to map the given LR MRI into HR. The training process of the generative model is independent of the MRI under-sampling process, ensuring the generalization of our method to many MRI SR problems with different input

---

Supplementary Information The online version contains supplementary material available at https://doi.org/10.1007/978-3-031-43999-5_42.

measurements. We validate our method on over 100 brain T1w MRIs from the IXI dataset. Our method can demonstrate that powerful priors given by LDM can be used for MRI reconstruction. Our source code is available online: https://github.com/BioMedAI-UCSC/InverseSR.

Keywords: MRI Super-Resolution · Latent Diffusion Model · Inverse Problem · Optimization Method

1   Introduction

End-to-end convolutional neural networks (CNNs) have shown remarkable performance compared to classical algorithms [14] on MRI SR. Deep CNNs have been widely applied in a variety of MRI SR situations; for instance, slice imputation on the brain, liver and prostate MRI [29] and brain MRI SR reconstruction on scaling factors ×2, ×3, ×4 [32]. Several techniques based on deep CNNs have been proposed to improve performance, such as densely connected networks [6], adversarial networks [5], and attention network [32]. However, their supervised training requires paired images, which necessitates re-training every time there is a shift in the input distribution [4,16]. As a result, such methods are unsuitable for MRI SR, as it is challenging to obtain paired training data that cover the variability in acquisition protocols and resolution of clinical brain MRI scans across institutions [14].

Building image priors through generative models has recently become a popular approach in the field of image SR, for both computer vision [1,2,7,17,19] as well as medical imaging [18,25], as they do not require re-training in the presence of several types of input distribution shifts. While these methods have shown promise in MRI SR, they have so far been limited to 2D slices [18,25], rendering them unsuitable for 3D brain MRIs slice imputation.

In this study, we propose solving the MRI SR problem by building powerful, 3D-native image priors through a recently proposed HR image generative model, the latent diffusion model (LDM) [21,22]. We solve the inverse problem by finding the optimal latent code z in the latent space of the pre-trained generative model, which could restore a given LR MRI I, using a known corruption function f. In this study, we focus on slice imputation, yet our method could be applied to other medical image SR problems by implementing different corruption functions f. We proposed two novel strategies for MRI SR: Inverse(LDM), which additionally inverts the input image through the deterministic DDIM model, and InverseSR(Decoder) which inverts the input image through the corruption function f and through the decoder $\mathcal{D}$ of the LDM model. We found that for large sparsity, InverseSR(LDM) had a better performance, while for low sparsity, InverseSR(Decoder) performed best. While the LDM model was trained on UK BioBank, we demonstrate our methods on an external dataset (IXI) which was inaccessible to the pre-trained generative model. Both quantitative and qualitative results show that our method achieves significantly better performance compared to two other baseline models. Furthermore, our method can also be applied to tumour/lesion filling by creating tumour/lesion shape masks.

1.1   Related Work

MRI Super-Resolution. End-to-end deep training [27,29,32] has been proposed recently for MRI SR, which has achieved superior results compared to classical methods. However, these methods require paired data to train, which is hard to acquire because of the large variability present in clinical MRIs [14,23]. To circumvent this limitation, several unsupervised methods have been proposed without requiring access to HR scans [3,8,14]. Dalca et al. [8] proposed a gaussian mixture model for sparse image patches. Brudfors et al. [3] presented an algorithm which could take advantage of multimodal MRI. Iglesias et al. [14] introduced a method to train a CNN for MRI SR on any given combination of contrasts, resolutions and orientations.

Solving Inverse Problems Using Generative Models. A common way to solve the inverse problem using an LDM is to use the encoder $\mathcal{E}$ to first encode the given image x into the latent space $z_0 = \mathcal{E}(x)$ [10,12,20], followed by DDIM (Denoising Diffusion Implicit Models) Inversion [9,24] to encode z_0 into the noise latent code z_T [20]. However, this approach does not work for low-resolution images, because the encoder $\mathcal{E}$ has only been trained on high-resolution images.

Our work is also similar to the optimization-based generative adversarial network (GAN) inversion approach [30], trying to find the optimal latent representation z^* in the latent space of GAN, which could be mapped to represent the given image $x \approx G(z^*)$. More recent works [7,10,13,17,25] have used diffusion models for inverse problems due to their superior performance. However, all these methods require the diffusion model to operate directly in the image space, which for large image resolutions can become GPU-memory intensive.

2   Methods

3D Brain Latent Diffusion Models. We leverage a state-of-the-art LDM [21] to create high-quality priors for 3D brain MRIs. There are two components in an LDM: an autoencoder and a diffusion model [22]. An encoder $\mathcal{E}$ maps each high-resolution T1w brain MRI $x \sim p_{data}(x)$ into a latent vector $z_0 = \mathcal{E}(x)$ of size $20 \times 28 \times 20$. The decoder $\mathcal{D}$ is trained to map the latent vectors z_0 back into the MRI image domain x. The autoencoder was trained on 31,740 T1w MRIs from the UK Biobank [26] using a combination of an L1 loss, a perceptual loss [31], a patch-based adversarial loss [11] and a KL regularization term in the latent space. The autoencoder was trained on pre-processed MRIs using UniRes [3] into a common MNI space with a voxel size of $1\,\text{mm}^3$ and was then kept unchanged during the LDM training. The latent representations of the T1w brain MRIs were then used to train the LDM. A conditional U-Net ϵ_θ was then trained to predict the artificial noise by the following objective:

$$\theta^* = \arg\min_\theta \mathbb{E}_{z\sim\mathcal{E}(x),\epsilon\sim\mathcal{N}(0,1),t}||\epsilon - \epsilon_\theta(z_t,t,\mathcal{C})||_2^2 \tag{1}$$

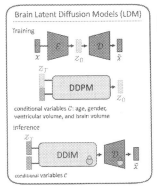

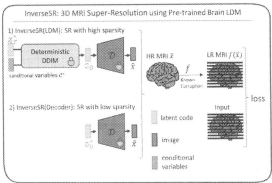

Fig. 1. (Left) The Brain LDM has two stages of training process. First, an autoencoder is pre-trained to map T1w brain MRIs into a latent code $z_0 = \mathcal{E}(x)$. Then, a diffusion model is trained to generate z_0 latents from this learned latent space. During inference, DDIM has been applied to reduce the sampling step with little performance drop. (Right) We proposed two methods to handle different scenarios of MRI SR, based on the architecture of brain LDM: 1) InverseSR(LDM): for SR with high sparsity, we optimize the latent code z_T^* and associated conditional variables $\mathcal{C}^*$ using deterministic DDIM and decoder $\mathcal{D}$ to map the latent code into brain MRI. 2) InverseSR(Decoder): for SR with low sparsity, we optimize the z_0^* which only use the decoder $\mathcal{D}$ to map the latent code into brain MRI.

DDIM [24] has been used in brain LDM to replace the denoising diffusion probabilistic models (DDPM) during inference to reduce the number of reverse steps with minimal performance loss [21,24]. This network ε_θ is conditioned on four conditional variables $\mathcal{C}$: *age, gender, ventricular volume* and *brain volume*, which are all introduced by cross-attention layers [22]. Gender is a binary variable, while the rest of the covariates are scaled to $[0, 1]$. Finally, the pre-trained decoder maps the latent vector into an HR MRI $\tilde{x} = \mathcal{D}(z_0)$. The architecture of the brain LDM can be found in Fig. 1.

Deterministic DDIM Sampling. In order to obtain a latent representation z_T capable of reconstructing a given noisy sample into a high-resolution image, we employ deterministic DDIM sampling [24]:

$$z_{t-1} = \sqrt{\alpha_{t-1}}\left(\frac{z_t - \sqrt{1-\alpha_t}\cdot\epsilon_\theta(z_t,\mathcal{C},t)}{\sqrt{\alpha_t}}\right) + \sqrt{1-\alpha_{t-1}}\cdot\epsilon_\theta(z_t,\mathcal{C},t) \quad (2)$$

where $\alpha_{1:T} \in (0,1]^T$ is a time-dependent decreasing sequence, $\frac{z_t-\sqrt{1-\alpha_t}\cdot\epsilon_\theta(z_t,\mathcal{C},t)}{\sqrt{\alpha_t}}$ represents the "predicted x_0", and $\sqrt{1-\alpha_{t-1}}\cdot\epsilon_\theta(z_t,\mathcal{C},t)$ can be understood as the "direction pointing to x_t" [24].

Corruption Function f. We assume a corruption function f known *a-priori* that is applied on the HR image $\tilde{x}$ obtained from the generative model, and compute the loss function based on the corrupted image $f \circ \tilde{x}$ and the given LR input image I. In clinical practice, a prevalent method for acquiring MR images is prioritizing high in-plane resolution while sacrificing through-plane resolution to expedite the acquisition process and reduce motion artifacts [33]. To account for this procedure, we introduce a corruption function that generates masks for non-acquired slices, enabling our method to in-paint the missing slices. For instance, on $1 \times 1 \times 4\,\mathrm{mm}^3$ undersampled volumes, we create masks for three slices every four slices on the generated HR $1 \times 1 \times 1\,\mathrm{mm}^3$ volumes.

Algorithm 1. InverseSR(LDM)

1: **Input**: Low-resolution MR image I.
2: **Output**: Optimized noise latent code z_T^* and conditional variables $\mathcal{C}^*$

3: Initialize z_T with gaussian noise from $N(0, \mathbb{I})$;
4: Initialize conditional variables $\mathcal{C} = 0.5$;
5: **for** $j = 0, \ldots, N-1$ **do**
6:     **for** $t = T, T-1, \ldots, 1$ **do**
7:         $z_{t-1} \leftarrow \sqrt{\alpha_{t-1}} \left(\frac{z_t - \sqrt{1-\alpha_t} \cdot \epsilon_\theta(z_t, \mathcal{C}, t)}{\sqrt{\alpha_t}} \right) + \sqrt{1-\alpha_{t-1}} \cdot \epsilon_\theta(z_t, \mathcal{C}, t)$;
8:     $L \leftarrow \lambda_{perc} L_{perc}(f \circ \mathcal{D}(z_0), I) + \lambda_{mae}\|f \circ \mathcal{D}(z_0) - I\|$;
9:     $\mathcal{C} \leftarrow \mathcal{C} - \alpha \nabla_\mathcal{C} L$;
10:     $z_T \leftarrow z_T - \alpha \nabla_{z_T} L$;
11: Set $z_T^* \leftarrow z_T$; $\mathcal{C}^* \leftarrow \mathcal{C}$;
12: **Return** $z_T^*, \mathcal{C}^*$;

InverseSR(LDM): In the case of high sparsity MRI SR, we optimize the noise latent code z_T^* and its associated conditional variables $\mathcal{C}^*$ to restore the HR image from the given LR input image I using the optimization method:

$$z_T^*, \mathcal{C}^* = \arg\min_{z_T, \mathcal{C}} \lambda_{perc} L_{perc}(f \circ \mathcal{D}(\mathrm{DDIM}(z_T, \mathcal{C}, T)), I) + $$
$$\lambda_{mae}\|f \circ \mathcal{D}(\mathrm{DDIM}(z_T, \mathcal{C}, T)) - I\| \tag{3}$$

where $\mathrm{DDIM}(z_T, \mathcal{C}, T)$ represents T deterministic DDIM sampling steps on the latent z_0 in Eq. 2. We follow the brain LDM model to use the perceptual loss L_{perc} and the L1 pixelwise loss. The loss function is computed on the corrupted image generated from the generative model and the given LR input. A detailed pseudocode description of this method can be found in Algorithm 1.

InverseSR(Decoder): For low sparsity MRI SR, we directly find the optimal latent code z_T^* using the decoder $\mathcal{D}$:

$$z_0^* = \arg\min_{z_0} \lambda_{perc} L_{perc}(f \circ \mathcal{D}(z_0), I) + \lambda_{mae}\|f \circ \mathcal{D}(z_0) - I\| \tag{4}$$

3   Experimental Design

Dataset for Validation: We use 100 HR T1 MRIs from the IXI dataset (http://brain-development.org/ixi-dataset/) to validate our method, after filtering out those scans where registration failed. We note that subjects in the IXI dataset are around 10 years younger on average than those in UK Biobank. The MRI scans from UK Biobank also had the faces masked out, while the scans from IXI did not. This caused the faces of our reconstructions to appear blurred.

Implementation: Conditional variables are all initialized to 0.5. Voxels in all input volumes are normalized to [0,1]. When sampling the pre-trained brain LDM with the DDIM sampler, we run $T = 46$ timesteps due to computational limitations on our hardware. For InverseSR(LDM), z_T is initialized with random gaussian noise. For InverseSR(Decoder), we compute the mean latent code $\bar{z}_0$ as $\bar{z}_0 = \sum_{i=1}^{S} \frac{1}{S}\text{DDIM}(z_T^i, \mathcal{C}, T)$ by first sampling $S = 10,000$ z_T^i samples from $N(0, \mathbb{I})$, then passing them through the DDIM model. $N = 600$ gradient descent steps are used for InverseSR(LDM) to guarantee converging (Algorithm 1, line 5). 600 optimization steps are also utilized in InverseSR(Decoder). We use the Adam optimizer with $\alpha = 0.07$, $\beta_1 = 0.9$ and $\beta_2 = 0.999$.

4   Results

Figure 2 shows the qualitative results on the coronal slices of SR from 4 and 8 mm axial scans. The advantage of our approach is clear compared to baseline methods because it is capable of restoring HR MRIs with smoothness even when the slice thickness is large (i.e., 8 mm). This is the case because the pre-trained

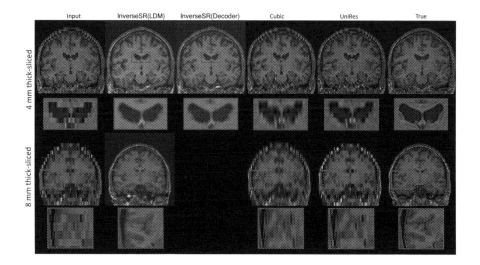

Fig. 2. Qualitative results of our approach (InverseSR) and the cubic and UniRes baselines on scans with 4 mm and 8 mm thickness.

LDM we use is able to build a powerful prior over the HR T1w MRI domain. Therefore, the generated images of our method are HR MRIs with smoothness in 3 directions: axial, sagittal and coronal, no matter how sparse the input images I are. Qualitative results of applying our method on tumour and lesion filling are available in the supplementary material.

Table 1 shows quantitative results on 100 HR T1 scans from the IXI dataset, which the brain LDM did not have access to during training. We investigated mean peak signal-to-noise ratio (PSNR), and structural similarity index measure (SSIM) [28] values and their corresponding standard deviation. We compare our method to cubic interpolation, as well as a similar unsupervised approach, UniRes [3]. We show our approach and the two compared methods on two different settings of slice imputation: 4 mm and 8 mm thick-sliced axial scans representing low sparsity and high sparsity LR MRIs, respectively. All the metrics are computed on a 3D volume around the brain of size $160 \times 224 \times 160$. For SR at 4 mm, InverseSR(Decoder) achieves the highest mean SSIM and PSNR scores among all compared methods, which are slightly higher than the scores for InverseSR(LDM). For SR at 8 mm, Inverse(LDM) achieves the highest mean SSIM and PSNR and lowest standard error than the two baseline methods, which could be attributed to the stronger prior learned by the DDIM model.

5   Limitations

One key limitation of our method is the need for large computational resources to perform the image reconstruction, in particular the long Markov chain of sampling steps required by the diffusion model to generate samples. An entire pass through the diffusion model (lines 6–8 in Algorithm 1) is required for every step in the gradient descent method. Another limitation of our method is that it is limited by the capacity and output heterogeneity of the LDM generator.

Table 1. Quantitative evaluation results (mean ± standard error) of our approach (InverseSR) and two baselines on 1 mm scans and corresponding SR counterpart - from 4 and 8 mm axial scans.

Slice Thickness	Methods	SSIM ↑	PSNR ↑
4 mm	InverseSR(LDM)	0.797 ± 0.037	28.59 ± 1.61
	InverseSR(Decoder)	**0.803 ± 0.030**	**29.64 ± 1.64**
	Cubic	0.760 ± 0.052	23.84 ± 2.36
	UniRes [3]	0.688 ± 0.079	21.49 ± 2.61
8 mm	InverseSR(LDM)	**0.754 ± 0.038**	**27.92 ± 1.60**
	Cubic	0.632 ± 0.067	21.80 ± 2.36
	UniRes [3]	0.633 ± 0.053	20.91 ± 2.29

6   Conclusions

In this study, we have developed an unsupervised technique for MRI super-resolution. We leverage a recent pre-trained Brain LDM [21] for building powerful image priors over T1w brain MRIs. Unlike end-to-end supervised approaches, which require retraining each time there is a distribution shift over the input, our method is capable of being adapted to different settings of MRI SR problems at test time. This feature is suitable for MRI SR since the acquisition protocols and resolution of clinical brain MRI exams vary across or even within institutions. We proposed two novel strategies for different settings of MRI SR: InverseSR(LDM) for low sparsity MRI and InverseSR(Decoder) for high sparsity MRI. We validated our method on 100 brain T1w MRIs from the IXI dataset through slice imputation using input scans of 4 and 8 mm slice thickness, and compared our method with cubic interpolation and UniRes [3].

Experimental results have shown that our approach achieves superior performance compared to the unsupervised baselines, and could create smooth HR images with fine detail even on an external dataset (IXI). Experiments in this paper focus on slice imputation, but our method could be adapted to other MRI under-sampling problems by implementing different corruption functions f. For instance, for reconstructing k-space under-sampled MR images, a new corruption function could be designed by first converting the HR image into k-space, then masking a chosen set of k-space measurements, and then converting back to image space. Instead of estimating a single image, future work could also estimate a distribution of reconstructed images through either variational inference (like the BRGM model [18]) or through sampling methods such as Markov Chain Monte Carlo (MCMC) or Langevin dynamics [15].

Acknowledgements. This work was fund by an NSERC Discovery Grant to JL. Funding was also provided by a Nova Scotia Graduate Scholarship and a StFX Graduate Scholarship to JW. Computational resources were provided by Compute Canada.

References

1. Abdal, R., Qin, Y., Wonka, P.: Image2StyleGAN: how to embed images into the StyleGAN latent space? In: Proceedings of the IEEE/CVF International Conference on Computer Vision, pp. 4432–4441 (2019)
2. Bora, A., Jalal, A., Price, E., Dimakis, A.G.: Compressed sensing using generative models. In: International Conference on Machine Learning. PMLR (2017)
3. Brudfors, M., Balbastre, Y., Nachev, P., Ashburner, J.: A tool for super-resolving multimodal clinical MRI. arXiv preprint arXiv:1909.01140 (2019)
4. Chen, H., Dou, Q., Yu, L., Qin, J., Heng, P.A.: VoxResNet: deep voxelwise residual networks for brain segmentation from 3D MR images. NeuroImage **170**, 446–455 (2018)
5. Chen, Y., Shi, F., Christodoulou, A.G., Xie, Y., Zhou, Z., Li, D.: Efficient and accurate MRI super-resolution using a generative adversarial network and 3D multilevel densely connected network. In: Frangi, A.F., Schnabel, J.A., Davatzikos, C., Alberola-López, C., Fichtinger, G. (eds.) MICCAI 2018. LNCS, vol. 11070, pp. 91–99. Springer, Cham (2018). https://doi.org/10.1007/978-3-030-00928-1_11

6. Chen, Y., Xie, Y., Zhou, Z., Shi, F., Christodoulou, A.G., Li, D.: Brain MRI super resolution using 3D deep densely connected neural networks. In: 2018 IEEE 15th International Symposium on Biomedical Imaging (ISBI 2018), pp. 739–742 (2018)
7. Chung, H., Kim, J., Mccann, M.T., Klasky, M.L., Ye, J.C.: Diffusion posterior sampling for general noisy inverse problems. arXiv preprint arXiv:2209.14687 (2022)
8. Dalca, A.V., Bouman, K.L., Freeman, W.T., Rost, N.S., Sabuncu, M.R., Golland, P.: Medical image imputation from image collections. IEEE Trans. Med. Imaging **38**(2), 504–514 (2018)
9. Dhariwal, P., Nichol, A.: Diffusion models beat GANs on image synthesis. Adv. Neural. Inf. Process. Syst. **34**, 8780–8794 (2021)
10. Elarabawy, A., Kamath, H., Denton, S.: Direct inversion: optimization-free text-driven real image editing with diffusion models. arXiv:2211.07825 (2022)
11. Esser, P., Rombach, R., Ommer, B.: Taming transformers for high-resolution image synthesis. In: Proceedings of the IEEE/CVF Conference on Computer Vision and Pattern Recognition, pp. 12873–12883 (2021)
12. Gal, R., et al.: An image is worth one word: personalizing text-to-image generation using textual inversion. In: ICLR (2023)
13. Hertz, A., Mokady, R., Tenenbaum, J., Aberman, K., Pritch, Y., Cohen-Or, D.: Prompt-to-prompt image editing with cross attention control. In: International Conference on Learning Representations (2023)
14. Iglesias, J.E., et al.: Joint super-resolution and synthesis of 1 mm isotropic MP-RAGE volumes from clinical MRI exams with scans of different orientation, resolution and contrast. Neuroimage **237**, 118206 (2021)
15. Jalal, A., Arvinte, M., Daras, G., Price, E., Dimakis, A.G., Tamir, J.: Robust compressed sensing MRI with deep generative priors. Adv. Neural. Inf. Process. Syst. **34**, 14938–14954 (2021)
16. Kamnitsas, K., et al.: Efficient multi-scale 3D CNN with fully connected CRF for accurate brain lesion segmentation. Med. Image Anal. **36**, 61–78 (2017)
17. Lugmayr, A., Danelljan, M., Romero, A., Yu, F., Timofte, R., Van Gool, L.: Repaint: inpainting using denoising diffusion probabilistic models. In: Proceedings of the IEEE/CVF Conference on Computer Vision and Pattern Recognition, pp. 11461–11471 (2022)
18. Marinescu, R.V., Moyer, D., Golland, P.: Bayesian image reconstruction using deep generative models. In: NeurIPS 2021 Workshop on Deep Generative Models and Downstream Applications (2021)
19. Menon, S., Damian, A., Hu, S., Ravi, N., Rudin, C.: Pulse: self-supervised photo upsampling via latent space exploration of generative models. In: Proceedings of the IEEE/CVF Conference on Computer Vision and Pattern Recognition (2020)
20. Mokady, R., Hertz, A., Aberman, K., Pritch, Y., Cohen-Or, D.: Null-text inversion for editing real images using guided diffusion models. arXiv:2211.09794 (2022)
21. Pinaya, W.H.L., et al.: Brain imaging generation with latent diffusion models. In: Mukhopadhyay, A., Oksuz, I., Engelhardt, S., Zhu, D., Yuan, Y. (eds.) DGM4MICCAI 2022. LNCS, vol. 13609, pp. 117–126. Springer, Cham (2022). https://doi.org/10.1007/978-3-031-18576-2_12
22. Rombach, R., Blattmann, A., Lorenz, D., Esser, P., Ommer, B.: High-resolution image synthesis with latent diffusion models. In: Proceedings of the IEEE/CVF Conference on Computer Vision and Pattern Recognition, pp. 10684–10695 (2022)
23. Sander, J., de Vos, B.D., Išgum, I.: Autoencoding low-resolution MRI for semantically smooth interpolation of anisotropic MRI. Med. Image Anal. **78**, 102393 (2022)

24. Song, J., Meng, C., Ermon, S.: Denoising diffusion implicit models. In: International Conference on Learning Representations (2020)
25. Song, Y., Shen, L., Xing, L., Ermon, S.: Solving inverse problems in medical imaging with score-based generative models. arXiv preprint arXiv:2111.08005 (2021)
26. Sudlow, C., et al.: UK biobank: an open access resource for identifying the causes of a wide range of complex diseases of middle and old age. PLOS Med. **12**(3), 1–10 (2015)
27. Wang, J., Chen, Y., Wu, Y., Shi, J., Gee, J.: Enhanced generative adversarial network for 3D brain MRI super-resolution. In: 2020 IEEE Winter Conference on Applications of Computer Vision (WACV), pp. 3616–3625 (2020)
28. Wang, Z., Bovik, A.C., Sheikh, H.R., Simoncelli, E.P.: Image quality assessment: from error visibility to structural similarity. IEEE Trans. Image Process. **13**(4), 600–612 (2004)
29. Wu, Z., Wei, J., Wang, J., Li, R.: Slice imputation: multiple intermediate slices interpolation for anisotropic 3D medical image segmentation. Comput. Biol. Med. **147**(C), 105667 (2022)
30. Xia, W., Zhang, Y., Yang, Y., Xue, J.H., Zhou, B., Yang, M.H.: Gan inversion: a survey. IEEE Trans. Pattern Anal. Mach. Intell. **45**, 3121–3138 (2021)
31. Zhang, R., Isola, P., Efros, A.A., Shechtman, E., Wang, O.: The unreasonable effectiveness of deep features as a perceptual metric. In: Proceedings of the IEEE Conference on Computer Vision and Pattern Recognition, pp. 586–595 (2018)
32. Zhang, Y., Li, K., Li, K., Fu, Y.: MR image super-resolution with squeeze and excitation reasoning attention network. In: Proceedings of the IEEE/CVF Conference on Computer Vision and Pattern Recognition, pp. 13425–13434 (2021)
33. Zhao, C., Dewey, B.E., Pham, D.L., Calabresi, P.A., Reich, D.S., Prince, J.L.: Smore: a self-supervised anti-aliasing and super-resolution algorithm for MRI using deep learning. IEEE Trans. Med. Imaging **40**(3), 805–817 (2020)

Unified Brain MR-Ultrasound Synthesis Using Multi-modal Hierarchical Representations

Reuben Dorent[1]([✉]), Nazim Haouchine[1], Fryderyk Kogl[1], Samuel Joutard[2], Parikshit Juvekar[1], Erickson Torio[1], Alexandra J. Golby[1], Sebastien Ourselin[2], Sarah Frisken[1], Tom Vercauteren[2], Tina Kapur[1], and William M. Wells III[1,3]

[1] Harvard Medical School, Brigham and Women's Hospital, Boston, MA, USA
rdorent@bwh.harvard.edu
[2] King's College London, London, UK
[3] Massachusetts Institute of Technology, Cambridge, MA, USA

Abstract. We introduce MHVAE, a deep hierarchical variational auto-encoder (VAE) that synthesizes missing images from various modalities. Extending multi-modal VAEs with a hierarchical latent structure, we introduce a probabilistic formulation for fusing multi-modal images in a common latent representation while having the flexibility to handle incomplete image sets as input. Moreover, adversarial learning is employed to generate sharper images. Extensive experiments are performed on the challenging problem of joint intra-operative ultrasound (iUS) and Magnetic Resonance (MR) synthesis. Our model outperformed multi-modal VAEs, conditional GANs, and the current state-of-the-art unified method (ResViT) for synthesizing missing images, demonstrating the advantage of using a hierarchical latent representation and a principled probabilistic fusion operation. Our code is publicly available (https://github.com/ReubenDo/MHVAE).

Keywords: Variational Auto-Encoder · Ultrasound · Brain Resection · Image Synthesis

1 Introduction

Medical imaging is essential during diagnosis, surgical planning, surgical guidance, and follow-up for treating brain pathology. Images from multiple modalities are typically acquired to distinguish clinical targets from surrounding tissues. For example, intra-operative ultrasound (iUS) imaging and Magnetic Resonance Imaging (MRI) capture complementary characteristics of brain tissues that can be used to guide brain tumor resection. However, as noted in [30], multi-modal

Supplementary Information The online version contains supplementary material available at https://doi.org/10.1007/978-3-031-43999-5_43.

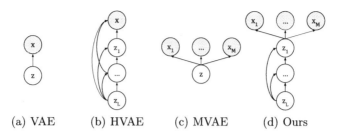

(a) VAE   (b) HVAE   (c) MVAE   (d) Ours

Fig. 1. Graphical models of: (a) variational auto-encoder (VAE); (b) hierarchical VAE (HVAE); (c) Multi-modal VAE (MVAE); (d) Multi-Modal Hiearchical VAE (Ours).

data is *expensive* and *sparse*, typically leading to incomplete sets of images. For example, the prohibitive cost of intra-operative MRI (iMRI) scanners often hampers the acquisition of iMRI during surgical procedures. Conversely, iUS is an affordable tool but has been perceived as difficult to read compared to iMRI [5]. Consequently, there is growing interest in synthesizing missing images from a subset of available images for enhanced visualization and clinical training.

Medical image synthesis aims to predict missing images given available images. Deep-learning based methods have reached the highest level of performance [29], including conditional generative adversarial (GAN) models [6,14,15,21] and conditional variational auto-encoders [3]. However, a key limitation of these techniques is that they must be trained for each subset of available images.

To tackle this challenge, unified approaches have been proposed. These approaches are designed to have the flexibility to handle incomplete image sets as input, improving practicality as only one network is used for generating missing images. To handle partial inputs, some studies proposed to fill missing images with arbitrary values [4,17,18,24]. Alternatively, other work aim at creating a common feature space that encodes shared information from different modalities. Feature representations are extracted independently for each modality. Then, arithmetic operations (e.g., mean [7,11,28], max [2] or a combination of sum, product and max [32]) are used to fuse these feature representations. However, these operations do not force the network to learn a shared latent representation of multi-modal data and lack theoretical foundations. In contrast, Multi-modal Variational Auto-Encoders (MVAEs) provide a principled probabilistic fusion operation to create a common representation space [8,30]. In MVAEs, the common representation space is low-dimensional (e.g., $\mathbb{R}^{256}$), which usually leads to blurry synthetic images. In contrast, hierarchical VAEs (HVAEs) [19,22,26,27] allow for learning complex latent representations by using a hierarchical latent structure, where the coarsest latent variable (z_L) represents global features, as in MVAEs, while the finer variables capture local characteristics. However, HVAEs have not yet been extended to multi-modal settings to synthesize missing images.

In this work, we introduce Multi-Modal Hierarchical Latent Representation VAE (MHVAE), the first multi-modal VAE approach with a hierarchical latent representation for unified medical image synthesis. Our contribution is four-fold.

First, we integrate a hierarchical latent representation into the multi-modal variational setting to improve the expressiveness of the model. Second, we propose a principled fusion operation derived from a probabilistic formulation to support missing modalities, thereby enabling image synthesis. Third, adversarial learning is employed to generate realistic image synthesis. Finally, experiments on the challenging problem of iUS and MR synthesis demonstrate the effectiveness of the proposed approach, enabling the synthesis of high-quality images while establishing a mathematically grounded formulation for unified image synthesis and outperforming non-unified GAN-based approaches and the state-of-the-art method for unified multi-modal medical image synthesis.

2   Background

Variational Auto-Encoders (VAEs). The goal of VAEs [16] is to train a generative model in the form of $p(x, z) = p(z)p(x|z)$ where $p(z)$ is a prior distribution (e.g. isotropic Normal distribution) over latent variables $z \in \mathbb{R}^H$ and where $p_\theta(x|z)$ is a decoder parameterized by θ that reconstructs data $x \in \mathbb{R}^N$ given z. The latent space dimension H is typically much lower than the image space dimension N, i.e. $H \ll N$. The training goal with respect to θ is to maximize the marginal likelihood of the data $p_\theta(x)$ (the "evidence"); however since the true posterior $p_\theta(z|x)$ is in general intractable, the variational evidence lower bound (ELBO) is instead optimized. The ELBO $\mathcal{L}_{\text{VAE}}(x; \theta, \phi)$ is defined by introducing an approximate posterior $q_\phi(z|x)$ with parameters ϕ:

$$\mathcal{L}_{\text{VAE}}(x; \theta, \phi) := \mathbb{E}_{q_\phi(z|x)}[\log(p_\theta(x|z))] - \text{KL}[q_\phi(z|x)||p(z)] , \qquad (1)$$

where $\text{KL}[q||p]$ is the Kullback-Leibler divergence between distributions q and p.

Multi-modal Variational Auto-Encoders (MVAE). Multi-modal VAEs [8, 25, 30] introduced a principled probabilistic formulation to support missing data at training and inference time. Multi-modal VAEs assume that M paired images $x = (x_1, ..., x_M) \in \mathbb{R}^{M \times N}$ are conditionally independent given a shared representation z as highlighted in Fig. 1, i.e. $p_\theta(x|z) = \prod_{i=1}^{M} p(x_i|z)$.

Instead of training one single variational network $q_\phi(z|x)$ that requires all images to be presented at all times, MVAEs factorize the approximate posterior as a combination of unimodal variational posteriors $(q_\phi(z|x_i))_{i=1}^{M}$. Given any subset of modalities $\pi \subseteq \{1, ..., M\}$, MVAEs have the flexibility to approximate the π-marginal posteriors $p(z|(x_i)_{i \in \pi})$ using the $|\pi|$ unimodal variational posteriors $(q_\phi(z|x_i))_{i \in \pi}$. MVAE [30] and U-HVED [8] factorize the π-marginal variational posterior as a product-of-experts (PoE), i.e.:

$$q_\phi^{\text{PoE}}(z|x_\pi) = p(z) \prod_{i \in \pi} q_\phi(z|x_i) . \qquad (2)$$

3   Methods

In this paper, we propose a deep multi-modal hierarchical VAE called MHVAE that synthesizes missing images from available images. MHVAE's design focuses on tackling three challenges: (i) improving expressiveness of VAEs and MVAEs using a hierarchical latent representation; (ii) parametrizing the variational posterior to handle missing modalities; (iii) synthesizing realistic images.

3.1   Hierarchical Latent representation

Let $x = (x_i)_{i=1}^M \in \mathbb{R}^{M \times N}$ be a complete set of paired (i.e. co-registered) images of different modalities where M is the total number of image modalities and N the number of pixels (e.g. $M = 2$ for T_2 MRI and iUS synthesis). The images x_i are assumed to be conditionally independent given a latent variable z. Then, the conditional distribution $p_\theta(x|z)$ parameterized by θ can be written as:

$$p_\theta(x|z) = \prod_{i=1}^M p_\theta(x_i|z) . \tag{3}$$

Given that VAEs and MVAEs typically produce blurry images, we propose to use a hierarchical representation of the latent variable z to increase the expressiveness the model as in HVAEs [19,22,26,27]. Specifically, the latent variable z is partitioned into disjoint groups, as shown in Fig. 1 i.e. $z = \{z_1, ... z_L\}$, where L is the number of groups. The prior $p(z)$ is then represented by:

$$p_\theta(z) = p(z_L) \prod_{l=1}^{L-1} p_{\theta_l}(z_l|z_{>l}) , \tag{4}$$

where $p(z_L) = \mathcal{N}(z_L; 0, I)$ is an isotropic Normal prior distribution and the conditional prior distributions $p_{\theta_l}(z_l|z_{>l})$ are factorized Normal distributions with diagonal covariance parameterized using neural networks, i.e. $p_{\theta_l}(z_l|z_{>l}) = \mathcal{N}(z_l; \mu_{\theta_l}(z_{>l}), D_{\theta_l}(z_{>l}))$. Note that the dimension of the finest latent variable $z_1 \in \mathbb{R}^{H_1}$ is similar to number of pixels, i.e. $H_1 = \mathcal{O}(N)$ and the dimension of the latent representation exponentially decreases with the depth, i.e. $H_L \ll H_1$.

Reusing Eq. 1, the evidence $\log(p_\theta(x))$ is lower-bounded by the tractable variational ELBO $\mathcal{L}_{\text{MHVAE}}^{\text{ELBO}}(x; \theta, \phi)$:

$$\mathcal{L}_{\text{MHVAE}}^{\text{ELBO}}(x; \theta, \phi) = \sum_{i=1}^M \mathbb{E}_{q_\phi(z|x)}[\log(p_\theta(x_i|z))] - \text{KL}\left[q_\phi(z_L|x)||p(z_L)\right]$$
$$- \sum_{l=1}^{L-1} \mathbb{E}_{q_\phi(z_{>l}|x)}\left[\text{KL}[q_\phi(z_l|x, z_{>l})||p_\theta(z_l|z_{>l})]\right] \tag{5}$$

where $q_\phi(z|x) = \prod_{l=1}^L q_\phi(z_l|x, z_{>l})$ is a variational posterior that approximates the intractable true posterior $p_\theta(z|x)$.

3.2 Variational Posterior's Parametrization for Incomplete Inputs

To synthesize missing images, the variational posterior $(q_\phi(z_l|x, z_{>l}))_{l=1}^L$ should handle missing images. We propose to parameterize it as a combination of uni-modal variational posteriors. Similarly to MVAEs, for any set $\pi \subseteq \{1, ..., M\}$ of images, the conditional posterior distribution at the coarsest level L is expressed

$$q_{\phi_L}^{\mathrm{PoE}}(z_L|x_\pi) = p(z_L) \prod_{i \in \pi} q_{\phi_L^i}(z|x_i) . \tag{6}$$

where $p(z_L) = \mathcal{N}(z_L; 0, I)$ is an isotropic Normal prior distribution and $q_{\phi_L}(z|x_i)$ is a Normal distribution with diagonal covariance parameterized using CNNs.

For the other levels $l \in \{1, .., L-1\}$, we similarly propose to express the conditional variational posterior distributions as a product-of-experts:

$$q_{\phi_l, \theta_l}^{\mathrm{PoE}}(z_l|x_\pi, z_{>l}) = p_{\theta_l}(z_l|z_{>l}) \prod_{i \in \pi} q_{\phi_l^i}(z_l|x_i, z_{>l}) \tag{7}$$

where $q_{\phi_l^i}(z_l|x_i, z_{>l})$ is a Normal distribution with diagonal covariance parameterized using CNNs, i.e. $q_{\phi_l^i}(z_l|x_i, z_{>l}) = \mathcal{N}(z_l; \mu_{\phi_l^i}(x_i, z_{>l}); D_{\phi_l^i}(x_i, z_{>l}))$.

This formulation allows for a principled operation to fuse content information from available images while having the flexibility to handle missing ones. Indeed, at each level $l \in \{1, ..., L\}$, the conditional variational distributions $q_{\phi_l, \theta_l}^{\mathrm{PoE}}(z_l|x_\pi, z_{>l})$ are Normal distributions with mean $\mu_{\phi_l, \theta_l}(x_\pi, z_{>l})$ and diagonal covariance $D_{\phi_l, \theta_l}(x_\pi, z_{>l})$ expressed in closed-form solution [12] as:

$$\begin{cases} D_{\phi_l, \theta_l}(x_\pi, z_{>l}) = \left(D_{\theta_l}^{-1}(z_{>l}) + \sum_{i \in \pi} D_{\phi_l^i}^{-1}(x_i, z_{>l}) \right)^{-1} \\ \mu_{\phi_l, \theta_l}(x_\pi, z_{>l}) = D_{\phi_l, \theta_l}^{-1}(x_\pi, z_{>l}) \left(D_{\theta_l}^{-1}(z_{>l})\mu_{\theta_l}(z_{>l}) + \sum_{i \in \pi} D_{\phi_l^i}^{-1}(x_i, z_{>l})\mu_{\phi_l^i}(x_i, z_{>l}) \right) \end{cases}$$

with $D_{\theta_L}(z_{>L}) = I$ and $\mu_{\theta_L}(z_{>L}) = 0$.

3.3 Optimization Strategy for Image Synthesis

The joint reconstruction and synthesis optimization goal is to maximize the expected evidence $\mathbb{E}_{x \sim p_{\mathrm{data}}}[\log(p(x))]$. As the ELBO defined in Eq. 5 is valid for any approximate distribution q, the evidence, $\log(p_\theta(x))$, is in particular lower-bounded by the following subset-specific ELBO for any subset of images π:

$$\mathcal{L}_{\mathrm{MAVAE}}^{\mathrm{ELBO}}(x_\pi; \theta, \phi) = \underbrace{\sum_{i=1}^M \mathbb{E}_{q_\phi(z|x_\pi)}[\log(p_\theta(x_i|z_1))]}_{\mathrm{reconstruction}} - \mathrm{KL}\left[q_{\phi_L}(z_L|x_\pi)||p(z_L)\right]$$

$$- \sum_{l=1}^{L-1} \mathbb{E}_{q_{\phi_l, \theta_l}(z_{>l}|x_\pi)}\left[\mathrm{KL}[q_{\phi_l, \theta_l}(z_l|x_\pi, z_{>l})||p_{\theta_l}(z_l|z_{>l})]\right] . \tag{8}$$

Hence, the expected evidence $\mathbb{E}_{x \sim p_{\text{data}}}\left[\log(p(x))\right]$ is lower-bounded by the average of the subset-specific ELBO, i.e.:

$$\mathcal{L}_{\text{MHVAE}} := \frac{1}{|\mathcal{P}|} \sum_{\pi \in \mathcal{P}} \mathcal{L}_{\text{MAVAE}}^{\text{ELBO}}(x_\pi; \theta, \phi) \ . \tag{9}$$

Consequently, we propose to average all the subset-specific losses at each training iteration. The image decoding distributions are modelled as Normal with variance σ, i.e. $p_\theta(x_i|z_1) = \mathcal{N}(x_i; \mu_i(z_1), \sigma I)$, leading to reconstruction losses $-\log(p_\theta(x_i|z_1))$, which are proportional to $||x_i - \mu_i(z_1)||^2$. To generate sharper images, the L_2 loss is replaced by a combination of L_1 loss and GAN loss via a PatchGAN discriminator [14]. Moreover, the expected KL divergences are estimated with one sample as in [19]. Finally, the loss associated with the subset-specific ELBOs Eq. (9) is:

$$L = \sum_{i=1}^{N} (\lambda_{L_1} L_1(\mu_i, x_i) + \lambda_{\text{GAN}} L_{\text{GAN}}(\mu_i)) + \text{KL} \ .$$

Following standard practices [4,14], images are normalized in $[-1, 1]$ and the weights of the L_1 and GAN losses are set to $\lambda_{L_1} = 100$ and $\lambda_{\text{GAN}} = 1$.

4   Experiments

In this section, we report experiments conducted on the challenging problem of MR and iUS image synthesis.

Data. We evaluated our method on a dataset of 66 consecutive adult patients with brain gliomas who were surgically treated at the Brigham and Women's hospital, Boston USA, where both pre-operative 3D T2-SPACE and pre-dural opening intraoperative US (iUS) reconstructed from a tracked handheld 2D probe were acquired. The data will be released on TCIA in 2023. 3D T2-SPACE scans were affinely registered with the pre-dura iUS using NiftyReg [20] following the pipeline described in [10]. Three neurological experts manually checked registration outputs. The dataset was randomly split into a training set (N = 56) and a testing set (N = 10). Images were resampled to an isotropic 0.5 mm resolution, padded for an in-plane matrix of (192, 192), and normalized in $[-1, 1]$.

Implementation Details. Since raw brain ultrasound images are typically 2D, we employed a 2D U-Net-based architecture. The spatial resolution and the feature dimension of the coarsest latent variable (z_L) were set to 1×1 and 256. The spatial and feature dimensions are respectively doubled and halved after each level to reach a feature representation of dimension 8 for each pixel, i.e. $z_1 \in \mathbb{R}^{196 \times 196 \times 8}$ and $z_L \in \mathbb{R}^{1 \times 1 \times 256}$. This leads to 7 latent variable levels, i.e. $L = 7$. Following state-of-the-art NVAE architecture [27], residual cells

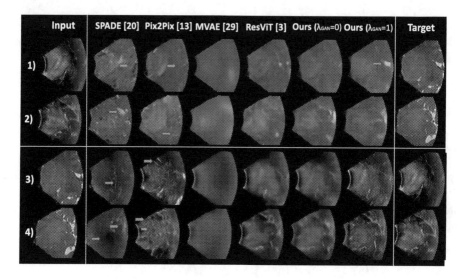

Fig. 2. Examples of image synthesis (rows 1 and 2: iUS → T_2; rows 3 and 4: T_2 → iUS) using SPADE [21], Pix2Pix [14], MVAE [30], ResViT [4] and MHVAE (ours) without and with GAN loss. As highlighted by the arrows, our approach better preserves anatomy compared to GAN-based approach and produces more realistic approach than the transformer-based approach (ResViT).

for the encoder and decoder from MobileNetV2 [23] are used with Squeeze and Excitation [13] and Swish activation. The image decoders $(\mu_i)_{i=1}^M$ correspond to 5 ResNet blocks. Following state-of-the-art bidirectional inference architectures [19,27], the representations extracted in the contracting path (from x_i to $(z_l)_l$) and the expansive path (from z_L to x_i and $(z_l)_{l<L}$) are partially shared. Models are trained for 1000 epochs with a batch size of 16. To improve convergence, λ_{GAN} is set to 0 for the first 800 epochs. Network architecture is presented in Appendix, and the code is available at https://github.com/ReubenDo/MHVAE.

Evaluation. Since paired data was available for evaluation, standard supervised evaluation metrics are employed: PSNR (Peak Signal-to-Noise Ratio), SSIM (Structural Similarity), and LPIPS [31] (Learned Perceptual Image Patch Similarity). Quantitative results are presented in Table 1, and qualitative results are shown in Fig. 2. Wilcoxon signed rank tests ($p < 0.01$) were performed.

Ablation Study. To quantify the importance of each component of our approach, we conducted an ablation study. First, our model (MHVAE) was compared with MVAE, the non-hierarchical multi-modal VAE described in [30]. It can be observed in Table 1 that MHVAE (ours) significantly outperformed MVAE. This highlights the benefits of introducing a hierarchy in the latent representation. As shown in Fig. 2, MVAE generated blurry images, while our approach produced sharp and detailed synthetic images. Second, the impact of the GAN loss was evaluated by comparing our model with ($\lambda_{GAN} = 0$) and without ($\lambda_{GAN} = 1$) the adversarial loss. Both models performed similarly in terms of evaluation metrics. However, as highlighted in Fig. 2, adding the GAN loss led to more realistic textures with characteristic iUS speckles on synthetic iUS. Finally, the image similarity between the target and reconstructed images (i.e., target image used as input) was excellent, as highlighted in Table 1. This shows that the learned latent representations preserved the content information from input modalities.

Table 1. Comparison against the state-of-the-art conditional GAN models for image synthesis. Available modalities are denoted by •, the missing ones by ∘. Mean and standard deviation values are presented. * denotes significant improvement provided by a Wilcoxon test ($p < 0.01$). Arrows indicate favorable direction of each metric.

	Input		iUS			T_2		
	iUS	T_2	PSNR(dB)↑	SSIM(%)↑	LPIPS(%)↓	PSNR(dB)↑	SSIM(%)↑	LPIPS(%)↓
MHVAE ($\lambda_{GAN} = 0$)	•	•	33.15 (2.48)	91.3 (3.5)	6.3 (2.3)	36.38 (2.40)	95.3 (1.9)	2.2 (0.8)
MHVAE ($\lambda_{GAN} = 1$)	•	•	31.54 (2.62)	89.1 (4.3)	7.1 (2.6)	34.35 (2.67)	93.6 (2.7)	2.8 (1.2)
Pix2Pix [14] $T_2 \to$ iUS	∘	•	20.31 (3.78)	70.2 (12.0)	19.8 (5.7)	×	×	×
SPADE [21] $T_2 \to$ iUS	∘	•	20.30 (3.62)	70.1 (12.1)	21.5 (6.9)	×	×	×
MVAE [30]	∘	•	21.21 (4.20)	73.5 (10.9)	26.9 (10.5)	23.23 (4.55)	83.4 (8.1)	21.4 (9.0)
ResViT [4]	∘	•	21.22 (3.10)	75.2* (9.7)	24.0 (7.5)	37.14 (5.94)	99.1 (0.9)	1.0 (0.5)
MHVAE ($\lambda_{GAN} = 0$)	∘	•	21.87* (4.06)	74.9 (10.4)	24.2 (9.1)	36.41 (2.13)	95.5 (1.8)	7.2 (3.0)
MHVAE ($\lambda_{GAN} = 1$)	∘	•	21.26 (3.93)	71.9 (11.4)	19.0* (7.6)	34.94 (2.27)	94.4 (2.3)	7.6 (3.2)
Pix2Pix [14] iUS $\to T_2$	•	∘	×	×	×	21.01 (3.70)	77.9 (9.2)	17.4 (4.7)
SPADE [21] iUS $\to T_2$	•	∘	×	×	×	20.12 (3.20)	74.3 (8.5)	18.6 (3.8)
MVAE [30]	•	∘	23.02 (4.12)	75.3 (10.4)	25.5 (9.9)	21.70 (4.60)	82.6 (8.2)	21.7 (9.1)
ResViT [4]	•	∘	35.09 (3.96)	97.6 (1.0)	3.5 (1.2)	21.70 (3.40)	82.8* (7.6)	18.9 (6.8)
MHVAE ($\lambda_{GAN} = 0$)	•	∘	33.07 (2.34)	91.3 (3.4)	13.2 (4.8)	22.16* (4.13)	82.8* (8.0)	18.3 (7.6)
MHVAE ($\lambda_{GAN} = 1$)	•	∘	31.58 (2.26)	90.8 (3.6)	12.0 (4.4)	22.12 (4.28)	81.7 (8.2)	17.4* (7.3))

State-of-the-Art Comparison. To evaluate the performance of our model (MHVAE) against existing image synthesis frameworks, we compared it to two state-of-the-art GAN-based conditional image synthesis methods: Pix2Pix [14] and SPADE [21]. These models have especially been used as synthesis backbones in previous MR/iUS synthesis studies [6,15]. Results in Table 1 show that our approach statistically outperformed these GAN methods with and without adversarial learning. As shown in Fig. 2, these conditional GANs produced realistic images but did not preserve the brain anatomy. Given that these models are not unified, Pix2Pix and SPADE must be trained for each synthesis direction ($T_2 \to$ iUS and iUS $\to T_2$). In contrast, MHVAE is a unified approach where one model is trained for both synthesis directions, improving inference practicality without a drop in performance. Finally, we compared our approach with ResViT [4], a transformer-based method that is the current state-of-the-art for unified multi-modal medical image synthesis. Our approach outperformed or reached similar performance depending on the metric. In particular, as shown in Fig. 2 and in Table 1 for the perceptual LPIPS metric, our GAN model synthesizes images that are visually more similar to the target images. Finally, our approach demonstrates significantly lighter computational demands when compared to the current SOTA unified image synthesis framework (ResViT), both in terms of time complexity (8G MACs vs. 487G MACs) and model size (10M vs. 293M parameters). Compared to MVAEs, our hierarchical multi-modal approach only incurs a marginal increase in time complexity (19%) and model size (4%). Overall, this set of experiments demonstrates that variational auto-encoders with hierarchical latent representations, which offer a principled formulation for fusing multi-modal images in a shared latent representation, are effective for image synthesis.

5 Discussion and Conclusion

Other Potential Applications. The current framework enables the generation of iUS data using T_2 MRI data. Since image delineation is much more efficient on MRI than on US, annotations performed on MRI could be used to train a segmentation network on pseudo-iUS data, as performed by the top-performing teams in the crossMoDA challenge [9]. For example, synthetic ultrasound images could be generated from the BraTS dataset [1], the largest collection of annotated brain tumor MR scans. Qualitative results shown in Appendix demonstrate the ability of our approach to generalize well to T_2 imaging from BraTS. Finally, the synthetic images could be used for improved iUS and T_2 image registration.

Conclusion and Future Work. We introduced a multi-modal hierarchical variational auto-encoder to perform unified MR/iUS synthesis. By approximating the true posterior using a combination of unimodal approximates and optimizing the ELBO with multi-modal and uni-modal examples, MHVAE demonstrated state-of-the-art performance on the challenging problem of iUS and MR synthesis. Future work will investigate synthesizing additional imaging modalities such as CT and other MR sequences.

Acknowledgement. This work was supported by the National Institutes of Health (R01EB032387, R01EB027134, P41EB028741, R03EB032050), the McMahon Family Brain Tumor Research Fund and by core funding from the Wellcome/EPSRC [WT203148/Z/16/Z; NS/A000049/1]. For the purpose of open access, the authors have applied a CC BY public copyright licence to any Author Accepted Manuscript version arising from this submission.

References

1. Bakas, S., et al.: Identifying the best machine learning algorithms for brain tumor segmentation, progression assessment, and overall survival prediction in the brats challenge (2019)
2. Chartsias, A., Joyce, T., Giuffrida, M.V., Tsaftaris, S.A.: Multimodal mr synthesis via modality-invariant latent representation. IEEE Trans. Med. Imaging **37**(3), 803–814 (2017)

3. Chartsias, A., et al.: Disentangled representation learning in cardiac image analysis. Med. Image Anal. **58**, 101535 (2019)
4. Dalmaz, O., Yurt, M., Çukur, T.: Resvit: residual vision transformers for multimodal medical image synthesis. IEEE Trans. Med. Imaging **41**(10), 2598–2614 (2022). https://doi.org/10.1109/TMI.2022.3167808
5. Dixon, L., Lim, A., Grech-Sollars, M., Nandi, D., Camp, S.: Intraoperative ultrasound in brain tumor surgery: a review and implementation guide. Neurosurg. Rev. **45**(4), 2503–2515 (2022)
6. Donnez, M., Carton, F.X., Le Lann, F., De Schlichting, E., Chabanas, M.: Realistic synthesis of brain tumor resection ultrasound images with a generative adversarial network. In: Medical Imaging 2021: Image-Guided Procedures, Robotic Interventions, and Modeling, vol. 11598, pp. 637–642. SPIE (2021)
7. Dorent, R., et al.: Learning joint segmentation of tissues and brain lesions from task-specific hetero-modal domain-shifted datasets. Med. Image Anal. **67**, 101862 (2021)
8. Dorent, R., Joutard, S., Modat, M., Ourselin, S., Vercauteren, T.: Hetero-modal variational encoder-decoder for joint modality completion and segmentation. In: Shen, D., et al. (eds.) MICCAI 2019. LNCS, vol. 11765, pp. 74–82. Springer, Cham (2019). https://doi.org/10.1007/978-3-030-32245-8_9
9. Dorent, R., Kujawa, A., Ivory, M., Bakas, S., Rieke, N., et al.: Crossmoda 2021 challenge: benchmark of cross-modality domain adaptation techniques for vestibular schwannoma and cochlea segmentation. Med. Image Anal. **83**, 102628 (2023)
10. Drobny, D., Vercauteren, T., Ourselin, S., Modat, M.: Registration of MRI and iUS data to compensate brain shift using a symmetric block-matching based approach. In: CuRIOUS (2018)
11. Havaei, M., Guizard, N., Chapados, N., Bengio, Y.: HeMIS: hetero-modal image segmentation. In: Ourselin, S., Joskowicz, L., Sabuncu, M.R., Unal, G., Wells, W. (eds.) MICCAI 2016. LNCS, vol. 9901, pp. 469–477. Springer, Cham (2016). https://doi.org/10.1007/978-3-319-46723-8_54
12. Hernández-Lobato, J.M., et al.: Balancing flexibility and robustness in machine learning: semi-parametric methods and sparse linear models. Appendix C.2 (2010)
13. Hu, J., Shen, L., Sun, G.: Squeeze-and-excitation networks. In: CVPR (2018)
14. Isola, P., Zhu, J.Y., Zhou, T., Efros, A.A.: Image-To-Image translation with conditional adversarial networks. In: CVPR (2017)
15. Jiao, J., Namburete, A.I.L., Papageorghiou, A.T., Noble, J.A.: Self-supervised ultrasound to mri fetal brain image synthesis. IEEE Trans. Med. Imaging **39**(12), 4413–4424 (2020). https://doi.org/10.1109/TMI.2020.3018560
16. Kingma, D.P., Welling, M.: Auto-encoding variational Bayes. In: ICLR (2014)
17. Lee, D., Kim, J., Moon, W.J., Ye, J.C.: Collagan: collaborative gan for missing image data imputation. In: Proceedings of the IEEE/CVF Conference on Computer Vision and Pattern Recognition, pp. 2487–2496 (2019)
18. Li, H., et al.: DiamondGAN: unified multi-modal generative adversarial networks for MRI sequences synthesis. In: Shen, D., et al. (eds.) MICCAI 2019. LNCS, vol. 11767, pp. 795–803. Springer, Cham (2019). https://doi.org/10.1007/978-3-030-32251-9_87
19. Maaløe, L., Fraccaro, M., Liévin, V., Winther, O.: BIVA: a very deep hierarchy of latent variables for generative modeling. In: NeurIPS (2019)
20. Modat, M., et al.: Fast free-form deformation using graphics processing units. Comput. Methods Prog. Biomed. **98**, 278–224 (2010)
21. Park, T., Liu, M.Y., Wang, T.C., Zhu, J.Y.: Semantic image synthesis with spatially-adaptive normalization. In: CVPR (2019)

22. Ranganath, R., Tran, D., Blei, D.: Hierarchical variational models. In: ICML (2016)
23. Sandler, M., Howard, A., Zhu, M., Zhmoginov, A., Chen, L.C.: MobileNetV2: inverted residuals and linear bottlenecks. In: CVPR (2018)
24. Sharma, A., Hamarneh, G.: Missing MRI pulse sequence synthesis using multi-modal generative adversarial network. IEEE Trans. Med. Imaging **39**(4), 1170–1183 (2020)
25. Shi, Y., Paige, B., Torr, P., et al.: Variational mixture-of-experts autoencoders for multi-modal deep generative models. In: NeurIPS, vol. 32 (2019)
26. Sønderby, C.K., Raiko, T., Maaløe, L., Sønderby, S.K., Winther, O.: Ladder variational autoencoders. In: NeurIPS (2016)
27. Vahdat, A., Kautz, J.: NVAE: a deep hierarchical variational autoencoder. In: NeurIPS, vol. 33 (2020)
28. Varsavsky, T., Eaton-Rosen, Z., Sudre, C.H., Nachev, P., Cardoso, M.J.: PIMMS: permutation invariant multi-modal segmentation. In: Stoyanov, D., et al. (eds.) DLMIA/ML-CDS -2018. LNCS, vol. 11045, pp. 201–209. Springer, Cham (2018). https://doi.org/10.1007/978-3-030-00889-5_23
29. Wang, T., et al.: A review on medical imaging synthesis using deep learning and its clinical applications. J. Appl. Clin. Med. Phys. **22**(1), 11–36 (2021)
30. Wu, M., Goodman, N.: Multimodal generative models for scalable weakly-supervised learning. In: NeurIPS, vol. 31 (2018)
31. Zhang, R., Isola, P., Efros, A.A., Shechtman, E., Wang, O.: The unreasonable effectiveness of deep features as a perceptual metric. In: CVPR (2018)
32. Zhou, T., Fu, H., Chen, G., Shen, J., Shao, L.: Hi-net: hybrid-fusion network for multi-modal mr image synthesis. IEEE Trans. Med. Imaging **39**(9), 2772–2781 (2020)

S3M: Scalable Statistical Shape Modeling Through Unsupervised Correspondences

Lennart Bastian[✉], Alexander Baumann, Emily Hoppe, Vincent Bürgin,
Ha Young Kim, Mahdi Saleh, Benjamin Busam, and Nassir Navab

Computer Aided Medical Procedures, Technical University Munich,
Munich, Germany
lennart.bastian@tum.de

Abstract. Statistical shape models (SSMs) are an established way to represent the anatomy of a population with various clinically relevant applications. However, they typically require domain expertise, and labor-intensive landmark annotations to construct. We address these shortcomings by proposing an unsupervised method that leverages deep geometric features and functional correspondences to simultaneously learn local and global shape structures across population anatomies. Our pipeline significantly improves unsupervised correspondence estimation for SSMs compared to baseline methods, even on highly irregular surface topologies. We demonstrate this for two different anatomical structures: the thyroid and a multi-chamber heart dataset. Furthermore, our method is robust enough to learn from noisy neural network predictions, potentially enabling scaling SSMs to larger patient populations without manual segmentation annotation. The code is publically available at:
https://github.com/alexanderbaumann99/S3M

Keywords: Statistical Shape Modeling · Unsupervised
Correspondence Estimation · Geometric Deep Learning

1 Introduction

Statistical shape models (SSMs) are a powerful tool to characterize anatomical variations across a population. They have been widely used in medical image analysis and computational anatomy to represent organ structures, with numerous clinically relevant applications such as clustering, classification, and shape regression [2,7,21]. SSMs are generally represented by point-wise correspondences between shapes [4,13], or deformation fields to a pre-defined template [14,28]. Despite the existence of implicit models [2], abstracting shape correspondences in the form of linear point distribution models (PDM) constitutes an appealing and interpretable

---

Lennart Bastian and Alexander Baumann contributed equally to this work.

Supplementary Information The online version contains supplementary material available at https://doi.org/10.1007/978-3-031-43999-5_44.

way to represent shape distributions [3]. Furthermore, many implicit models still rely on correspondence annotations during training [2,7].

Creating SSMs is cumbersome and intricate, as significant manual human annotation is necessary. Domain experts typically first segment images in 3D. The labeled 3D organ surfaces must then be aligned and brought into correspondence, typically achieved through deformable image registration methods using manual landmark annotations [6]. This is labor-intensive and error-prone, potentially inducing bias into downstream SSMs and applications [39].

Unsupervised methods have been proposed to estimate correspondences for SSMs [4,11,24]. However, they typically require precisely segmented and smooth surfaces to generate accurate inter-organ correspondences. ShapeWorks has been established to produce high-quality correspondences on several organs such as femurs or left atria [1,3,7]. However, as domain experts carefully curate most medical datasets, the robustness of such methods has not yet been thoroughly evaluated concerning label noise and segmentation inaccuracies. The main obstacles that prevent scaling SSMs to larger patient populations are unsupervised correspondence methods that can handle topological variations in noisy annotations, such as those produced by inexperienced annotators or predictions from deep neural networks. Robust methods to deal with these obstacles are required.

To address these challenges, we propose S3M, which leverages unsupervised deep geometric features while incorporating a global shape structure. Geometric Deep Learning (GDL) provides techniques to process 3D shapes and geometries, which are robust to noise, 3D rotations, and global deformations. We utilize graph neural networks (GNN) and functional mappings to establish dense surface correspondences of samples without supervision. This approach has significant clinical implications as it enables automatically representing anatomical shapes across large patient populations without requiring manual expert landmark annotations. We demonstrate that our proposed method creates objectively superior SSMs from shapes with noisy surface topologies. Moreover, it accurately corresponds regions of complex anatomies with mesh bifurcations such as the heart, which could ease the modeling of inter-organ relations [12].

Our contributions can be summarized as follows:

- We propose a novel unsupervised correspondence method for SSM curation based on geometric deep learning and functional correspondence.
- S3M exhibits superior performance on two challenging anatomies: thyroid and heart. It generates objectively more suitable SSMs from noisy thyroid labels and challenging multi-chamber heart reconstructions.
- To pave the way for unsupervised SSM generation in other medical domains, we open-source the code of our pipeline.

2   Related Work

Point Distribution Models. A population of shapes must be brought into correspondence to construct a PDM. This has been traditionally achieved through pair-wise registration methods [19,21]. However, pairwise approaches can admit

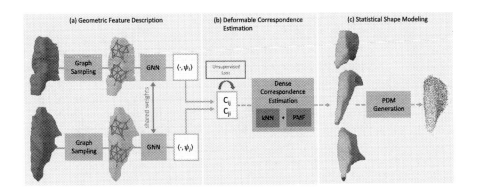

Fig. 1. Our proposed method for unsupervised SSM curation. (a) We use a Siamese GNN as shape descriptor and project the extracted features onto the LBO eigenfunctions ψ to obtain spectral representations. (b) From these, we optimize a functional mapping between pairs of shapes with an unsupervised loss. Gradients are backpropagated to the geometric descriptors. (c) During inference, the dense correspondences are estimated between pairs of shapes based on the learned population parameters, which are then used to construct an SSM.

bias as they neglect the population during correspondence generation [19]. More recently, group-wise optimization methods such as ShapeWorks [11] have been adopted as they jointly optimize over a cohort, overcoming such biases. They demonstrate superior prediction of clinically relevant anatomical variations [19]. Furthermore, generic models that can perform well across various organs are sought after.

Graph Neural Networks (GNNs) have been used to enable structural feature extraction through message passing. They constitute a powerful tool to process 3D data and extract geometric features [32,33,38] which can be useful for disease prediction [23,30]. Other medical applications involve brain cortex mesh reconstruction [8] and 3D organ-at-risk (ORA) segmentation [22]. We use GNNs for deformable 3D organ shapes and learn to estimate dense correspondences in the presence of noise and anatomical variations.

Functional Correspondence. Functional maps abstract the notion of point-to-point correspondences by corresponding arbitrary functions, such as texture or curvature, across shapes. They are extensively used to estimate dense correspondences across deformable shapes [29] and can be incorporated into learning frameworks [17,26]. These methods are typically evaluated on synthetic meshes with dense annotations and limited variable surface topology. In contrast, medical shapes exhibit higher variability, requiring robust surface representation for reliable correspondence matching. More recently, unsupervised functional correspondence models have been proposed [10,31]. These methods demonstrate strong performance on synthetic data without manual correspondence annotations. They extract features from the surface geometry using hand-crafted descriptors such as SHOT [36], wave-kernel signatures (WKS) [5] or heat-kernel signatures (HKS) [9]. The extracted features are then typically refined and projected onto the Laplace-Beltrami Operator (LBO) eigenfunctions [29]. μMatch [24] recently leveraged such an unsupervised approach in the medical domain. They employ handcrafted features to extract representations from shapes with a relatively smooth surface topology; however, they fail for shapes with high degrees of surface noise or label inconsistencies. To scale SSM curation to larger datasets encompassing population variance, our method must be robust to a more variable and complex surface topology.

3   Method

In the following, we propose a method to establish an SSM as a Point Distribution Model (PDM), illustrated in Fig. 1. Robust local features from the surface mesh are extracted using GNNs. These features are then projected onto the truncated eigenspace of the Laplace-Beltrami Operator using $m = 20$ eigenfunctions [29]. We perform post-processing with a Product Manifold Filter (PMF) [37] to obtain bijective correspondences for SSM generation. The shape model is subsequently created by aggregating correspondences across a dataset of predicted correspondences using the eigendecomposition of the covariance matrix.

Geometric Feature Description. Handcrafted descriptors [5,9,36] are unable to represent the complex surface topology of medical data adequately. To cope with surface artifacts and irregular morphologies, we use a graph-based descriptor [32]. We first extract a surface mesh from a 3D volumetric grid using marching cubes [27]. Graph nodes are defined as the mesh's vertices; edges are obtained using a k-nearest neighbor search with $k = 10$. Node features are given by spatial xyz-coordinates. The graph is fed into three topology adaptive layers [18] using graph convolutions with a specific number of hops to define the number of nodes a message is passed to. Increasing the number of hops (we use 1, 2, 3 hops per layer, respectively) increases the receptive field, incorporating features from more distant nodes. Finally, features pass a linear layer before being projected onto the Laplace-Beltrami eigenfunctions.

Deformable Correspondence Estimation. PDMs require correspondences between samples to model the statistical distribution of the organ. Inspired by methods for geometric shape correspondence, we propose to estimate a functional mapping T to correspond high-level semantics from two input shapes, X_i and X_j. The LBO extends the Laplace operator to Riemannian manifolds, capturing intrinsic characteristics of the shape independent of its position and orientation in Euclidean space. It can be efficiently computed on a surface mesh using, for example, the cotangent weight discretization scheme [15]. This results in a matrix representation of the LBO from which one can then calculate the associated

Fig. 2. PDM results from the proposed method S3M on the thyroid dataset [25]. The top row depicts the PDM generated from manual annotations and the bottom row from network pseudo-labels. From left to right, we depict $-3\sqrt{\lambda_1}$, $-3\sqrt{\lambda_2}$, the mean shape, $+3\sqrt{\lambda_2}$, $+3\sqrt{\lambda_1}$, with λ_i the eigenmode corresponding to the i-th largest eigenvalue. Similar colors indicate corresponding regions predicted by the model.

eigenfunctions $\psi_i \in \mathbf{R}^{n \times m}$ for a shape $X_i \in \mathbf{R}^{n \times 3}$. Given a feature vector D_i extracted from a surface mesh of shape X_i and a neural network T_ϕ, we can approximate a functional mapping between shapes by solving the following optimization problem:

$$\min_{\phi} \sum_{(X_i, X_j)} \mathcal{L}(C_{ij}, C_{ji}) \quad \text{where} \quad C_{ij} = \arg\min_{C} \| C A_{T_\phi(D_i)} - A_{T_\phi(D_j)} \| \quad (1)$$

Here, $A_{T_\phi(D_i)} \in \mathbf{R}^{m \times m}$ denotes the transformed descriptor, written in the basis of the LBO eigenfunctions of shape X_i and $C_{ij} \in \mathbf{R}^{m \times m}$ represents the optimal functional mapping from the descriptor space of X_i to the one of X_j. Inspired by existing works on shape correspondence [10,31,34], our loss function enforces four separate characteristics on the learned functional mapping, including bijectivity, orthogonality, and isometric properties. We refer to the supplementary materials for the complete definition. Notably, none of these losses uses ground truth correspondences, making the entire process unsupervised.

Training and Inference. During training, two shapes are sampled from the dataset, and the pipeline is optimized with Eq. 1. We increase model robustness by augmenting with rotations and small surface deformations. The point cloud is sub-sampled in each training iteration using farthest point sampling with random initialization. During inference, our model predicts pairwise correspondences. To accumulate these over an entire dataset of N shapes, we choose a template shape $X_T = \arg\min_{X_i} \sum_{j=0}^{N} \mathbb{1}_{i \neq j} \mathcal{L}(C_{ij}, C_{ji})$ as the instance with the lowest average loss to all other shapes in the dataset. As in [29], we extract

(a) Dataset Sample   (b) Shapeworks   (c) μMatch   (d) SURFMNet   (e) Ours

Fig. 3. Qualitative Analysis of Whole-Heart SSMs. A sample from the heart dataset (a). Composition 1 (right ventricle) is denoted in red. Composition 2 (both ventricles and atria) combines red and green regions. Composition 3 additionally incorporates the vessels, denoted in blue The predicted SSM mean shapes for composition 3 are portrayed for ShapeWorks (b), μMatch (c), SURFMNet (d), and S3M (e).

point-to-point correspondences between the template X_T and another shape X_i by matching the transformed LBO eigenfunctions of X_i, namely $C_{iT}\psi_i$, with the LBO eigenfunctions of the template ψ_T using nearest neighbors. As PDMs require bijective correspondences, we subsequently post-process the results with PMF [37].

Statistical Shape Modeling. We use the PDM [13] as the underlying method of the shape model. It takes input points of the form $X \in \mathbf{R}^{N \times d}$, where N, d are the number of shape samples and coordinates per shape, respectively. It returns a multi-variate normal distribution. In our case, $d = 3n$ given each shape has n points. We calculate the mean shape $\bar{X}$ and the empirical covariance matrix $S = \mathrm{cov}(\mathrm{X})$ over the N samples [13]. Since S has rank $N - 1$, it has $N - 1$ real eigenvectors v_j with eigenvalues λ_j. If we consider the sum

$$s = \bar{X} + \sum_{j=1}^{N-1} \alpha_j \lambda_j v_j, \qquad \alpha_j \sim \mathcal{N}(0, 1) \tag{2}$$

then $s \sim \mathcal{N}\left(\bar{X}, S\right)$, which is the desired distribution of the model. For the above, the points must be in correspondence across the samples. We thus use the correspondences generated in Sect. 3 to construct the PDM.

4   Experiments

All experiments are carried out using two publicly available datasets: thyroid ultrasound scans and heart MRI acquisitions. Our model is implemented in PyTorch 1.12 using CUDA 11.6. Training takes between 2.5–3 h on an Nvidia A40 GPU and inference about 0.71 s for a pair of shapes. We use publically available implementations for all baseline methods.

Thyroid Dataset (SegThy) [25]. The dataset comprises 3D freehand US scans of healthy thyroids from 32 volunteers aged 24–39. For each volunteer, three physicians acquired three scans each. Ultrasound scans generally exhibit noise induced by physical properties such as phase aberrations and attenuation. This leads to label inconsistencies or topological irregularities that pose a challenge for shape modeling (see Fig. 2). US sweeps were compounded to a 3D grid of resolution $0.12 \times 0.12 \times 0.12\,\mathrm{mm}^3$. A single scan from each of the 16 volunteers was manually annotated by experts (ground truth) and used to train QuickNAT [25]. The remaining scans were pseudo-labeled through QuickNAT segmentation predictions exhibiting moderate degrees of noise and inaccuracies (dice score of 0.94 [25]). We divide the dataset into manual and pseudo-label predictions and evaluate them separately. The pseudo-label experiment evaluates the model's performance under topological noise and inaccuracies and is limited to 100 scans due to ShapeWorks memory constraints [20]. We extract a surface mesh for each scan using marching cubes and subsample the meshes to 5000 vertices.

Heart Dataset [35]. The data constitutes 30 MRI scans from a single cardiac phase of the heart. Each image has a voxel resolution of $1.25 \times 1.25 \times 2.7\,\mathrm{mm}^3$. Segmentation is carried out using an automated method, with subsequent manual corrections by domain experts. Labels are provided for: the right/left ventricle, right/left atrium, aorta, and pulmonary artery. We evaluate the capability of the models to reconstruct complex organs using three hierarchical compositions of the heart chambers. *Composition 1* consists of the right ventricle, *Composition 2* of the left and right atrium and ventricle, and *Composition 3* of the whole heart, including the aorta and pulmonary artery.

Table 1. SSM quality metrics for the Thyroid dataset [25]

Metrics	Ground-Truth Segmentation		Network Pseudo-Labels	
	Generality [mm] ↓	Specificity [mm] ↓	Generality [mm] ↓	Specificity [mm] ↓
SURFMNet	2.20 ± 0.20	3.20 ± 0.29	–	–
μMatch	1.92 ± 0.07	2.81 ± 0.18	1.90 ± 0.08	2.84 ± 0.10
Shapeworks	1.94 ± 0.27	1.81 ± 0.06	1.60 ± 0.04	1.75 ± 0.07
S3M (Ours)	$\mathbf{1.25 \pm 0.11}$	$\mathbf{1.59 \pm 0.06}$	$\mathbf{0.95 \pm 0.07}$	1.84 ± 0.08

SSM Evaluation. We compare Shapeworks [11], μMatch [24] and SURFMNet [31], with S3M. A four-fold cross-validation is employed. SURFMNet, μMatch, and S3M are trained on the training folds and correspondences are predicted on the training and validation set. Since the particle-based optimization of Shapeworks does not generalize to unseen data, it uses all folds for correspondence estimation. The SSM is built using correspondences from the training set, and evaluated with respect to two standardized metrics: generality and specificity [16]. For generality, we measure how well the SSM can represent unseen instances from the fourth fold through the Chamfer distance between the original shape and its SSM reconstruction. Specificity indicates how well random samples from the SSM represent the training data. We sample from the PDM 1000 times and calculate each sample's minimum Chamfer distance to the training shapes. Generalization and specificity are reported in mm. Numbers in bold indicate statistically significant results by a one-sided t-test ($p < 0.05$).

5   Results and Discussion

Experiment 1: Thyroid SSM. Table 1 depicts the performance for all methods on the thyroid shapes. SURFMNet results for the thyroid pseudo-labels are omitted, as the method did not converge. Consistent trends can be observed across all methods for both sets of thyroid labels. The two existing functional map-based methods were outperformed by Shapeworks, while the proposed S3M exceeded the latter's scores. The descriptor is the most significant difference between the three learned functional map methods. Hand-crafted shape descriptors like SHOT and a simple fully-connected residual network architecture do not adequately represent thyroids' noisy and heterogeneous shapes.

Our proposed method significantly outperformed Shapeworks in all metrics except the specificity of pseudo-labeled thyroids, where the results are not statistically significant. This was despite the advantage of optimizing correspondences across all shapes in training and validation. S3M can better cope with topological noise and generalizes to unseen samples, demonstrating potential in scaling SSM generation to larger datasets. Furthermore, it does not suffer from increasing memory requirements with the number of samples.

Table 2. Whole Heart Statistical Shape Modeling

Metrics	Composition 1		Composition 2		Composition 3	
	Generality [mm] ↓	Specificity [mm] ↓	Generality [mm] ↓	Specificity [mm] ↓	Generality [mm] ↓	Specificity [mm] ↓
SURFMNet	1.10 ± 0.39	1.79 ± 0.87	1.37 ± 0.16	2.00 ± 0.30	1.71 ± 0.17	2.54 ± 0.36
μMatch	3.39 ± 0.26	4.65 ± 0.16	2.82 ± 0.14	4.81 ± 0.41	3.30 ± 0.07	5.80 ± 0.08
ShapeWorks	0.89 ± 0.08	1.40 ± 0.04	2.57 ± 0.17	3.60 ± 0.06	3.07 ± 0.19	4.99 ± 0.22
S3M (Ours)	0.85 ± 0.07	1.30 ± 0.01	1.30 ± 0.13	1.72 ± 0.05	1.63 ± 0.17	2.14 ± 0.07

Experiment 2: Whole Heart SSM. Table 2 depicts the results of the different models on the three heart chamber compositions as previously defined. For the single-organ right atrium (composition 1), our proposed method fares comparably to ShapeWorks. For the more complex compositions 2 and 3, we observe larger increases in generalization and specificity for Shapeworks. μMatch fails to create a convincing SSM for any heart composition. Interestingly, SURFMNet can represent the more complex compositions 2 and 3 better than ShapeWorks, showing the strength of functional maps at representing complex high-level structures. S3M still exceeded the performance of SURFMNet, possibly due to the graph descriptor being better able to represent the surface topology.

From the qualitative results in Fig. 3, it becomes more apparent that Shape-Works does not generate an adequate SSM for the more complex compositions. This further supports our proposed method's ability to learn correspondences for intricate and complex surface topologies, even consisting of meshes with bifurcations. The flexibility of our surface representation enables unsupervised correspondence estimation from multiple hierarchical sub-shapes, which is invaluable in multi-organ modeling such as for the heart [6,12].

Experiment 3: Thyroid Pseudo-label Generalization
To further highlight the proposed methods' robustness to network-generated segmentation labels, we additionally measure the reconstruction ability of SSMs created from pseudo-labels on manually annotated thyroid labels under the Chamfer distance (Ours: 1.05 ± 0.10 mm, Shapeworks: 1.84 ± 0.40 mm). Notably, the proposed PDM on pseudo-labels generalizes better than the SSM built on few manual labels (1.25 ± 0.11 mm; see Table 1), suggesting that more data can improve the SSM even if the labels are inaccurate. This is further supported by differences in shape (suppl. Figure 1); the SSM's mean shape generated from pseudo-labels approximates the mean shapes on GT labels (and thus, the true organ shape) more closely.

6 Conclusion

We present an unsupervised approach for learning correspondences between shapes that exhibit noisy and irregular surface topologies. Our method leverages the strengths of geometric feature extractors to learn the intricacies of organ surfaces, as well as high-level functional bases of the Laplace-Beltrami operator to capture more extensive organ semantics. S3M outperforms existing methods on both manual labels, and label predictions from a network, demonstrating the potential to scale existing SSM pipelines to datasets that encompass more substantial population variance without additional annotation burden. Finally, we show that our model has the potential to learn correspondences between complex multi-organ shape hierarchies such as chambers of the heart, which would ease the manual burden of SSM curation for structures that currently still require meticulous manual landmark annotations.

References

1. Adams, J., Bhalodia, R., Elhabian, S.: Uncertain-DeepSSM: from images to probabilistic shape models. In: Reuter, M., Wachinger, C., Lombaert, H., Paniagua, B., Goksel, O., Rekik, I. (eds.) ShapeMI 2020. LNCS, vol. 12474, pp. 57–72. Springer, Cham (2020). https://doi.org/10.1007/978-3-030-61056-2_5
2. Adams, J., Elhabian, S.: From images to probabilistic anatomical shapes: a deep variational bottleneck approach. In: Wang, L., Dou, Q., Fletcher, P.T., Speidel, S., Li, S. (eds.) MICCAI 2022. LNCS, vol. 13432, pp. 474–484. Springer, Heidelberg (2022). https://doi.org/10.1007/978-3-031-16434-7_46

3. Adams, J., Khan, N., Morris, A., Elhabian, S.: Spatiotemporal cardiac statistical shape modeling: a data-driven approach. In: Camara, O., et al. (eds.) STACOM MICCAI 2022, vol. 13593, pp. 143–156. Springer, Heidelberg (2023). https://doi.org/10.1007/978-3-031-23443-9_14

4. Agrawal, P., Whitaker, R.T., Elhabian, S.Y.: Learning deep features for shape correspondence with domain invariance. arXiv preprint arXiv:2102.10493 (2021)

5. Aubry, M., Schlickewei, U., Cremers, D.: The wave kernel signature: a quantum mechanical approach to shape analysis. In: ICCV Workshops (2011)

6. Banerjee, A., Zacur, E., Choudhury, R.P., Grau, V.: Automated 3d whole-heart mesh reconstruction from 2d cine mr slices using statistical shape model. In: 2022 IEEE EMBS, pp. 1702–1706. IEEE (2022)

7. Bhalodia, R., Elhabian, S., Adams, J., Tao, W., Kavan, L., Whitaker, R.: Deepssm: a blueprint for image-to-shape deep learning models. arXiv preprint arXiv:2110.07152 (2021)

8. Bongratz, F., Rickmann, A.M., Pölsterl, S., Wachinger, C.: Vox2cortex: fast explicit reconstruction of cortical surfaces from 3d MRI scans with geometric deep neural networks. In: CVPR 2022, pp. 20773–20783 (2022)

9. Bronstein, M.M., Kokkinos, I.: Scale-invariant heat kernel signatures for non-rigid shape recognition. In: CVPR 2010, pp. 1704–1711. IEEE (2010)

10. Cao, D., Bernard, F.: Unsupervised deep multi-shape matching. In: Avidan, S., Brostow, G., Cisse, M., Farinella, G.M., Hassner, T. (eds.) ECCV 2022. LNCS, vol. 13663, pp. 55–71. Springer, Heidelberg (2022). https://doi.org/10.1007/978-3-031-20062-5_4

11. Cates, J., Elhabian, S., Whitaker, R.: ShapeWorks. In: Statistical Shape and Deformation Analysis, pp. 257–298. Elsevier (2017)

12. Cerrolaza, J.J., et al.: Computational anatomy for multi-organ analysis in medical imaging: a review. Med. Image Anal. **56**, 44–67 (2019)

13. Cootes, T.F., Taylor, C.J., Cooper, D.H., Graham, J.: Training models of shape from sets of examples. In: BMVC92 (1992)

14. Cootes, T.F., Twining, C.J., Babalola, K.O., Taylor, C.J.: Diffeomorphic statistical shape models. Image Vision Comput. **26**(3), 326–332 (2008)

15. Crane, K.: Discrete differential geometry: an applied introduction. Not. AMS Commun. **7**, 1153–1159 (2018)

16. Davies, R.H.: Learning Shape: Optimal Models for Analysing Natural Variability. The University of Manchester (United Kingdom) (2002)

17. Donati, N., Sharma, A., Ovsjanikov, M.: Deep geometric functional maps: robust feature learning for shape correspondence. In: CVPR 2020, pp. 8592–8601 (2020)

18. Du, J., Zhang, S., Wu, G., Moura, J.M., Kar, S.: Topology adaptive graph convolutional networks. arXiv preprint arXiv:1710.10370 (2017)

19. Goparaju, A., Iyer, K., Bone, A., Hu, N., Henninger, H.B., et al.: Benchmarking off-the-shelf statistical shape modeling tools in clinical applications. Med. Image Anal. **76**, 102271 (2022)

20. Gutiérrez-Becker, B., Wachinger, C.: Learning a conditional generative model for anatomical shape analysis. In: Chung, A.C.S., Gee, J.C., Yushkevich, P.A., Bao, S. (eds.) IPMI 2019. LNCS, vol. 11492, pp. 505–516. Springer, Cham (2019). https://doi.org/10.1007/978-3-030-20351-1_39

21. Heimann, T., Meinzer, H.P.: Statistical shape models for 3D medical image segmentation: a review. Med. Image Anal. **13**(4), 543–563 (2009)

22. Henderson, E.G., Green, A.F., van Herk, M., Vasquez Osorio, E.M.: Automatic identification of segmentation errors for radiotherapy using geometric learning. In:

Wang, L., Dou, Q., Fletcher, P.T., Speidel, S., Li, S. (eds.) MICCAI 2022. LNCS, vol. 13435, pp. 319–329. Springer, Heidelberg (2022). https://doi.org/10.1007/978-3-031-16443-9_31

23. Kazi, A., et al.: InceptionGCN: receptive field aware graph convolutional network for disease prediction. In: Chung, A.C.S., Gee, J.C., Yushkevich, P.A., Bao, S. (eds.) IPMI 2019. LNCS, vol. 11492, pp. 73–85. Springer, Cham (2019). https://doi.org/10.1007/978-3-030-20351-1_6

24. Klatzow, J., Dalmasso, G., Martínez-Abadías, N., Sharpe, J., Uhlmann, V.: μMatch: 3D shape correspondence for biological image data, vol. 4 (2022)

25. Krönke, M., Eilers, C., Dimova, D., Köhler, M., Buschner, G., et al.: Tracked 3d ultrasound and deep neural network-based thyroid segmentation reduce interobserver variability in thyroid volumetry. Plos One **17**(7) (2022)

26. Litany, O., Remez, T., Rodola, E., Bronstein, A., Bronstein, M.: Deep functional maps: structured prediction for dense shape correspondence. In: ICCV, pp. 5659–5667 (2017)

27. Lorensen, W.E., Cline, H.E.: Marching cubes: a high resolution 3d surface construction algorithm. ACM Siggraph Comput. Graph. **21**(4), 163–169 (1987)

28. Lüdke, D., Amiranashvili, T., Ambellan, F., Ezhov, I., Menze, B.H., Zachow, S.: Landmark-free statistical shape modeling via neural flow deformations. In: Wang, L., Dou, Q., Fletcher, P.T., Speidel, S., Li, S. (eds.) MICCAI 2022. LNCS, vol. 13432, pp. 453–463. Springer, Heidelberg (2022). https://doi.org/10.1007/978-3-031-16434-7_44

29. Ovsjanikov, M., Ben-Chen, M., Solomon, J., Butscher, A., Guibas, L.: Functional maps: a flexible representation of maps between shapes. ACM ToG **31**(4) (2012)

30. Parisot, S., et al.: Spectral graph convolutions for population-based disease prediction. In: Descoteaux, M., Maier-Hein, L., Franz, A., Jannin, P., Collins, D.L., Duchesne, S. (eds.) MICCAI 2017. LNCS, vol. 10435, pp. 177–185. Springer, Cham (2017). https://doi.org/10.1007/978-3-319-66179-7_21

31. Roufosse, J.M., Sharma, A., Ovsjanikov, M.: Unsupervised deep learning for structured shape matching. In: ICCV (2019)

32. Saleh, M., Dehghani, S., Busam, B., Navab, N., Tombari, F.: Graphite: graph-induced feature extraction for point cloud registration. In: 2020 3DV, pp. 241–251. IEEE (2020)

33. Saleh, M., Wu, S.C., Cosmo, L., Navab, N., Busam, B., Tombari, F.: Bending graphs: hierarchical shape matching using gated optimal transport. In: CVPR 2022, pp. 11757–11767 (2022)

34. Sharma, A., Ovsjanikov, M.: Weakly supervised deep functional maps for shape matching. NeurIPS **33**, 19264–19275 (2020)

35. Tobon-Gomez, C., et al.: Benchmark for algorithms segmenting the left atrium from 3d ct and mri datasets. IEEE Trans. Med. Imaging **34**(7), 1460–1473 (2015)

36. Tombari, F., Salti, S., Di Stefano, L.: Unique signatures of histograms for local surface description. In: Daniilidis, K., Maragos, P., Paragios, N. (eds.) ECCV 2010. LNCS, vol. 6313, pp. 356–369. Springer, Heidelberg (2010). https://doi.org/10.1007/978-3-642-15558-1_26

37. Vestner, M., Litman, R., Rodola, E., Bronstein, A., Cremers, D.: Product manifold filter: non-rigid shape correspondence via kernel density estimation in the product space. In: CVPR (2017)

38. Wang, Y., Sun, Y., Liu, Z., Sarma, S.E., Bronstein, M.M., Solomon, J.M.: Dynamic graph cnn for learning on point clouds. Acm ToG **38**(5), 1–12 (2019)

39. Zhang, L., et al.: Disentangling human error from ground truth in segmentation of medical images. NeurIPS **33**, 15750–15762 (2020)

RESToring Clarity: Unpaired Retina Image Enhancement Using Scattering Transform

Ellen Jieun Oh[1], Yechan Hwang[1], Yubin Han[1], Taegeun Choi[2],
Geunyoung Lee[2], and Won Hwa Kim[1(✉)]

[1] Pohang University of Science and Technology (POSTECH), Pohang, South Korea
{jieunoh,yechan99,yubin,wonhwa}@postech.ac.kr
[2] Mediwhale, Seoul, South Korea

Abstract. Retina images are non-invasive and highly effective in the diagnosis of various diseases such as cardiovascular and ophthalmological diseases. Accurate diagnosis depends on the quality of the retina images, however, obtaining high-quality images can be challenging due to various factors, such as noise, artifacts, and eye movement. Methods for enhancing retina images are therefore in high demand for clinical purposes, yet the problem remains challenging as there is a natural trade-off between preserving anatomical details (e.g., vessels) and increasing overall image quality other than the content in it. Moreover, training an enhancement model often requires paired images that map low-quality images to high-quality images, which may not be available in practice. In this regime, we propose a novel Retina image Enhancement framework using Scattering Transform (REST). REST uses unpaired retina image sets and does not require prior knowledge of the degraded factors. The generator in REST enhances retina images by utilizing the Anatomy Preserving Branch (APB) and the Tone Transferring Branch (TTB) with different roles. Our model successfully enhances low-quality retina images demonstrating commendable results on two independent datasets.

1   Introduction

Microvasculature and neural tissue in the retina can be directly and non-invasively visualized in vivo [19], and retina images are used for the diagnosis of various diseases including cardiovascular diseases [18], Alzheimer's disease [23], and ophthalmological diseases such as glaucoma and cataract [9]. Therefore, screening diseases using retina images has significant advantages over painful invasive methods such as blood tests or biopsies. However, the screening accuracy highly depends on the quality of the image [21] which can be compromised by various factors such as noise, low contrast, uneven illumination, blurriness, camera model, and experience level of the clinician taking the image [6,16,17,22].

Supplementary Information The online version contains supplementary material available at https://doi.org/10.1007/978-3-031-43999-5_45.

H. Greenspan et al. (Eds.): MICCAI 2023, LNCS 14229, pp. 470–480, 2023.
https://doi.org/10.1007/978-3-031-43999-5_45

These issues make it difficult to discriminate important indicators and biomarkers, leading to the failure of early detection of diseases and proper intervention. Therefore, image enhancement of low-quality retina images is highly necessary. However, retina image enhancement is a challenging task due to several reasons. Intuitively, it requires paired low and high-quality retina images to learn to map low-quality images to their high-quality ones and such data are difficult to acquire in practice [1]. Moreover, enhancing images according to signal-to-noise ratio might discard important anatomical structures, e.g., the shape of the optic disc, vessels, and disease-related features, that are critical for disease screening.

To deal with the issues above, a structure-preserving guided retina image filter (SGRIF) was proposed to enhance unpaired retina images with prior knowledge of clouding effect [4]. While dealing with unpaired data, it is biased with de-clouding the artifact caused by the cataract. Several parametric methods were recently proposed including [25] and [24] with simple CycleGAN [26] structure which require authentic unpaired data only. Although they can map the tone of the input image such as color, contrast, and illuminance, to those of high-quality images, their ability to preserve crucial anatomical structures that are essential for accurate image examination [6] remains limited. Later, ISE-CRET [5] and PCENet (Pyramid Constraint Enhancement Network) [13] were proposed, which utilize supervised learning with synthesized low-quality images and authentic high-quality image pairs. These models preserve detailed features effectively, however, generating degraded images with priors remains a labor-intensive process, and trained models are specific to the dataset in question.

In this context, we propose an unpaired Retina image Enhancement with Scattering Transform (REST) which preserves the anatomical structure (e.g., vessels, optic disc, and cup) and maps the tone (e.g., color and illuminance) effectively by utilizing an unpaired dataset. Our model contains a genuine generator that includes two branches: the Anatomy Preserving Branch (APB) and the Tone Transferring Branch (TTB). The APB incorporates scattering transform, which effectively captures anatomic structures and the TTB employs a multi-layer convolution to refine the tone of the image as that of high-quality images. Constructing a cyclic architecture [26] with the proposed generators and discriminators for consistency regularization, the REST successfully learns how to enhance the low-quality retina images with the following contributions: **1)** REST only uses authentic unpaired data and does not require any prior knowledge, **2)** Successful preservation of anatomic structures is achieved with scattering transform, **3)** REST is extensively evaluated qualitatively and quantitatively on two independent datasets. Experiments on UK Biobank and EyeQ [6] datasets demonstrate that the REST can adequately enhance retina images by restoring dark and uncertain regions without compromising anatomical structures.

2   Related Works

The SGRIF proposed in [4] is a non-parametric method that utilizes a filter with degradation equations describing clouding effects caused by cataracts on the retina image. SGRIF pose and tackles the enhancement problem as a dehazing problem in computer vision. Using the degradation equation, global structure transfer filter, and global edge-preserving smoothing filter, SGRIF de-clouds the retina image. Although this method does not require image pairs, its performance depends on prior functions for degradation which causes a lack of generalizability. Parametric methods, especially those based on deep learning, perform better at generalization by identifying and improving low-quality factors in input images with the parameters. Simple CycleGAN-based models, such as [25] and [24], use only the unpaired images and do not require prior knowledge but face challenges in preserving detailed information. To address this limitation, recent approaches synthesize low-quality images paired with authentic high-quality images for supervised learning, as proposed in [5,12,13]. Although such methods preserve structural information, it is prone to being dataset-specific by the degrading model, and synthesizing low-quality images is time-consuming due to the difficulty in designing an accurate degradation model.

3   Methods

We propose a generative framework for unpaired retina image enhancement, which translates low-quality retina images X to high-quality images Y through two separate branches: APB and TTB. As illustrated in Fig. 1, an input image is simultaneously fed into both branches with encoder-decoder structures, and their outputs are combined with a kernel to obtain an enhanced image.

3.1   Anatomy Preserving Branch

In order to preserve fine anatomical details, such as the shape of an optic disc, an optic cup, and vessels, the model must accurately capture high-frequency components, i.e., edges. To extract these high-frequency components, we designed the APB, which employs a wavelet scattering transform over multiple encoding layers. Wavelet scattering transform extracts high-frequency factors and invariants with respect to translation and rotation [3] by comprising wavelet filters Ψ with the set of frequency and phase indices Λ and J number of scales, a low pass filter Φ, and a modulus operator [2,11]. Given an image x as an input, the scattering transform consisting of one layer, results in 0-th and first-order scattering coefficients, S_J^0 and S_J^1, as

$$S_J x = [S_J^0 x, S_J^1 x] = [x * \Phi, |x * \Psi_\lambda| * \Phi], \lambda \in \Lambda \tag{1}$$

The 0-th order scattering coefficient $S_J^0 x$ is computed by convolving ($*$) x with Φ. The first-order coefficients set $S_J^1 x$ is obtained by convolving x with a set of wavelet filters Ψ_λ and modulus operation followed by a low pass filter Φ. Since the wavelet filter sets Ψ_λ have diverse frequencies and angles (i.e., Λ) within 2D image space as well as scales (i.e., J), the filters ensure the scattering transform to capture the structural information with respect to the frequency and direction [3].

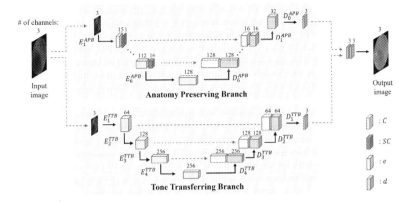

Fig. 1. The architecture of the generator in REST with APB (top) and TTB (bottom). Input images are fed into two branches and the two outputs are concatenated and convolved by 1×1 kernel to generate the output images. The symbols on the legend will be explained in Sect. 3.1 and Sect. 3.2.

The design of APB with scattering transform can be seen in Fig. 1 (top). First, the input image, denoted as x, is processed through a sequence of N encoding operations E_i^{APB}, where i denotes the layer index as

$$e_i^{APB} = \begin{cases} x, & i = 0 \\ E_i^{APB}(e_{i-1}^{APB}) = C_i \| SC_i, & 0 < i < N \\ E_i^{APB}(e_{i-1}^{APB}) = C_i, & i = N \end{cases} \quad (2)$$

where,

$$C_i = e_{i-1}^{APB} * k_{Ei}^{APB}, \quad SC_i = (S_1 e_{i-1}^{APB}) * k_{1\times 1}. \quad (3)$$

The input of each layer e_{i-1}^{APB} undergoes both the scattering transform S_1 and convolution with a trainable kernel $k_{1\times 1}$ and k_{Ei}^{APB}, except for the last layer in the encoding process. The C_i and SC_i are concatenated ($\|$) to yield an input for a subsequent layer. In the final layer, the input undergoes only convolution with a kernel k_{Ei}^{APB}.

After the completing N encoding processes, the decoding process, denoted as D_i^{APB}, commences. The final encoded feature map of the encoder, e_N^{APB}, undergoes $N + 1$ decoding processes. The decoding process is composed of a resizing $R(\cdot)$ and convolution with trainable kernel k_{Di}^{APB} as

$$d_i^{APB} = \begin{cases} D_i^{APB}(e_i^{APB}) = R(e_i^{APB}) * k_{Di}^{APB}, & i = N \\ D_i^{APB}(d_{i+1}^{APB}, e_i^{APB}) = R(d_{i+1}^{APB} \| e_i^{APB}) * k_{Di}^{APB}, & 0 < i < N \\ D_i^{APB}(d_{i+1}^{APB}) = d_{i+1}^{APB} * k_{Di}^{APB}, & i = 0. \end{cases} \quad (4)$$

In the upscaling phase ($0 < i \leq N$), a combination of resizing and convolution techniques is implemented instead of transpose convolution. This approach prevents the occurrence of checkerboard artifacts, which can arise from uneven overlap during the transpose convolution [14]. The artifacts are particularly problematic in the APB since the APB deals with high-frequency features which are easily affected them. The use of a combination of resizing and convolution techniques prevents the occurrence of these artifacts and ultimately produces clearer images. More specifically, we utilized interpolation with a factor of 2 and convolution with the k_{Di}^{APB} kernel. Additionally, to preserve anatomical structures such as vessels and an optic disc of the input image, we introduce skip connections [20] between the encoder and decoder. These connections involve the concatenation of the output of each encoder layer, e_i ($0 < i < N$), with the corresponding decoder layer's input, as illustrated in Fig. 1 (top).

3.2 Tone Transferring Branch

As the quality of a retina image depends on the tone of the image such as color, illumination, and contrast in addition to anatomical structures [6], it is crucial to generate synthetic images that have a similar tone to high-quality images. To ensure that the synthetic images resemble the tone of high-quality images, we designed TTB, which consists of multiple convolutional layers with a U-Net [20] architecture as in Fig. 1 (bottom). Similar to the APB, the input x is subject to a sequence of N encoding and decoding operations, denoted as E_i^{TTB} and D_i^{TTB}, respectively. For the encoding layers E_i^{TTB}, the input e_{i-1}^{TTB} is convolved with the trainable kernel k_i^{TTB} as

$$e_i^{TTB} = \begin{cases} x, & i = 0 \\ E_i^{TTB}(e_{i-1}^{TTB}) = e_{i-1}^{TTB} * k_i^{TTB}, & 0 < i \leq N. \end{cases} \quad (5)$$

After the encoding process, the N sequence of decoding operations D_i^{TTB} commences with the layer $i = N$. In the first decoding layer, the output of the encoder e_N^{TTB} is up-scaled using the transposed convolution operator ($\circledast$). For subsequent layers ($1 \leq i < N$), the output of the previous decoding layer d_{i+1}^{TTB} and the corresponding output of the encoding layer e_i^{TTB} are concatenated processed through the transposed convolution operator as

$$d_i^{TTB} = \begin{cases} D_i^{TTB}(e_i^{TTB}) = e_i^{TTB} \circledast k_i^{TTB}, & i = N \\ D_i^{TTB}(d_{i+1}^{TTB}, e_i^{TTB}) = (d_{i+1}^{TTB} \| e_i^{TTB}) \circledast k_i^{TTB}, & 1 \leq i < N. \end{cases} \quad (6)$$

Along with our objective function containing GAN Loss [7], which will be illustrated in the next section, Sect. 3.3, the sequence of convolution and transposed convolution layers encourage the distribution of the generated image, including the tone of the image, to match that of the high-quality images [26]. Moreover, skip connections that provide direct information from the encoder to the decoder constrain the branch from modifying the input image beyond recognition [20].

3.3 Loss Function for Unpaired Image Enhancement

The popular adversarial training architecture for unpaired image translation [26] is devised to train an enhancement model G comprised of APB and TTB, from Sect. 3.1 and Sect. 3.2, for synthesizing high-quality images Y from low-quality

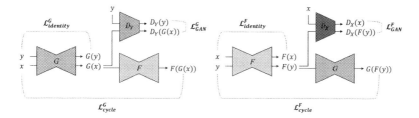

Fig. 2. The overall architecture of the REST is composed of cycles of two generators: G, F, and two discriminators: D_X, D_Y.

images X. As seen in Fig. 2, the overall architecture is composed of two generators $G{:}X \to Y$ and $F{:}Y \to X$, and two discriminators D_X and D_Y which discriminate authentic images and synthetic images of low-quality and high-quality respectively. The two generators share the same architecture as Fig. 1 and the two discriminators are also identically modeled with multiple convolution layers. To successfully train the enhancement model $G{:}X \to Y$, we composed our loss function $\mathcal{L}$ with GAN losses $\mathcal{L}_{GAN}^{G}, \mathcal{L}_{GAN}^{F}$, cycle consistency losses $\mathcal{L}_{cycle}^{G}, \mathcal{L}_{cycle}^{F}$ and identity losses $\mathcal{L}_{identity}^{G}$, $\mathcal{L}_{identity}^{F}$ [26] as follows:

$$\mathcal{L}(G, F, D_X, D_Y) = \mathcal{L}_{GAN}^{G} + \mathcal{L}_{GAN}^{F} + \mathcal{L}_{cycle} + \mathcal{L}_{identity}. \tag{7}$$

To ensure that generated images look like genuine images, the GAN losses $\mathcal{L}_{GAN}^{G}, \mathcal{L}_{GAN}^{F}$ are defined as [7]

$$\begin{aligned} \mathcal{L}_{GAN}^{G} &= \mathbb{E}_{y \in Y}[||D_Y(y)||_2^2] + \mathbb{E}_{x \in X}[||1 - D_Y(G(x))||_2^2], \\ \mathcal{L}_{GAN}^{F} &= \mathbb{E}_{x \in X}[||D_X(x)||_2^2] + \mathbb{E}_{y \in Y}[||(1 - D_X(F(y))||_2^2]. \end{aligned} \tag{8}$$

To preserve the contents of the input image, we aim to make $F(G(x)) \approx x$ and $G(F(y)) \approx y$ with the cycle consistency losses as [26]

$$\mathcal{L}_{cycle}^{G} = \lambda_{c1}\mathbb{E}_{x \in X}[||F(G(x)) - x||_1], \quad \mathcal{L}_{cycle}^{F} = \lambda_{c2}\mathbb{E}_{y \in Y}[||G(F(y)) - y||_1] \tag{9}$$

where λ_{c1} and λ_{c2} are hyperparameters. To make sure that the generators avoid translation when there is no need, i.e., a high-quality image as an input to G should lead to a high-quality image without changes, identity losses are defined as [26]

$$\mathcal{L}_{identity}^{G} = \lambda_{i1}\mathbb{E}_{y \in Y}[||G(y) - y||_1], \quad \mathcal{L}_{identity}^{F} = \lambda_{i2}\mathbb{E}_{x \in X}[||F(x) - x||_1] \tag{10}$$

with the hyperparameters λ_{i1} and λ_{i2}. With the aforementioned losses, we aim to obtain generators G^*, F^* that minimize the loss function while discriminators D_X^*, D_Y^* that maximize it as

$$G^*, F^* D_X^*, D_Y^* = \arg \max_{D_X, D_Y} \min_{G,F} \mathcal{L}(G, F, D_X, D_Y). \tag{11}$$

After training, G^* is utilized at the inference stage to enhance retina images.

4 Experiments

4.1 Datasets and Experimental Setup

Datsets. From **UKB**, 2000 images, i.e., 1000 high-quality and 1000 low-quality images respectively, were randomly sampled and split into train and test sets by a ratio of 2:1. For the **EyeQ**, we utilized 23252 labeled images, i.e., 6434 low-quality images labeled 'usable' and 16818 high-quality images labeled 'good' [6] and followed the data splitting protocol into the train, validation, and test sets as in [5]. As ISECRET and PCENet require low-quality pairs and a mask for every image during training, degraded pairs were generated between high-quality ones and masks with the given degrading and masking methods in [5,13].

Table 1. FIQA, WFQA, the number (#) of parameters, usage of paired data, and training samples (# of data) for UKB (left) and EyeQ (right) are compared.

	Parameters	Pair	Samples	FIQA	WFQA
Real high	-	-	-	0.989	1.978
Real low	-	-	-	0.051	0.141
CycleGAN (ResNet) [7,8]	28.29 M	-	1332	0.926 ± 0.04	1.853 ± 0.07
CycleGAN (U-Net) [7,20]	114.35 M	-	1332	0.893 ± 0.04	1.765 ± 0.09
ISECRET [5]	15.36 M	✓	1998	0.898 ± 0.02	1.765 ± 0.09
PCENet [13]	26.65 M	✓	1332	0.389 ± 0.06	0.782 ± 0.12
REST (ours)	16.35 M	-	1332	**0.940 ± 0.01**	**1.881 ± 0.03**
REST w/o APB (ours)	13.70 M	-	1332	0.920 ± 0.02	1.843 ± 0.54
REST w/o TTB (ours)	8.18 M	-	1332	0.223 ± 0.05	0.454± 0.09

	Pair	FIQA
Real high	-	0.980
Real low	-	0.198
cGAN [10]	-	0.304
CycleGAN [7]	-	0.890
CutGAN [15]	-	0.919
ISECRET [5]	✓	**0.937**
PCENet [13]	✓	0.894
REST (ours)	-	0.936

Evaluation Setup. To quantitatively evaluate the quality of the generated retina images, Fundus Image Quality Assessment (FIQA) [6] and Weighted FIQA (WFQA) [13] scores were adopted. For FIQA, a Multiple Color-space Fusion Network (MCF-Net) [6] which labels each input image of the retina as 'Good', 'Usable', or 'Reject' is utilized. With the output of the MCF-Net, the FIQA score is calculated. The FIQA is calculated as the ratio of the number of images labeled as 'Good' to the total number of inputs, while the WFQA is determined by the ratio of the weighted sum of the number of images labeled as 'Good' and 'Usable' to the total number of inputs, with weights of 2 and 1 assigned to 'Good' and 'Usable' respectively. For evaluation, the FIQA and WFQA scores were calculated by inputting output images from enhancement models to pretrained MCF-Net. For **UKB** data, both quantitative and qualitative evaluations were done for our model (REST) and four baseline models, i.e., two traditional unpaired image translation models, CycleGANs with ResNet and U-Net [8,20,26] and two latest models designed for unpaired retina image enhancement, ISECRET [5] and PCENet [13]. For **EyeQ** data, experiments were done for REST and PCENet following the settings in [5], and results are compared with traditional paired image translation model cGAN [15] and unpaired image translation models, CycleGAN [26], CutGAN [15], and ISECRET [5], reported in [5].

Implmenemtation. The batch size was set to 4 and the Adam optimizer was adopted with the initial learning rate of 0.0002 linearly decaying. For scattering transform, a Morlet wavelet was used with eight different angles within 2D image space and J was set to 1. A 2D Gabor filter was used as a low-pass filter. Parameters in losses, i.e., $\lambda_{c1}, \lambda_{c2}, \lambda_{i1}$ and λ_{i2}, were set to 10, 10, 5 and 5 respectively.

Input image	CycleGAN	CycleGAN	ISECRET	PCENet	REST
(authentic lq)	w/ResNet [7, 8]	w/U-Net [7, 20]	[5]	[13]	(ours)

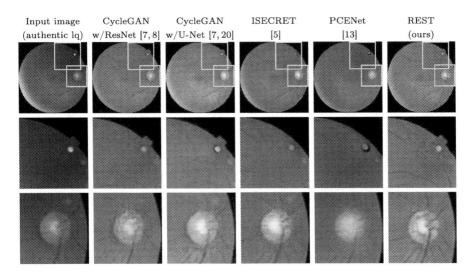

Fig. 3. Examples of the low-quality input image and the outputs of different models. The zoomed-in figures from yellow and blue boxes (2nd and 3rd rows) show that our model REST better preserves the anatomical structures (vessels, optic disc, and cup) and transfers the tone. (Color figure online)

4.2 Results

Regarding the experiments on **UKB**, the result of the qualitative evaluation is illustrated in Table 1 (left). In the Table 1, (1) the number of parameters, (2) the requirement for the paired data while training, (3) the total number of images used for training, and (4) the FIQA score and WFQA score for each model are illustrated with the best score denoted in bold, and the second-best in underlined. REST achieved the best FIQA with the smallest standard deviation. Compared with the second-best model, excluding the ablation study, cycleGAN with Resnet backbone, the FIQA score of REST is 0.014 (1.4%p) higher with a lot smaller number of total parameters. Qualitative evaluation on a UKB sample is illustrated in Fig. 3. In Fig. 3, the low-quality input image and synthesized high-quality images by each model are shown sequentially. The second and third rows in Fig. 3 show the details of the image marked with a box in the first row. As can be seen in Fig. 3, REST preserved the anatomical structures such as vessels both in and out of the optic disc and the shape of the optic disc and cup better than the baselines. The overall tone of the image was also successfully transferred to make the contrast and the illuminance become even.

For the experiment on the **EyeQ** dataset, the result of the qualitative evaluation is illustrated in Table 1 (right). REST achieved the second-best FIQA score slightly behind by only 0.001p from ISECRET. However, notice that ISECRET requires synthetic paired images while REST uses only unpaired images. As the low-quality images utilized in the EyeQ experiment, i.e., labeled as 'Usable', are already deemed to be of relatively high quality, the capability of REST that shows superior preservation of anatomical structure in contaminated images was not sufficiently manifested as in the UKB experiment. Still, the results are highly competitive and demonstrate potentials to be used for real applications.

5   Conclusion

In the study, we proposed a novel approach to unpaired retina image enhancement, i.e., REST. While using only the authentic unpaired images, the proposed method effectively preserves anatomical structures during the enhancement process. This is done by utilizing APB for preserving the details by the scattering transform and TTB for transferring the tone of the images. Notably, REST has demonstrated commendable performance on both two different datasets.

Acknowledgement. This research was supported by NRF-2022R1A2C2092336 (50%), IITP-2022-0-00290 (30%), IITP-2019-0-01906 (AI Graduate Program at POSTECH, 10%), and IITP-2022-2020-0-01461 (ITRC, 10%) funded by MSIT in South Korea.

References

1. Ancuti, C.O., Ancuti, C., Vasluianu, F.A., Timofte, R.: Ntire 2021 nonhomogeneous dehazing challenge report. In: CVPR, pp. 627–646 (2021)
2. Andreux, M., Angles, T., Exarchakisgeo, G., et al.: Kymatio: scattering transforms in python. J. Mach. Learn. Res. **21**(1), 2256–2261 (2020)
3. Bruna, J., Mallat, S.: Invariant scattering convolution networks. IEEE Trans. Pattern Anal. Mach. Intell. **35**(8), 1872–1886 (2013)
4. Cheng, J., Li, Z., Gu, Z., Fu, H., Wong, D.W.K., Liu, J.: Structure-preserving guided retinal image filtering and its application for optic disk analysis. IEEE Trans. Med. Imaging **37**(11), 2536–2546 (2018)
5. Cheng, P., Lin, L., Huang, Y., Lyu, J., Tang, X.: I-SECRET: importance-guided fundus image enhancement via semi-supervised contrastive constraining. In: de Bruijne, M., et al. (eds.) MICCAI 2021. LNCS, vol. 12908, pp. 87–96. Springer, Cham (2021). https://doi.org/10.1007/978-3-030-87237-3_9

6. Fu, H., Wang, B., Shen, J., Cui, S., Xu, Y., Liu, J., Shao, L.: Evaluation of retinal image quality assessment networks in different color-spaces. In: Shen, D., et al. (eds.) MICCAI 2019. LNCS, vol. 11764, pp. 48–56. Springer, Cham (2019). https://doi.org/10.1007/978-3-030-32239-7_6

7. Goodfellow, I., Pouget-Abadie, J., Mirza, M., et al.: Generative adversarial networks. Commun. ACM **63**(11), 139–144 (2020)

8. He, K., Zhang, X., Ren, S., Sun, J.: Deep residual learning for image recognition. In: CVPR, pp. 770–778 (2016)

9. Huang, Y., et al.: Automated hemorrhage detection from coarsely annotated fundus images in diabetic retinopathy. In: ISBI, pp. 1369–1372. IEEE (2020)

10. Isola, P., Zhu, J.Y., Zhou, T., Efros, A.A.: Image-to-image translation with conditional adversarial networks. In: CVPR, pp. 1125–1134 (2017)

11. Kim, W.H., Ravi, S.N., Johnson, S.C., et al.: On statistical analysis of neuroimages with imperfect registration. In: ICCV, pp. 666–674 (2015)

12. Li, H., Liu, H., Fu, H., et al.: Structure-consistent restoration network for cataract fundus image enhancement. In: Wang, L., Dou, Q., Fletcher, P.T., Speidel, S., Li, S. (eds.) MICCAI 2022. LNCS, vol. 13432, pp. 487–496. Springer, Heidelberg (2022). https://doi.org/10.1007/978-3-031-16434-7_47

13. Liu, H., Li, H., Fu, H., et al.: Degradation-invariant enhancement of fundus images via pyramid constraint network. In: Wang, L., Dou, Q., Fletcher, P.T., Speidel, S., Li, S. (eds.) MICCAI 2022. LNCS, vol. 13432, pp. 507–516. Springer, Heidelberg (2022). https://doi.org/10.1007/978-3-031-16434-7_49

14. Odena, A., Dumoulin, V., Olah, C.: Deconvolution and checkerboard artifacts. Distill **1**(10), e3 (2016)

15. Park, T., Efros, A.A., Zhang, R., Zhu, J.-Y.: Contrastive learning for unpaired image-to-image translation. In: Vedaldi, A., Bischof, H., Brox, T., Frahm, J.-M. (eds.) ECCV 2020. LNCS, vol. 12354, pp. 319–345. Springer, Cham (2020). https://doi.org/10.1007/978-3-030-58545-7_19

16. Pérez, A.D., Perdomo, O., Rios, H., Rodríguez, F., González, F.A.: A conditional generative adversarial network-based method for eye fundus image quality enhancement. In: Fu, H., Garvin, M.K., MacGillivray, T., Xu, Y., Zheng, Y. (eds.) OMIA 2020. LNCS, vol. 12069, pp. 185–194. Springer, Cham (2020). https://doi.org/10.1007/978-3-030-63419-3_19

17. Raj, A., Tiwari, A.K., Martini, M.G.: Fundus image quality assessment: survey, challenges, and future scope. IET Image Process. **13**(8), 1211–1224 (2019)

18. Rim, T.H., Lee, C.J., Tham, Y.C., et al.: Deep-learning-based cardiovascular risk stratification using coronary artery calcium scores predicted from retinal photographs. Lancet Dig. Health **3**(5), e306–e316 (2021)

19. Rim, T.H., Lee, G., Kim, Y., Tham, Y.C., et al.: Prediction of systemic biomarkers from retinal photographs: development and validation of deep-learning algorithms. Lancet Dig. Health **2**(10), e526–e536 (2020)

20. Ronneberger, O., Fischer, P., Brox, T.: U-Net: convolutional networks for biomedical image segmentation. In: Navab, N., Hornegger, J., Wells, W.M., Frangi, A.F. (eds.) MICCAI 2015. LNCS, vol. 9351, pp. 234–241. Springer, Cham (2015). https://doi.org/10.1007/978-3-319-24574-4_28

21. Şevik, U., Köse, C., Berber, T., Erdöl, H.: Identification of suitable fundus images using automated quality assessment methods. J. Biomed. Opt. **19**(4), 046006–046006 (2014)

22. Shen, Z., Fu, H., Shen, J., Shao, L.: Modeling and enhancing low-quality retinal fundus images. IEEE Trans. Med. Imaging **40**(3), 996–1006 (2020)

23. Snyder, P.J., Alber, J., Alt, C., et al.: Retinal imaging in Alzheimer's and neurode-generative diseases. Alzheimer's Dementia **17**(1), 103–111 (2021)
24. You, Q., Wan, C., Sun, J., Shen, J., Ye, H., Yu, Q.: Fundus image enhancement method based on cyclegan. In: EMBC, pp. 4500–4503. IEEE (2019)
25. Zhao, H., Yang, B., Cao, L., Li, H.: Data-driven enhancement of blurry retinal images via generative adversarial networks. In: Shen, D., et al. (eds.) MICCAI 2019. LNCS, vol. 11764, pp. 75–83. Springer, Cham (2019). https://doi.org/10.1007/978-3-030-32239-7_9
26. Zhu, J.Y., Park, T., Isola, P., Efros, A.A.: Unpaired image-to-image translation using cycle-consistent adversarial networks. In: ICCV, pp. 2223–2232 (2017)

Noise2Aliasing: Unsupervised Deep Learning for View Aliasing and Noise Reduction in 4DCBCT

Samuele Papa[1,2]([✉]), Efstratios Gavves[1], and Jan-Jakob Sonke[2]

[1] University of Amsterdam, Amsterdam, The Netherlands
s.papa@nki.nl
[2] The Netherlands Cancer Institute, Amsterdam, The Netherlands

Abstract. Respiratory Correlated Cone Beam Computed Tomography (4DCBCT) is a technique used to address respiratory motion artifacts that affect reconstruction quality, especially for the thorax and upper-abdomen. 4DCBCT sorts the acquired projection images in multiple respiratory correlated bins. This technique results in the emergence of aliasing artifacts caused by the low number of projection images per bin, which severely impacts the image quality and limits downstream use. Previous attempts to address this problem relied on traditional algorithms, while only recently deep learning techniques are being employed.

In this work, we propose Noise2Aliasing, which reduces both view-aliasing and statistical noise present in 4DCBCT scans. Using a fundamental property of the FDK reconstruction algorithm, and prior results from the literature, we prove mathematically the ability of the method to work and specify the underlying assumptions.

We apply the method to a public dataset and to an in-house dataset and show that it matches the performance of a supervised approach and outperforms it when measurement noise is present in the data.

Keywords: Medical Imaging · Adaptive Radiotherapy · 4DCBCT · Deep Learning · Unsupervised Learning

1 Introduction

Radiotherapy (RT) is one of the cornerstones of cancer patients. It utilizes ionizing radiation to eradicate all cells of a tumor. The total radiation dose is typically divided over 3–30 daily fractions to optimize its effect. As the surrounding normal tissue is also sensitive to radiation, highly accurate delivery is vital. Image guided RT (IGRT) is a technique to capture the anatomy of the day using in room imaging in order to align the treatment beam with the tumor location [1]. Cone Beam CT (CBCT) is the most widely used imaging modality for IGRT.

Supplementary Information The online version contains supplementary material available at https://doi.org/10.1007/978-3-031-43999-5_46.

A major challenge especially for CBCT imaging of the thorax and upper-abdomen is the respiratory motion that introduces blurring of the anatomy, reducing the localization accuracy and the sharpness of the image.

A technique used to alleviate motion artifacts is Respiratory Correlated CBCT (4DCBCT) [16]. From the projections, it is possible to extract a respiratory signal [12], which indicates the position of the organs within the patient during breathing. With this, subsets of the projections can be defined to create reconstructions that resolve the motion. However, only 20 to 60 respiratory periods are imaged. This limits the number of projections available and results in view-aliasing [16]. Additionally, the projections are affected by stochastic measurement noise caused by the finite imaging dose used, which further degrades the quality of the reconstruction even when all projections are used.

Several traditional methods based on iterative reconstruction algorithms and motion compensation techniques are used to reduce view-aliasing in 4DCBCTs [7,10,11,14,15]. Although effective, these methods suffer from motion modeling uncertainty and prolonged reconstruction times.

Deep learning has been proposed as a way to address view-aliasing with accelerated reconstruction [6]. However, the method cannot reduce measurement noise because it is still present in the images used as targets during training.

A different method, called Noise2Inverse, uses an unsupervised approach to reduce measurement noise in the traditional CT setting [4]. There are two ways to apply it to 4DCBCT and both fail to reduce stochastic noise effectively. The first is to apply Noise2Inverse to each respiratory-correlated reconstruction. In this case, the method will struggle because of the very low number of projections that are available. The second is to apply Noise2Inverse directly to all the projections. In this case, the motion artifacts that blur the image will appear again, as Noise2Inverse requires averaging the sub-reconstructions to obtain a clean reconstruction.

We propose **Noise2Aliasing** to address these limitations. The method can be used to *provably* train models to reduce both *view-aliasing artifacts* and *stochastic noise* from 4DCBCTs in an *unsupervised* way. Training deep learning models for medical applications often needs new data. This was not the case for Noise2Aliasing, and historical clinical data sufficed for training.

We validated our method on publicly available data [15] against a supervised approach [6] and applied it to an internal clinical dataset of 30 lung cancer patients. We explore different dataset sizes to understand their effects on the reconstructed images.

2   Theoretical Background

In this section, we will introduce the concepts and the notation necessary to understand the method and the choices made during implementation.

Unsupervised noise reduction with Noise2Noise. Given input-target pairs $x, y \in \mathbb{R}$ we can define the regression problem in the one-dimensional setting as finding

$f^* : \mathbb{R} \to \mathbb{R}$ which satisfies the following:

$$f^* = \arg \min_f \mathbb{E}_{x,y} \left[\|f(x) - y\|_2^2 \right], \tag{1}$$

which can be minimized point-wise [3], yielding:

$$f^*(x) = \mathbb{E}_{y|x} \left[y|x \right]. \tag{2}$$

In Noise2Noise [5], input-target pairs are two samples of the same image that only differ because of some independent mean-zero noise $(x + \delta_1, x + \delta_2)$ with $\mathbb{E}_{\delta_2} \left[x + \delta_2 | x + \delta_1 \right] = x$. Then f^* will recover the input image without any noise:

$$f^*(x + \delta_1) = \mathbb{E}_{\delta_2} \left[x + \delta_2 | x + \delta_1 \right] = x. \tag{3}$$

Denoising for Tomography with Noise2Inverse. During a CT scan, a volume $\mathbf{x}$ is imaged by acquiring projections $\mathbf{y} = A\mathbf{x}$ using an x-ray source and a detector placed on the opposite side of the volume. The projections can then be used by an algorithm that computes a linear operator R to obtain an approximation of the original distribution of x-ray attenuation coefficients $\hat{x} = Ry$. The algorithm can also operate on a subset of the projections. Let $\mathcal{J} = \{1, 2, \dots\}$ be the set of all projections and $J \subset \mathcal{J}$, then $\hat{x}_J = R_J y_J$ is the reconstruction obtained using only projections y_J. Let us now assume that the projections have some mean-zero noise $\tilde{y}_i = y_i + \epsilon$ with $\mathbb{E}_\epsilon (\tilde{y}_i) = y_i$. Then, in Noise2Inverse [4] the results from Noise2Noise are extended to find a function f^* which removes projection noise when trained using noisy reconstructions $\tilde{x}_J = R_J \tilde{y}_J = R_J y_J + R_J \epsilon = \hat{x}_J + R_J \epsilon$ and the expected MSE as loss function. In particular, they find that the loss function can be decomposed in the following way:

$$\mathcal{L} = \mathbb{E} \| f(\tilde{x}_{J'}) - \tilde{x}_J \|_2^2 = \mathbb{E} \| f(\tilde{x}_{J'}) - \hat{x}_J \|_2^2 + \mathbb{E} \| \hat{x}_J - \tilde{x}_J \|_2^2, \tag{4}$$

where J is a random variable that picks subsets of projections at random and J' is its complementary.

Given Eq. 2, we observe that function f^* which minimizes $\mathcal{L}$ is:

$$f^*(\tilde{x}_{J'}) = \mathbb{E}_J(\hat{x}_J | \tilde{x}_{J'}). \tag{5}$$

When using reconstructions from a subset of noisy projections as input and reconstructions from their complementary as its output, a neural network will learn to predict the expected reconstruction without the noise.

Property of Expectation over Subsets of Projections Using FDK. Now let J be a random variable that selects subsets of projections $J \subset \mathcal{J}$ at random such that each projection is selected at least once. Define $R_J : \mathbb{R}^{D_d \times |J|} \to \mathbb{R}^{D_v}$ to be the FDK reconstruction algorithm [2] that reconstructs a volume of dimensionality D_v from projections J each with dimensionality D_d (geometrical details on the exact setup are not relevant). The FDK uses, as its fundamental step, the dual

Radon transform [9], which is a weighted summation that can be written as an expectation. Then, the following holds:

$$\hat{x} = R_{\mathcal{J}} y = \mathbb{E}_{J \sim \mathsf{J}} [R_J y_J] = \mathbb{E}_{J \sim \mathsf{J}} [\hat{x}_J]. \tag{6}$$

3 Noise2Aliasing

Here, we propose Noise2Aliasing, an unsupervised method capable of reducing both view-aliasing and projection noise in 4DCBCTs. At the core of this method is the following proposition.

Proposition. Given the projection set $\mathcal{J} = \{1, 2, \dots\}$, the FDK reconstruction algorithm R, and the noisy projections $\tilde{y} = Ax + \epsilon$ with ϵ mean-zero element-wise independent noise. Let $\mathsf{J}_1, \mathsf{J}_2$ be two random variables that pick different subsets at random belonging to a partition of $\mathcal{J}$, and $(\tilde{x}_{\mathsf{J}_1} = R_{\mathsf{J}_1} \tilde{y}_{\mathsf{J}_1}, \tilde{x}_{\mathsf{J}_2} = R_{\mathsf{J}_2} \tilde{y}_{\mathsf{J}_2}) \in \mathcal{D}$ be the input-target pairs in dataset $\mathcal{D}$ of reconstructions using disjoint subsets of noisy projections. Let $\mathcal{L}$ be the expected MSE over $\mathcal{D}$ with respect to a function $f : \mathbb{R}^{D_v} \to \mathbb{R}^{D_v}$ and the previously-described input-target pairs. Then, we find that the function f^* that minimizes $\mathcal{L}$ for any given $J \in \mathcal{J}$ will reconstruct the volume using all the projections and remove the noise ϵ:

$$f^*(\tilde{x}_J) = \hat{x}. \tag{7}$$

Proof. The loss function $\mathcal{L}$ is defined in the following way:

$$\mathcal{L} = \mathbb{E}_{\mathcal{D}} \| f(\tilde{x}_{\mathsf{J}_2}) - \tilde{x}_{\mathsf{J}_1} \|_2^2. \tag{8}$$

Additionally, $\mathsf{J}_1, \mathsf{J}_2$ are disjoint, the noise is mean-zero element-wise, and we are using the FDK reconstruction algorithm which defines a linear operator R. These allow us to use Eq. 5 to find that the function f^* that minimizes $\mathcal{L}$ is the following:

$$f^*(\tilde{x}_J) = \mathbb{E}_{\mathsf{J}_1, \mathsf{J}_2}(\hat{x}_{\mathsf{J}_1} | \tilde{x}_{\mathsf{J}_2} = \tilde{x}_J). \tag{9}$$

This is sufficient to reduce stochastic noise but we need to further manipulate this expression to address view aliasing. Simplifying notation and using the properties of conditional expectations, we can write:

$$f^*(z) = \mathbb{E}_{j_1 \sim \mathsf{J}_1} [\mathbb{E}_{j_2 \sim \mathsf{J}_2}(\hat{x}_{j_1} | \tilde{x}_{j_2} = z)], \tag{10}$$

now assume that $\hat{x}_{j_1}$ is the clean reconstruction that is consistent with the observed noisy reconstruction z obtained from each disjoint subset j_2, then:

$$f^*(z) = \mathbb{E}_{j_1 \sim \mathsf{J}_1}(\hat{x}_{j_1}). \tag{11}$$

Finally, we use the property of the FDK from Eq. 6:

$$f^*(z) = \mathbb{E}_{j_1 \sim \mathsf{J}_1}(\hat{x}_{j_1}) = \hat{x}. \tag{12}$$

□

3.1   Design Choices Based on the Proposition

The proposition guided the choice of reconstruction method to be FDK and the design of the subset selection method from considerations that are now explained.

Equation 12 holds true only when the *same* underlying *clean* reconstruction $\hat{x}$ can be determined from the noisy reconstruction using any subset from a partition of the projections $\mathcal{J}$. This means that, in our dataset, we should have at our disposal reconstructions of the same underlying volume x using disjoint subsets of projections. In 4DCBCTs this is not the case, as separate respiratory phases are being reconstructed, where the organs are in different positions. We can address this problem by carefully choosing subsets of projections that result in *respiratory-uncorrelated* reconstructions. The reconstructions will display organs in their average position and, therefore, have the same underlying structure. When the projections are selected with the same sampling pattern as the one used in respiratory-correlated reconstructions, then the view-aliasing artifacts display will have the same pattern as the ones present in the 4DCBCTs.

Compared to previous work, to obtain the additional effect of reducing projection noise, the *respiratory-uncorrelated* reconstructions must use non-overlapping subsets of projections. Coincidentally, a previously proposed subset selection method utilized for supervised aliasing reduction fits all these requirements and will, therefore, be used in this work [4].

4   Experiments

First, we used the SPARE Varian dataset to study whether Noise2Aliasing can *match* the performance of the supervised baseline and if it can *outperform* it when adding noise to the projections. Then, we use the internal dataset to explore the *requirements* for the method to be applied to an existing clinical dataset. These required around 64 GPU days on NVIDIA A100 GPUs.

Training of the model is done on 2D slices. The projections obtained during a scan are sub-sampled according to the pseudo-average subset selection method described in [6] and then used to obtain 3D reconstructions. In Noise2Aliasing these are used for both input and target during training. Given two volumes (x, y), the training pairs $(x_{i^{(k)}}, y_{i^{(k)}})$ are the same i-th slice along the k-th dimension of each volume chosen to be the axial plane.

The Datasets used in this study are two:

1. The SPARE Varian dataset was used to provide performance results on publicly available patient data. To more closely resemble normal respiratory motion per projection image, the 8 min scan has been used from each patient (five such scans are available in the dataset). Training is performed over 4 patients while 1 patient is used as a test set. The hyperparameters are optimized over the training dataset.

2. An internal dataset (IRB approved) of 30 lung cancer patients' 4DCBCTs from 2020 to 2022, originally used for IGRT, with 25 patients for training and 5 patients for testing. The scans are 4 min 205° scans with 120keV source and 512×512 sized detector, using Elekta LINACs. The data were anonymized prior to analysis.

Projection Noise was added using the Poisson distribution to the SPARE Varian dataset to evaluate the ability of the unsupervised method to reduce it. Given a projected value of p and a photon count π (chosen to be 2500), the rate of the Poisson distribution is defined as πe^{-p} and given a sample q from this distribution, then the new projected value is $\tilde{p} = -\log\left(\frac{q}{\pi}\right)$.

The Architecture used in this work is the Mixed Scale Dense CNN (MSD) [8], the most successful architecture from Noise2Inverse [4]. The MSD makes use of dilated convolutions to process features at all scales of the image. We use the MSD with depth 200 and width 1, Adam optimizer, MSE loss, a batch size of 16, and a learning rate of 0.0001.

The Baselines we compare against are two. The first is the traditional FDK obtained using RTK [13]. The second is the supervised approach proposed by [6], where we replace the model with the MSD, for a fair comparison. In the supervised approach, the model is trained by using as input reconstructions obtained from subsets defined with pseudo-average subset selection while the targets use all of the projections available.

The Metrics used in this work are the Root Mean Squared Error (RMSE), Peak Signal-to-Noise Ratio (PSNR), and Structural Similarity Index Measure (SSIM) [17] All the metrics are defined between the output of the neural network and a 3D (CB)CT scan. For the SPARE Varian dataset, we use the ROIs defined provided [15] and used the 3D reconstruction using all the projections available as a ground truth. For the internal dataset, we deformed the planning CT to each of the phases reconstructed using the FDK algorithm and evaluate the metric over only the 4DCBCT volume boundaries.

5   Results and Discussion

SPARE Varian. Inference speed with the NVIDIA A100 GPU averages 600ms per volume made of 220 slices. From the qualitative evaluation of the methods in Fig. 1, Noise2Aliasing matches the visual quality of the supervised approach on the low-noise dataset on both soft tissue and bones. The metrics in Table 1 show mean and standard deviation across all phases for a single patient. In the low-noise setting, both supervised and Noise2Aliasing outperform FDK with very similar results, often within a single standard deviation.

Noise2Aliasing successfully matches the performance of the supervised baseline.

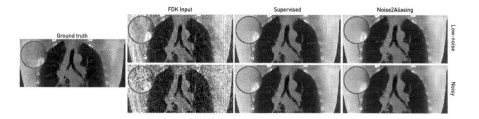

Fig. 1. Qualitative comparison between methods using coronal view of the patient in the test set. Noise2Aliasing and the Supervised method produce very similar images in the low-noise case. With noisy data, the supervised method tends to re-create the noise seen during training.

Table 1. Metrics for the comparison between FDK, Supervised method, and Noise2Aliasing (N2A). Values are *mean* and *std* computed across all phases of patient 1 of the SPARE Varian dataset. The Planning Target Volume (PTV) ROI is less affected by noise compared to the whole *Body*, which is what causes the supervised model to outperform N2A in terms of PSNR and RMSE.

	SSIM ↑ $(\times 10^{-2})$		**PSNR** ↑		**RMSE** ↓ $(\times 10^{-3})$	
Noisy	*Body*	*PTV*	*Body*	*PTV*	*Body*	*PTV*
FDK	12.99 ± 2.1	25.31 ± 4.2	14.66 ± 1.0	13.83 ± 1.0	6.70 ± 0.8	5.51 ± 0.6
Supervised	59.76 ± 2.1	72.61 ± 2.7	22.20 ± 0.2	$\mathbf{20.59 \pm 0.4}$	2.80 ± 0.1	$\mathbf{2.52 \pm 0.1}$
N2A	$\mathbf{64.90 \pm 0.8}$	$\mathbf{76.33 \pm 1.4}$	$\mathbf{22.31 \pm 0.2}$	20.41 ± 0.4	$\mathbf{2.76 \pm 0.1}$	2.57 ± 0.1
Low-Noise						
FDK	41.75 ± 2.1	56.77 ± 4.2	20.86 ± 1.0	19.09 ± 1.0	3.27 ± 0.8	2.99 ± 0.6
Supervised	$\mathbf{67.49 \pm 0.8}$	$\mathbf{79.54 \pm 2.3}$	$\mathbf{22.68 \pm 0.2}$	$\mathbf{20.92 \pm 0.5}$	$\mathbf{2.65 \pm 0.1}$	$\mathbf{2.43 \pm 0.1}$
N2A	67.13 ± 0.7	79.52 ± 2.0	22.50 ± 0.2	20.79 ± 0.5	2.70 ± 0.1	2.46 ± 0.1

Noisy SPARE Varian. From Fig. 1 and Table 1, the supervised approach reproduces the noise that was seen during training, while Noise2Aliasing manages to remove it consistently, outperforming the supervised approach, especially in the soft tissue area around the lungs, where the noise affects attenuation coefficients the most.

Noise2Aliasing is capable of reducing the artifacts present in reconstructions caused by stochastic noise in the projections used, outperforming the supervised baseline.

Internal Dataset. Noise2Alisting trained on 25 patients and tested on 5 achieved mean PSNR of 35.24 and SSIM of 0.91, while the clinical method achieved mean PSNR of 29.97 and 0.74 SSIM with p-value of 0.048 for the PSNR and 0.0015 for the SSIM, so Noise2Aliasing was significantly better according to both metrics. Additionally, from Fig. 3 we can see how the breathing extent is matched with sharp reconstruction of the diaphragm. Overall, using more patients results in better noise reduction and sharper reconstructions (see Fig. 2),

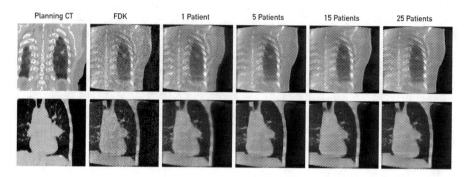

Fig. 2. Reconstruction using Noise2Aliasing with different-sized datasets. With fewer patients, the model is more conservative and tends to keep more noise, but also smudges the interface between tissues and bones. With more patients, more of the view-aliasing is addressed, and the reconstruction is sharper, however, a few small anatomical structures tend to be suppressed by the model.

especially between fat tissue and skin and around the bones. However, the model also tends to remove small anatomical structures as high-frequency objects that cannot be distinguished from the noise.

When applied to a clinical dataset, Noise2Aliasing benefits from more patients being included in the dataset, however, qualitatively good performance is already achieved with 5 patients. No additional data collection was required and the method can be applied without major changes to the current clinical practice.

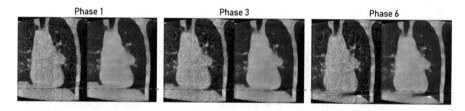

Fig. 3. Motion extent is accurately resolved by Noise2Aliasing when using 25 patients. On the left is the FDK, while on the right is the output of the model.

6   Conclusion

We have presented Noise2Aliasing, a method to provably remove both view-aliasing and stochastic projection noise from 4DCBCTs using an unsupervised deep learning method. We have empirically demonstrated its performance on a publicly available dataset and on an internal clinical dataset. Noise2Aliasing outperforms a supervised approach when stochastic noise is present in the projections and matches its performance on a popular benchmark. Noise2Aliasing can be trained on existing historical datasets and does not require changing current clinical practices. The method removes noise more reliably when the dataset size is increased, however further analysis is required to establish a good quantitative measurement of this phenomenon. As future work, we plan to study Noise2Aliasing in the presence of changes in the breathing frequency and amplitude between patients and during a scan.

Acknowledgements and Disclosures. We thank Celia Juan de la Cruz, Nikita Moriakov, Xander Staal, and Jonathan Mason for helping during the development of this work.

This work was funded by ROV with grant number PPS2102 and Elekta Oncology AB and was supported by an institutional grant of the Dutch Cancer Society and the Dutch Ministry of Health.

References

1. Dawson, L.A., Jaffray, D.A.: Advances in image-guided radiation therapy. J. Clin. Oncol. **25**(8), 938–946 (2007)
2. Feldkamp, L.A., Davis, L.C., Kress, J.W.: Practical cone-beam algorithm. Josa a **1**(6), 612–619 (1984)
3. Hastie, T., Tibshirani, R., Friedman, J.H., Friedman, J.H.: The Elements of Statistical Learning: Data Mining, Inference, and Prediction, vol. 2. Springer, New York (2009)
4. Hendriksen, A.A., Pelt, D.M., Batenburg, K.J.: Noise2Inverse: self-supervised deep convolutional denoising for tomography. IEEE Trans. Comput. Imaging **6**, 1320–1335 (2020). https://doi.org/10.1109/TCI.2020.3019647
5. Lehtinen, J., et al.: Noise2Noise: learning image restoration without clean data. In: Proceedings of the 35th International Conference on Machine Learning, pp. 2965–2974. PMLR, July 2018
6. Madesta, F., Sentker, T., Gauer, T., Werner, R.: Self-contained deep learning-based boosting of 4D cone-beam CT reconstruction. Med. Phys. **47**(11), 5619–5631 (2020). https://doi.org/10.1002/mp.14441
7. Mory, C., Janssens, G., Rit, S.: Motion-aware temporal regularization for improved 4d cone-beam computed tomography. Phys. Med. Biol. **61**(18), 6856 (2016)
8. Pelt, D.M., Sethian, J.A.: A mixed-scale dense convolutional neural network for image analysis. Proc. Natl. Acad. Sci. **115**(2), 254–259 (2018)
9. Quinto, E.T.: An introduction to x-ray tomography and radon transforms. In: Proceedings of Symposia in Applied Mathematics, vol. 63, p. 1 (2006)
10. Ren, L., et al.: A novel digital tomosynthesis (DTS) reconstruction method using a deformation field map. Med. Phys. **35**(7Part1), 3110–3115 (2008)
11. Riblett, M.J., Christensen, G.E., Weiss, E., Hugo, G.D.: Data-driven respiratory motion compensation for four-dimensional cone-beam computed tomography (4D-CBCT) using GroupWise deformable registration. Med. Phys. **45**(10), 4471–4482 (2018)
12. Rit, S., van Herk, M., Zijp, L., Sonke, J.J.: Quantification of the variability of diaphragm motion and implications for treatment margin construction. Int. J. Radiat. Oncol. * Biol.* Phys. **82**(3), e399–e407 (2012)

13. Rit, S., Oliva, M.V., Brousmiche, S., Labarbe, R., Sarrut, D., Sharp, G.C.: The reconstruction toolkit (RTK), an open-source cone-beam CT reconstruction toolkit based on the insight toolkit (ITK). J. Phys. Conf. Ser. **489**, 012079 (2014)
14. Rit, S., Wolthaus, J.W., van Herk, M., Sonke, J.J.: On-the-fly motion-compensated cone-beam CT using an a priori model of the respiratory motion. Med. Phys. **36**(6Part1), 2283–2296 (2009)
15. Shieh, C.C., et al.: Spare: sparse-view reconstruction challenge for 4d cone-beam CT from a 1-min scan. Med. Phys. **46**(9), 3799–3811 (2019)
16. Sonke, J.J., Zijp, L., Remeijer, P., van Herk, M.: Respiratory correlated cone beam CT: respiratory correlated cone beam CT. Med. Phys. **32**(4), 1176–1186 (2005). https://doi.org/10.1118/1.1869074
17. Wang, Z., Bovik, A.C., Sheikh, H.R., Simoncelli, E.P.: Image quality assessment: from error visibility to structural similarity. IEEE Trans. Image Process. **13**(4), 600–612 (2004)

Self-supervised MRI Reconstruction with Unrolled Diffusion Models

Yilmaz Korkmaz[1]([✉]), Tolga Cukur[2,3], and Vishal M. Patel[1]

[1] Johns Hopkins University, Baltimore, MD, USA
{ykorkma1,vpatel36}@jhu.edu
[2] Bilkent University, Ankara, Turkey
cukur@ee.bilkent.edu.tr
[3] National Magnetic Resonance Research Center (UMRAM), Ankara, Turkey

Abstract. Magnetic Resonance Imaging (MRI) produces excellent soft tissue contrast, albeit it is an inherently slow imaging modality. Promising deep learning methods have recently been proposed to reconstruct accelerated MRI scans. However, existing methods still suffer from various limitations regarding image fidelity, contextual sensitivity, and reliance on fully-sampled acquisitions for model training. To comprehensively address these limitations, we propose a novel self-supervised deep reconstruction model, named Self-Supervised Diffusion Reconstruction (SSDiffRecon). SSDiffRecon expresses a conditional diffusion process as an unrolled architecture that interleaves cross-attention transformers for reverse diffusion steps with data-consistency blocks for physics-driven processing. Unlike recent diffusion methods for MRI reconstruction, a self-supervision strategy is adopted to train SSDiffRecon using only undersampled k-space data. Comprehensive experiments on public brain MR datasets demonstrates the superiority of SSDiffRecon against state-of-the-art supervised, and self-supervised baselines in terms of reconstruction speed and quality. Implementation will be available at https://github.com/yilmazkorkmaz1/SSDiffRecon.

Keywords: Magnetic Resonance Imaging · Self-Supervised Learning · Cross-Attention · Transformers · Accelerated MRI

1 Introduction

Magnetic Resonance Imaging (MRI) is one of the most widely used imaging modalities due to its excellent soft tissue contrast, but it has prolonged and costly scan sessions. Therefore, accelerated MRI methods are needed to improve its clinical utilization. Acceleration through undersampled acquisitions of a subset of k-space samples (i.e., Fourier domain coefficients) results in aliasing artifacts [8,17]. Many promising deep-learning methods have been proposed to reconstruct

Supplementary Information The online version contains supplementary material available at https://doi.org/10.1007/978-3-031-43999-5_47.

images by suppressing aliasing artifacts [1, 2, 7, 9, 11, 15, 16, 18, 22, 24, 27, 29]. However, many existing methods are limited by suboptimal capture of the data distribution, poor contextual sensitivity, and reliance on fully-sampled acquisitions for model training [7, 13, 23].

A recently emergent framework for learning data distributions in computer vision is based on diffusion models [10, 19]. Several recent studies have considered diffusion-based MRI reconstructions, where either an unconditional or a conditional diffusion model is trained to generate images and reconstruction is achieved by later injecting data-consistency projections in between diffusion steps during inference [3, 4, 6, 20, 25]. While promising results have been reported, these diffusion methods can show limited reliability due to omission of physical constraints during training, and undesirable reliance on fully-sampled images. There is a more recent work that tried to mitigate fully-sampled data needs by Cui et al. [5]. In this work authors proposed a two-staged training strategy where a Bayesian network is used to learn the fully-sampled data distribution to train a score model which is then used for conditional sampling. Our model differs from this approach since we trained it end-to-end without allowing error propagation from distinct training sessions.

To overcome mentioned limitations, we propose a novel self-supervised accelerated MRI reconstruction method, called SSDiffRecon. SSDiffRecon leverages a conditional diffusion model that interleaves linear-complexity cross-attention transformer blocks for denoising with data-consistency projections for fidelity to physical constraints. It further adopts self-supervised learning by prediction of masked-out k-space samples in undersampled acquisitions. SSDiffRecon achieves on par performance with supervised baselines while outperforming self-supervised baselines in terms of inference speed and image fidelity.

2   Background

2.1   Accelerated MRI Reconstruction

Acceleration in MRI is achieved via undersampling the acquisitions in the Fourier domain as follows

$$F_p C I = y_p, \qquad (1)$$

where F_p is the partial Fourier operator, C denotes coil sensitivity maps, I is the MR image and y_p is partially acquired k-space data. Reconstruction of fully sampled target MR image I from y_p is an ill-posed problem since the number of unknowns are higher than the number of equations. Supervised deep learning methods try to solve this ill-posed problem using prior knowledge gathered in the offline training sessions as follows

$$\widehat{I} = \operatorname*{argmin}_{I} \frac{1}{2} \|y_p - F_p C I\|^2 + \lambda(I), \qquad (2)$$

where $\widehat{I}$ is the reconstruction, and $\lambda(I)$ is the prior knowledge-guided regularization term. In supervised reconstruction frameworks, prior knowledge is induced from underlying mapping between under- and fully sampled acquisitions.

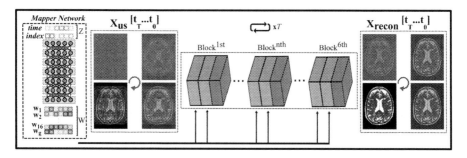

Fig. 1. Overall network architecture. SSDiffRecon utilizes an unrolled physics-guided network as a denoiser in the diffusion process while allowing time index guidance through the Mapper Network via cross-attention transformer layers (shown in green). After two transformer layers, it performs data-consistency (shown in orange). Corresponding noisy input under-sampled and denoised reconstructed images are shown during training for different time indexes descending in the direction of circular arrow. (Color figure online)

2.2 Denoising Diffusion Models

In diffusion models [10], Gaussian noise is progressively mapped on the data via a forward noising process

$$q\left(\mathbf{x}_t \mid \mathbf{x}_{t-1}\right) = \mathcal{N}\left(\mathbf{x}_t; \sqrt{1 - \beta_t}\mathbf{x}_{t-1}, \beta_t\mathbf{I}\right), \tag{3}$$

where β_t refers to the fixed variance schedule. After a sufficient number of forward diffusion steps (T), x_t follows a Gaussian distribution. Then, the backward diffusion process is deployed to gradually denoise x_T to get x_0 using a deep neural network as a denoiser as follows

$$p_\theta\left(x_{t-1} \mid x_t\right) = \mathcal{N}\left(x_{t-1}; \epsilon_\theta\left(x_t, t\right), \sigma_t^2\mathbf{I}\right), \tag{4}$$

where $\sigma_t^2 = \tilde{\beta}_t = \frac{1-\bar{\alpha}_{t-1}}{1-\bar{\alpha}_t}\beta_t$ and ϵ_θ represents the denoising neural network parametrized during backward diffusion and trained using the following loss [10]

$$L(\theta) = \mathbb{E}_{t, \mathbf{x}_0, \epsilon}\left[\left\|\epsilon - \epsilon_\theta\left(\sqrt{\bar{\alpha}_t}\mathbf{x}_0 + \sqrt{1 - \bar{\alpha}_t}\epsilon, t\right)\right\|^2\right], \tag{5}$$

where $\bar{\alpha}_t = \prod_{m=1}^t \alpha_m$, $\alpha_t = 1 - \beta_t$ and $\epsilon \sim \mathcal{N}(0, I)$.

3   SSDiffRecon

In SSDiffRecon, we utilize a conditional diffusion probabilistic model to reconstruct fully-sampled MR images given undersampled acquisitions as input. The reverse diffusion steps are parametrized using an unrolled transformer architecture as shown in Fig. 1. To improve adaptation across time steps in the diffusion process, we inject the time-index t via cross-attention transformers as opposed to the original DDPM models that add time embeddings as a bias term. In what follows we describe the training and inference procedures of SSDiffRecon.

Self-Supervised Training: For self-supervised learning, we adopt a k-space masking strategy for diffusion models [26] as follows

$$L(\theta) = \|\mathcal{M}_l \odot \mathcal{F}(Cx_{us}) - \mathcal{M}_l \odot \mathcal{F}(Cx_{recon}^t)\|_1, \tag{6}$$

where $\|\cdot\|_1$ denotes the $L1$-norm, $\mathcal{F}$ denotes 2D Fourier Transform, C are coil sensitivities, x_{us} is the image derived from undersampled acquisitions, and $\mathcal{M}_l$ is the random sub-mask within the main undersampling mask $\mathcal{M}$. Here x_{recon}^t is the output of the unrolled denoiser network (R_θ) at time instant $t \in \{T, T-1, ..., 0\}$

$$x_{recon}^t = R_\theta(x_{us'}^t, x_{us}, \mathcal{M}_p, C, t), \quad x_{us'}^t = \sqrt{\bar{\alpha}_t}x_{us} + \sqrt{1 - \bar{\alpha}_t}\epsilon, \tag{7}$$

where $\mathcal{M}_p$ is the sub-mask of the remaining points in $\mathcal{M}$ after excluding $\mathcal{M}_l$.

Inference: To speed up image sampling, inference starts with zero-filled Fourier reconstruction of the undersampled acquisitions as opposed to a pure noise sample. Conditional diffusion sampling is then performed with the trained diffusion model that iterates through cross-attention transformers for denoising and data-consistency projections. For gradual denoising, we introduce a descending random noise onto the undersampled data within data-consistency layers. Accordingly, the reverse diffusion step at time-index t is given as

$$x_{recon}^{t-1} = R_\theta(x_{recon}^t, x_{us''}^t, \mathcal{M}_p, C, t) + \sigma_t z, \quad x_{us''}^t = \sqrt{\bar{\alpha}_t}x_{us} + \sqrt{1 - \bar{\alpha}_t}\epsilon_{low}, \tag{8}$$

where $\epsilon_{low} \sim \mathcal{N}(0, 0.1I)$ and $z \sim \mathcal{N}(0, I)$.

Unrolled Denoising Network $R_\theta(.)$: SSDiffRecon deploys an unrolled physics-guided denoiser in the diffusion process instead of UNET as is used in [10]. Our denoiser network consists of the following two fundamental structures as shown in Fig. 1. The entire network is trained end-to-end.

1. Mapper Network
2. Unrolled Denoising Blocks

Mapper Network: Mapper network is trained to generate local and global latent variables (w_l and w_g, respectively) that control the fine and global features in the generated images via cross-attention and instance modulation detailed in later sections. The mapper network is taking time index of the diffusion and extracted label of undersampled image (i.e., undersampling rate and target contrast in multiple contrast dataset) as input and built with 12 fully-connected layers each with 32 neurons.

Unrolled Denoising Blocks: Each denoising block consists of cross-attention and data-consistency layers sequentially. Let the input of the jth denoising block at time instant t be $x_{in,j}^t \in \mathbb{R}^{(h \times w) \times n}$, where h and w denote the height and width of the image, and n denotes the number of feature channels. For the first denoising block $n = 2$ and ($x_{in,j}^t = x_{us'}^t$). First, input is modulated with the affine-transformed global latent variable ($w_g \in \mathbb{R}^{32}$) via modulated-convolution adopted from [12]. Assuming that the modulated-convolution kernel is given as β_j, this operation is expressed as follows

$$x^t_{output,j} = \begin{bmatrix} \sum_m x^{t,m}_{in,j} \circledast \beta^{m,1}_j \\ \vdots \\ \sum_m x^{t,m}_{in,j} \circledast \beta^{m,v}_j \end{bmatrix}, \tag{9}$$

where $\beta^{u,v}_j \in \mathbb{R}^{3\times3}$ is the convolution kernel for the uth input channel and the vth output channel, and m is the channel index. Then, the output of modulated convolution goes into the cross-attention transformer where the attention map att^t_j is calculated using local latent variables w^t_l at time index t as follows

$$att^t_j = softmax\left(\frac{Q_j(x^t_{output,j} + P.E.)K_j(w^t_l + P.E.)^T}{\sqrt{n}}\right)V_j(w^t_l), \tag{10}$$

where $Q_j(.)$, $K_j(.)$, $V_j(.)$ are queries, keys and values, respectively where each function represents a dense layer with input inside the parenthesis, and $P.E.$ is the positional encoding. Then, $x^t_{output,j}$ is normalized to zero-mean unit variance and scaled with a learned projection of the attention maps att^t_j as follows

$$x^t_{output,j} = \alpha_j(att_j) \odot \left(\frac{x^t_{output,j} - \mu(x^t_{output,j})}{\sigma(x^t_{output,j})}\right), \tag{11}$$

where $\alpha_j(.)$ is the learned scale parameter. After repeating the sequence of cross-attention layer twice, lastly the data-consistency is performed. To perform data-consistency the number of channels in $x^t_{output,j}$ is decreased to 2 with an additional convolution layer. Then, 2-channel images are converted, where channels represent real and imaginary components, to complex and data-consistency is applied as follows

$$x^t_{output,j} = \mathcal{F}^{-1}\{\mathcal{F}(Cx^t_{output,j}) \odot (1 - \mathcal{M}_p) + \mathcal{F}(Cx^t_{us''}) \odot \mathcal{M}_p\}, \tag{12}$$

where $\mathcal{F}^{-1}$ represents the inverse 2D Fourier transform and $x^t_{us''} = x_{us}$ during training. Then, using another extra convolution, the number of feature maps are increased to n again for the next denoising block.

Implementation Details : Adam optimizer is used for self-supervised training with $\beta = (0.9, 0.999)$ and learning rate 0.002. Default noise schedule paramaters are taken from [10]. 1000 forward and 5 reverse diffusion steps are used for training and inference respectively with batch size equals to 1. $\mathcal{M}_l$ are sampled from $\mathcal{M}$ using uniform distribution by collecting 5% of acquired points. We used network snapshots at 445K and 654K steps which corresponds to 28th and 109th epochs for IXI and fastMRI datasets respectively. A single NVIDIA RTX A5000 gpu is used for training and inference.

4   Experimental Results

4.1   Datasets

Experiments are performed using the following multi-coil and single-coil brain MRI datasets:

1. **fastMRI**: Reconstruction performance illustrated in multi-coil brain MRI dataset [14], 100 subjects are used for training, 10 for validation and 40 for testing. Data from multiple sites are included with no common protocol. T_1-, T_2- and Flair-weighted acquisitions are considered. GCC [28] is used to decrease the number of coils to 5 to reduce the computational complexity.
2. **IXI**: Reconstruction performance illustrated in single-coil brain MRI data from IXI (http://brain-development.org/ixi-dataset/). T_1-, T_2- and PD-weighted acquisitions are considered. In IXI, 25 subjects are used for training, 5 for validation and 10 for testing.

Acquisitions are retrospectively undersampled using variable-density masks. Undersampling masks are generated based on a 2D Gaussian distribution with variance adjusted to obtain acceleration rates of $R = [4, 8]$.

4.2   Competing Methods

We compare the performance of SSDiffRecon with the following supervised and self-supervised baselines:

1. **DDPM**: Supervised diffusion-based reconstruction baseline. DDPM is trained with fully sampled MR images and follows a novel k-space sampling approach during inference introduced by Peng et al. [20]. 1000 forward and backward diffusion steps are used in training and inference respectively.
2. **self-DDPM**: Self-supervised diffusion-based reconstruction baseline. Self-DDPM is trained using only under-sampled MRI acquisitions. Other than training, the inference procedure is identical to the DDPM.
3. **D5C5**: Supervised model-based reconstruction baseline. D5C5 is trained using under- and fully sampled paired MR images. Network architecture and training loss are adopted from [21].
4. **self-D5C5**: Self-supervised model-based reconstruction baseline. Self-D5C5 is trained using the self-supervision approach introduced in [26] using under-sampled acquisitions. The hyperparameters and network architecture are the same as in D5C5.
5. **RGAN**: CNN-based reconstruction baseline. RGAN is trained using paired under- and fully sampled MR images. Network architecture and hyperparameters are adapted from [7].
6. **self-RGAN**: Self-supervised CNN-based reconstruction baseline. Self-RGAN is trained using the self-supervision loss in [26] using only under-sampled images. Network architecture and other hyperparameters are identical to RGAN.

4.3   Experiments

We compared the reconstruction performance using Peak-Signal-to-Noise-Ratio (PSNR, dB) and Structural-Similarity-Index (SSIM, %) between reconstructions and the ground truth images. Hyperparameter selection for each method is performed via cross-validation to maximize PSNR.

Ablation Experiments. We perform the following four ablation experiments to show the relative effect of each component in the model on the reconstruction quality as well as the effect of self-supervision in Table 1.

1. Supervised: Supervised training of SSDiffRecon using paired under- and fully sampled MR images and pixel-wise loss is performed. Other than training, inference sampling procedures are the same as the SSDiffRecon.
2. UNET: Original UNET architecture in DDPM [10] is trained with the same self-supervised loss as in SSDiffRecon. Other than the denoising network architecture, the training and inference procedures are not changed.
3. Without TR: SSDiffRecon without cross-attention transformer layers is trained and tested. This model only consists of data-consistency and CNN layers. Other than the network, training and inference procedures are not changed.
4. Without DC: SSDiffRecon without the data-consistency layers is trained and tested. This model does not utilize data-consistency but the other training and inference details are the same as the SSDiffRecon.

Table 1. Ablation results as avaraged across whole fastMRI test set.

	SSDiffRecon		Supervised		UNET		Without TR		Without DC	
	PSNR	SSIM	PSNR	SSIM	PSNR	SSIM	PSNR	SSIM	PSNR	SSIM
fastMRI	35.9	**94.1**	**36.3**	93.2	26.9	84.7	35.1	93.8	26.4	66.4

Table 2. Reconstruction performance on the IXI dataset for R = 4 and 8.

	DDPM		D5C5		RGAN		self-DDPM		Self-D5C5		self-RGAN		SSDiffRecon	
	PSNR	SSIM	PSNR	SSIM	PSNR	SSIM	PSNR	SSIM	PSNR	SSIM	PSNR	SSIM	PSNR	SSIM
T_1-4x	39.4	98.8	37.1	97.1	36.8	97.6	33.6	92.7	39.3	98.4	36.9	97.9	**42.3**	**99.3**
T_1-8x	33.0	96.8	30.9	94.0	32.3	95.4	30.1	90.2	32.7	96.0	31.9	95.9	**34.6**	**97.9**
T_2-4x	41.8	98.5	38.9	95.0	38.5	96.6	35.4	88.1	40.2	97.4	39.0	96.9	**45.9**	**99.1**
T_2-8x	36.2	96.2	34.2	91.6	34.6	94.3	32.9	84.9	34.6	94.5	34.9	94.6	**39.3**	**98.0**
PD-4x	37.1	98.5	36.5	94.7	35.6	96.5	32.5	88.4	**39.4**	98.5	36.3	96.8	38.4	**99.0**
PD-8x	32.2	96.1	30.1	90.5	31.9	93.9	29.8	85.0	32.3	94.5	31.8	94.4	**33.3**	**97.8**

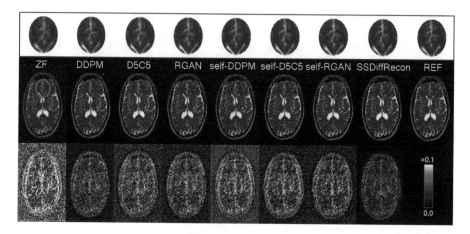

Fig. 2. Reconstructions of T_2- weighted images from the IXI dataset, along with the zoomed-in regions on the top and the corresponding error maps underneath.

5   Results

The results are presented in two clusters using a single figure and table for each dataset; fastMRI results are presented in Fig. 3 and Table 3, and IXI results are presented in Fig. 2 and Table 2. The best performed method in each test case is marked in bold in the tables. SSDiffRecon yields 2.55 dB more average PSNR and %1.96 SSIM than the second best self-supervised baseline in IXI, while performing 0.4 dB better in terms of PSNR and %0.25 in terms of SSIM on fastMRI. Visually, it captured most of the high frequency details while other self-supervised reconstructions suffer from either high noise or blurring artifact. Moreover, visual quality of reconstructions is either very close or better than supervised methods as be seen in the figures. It is also important to note that SSDiffRecon is performing only five backward diffusion steps while regular DDPM perform thousand diffusion steps for an equivalent reconstruction performance.

Table 3. Reconstruction performance on the fastMRI dataset for R = 4 and 8.

	DDPM		D5C5		RGAN		self-DDPM		self-D5C5		Self-RGAN		SSDiffRecon	
	PSNR	SSIM	PSNR	SSIM	PSNR	SSIM	PSNR	SSIM	PSNR	SSIM	PSNR	SSIM	PSNR	SSIM
T_1-4x	**40.2**	95.3	39.3	94.8	39.6	95.5	38.4	95.8	38.0	93.1	38.3	95.0	40.1	**96.5**
T_1-8x	**36.2**	91.7	35.6	92.6	36.0	92.7	35.4	93.3	34.8	90.5	35.0	92.4	35.1	**93.3**
T_2-4x	**38.2**	96.0	37.5	96.2	36.8	95.8	36.8	96.0	37.1	95.6	36.8	95.6	37.7	**96.6**
T_2-8x	**34.5**	93.0	34.3	94.0	33.8	93.1	34.3	**94.2**	33.8	93.0	34.0	93.4	33.7	93.8
Flair-4x	36.8	93.5	36.2	93.3	35.1	92.8	35.7	94.1	35.4	91.1	35.3	92.1	**36.9**	94.6
Flair-8x	**33.1**	87.8	32.7	89.1	32.0	88.1	32.5	89.6	32.2	86.3	32.1	87.8	32.1	**89.7**

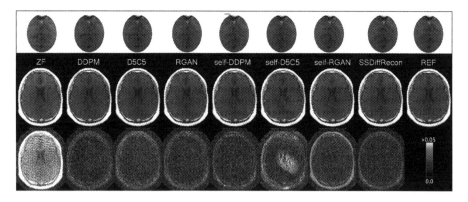

Fig. 3. Reconstructions of T_1- weighted images from fastMRI, along with the zoomed-in regions on the top and the corresponding error maps underneath.

6 Conclusion

We proposed a novel diffusion-based unrolled architecture for accelerated MRI reconstruction. Our model performs better than self-supervised baselines in a relatively short inference time while performing on-par with the supervised reconstruction methods. Inference time and model complexity analyses are presented in the supplementary materials.

Acknowledgement. This work was supported by NIH R01 grant R01CA276221 and TUBITAK 1001 grant 121E488.

References

1. Aggarwal, H.K., Mani, M.P., Jacob, M.: MoDL: model-based deep learning architecture for inverse problems. IEEE Trans. Med. Imaging **38**(2), 394–405 (2019)
2. Bakker, T., Muckley, M., Romero-Soriano, A., Drozdzal, M., Pineda, L.: On learning adaptive acquisition policies for undersampled multi-coil MRI reconstruction. arXiv preprint arXiv:2203.16392 (2022)
3. Cao, C., Cui, Z.X., Liu, S., Liang, D., Zhu, Y.: High-frequency space diffusion models for accelerated mri. arXiv preprint arXiv:2208.05481 (2022)
4. Cao, Y., Wang, L., Zhang, J., Xia, H., Yang, F., Zhu, Y.: Accelerating multi-echo MRI in k-space with complex-valued diffusion probabilistic model. In: 2022 16th IEEE International Conference on Signal Processing (ICSP), vol. 1, pp. 479–484. IEEE (2022)
5. Cui, Z.X., et al.: Self-score: Self-supervised learning on score-based models for MRI reconstruction. arXiv preprint arXiv:2209.00835 (2022)
6. Dar, S.U., et al.: Adaptive diffusion priors for accelerated MRI reconstruction. arXiv preprint arXiv:2207.05876 (2022)
7. Dar, S.U., Yurt, M., Shahdloo, M., Ildız, M.E., Tınaz, B., Çukur, T.: Prior-guided image reconstruction for accelerated multi-contrast MRI via generative adversarial networks. IEEE J. Sel. Top. Signal Process. **14**(6), 1072–1087 (2020)

8. Haldar, J.P., Hernando, D., Liang, Z.P.: Compressed-sensing MRI with random encoding. IEEE Trans. Med. Imaging **30**(4), 893–903 (2010)
9. Hammernik, K., Pan, J., Rueckert, D., Küstner, T.: Motion-guided physics-based learning for cardiac MRI reconstruction. In: 2021 55th Asilomar Conference on Signals, Systems, and Computers, pp. 900–907. IEEE (2021)
10. Ho, J., Jain, A., Abbeel, P.: Denoising diffusion probabilistic models. Adv. Neural. Inf. Process. Syst. **33**, 6840–6851 (2020)
11. Huang, W., et al.: Rethinking the optimization process for self-supervised model-driven MRI reconstruction. arXiv preprint arXiv:2203.09724 (2022)
12. Karras, T., Laine, S., Aittala, M., Hellsten, J., Lehtinen, J., Aila, T.: Analyzing and improving the image quality of StyleGAN. In: Proceedings of the IEEE/CVF Conference on Computer Vision and Pattern Recognition (CVPR), pp. 8107–8116 (2020)
13. Knoll, F., Hammernik, K., Kobler, E., Pock, T., Recht, M.P., Sodickson, D.K.: Assessment of the generalization of learned image reconstruction and the potential for transfer learning. Magn. Reson. Med. **81**(1), 116–128 (2019)
14. Knoll, F., et al.: fastMRI: a publicly available raw k-space and DICOM dataset of knee images for accelerated MR image reconstruction using machine learning. Radiol. Artif. Intell. **2**(1), e190007 (2020)
15. Kwon, K., Kim, D., Park, H.: A parallel MR imaging method using multilayer perceptron. Med. Phys. **44**(12), 6209–6224 (2017). https://doi.org/10.1002/mp.12600
16. Lee, D., Yoo, J., Tak, S., Ye, J.C.: Deep residual learning for accelerated MRI using magnitude and phase networks. IEEE Trans. Biomed. Eng. **65**(9), 1985–1995 (2018)
17. Lustig, M., Donoho, D., Pauly, J.M.: Sparse MRI: the application of compressed sensing for rapid MR imaging. Magn. Resonan. Med. Off. J. Int. Soc. Magn. Resonan. Med. **58**(6), 1182–1195 (2007)
18. Mardani, M., et al.: Deep generative adversarial neural networks for compressive sensing MRI. IEEE Trans. Med. Imaging **38**(1), 167–179 (2019)
19. Nichol, A.Q., Dhariwal, P.: Improved denoising diffusion probabilistic models. In: International Conference on Machine Learning, pp. 8162–8171. PMLR (2021)
20. Peng, C., Guo, P., Zhou, S.K., Patel, V.M., Chellappa, R.: Towards performant and reliable undersampled MR reconstruction via diffusion model sampling. In: Wang, L., Dou, Q., Fletcher, P.T., Speidel, S., Li, S. (eds.) Medical Image Computing and Computer Assisted Intervention. MICCAI 2022. LNCS, vol. 13436, pp. 623–633. Springer, Cham (2022). https://doi.org/10.1007/978-3-031-16446-0_59
21. Qin, C., Schlemper, J., Caballero, J., Price, A.N., Hajnal, J.V., Rueckert, D.: Convolutional recurrent neural networks for dynamic MR image reconstruction. IEEE Trans. Med. Imaging **38**(1), 280–290 (2018)
22. Schlemper, J., Caballero, J., Hajnal, J.V., Price, A., Rueckert, D.: A deep cascade of convolutional neural networks for MR image reconstruction. In: International Conference on Information Processing in Medical Imaging, pp. 647–658 (2017)
23. Sriram, A., Zbontar, J., Murrell, T., Zitnick, C.L., Defazio, A., Sodickson, D.K.: GrappaNet: combining parallel imaging with deep learning for multi-coil MRI reconstruction. In: Proceedings of the IEEE/CVF Conference on Computer Vision and Pattern Recognition (CVPR), pp. 14303–14310, June 2020
24. Wang, S., et al.: Accelerating magnetic resonance imaging via deep learning. In: IEEE 13th International Symposium on Biomedical Imaging (ISBI), pp. 514–517 (2016). https://doi.org/10.1109/ISBI.2016.7493320

25. Xie, Y., Li, Q.: Measurement-conditioned denoising diffusion probabilistic model for under-sampled medical image reconstruction. In: Wang, L., Dou, Q., Fletcher, P.T., Speidel, S., Li, S. (eds.) Medical Image Computing and Computer Assisted Intervention. MICCAI 2022. LNCS, vol. 13436, pp. pp. 655–664. Springer, Cham (2022). https://doi.org/10.1007/978-3-031-16446-0_62
26. Yaman, B., Hosseini, S.A.H., Moeller, S., Ellermann, J., Uğurbil, K., Akçakaya, M.: Self-supervised learning of physics-guided reconstruction neural networks without fully sampled reference data. Magn. Reson. Med. 84(6), 3172–3191 (2020)
27. Yu, S., et al.: DAGAN: deep de-aliasing generative adversarial networks for fast compressed sensing MRI reconstruction. IEEE Trans. Med. Imaging 37(6), 1310–1321 (2018)
28. Zhang, T., Pauly, J.M., Vasanawala, S.S., Lustig, M.: Coil compression for accelerated imaging with cartesian sampling. Magn. Reson. Med. 69(2), 571–582 (2013)
29. Zhu, B., Liu, J.Z., Rosen, B.R., Rosen, M.S.: Image reconstruction by domain transform manifold learning. Nature 555(7697), 487–492 (2018)

LightNeuS: Neural Surface Reconstruction in Endoscopy Using Illumination Decline

Víctor M. Batlle[1]([✉]), José M. M. Montiel[1], Pascal Fua[2], and Juan D. Tardós[1]

[1] Inst. Investigación en Ingeniería de Aragón, I3A, Universidad de Zaragoza, Zaragoza, Spain
`{vmbatlle,josemari,tardos}@unizar.es`
[2] CVLab, École Polytechnique Fédérale de Lausanne, Lausanne, Switzerland
`pascal.fua@epfl.ch`

Abstract. We propose a new approach to 3D reconstruction from sequences of images acquired by monocular endoscopes. It is based on two key insights. First, endoluminal cavities are watertight, a property naturally enforced by modeling them in terms of a signed distance function. Second, the scene illumination is variable. It comes from the endoscope's light sources and decays with the inverse of the squared distance to the surface. To exploit these insights, we build on NeuS [25], a neural implicit surface reconstruction technique with an outstanding capability to learn appearance and a SDF surface model from multiple views, but currently limited to scenes with static illumination. To remove this limitation and exploit the relation between pixel brightness and depth, we modify the NeuS architecture to explicitly account for it and introduce a calibrated photometric model of the endoscope's camera and light source.

Our method is the first one to produce watertight reconstructions of whole colon sections. We demonstrate excellent accuracy on phantom imagery. Remarkably, the watertight prior combined with illumination decline, allows to complete the reconstruction of unseen portions of the surface with acceptable accuracy, paving the way to automatic quality assessment of cancer screening explorations, measuring the global percentage of observed mucosa.

Keywords: Reconstruction · Photometric multi-view · Endoscopy

1 Introduction

Colorectal cancer (CRC) is the third most commonly diagnosed cancer and is the second most common cause of cancer death [23]. Early detection is crucial for a good prognosis. Despite the existence of other techniques, such as virtual colonoscopy (VC), optical colonoscopy (OC) remains the gold standard for colonoscopy screening and the removal of precursor lesions. Unfortunately, we do

---

Supplementary Information The online version contains supplementary material available at https://doi.org/10.1007/978-3-031-43999-5_48.

H. Greenspan et al. (Eds.): MICCAI 2023, LNCS 14229, pp. 502–512, 2023.
https://doi.org/10.1007/978-3-031-43999-5_48

not yet have the ability to reconstruct densely the 3D shape of large sections of the colon. This would usher exciting new developments, such as post-intervention diagnosis, measuring polyps and stenosis, and automatically evaluating exploration thoroughness in terms of the surface percentage that has been observed.

This is the problem we address here. It has been shown that the colon 3D shape can be estimated from single images acquired during human colonoscopies [3]. However, to model large sections of it while increasing the reconstruction accuracy, multiple images must be used. As most endoscopes contain a single camera, the natural way to do this is to use video sequences acquired by these cameras in the manner of structure-from-motion algorithms. An important first step in that direction is to register the images from the sequences. This can now be done reliably using either batch [21] or SLAM techniques [8]. Unfortunately, this solves only half the problem because these techniques provide very sparse reconstructions and going from there to dense ones remains an open problem. And occlusions, specularities, varying albedos, and specificities of endoscopic lighting make it a challenging one.

To overcome these difficulties, we rely on two properties of endoscopic images:

- Endoluminal cavities such as the gastrointestinal tract, and in particular the human colon, are watertight surfaces. To account for this, we represent its surface in terms of a signed distance function (SDF), which by its very nature presents continuous watertight surfaces.
- In endoscopy the light source is co-located with the camera. It illuminates a dark scene and is always close to the surface. As a result, the irradiance decreases rapidly with distance t from camera to surface; more specifically it is a function of $1/t^2$. In other words, there is a strong correlation between light and depth, which remains unexploited to date.

To take advantage of these specificities, we build on the success of Neural implicit Surfaces (NeuS) [25] that have been shown to be highly effective at deriving surface 3D models from sets of registered images. As the Neural Radiance Fields (NeRFs) [15] that inspired them, they were designed to operate on regular images taken around a scene, sampling fairly regularly the set of possible viewing directions. Furthermore, the lighting is assumed to be static and distant so that the brightness of a pixel and its distance to the camera are unrelated. Unfortunately, none of these conditions hold in endoscopies. The camera is inside a cavity (in the colon, a roughly cylindrical tunnel) that limits viewing directions. The light source is co-located with the camera and close to the surface, which results in a strong correlation between pixel brightness and distance to the camera. In this paper, we show that, far from being a handicap, this correlation is a key information for neural network self-supervision.

NeuS training selects a pixel from an image and samples points along its projecting ray. However, the network is agnostic to the sampling distance. In LightNeuS, we explicitly feed to the renderer the distance of each one of these sampled points to the light source, as shown in Fig. 1. Hence, the renderer can exploit the inverse-square illumination decline. We also introduce and calibrate a photometric model for the endoscope light and camera, so that the inverse

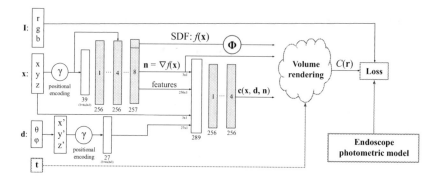

Fig. 1. From NeuS to LightNeuS. The original NeuS architecture is depicted by the black arrows. In LightNeuS, when training the network with a sampled point, we provide the sampling distance t to the renderer, that takes into account illumination decline. We also incorporate a calibrated photometric endoscope model that is used to correctly compute the photometric loss. The changes are shown in red. (Color figure online)

square law discussed above actually holds. Together, these two changes make the minimization problem better posed and the automatic depth estimation more reliable.

Our results show that exploiting the illumination is key to unlocking implicit neural surface reconstruction in endoscopy. It delivers accuracies in the range of 3 mm, whereas an unmodified NeuS is either 5 times less accurate or even fails to reconstruct any surface at all. Earlier methods [3] have reported similar accuracies but only on very few synthetic images and on short sections of the colon. By contrast, we can handle much longer ones and provide a broad evaluation in a real dataset (C3VD) over multiple sequences. This makes us the first to show accurate results of extended 3D watertight surfaces from monocular endoscopy images.

2   Related Works

3D Reconstruction from Endoscopic Images. It can help with the effective localization of lesions, such as polyps and adenomas, by providing a complete representation of the observed surface. Unfortunately, many state-of the-art SLAM techniques based on feature matching [5] or direct methods [6,7] are impractical for dense endoscopic reconstruction due to the lack of texture and the inconsistent lighting that moves along with the camera. Nevertheless, sparse reconstructions by classical Structure-from-Motion (SfM) algorithms can be good starting points for refinement and densification based on Shape-from-Shading (SfS) [24,28]. However, classical multi-view and SfS methods require strong suboptimal priors on colon surface shape and reflectance.

In monocular dense reconstructions, it is common practice to encode shape priors in terms of smooth rigid surfaces [14,17,20]. Recently, [22] proposes a tubular topology prior for NRSfM aimed to process endoluminal cavities where these tubular shapes are prevalent. In contrast, for the same environments, we propose the watertight prior coded by implicit SDF representations.

Recent methods for dense reconstruction rely on neural networks to predict per-pixel depth in the 2D space of each image and fuse the depth maps by using multi-view stereo (MVS) [2] or a SLAM pipeline [12,13]. However, holes in the reconstruction appear due to failures in triangulation and inaccurate depth estimation or in areas not observed in any image. Wang et al. [27] show the potential of neural rendering in reconstruction from medical images, although they use a binocular static camera with fixed light source, which is not feasible in endoluminal endoscopy. Unfortunately, most of the previous 3D methods do not provide code [14,22], are not evaluated in biomedical settings [17,20], or do not report reconstruction accuracy [12,13].

Neural Radiance Fields (NeRFs) were first proposed to reconstruct novel views of non-Lambertian objects [15]. This method provides an *implicit neural representation* of a scene in terms of local densities and associated colors. In effect, the scene representation is stored in the weights of a neural network, usually a multilayer perceptron (MLP), that learns its shape and reflectance for any coordinate and viewing direction. NeRFs use volume rendering [9], based on ray-tracing from multiple camera positions. The volume density $\sigma(\mathbf{x})$ can be interpreted as the differential probability of a ray terminating at an infinitesimal particle at location $\mathbf{x}$. The expected color $C(\mathbf{r})$ of the pixel with camera ray $\mathbf{r}(t) = \mathbf{o} + t\mathbf{d}$ is the integration of the radiance emitted by the field at every traveled distance t from near to far bounds t_n and t_f, such that

$$C(\mathbf{r}) = \int_{t_n}^{t_f} T(t)\,\sigma(\mathbf{r}(t))\,\mathbf{c}(\mathbf{r}(t),\mathbf{d})\,\mathrm{d}t \quad \text{where } T(t) = \exp\left(-\int_{t_n}^{t}\sigma(\mathbf{r}(s))\,\mathrm{d}s\right) \quad (1)$$

where $\mathbf{c}$ stands for the color. The function T denotes the accumulated transmittance along the ray from t_n to t, that is the probability that the ray travels from t_n to t without hitting any other particle. The authors propose two MLPs to estimate the volume density function $\sigma : \mathbf{x} \to [0,1]$ and the directional emitted color function $\mathbf{c} : (\mathbf{x},\mathbf{d}) \to [0,1]^3$, so the density of a point does not depend on the viewing direction $\mathbf{d}$, but the color does. This allows them to model non-Lambertian reflectance. In addition, they propose a positional encoding for location $\mathbf{x}$ and direction $\mathbf{d}$, which allows high-frequency details in the reconstruction.

Neural Implicit Surfaces (NeuS) were introduced in [25] to improve the quality of NeRF representation modelling watertight surfaces. For that, the volume density σ is computed so as to be maximal at the zero-crossings of a signed distance function (SDF) f:

$$\sigma(\mathbf{r}(t)) = \max\left(\frac{\Phi'_s(f(\mathbf{r}(t)))}{\Phi_s(f(\mathbf{r}(t)))}, 0\right) \quad \text{where } \Phi_s(x) = \frac{1}{1 + e^{-sx}} \quad (2)$$

The SDF formulation makes it possible to estimate the surface normal as $\mathbf{n} = \nabla f(\mathbf{x})$. The reflectance of a material is usually determined as a function of the incoming and outgoing light directions with respect to the surface normal. Therefore, the normal is added as an input to the MLP that estimates color $\mathbf{c} : (\mathbf{x}, \mathbf{d}, \mathbf{n})$, as shown in Fig. 1.

3 LightNeuS

In this section, we present the key contributions that make *LightNeuS* a neural implicit reconstruction method suitable for endoscopy in endoluminal cavities. In this context, the light source is located next to the camera and moves with it. Furthermore, it is close to the surfaces to be modeled. As a result, for any surface point $\mathbf{x} = \mathbf{o} + t\mathbf{d}$, the irradiance decreases with the square of the distance to the camera t. Hence, we can write the color of the corresponding pixel as [3]:

$$\mathcal{I}(\mathbf{x}) = \left(\frac{L_e}{t^2} \text{ BRDF}(\mathbf{x}, \mathbf{d}) \ \cos{(\theta)} \ g \right)^{1/\gamma} \tag{3}$$

where L_e is the radiance emitted by the light source to the surface point, that was modeled and calibrated in the EndoMapper dataset [1] according to the SLS model from [16]. The bidirectional reflectance distribution function (BRDF) determines how much light is reflected to the camera, and the cosine term $\cos{(\theta)} = -\mathbf{d} \cdot \mathbf{n}$ weights the incoming radiance with respect to the surface normal $\mathbf{n}$. Equation (3) also takes into account the camera gain g and gamma correction γ.

3.1 Using Illumination Decline as a Depth Cue

The NeuS formulation of Sect. 2 assumes distant and fixed lighting. However, in endoscopy inverse-square light decline is significant, as quantified in Eq. (3). Accounting for this is done by modifying the original NeuS formulation as follows. Figure 1 depicts the original NeuS network in black. It uses a SDF network—shown in orange—to estimate a view-independent geometry and only the final RGB color depends on the viewing·direction $\mathbf{d}$. It is estimated by the network shown in green. Thus, this second network $\mathbf{c}(\mathbf{x}, \mathbf{d}, \mathbf{n})$ may learn to model non-Lambertian BRDF$(\mathbf{x}, \mathbf{d})$, including specular highlights, and the cosine term of Eq. (3). However, if the distance t from the light to the point $\mathbf{x}$ is not provided to the color network, the $1/t^2$ dependency cannot be learned, and surface reconstruction will fail. Our key insight is to explicitly supply this distance as input to the volume rendering algorithm, as shown in red in Fig. 1 and reformulate Eq. (1) as

$$C(\mathbf{r}) = \int_{t_n}^{t_f} T(t) \ \sigma(\mathbf{r}(t)) \ \frac{\mathbf{c}(\mathbf{r}(t), \mathbf{d}, \mathbf{n})}{t^2} \ dt \tag{4}$$

This conceptually simple change, using illumination decline while training, unlocks all the power of neural surface reconstruction in endoscopy.

3.2   Endoscope Photometric Model

Apart from illumination decline, there are several significant differences between the images captured by endoscopes and those conventionally used to train NeRFs and NeuS: fish-eye lenses, strong vignetting, uneven scene illumination, and post-processing.

Endoscopes use fisheye lenses to cover a wide field of view, usually close to $170°$. These lenses produce strong deformations, making it unwise to use the standard pinhole camera model. Instead, specific models [10,19] must be used. Hence, we also modified the original NeuS implementation to support these models.

The light sources of endoscopes behave like spotlights. In other words, they do not emit with the same intensity in all directions, so L_e in Eq. (3) is not constant for all image pixels. This effect is similar to the vignetting effect caused by conventional lenses, that is aggravated in fisheye lenses. Fortunately, they can be accurately calibrated [1,16] and compensated for.

The post-processing software of medical endoscopes is designed to always display well-exposed images, so that physicians can see details correctly. An adaptive gain factor g is applied by the endoscope's internal logic and gamma correction is also used to adapt to non-linear human vision, achieving better contrast perception in mid tones and dark areas. Endoscope manufacturers know the post-processing logic of their devices, but this information is proprietary and not available to users. Again, gamma correction can be calibrated assuming it is constant [3], and the gain change between successive images can be estimated, for example, by sparse feature matching.

All these factors must be taken into account during network training. Thus, our photometric loss is computed using a normalized image:

$$I' = \left(\frac{I^\gamma}{L_e g} \right)^{1/\gamma} \tag{5}$$

4   Experiments

We validate our method on the C3VD dataset [4], which covers all different sections of the colon anatomy in 22 video sequences. This dataset contains sequences recorded with a medical video colonoscope, Olympus Evis Exera III CF-HQ190L. The images were recorded inside a *phantom*, a model of a human colon made of silicone. The intrinsic camera parameters are provided. The camera extrinsics for each frame are estimated by 2D-3D registration against the known 3D model. In an operational setting, we could use a structure-from-motion approach such as COLMAP [21] or a SLAM technique such as [8], which have been shown to work well in endoscopic settings. The gain values were easily estimated from the dataset itself. For vignetting, we use the calibration obtained from a colonoscope of the same brand and series from the EndoMapper dataset [1].

During training, we follow the NeuS paper approach of using a few informative frames per scene, as separated as possible, by sampling each video uniformly.

Table 1. Reconstruction error [mm] on the C3VD dataset. <u>Sur</u>veyed: points seen at least once. <u>Ext</u>ended: points within 20 mm of a visible point. Anatomical regions: <u>C</u>ecum, <u>D</u>escending, <u>S</u>igmoid and <u>T</u>ransverse. For NeuS, we provide two sets of numbers because the optimization failed on the other sections. In *italics* we mark the sequences where the camera moves less than 1 cm yielding higher errors.

		NeuS		LightNeuS (ours)										
	Sequence	C1a	C4b	C1a	C1b	C2a	C2b	C2c	C3a	C4a	C4b	D4a	S1a	S2a
Sur.	MedAE	4.53	10.6	0.95	4.85	1.40	3.26	2.57	1.12	1.90	1.41	2.66	4.23	1.19
	MAE	5.07	10.6	1.48	5.11	1.54	3.65	3.00	2.54	2.14	1.63	3.26	4.33	1.89
	RMSE	6.40	11.6	2.01	5.63	1.87	4.39	3.74	5.49	2.92	2.10	4.08	4.96	2.78
Ext.	MedAE	4.68	5.35	0.83	4.89	1.41	3.32	2.54	1.27	1.91	1.45	4.50	4.01	1.40
	MAE	6.24	6.74	1.26	5.10	1.56	3.70	3.01	3.83	2.18	1.72	6.61	4.19	2.36
	RMSE	8.77	8.56	1.72	5.60	1.90	4.42	3.77	7.96	2.95	2.20	9.32	4.87	3.96

LightNeuS (ours)												
S3a	S3b	T1a	T1b	T2a	T2b	T4a	Mean	T2c	T3a	T3b	T4b	Mean
2.57	3.63	3.43	2.33	2.24	2.16	1.15	**2.39**	*5.07*	*6.39*	*11.0*	*1.75*	***6.04***
2.68	4.16	3.47	2.72	2.28	2.30	2.31	**2.80**	*5.45*	*8.65*	*12.1*	*6.70*	***8.23***
3.18	4.81	4.07	3.34	2.58	2.70	3.79	**3.58**	*6.48*	*10.7*	*14.4*	*11.3*	***10.7***
2.87	3.54	3.38	2.69	2.19	2.12	1.29	**2.53**	*4.44*	*6.54*	*13.6*	*8.00*	***8.16***
3.27	4.64	3.31	3.21	2.22	2.28	2.22	**3.15**	*5.36*	*8.10*	*14.1*	*10.4*	***9.47***
4.04	6.10	3.86	3.96	2.55	2.69	3.32	**4.18**	*6.78*	*9.94*	*15.9*	*13.9*	***11.6***

For each sequence, we train both the vanilla NeuS and our LighNeuS using 20 frames each time. They are extracted uniformly over the duration of the video. We use the same batch size and number of iterations as in the original NeuS paper, 512 and 300k respectively. Once the network is trained, we can extract triangulated meshes from the reconstruction. Since the C3VD dataset comprises a ground-truth triangle mesh, we compute point-to-triangle distances from all the vertices in the reconstruction to the closest ground-truth triangle.

In the first rows of Table 1, we report median (MedAE), mean (MAE), and root mean square (RMSE) values of these distances for all vertices seen in at least one image. Columns show the result for 22 sequences. We note 18 sequences where the camera moved at least 1 cm, and the reconstruction yielded a mean error of 2.80 mm. The other four smaller trajectories (<1 cm) lack parallax and the mean error is higher (8.23 mm).

This is in the range of reported accuracy in the literature for monocular dense non-watertight depth estimation, 1.1 mm in [14] for high parallax geometry in laparoscopy, which is a much more favorable geometry than the one we have here, or 0.85 mm for the significantly smaller-size cavities of endoscopic endonasal surgery (ESS) [11].

In contrast, vanilla NeuS assumes constant illumination. The strong light changes typical of endoscopy fatally mislead the method. We only report numerical results of NeuS in two sequences because in all the rest, the SDF diverges and ends up blown out of the rendering volume, giving no result at all.

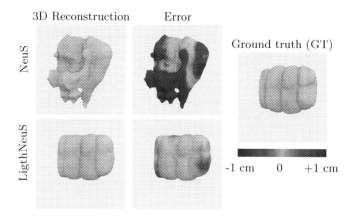

Fig. 2. Benefits of illumination decline. Result on the *"Cecum 1 a"* sequence. **Top:** The NeuS reconstruction exhibits multiple artifacts that make it unusable. **Bottom:** Our reconstruction is much closer to the ground truth shape. The error is shown in blue if the reconstruction is inside the surface, and in red otherwise. A fully saturated red or blue denotes an error of more than 1 cm and grey denotes no error at all.

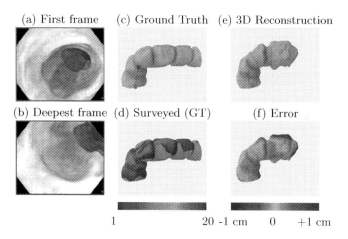

Fig. 3. Reconstructing partially observed regions. Results on *"Transcending 4 a"* sequence. The camera performs a short trajectory from (a) to (b). In (c) we represent both frames and intermediate camera poses. (d) Number of frames seeing each surface point, with GT unobserved areas shown in gray. (e) We managed to reconstruct a curved section of the colon. (f) Our method plausibly estimates the wall of the colon at the right of camera (b), although it was never seen in the images.

We provide a qualitative result in Fig. 2 and additional ones in the supplementary material. Note that the watertight prior inherent to an SDF allows the network to hallucinate unseen areas. Remarkably, these unsurveyed areas continue the tubular shape of the colon and we found them to be mostly accurate when compared to the ground truth. For example, the curved areas of the colon where a wall is occluded behind the corner of the curve is reconstructed, as shown in Fig. 3. This ability to "fill in" observation gaps may be useful in providing the endoscopist with an estimate of the percentage of unsurveyed area during a procedure.

We hypothesize that this desirable behavior stems from the fact that the network learns an empirical shape prior from the observed anatomy of the colon. However, we don't expect this behavior to hold for distant unseen parts, but only for regions closer than 20 mm to one observation. In the last rows of Table 1, we compute accuracy metrics for this *extended* region. It includes not only surveyed areas, but also neighboring areas that were not observed.

5   Conclusion

We have presented a method for 3D dense multi-view reconstruction from endo-scopic images. We are the first to show that neural radiance fields can be used to obtain accurate dense reconstructions of colon sections of significant length. At the heart of our approach, is exploiting the correlation between depth and brightness. We have observed that, without it, neural reconstruction fails.

The current method could be used offline for post-exploration coverage analysis and endoscopist training. But real-time performance could be achieved in the future as the new NeuS2 [26] converges in minutes, enabling automatic coverage reporting. Similar to other reconstruction methods, for now our approach works in areas of the colon where there is little deformation. Several sub-maps of non-deformed areas can be created if necessary. However, this limitation could be overcome by adopting the deformable NeRFs formalism [18].

Acknowledgement. This work was supported by EU-H2020 grant 863146: ENDO MAPPER, Spanish government grants PID2021-127685NB-I00 and FPU20/06782 and by Aragón government grant DGA_T45-17R.

References

1. Azagra, P., et al.: EndoMapper dataset of complete calibrated endoscopy procedures. arXiv:2204.14240 (2022)
2. Bae, G., Budvytis, I., Yeung, C.-K., Cipolla, R.: Deep multi-view stereo for dense 3D reconstruction from monocular endoscopic video. In: Martel, A.L., et al. (eds.) MICCAI 2020. LNCS, vol. 12263, pp. 774–783. Springer, Cham (2020). https://doi.org/10.1007/978-3-030-59716-0_74
3. Batlle, V.M., Montiel, J.M.M., Tardós, J.D.: Photometric single-view dense 3D reconstruction in endoscopy. In: IEEE/RSJ International Conference on Intelligent Robots and Systems (IROS), pp. 4904–4910 (2022)
4. Bobrow, T.L., Golhar, M., Vijayan, R., Akshintala, V.S., Garcia, J.R., Durr, N.J.: Colonoscopy 3D video dataset with paired depth from 2D–3D registration. arXiv:2206.08903 (2022)

5. Campos, C., Elvira, R., Gómez-Rodríguez, J.J., Montiel, J.M.M., Tardós, J.D.: ORB-SLAM3: an accurate open-source library for visual, visual-inertial, and multimap SLAM. IEEE Trans. Rob. **37**(6), 1874–1890 (2021)
6. Engel, J., Koltun, V., Cremers, D.: Direct sparse odometry. IEEE Trans. Pattern Anal. Mach. Intell. **40**(3), 611–625 (2018)
7. Engel, J., Schöps, T., Cremers, D.: LSD-SLAM: large-scale direct monocular SLAM. In: Fleet, D., Pajdla, T., Schiele, B., Tuytelaars, T. (eds.) ECCV 2014. LNCS, vol. 8690, pp. 834–849. Springer, Cham (2014). https://doi.org/10.1007/978-3-319-10605-2_54
8. Gómez-Rodríguez, J.J., Lamarca, J., Morlana, J., Tardós, J.D., Montiel, J.M.M.: SD-DefSLAM: Semi-direct monocular SLAM for deformable and intracorporeal scenes. In: IEEE Int. Conf. on Robotics and Automation (ICRA). pp. 5170–5177 (2021)
9. Kajiya, J.T., Von Herzen, B.P.: Ray tracing volume densities. SIGGRAPH Comput. Graph. 18(3), 165–174 (jan 1984)
10. Kannala, J., Brandt, S.: A generic camera model and calibration method for conventional, wide-angle, and fish-eye lenses. IEEE Trans. Pattern Anal. Mach. Intell. **28**(8), 1335–1340 (2006)
11. Liu, X., Li, Z., Ishii, M., Hager, G.D., Taylor, R.H., Unberath, M.: Sage: Slam with appearance and geometry prior for endoscopy. In: IEEE Int. Conf. on Robotics and Automation (ICRA). pp. 5587–5593 (2022)
12. Ma, R., Wang, R., Pizer, S., Rosenman, J., McGill, S.K., Frahm, J.M.: Real-time 3D reconstruction of colonoscopic surfaces for determining missing regions. In: Int. Conf. on Medical Image Computing and Computer Assisted Intervention (MICCAI). pp. 573–582 (2019)
13. Ma, R., Wang, R., Zhang, Y., Pizer, S., McGill, S.K., Rosenman, J., Frahm, J.M.: RNNSLAM: Reconstructing the 3D colon to visualize missing regions during a colonoscopy. Med. Image Anal. **72**, 102100 (2021)
14. Mahmoud, N., Collins, T., Hostettler, A., Soler, L., Doignon, C., Montiel, J.M.M.: Live tracking and dense reconstruction for handheld monocular endoscopy. IEEE Trans. Med. Imaging **38**(1), 79–89 (2019)
15. Mildenhall, B., Srinivasan, P.P., Tancik, M., Barron, J.T., Ramamoorthi, R., Ng, R.: NeRF: representing scenes as neural radiance fields for view synthesis. Commun. ACM. **65**(1), 99–106 (2021)
16. Modrzejewski, R., Collins, T., Hostettler, A., Marescaux, J., Bartoli, A.: Light modelling and calibration in laparoscopy. Int. J. Comput. Assist. Radiol. Surg. **15**(5), 859–866 (2020)
17. Newcombe, R.A., Lovegrove, S.J., Davison, A.J.: DTAM: dense tracking and mapping in real-time. In: IEEE International Conference on Computer Vision (ICCV), pp. 2320–2327 (2011)
18. Park, K., et al.: Nerfies: deformable neural radiance fields. In: IEEE/CVF International Conference on Computer Vision (ICCV), pp. 5865–5874 (2021)
19. Scaramuzza, D., Martinelli, A., Siegwart, R.: A toolbox for easily calibrating omnidirectional cameras. In: IEEE/RJS International Conference on Intelligent Robots and Systems (IROS), pp. 5695–5701 (2006)
20. Schönberger, J.L., Zheng, E., Frahm, J.-M., Pollefeys, M.: Pixelwise view selection for unstructured multi-view stereo. In: Leibe, B., Matas, J., Sebe, N., Welling, M. (eds.) ECCV 2016. LNCS, vol. 9907, pp. 501–518. Springer, Cham (2016). https://doi.org/10.1007/978-3-319-46487-9_31
21. Schönberger, J.L., Frahm, J.M.: Structure-from-motion revisited. In: IEEE Conference on Computer Vision and Pattern Recognition (CVPR) (2016)

22. Sengupta, A., Bartoli, A.: Colonoscopic 3D reconstruction by tubular non-rigid structure-from-motion. Int. J. Comput. Assist. Radiol. Surg. **16**(7), 1237–1241 (2021)
23. Sung, H., et al.: Global cancer statistics 2020: Globocan estimates of incidence and mortality worldwide for 36 cancers in 185 countries. CA: Cancer J. Clin. **71**(3), 209–249 (2021)
24. Tokgozoglu, H.N., Meisner, E.M., Kazhdan, M., Hager, G.D.: Color-based hybrid reconstruction for endoscopy. In: IEEE Conference on Computer Vision and Pattern Recognition (CVPR) Workshops, pp. 8–15 (2012)
25. Wang, P., Liu, L., Liu, Y., Theobalt, C., Komura, T., Wang, W.: NeuS: learning neural implicit surfaces by volume rendering for multi-view reconstruction. In: Advances in Neural Information Processing Systems, vol. 34, pp. 27171–27183 (2021)
26. Wang, Y., Han, Q., Habermann, M., Daniilidis, K., Theobalt, C., Liu, L.: NeuS2: fast learning of neural implicit surfaces for multi-view reconstruction. arXiv:2212.05231 (2022)
27. Wang, Y., Long, Y., Fan, S.H., Dou, Q.: Neural rendering for stereo 3D reconstruction of deformable tissues in robotic surgery. In: Wang, L., Dou, Q., Fletcher, P.T., Speidel, S., Li, S. (eds.) Medical Image Computing and Computer Assisted Intervention. MICCAI 2022. LNCS, vol. 13437, pp. 431–441. Springer, Cham (2022). https://doi.org/10.1007/978-3-031-16449-1_41
28. Zhao, Q., Price, T., Pizer, S., Niethammer, M., Alterovitz, R., Rosenman, J.: The Endoscopogram: a 3D model reconstructed from endoscopic video frames. In: Ourselin, S., Joskowicz, L., Sabuncu, M.R., Unal, G., Wells, W. (eds.) MICCAI 2016. LNCS, vol. 9900, pp. 439–447. Springer, Cham (2016). https://doi.org/10.1007/978-3-319-46720-7_51

Reflectance Mode Fluorescence Optical Tomography with Consumer-Grade Cameras

Mykhaylo Zayats[1]([✉]), Christopher Hansen[2], Ronan Cahill[3],
Gareth Gallagher[3], Ra'ed Malallah[3,4], Amit Joshi[2], and Sergiy Zhuk[1]

[1] IBM Research - Europe, Dublin, Ireland
mykhaylo.zayats1@ibm.com, sergiy.zhuk@ie.ibm.com
[2] Department of Biomedical Engineering, Medical College of Wisconsin,
Milwaukee, WI, USA
{chhansen,ajoshi}@mcw.edu
[3] Centre for Precision Surgery, School of Medicine, University College Dublin,
Dublin, Ireland
{ronan.cahill,gareth.gallagher,raed.malallah}@ucd.ie
[4] Physics Department, Faculty of Science, University of Basrah, Garmat Ali, Basra,
Iraq

Abstract. Efficient algorithms for solving inverse optical tomography problems with noisy and sparse measurements are a major challenge for near-infrared fluorescence guided surgery. To address that challenge, we propose an Incremental Fluorescent Target Reconstruction scheme based on the recent advances in convex optimization and sparse regularization. We demonstrate the efficacy of the proposed scheme on continuous wave reflectance mode boundary measurements of emission fluence from a 3D fluorophore target immersed in a tissue like media and acquired by an inexpensive consumer-grade camera.

Keywords: near-infrared imaging · diffuse optical tomography

1 Introduction

Near-infrared (NIR) fluorescence imaging can allow the detection of fluorophores up to 4 cm depth in tissue [11]. Recently, with the availability of clinically approved NIR fluorophores such as indocyanine green or ICG, fluorescence imaging is increasingly being employed for intra-operative guidance during surgically excision of malignant tumors and lymph nodes [6,15,16]. Fluorescence imaging is also a workhorse for small animal or preclinical research with multiple commercial devices utilizing sensitive front or back-illuminated and cooled CCD camera detectors available at prices ranging from 250–600K USD [9,14].

M. Zayats, C. Hansen—These authors contributed equally.

Supplementary Information The online version contains supplementary material available at https://doi.org/10.1007/978-3-031-43999-5_49.

A majority of fluorescence imaging applications including Fluorescence Guided Surgery (FGS) rely upon visible 2D surface imaging [5,8,17,20] while reconstruction of the invisible 3D target in tissue is not widely used for reflectance mode imaging despite a large number of publications in 3D fluorescence diffuse optical tomography (FDOT) since early 1990s [1,4,10,18,19]. The primary cause of this impasse is the ill-posedness of the mathematical inverse problem underlying the 3D reconstruction of the target in tissue from boundary measurements.

The prime motivation of our work is to enable an efficient 3D tumor shape reconstruction for FGS in an operating room environment, where we do not have full control of the ambient light and we cannot rely on sophisticated time or frequency domain imaging instrumentation and setup. In these situations, one has to use clinical cameras producing rapid Continuous Wave (CW) fluorescence boundary measurements [19] in reflectance mode (i.e., the transmission of the light through the domain is not measured), and with low signal-to-noise ratio which further exacerbates the ill-posedness of FDOT problem. The standard approach for solving FDOT problem with CW measurements is based on Born approximation which works well in the case of a small compared to the computational domain target and a very large number of reflectance-transmission type measurements made by "slow in acquisition" light sources and detector arrays of highly sensitive cooled CCD cameras or photomultiplier tube arrays collecting both reflected and transmitted light [18]. None of these is suitable for FGS settings where time is limited, just a few reflectance mode CW-measurements are available, and the target can be large compared to the imaged domain.

We propose an Incremental Fluorescent Target Reconstruction (IFTR) scheme, based on the recent advances in quadratic and conic convex optimization and sparse regularization, which can recover a relatively large 3D target in tissue-like media. In our experiments, IFTR scheme demonstrates accurate reconstruction of 3D targets from reflectance mode CW-measurements collected at the top surface of the domain. To our best knowledge, this is the first report where the 3D shape of tumor-like target has been recovered from reflectance mode steady-state CW measurements. Previously such results were reported in FDOT literature only for time-consuming frequency-domain or time-domain measurements [12] where photon path-length information is available. Moreover, the data is acquired almost instantly by an inexpensive (<100 Euros) camera with flexible fiber-optics making it suitable for endoscopic FGS in contrast to the standard slow in acquisition frequency-based measurements obtained by expensive (USD100K+ range) stationary cameras. Lastly, IFTR scheme is implemented using FEniCS [3], a high-level Python package for FEM discretization of the physical model, and CVXPY [2, 7], a convex optimization package making this method easy to reuse/adjust for a different setup. The code and data produced for this work are released as an open source at https://github.com/IBM/DOT.

2   Methods

Figure 1 describes the setup representing a typical surgical field while excising tumors. We simulate the provision of 3D surgical guidance via a flexible endo-

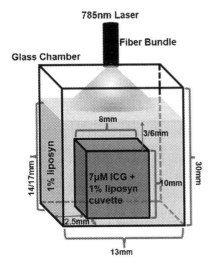

Fig. 1. Schematics of the laboratory experiment setting.

Fig. 2. Depiction of the laboratory experiment setting.

scope type fluorescence imager. For such provision we need to solve the following FDOT problem: estimate the spatial shape χ of the ICG tagged tumor target (the cube in green) within the tissue domain $\Omega \in R^3$ (the area in grey) from measurements y. Measurements are obtained by illuminating the tissue domain with NIR light at ICG peak excitation wavelength via the expanded beam from endoscope fiber bundle, and then measuring the light emitted by the tumor like target diffusing to the top face of the phantom surface $\partial\Omega_{obs}$, by the fiber bundle with suitable emission filter which is coupled to a camera at the backend.

In this section we briefly describe the mathematical formulation of the FDOT problem and introduce the IFTR scheme for solving it.

Forward and Inverse Problems. Photon propagation in tissue-like media is described by a coupled system of elliptic Partial Differential Equations (PDEs) for determining photon fluence ϕ (W/cm^2) at excitation and fluorescence emission wavelengths through out the domain. Wavelength and space dependent absorption and scattering coefficients and fluorophore properties comprise the coefficients of this PDE system (see Appendix A). The discretization of coupled diffusion PDEs is obtained by applying a standard FEM methodology [13]: domain Ω is covered by a uniform grid comprised of N nodes $\{x_i\}_{i=1}^N$; each function ϕ is approximated by a vector $\phi \in \mathbb{R}^N$ with components $\phi_i = \phi(x_i)$; PDEs are approximated using weak formulations incorporating boundary conditions. This results in a system of algebraic equations:

$$\begin{cases} S_x(\chi)\phi_x + M \odot \chi\phi_x = f \\ S_m\phi_m = M \odot \chi\phi_x \end{cases} \tag{1}$$

where the first equation describes the excitation photon fluence $\boldsymbol{\phi}_x \in \mathbb{R}^N$, and the second describes photon emission fluence $\boldsymbol{\phi}_m \in \mathbb{R}^N$; subscripts x and m indicate excitation and emission respectively. Vectors $\boldsymbol{f}, \boldsymbol{\chi} \in \mathbb{R}^N$ are the source of excitation light and target's shape indicator, i.e., a binary vector such that $\chi_i = 1$ if x_i belongs to the target and 0 otherwise. $S_{x/m}(\cdot) \in \mathbb{R}^{N \times N}$ are the stiffness matrices obtained by discretizing the diffusion terms of excitation/emission PDEs respectively and additionally S_x depends on $\boldsymbol{\chi}$. $M \in \mathbb{R}^{N \times N}$ is the mass matrix and $\odot$ denotes Hadamard (elementwise) product such that $M \odot \boldsymbol{\chi} = M \mathrm{diag}(\boldsymbol{\chi})$ and $M \odot \boldsymbol{\chi} \boldsymbol{\phi}_x = M \odot \boldsymbol{\phi}_x \boldsymbol{\chi}$. Finally, vector of measurements $\boldsymbol{y} \in \mathbb{R}^K$ is related to the emission fluence $\boldsymbol{\phi}_m$ as follows

$$\boldsymbol{y} = T \boldsymbol{\phi}_m \tag{2}$$

Here $T \in \mathbb{R}^{K \times N}$ is a binary matrix that selects components of $\boldsymbol{\phi}_m$ corresponding to the observed grid nodes and K is a number of observed nodes.

In the following if target indicator $\boldsymbol{\chi}$ is given then the system (1) is referred to as the forward FDOT problem to compute unknown excitation and emission fluence $\boldsymbol{\phi}_x$, $\boldsymbol{\phi}_m$. If vector $\boldsymbol{\chi}$ is unknown but measurements of emission fluence are present then the system (1)–(2) is referred to as the FDOT inverse problem.

Search Space Regularization. In what follows we propose an algorithm that estimates target's indicator $\boldsymbol{\chi}$ from data $\boldsymbol{y}$, i.e. solves the inverse FDOT problem. To reduce the ill-posedness of the inverse problem (1)–(2) we introduce several regularization schemes. These regularizations describe prior knowledge about the desired solution $\boldsymbol{\chi}$ and thus reduce the search space of admissible targets.

The first regularization represents an assumption that the correct $\boldsymbol{\chi}$ is a binary vector. Since binary constraints are not convex, we adopt a more relaxed condition on $\boldsymbol{\chi}$ referred to as the *box constraints*: $0 \leq \boldsymbol{\chi} \leq 1$.

The second regularization describes the piece-wise constant structure of the indicator $\boldsymbol{\chi}$, and is referred to as the *piece-wise total variation* (PTV). It is obtained by extending the notion of total variation which has been successfully applied in optical tomography. To this end, assume $m(j)$, $n(j)$, and $j \in \mathcal{I}$ are indices corresponding to the j-th pair of neighboring nodes. Let the domain Ω be split into N_{ptv} non-overlapping subdomains, e.g., cuboids, and the index $\mathcal{I}$ is correspondingly split into non-overlapping sub-indices $\mathcal{I}_i$, $i = 1, \ldots, N_{ptv}$ of nodes pairs that belong to Ω_i. PTV is obtained as a sum of total variations computed using sub-indices $\mathcal{I}_i$:

$$v(\boldsymbol{\chi}) = \sum_{i=1}^{N_{ptv}} \sum_{j \in \mathcal{I}_i} |\chi_{m(j)} - \chi_{n(j)}| = \|V \boldsymbol{\chi}\|_1 \tag{3}$$

and is also written in a matrix form assuming matrix V encodes subtraction across node pairs across all sub-indices.

The third regularization aims to reduce a null space of the inverse problem in the boundary layer of a thickness ϵ, reflecting the assumption that the target is under the surface. It is referred to as the *boundary regularization* and is defined as $W\boldsymbol{\chi} = 0$ where W selects components of $\boldsymbol{\chi}$ that belong to the boundary layer.

Finally, the fourth regularization referred to as the *minimum volume regularization* requires that $\boldsymbol{\chi}$ has at least m_0 non-zero components: $\mathbb{1}^T\boldsymbol{\chi} \geq m_0$.

Optimization Framework. In this subsection we present an *Incremental Fluorescent Target Reconstruction* (IFTR) scheme solving the inverse problem (1)–(2). Noting that the nonlinearity of the inverse problem stems from the fact that $\boldsymbol{\chi} \odot \boldsymbol{\phi}_x$ is a bi-linear vector function the IFTR scheme employs the following splitting method: (i) for $n = 0, 1, \ldots$ fix $\boldsymbol{\chi}^n$ and compute $\boldsymbol{\phi}_x^n$ as the unique solution of linear excitation equation:

$$\left(S_x(\boldsymbol{\chi}^n) + M \odot \boldsymbol{\chi}^n\right)\boldsymbol{\phi}_x^n = \boldsymbol{f} \tag{4}$$

then (ii) fix the obtained $\boldsymbol{\phi}_x^n$ and compute $\boldsymbol{\chi}^{n+1}$ as the unique solution of one of the 3 convex optimization problems:

Variant I. This variant relies upon direct inversion of the emission equation matrix S_m^{-1} to find $\boldsymbol{\chi}^{n+1}$:

$$\boldsymbol{\chi}^{n+1} = \underset{\boldsymbol{\chi}}{\mathrm{argmin}} \quad \|\boldsymbol{y} - TS_m^{-1}M \odot \boldsymbol{\phi}_x^n\boldsymbol{\chi}\|_2^p/\|\boldsymbol{y}\|_2^p + \|V\boldsymbol{\chi}\|_1$$
$$\text{s.t.} \quad W\boldsymbol{\chi} = 0, \ \mathbb{1}^T\boldsymbol{\chi} \geq m_0, \ 0 \leq \boldsymbol{\chi} \leq 1 \tag{5}$$

Variant II. This variant imposes the emission equation as an inequality constraint:

$$\boldsymbol{\chi}^{n+1}, \boldsymbol{\phi}_m^{n+1} = \underset{\boldsymbol{\chi}, \boldsymbol{\phi}_m}{\mathrm{argmin}} \quad \|\boldsymbol{y} - T\boldsymbol{\phi}_m\|_2^p/\|\boldsymbol{y}\|_2^p + \|V\boldsymbol{\chi}\|_1$$
$$\text{s.t.} \quad \langle S_m\boldsymbol{\phi}_m - M \odot \boldsymbol{\phi}_x^n\boldsymbol{\chi}\rangle_p \leq E_m \tag{6}$$
$$W\boldsymbol{\chi} = 0, \ \mathbb{1}^T\boldsymbol{\chi} \geq m_0, \ 0 \leq \boldsymbol{\chi} \leq 1$$

Here, depending on the value of p we take $\langle\boldsymbol{x}\rangle_1 = \|\boldsymbol{x}\|_2$ or $\langle\boldsymbol{x}\rangle_2 = \mathbb{1}^T\boldsymbol{x}$ and E_m is a parameter defining emission equation constraint tolerance.

Variant III. This variant uses the emission equation as a term of the loss function:

$$\boldsymbol{\chi}^{n+1}, \boldsymbol{\phi}_m^{n+1} = \underset{\boldsymbol{\chi}, \boldsymbol{\phi}_m}{\mathrm{argmin}} \quad \|\boldsymbol{y} - T\boldsymbol{\phi}_m\|_2^p/\|\boldsymbol{y}\|_2^p + \|S_m\boldsymbol{\phi}_m - M \odot \boldsymbol{\phi}_x^n\boldsymbol{\chi}\|_2^p + \|V\boldsymbol{\chi}\|_1$$
$$\text{s.t.} \quad W\boldsymbol{\chi} = 0, \ \mathbb{1}^T\boldsymbol{\chi} \geq m_0, \ 0 \leq \boldsymbol{\chi} \leq 1 \tag{7}$$

We note that all the three variants depend on parameter $p = 1, 2$ which defines the type of optimization problem that should be solved: i) if $p = 1$ we get conic optimization problems of the loss function in the form $\|\cdot\|_2$ which would be treated as conic constraints); ii) if $p = 2$ we get quadratic optimization problems.

To get a good initial guess for $\boldsymbol{\chi}^0$ we borrow from the *Born approximation* which suggests that excitation field $\boldsymbol{\phi}_x$ can be approximated by the background excitation obtained by solving excitation equation with no ICG, i.e., $\boldsymbol{\chi}^0 = 0$.

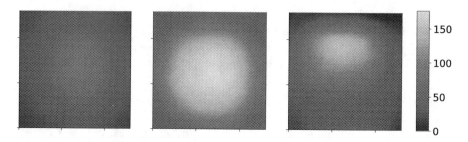

Fig. 3. Emission fluence measured during the laboratory experiment collected on: the top surface from 6 mm deep target (left panel), the top surface from 3 mm deep target (middle panel), the side face $x = 0$ from 3 mm deep target (right panel) and passed through the median filter.

Iterating this splitting method for $n = 0, 1, 2, \dots$ we obtain a sequence of updates χ^n that converge into a vicinity of the true χ provided data is "representative enough". We conclude the presentation of IFTR scheme with stopping criteria of the iterative process. For this we use a standard Dice coefficient $d(\cdot, \cdot)$ and a binary projector $b(\cdot)$: the scheme stops once the following condition is met

$$d(b(\chi^{n-1}), b(\chi^n)) = 1, \quad [b(\boldsymbol{x})]_i = \begin{cases} 1, & \text{if } \boldsymbol{x}_i \geq 0.5 \\ 0, & \text{if } \boldsymbol{x}_i < 0.5 \end{cases}, \quad d(\boldsymbol{a}, \boldsymbol{b}) = \frac{2|\boldsymbol{a} \cap \boldsymbol{b}|}{|\boldsymbol{a}| + |\boldsymbol{b}|} \quad (8)$$

3   Data Collection

To validate the IFTR scheme, we performed an experiment capturing the essential elements of FGS applications. Figure 1 describes the experiment setup. The tissue phantom was composed of a $13 \times 13 \times 30$ mm (inner dimensions) glass box filled with a 1% liposyn solution, which is a fat emulsion with scattering absorption properties mimicking human soft tissue [12]. The fluorescent target used was a $8 \times 8 \times 8$ mm (inner dimensions) acrylic spectrophotometry cuvette filled with a 5% BSA, 1% liposyn, 7μM ICG solution.

Figure 2 depicts the imaging system consisting of relatively inexpensive components: a Raspberry Pi Computer (4B/2GB), 12MP RGB camera (Raspberry Pi, SC0261) with IR filter removed, 16 mm telephoto lens (Raspberry Pi, SC0123), 700–800 nm band stop filter(Midwest Optical Sytems, DB850-25.4), 785 nm laser (Roithner Lasertechnik, RLTMDL-785-300-5) as the excitation source, and a polyscope fiber bundle (PolyDiagnost, PD-PS-0095). The detector and lens are approximately €100 combined, with just 8 bits of dynamic range. This is in stark contrast to the ultra-sensitive 16-bit scientific cameras priced an order of magnitude more used in other studies.

We collected 3 sets of experimental measurements (see Fig. 3): i) $\boldsymbol{y}_{top}^{6\,\text{mm}}$ is emission fluence collected on the top surface from the target immersed 6 mm under the top surface of tissue phantom; ii) $\boldsymbol{y}_{top}^{3\,\text{mm}}$ is emission fluence from the target immersed 3 mm under the top surface; and iii) $\boldsymbol{y}_{top\&side}^{3\,\text{mm}}$ is also emission

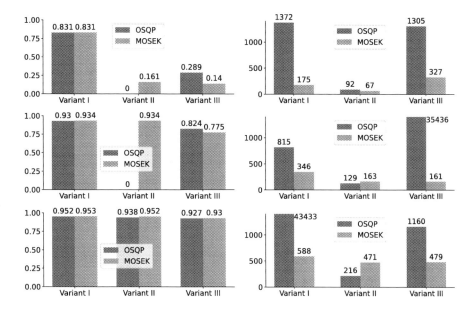

Fig. 4. Dice coefficient (left column) and execution time (right column) of experiments with 3 sets of boundary measurements: 1) $\boldsymbol{y}_{top}^{6\,mm}$ (top row); 2) $\boldsymbol{y}_{top}^{3\,mm}$ (middle row) and 3) $\boldsymbol{y}_{top\&side}^{3\,mm}$ (bottom row).

fluence from the 3 mm deep target which additionally contains measurements taken on a side face ($x = 0$) and then reflected to the other 3 side faces. The raw data was median filtered and re-scaled since the camera we used does not measure fluence in physical units. The re-scaling factor was found by taking the largest value of per pixel division of $\boldsymbol{y}_{top}^{3\,mm}$ image by the emission fluence computed as a solution of forward FDOT problem with known true target $\boldsymbol{\chi}_{true}$ and taken at the top face. It was computed to be 748.4.

4   Experimental Validation

Performance of the proposed IFTR scheme is characterized by a set of numerical experiments. IFTR scheme was implemented using the FEniCS package for FEM matrices computation and CVXPY for the construction of the loss functions and constraints. Additionally CVXPY provides a common interface to various state-of-the-art optimization solvers making it very easy to switch between them. Although we tested IFTR with 4 commonly used solvers: OSQP, SCS and ECOS (distributed together with CVXPY) and MOSEK (required an additional installation) we report results for the solvers that performed the best. Thus, for the quadratic optimization formulation we selected OSQP and for the conic optimization formulation we selected MOSEK. The resulting configurations were compared in terms of Dice coefficient $d(\boldsymbol{\chi}_{est}, \boldsymbol{\chi}_{true})$ comparing estimated target and the true target as well as execution time.

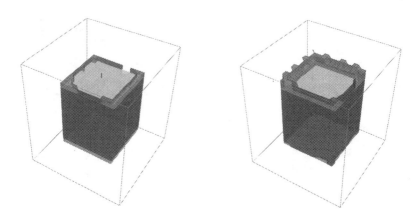

Fig. 5. Reconstructed targets (green) by: IFTR variant I with quadratic OSQP solver from $\boldsymbol{y}_{top}^{6\,mm}$ (left panel); and IFTR variant II with conic MOSEK solver from $\boldsymbol{y}_{top}^{3\,mm}$ (right panel) plotted over the true target (white). (Color figure online)

The results of the performed experiments are summarised in Fig. 4 where the left column of panels presents Dice coefficients and the right column of panels presents respective execution time. Each row of panels in Fig. 4 corresponds (from the top to the bottom) to the experiment using one of three measurements vectors: 1) $\boldsymbol{y}_{top}^{6\,mm}$; 2) $\boldsymbol{y}_{top}^{3\,mm}$ and 3) $\boldsymbol{y}_{top\&side}^{3\,mm}$.

The first experiment is the most challenging as it recovers the target 6 mm deep under the surface. Yet, variant I of IFTR obtains good reconstruction with both quadratic OSQP and conic MOSEK solvers for which Dice score reaches value of 0.831. Good quality reconstruction is indeed confirmed on the left panel in Fig. 5 depicting target recovered by IFTR variant I with quadratic OSQP solver and plotted over the true target. The second experiment recovers the target 3 mm deep which is easier and thus more IFTR variants are capable of obtaining good reconstructions as suggested on the left panel in the middle row in Fig. 4. Additionally, the right panel in Fig. 5 depicts the target reconstructed by IFTR variant II with conic MOSEK solver. We stress that both of these experiments employ reflectance-mode measurements suggesting the proposed IFTR scheme is promising for adoption in FGS-related applications.

The third experiment demonstrates the consistency of IFTR scheme: adding side measurements allows all variants to obtain good reconstructions and further increases Dice coefficients. This, however, comes at a price of increased computational demands, particularly for variant I solved with quadratic OSQP solver. The performed experiments reveal that it is difficult to pick a single winning configuration of IFTR scheme but there are several considerations: i) variant I provides the lowest errors but is the slowest variant with MOSEK solver has been consistently faster than OSQP; ii) variant II is the fastest variant but it is more sensitive to the amount of measurement compared to others; iii) variant III is less sensitive to the amount of measurements compared to Variant II has similar execution time but is less accurate.

We also note that IFTR scheme is robust with respect to PTV regularization parameter. This was achieved by scaling the data misfit and PTV term to similar magnitude: we normalised the misfit term by the norm of the observations vector and rescaled PTV term by the number of subdomains and each local total variation weight by the number of nodes in that subdomain. The robustness to regularization parameter choice was confirmed by our experiments with several different values of such parameter. Another relevant consideration is that PTV impacts the loss function in a different way compared to a standard L1 or L2 regularization: the latter has the unique global minimizer (0-vector) while the former has many global minimizers and IFTR benefits from this.

5   Conclusions

In this work we proposed novel IFTR scheme for solving FDOT problem. It performs a splitting of the bi-linearity of the original non-convex problem into a sequence of convex ones. Additionally, IFTR restricts the search space by a set of regularizers promoting piece-wise constant structure of target's indicator function which in turn allows to recover fluorescent targets from only the reflectance mode CW measurements collected by a consumer grade camera.

Although the scheme was tested using proof-of-concept experimental data and cubical shape target the method is general and depending on mesh discretization level, scalable to arbitrary domain and target shapes. Thus, the obtained results suggest strong potential for adoption of IFTR scheme in FGS related applications.

A   Continuous formulation of the FDOT problem

Near-infrared photon propagation in tissue like media is described by the following coupled system of PDEs:

$$-\nabla \cdot (D_x \nabla \phi_x) + k_x \phi_x = 0 \tag{9}$$

$$-\nabla \cdot (D_m \nabla \phi_m) + k_m \phi_m = \Gamma \mu_{axf} \phi_x \tag{10}$$

Here (9) is the excitation equation describing the excitation (at wavelength 785 nm) photon fluence $\phi_x, W/\text{cm}^2$, and (10) – emission equation describing emission (at wavelength 830 nm) photon fluence $\phi_m, W/\text{cm}^2$, subscripts x and m indicate excitation and emission respectively. The parameters of those equations are taken according to the laboratory experiment setup. $\Gamma = 0.016$ is a constant representing the dimensionless quantum efficiency of ICG fluorescence emission. $D_{x/m}, \text{cm}$ and $k_{x/m}, \text{cm}^{-1}$ refer to coefficients in excitation and emission equations, which determine light scattering and absorption properties of tissues:

$$D_{x/m} = \frac{1}{3\left(\mu_{ax/mi} + \mu_{ax/mf} + \mu'_{sx/m}\right)}, \qquad k_{x/m} = \mu_{ax/mi} + \mu_{ax/mf} \tag{11}$$

where $\mu'_{sx} = 9.84\,\mathrm{cm}^{-1}$ and $\mu'_{sm} = 9.84\,\mathrm{cm}^{-1}$ are the scattering coefficients of Liposyn at excitation and emission wavelength respectively; $\mu_{axi} = 0.023\,\mathrm{cm}^{-1}$ and $\mu_{ami} = 0.0289\,\mathrm{cm}^{-1}$ are the absorption coefficients of Liposyn at excitation and emission wavelengths; $\mu_{axf} = \mu_{ICG}\chi$, cm^{-1} is the absorption coefficient of the unknown ICG-tagged target and thus depends on the target's shape modelled by an indicator function χ and ICG absorption coefficient at excitation wavelength $\mu_{ICG} = 3.5\,\mathrm{cm}^{-1}$; $\mu_{amf} = 0\,\mathrm{cm}^{-1}$ as we assume there is no self-absorption of ICG fluorescence emission at the concentration ranges employed in this work and for practical applications [12].

The system (9)–(10) is complemented by Robin-type boundary conditions modelling the excitation source applied at the surface of the domain Ω:

$$\gamma\phi_x + 2D_x\frac{\partial\phi_x}{\partial\mathbf{n}} + S = 0 \tag{12}$$

$$\gamma\phi_m + 2D_m\frac{\partial\phi_m}{\partial\mathbf{n}} = 0 \tag{13}$$

where $\gamma = 2.5156$ – dimensionless constant depending on the optical reflective index mismatch at the boundary.

References

1. Abascal, J.J., et al.: Fluorescence diffuse optical tomography using the split Bregman method. Med. Phys. **38**(11), 6275–6284 (2011)
2. Agrawal, A., Verschueren, R., Diamond, S., Boyd, S.: A rewriting system for convex optimization problems. J. Control Decis. **5**(1), 42–60 (2018)
3. Alnaes, M.S., et al.: The FEniCS project version 1.5. Arch. Numer. Softw. **3**, 1–15 (2015)
4. Arridge, S.R., Schotland, J.C.: Optical tomography: forward and inverse problems. Inverse Prob. **25**(12), 123010 (2009)
5. Cahill, R.A., et al.: Artificial intelligence indocyanine green (ICG) perfusion for colorectal cancer intra-operative tissue classification. Br. J. Surg. **108**(1), 5–9 (2021)
6. Cho, S.S., Salinas, R., Lee, J.Y.: Indocyanine-green for fluorescence-guided surgery of brain tumors: evidence, techniques, and practical experience. Front. Surg. **6**, 11 (2019)
7. Diamond, S., Boyd, S.: CVXPY: a python-embedded modeling language for convex optimization. J. Mach. Learn. Res. **17**(83), 1–5 (2016)
8. Epperlein, J., et al.: Practical perfusion quantification in multispectral endoscopic video: Using the minutes after ICG administration to assess tissue pathology. In: AMIA Annual Symposium Proceedings (2021)
9. Graves, E., Weissleder, R., Ntziachristos, V.: Fluorescence molecular imaging of small animal tumor models. Curr. Mol. Med. **4**(4), 419–430 (2004)
10. Hoshi, Y., Yamada, Y.: Overview of diffuse optical tomography and its clinical applications. J. Biomed. Opt. **21**(9), 091312–091312 (2016)
11. Houston, J.P., Thompson, A.B., Gurfinkel, M., Sevick-Muraca, E.M.: Sensitivity and depth penetration of continuous wave versus frequency-domain photon migration near-infrared fluorescence contrast-enhanced imaging. Photochem. Photobiol. **77**(4), 420–430 (2003)

12. Joshi, A., Bangerth, W., Hwang, K., Rasmussen, J.C., Sevick-Muraca, E.M.: Fully adaptive fem based fluorescence optical tomography from time-dependent measurements with area illumination and detection. Med. Phys. **33**(5), 1299–1310 (2006). https://doi.org/10.1118/1.2190330, https://aapm.onlinelibrary.wiley.com/doi/abs/10.1118/1.2190330
13. Langtangen, H.P., Logg, A.: Solving PDEs in Python. Springer, Cham (2016). https://doi.org/10.1007/978-3-319-52462-7
14. Leblond, F., Davis, S.C., Valdés, P.A., Pogue, B.W.: Pre-clinical whole-body fluorescence imaging: review of instruments, methods and applications. J. Photochem. Photobiol. B **98**(1), 77–94 (2010)
15. Low, P.S., Singhal, S., Srinivasarao, M.: Fluorescence-guided surgery of cancer: applications, tools and perspectives. Curr. Opin. Chem. Biol. **45**, 64–72 (2018)
16. Nagaya, T., Nakamura, Y.A., Choyke, P.L., Kobayashi, H.: Fluorescence-guided surgery. Front. Oncol. **7**, 314 (2017)
17. Shafiee, S., et al.: Dynamic NIR fluorescence imaging and machine learning framework for stratifying high vs low notch-dll4 expressing host microenvironment in triple-negative breast cancer. Cancers **15**(5), 1460 (2023)
18. Stuker, F., Ripoll, J., Rudin, M.: Fluorescence molecular tomography: principles and potential for pharmaceutical research. Pharmaceutics **3**(2), 229–274 (2011)
19. Yamada, Y., Okawa, S.: Diffuse optical tomography: present status and its future. Opt. Rev. **21**(3), 185–205 (2014)
20. Zhuk, S., et al.: Perfusion quantification from endoscopic videos: learning to read tumor signatures. In: Martel, A.L., et al. (eds.) MICCAI 2020. LNCS, vol. 12263, pp. 711–721. Springer, Cham (2020). https://doi.org/10.1007/978-3-030-59716-0_68

Solving Low-Dose CT Reconstruction via GAN with Local Coherence

Wenjie Liu[ID] and Hu Ding[✉][ID]

School of Computer Science and Technology, University of Science and Technology of China, Hefei, Anhui, China
lwj1217@mail.ustc.edu.cn, huding@ustc.edu.cn

Abstract. The Computed Tomography (CT) for diagnosis of lesions in human internal organs is one of the most fundamental topics in medical imaging. Low-dose CT, which offers reduced radiation exposure, is preferred over standard-dose CT, and therefore its reconstruction approaches have been extensively studied. However, current low-dose CT reconstruction techniques mainly rely on model-based methods or deep-learning-based techniques, which often ignore the coherence and smoothness for sequential CT slices. To address this issue, we propose a novel approach using generative adversarial networks (GANs) with enhanced local coherence. The proposed method can capture the local coherence of adjacent images by optical flow, which yields significant improvements in the precision and stability of the constructed images. We evaluate our proposed method on real datasets and the experimental results suggest that it can outperform existing state-of-the-art reconstruction approaches significantly.

Keywords: CT reconstruction · Low-dose · Generative adversarial networks · Local coherence · Optical flow

1 Introduction

Computed Tomography (CT) is one of the most widely used technologies in medical imaging, which can assist doctors for diagnosing the lesions in human internal organs. Due to harmful radiation exposure of standard-dose CT, the low dose CT is more preferable in clinical application [4,6,34]. However, when the dose is low together with the issues like sparse-view or limited angles, it becomes quite challenging to reconstruct high-quality CT images. The high-quality CT images are important to improve the performance of diagnosis in clinic [27]. In mathematics, we model the CT imaging as the following procedure

$$\mathbf{y} = \mathcal{T}(\mathbf{x^r}) + \delta, \tag{1}$$

Supplementary Information The online version contains supplementary material available at https://doi.org/10.1007/978-3-031-43999-5_50.

where $\mathbf{x^r} \in \mathbb{R}^d$ denotes the **unknown** ground-truth picture, $\mathbf{y} \in \mathbb{R}^m$ denotes the received measurement, and δ is the noise. The function $\mathcal{T}$ represents the forward operator that is analogous to the Radon transform, which is widely used in medical imaging [23,28]. The problem of CT reconstruction is to recover $\mathbf{x^r}$ from the received $\mathbf{y}$.

Solving the inverse problem of (1) is often very challenging if there is no any additional information. If the forward operator $\mathcal{T}$ is well-posed and δ is neglectable, we know that an approximate $\mathbf{x^r}$ can be easily obtained by directly computing $\mathcal{T}^{-1}(\mathbf{y})$. However, $\mathcal{T}$ is often ill-posed, which means the inverse function $\mathcal{T}^{-1}$ does not exist and the inverse problem of (1) may have multiple solutions. Moreover, when the CT imaging is low-dose, the filter backward projection (FBP) [11] can produce serious detrimental artifact. Therefore, most of existing approaches usually incorporate some prior knowledge during the reconstruction [14,17,26]. For example, a commonly used method is based on regularization:

$$\mathbf{x} = argmin_{\mathbf{x}} \|\mathcal{T}(\mathbf{x}) - \mathbf{y}\|_p + \lambda \mathcal{R}(\mathbf{x}), \tag{2}$$

where $\| \cdot \|_p$ denotes the p-norm and $\mathcal{R}(\mathbf{x})$ denotes the penalty item from some prior knowledge.

In the past years, a number of methods have been proposed for designing the regularization $\mathcal{R}$. The traditional model-based algorithms, e.g., the ones using total variation [3,26], usually apply the sparse gradient assumptions and run an iterative algorithm to learn the regularizers [12,18,24,29]. Another popular line for learning the regularizers comes from deep learning [13,17]; the advantage of the deep learning methods is that they can achieve an end-to-end recovery of the true image $\mathbf{x^r}$ from the measurement $\mathbf{y}$ [1,21]. Recent researches reveal that convolutional neural networks (CNNs) are quite effective for image denoising, e.g., the CNN based algorithms [10,34] can directly learn the reconstructed mapping from initial measurement reconstructions (e.g., FBP) to the ground-truth images. The dual-domain network that combines the sinograms with reconstructed low-dose CT images were also proposed to enhance the generalizability [15,30].

A major drawback of the aforementioned reconstruction methods is that they deal with the input CT 2D slices independently (note that the goal of CT reconstruction is to build the 3D model of the organ). Namely, the neighborhood correlations among the 2D slices are often ignored, which may affect the reconstruction performance in practice. In the field of computer vision, "optical flow" is a common technique for tracking the motion of object between consecutive frames, which has been applied to many different tasks like video generation [35], prediction of next frames [22] and super resolution synthesis [5,31]. To estimate the optical flow field, existing approaches include the traditional brightness gradient methods [2] and the deep networks [7]. The idea of optical flow has also been used for tracking the organs movement in medical imaging [16,20,33]. However, to the best of our knowledge, there is no work considering GANs with using optical flow to capture neighbor slices coherence for low dose 3D CT reconstruction.

In this paper, we propose a novel optical flow based generative adversarial network for 3D CT reconstruction. Our intuition is as follows. When a patient is located in a CT equipment, a set of consecutive cross-sectional images are generated. If the vertical axial sampling space of transverse planes is small, the corresponding CT slices should be highly similar. So we apply optical flow, though there exist several technical issues waiting to solve for the design and implementation, to capture the local coherence of adjacent CT images for reducing the artifacts in low-dose CT reconstruction. Our contributions are summarized below:

1. We introduce the "local coherence" by characterizing the correlation of consecutive CT images, which plays a key role for suppressing the artifacts.
2. Together with the local coherence, our proposed generative adversarial networks (GANs) can yield significant improvement for texture quality and stability of the reconstructed images.
3. To illustrate the efficiency of our proposed approach, we conduct rigorous experiments on several real clinical datasets; the experimental results reveal the advantages of our approach over several state-of-the-art CT reconstruction methods.

2   Preliminaries

In this section, we briefly review the framework of the ordinary generative adversarial network, and also introduce the local coherence of CT slices.

Generative Adversarial Network. Traditional generative adversarial network [8] consists of two main modules, a generator and a discriminator. The generator $\mathcal{G}$ is a mapping from a latent-space Gaussian distribution $\mathbb{P}_Z$ to the synthetic sample distribution $\mathbb{P}_{X_G}$, which is expected to be close to the real sample distribution $\mathbb{P}_X$. On the other hand, the discriminator $\mathcal{D}$ aims to maximize the distance between the distributions $\mathbb{P}_{X_G}$ and $\mathbb{P}_X$. The game between the generator and discriminator actually is an adversarial process, where the overall optimization objective follows a min-max principle:

$$\min_{\mathcal{G}} \max_{\mathcal{D}} \mathbb{E}_{\mathbf{x^r} \sim \mathbb{P}_X, \mathbf{z} \sim \mathbb{P}_Z} (\log(\mathcal{D}(\mathbf{x^r})) + \log(1 - \mathcal{D}(\mathcal{G}(\mathbf{z})))). \tag{3}$$

Local Coherence. As mentioned in Sect. 1, optical flow can capture the temporal coherence of object movements, which plays a crucial role in many video-related tasks. More specifically, the optical flow refers to the instantaneous velocity of pixels of moving objects on consecutive frames over a short period of time [2]. The main idea relies on the practical assumptions that the brightness of the object more likely remains stable across consecutive frames, and the brightness of the pixels in a local region are usually changed consistently [9].

Based on these assumptions, the brightness of optical flow can be described by the following equation:

$$\nabla I_w \cdot v_w + \nabla I_h \cdot v_h + \nabla I_t = 0, \tag{4}$$

where $v = (v_w, v_h)$ represents the optical flow of the position (w, h) in the image. $\nabla I = (\nabla I_w; \nabla I_h)$ denotes spatial gradients of image brightness, and ∇I_t denotes the temporal partial derivative of the corresponding region.

Following the Eq. (4), we consider the question that whether the optical flow idea can be applied to 3D CT reconstruction. In practice, the brightness of adjacent CT images often has very tiny difference, due to the inherent continuity and structural integrity of human body. Therefore, we introduce the "local coherence" that indicates the correlation between adjacent images of a tissue. Namely, adjacent CT images often exhibit significant similarities within a certain local range along the vertical axis of the human body. Due to the local coherence, the noticeable variations observed in CT slices within the local range often occur at the edges of organs. We can substitute the temporal partial derivative ∇I_t by the vertical axial partial derivative ∇I_z in the Eq. (4), where "z" indicates the index of the vertical axis. As illustrated in Fig. 1, the local coherence can be captured by the optical flow between adjacent CT slices.

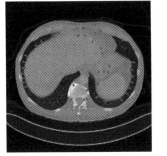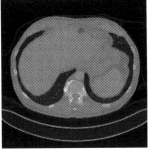

Fig. 1. The optical flow between two adjacent CT slices. The scanning window of X-ray slides from the position of the left image to the position of the right image. The directions and lengths of the red arrows represent the optical flow field. The left and right images share the local coherence and thus the optical flows are small. (Color figure online)

3   GANs with Local Coherence

In this section, we introduce our low-dose CT image generation framework with local coherence in detail.

The Framework of Our Network. The proposed framework comprises three components, including a generator $\mathcal{G}$, a discriminator $\mathcal{D}$ and an optical flow estimator $\mathcal{F}$. The generator is the core component, and the flow estimator provides auxiliary warping images for the generation process.

Suppose we have a sequence of measurements $\mathbf{y_1}, \mathbf{y_2}, \cdots, \mathbf{y_n}$; for each $\mathbf{y_i}$, $1 \leq i \leq n$, we want to reconstruct its ground truth image $\mathbf{x_i^r}$ as the Eq. (1). Before performing the reconstruction in the generator $\mathcal{G}$, we apply some prior knowledge in physics and run filter backward projection on the measurement $\mathbf{y_i}$ in Eq. (1) to obtain an initial recovery solution $\mathbf{s_i}$. Usually $\mathbf{s_i}$ contains significant noise comparing with the ground truth $\mathbf{x_i^r}$. Then the network has two input components, i.e., the initial backward projected image $\mathbf{s_i}$ that serves as an approximation of the ground truth $\mathbf{x_i^r}$, and a set of neighbor CT slices $\mathcal{N}(\mathbf{s_i}) = \{\mathbf{s_{i-1}}, \mathbf{s_{i+1}}\}^1$ for preserving the local coherence. The overall structure of our framework is shown in Fig. 2. Below, we introduce the three key parts of our framework separately.

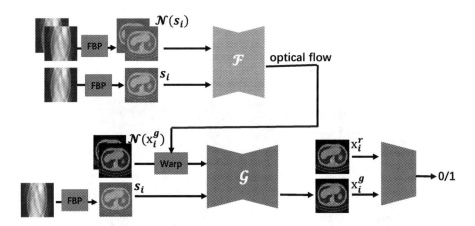

Fig. 2. The framework of our generate adversarial network with local coherence for CT reconstruction.

Optical Flow Estimator. The optical flow $\mathcal{F}(\mathcal{N}(\mathbf{s_i}), \mathbf{s_i})$ denotes the brightness changes of pixels from $\mathcal{N}(\mathbf{s_i})$ to $\mathbf{s_i}$, where it captures their local coherence. The estimator is derived by the network architecture of FlowNet [7]. The FlowNet is an autoencoder architecture with extraction of features of two input frames to learn the corresponding flow, which is consist of 6 (de)convolutional layers for both encoder and decoder.

Discriminator. The discriminator $\mathcal{D}$ assigns the label "1" to real standard-dose CT images and "0" to generated images. The goal of $\mathcal{D}$ is to maximize the separation between the distributions of real images and generated images:

---
^{1} If $i = 1$, $\mathcal{N}(\mathbf{s_i}) = \{\mathbf{s_2}\}$; if $i = n$, $\mathcal{N}(\mathbf{s_i}) = \{\mathbf{s_{n-1}}\}$.

$$\mathcal{L}_{\mathcal{D}} = \sum_{i=1}^{n} -(\log(\mathcal{D}(\mathbf{x}_i^r)) + \log(1 - \mathcal{D}(\mathbf{x}_i^g))), \qquad (5)$$

where $\mathbf{x}_i^g$ is the image generated by $\mathcal{G}$ (the formal definition for $\mathbf{x}_i^g$ will be introduced below). The discriminator includes 3 residual blocks, with 4 convolutional layers in each residual block.

Generator. We use the generator $\mathcal{G}$ to reconstruct the high-quality CT image for the ground truth $\mathbf{x}_i^r$ from the low-dose image $\mathbf{s}_i$. The generated image is obtained by

$$\begin{aligned} \mathbf{x}_i^g &= \mathcal{G}(\mathbf{s}_i, \mathcal{W}(\mathcal{N}(\mathbf{x}_i^g))); \\ \mathcal{N}(\mathbf{x}_i^g) &= \mathcal{G}(\mathcal{N}(\mathbf{s}_i)), \end{aligned} \qquad (6)$$

where $\mathcal{W}(\cdot)$ is the warping operator. Before generating $\mathbf{x}_i^g$, $\mathcal{N}(\mathbf{x}_i^g)$ is reconstructed from $\mathcal{N}(\mathbf{s}_i)$ by the generator without considering local coherence. Subsequently, according to the optical flow $\mathcal{F}(\mathcal{N}(\mathbf{s}_i), \mathbf{s}_i)$, we warp the reconstructed images $\mathcal{N}(\mathbf{x}_i^g)$ to align with the current slice by adjusting the brightness values. The warping operator $\mathcal{W}$ utilizes bi-linear interpolation to obtain $\mathcal{W}(\mathcal{N}(\mathbf{x}_i^g))$, which enables the model to capture subtle variations in the tissue from the generated $\mathcal{N}(\mathbf{x}_i^g)$; also, the warping operator can reduce the influence of artifacts for the reconstruction. Finally, $\mathbf{x}_i^g$ is generated by combining $\mathbf{s}_i$ and $\mathcal{W}(\mathcal{N}(\mathbf{x}_i^g))$. Since $\mathbf{x}_i^r$ is our target for reconstruction in the i-th batch, we consider the difference between $\mathbf{x}_i^g$ and $\mathbf{x}_i^r$ in the loss. Our generator is mainly based on the network architecture of Unet [25]. Partly inspired by the loss in [5], the optimization objective of the generator G comprises three items with the coefficients $\lambda_{\text{pix}}, \lambda_{\text{adv}}, \lambda_{\text{per}} \in (0, 1)$:

$$\mathcal{L}_{\mathcal{G}} = \lambda_{\text{pix}} \mathcal{L}_{\text{pixel}} + \lambda_{\text{adv}} \mathcal{L}_{\text{adv}} + \lambda_{\text{per}} \mathcal{L}_{\text{percept}}. \qquad (7)$$

In (7), "$\mathcal{L}_{\text{pixel}}$" is the loss measuring the pixel-wise mean square error of the generated image $\mathbf{x}_i^g$ with respect to the ground-truth $\mathbf{x}_i^r$. "$\mathcal{L}_{\text{adv}}$" represents the adversarial loss of the discriminator $\mathcal{D}$, which is designed to minimize the distance between the generated standard-dose CT image distribution $\mathbb{P}_{X_G}$ and the real standard-dose CT image distribution $\mathbb{P}_X$. "$\mathcal{L}_{\text{percept}}$" denotes the perceptual loss, which quantifies the dissimilarity between the feature maps of $\mathbf{x}_i^r$ and $\mathbf{x}_i^g$; the feature maps denote the feature representation extracted from the hidden layers in the discriminator $\mathcal{D}$ (suppose there are t hidden layers):

$$\mathcal{L}_{percept} = \sum_{i=1}^{n} \sum_{j=1}^{t} \|\mathcal{D}_j(\mathbf{x}_i^r) - \mathcal{D}_j(\mathbf{x}_i^g)\|_1 \qquad (8)$$

where $\mathcal{D}_j(\cdot)$ refers to the feature extraction performed on the j-th hidden layer. Through capturing the high frequency differences in CT images, $\mathcal{L}_{percept}$ can enhance the sharpness for edges and increase the contrast for the reconstructed images. $\mathcal{L}_{\text{pixel}}$ and $\mathcal{L}_{\text{adv}}$ are designed to recover global structure, and $\mathcal{L}_{percept}$ is utilized to incorporate additional texture details into the reconstruction process.

4   Experiment

Datasets. First, our proposed approaches are evaluated on the "Mayo-Clinic low-dose CT Grand Challenge" (Mayo-Clinic) dataset of lung CT images [19]. The dataset contains 2250 two dimensional slices from 9 patients for training, and the remaining 128 slices from 1 patient are reserved for testing. The low-dose measurements are simulated by parallel-beam X-ray with 200 (or 150) uniform views, i.e., $N_v = 200$ (or $N_v = 150$), and 400 (or 300) detectors, i.e., $N_d = 400$ (or $N_d = 300$). In order to further verify the denoising ability of our approaches, we add the Gaussian noise with standard deviation $\sigma = 2.0$ to the sinograms after X-ray projection in 50% of the experiments. To evaluate the generalization of our model, we also consider another dataset RIDER with non-small cell lung cancer under two CT scans [36] for testing. We randomly select 4 patients with 1827 slices from the dataset. The simulation process is identical to that of Mayo-Clinic. The proposed networks were implemented in the PyTorch framework and trained on Nvidia 3090 GPU with 100 epochs.

Baselines and Evaluation Metrics. We consider several existing popular algorithms for comparison. (1) **FBP** [11]: the classical filter backward projection on low-dose sinograms. (2) **FBPConvNet** [10]: a direct inversion network followed by the CNN after initial **FBP** reconstruction. (3) **LPD** [1]: a deep learning method based on proximal primal-dual optimization. (4) **UAR** [21]: an end-to-end reconstruction method based on learning unrolled reconstruction operators and adversarial regularizers. Our proposed method is denoted by **GAN-LC**. We set $\lambda_{\texttt{pix}} = 1.0, \lambda_{\texttt{adv}} = 0.01$ and $\lambda_{\texttt{per}} = 1.0$ for the optimization objective in Eq. (7) during our training process. Following most of the previous articles on 3D CT reconstruction, we evaluate the experimental performance by two metrics: the peak signal-to-noise ratio (PSNR) and the structural similarity index (SSIM) [32]. PSNR measures the pixel differences of two images, which is negatively correlated with mean square error. SSIM measures the structure similarity between two images, which is related to the variances of the input images. For both two measures, the higher the better.

Results. Table 1 presents the results on the Mayo-Clinic dataset, where the first row represents different parameter settings (i.e., the number of uniform views N_v, the number of detectors N_d and the standard deviation of Gaussian noise σ) for simulating low-dose sinograms. Our proposed approach **GAN-LC** consistently outperforms the baselines under almost all the low-dose parameter settings. The methods **FBP** and **UAR** are very sensitive to noise; the performance of **LPD** is relatively stable but with low reconstruction accuracy. **FBPConvNet** has

Table 1. Experimental results for Mayo-Clinic dataset. The value in first row of the table represents N_v, N_d and σ for simulating low-dose sinograms, respectively.

Sinograms	200, 400, 0.0		200, 400, 2.0		150, 300, 0.0		150, 300, 2.0	
	PSNR	SSIM	PSNR	SSIM	PSNR	SSIM	PSNR	SSIM
FBP	26.449	0.721	13.517	0.191	21.460	0.616	12.593	0.168
FBPConvNet	38.213	0.918	30.148	0.743	35.263	0.869	29.095	0.723
LPD	28.050	0.844	28.357	0.794	28.376	0.826	27.409	**0.801**
UAR	33.248	0.902	22.048	0.272	29.829	0.848	21.227	0.238
GAN-LC	**39.548**	**0.950**	**32.437**	**0.819**	**36.542**	**0.899**	**31.586**	0.725

Table 2. Experimental results for RIDER dataset. The value in first row of the table represents N_v, N_d and σ for simulating low-dose sinograms, respectively.

Sinograms	200, 400, 0.0		200, 400, 2.0		150, 300, 0.0		150, 300, 2.0	
	PSNR	SSIM	PSNR	SSIM	PSNR	SSIM	PSNR	SSIM
FBP	21.398	0.647	15.609	0.233	19.49	0.597	14.845	0.203
FBPConvNet	27.256	0.671	19.520	0.444	27.504	0.650	18.517	0.431
LPD	22.341	0.615	12.196	0.466	22.172	0.556	12.215	0.455
UAR	24.915	0.667	20.943	0.207	21.136	0.557	**19.873**	0.176
GAN-LC	**28.861**	**0.721**	**22.624**	**0.517**	**29.171**	**0.705**	19.607	**0.470**

a similar increasing trend with our approach across different settings but has worse reconstruction quality. To evaluate the stability and generalization of our model and the baselines trained on Mayo-Clinic dataset, we also test them on the RIDER dataset. The results are shown in Table 2. Due to the bias in the datasets collected from different facilities, the performances of all the models are declined to some extents. But our proposed approach still outperforms the other models for most testing cases.

To illustrate the reconstruction performances more clearly, we also show the reconstruction results for testing images in Fig. 3. We can see that our network can reconstruct the CT image with higher quality. Due to the space limit, the experimental results of different views N_v and more visualized results are placed in our supplementary material.

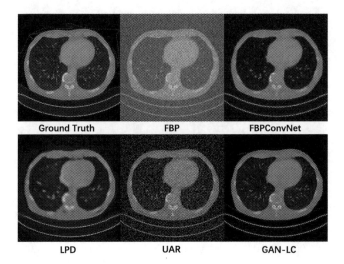

Fig. 3. Reconstruction results on Mayo-Clinic dataset. The sparse view setting of sinograms is $N_v = 200$, $N_d = 400$ and $\sigma = 2.0$. "Ground Truth" is the standard-dose CT image.

5   Conclusion

In this paper, we propose a novel approach for low-dose CT reconstruction using generative adversarial networks with local coherence. By considering the inherent continuity of human body, local coherence can be captured through optical flow, which is small deformations and structural differences between consecutive CT slices. The experimental results on real datasets demonstrate the advantages of our proposed network over several popular approaches. In future, we will evaluate our network on real-world CT images from local hospital and use the reconstructed images to support doctors for the diagnosis and recognition of lung nodules. Our code is publicly available at https://github.com/lwjie595/GANLC.

Acknowledgements. The research of this work was supported in part by National Key R&D program of China through grant 2021YFA1000900, the NSFC throught Grant 62272432, and the Provincial NSF of Anhui through grant 2208085MF163.

References

1. Adler, J., Öktem, O.: Learned primal-dual reconstruction. IEEE Trans. Med. Imaging **37**(6), 1322–1332 (2018)
2. Beauchemin, S.S., Barron, J.L.: The computation of optical flow. ACM Comput. Surv. (CSUR) **27**(3), 433–466 (1995)
3. Chambolle, A.: An algorithm for total variation minimization and applications. J. Math. Imaging Vis. **20**(1), 89–97 (2004)
4. Chen, H., et al.: Low-dose CT via convolutional neural network. Biomed. Opt. Express **8**(2), 679–694 (2017)

5. Chu, M., Xie, Y., Mayer, J., Leal-Taixé, L., Thuerey, N.: Learning temporal coherence via self-supervision for GAN-based video generation. ACM Trans. Graph. (TOG) **39**(4), 75-1 (2020)

6. Ding, Q., Nan, Y., Gao, H., Ji, H.: Deep learning with adaptive hyper-parameters for low-dose CT image reconstruction. IEEE Trans. Comput. Imaging **7**, 648–660 (2021)

7. Dosovitskiy, A., et al.: Flownet: learning optical flow with convolutional networks. In: Proceedings of the IEEE International Conference on Computer Vision, pp. 2758–2766 (2015)

8. Goodfellow, I.J., et al.: Generative adversarial networks. CoRR abs/1406.2661 (2014). https://arxiv.org/abs/1406.2661

9. Horn, B.K., Schunck, B.G.: Determining optical flow. Artif. Intell. **17**(1–3), 185–203 (1981)

10. Jin, K.H., McCann, M.T., Froustey, E., Unser, M.: Deep convolutional neural network for inverse problems in imaging. IEEE Trans. Image Process. **26**(9), 4509–4522 (2017)

11. Kak, A.C., Slaney, M.: Principles of computerized tomographic imaging. SIAM (2001)

12. Knoll, F., Bredies, K., Pock, T., Stollberger, R.: Second order total generalized variation (TGV) for MRI. Magn. Reson. Med. **65**(2), 480–491 (2011)

13. Kobler, E., Effland, A., Kunisch, K., Pock, T.: Total deep variation for linear inverse problems. In: Proceedings of the IEEE/CVF Conference on Computer Vision and Pattern Recognition, pp. 7549–7558 (2020)

14. Li, H., Schwab, J., Antholzer, S., Haltmeier, M.: NETT: solving inverse problems with deep neural networks. Inverse Prob. **36**(6), 065005 (2020)

15. Lin, W.A., et al.: Dudonet: dual domain network for CT metal artifact reduction. In: Proceedings of the IEEE/CVF Conference on Computer Vision and Pattern Recognition, pp. 10512–10521 (2019)

16. Liu, H., Lin, Y., Ibragimov, B., Zhang, C.: Low dose 4D-CT super-resolution reconstruction via inter-plane motion estimation based on optical flow. Biomed. Signal Process. Control **62**, 102085 (2020)

17. Lunz, S., Öktem, O., Schönlieb, C.B.: Adversarial regularizers in inverse problems. In: Advances in Neural Information Processing Systems, vol. 31 (2018)

18. McCann, M.T., Nilchian, M., Stampanoni, M., Unser, M.: Fast 3D reconstruction method for differential phase contrast X-ray CT. Opt. Express **24**(13), 14564–14581 (2016)

19. McCollough, C.: TU-FG-207A-04: overview of the low dose CT grand challenge. Med. Phys. **43**(6Part35), 3759–3760 (2016)

20. Mira, C., Moya-Albor, E., Escalante-Ramírez, B., Olveres, J., Brieva, J., Vallejo, E.: 3D hermite transform optical flow estimation in left ventricle CT sequences. Sensors **20**(3), 595 (2020)

21. Mukherjee, S., Carioni, M., Öktem, O., Schönlieb, C.B.: End-to-end reconstruction meets data-driven regularization for inverse problems. Adv. Neural. Inf. Process. Syst. **34**, 21413–21425 (2021)

22. Patraucean, V., Handa, A., Cipolla, R.: Spatio-temporal video autoencoder with differentiable memory. arXiv preprint arXiv:1511.06309 (2015)

23. Ramm, A.G., Katsevich, A.I.: The Radon Transform and Local Tomography. CRC Press, Boca Raton (2020)

24. Romano, Y., Elad, M., Milanfar, P.: The little engine that could: regularization by denoising (RED). SIAM J. Imag. Sci. **10**(4), 1804–1844 (2017)

25. Ronneberger, O., Fischer, P., Brox, T.: U-Net: convolutional networks for biomedical image segmentation. In: Navab, N., Hornegger, J., Wells, W.M., Frangi, A.F. (eds.) MICCAI 2015. LNCS, vol. 9351, pp. 234–241. Springer, Cham (2015). https://doi.org/10.1007/978-3-319-24574-4_28

26. Rudin, L.I., Osher, S., Fatemi, E.: Nonlinear total variation based noise removal algorithms. Physica D **60**(1–4), 259–268 (1992)

27. Sori, W.J., Feng, J., Godana, A.W., Liu, S., Gelmecha, D.J.: DFD-Net: lung cancer detection from denoised CT scan image using deep learning. Front. Comp. Sci. **15**, 1–13 (2021)

28. Toft, P.: The radon transform. Theory and Implementation (Ph.D. dissertation), Technical University of Denmark, Copenhagen (1996)

29. Venkatakrishnan, S.V., Bouman, C.A., Wohlberg, B.: Plug-and-play priors for model based reconstruction. In: 2013 IEEE Global Conference on Signal and Information Processing, pp. 945–948. IEEE (2013)

30. Wang, C., Shang, K., Zhang, H., Li, Q., Zhou, S.K.: DuDoTrans: dual-domain transformer for sparse-view CT reconstruction. In: Haq, N., Johnson, P., Maier, A., Qin, C., Würfl, T., Yoo, J. (eds.) MLMIR 2022. LNCS, vol. 13587, pp. 84–94. Springer, Cham (2022). https://doi.org/10.1007/978-3-031-17247-2_9

31. Wang, T.C., et al.: Video-to-video synthesis. arXiv preprint arXiv:1808.06601 (2018)

32. Wang, Z., Bovik, A.C., Sheikh, H.R., Simoncelli, E.P.: Image quality assessment: from error visibility to structural similarity. IEEE Trans. Image Process. **13**(4), 600–612 (2004)

33. Weng, N., Yang, Y.H., Pierson, R.: Three-dimensional surface reconstruction using optical flow for medical imaging. IEEE Trans. Med. Imaging **16**(5), 630–641 (1997)

34. Wolterink, J.M., Leiner, T., Viergever, M.A., Išgum, I.: Generative adversarial networks for noise reduction in low-dose CT. IEEE Trans. Med. Imaging **36**(12), 2536–2545 (2017)

35. Xue, T., Wu, J., Bouman, K., Freeman, B.: Visual dynamics: probabilistic future frame synthesis via cross convolutional networks. In: Advances in Neural Information Processing Systems, vol. 29 (2016)

36. Zhao, B., et al.: Evaluating variability in tumor measurements from same-day repeat CT scans of patients with non-small cell lung cancer. Radiology **252**(1), 263–272 (2009)

Image Registration

Co-learning Semantic-Aware Unsupervised Segmentation for Pathological Image Registration

Yang Liu and Shi Gu[✉]

University of Electronic Science and Technology of China, Chengdu, China
gus@uestc.edu.cn

Abstract. The registration of pathological images plays an important role in medical applications. Despite its significance, most researchers in this field primarily focus on the registration of normal tissue into normal tissue. The negative impact of focal tissue, such as the loss of spatial correspondence information and the abnormal distortion of tissue, are rarely considered. In this paper, we propose a novel unsupervised approach for pathological image registration by incorporating segmentation and inpainting. The registration, segmentation, and inpainting modules are trained simultaneously in a co-learning manner so that the segmentation of the focal area and the registration of inpainted pairs can improve collaboratively. Overall, the registration of pathological images is achieved in a completely unsupervised learning framework. Experimental results on multiple datasets, including Magnetic Resonance Imaging (MRI) of T1 sequences, demonstrate the efficacy of our proposed method. Our results show that our method can accurately achieve the registration of pathological images and identify lesions even in challenging imaging modalities. Our unsupervised approach offers a promising solution for the efficient and cost-effective registration of pathological images. Our code is available at https://github.com/brain-intelligence-lab/GIRNet.

Keywords: Unsupervised · Collaborative Learning · Registration · Segmentation · Pathological Image

1 Introduction

Image registration has been widely studied in both academia and industry over the past two decades. In general, the goal of deformable image registration is to estimate a suitable nonlinear transformation that overlaps the pair of images with corresponding spatial relationships [4, 22]. This goal is usually achieved by minimizing a well-defined similarity score. However, these methods often assume that there is no spatial non-correspondence between the two images. In the field

Supplementary Information The online version contains supplementary material available at https://doi.org/10.1007/978-3-031-43999-5_51.

of medical image analysis, this assumption is often not valid, particularly in cases such as pathology image to atlas registration or pre-operative and post-operative longitudinal registration. Direct registration of pathology images without taking into account the impact of focal tissue can result in missed pixel-level correspondence and large registration errors.

A variety of approaches have been proposed to handle the non-correspondence problem in medical image registration. These methods can be roughly divided into three main categories: *1) Cost function masking.* The authors of [5,17] used the segmentation of the non-corresponding regions to mask the image similarity measure in optimization. *2) Converting pathological image to normal appearance.* This class of approaches aims to replace or reconstruct the focal area as normal tissue to guide the registration either through low-rank and sparse image decomposition [10,11] or generative models [23]. *3) Non-correspondence detection via intensity criteria.* This category of methods can be formulated as joint segmentation and registration to detect non-corresponding regions during the registration process [6,7]. Although these approaches partially handle the issue of non-correspondence in the registration, they still have some serious shortcomings. The cost function masking and image conversion approaches require ground truth or accurate labels during registration and may decrease the alignment accuracy when the focal area is large. The non-correspondence detection approach, which typically relies on a sophisticated designed loss function, is very sensitive to the dataset [1] and difficult to find a set of unified parameters.

Therefore, to effectively address the non-correspondence problem in registering pathology images, it is necessary to incorporate both a data-independent segmentation module and a modality-adaptive inpainting module into the registration pipeline. To bridge this gap, we introduce the semantic information of the category based on [21,24]. It employs the non-correspondence in registration to achieve accurate segmentation of the lesion region and uses the segmented mask to reconstruct the lesion area and guide the registration. In this paper, we address the challenge of large alignment errors due to the loss of spatial correspondence in processing pathological images. To overcome this challenge, we propose **GIRNet**, a tri-net collaborative learning framework that simultaneously updates the segmentation, inpainting, and registration networks. The segmentation network minimizes the mutual information between the lesion and normal tissue based on the semantic information introduced by the registration network, allowing for accurate segmentation of regions with missing spatial correspondence. The registration network, in turn, weakens the adverse effects of the lesions based on the mask generated by the segmentation network. To the best of our knowledge, this is the first work to apply an unsupervised segmentation method based on minimal mutual information (**MMI**) to pathological image registration, with simultaneous training of segmentation and registration. Our work makes the following key contributions.

- We propose a collaborative learning method for the simultaneous optimization of registration, segmentation, and inpainting networks.

- We show the effectiveness of using mutual information minimization in an unsupervised manner for pathological image segmentation and registration by incorporating semantic information through the registration process.
- We perform a series of experiments to validate our method's superiority in accurately finding lesions and effectively registering pathological images.

2   Method

Our proposed framework (Fig. 1) involves three modules: a register denoted by ψ, a segmenter denoted by θ, and an inpainter denoted by ϕ. The three modules are trained in a co-learning manner to enable the registration aware of semantic information. Importantly, our proposed training procedure is fully unsupervised which does not require any labeled data for training the network.

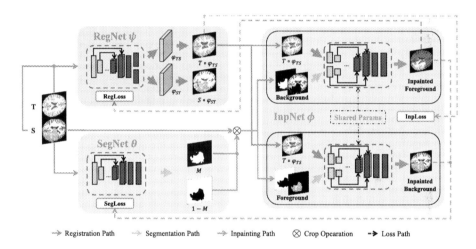

Fig. 1. The proposed tri-modules collaborative learning framework for medical image analysis includes RegNet, SegNet, and InpNet to achieve accurate image registration and segmentation through the optimization of semantic-informed mutual information.

2.1   Collaborative Optimization

The most critical problem in pathological image registration is identifying and dealing with the lesion area. If we naively register a source pathological image S to a template T without caring about the lesion boundary, the deformation field near the boundary would be uncontrollable because a healthy template does not have a lesion. A possible approach here is to initialize an inflating boundary containing the lesion area, followed by calculating the registration loss either outside of the boundary only or based on a modified S that is inpainted within the given boundary. However, the registration error has no sensitivity to the location of the inflated boundary as long as it is larger than the real one. On the other hand, if we compared the inpainted image and the pathological image S within the boundary only, we can notice that their dissimilarity increases when the boundary shrinks as the inpainting algorithm only generates healthy parts. This mechanism can then induce a segmentation module that segments the lesion as the foreground and the remaining as the background, which iteratively serves as the input mask for the inpainting module. Further, as the registration loss is calculated based on the registered inpainted image and the target image, the registration provides a regularization for the inpainting module such that the inpainting is specialized to facilitate the registration.

Specially for the input and output of the three modules, RegNet takes images S and T as input and generates the deformation field from S to T and T to S as φ_{ST} and φ_{TS} respectively. InpNet takes the background (foreground) cropped by SegNet and image $T \circ \varphi_{TS}$ warped by RegNet as input and outputs foreground (background) with a normal appearance. SegNet takes the pathology image S as input and employs the normal foreground and background inpainted by InpNet to segment the lesion region based on MMI. SegNet and InpNet are actually in an adversarial relationship. Through this joint optimization approach, the three networks collectively work to achieve registration and segmentation of pathological images under entirely unsupervised conditions, without being limited by the specific network structure. For the sake of simplicity, we employ a Unet-like [20] basic structure without any normalization layer.

2.2   Network Modules

RegNet. The primary objective of registration is to generate a deformation field that minimizes the dissimilarity between the source image (S) and the template image (T). The deformation is usually required to satisfy constraints like smoothness and even diffeomorphism. In terms of pathological image registration, the deformation field is only valid off the lesion area. Thus the registration loss should be calculated on the normal area only. Suppose that the lesion area is already obtained as $\theta(S)$ and inpainted with normal tissue, the registration loss can then be formulated as

$$\mathcal{L}_{reg} = \min_{\psi} \{ \mathcal{L}_{sym}(\phi(S \cdot \overline{\theta(S)}|T \circ \varphi_{TS}) \circ \varphi_{ST}, T)$$
$$+ \mathcal{L}_{sym}(T \circ \varphi_{TS}, \phi(S \cdot \overline{\theta(S)}|T \circ \varphi_{TS})) \} \tag{1}$$

where $\varphi_{ST} = \psi(S, T)$, $\varphi_{TS} = \psi(T, S)$ are the deformation fields that warp $S \to T$ and $T \to S$ respectively. The symbol $\cdot$ denotes element-wise multiplication. Furthermore, $\mathcal{L}_{sym}$ denotes the registration loss of SymNet [14], which aims to balance the losses of orientation consistency, regularization and magnitude.

SegNet. Minimal Mutual Information (MMI) is a typically used unsupervised segmentation method that distinguishes foreground from background. However, for a pathological image, the lesion regions often have a similar intensity to normal tissues near the boundary, which prevents the MMI from accurate segmentation without the semantic information. To address this limitation, we warp a healthy image T onto a pathology image S using a deformation field $\varphi_{TS} = \psi(T, S)$. This process maximizes the mutual information between corresponding regions of the two images and minimizes that of non-corresponding regions, thereby facilitating accessible lesion segmentation with MMI. Let $\Omega \in \mathbb{R}$ denote the image domain, M denote the mask, $F_\theta = \Omega \cdot M$ and $B_\theta = \Omega \cdot \overline{M}$ denote the foreground and background, where $\overline{M} = 1 - M, M \in \{0,1\}$. Regarding a pathological image S, when the background (normal) is given, the inpainted foreground (normal) will be different from the true foreground (lesion). When the foreground (lesion) is given, the inpainted background will remain the same as the background (normal). Thus we can formulate the adversarial loss of unsupervised segmentation as

$$\mathcal{L}_{seg} = \max_\theta \min_\phi \left\{ \frac{\mathbb{E}\{\theta(S) \cdot \mathcal{D}[S, \phi(S \cdot \overline{\theta(S)}|T \circ \varphi_{TS})]\}}{\mathbb{E}\|\theta(S)\|} \\ - \frac{\mathbb{E}\{\overline{\theta(S)} \cdot \mathcal{D}[S, \phi(S \cdot \theta(S)|T \circ \varphi_{TS})]\}}{\mathbb{E}\|\overline{\theta(S)}\|} \right\}, \tag{2}$$

where $\mathcal{D}$ is the distance function given by localized normalized cross-correlation (**LNCC**) [3]. Appendix A provides a detailed derivation.

InpNet. Let M denote the mask and φ_{TS} denote the deformation field from T to S. To handle the potential domain differences between the masked image $S \cdot M$ and the aligned image $T \circ \varphi_{TS}$, InpNet employs two encoders. The adversarial loss function of InpNet is represented as $\mathcal{L}_{MI}$. To incorporate semantic information, we include an additional similarity term $\mathcal{L}_{sim}$ that prevents InpNet from focusing too heavily on the foreground (lesion) and encourages it to produce healthy tissue. The proposed loss function $\mathcal{L}_{inp}$ is then formulated as the combination of mutual information loss defined through the normalized correlation coefficient (NCC) and similarity loss through the mean squared error (MSE):

$$\mathcal{L}_{inp} = \mathcal{L}_{MI} + \lambda \mathcal{L}_{sim}, \tag{3}$$

with

$$\mathcal{L}_{MI} = \mathcal{L}_{NCC}(S, \phi(S \cdot \overline{\theta(S)}|T \circ \varphi_{TS})) + \mathcal{L}_{NCC}(S, \phi(S \cdot \theta(S)|T \circ \varphi_{TS})),$$
$$\mathcal{L}_{sim} = \mathcal{L}_{MSE}(T^M, \phi(S \cdot \overline{\theta(S)}|T \circ \varphi_{TS})) + \mathcal{L}_{MSE}(T^M, \phi(S \cdot \theta(S)|T \circ \varphi_{TS})), \tag{4}$$

where λ represents the weight that balances the contributions of mutual information loss and similarity loss, and T^M denotes image T after histogram matching. We modify the histogram of $T \circ \varphi_{TS}$ to be similar to that of S in order to mitigate the effects of domain differences.

3   Experiments

Our experimental design focuses on two common clinical tasks: atlas-based registration, which involves warping pathology images to a standard atlas template, and longitudinal registration, which involves registering pre-operative images to post-operative images for the purpose of tracking changes over time.

Dataset and Pre-processing. For our study, we selected the ICBM 152 Nonlinear Symmetric template as our atlas [9]. We reoriented all MRI scans of the T1 sequence to the RAS orientation with a resolution of $1\,\mathrm{mm} \times 1\,\mathrm{mm} \times 1\,\mathrm{mm}$ and align the images to atlas using the *mri_robust_register* tool in FreeSurfer [19]. We then cropped the resulting MRI scans to a size of $160 \times 192 \times 144$, without any image augmentation. To evaluate our approach, we employed a 5-fold cross-validation method and divided our data into training and test sets in an 8:2 ratio.

3D Brain MRI. OASIS-1 [12] includes 416 cross-sectional MRI scans from individuals aged 18 to 96, with 100 of them diagnosed with mild to moderate Alzheimer's disease. BraTS2020 [13] provides 369 expert-labeled pre-operative MRI scans of glioblastomas and low-grade gliomas, acquired from multiple institutions for routine clinical use.

3D Pseudo Brain MRI. To evaluate the performance of atlas-based registration, it is essential to have the correct mapping of pathological regions to healthy brain regions. To create such a mapping, we created a pseudo dataset by utilizing images from the OASIS-1 and BraTS2020. From the resulting t1 sequences, a pseudo dataset of 300 images was randomly selected for further analysis. Appendix B provides a detailed process for creating the pseudo dataset.

Real Data with Landmarks. BraTS-Reg 2022 [2] provides extensive annotations of landmarks points within both the pre-operative and the follow-up scans that have been generated by clinical experts. A total of 140 images are provided, of which 112 are for training, and 28 for testing.

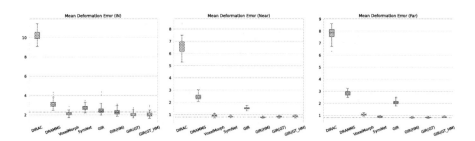

Fig. 2. Boxplots of mean deformation errors with respect to the gold standard deformations in three different regions on the pseudo dataset. Left to right: in tumor, near tumor and far from tumor.

Comparison to Pathology Registration. We compared our method (GIR-Net) with competitive algorithms: 1) three cutting-edge deep learning-based unsupervised deformable registration approaches: VoxelMorph [3], VoxelMorph-DF [8] and SymNet [14]. 2) two unsupervised deformable registration methods for pathological images: DRAMMS [18] and DIRAC [16]. DRAMMS is an optimization-based method that reduces the impact of non-corresponding regions. DIRAC jointly estimates regions with absent correspondence and bidirectional deformation fields and ranked first in the BraTSReg2022 challenge.

Atlas-Based Registration. After creating the pseudo dataset, we warped brain MR images without tumors to the atlas and used the resulting deformation field as the gold standard for evaluation. We then evaluated the mean deformation error (MDE) [10], which is calculated as the average Euclidean distance between the coordinates of the deformation field and the gold standard within specific regions of interest. These regions include: 1) the tumor region. 2) the normal region near the tumor (within 30 voxels). 3) the normal region far from the tumor (over 30 voxels but within brain tissue). Our results, presented in Fig. 2, show that our method with histogram matching (HM) outperforms other methods in all three regions, particularly in the normal regions (near and far). By utilizing HM, our network achieves an MDE of less than 1 mm compared to the gold standard deformations. These results demonstrate the effectiveness of our method in differentiating the impact of pathology in atlas-based registration tasks. Specifically, DIRAC is unable to eliminate the influence of domain differences and resulting in the largest registration error among the evaluated methods.

Longitudinal Registration. To perform the longitudinal registration task, we registered each pre-operative scan to the corresponding follow-up scan of the same patient and measured the mean target registration error (TRE) of the paired landmarks using the resulting deformation field. For this purpose, we leveraged SegNet, trained on BraTS2020, to segment the tumor of BraT-SReg2022 and separated the landmarks into two regions: near tumor and far from tumor. Figure 3 shows the mean TRE for the various registration approaches.

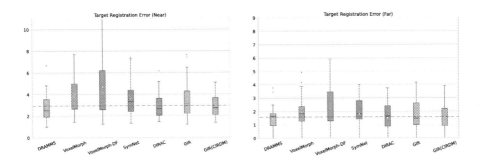

Fig. 3. Boxplots of the average target registration error (TRE) in two different regions: near tumor (left) and far from tumor (right).

In our proposed framework, we replaced RegNet with CIR-DM [15] (denoted as GIR(CIRDM)) without the need for supervised training or pretraining, and achieved comparable performance with the state-of-the-art method DIRAC. Moreover, our GIR approach outperforms other deep learning-based methods and achieved accurate segmentation of pathological images.

To quantitatively evaluate the segmentation capability of our proposed framework, we compared its performance with other unsupervised segmentation techniques methods, including unsupervised clustering toolbox AUCseg [25], joint non-correspondence segmentation and registration method NCRNet [1], and DIRAC. We used the mean Dice similarity coefficient (DSC) to evaluate the similarity between predicted masks and the ground truth. As shown in Table 1, AUCseg fails to detect the lesion in T1 scans. Our proposed framework achieved the highest DSC result of 0.83, following post-processing.

Ablation Study. We compared the performance of the InpNet trained with histogram matching (HM) and the SegNet trained with ground truth masks (Supervised). The results, shown in Table 1 and Fig. 2, demonstrate that domain differences between S and T have a significant effect on segmentation accuracy (without HM), leading to lower registration quality overall. Additionally, Fig. 4 shows an example of a pseudo image. We reconstructed the spatial correspondence by first using SegNet to localize the lesion and then using InpNet to inpaint it with the normal appearance.

Table 1. Average Dice Similarity Coefficients (DSCs) of Various Model Segmentation Results, Including GIRNet using different techniques: Histogram Matching (HM), Training with ground truth (Supervised), Mask binarized by threshold 0.5 (TH), and Post-processed by random walker algorithm (PP).

Dataset	AUCseg	NCRNet	DIRAC	GIRNet			
				TH	HM+TH	HM+PP	Supervised
Pesudo	0.095(±0.007)	0.201	0.18	0.254(±0.03)	0.744(±0.02)	0.831(±0.013)	0.921(±0.001)
BraTS2020	0.088(±0.010)	0.191	0.187	0.287(±0.01)	0.588(±0.014)	0.611(±0.012)	0.746(±0.02)

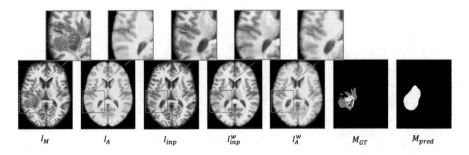

I_M I_A I_{inp} I_{inp}^W I_A^W M_{GT} M_{pred}

Fig. 4. Registration and segmentation results for Pseudo dataset. The 7 columns show: 1) the moving image; 2) the atlas; 3) the inpainted image; 4) the warped inpainted image; 5) the warped atlas image; 6) the ground truth mask 7) the predicted mask.

4   Conclusion

In this paper, we proposed a novel tri-net framework for joint image registration and unsupervised segmentation in medical imaging based on mutual information minimization in collaborative learning. Our experiments demonstrate that the proposed framework is effective for both atlas-based and longitudinal pathology image registration. We also observed that the accuracy of the segmentation network is significantly influenced by the quality of the inpainting, which, in turn, affects the registration outcome. In the future, our research will focus on enhancing the performance of InpNet to address domain differences better to improve the registration results.

References

1. Andresen, J., Kepp, T., Ehrhardt, J., Burchard, C., Roider, J., Handels, H.: Deep learning-based simultaneous registration and unsupervised non-correspondence segmentation of medical images with pathologies. Int. J. Comput. Assist. Radiol. Surg. **17**(4), 699–710 (2022). https://doi.org/10.1007/s11548-022-02577-4
2. Baheti, B., et al.: The brain tumor sequence registration challenge: establishing correspondence between pre-operative and follow-up MRI scans of diffuse glioma patients. arXiv preprint arXiv:2112.06979 (2021)
3. Balakrishnan, G., Zhao, A., Sabuncu, M.R., Guttag, J., Dalca, A.V.: Voxelmorph: a learning framework for deformable medical image registration. IEEE Trans. Med. Imaging **38**(8), 1788–1800 (2019)
4. Beg, M.F., Miller, M.I., Trouvé, A., Younes, L.: Computing large deformation metric mappings via geodesic flows of diffeomorphisms. Int. J. Comput. Vision **61**, 139–157 (2005)
5. Brett, M., Leff, A.P., Rorden, C., Ashburner, J.: Spatial normalization of brain images with focal lesions using cost function masking. Neuroimage **14**(2), 486–500 (2001)
6. Chen, K., Derksen, A., Heldmann, S., Hallmann, M., Berkels, B.: Deformable image registration with automatic non-correspondence detection. In: Aujol, J.-F., Nikolova, M., Papadakis, N. (eds.) SSVM 2015. LNCS, vol. 9087, pp. 360–371. Springer, Cham (2015). https://doi.org/10.1007/978-3-319-18461-6_29
7. Chitphakdithai, N., Duncan, J.S.: Non-rigid registration with missing correspondences in preoperative and postresection brain images. In: Jiang, T., Navab, N., Pluim, J.P.W., Viergever, M.A. (eds.) MICCAI 2010. LNCS, vol. 6361, pp. 367–374. Springer, Heidelberg (2010). https://doi.org/10.1007/978-3-642-15705-9_45
8. Dalca, A.V., Balakrishnan, G., Guttag, J., Sabuncu, M.R.: Unsupervised learning for fast probabilistic diffeomorphic registration. In: Frangi, A.F., Schnabel, J.A., Davatzikos, C., Alberola-López, C., Fichtinger, G. (eds.) MICCAI 2018. LNCS, vol. 11070, pp. 729–738. Springer, Cham (2018). https://doi.org/10.1007/978-3-030-00928-1_82
9. Fonov, V.S., Evans, A.C., McKinstry, R.C., Almli, C.R., Collins, D.: Unbiased nonlinear average age-appropriate brain templates from birth to adulthood. NeuroImage (47), S102 (2009)
10. Han, X., Yang, X., Aylward, S., Kwitt, R., Niethammer, M.: Efficient registration of pathological images: a joint PCA/image-reconstruction approach. In: 2017 IEEE

14th International Symposium on Biomedical Imaging (ISBI 2017), pp. 10–14. IEEE (2017)

11. Liu, X., Niethammer, M., Kwitt, R., Singh, N., McCormick, M., Aylward, S.: Low-rank atlas image analyses in the presence of pathologies. IEEE Trans. Med. Imaging **34**(12), 2583–2591 (2015)

12. Marcus, D.S., Wang, T.H., Parker, J., Csernansky, J.G., Morris, J.C., Buckner, R.L.: Open access series of imaging studies (OASIS): cross-sectional MRI data in young, middle aged, nondemented, and demented older adults. J. Cogn. Neurosci. **19**(9), 1498–1507 (2007)

13. Menze, B.H., et al.: The multimodal brain tumor image segmentation benchmark (BRATS). IEEE Trans. Med. Imaging **34**(10), 1993–2024 (2014)

14. Mok, T.C., Chung, A.: Fast symmetric diffeomorphic image registration with convolutional neural networks. In: Proceedings of the IEEE/CVF Conference on Computer Vision and Pattern Recognition, pp. 4644–4653 (2020)

15. Mok, T.C.W., Chung, A.C.S.: Conditional deformable image registration with convolutional neural network. In: de Bruijne, M., et al. (eds.) MICCAI 2021. LNCS, vol. 12904, pp. 35–45. Springer, Cham (2021). https://doi.org/10.1007/978-3-030-87202-1_4

16. Mok, T.C., Chung, A.C.: Unsupervised deformable image registration with absent correspondences in pre-operative and post-recurrence brain tumor MRI scans. In: Wang, L., Dou, Q., Fletcher, P.T., Speidel, S., Li, S. (eds.) MICCAI 2022. LNCS, vol. 13436, pp. 25–35. Springer, Cham (2022). https://doi.org/10.1007/978-3-031-16446-0_3

17. Nachev, P., Coulthard, E., Jäger, H.R., Kennard, C., Husain, M.: Enantiomorphic normalization of focally lesioned brains. Neuroimage **39**(3), 1215–1226 (2008)

18. Ou, Y., Sotiras, A., Paragios, N., Davatzikos, C.: Dramms: deformable registration via attribute matching and mutual-saliency weighting. Med. Image Anal. **15**(4), 622–639 (2011)

19. Reuter, M., Rosas, H.D., Fischl, B.: Highly accurate inverse consistent registration: a robust approach. Neuroimage **53**(4), 1181–1196 (2010)

20. Ronneberger, O., Fischer, P., Brox, T.: U-Net: convolutional networks for biomedical image segmentation. In: Navab, N., Hornegger, J., Wells, W.M., Frangi, A.F. (eds.) MICCAI 2015. LNCS, vol. 9351, pp. 234–241. Springer, Cham (2015). https://doi.org/10.1007/978-3-319-24574-4_28

21. Savarese, P., Kim, S.S., Maire, M., Shakhnarovich, G., McAllester, D.: Information-theoretic segmentation by inpainting error maximization. In: Proceedings of the IEEE/CVF Conference on Computer Vision and Pattern Recognition, pp. 4029–4039 (2021)

22. Wu, Y., Jiahao, T.Z., Wang, J., Yushkevich, P.A., Hsieh, M.A., Gee, J.C.: Nodeo: a neural ordinary differential equation based optimization framework for deformable image registration. In: Proceedings of the IEEE/CVF Conference on Computer Vision and Pattern Recognition, pp. 20804–20813 (2022)

23. Yang, X., Han, X., Park, E., Aylward, S., Kwitt, R., Niethammer, M.: Registration of pathological images. In: Tsaftaris, S.A., Gooya, A., Frangi, A.F., Prince, J.L. (eds.) SASHIMI 2016. LNCS, vol. 9968, pp. 97–107. Springer, Cham (2016). https://doi.org/10.1007/978-3-319-46630-9_10

24. Yang, Y., Loquercio, A., Scaramuzza, D., Soatto, S.: Unsupervised moving object detection via contextual information separation. In: Proceedings of the IEEE/CVF Conference on Computer Vision and Pattern Recognition, pp. 879–888 (2019)
25. Zhao, B., Ren, Y., Yu, Z., Yu, J., Peng, T., Zhang, X.Y.: Aucseg: an automatically unsupervised clustering toolbox for 3D-segmentation of high-grade gliomas in multi-parametric MR images. Front. Oncol. **11**, 679952 (2021)

H²GM: A Hierarchical Hypergraph Matching Framework for Brain Landmark Alignment

Zhibin He[1], Wuyang Li[3], Tuo Zhang[1(✉)], and Yixuan Yuan[2(✉)]

[1] School of Automation, Northwestern Polytechnical University, Xi'an, China
`tuozhang@nwpu.edu.cn`
[2] Department of Electronic Engineering, The Chinese University of Hong Kong, Shatin, Hong Kong SAR, China
`yxyuan@ee.cuhk.edu.hk`
[3] Department of Electrical Engineering, City University of Hong Kong, Kowloon, Hong Kong SAR, China

Abstract. Gyral hinges (GHs) are novel brain gyrus landmarks, and their precise alignment is crucial for understanding the relationship between brain structure and function. However, accurate and robust GH alignment is challenging due to the massive cortical morphological variations of GHs between subjects. Previous studies typically construct a single-scale graph to model the GHs relations and deploy the graph matching algorithms for GH alignment but suffer from two overlooked deficiencies. First, they consider only pairwise relations between GHs, ignoring that their relations are highly complex. Second, they only consider the point scale for graph-based GH alignment, which introduces several alignment errors on small-scaled regions. To overcome these deficiencies, we propose a Hierarchical HyperGraph Matching (H²GM) framework for GH alignment, consisting of a Multi-scale Hypergraph Establishment (MsHE) module, a Multi-scale Hypergraph Matching (MsHM) module, and an Inter-Scale Consistency (ISC) constraint. Specifically, the MsHE module constructs multi-scale hypergraphs by utilizing abundant biological evidence and models high-order relations between GHs at different scales. The MsHM module matches hypergraph pairs at each scale to entangle a robust GH alignment with multi-scale high-order cues. And the ISC constraint incorporates inter-scale semantic consistency to encourage the agreement of multi-scale knowledge. Experimental results demonstrate that the H²GM improves GH alignment remarkably and outperforms state-of-the-art methods. The code is available at here.

Keywords: Graph matching · Point cloud registration · Hypergraph

Z. He—This work was done when Zhibin He was a visiting student at the Department of Electronic Engineering, The Chinese University of Hong Kong.

1   Introduction

The ultimate goal of entire cortex registration (ECR) is to achieve functional region alignment across subjects. However, existing software often sacrifices the local functional alignment accuracy to accomplish ECR [6]. Previous studies have identified a novel brain gyrus landmark, termed the gyral hinge (GH) [10,13,30]. It has been demonstrated that the precise GH alignment over the whole brain is critical for understanding the relationship between brain structure and function [16]. However, achieving accurate and robust GH alignment is challenging due to the massive cortical morphological variations in GHs between subjects [9,16].

To address this issue, existing works have introduced the single-scale graph structure to represent and align GHs [16,29,31]. As shown in Fig. 1(a), they model GHs as graph nodes, formulate graph edges with the white matter fiber connections between GHs, and deploy the graph matching algorithms to align GHs [29,31]. Despite the great successes, existing works suffer from two overlooked deficiencies. First, they consider only pairwise relations between GHs to construct graphs that use second-order edges (the one-hop connection between two GHs), without considering underlying complex (e.g., high-order) relations among more than two GHs [23,28]. For example, the high-order relations of GHs within the same brain region cannot be well modeled only in low-order cues [5]. Hence, we are committed to going beyond the second-order relations and seeking a more effective graph structure, hypergraph, to model such high-order relations [2,15,26,32]. Second, existing works focus only on a single-scale, *i.e.*, the point scale [16,29,31], ignoring the hierarchy of brain structure and function at multiple scales [1]. Only considering the single-scale knowledge is insufficient to capture the multi-scale dependence within graphs, inevitably introducing several alignment errors on small-scaled regions [3]. Hence, we aim to introduce multiscale knowledge in hypergraph matching (Fig. 1(b)), exploring more effective message propagation with hierarchical hypergraph learning.

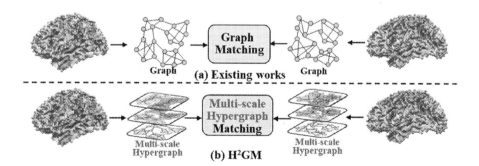

Fig. 1. Illustration of the GH alignment. (a) Previous works use graph structure to represent and align GHs at a single scale. (b) The proposed H²GM introduces multiscale hypergraph matching for a better GH alignment.

To overcome the aforementioned challenges, we propose a Hierarchical Hyper-Graph Matching (H²GM) framework for GH alignment, which consists of a Multi-scale Hypergraph Establishment (MsHE), a Multi-scale Hypergraph Matching (MsHM), and an Inter-Scale Consistency (ISC) constraint. Specifically, in the MsHE module, we construct multi-scale hypergraphs with three hierarchical scales (Five Lobes Atlas, DK Atlas [4], GH) and establish correspondences for both inter-scale and inter-subject hypergraphs. As for the MsHM module, we match the hypergraph pairs at each scale to entangle a robust GH alignment with multi-scale high-order cues. Finally, ISC incorporates inter-scale semantic consistency to encourage the agreement of multi-scale knowledge. Experimental results demonstrate that our framework enables more effective GH alignment. In summary, the main contributions are as follows: (1) We propose a H²GM framework for GH alignment. To the best of our knowledge, this work is the first attempt to leverage multi-scale hypergraphs to align the brain landmark. (2) We design a MsHE module to extract the high-order relations among GHs at different scales and a MsHM module to propagate GHs features to align GHs. Moreover, the feature distribution of GH optimized by the inter-scale semantic consistency further improves the alignment accuracy. (3) Extensive experiments verify that the proposed matching framework improves the GH alignment remarkably and outperforms state-of-the-art methods.

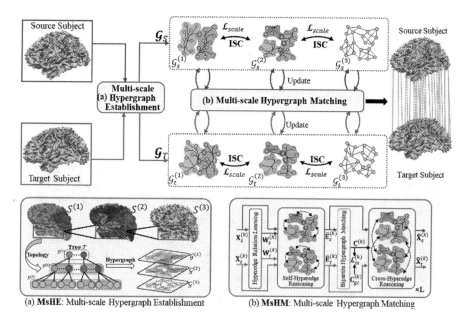

Fig. 2. Overview of the proposed H²GM. ISC represents inter-scale consistency.

2   Method

The overview of our proposed framework is shown in Fig. 2, which contains two parts. Given two subjects MRI, we extract GHs and establish multi-scale hypergraphs $\mathcal{G}_{s/t}^{(k)} = \left\{ \mathbf{H}_{s/t}^{(k)}, \mathbf{X}_{s/t}^{(k)}, \mathbf{W}_{s/t}^{(k)} \right\}$ in the MsHE module (Fig. 2(a)), illustrated in Sect. 2.1. Then we perform multi-scale GH alignment in the MsHM module (Fig. 2(b)) to obtain multi-scale correspondence matrix $\mathbf{C}^{(k)}$ (Sect. 2.2). The $\mathbf{C}^{(k)}$ indicating the correspondence between two hypergraphs, i.e., $\mathbf{C}_{i,j}^{(k)} = 1$ means the i-th GH in source matches to the j-th GH in target at the k scale. The GH alignment can be formulated as the hypergraph matching problem:

$$\min_{\theta} g(\theta) = \min_{\theta} \sum_{k=1}^{K} \left(\mathbf{H}_s^{(k)\,T} \mathbf{X}_s^{(k)} - \mathbf{C}^{(k)} \mathbf{H}_t^{(k)\,T} \mathbf{X}_t^{(k)} \mathbf{C}^{(k)\,T} \right), \tag{1}$$

where θ represents the neural network parameters, the $\mathbf{H}$ is the incidence matrix of hypergraph [5], the $\mathbf{X}$ is the GHs' features, and the k is the number of scale. We use a neural network to predict $\mathbf{C}^{(k)}$. The details will be introduced below.

2.1   Multi-scale Hypergraph Establishment (MsHE)

As shown in Fig. 2(a), we utilize the Five Lobes Atlas, DK Atlas [4], and GH as three scales to model the inter-scale relations through a tree structure $\mathcal{T}$ [3]. In the tree structure $\mathcal{T}$, the larger-scale brain regions are parent nodes, and the smaller-scale brain regions or GHs are child nodes. As for the multi-scale hypergraph incidence matrix $\mathbf{H}_{s/t}^{(k)}$, we capture the topological relations among brain atlases to serve as the incidence matrix. The element $\mathbf{H}_{i,j}^{(k)}$ indicates that the i-th GH is included in the j-th hyperedge (brain region or GH). Notably, when $k=3$, the row sum of $\mathbf{H}_{s/t}^{(k)}$ is 1 and $\mathbf{H}_{s/t}^{(k)}$ is in the diagonal format. As for the hypergraph nodes $\mathbf{V}_{s/t}$, i.e., GHs, we capture various cues in the subject to serve as the raw node feature. Specifically, considering the limited representation of the single vertex knowledge, we expand each GH on the surface by two rings, resulting in a total of 19 vertices as the GHs' raw features. Each vertex features contain three-dimensional coordinates ($\times 3$), normal vector coordinates ($\times 3$), curvature ($\times 1$), convexity ($\times 1$), and cortical thickness ($\times 1$). Then, the aforementioned features are contacted to obtain the raw node features $\mathbf{V}_{s/t}^{(k)} \in \mathbb{R}^{m/n \times d}$, where m/n represent the number of source/target subject GHs, respectively. After that, the raw features are sent to Dynamic Graph CNN [25] (DGCNN) to extract GHs descriptors $\mathbf{X}_{s/t}^{(k)} \in \mathbb{R}^{m/n \times D}$ as the final node representation. Notably, the parameters of the DGCNN are shared across all scales. For the hyperedge weight matrix $\mathbf{W}_{s/t}^{(k)}$, we propose to use a diagonal metric with one-value entries as the initialization, which will be optimized in the following hyperedge relation learning module. Finally, we obtain the multi-scale hypergraphs of source/target subjects as $\mathcal{G}_{s/t}^{(k)} = \left\{ \mathbf{H}_{s/t}^{(k)}, \mathbf{X}_{s/t}^{(k)}, \mathbf{W}_{s/t}^{(k)} \right\}$. The proposed multi-scale hypergraphs can well model significant surface morphological information and multi-scale GH relations for better GH alignment.

2.2   Multi-scale Hypergraph Matching (MsHM)

As shown in Fig. 2(b), the hyperedge features at k scale are computed as $\mathbf{E}_{s/t}^{(k)} = \mathbf{H}_{s/t}^{(k)T}\mathbf{X}_{s/t}^{(k)}$, and then sent to MsHM to solve the multi-scale matching with four sub-stages, including hyperedge relation learning, self-hyperedge reasoning, bipartite hypergraph matching, and cross-hyperedge reasoning.

Hyperedge Relation Learning: To dynamically learn hyperedge relation with better structural information, we propose hyperedge relation learning to model the high-order relation among hyperedges, instead of directly using the handcraft hyperedge weight metric $\mathbf{W}_{s/t}$ [5,15]. The layer utilizes transformer [7,22], which comprises several encoder-decoder layers, to generate hyperedge soft edges. Specifically, the transformer takes hyperedge features $\mathbf{E}_{s/t}$ as input and encodes them into embedding features. The inner product of these embedded features is then passed through a softmax function to generate the hyperedge weight matrices. This process can be expressed as follows:

$$\mathbf{W}_{s/t}^{(k)} = softmax((f_{emb}(\mathbf{E}_{s/t}^{(k)}))^T, f_{emb}(\mathbf{E}_{s/t}^{(k)}), \tag{2}$$

where f_{emb} is a transformer-based feature embedding function.

Self-hyperedge Reasoning: We utilize a self-hyperedge reasoning network to capture the self-correlation of hyperedge features. Propagating features achieve this by hyperedge weight matrix within each hypergraph. It can be written as:

$$\bar{\mathbf{E}}_{s/t}^{(k)} = \sigma\left(\mathbf{W}_{s/t}^{(k)}\mathbf{E}_{s/t}^{(k)}\mathbf{\Theta}_{s/t}^{(k)}\right), \tag{3}$$

where $\mathbf{\Theta}_{s/t}^{(k)} \in \mathbb{R}^{D \times D}$ denotes the learnable parameters of self-hyperedge reasoning network. And $\sigma(\cdot)$ is the nonlinear activation function.

Bipartite Hypergraph Matching: We use bipartite matching to determine a soft correspondence matrix between two subjects, which is achieved using the following expression: $\mathbf{C}^{(k)} = \text{Gbm}\left(\bar{\mathbf{E}}_s^{(k)}, \bar{\mathbf{E}}_t^{(k)}\right)$, where Gbm consists of an Affinity layer, an Instance normalization layer, a quadratic constrain (QC) layer, and a Sinkhorn layer. Initially, the affinity matrix is computed as $\mathbf{A}^{(k)} = \bar{\mathbf{E}}_s^{(k)}\mathbf{M}^{(k)}\bar{\mathbf{E}}_t^{(k)}$, where $\mathbf{M}^{(k)} \in \mathbb{R}^{D \times D}$ is the learnable parameter matrix in the affinity layer. Next, we apply instance normalization [20] to transform $\mathbf{A}^{(k)}$ into a matrix with positive elements within finite values. We then introduce QC to minimize the structural difference of matched GH pairs [8,14]. For unmatched GHs, we add an additional row and column of ones to the output of the QC layer matrix, which is then processed through the Sinkhorn layer [19] to obtain a double-stochastic affinity matrix with maximum iteration optimization. After deleting the added row and column, we obtain a soft assignment matrix $\mathbf{C}^{(k)}$. Finally, we use cross-entropy-type loss functions to compute the linear assignment cost between the ground truth and the soft assignment matrix, which is defined as follows:

$$\mathcal{L}_m = \sum_{k=1}^{K} \left(\sum_{l=1}^{L} \left(\mathbf{C}_{gt}^{(k)} \log \left(\mathbf{C}_l^{(k)} \right) + \left(1 - \mathbf{C}_{gt}^{(k)} \right) \log \left(1 - \mathbf{C}_l^{(k)} \right) \right) \right), \quad (4)$$

where $\mathbf{C}_{gt}^{(k)} \in \{0,1\}, \mathbf{C}_l^{(k)} \in [0,1]$, and this loss can optimize the Eq. 1. The k-scale soft assignment matrix of the MsHM l-th layer output is $\mathbf{C}_l^{(k)}$.

Cross-Hyperedge Reasoning: We further enhance the hyperedge features by exploring cross-correlation through cross-hyperedge reasoning. Different from self-hyperedge reasoning, the proposed cross-hyperedge reasoning enables subject-aware message propagation, facilitating effective interaction between subjects. The more similar a pair of hyperedges is between two subjects, the better features will be aggregated in better alignment. It can be written as:

$$\hat{\mathbf{E}}_{s/t}^{(k)} = f_{cross} \left(\bar{\mathbf{E}}_{s/t}^{(k)}, \mathbf{C}^{(k)} \bar{\mathbf{E}}_{t/s}^{(k)} \right), \quad (5)$$

where f_{cross} consists of a feature concatenate and a fully connected layer. Finally, the new GHs features with a symmetric normalization can be written as:

$$\mathbf{X}_{s/t}^{(k)} = \sigma((\mathbf{D}_{s/t}^{(k)})^{-1} \mathbf{H}_{s/t}^{(k)} \mathbf{W}_{s/t}^{(k)} (\mathbf{B}_{s/t}^{(k)})^{-1} \hat{\mathbf{E}}_{s/t}^{(k)} \mathbf{\Phi}_{s/t}^{(k)}), \quad (6)$$

where $\sigma(\cdot)$ is the nonlinear activation function. $\mathbf{D}_{s/t}^{(k)}$ and $\mathbf{B}_{s/t}^{(k)}$ are the diagonal node degree matrix and hyperedge degree matrix, respectively. $\mathbf{\Phi} \in \mathbb{R}^{D \times D}$ denotes the learnable parameters of cross-hyperedge reasoning network.

2.3   Inter-Scale Consistency (ISC)

To avoid the potential disagreement among different scales, it is critical to introduce the consistency constraint across scales, which achieves collaborative optimization in different scales and prevents sub-optimal alignment. Specifically, we propose a novel ISC mechanism that leverages the local properties of the tree structure. For a parent hyperedge e_p and its child hyperedges $\{e_c^q\}_{q=1}^{Q}$ (Q is the number of children nodes belonging to the same parent node.), which represent the same brain region at different scales, we enforce semantic-level consistency between the child hyperedges and their parent hyperedge to facility the reliable model learning, which is denoted as follows:

$$\mathcal{L}_{scale} = \sum_{\forall \mathcal{T}_{s/t}} \left\| e_p - \sum_{q=1}^{Q} e_c^q \right\|^2, \quad (7)$$

where $\mathcal{L}_{scale}$ denotes the loss function that enforces inter-scale semantic consistency among source/target subjects tree structure $\mathcal{T}_{s/t}$. By incorporating this loss, the model learns to maintain consistent semantic information across different scales for hyperedges corresponding to the same brain region.

2.4   Model Optimization

In the training of this work, we introduce a hyperparameter to add up $\mathcal{L}_m$ and $\mathcal{L}_{scale}$. Then, the overall train loss for the GH alignment model is denoted as:

$$\mathcal{L} = \mathcal{L}_m + \beta\mathcal{L}_{scale}, \tag{8}$$

where β is a hyperparameter to control the intensity. In testing, we utilize the Hungarian algorithm [12] to convert the soft assignment matrix into a binary matrix. Subsequently, the rows and columns of the binary matrix with a value of 1 represent the GHs correspondences between source and target subjects.

Table 1. Comparison with state-of-the-art methods and the ablation studies on the GH alignment dataset. The best performance is highlighted in bold.

Method	Accuracy (%)	Correlation	MGE
Ground Truth	100	39.84 ± 3.83	0
SurfReg [6]	\	27.67 ± 4.93	\
HNN (AAAI2019) [5]	77.10 ± 8.53	37.01 ± 3.96	6.7
PCA-GM (ICCV2019) [24]	74.32 ± 8.55	35.50 ± 4.07	10.1
RGM (CVPR2021) [7]	74.71 ± 8.67	35.85 ± 4.03	9.7
QC-DGM (CVPR2021) [8]	71.83 ± 9.74	34.72 ± 4.26	11.2
REGTR (CVPR2022) [27]	75.30 ± 8.31	36.30 ± 3.97	8.8
Sigma++ (TPAMI2023) [15]	77.01 ± 8.41	36.98 ± 3.96	6.7
w/o ISC	77.66 ± 8.27	37.83 ± 3.93	6.2
w/o Multi-scale	74.81 ± 8.66	35.92 ± 4.02	9.6
w/o Hyperedge Relation Learning	77.43 ± 8.28	37.79 ± 3.93	6.5
Our (H^2GM)	$\mathbf{78.03 \pm 8.21}$	$\mathbf{37.99 \pm 3.91}$	**5.8**

3   Experiments and Results

3.1   Experimental Setup

Dataset: We evaluate our proposed framework effectiveness by conducting experiments on the GH alignment dataset, which includes the ground truth of GH correspondences across 250 subjects. Each GH contains brain atlas information, various morphological features from T1-weighted (T1w) MRI, and task activation vectors obtained from the task fMRI (tfMRI). The Five Lobe Atlas consists of 10 brain regions, while the DK Atlas includes 66 brain regions. The T1w MRI and tfMRI are acquired from the WU-Minn Human Connectome Project (HCP) consortium [21], with written informed consent obtained from HCP participants and relevant institutional review boards approving the study.

Evaluation: In this study, we adopt the accuracy, the correlation ($\times 10^{-2}$), and the mean geodesic errors [17] (MGE, in centimeters) as evaluation metrics for our proposed model. Specifically, accuracy is computed as the average rate of correct GH alignments across all subjects. Furthermore, we use correlation to measure the similarity between the activation vectors of the aligned GH pairs across all subjects. The higher the value of correlation, the higher the confidence that the two GHs align correctly. To evaluate the performance of our model, we compare it against several state-of-the-art learning-based point cloud registration methods, including RGM [7], QC-DGM [8], and REGTR [27], as well as several learning-based graph matching methods, including HNN [5], PCA-GM [24], and Sigma++ [15], and the traditional surface registration method, SurfReg [6]. Notably, we exclude comparisons with non-open-source methods such as [16,29,31].

Implementation Details To train our model, we utilize the Adam optimizer [11] with 3×10^{-5} learning rate, 5×10^{-4} weight decay, 1 batch size, and 40 epochs. We use $L = 3$ layers of the MsHM and performe 20 Sinkhorn iterations. We set the hyperparameter $\beta = 10$, the raw feature channel $d = 171$, and the GH descriptor channel $D = 2048$. We use 200 subjects as the training set and 50 subjects as the test set. We implement our network using the PyTorch library [18].

Table 2. Sensitivity analysis on the proposed H^{2}GM

Weights	Accuracy (%)	Correlation	MGE
$\beta = 0.1$	77.96 ± 8.22	37.94 ± 3.91	6.3
$\beta = 1$	77.71 ± 8.24	37.90 ± 3.91	5.9
$\beta = 10$	$\mathbf{78.03 \pm 8.21}$	$\mathbf{37.99 \pm 3.91}$	$\mathbf{5.8}$
$\beta = 100$	76.90 ± 8.30	36.92 ± 3.97	6.9

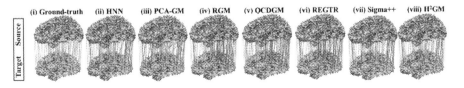

Fig. 3. Visualization of GH alignment results. The green and red lines represent the correct and incorrect alignments, respectively. (Color figure online)

3.2   Experimental Results

Comparison with State-of-the-Arts: We present the comparison results in Table 1. The accuracy, correlation, and MGE of H^2GM achieves $78.03 \pm 8.21\%$, $(37.99 \pm 3.91) \times 10^{-2}$, and 5.8, respectively, outperforming existing works by a large margin. Compared with learning-based approaches, showing our advantages over existing works. Besides, compared to other works, the highest correlation and the lowest MGE obtained by our proposed H^2GM demonstrates that our proposed framework provides the most precise GH alignment.

Ablation Studies: Table 1 displays ablation studies that verify the efficacy of each module of the proposed H^2GM. Our results demonstrate that inter-scale semantic consistency enhances the effectiveness of our approach, as evidenced by the removal of ISC alignment accuracy $77.66 \pm 8.27\%$. Introducing multi-scale atlases improves alignment accuracy by reducing error tolerance, as indicated by removing the two scales' brain atlas accuracy $74.81 \pm 8.66\%$. Removing the hyperedge relation learning accuracy $77.43 \pm 8.28\%$ suggests that the hyperedge structure-aware message is instrumental in improving alignment accuracy.

Sensitivity Analysis: We conduct experiments with varying hyperparameters to investigate the sensitivity of β in Eq. 8 and record the results in Table 2. Our findings indicate that the highest alignment accuracy is achieved when $\beta = 10$. However, a β setting that is too small leads to a slight performance decline due to insufficient inter-scale semantic consistency. Conversely, a β setting that is too large also leads to a performance decline.

Qualitative Analysis: Figure 3 presents the visualization results of GH alignment. Randomly selected pairs of subjects show that our method achieves the highest accuracy compared to state-of-the-art methods. These findings suggest that exploring higher-order relations between GHs can optimize GH alignment.

4   Conclusion

In this paper, we propose a novel framework H^2GM for brain landmark alignment. Specifically, H^2GM consists of a MsHE module for constructing the multi-scale hypergraphs, a MsHM module for matching them, and ISC for incorporating the semantic consistency among scales. Experimental results demonstrate that our proposed H^2GM outperforms existing approaches significantly.

Acknowledgment. This work was supported in part by the Innovation and Technology Commission-Innovation and Technology Fund ITS/100/20, in part by the National Natural Science Foundation of China [62001410, 31671005, 31971288, and U1801265], and in part by the Innovation Foundation for Doctor Dissertation of Northwestern Polytechnical University [CX2022052].

References

1. Avena-Koenigsberger, A., Misic, B., Sporns, O.: Communication dynamics in complex brain networks. Nat. Rev. Neurosci. **19**(1), 17–33 (2018)
2. Bai, S., Zhang, F., Torr, P.H.: Hypergraph convolution and hypergraph attention. Pattern Recogn. **110**, 107637 (2021)
3. Chen, Z., Zhang, J., Che, S., Huang, J., Han, X., Yuan, Y.: Diagnose like a pathologist: weakly-supervised pathologist-tree network for slide-level immunohistochemical scoring. In: Proceedings of the AAAI Conference on Artificial Intelligence, vol. 35, pp. 47–54 (2021)
4. Desikan, R.S., et al.: An automated labeling system for subdividing the human cerebral cortex on MRI scans into gyral based regions of interest. Neuroimage **31**(3), 968–980 (2006)
5. Feng, Y., You, H., Zhang, Z., Ji, R., Gao, Y.: Hypergraph neural networks. In: Proceedings of the AAAI Conference on Artificial Intelligence, vol. 33, pp. 3558–3565 (2019)
6. Fischl, B., Sereno, M.I., Tootell, R.B., Dale, A.M.: High-resolution intersubject averaging and a coordinate system for the cortical surface. Hum. Brain Mapp. **8**(4), 272–284 (1999)
7. Fu, K., Liu, S., Luo, X., Wang, M.: Robust point cloud registration framework based on deep graph matching. In: Proceedings of the IEEE/CVF Conference on Computer Vision and Pattern Recognition, pp. 8893–8902 (2021)
8. Gao, Q., Wang, F., Xue, N., Yu, J.G., Xia, G.S.: Deep graph matching under quadratic constraint. In: Proceedings of the IEEE/CVF Conference on Computer Vision and Pattern Recognition, pp. 5069–5078 (2021)
9. He, H., Razlighi, Q.R.: Landmark-guided region-based spatial normalization for functional magnetic resonance imaging. Hum. Brain Mapp. **43**(11), 3524–3544 (2022)
10. He, Z., et al.: Gyral hinges account for the highest cost and the highest communication capacity in a corticocortical network. Cereb. Cortex **32**(16), 3359–3376 (2022)
11. Kingma, D.P., Ba, J.: Adam: a method for stochastic optimization. arXiv preprint arXiv:1412.6980 (2014)
12. Kuhn, H.W.: The Hungarian method for the assignment problem. Nav. Res. Logist. Q. **2**(1–2), 83–97 (1955)
13. Li, K., et al.: Gyral folding pattern analysis via surface profiling. Neuroimage **52**(4), 1202–1214 (2010)
14. Li, W., Liu, X., Yuan, Y.: Sigma: semantic-complete graph matching for domain adaptive object detection. In: Proceedings of the IEEE/CVF Conference on Computer Vision and Pattern Recognition, pp. 5291–5300 (2022)
15. Li, W., Liu, X., Yuan, Y.: SIGMA++: improved semantic-complete graph matching for domain adaptive object detection. IEEE Trans. Pattern Anal. Mach. Intell. (2023)
16. Li, X., et al.: Commonly preserved and species-specific gyral folding patterns across primate brains. Brain Struct. Funct. **222**, 2127–2141 (2017)
17. Litany, O., Remez, T., Rodola, E., Bronstein, A., Bronstein, M.: Deep functional maps: structured prediction for dense shape correspondence. In: Proceedings of the IEEE International Conference on Computer Vision, pp. 5659–5667 (2017)
18. Paszke, A., et al.: Pytorch: an imperative style, high-performance deep learning library. In: Advances in Neural Information Processing Systems, vol. 32 (2019)

19. Sinkhorn, R.: A relationship between arbitrary positive matrices and doubly stochastic matrices. Ann. Math. Stat. **35**(2), 876–879 (1964)
20. Ulyanov, D., Vedaldi, A., Lempitsky, V.: Instance normalization: the missing ingredient for fast stylization. arXiv preprint arXiv:1607.08022 (2016)
21. Van Essen, D.C., et al.: The WU-Minn human connectome project: an overview. Neuroimage **80**, 62–79 (2013)
22. Vaswani, A., et al.: Attention is all you need. In: Advances in Neural Information Processing Systems, vol. 30 (2017)
23. Wang, Q., et al.: Modeling functional difference between gyri and sulci within intrinsic connectivity networks. Cerebral Cortex **33**(4), 933–947 (2022)
24. Wang, R., Yan, J., Yang, X.: Learning combinatorial embedding networks for deep graph matching. In: Proceedings of the IEEE/CVF International Conference on Computer Vision, pp. 3056–3065 (2019)
25. Wang, Y., Sun, Y., Liu, Z., Sarma, S.E., Bronstein, M.M., Solomon, J.M.: Dynamic graph CNN for learning on point clouds. ACM Trans. Graph. (TOG) **38**(5), 1–12 (2019)
26. Xu, C., Li, M., Ni, Z., Zhang, Y., Chen, S.: Groupnet: multiscale hypergraph neural networks for trajectory prediction with relational reasoning. In: Proceedings of the IEEE/CVF Conference on Computer Vision and Pattern Recognition, pp. 6498–6507 (2022)
27. Yew, Z.J., Lee, G.H.: REGTR: end-to-end point cloud correspondences with transformers. In: Proceedings of the IEEE/CVF Conference on Computer Vision and Pattern Recognition, pp. 6677–6686 (2022)
28. Zhang, S., et al.: Gyral peaks: novel gyral landmarks in developing macaque brains. Hum. Brain Mapp. **43**(15), 4540–4555 (2022)
29. Zhang, T., et al.: Identifying cross-individual correspondences of 3-hinge gyri. Med. Image Anal. **63**, 101700 (2020)
30. Zhang, T., et al.: Cortical 3-hinges could serve as hubs in cortico-cortical connective network. Brain Imaging Behav. **14**(6), 2512–2529 (2020). https://doi.org/10.1007/s11682-019-00204-6
31. Zhang, T., et al.: Group-wise graph matching of cortical gyral hinges. In: Shen, D., et al. (eds.) MICCAI 2019. LNCS, vol. 11767, pp. 75–83. Springer, Cham (2019). https://doi.org/10.1007/978-3-030-32251-9_9
32. Zhang, Z., et al.: H2MN: graph similarity learning with hierarchical hypergraph matching networks. In: Proceedings of the 27th ACM SIGKDD Conference on Knowledge Discovery & Data Mining, pp. 2274–2284 (2021)

SAMConvex: Fast Discrete Optimization for CT Registration Using Self-supervised Anatomical Embedding and Correlation Pyramid

Zi Li[1,4(✉)], Lin Tian[1,2], Tony C. W. Mok[1,4], Xiaoyu Bai[1,4], Puyang Wang[1,4], Jia Ge[3], Jingren Zhou[1,4], Le Lu[1], Xianghua Ye[3], Ke Yan[1,4], and Dakai Jin[1]

[1] DAMO Academy, Alibaba Group, Hangzhou, China
alisonbrielee@gmail.com, lintian@cs.unc.edu
[2] University of North Carolina at Chapel Hill, Chapel Hill, USA
[3] The First Affiliated Hospital of College of Medicine, Zhejiang University, Hangzhou, China
[4] Hupan Lab, 310023 Hangzhou, China

Abstract. Estimating displacement vector field via a cost volume computed in the feature space has shown great success in image registration, but it suffers excessive computation burdens. Moreover, existing feature descriptors only extract local features incapable of representing the global semantic information, which is especially important for solving large transformations. To address the discussed issues, we propose SAMConvex, a fast coarse-to-fine discrete optimization method for CT registration that includes a decoupled convex optimization procedure to obtain deformation fields based on a self-supervised anatomical embedding (SAM) feature extractor that captures both local and global information. To be specific, SAMConvex extracts per-voxel features and builds 6D correlation volumes based on SAM features, and iteratively updates a flow field by performing lookups on the correlation volumes with a coarse-to-fine scheme. SAMConvex outperforms the state-of-the-art learning-based methods and optimization-based methods over two inter-patient registration datasets (Abdomen CT and HeadNeck CT) and one intra-patient registration dataset (Lung CT). Moreover, as an optimization-based method, SAMConvex only takes ∼2 s (∼5 s with instance optimization) for one paired images.

Keywords: Medical Image Registration · Large Deformation

Z. Li and L. Tian—Equal contribution.

Supplementary Information The online version contains supplementary material available at https://doi.org/10.1007/978-3-031-43999-5_53.

1   Introduction

Deformable image registration [21], a fundamental medical image analysis task, has traditionally been approached as a continuous optimization [1,2,9,19,24] problem over the space of dense displacement fields between image pairs. The iterative process always leads to inefficiency. Recent learning-based approaches that use a deep network to predict a displacement field [3,6,16–18,25,30], yield much faster runtime and have gained huge attention. However, they often struggle with versatile applicability due to the fact that they require training per registration task. Moreover, gathering enough data for the training is not a trivial task in practice. In addition, both optimization and learning-based methods rely on similarity measures computed over the intensity, which prevents the methods to utilize the anatomy correspondence. Several works [8,10] use feature descriptors that provide modality and contrast invariant information but they still can only represent local information and do not contain the global semantic information. Thus, they face challenges in settings with large deformations or complex anatomical differences (*e.g.*, inter-patient Abdomen).

To address this issue, we incorporate a Self-supervised Anatomical eMbedding (SAM) [27] into registration. SAM generates a unique semantic embedding for each voxel in CT that describes its anatomical structure, thus, providing semantically coherent information suitable for registration. SAME [15] enhances learning-based registration with SAM embeddings, but it suffers the applicability issue as the other learning-based methods even when the SAM embedding is pre-trained and ready to use out of the box for multiple anatomical regions.

Registration has also been formulated as a discrete optimization problem [5,7, 20,22,26] that employs a dense set of discrete displacements, called cost volume. The main challenge of this category of approach is the massive size of the search space, as millions of voxels exist in a typical 3D CT scan and each voxel in the moving scan can be reasonably paired with thousands of points in the other scan, leading to a high computational burden. To obtain fast registration with discrete optimization, Heinrich *et al.* [11] prunes the search space by constructing the cost volume only within the neighborhood of each voxel. However, the magnitude range of the deformation it can solve is limited by the size of the neighborhood window, leading to reliance on an accurate pre-alignment.

We propose SAMConvex, a coarse-to-fine discrete optimization method for CT registration. Specifically, SAMConvex presents two main components: (1) a *discriminative* feature extractor that encodes global and local embeddings for each voxel; (2) a *lightweight* correlation pyramid that constructs multi-scale 6D cost volume by taking the inner product of SAM embeddings. It enjoys the strengths of state-of-the-art accuracy (Sect. 3.2), good generalization (Sect. 3.2 and Sect. 3.3) and fast runtime (Sect. 3.2).

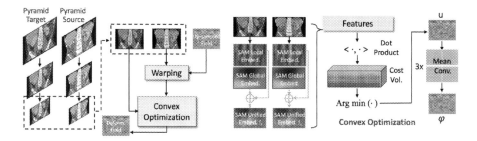

Fig. 1. Left: Image pyramid and refinement at one pyramid level by discrete registration with convex optimization. **Right**: the detail of convex optimization that consists of SA pyramid and hierarchical optimization at one pyramid level.

2 SAMConvex Method

2.1 Problem Formulation

Given a source image $I_s : \Omega_s \to \mathbb{R}$ and a target image $I_t : \Omega_t \to \mathbb{R}$ within a spatial domain $\Omega \subset \mathbb{R}^n$, our goal is to find the spatial transformation $\varphi^{-1} : \Omega_t \to \Omega_s$. We aim at minimizing the following energy:

$$\hat{u} = \arg\min_u E_\mathcal{D}(I_s \circ \varphi^{-1}, I_t) + \lambda E_\mathcal{R}(u), \tag{1}$$

where the transformation model is the displacement vector field (DVF) $\varphi^{-1} = Id + u$ with Id being the identity transformation, the $E_\mathcal{D}$ is the similarity term measuring the similarity between I_t and warped image $I_s \circ \varphi^{-1}$, $E_\mathcal{R} = ||\nabla u||_2^2$ is the diffusion regularizer that advocates the regularity of φ^{-1}. λ weights between the data matching term and the regularization term.

2.2 Decoupling to Convex Optimizations

We conduct the optimization scheme proposed in [4,11,23,29] and we give a brief review here. By introducing an extra term into Eq. 1

$$\hat{v}, \hat{u} = \arg\min_{v,u} E_\mathcal{D}(I_s \circ (Id + v), I_t) + \frac{1}{2\theta}||v - u||^2 + \lambda E_\mathcal{R}(u), \tag{2}$$

the optimization then can be decomposed into two sub-optimization problems[1]

$$\begin{cases} \hat{v} = \arg\min_v E_\mathcal{D}(I_s \circ (Id + v), I_t) + \dfrac{1}{2\theta}||v - \hat{u}||^2 + const, \\[2mm] \hat{u} = \arg\min_u \dfrac{1}{2\theta}||\hat{v} - u||^2 + \lambda E_\mathcal{R}(u) + const. \end{cases} \tag{3}$$

---

[1] For better illustration, we substitute $\lambda E_\mathcal{R}(u)$ and $E_\mathcal{D}(I_s \circ \varphi, I_t)$ with $const$ in the two equations in Eq. 3, respectively.

with each can be solved via global optimizations. By alternatively optimizing the two subproblems and progressively reducing θ during the alternative optimization, we get $\hat{v} \approx \hat{u}$ and obtain the solution of Eq. 1 as $\varphi^{-1} = Id + \hat{v}$. To be noted, the first problem in Eq. 3 can be solved point-wise because the spatial derivatives of v is not involved. We can search over the cost volume of each point to obtain the optimal solution of the first problem in Eq. 3. For the second problem, we are inspired by mean-field inference [14] and process u using average pooling.

2.3 $E_{\mathcal{D}}(\cdot, \cdot)$ Using Global and Local Semantic Embedding

Presumably, handling complex registration tasks rely on: (1) distinct features that are robust to inter-subject variation, organ deformation, contrast injection, and pathological changes, and (2) global/contextual information that can benefit the registration accuracy in complex deformation. To achieve these goals, we adopt self-supervised anatomical embedding (SAM) [27] based feature descriptor that encodes both global and local embeddings. The global embeddings memorize the 3D contextual information of body parts on a coarse level, while the local embeddings differentiate adjacent structures with similar appearances, as shown on the left of Fig. 1. The former helps the latter to focus on small regions to distinguish with fine-level features.

To be specific, given an image $I \in \mathbb{R}^{H \times W \times D}$, we utilize a pre-trained SAM model that outputs a global embedding $f_{global} \in \mathbb{R}^{128 \times \frac{H}{8} \times \frac{W}{8} \times \frac{D}{2}}$ and a local embedding $f_{local} \in \mathbb{R}^{128 \times \frac{H}{2} \times \frac{W}{2} \times \frac{D}{2}}$. We resample the global embedding with bilinear interpolation to match the shape of f_{global} to f_{local} and concatenate them to get $f_{SAM} \in \mathbb{R}^{256 \times \frac{H}{2} \times \frac{W}{2} \times \frac{D}{2}}$. We normalize f_{global} and f_{local} before concatenation and adopt the dot product $< \cdot, \cdot >$ as the similarity measure in the following part, with a higher value indicating better alignment.

With f_{SAM}, we construct $E_{\mathcal{D}}(\cdot, \cdot)$ as the cost volume in the SAM feature space within a neighborhood of $[-N, N]$, where N represents searching radius. Given two images I^0 and I^1, the resulting $E_{\mathcal{D}}(\cdot, \cdot)$ can then be formulated as

$$E_{\mathcal{D}}(I^0, I^1) = < f^0_{SAM}(x), f^1_{SAM}(x+d) > \tag{4}$$

where f^0_{SAM} and f^1_{SAM} are the SAM embedding of I^0 and I^1, respectively. x is any voxel in the image domain and d is the voxel displacement within the neighborhood. $E_{\mathcal{D}}(\cdot, \cdot)$ has the shape of $(2N+1) \times (2N+1) \times (2N+1) \times \frac{H}{2} \times \frac{W}{2} \times \frac{D}{2}$.

2.4 Coarse-to-Fine Optimization Strategy

Since $E_{\mathcal{D}}(\cdot, \cdot)$ is computed within the neighborhood, the magnitude of deformation is bounded by the size of the neighborhood. Our approach to addressing this issue is based on a surprisingly simple but effective observation: performing registration based on a coarse-to-fine scheme with a small search range at each level instead of a large search range at one resolution benefits to 1) significantly improve the efficiency such as low computation burden and fast running time; 2) enlarge receptive field and improve registration accuracy.

We first build an image pyramid of I_s and I_t. At each resolution, we warp the source image with the composed deformation computed from all the previous levels (starting from the coarsest resolution), transform the warped image and target image to the SAM space, and conduct the decoupling to convex optimizations strategy to obtain the deformation between the warped image and target at the current resolution. With such a coarse-to-fine strategy, we estimate a sequence of displacement fields from a starting identity transformation. The final field is computed via the composition of all the displacement fields [30].

To be noted, the cost volume at each level is computed by taking the inputs of $\{1, \frac{1}{2}, \frac{1}{4}\}$ resolution. Adding one coarser level consumes less computation than doubling the size of the neighborhood at the fine resolution but yields the same search range.

2.5   Implementation Detail

We conduct the registration in 3 levels of resolutions. At each level, we solve Eq. 3 via five alternative optimizations with $\frac{1}{2\theta} = \{0.003, 0.01, 0.03, 0.1, 0.3, 1\}$. To further improve the registration performance, we append a SAM-based instance optimization after the coarse-to-fine registration with a learning rate of 0.05 for 50 iterations. It solves an instance-specific optimization problem [3] in the SAM feature space. The optimization objective consists of similarity (dot product between SAM feature vectors on the highest resolution) and diffusion regularization terms. All the experiments are run on the CPU of Intel Xeon Platinum 8163 with 16 cores of 2.50 GHz and the GPU of NVIDIA Tesla V100. Code will be available at https://github.com/Alison-brie/SAMConvex.

3   Experiments

3.1   Datasets and Evaluation Metrics

We split Abdomen and HeadNeck into a training set and test set to accommodate the requirement of the training dataset of comparing learning-based methods. Lung dataset contains 35 CT pairs, which is not sufficient for developing learning-based methods. Hence, it is only used as a testing set for optimization-based methods. All the methods are evaluated on the test set. The SAM is pre-trained on NIH Lymph Nodes dataset [27]. All 3 datasets in this paper are not used for pre-training.

Inter-patient Task on Abdomen: The Abdomen CT dataset [12] contains 30 abdominal scans with 20 for training and 10 for testing. Each image has 13 manually labeled anatomical structures: spleen, right kidney, left kidney, gall bladder, esophagus, liver, stomach, aorta, inferior vena cava, portal and splenic vein, pancreas, left adrenal gland, and right adrenal gland. The images are resampled to the same voxel resolution of 2 mm and spatial dimensions of $192 \times 160 \times 256$.

Inter-patient Task on HeadNeck: The HeadNeck CT dataset [28] contains 72 subjects with 13 organs labeled. The manually labeled anatomical structures include the brainstem, left eye, right eye, left lens, right lens, optic chiasm, left optic nerve, right optic nerve, left parotid, right parotid, left and right temporo-mandibular joint, and spinal cord. We split the dataset into 52, 10, and 10 for training, validation, and test set. For testing, we construct 90 image pairs for registration, with each image, resampled to an isotropic resolution of 2 mm and cropped to $256 \times 128 \times 224$.

Intra-patient Task on Lung: The images are acquired as part of the radio-therapy planning process for the treatment of malignancies. We collect a lung 4D CT dataset containing 35 patients, each with inspiratory and expiratory breath-hold image pairs, and take this dataset as an extra test set. Each image has labeled malignancies, and we try to align two phases for motion estimation of malignancies. The images are resampled to a spacing of $2 \times 2 \times 2$ mm and cropped to $256 \times 256 \times 112$. All images are used as testing cases.

Table 1. Quantitative results of inter- and intra-subject registration. The subscript of each metric indicates the number of anatomical structures involved. ↑: higher is better, and ↓: lower is better. Initial: initial affine results without deformable registration. Ours indicates the coarse-to-fine registration and IO indicates the instance optimization.

Method	Abdomen CT			HeadNeck CT			Lung CT		
	DSC_{13} ↑	SDlogJ ↓	T_{test} ↓	DSC_{13} ↑	SDlogJ ↓	T_{test} ↓	DSC_1 ↑	SDlogJ ↓	T_{test} ↓
Initial	32.64	-	-	33.91	-	-	67.06	-	-
NiftyReg	34.98	0.22	186.54 s	57.66	0.21	168.39 s	74.59	0.02	85.50 s
Deeds	46.52	0.44	45.21 s	61.32	1.07	42.05 s	81.53	0.49	227.59 s
ConvexAdam	44.44	0.74	6.06 s	61.45	0.77	3.12 s	79.26	1.17	2.20 s
LapIRN	46.44	0.72	0.07 s	60.16	0.58	0.03 s	-	-	-
SAME	47.12	0.90	6.88 s	61.35	1.00	3.34 s	-	-	-
Ours	48.30	0.86	2.12 s	61.88	0.79	1.88 s	80.13	0.37	1.86 s
Ours + IO	**51.17**	0.86	5.14 s	**63.72**	0.88	4.59 s	**81.61**	0.21	4.34 s

Evaluation Metrics: We use the average Dice score (DSC) to evaluate the accuracy and compute the standard deviation of the logarithm of the Jacobian determinant (SDlogJ) to evaluate the plausibility of deformation fields, also comparing running time (T_{test}) on the same hardware.

3.2 Registration Results

We compare with five widely-used and top-performing deformable registration methods, including NiftyReg [24], Deeds [9], top-ranked ConvexAdam [20] in Learn2Reg Challenge [12], and SOTA learning-based methods LapIRN [18] and SAME [15]. All results are based on the SAM-affine [15] pre-alignment.

As shown in Table 1, SAMConvex outperforms the widely-used NiftyReg [24] across the three datasets with an average of **10.0%** Dice score improvement.

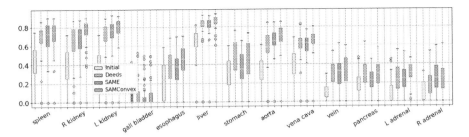

Fig. 2. Comparison of performance-top-three deformable registration methods on all organ groups on **Abdomen** dataset.

Compared with the best traditional optimization-based method Deeds [9], SAM-Convex performs better or comparatively on the three datasets with approximately **20–100** times faster runtime. We also see a better performance of SAMConvex over ConvexAdam [20]. To be noted, the performance gap between SAMConvex and ConvexAdam is greater on Abdomen CT than the other two datasets. This may be because the deformation is more complex in Abdomen and the anatomical differences between inter-subjects make the registration more challenging. With a coarse-to-fine strategy, SAMConvex can better bear these issues. Although LapIRN [18] has the fastest inference time, it has slightly inferior registration accuracy as compared to SAMConvex. Moreover, it needs training when being applied to a new dataset. Equipped with SAM, SAME achieves the overall 2nd best performance in two inter-patient registration tasks but with a notably higher folding rate overall. Moreover, it also requires individual training for a new dataset and struggles with the intra-patient lung registration task (small dataset and no available training data). In summary, our SAMConvex achieves the overall best performance as compared to other methods with a fast inference time and comparable deformation field smoothness.

To better understand the performance of different deformable registration methods, we display organ-specific results of the inter-patient registration task of abdomen CT in Fig. 2 where large deformations and complex anatomical differences exist. As shown, SAMConvex is consistently better than the second-best and third-best registration methods on most examined abdominal organs, in 11 out of 13 organs.

3.3 Ablation Study on SAMConvex

Loss Landscape of SAM. We explore the loss landscapes of SAM-global, SAM-local, SAM (SAM-global and SAM-local), and MIND via varying the transformation parameters. We conduct experiments on the abdomen dataset[2]. Figure 3 shows the comparison of landscapes when we vary the rotation along two axes. The loss landscape falls to flat quickly when the rotation is greater than $20°$ degrees and $40°$ degrees for SAM-local and MIND, respectively. The flattened

---

[2] Refer to Appendix for more experiment results.

area indicates where the similarity measure loses the capability to guide the registration because multiple transformation parameters yield the same similarity value. Compared with SAM-local and MIND, SAM-global does not have the flattened area within $[-60°, 60°]$, meaning it can give correct measure over a larger capture range. However, it does not have the fast converging area shown as a spike for the loss landscapes of MIND and SAM-local. When combining SAM-global and SAM-local together, SAM shows a greater capture range meanwhile fast convergence when the transformation is small.

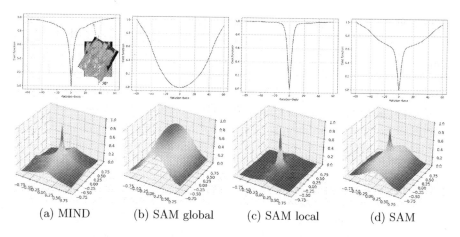

(a) MIND   (b) SAM global   (c) SAM local   (d) SAM

Fig. 3. Loss landscapes of MIND, SAM-global, SAM-local and SAM on **Abdomen** dataset with varying rotations along two axes.

Robustness to Pre-alignment. We study the robustness of SAMConvex over different affine pre-alignment. Elastix-Affine [13] and SAM-Affine are used as the pre-alignment and results are notated as $Initial_e$ and $Initial_a$ in Table 2, respectively. Compared to ConvexAdam, our method SAMConvex is less affected by the performance of the Affine pre-alignment. This is in line with our expectations. The global information contained in SAM provides $E_D(I_s \circ \varphi, I_t)$ a greater capture range than features that contain local information only. Thus, SAM-Convex can register images with larger transformation, leading to less reliance on the performance of the pre-alignment.

Ablation on Coarse-to-Fine. We study how the number of coarse-to-fine layers and how the size of the neighborhood window affect the registration result on the Abdomen dataset. From Table 3, we can conclude that performing registration with a small search range with a coarse-to-fine scheme instead of a cost volume with a large search range can help to improve registration accuracy and computation efficiency.

Table 2. ConvexAdam and SAMConvex on different affine pre-alignment.

Method	DSC$_{13}$ ↑	
	Initial$_e$	Initial$_a$
Initial	25.88	32.64
ConvexAdam	34.42	44.44
SAMConvex	45.38	48.30

Table 3. Ablation study on how different pyramid designs affect the performance. The subscript of Pym indicates the number of scales.

Method	DSC$_{13}$ ↑	SDlogJ ↓	T_{test} ↓
Pym$_1$ $N \in [3]$	40.51	0.59	1.69 s
Pym$_1$ $N \in [6]$	44.22	1.62	7.84 s
Pym$_2$ $N \in [3,3]$	47.80	0.76	2.03 s
Pym$_3$ $N \in [2,3,3]$	48.30	0.86	2.12 s

Discussion About the Differences with ConvexAdam [20]. Apart from the SAM extraction module, first, we introduce a cost volume pyramid to the convex optimization framework to reduce the intensive computation burdens. To be specific, with the coarse-to-fine strategy, one can sparsely search on a coarser level with a smaller search radius in each iteration, reaching the same search range as the complete search with less computational complexity. Second, we explicitly validate the robustness of the SAM feature against geometrical transformations and integrate the SAM feature into an instance-specific optimization pipeline. Our SAMConvex is less sensitive to local minimal, achieving superior performance with comparable running time. Ablation study on pyramid designs (Table 3) and leading accuracy on large deformation registration tasks (Table 1) further support our claims.

4   Conclusion

We present SAMConvex, a coarse-to-fine discrete optimization method for CT registration. It extracts per-voxel semantic global and local features and builds a series of lightweight 6D correlation volumes, and iteratively updates a flow field by performing lookups on the correlation volumes. The performance on two inter-patient and one intra-patient registration datasets demonstrates state-of-the-art accuracy, good generalization, and high computation efficiency of SAMConvex.

References

1. Ashburner, J.: A fast diffeomorphic image registration algorithm. Neuroimage **38**(1), 95–113 (2007)
2. Avants, B.B., Epstein, C.L., Grossman, M., Gee, J.C.: Symmetric diffeomorphic image registration with cross-correlation: evaluating automated labeling of elderly and neurodegenerative brain. Med. Image Anal. **12**(1), 26–41 (2008)
3. Balakrishnan, G., Zhao, A., Sabuncu, M.R., Guttag, J.V., Dalca, A.V.: Voxelmorph: a learning framework for deformable medical image registration. IEEE Trans. Med. Imaging **38**(8), 1788–1800 (2019)
4. Chambolle, A.: An algorithm for total variation minimization and applications. J. Math. Imaging Vis. **20**, 89–97 (2004)

5. Chen, Q., Koltun, V.: Full flow: optical flow estimation by global optimization over regular grids. In: Proceedings of the IEEE Conference on Computer Vision and Pattern Recognition, pp. 4706–4714 (2016)
6. Fan, X., et al.: Automated learning for deformable medical image registration by jointly optimizing network architectures and objective functions. arXiv:2203.06810 (2022)
7. Heinrich, M.P.: Closing the gap between deep and conventional image registration using probabilistic dense displacement networks. In: International Conference on Medical Image Computing and Computer-Assisted Intervention (2019)
8. Heinrich, M.P., et al.: MIND: modality independent neighbourhood descriptor for multi-modal deformable registration. Med. Image Anal. **16**(7), 1423–1435 (2012)
9. Heinrich, M.P., Jenkinson, M., Brady, S.M., Schnabel, J.A.: Globally optimal deformable registration on a minimum spanning tree using dense displacement sampling. In: Ayache, N., Delingette, H., Golland, P., Mori, K. (eds.) MICCAI 2012. LNCS, vol. 7512, pp. 115–122. Springer, Heidelberg (2012). https://doi.org/10.1007/978-3-642-33454-2_15
10. Heinrich, M.P., Jenkinson, M., Papiez, B.W., Brady, M., Schnabel, J.A.: Towards realtime multimodal fusion for image-guided interventions using self-similarities. In: Medical Image Computing and Computer-Assisted Intervention, vol. 8149, pp. 187–194 (2013)
11. Heinrich, M.P., Papież, B.W., Schnabel, J.A., Handels, H.: Non-parametric discrete registration with convex optimisation. In: Ourselin, S., Modat, M. (eds.) WBIR 2014. LNCS, vol. 8545, pp. 51–61. Springer, Cham (2014). https://doi.org/10.1007/978-3-319-08554-8_6
12. Hering, A., et al.: Learn2Reg: comprehensive multi-task medical image registration challenge, dataset and evaluation in the era of deep learning. IEEE Trans. Med. Imaging **42**(3), 697–712 (2022)
13. Klein, S., Staring, M., Murphy, K., Viergever, M.A., Pluim, J.P.W.: elastix: a toolbox for intensity-based medical image registration. IEEE Trans. Med. Imaging **29**(1), 196–205 (2010)
14. Krähenbühl, P., Koltun, V.: Efficient inference in fully connected CRFs with gaussian edge potentials. In: Advances in Neural Information Processing Systems, pp. 109–117 (2011)
15. Liu, F., et al.: SAME: deformable image registration based on self-supervised anatomical embeddings. In: de Bruijne, M., et al. (eds.) MICCAI 2021. LNCS, vol. 12904, pp. 87–97. Springer, Cham (2021). https://doi.org/10.1007/978-3-030-87202-1_9
16. Liu, R., Li, Z., Fan, X., Zhao, C., Huang, H., Luo, Z.: Learning deformable image registration from optimization: perspective, modules, bilevel training and beyond. IEEE Trans. Pattern Anal. Mach. Intell. **44**(11), 7688–7704 (2022)
17. Liu, R., Li, Z., Zhang, Y., Fan, X., Luo, Z.: Bi-level probabilistic feature learning for deformable image registration. In: Proceedings of the Twenty-Ninth International Joint Conference on Artificial Intelligence, pp. 723–730 (2020)
18. Mok, T.C.W., Chung, A.C.S.: Large deformation diffeomorphic image registration with laplacian pyramid networks. In: Martel, A.L., et al. (eds.) MICCAI 2020. LNCS, vol. 12263, pp. 211–221. Springer, Cham (2020). https://doi.org/10.1007/978-3-030-59716-0_21
19. Shen, D., Davatzikos, C.: Hammer: hierarchical attribute matching mechanism for elastic registration. IEEE Trans. Med. Imaging **21**(11), 1421–1439 (2002)

20. Siebert, H., Hansen, L., Heinrich, M.P.: Fast 3D registration with accurate optimisation and little learning for Learn2Reg 2021. In: Aubreville, M., Zimmerer, D., Heinrich, M. (eds.) MICCAI 2021. LNCS, vol. 13166, pp. 174–179. Springer, Cham (2022). https://doi.org/10.1007/978-3-030-97281-3_25

21. Sotiras, A., Davatzikos, C., Paragios, N.: Deformable medical image registration: a survey. IEEE Trans. Med. Imaging **32**(7), 1153–1190 (2013)

22. Steinbrücker, F., Pock, T., Cremers, D.: Large displacement optical flow computation withoutwarping. In: 2009 IEEE 12th International Conference on Computer Vision, pp. 1609–1614 (2009). https://doi.org/10.1109/ICCV.2009.5459364

23. Steinbrücker, F., Pock, T., Cremers, D.: Large displacement optical flow computation withoutwarping. In: 2009 IEEE 12th International Conference on Computer Vision, pp. 1609–1614. IEEE (2009)

24. Sun, W., Niessen, W.J., Klein, S.: Free-form deformation using lower-order b-spline for nonrigid image registration. In: Medical Image Computing and Computer Assisted Intervention, pp. 194–201 (2014)

25. Tian, L., Greer, H., Vialard, F.X., Kwitt, R., Estépar, R.S.J., Niethammer, M.: Gradicon: approximate diffeomorphisms via gradient inverse consistency. arXiv preprint arXiv:2206.05897 (2022)

26. Xu, J., Ranftl, R., Koltun, V.: Accurate optical flow via direct cost volume processing. In: 2017 IEEE Conference on Computer Vision and Pattern Recognition, pp. 5807–5815 (2017)

27. Yan, K., et al.: SAM: Self-supervised learning of pixel-wise anatomical embeddings in radiological images. IEEE Trans. Med. Imaging **41**(10), 2658–2669 (2022)

28. Ye, X., et al.: Comprehensive and clinically accurate head and neck cancer organs-at-risk delineation on a multi-institutional study. Nat. Commun. **13**(1), 6137 (2022)

29. Zach, C., Pock, T., Bischof, H.: A duality based approach for realtime TV-L^1 optical flow. In: Hamprecht, F.A., Schnörr, C., Jähne, B. (eds.) DAGM 2007. LNCS, vol. 4713, pp. 214–223. Springer, Heidelberg (2007). https://doi.org/10.1007/978-3-540-74936-3_22

30. Zhao, S., Lau, T., Luo, J., Eric, I., Chang, C., Xu, Y.: Unsupervised 3D end-to-end medical image registration with volume tweening network. IEEE J. Biomed. Health Inform. **24**(5), 1394–1404 (2019)

CycleSTTN: A Learning-Based Temporal Model for Specular Augmentation in Endoscopy

Rema Daher[1]([✉])[ID], O. León Barbed[2][ID], Ana C. Murillo[2][ID],
Francisco Vasconcelos[1][ID], and Danail Stoyanov[1][ID]

[1] University College London, London, UK
{rema.daher.20,danail.stoyanov}@ucl.ac.uk
[2] Universidad de Zaragoza, Zaragoza, Spain
acm@unizar.es

Abstract. Feature detection and matching is a computer vision problem that underpins different computer assisted techniques in endoscopy, including anatomy and lesion recognition, camera motion estimation, and 3D reconstruction. This problem is made extremely challenging due to the abundant presence of specular reflections. Most of the solutions proposed in the literature are based on filtering or masking out these regions as an additional processing step. There has been little investigation into explicitly learning robustness to such artefacts with single-step end-to-end training. In this paper, we propose an augmentation technique (CycleSTTN) that adds temporally consistent and realistic specularities to endoscopic videos. Such videos can act as ground truth data with known texture occluded behind the added specularities. We demonstrate that our image generation technique produces better results than a standard CycleGAN model. Additionally, we leverage this data augmentation to re-train a deep-learning based feature extractor (SuperPoint) and show that it improves. CycleSTTN code is made available here.

Keywords: Surgical Data Science · Surgical AI · Generative AI · Deep Learning · Endoscopy · Specularity

1 Introduction

During endoscopic procedures, such as colonoscopy, the camera light source produces abundant specular highlight reflections on the visualised anatomy. This is due to its very close proximity to the scene coupled with the presence of wet tissue. These reflections can occlude texture and produce salient artifacts, which may reduce the accuracy of surgical vision algorithms aiming at scene understanding, including depth estimation and 3D reconstruction [5,16]. To resolve

Supplementary Information The online version contains supplementary material available at https://doi.org/10.1007/978-3-031-43999-5_54.

these challenges, a simple solution is to detect and mask out these regions before performing any other downstream visual task. However, this approach comes with limitations. In the context of sparse feature point detection, simply discarding points falling on specularities results in excessive filtering, often leading to an insufficient number of detected features. As for dense estimation problems, such as pixel depth regression and optical flow, masking out specular regions makes interpolation necessary.

Alternatively, videos can be pre-processed to inpaint specular highlights with its hidden texture inferred from neighbouring frames [7]. This allows running other algorithms with reflection-free data. However, this adds a significant computational overhead and requires the processing of temporal frame windows that restrict online inference applications. Given that state-of-the-art feature detection methods are deep learning models, an appealing approach would be to learn robustness to reflections during training, as this would result in an single-step end-to-end inference model without any pre or post-processing overhead.

In this paper, we propose to learn robustness to specular reflections via data augmentation. We use a CycleGAN [24] methodology that takes advantage of a pre-trained specular highlight removal network in adversarial training. Our proposed generator network, based on STTN [7,23], performs video-to-video translation. In doing so, we create a cycle structure for STTN, that we call CycleSTTN. To demonstrate the effectiveness of our approach, we use it to improve the performance of the SuperPoint feature detector/descriptor. We combine the proposed method with that of [7] to add and remove specularities as data augmentation. The contributions of this paper can be summarised as:

- We propose the CycleSTTN training pipeline as an extension of STTN to a cyclic structure.
- We use CycleSTTN to train a model for synthetic generation of temporally consistent and realistic specularities in endoscopy videos. We compare results of our method against CycleGAN.
- We demonstrate CycleSTTN as a data augmentation technique that improves the performance of SuperPoint feature detector in endoscopy videos.

2   Related Work

Barbed et al. investigated the challenge of specular reflections when performing feature detection in endoscopy images [3]. As a solution, they design a loss function to explicitly avoid specular regions and focus on detecting points in the remaining parts of the image. However, this ignores the fact that features extracted from these points will still be contaminated by specularity pixels in their neighborhood. Following a different strategy, we propose data augmentation as a way to induce specularity robustness in both point detection and feature extraction.

Data augmentation for lighting conditions in surgical data has been explored extensively before. Classical techniques for modelling specular highlights include the use of a parametric model to add specularities and illumination to real images [1,13]. Another technique for augmentation uses synthetic data or phantoms, where different light sources and environment variables can be captured. However, synthetic data lacks real textures and artifacts. Thus many image-to-image translation techniques have been developed to add more realistic textures to synthetic data [15,17,21,22]. These methods rely on CycleGAN [24] for unsupervised learning and thus, are able to map from real to synthetic domain and vice-versa. This allows for the generation of real images with the same structure but different lighting, texture, and blurriness. To have more control over the augmentations, some methods add a controllable noise vector to the network input that modifies the image lighting, specular reflections, and texture. However, this vector does not have a physical meaning, and it is thus challenging to independently control the different environment variables by directly manipulating its values. To address this, [14] uses two separate noise inputs, one controlling texture and specular reflection and another controlling color and light. However, all these approaches still use multiple steps to finally create new real data, which might lead to loss of important information in the process.

Single-step approaches have also been developed that augment real data directly, but the generated images have different structures and thus, do not create paired data [9,20]. A CycleGAN model has been proposed to map from images with specularities to images without specularities [11] using manually labelled patches cropped from frames, however, this work focuses on specular highlight removal, and does not test the data augmentation capabilities of generating synthetic specularities. In [12] a classification model is used to categorize data for unpaired training of CycleGAN. From the output of CycleGAN, they generate a paired dataset, however, only quality metrics are used to filter out images in the generated paired dataset. Furthermore, most of these approaches are applicable to single frames and are not able to generate synthetic videos with temporally consistent specularities. While some other methods use a temporal component for endoscopic video augmentation [17,21], they do not have a single-step structure and have not been applied to generate/remove specular highlights.

3   Methods

We use the STTN model as our video-to-video translation architecture and T-PatchGAN [6] as the discriminator. STTN contains an encoder followed by a spatio-temporal transformer and a decoder [7,23]. While STTN was originally proposed to remove occlusions in videos, in this paper, we use two instances of this model, $STTN_R$ and $STTN_A$, to respectively remove and add specular occlusions. We start by pre-training these models separately using their respective discriminators D_R and D_A. Then, we continue training them simultaneously in an adversarial manner with a CycleGAN methodology. We denote the complete training pipeline as CycleSTTN (Fig. 1), which is divided into 3 sections:

1. ***Paired Dataset Generation*** We generate a dataset of paired videos with and without specularities. We train a generator for specularity removal fol-

lowing the same methodology as in [7]. This model is denoted as $(STTN_{R0},$
$D_{R0})$, where $STTN$ is the generator and D is the discriminator. For a set
of real endoscopic videos with specularities V_A and specularity masks M, we
run $STTN_{R0}$ to generate their inpainted counterparts without specularities
V_R (Fig. 1 - Step 1).

2. **$STTN_A$ Pre-training** Using the paired dataset (V_A, V_R) we train a new
model to add specularities. We denote it as $STTN_{A0}$ with D_{A0} as its discriminator. This is shown in Fig. 1 as step 2.

3. **$STTN_R, STTN_A$ Joint Training** By initializing with the models from
Step 1 and 2 $((STTN_{R0}, D_{R0})$ and $(STTN_{A0}, D_{A0}))$, we continue train-
ing $STTN_R$ and $STTN_A$ simultaneously in an adversarial manner with a
CycleGAN methodology. We denote the final removal and addition models
as $(STTN_{R1}, D_{R1})$ and $(STTN_{A1}, D_{A1})$, respectively. This is also shown in
Fig. 1 as step 3.

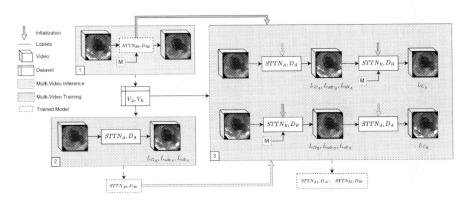

Fig. 1. CycleSTTN training pipeline with 3 main steps: 1 Paired Dataset Generation,
2 $STTN_A$ Pre-training, and 3 $STTN_R, STTN_A$ Joint Training.

3.1 Model Inputs

The STTN architecture receives as input a paired sequence of frames and masks.
Originally masks are meant to represent occluded image regions that should
be inpainted, however, in this work, we do not always use them in this way.
When training $STTN_R$ we define the mask inputs as regions to be inpainted (to
remove specularities). However, when training $STTN_A$, input masks are set to
1 for all pixels, since we do not want to enforce specific locations for specularity
generation; we want to encourage the model to learn these patterns from data.

3.2   Losses

The loss function of the T-PatchGAN discriminators are shown below, such that $\mathbb{E}$ is the expected value of the data distributions as done in [23]:

$$L_{D_R} = \mathbb{E}_{V_R \sim p(V_R)}[ReLU(1 - D_R(V_R))] + \mathbb{E}_{Fake'_R \sim p(Fake'_R)}[ReLU(1 + D_R(Fake'_R))] \tag{1}$$

$$L_{D_A} = \mathbb{E}_{V_A \sim p(V_A)}[ReLU(1 - D_A(V_A))] + \mathbb{E}_{Fake_A \sim p(Fake_A)}[ReLU(1 + D_A(Fake_A))] \tag{2}$$

where $Fake_A = STTN_A(V_R)$ represents fake videos with added specularities, and analogously $Fake_R = STTN_R(V_A)$ represents fake videos with removed specularities. Further, we also define $Fake'_R = M.Fake_R + V_A(1 - M)$ for the discriminator loss, where inpainted occluded regions from $Fake_R$ are overlaid over V_A. M denotes masks with 1 values in specular regions of V_A and 0 otherwise.

For the generators, an adversarial loss was used as done in [7]:

$$L_{adv_R} = -\mathbb{E}_{Fake'_R \sim p(Fake'_R)}[D_R(Fake'_R)]; \quad L_{adv_A} = -\mathbb{E}_{Fake_A \sim p(Fake_A)}[D_A(Fake_A)] \tag{3}$$

The identity loss was the only loss modified from the original STTN model [7]. The identity loss for $STTN_R$ and $STTN_A$ are:

$$L_{idt_R} = \|V_R - STTN_R(V_R)\|_1; \quad L_{idt_A} = \|V_A - Fake_A\|_1 \tag{4}$$

Here L_{idt_R} ensures that if a video does not have specularities, it would stay the same when fed into the model that removes specularities. Whereas, L_{idt_A} ensures predicted videos with specularities, $Fake_A$, resemble real specular videos V_A.

Finally, we added cycle loss terms:

$$L_{c_R} = \mathbb{E}_{V_A \sim p(V_A)}[\|V_A - STTN_A(Fake_R)\|_1]; \quad L_{c_A} = \mathbb{E}_{V_R \sim p(V_R)}[\|V_R - STTN_R(Fake_A)\|_1] \tag{5}$$

The total generator losses L_R and L_A for removing and adding specularities are shown below, such that the loss weights λ are all set to 1 except for λ_{adv}, which is set to 0.01 as advised by [23]:

$$L_R = \lambda_{adv}L_{adv_R} + \lambda_{idt}L_{idt_R} + \lambda_c L_{c_R}; \quad L_A = \lambda_{adv}L_{adv_A} + \lambda_{idt}L_{idt_A} + \lambda_c L_{c_A} \tag{6}$$

In summary, we adopted the original STTN model [7] and changed the training pipeline, model inputs, and losses as shown in Fig. 1. In particular, (a) the training pipeline was transformed into a multi-task one of adding and removing specularities, where $STTN_{R0}$ from [7] was used as an initialization model. (b) For $STTN_A$, specularity masks were removed from model inputs. (c) Identity losses and cycle losses were also added while masked based losses were removed.

4   Experiments and Results

4.1   Datasets and Parameters

To evaluate our pipeline, we use 373 videos from the Hyper Kvasir dataset [4] to generate our paired dataset (V_R, V_A) as described in Sect. 3 and Fig. 1. 343 video pairs were used for training and 30 for testing with an upper limit of 927 frames per video. Models were trained on NVIDIA A100-SXM4-40GB GPUs. We use the same training parameters as [7,23] with the exception of batch size, which we changed from 8 to 3 for training $(STTN_{A1}, STTN_{R1})$. CycleGAN models were trained with suggested parameters in CycleGAN's public repository[1], with the exception of batch size, which was changed from 1 to 3.

In our experimental analysis, we use our proposed models shown in Fig. 1 along with CycleGAN models, which use ResNet with 9 residual blocks as the generator. All these models are listed in Table 1. We note that even though $STTN_{R1}$ was trained with masks, it seems it was affected by the cycle loss and was only able to give decent results without a mask as input.

Table 1. Trained models used in our analysis input type and training iterations.

Proposed Models	Input	Iterations	CycleGAN Models	Identity Loss	Input	Iterations
$STTN_{R0}$	videos + masks	90,000	$(ResNet_{A0}, ResNet_{R0})$	Original [24]	videos	285,600
$STTN_{A0}$	videos	30,000	$(ResNet_{A1}, ResNet_{R1})$	L_{idt_A} - Eq. 4	videos	285,600
$(STTN_{A1}, STTN_{R1})$	videos	20,000				

4.2 Pseudo Evaluation Experiments

We input the pseudo ground truth V_R to our models, $STTN_{A0}$ and $STTN_{A1}$. We compare the output $Fake_A$ to real videos V_A. We conduct non-temporal testing by using single frame inputs, as opposed to video inputs, to demonstrate the temporal effect. We report the Peak Signal to Noise Ratio (PSNR), Structural Similarity Index (SSIM), and Mean Square Error (MSE) metrics.

We show visual results for different models in Fig. 2. When V_R (videos with no specularities) are used as input, our CycleSTTN model $STTN_{A1}$ shows the highest similarity to the ground truth (V_A). With V_A (videos with specularities) as input, $STTN_{A1}$ is able to add more realistic specularities that flow smoothly from one frame to another. CycleGAN based models were not able to add new specularities with V_A as input. For $ResNet_{A0}$, this is expected, due to the original CycleGAN identity loss. When changing the identity loss, $ResNet_{A1}$ only intensifies specularities and darkens the background texture. This was further validated through results shown in Table 2, where $STTN_{A1}$ has the best SSIM, PSNR and MSE values. We can also see that $STTN_{A1}$ results are only slightly

---

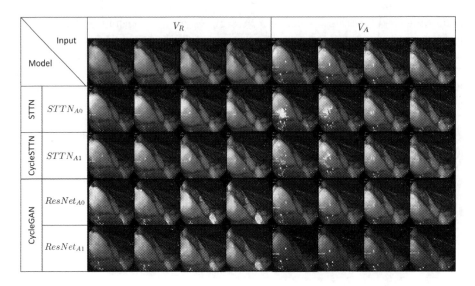

Fig. 2. Sample consecutive video frames from the model output using pseudo ground truth V_R as input in columns 1–4 and real videos V_A as input in columns 5–8.

Table 2. Mean SSIM, PSNR, and MSE values for model output videos using pseudo ground truth V_R as input. The output is compared to real videos V_A.

Method	Non-Temporal Testing						Temporal Testing					
	SSIM		PSNR		MSE		SSIM		PSNR		MSE	
	Mean	Std	Mean	Std	Mean	Std	Mean	Std	Mean	Std	Mean	Std
$STTN_{A0}$	0.802	0.040	26.20	2.31	231	98.0	0.808	0.039	26.33	2.37	226	91.9
$STTN_{A1}$	**0.826**	0.036	26.38	2.43	224	102.5	0.824	0.036	**26.49**	2.45	**219**	91.4
$ResNet_{A0}$	0.792	0.041	21.29	1.92	675	311	-	-	-	-	-	-
$ResNet_{A1}$	0.780	0.044	21.48	2.03	666	329	-	-	-	-	-	-

worse without temporal testing according to PSNR and MSE, yet slightly better according to SSIM. Thus, the temporal component only benefits training as a regularizer (i.e. our models outperforms CycleGAN based models). However, during inference, temporal testing is not necessary since it yields similar results to non-temporal testing.

The pseudo evaluation shows that generated specularities closely resemble real ones in appearance and location. While not fully enforcing physical realism, this augmentation improves upon traditional warping methods (Sect. 4.3).

4.3   Relative Pose Estimation Experiments

We use our models as data augmentation to re-train the feature detector proposed in [3], an adaptation of SuperPoint [8] to endoscopy. While [3] is originally trained with a specularity loss term that encourages the network to ignore specularity regions, we omit this term in our training. We want our models to be robust to specularities, rather than just avoiding them. We use the pre-trained model in [3] to generate data labels and initialize our models. SuperPoint is trained by generating a sparse set of point matches in an image and its warped version via homography. Augmentations already used by SuperPoint include traditional random brightness, contrast, Gaussian noise, speckle noise, blur, and shade. We use our models as augmentations to original and warped images separately by randomly choosing between (no augmentation, specularity addition, and specularity removal). SuperPoint models with various augmentations are listed in Table 3. These models are trained using 131 randomly chosen videos from V_A (with cropped boundaries) and 14 for validation for 25,000 iterations with the same training parameters as [3]. Temporal (non-temporal) refers to feeding an image and its warped version together (separately) to the augmentation model.

Table 3. Pose estimation analysis for 12 SuperPoint models each trained with different specular augmentations. Metrics are described in Sect. 4.3.

	Specularity Augmentation Models		Non-Temporal		Rot. error		Temporal		Rot. error	
	Addition	Removal	Matches	Precision	Mean	Median	Matches	Precision	Mean	Median
1	-	-	368.6	25.0	24.8	12.0	-	-	-	-
2	-	$STTN_{R0}$	537.4	29.2	21.4	10.1	538.2	29.2	21.2	9.8
3	$STTN_{A0}$	$STTN_{R0}$	402.1	27.8	21.7	10.7	404.2	27.5	22.0	10.6
4	$STTN_{A0}$	-	398.3	26.7	23.7	11.4	390.1	26.5	22.8	11.1
5	-	$STTN_{R1}$	373.3	26.0	22.6	10.3	542.8	28.6	23.4	11.1
6	$STTN_{A1}$	$STTN_{R1}$	360.3	27.3	21.3	9.6	548.5	28.7	23.1	11.0
7	$STTN_{A1}$	-	386.3	27.2	21.5	10.1	542.5	28.3	23.5	11.2
8	$STTN_{A1}$	$STTN_{R0}$	541.5	29.9	20.4	9.5	543.6	29.8	20.2	9.6
9	$STTN_{A0}$	$STTN_{R1}$	536.6	29.2	21.4	10.0	526.7	29.0	21.1	9.7
10	-	$ResNet_{R0}$	571.1	27.6	22.0	10.3				
11	$ResNet_{A0}$	$ResNet_{R0}$	531.0	26.8	22.5	10.6				
12	$ResNet_{A0}$	-	529.4	27.3	22.6	10.5				
13	-	$ResNet_{R1}$	571.6	28.2	22.7	10.5				
14	$ResNet_{A1}$	$ResNet_{R1}$	548.4	29.0	22.2	10.3				
15	$ResNet_{A1}$	-	527.3	29.0	22.4	10.5				

We evaluate the quality of point detections by using them to estimate relative camera motion in endoscopic sequences. We first apply brute-force matching of detected points in image pairs and then estimate motion via RANSAC [10]. The test data for this experiment is the same as in [3]. It includes 6 sequences (14191 frames) from the EndoMapper dataset [2] with a relative camera motion pseudo ground truth based on structure-from-motion (SFM: COLMAP [18,19]). Reported metrics include the precision of inlier points from RANSAC-estimated (threshold = 10 px) essential matrices as compared to inlier points using pseudo ground truth essential matrix (from COLMAP). To generate the pseudo ground truth inliers, the same distance metric used in RANSAC was applied to the pseudo ground truth essential matrix (from COLMAP). We also report the Rota-

R. Daher et al.

tion error, which is the geodesic angle in degrees between the RANSAC-based pose estimation and the pseudo ground truth pose (from COLMAP).

In Table 3, all specularity augmentations (models 2–15) improve Super-Point relative to not using them (model 1), which demonstrates their usefulness. Most STTN-based augmentations (models 2–9) produce better results than CycleGAN-based ones (models 10–15). Overall, the best performing augmentation is $(STTN_{A1}, STTN_{R0})$, showing the effectiveness of our system. This makes sense since the best removal and addition models are $STTN_{R0}$ and $STTN_{A1}$ according to the rotation error. However, it appears that non-temporal testing sometimes gives lower rotation errors than temporal testing. This could be due to the unrealistic nature of warped images used as consecutive frames. As also discussed in Sect. 4.2, temporal testing does not improve results, only temporal training does.

5   Conclusion

In conclusion, we introduce CycleSTTN, a temporal CycleGAN applied to generate temporally consistent and realistic specularities in endoscopy. Our model outperforms CycleGAN, as demonstrated by mean PSNR, MSE, and SSIM metrics using a pseudo ground truth dataset. We also observe a positive effect of our model as augmentation for training a feature extractor, resulting in improved inlier precision and rotation errors. However, our evaluation relies on SFM generated ground truth, different testing and training datasets, and indirect metrics, which may introduce some uncertainty. Nevertheless, augmentation shows great promise as an addition for training various endoscopic computer vision tasks to enhance performance and provide insights into the impact of specific artifacts.

Acknowledgments. This research was funded in part, by the Wellcome/EPSRC Centre for Interventional and Surgical Sciences (WEISS) [203145/Z/16/Z]; the Engineering and Physical Sciences Research Council (EPSRC) [EP/P027938/1, EP/R004080/1, EP/P012841/1]; the Royal Academy of Engineering Chair in Emerging Technologies Scheme; H2020 FET (GA863146); and the UCL Centre for Digital Innovation through the Amazon Web Services (AWS) Doctoral Scholarship in Digital Innovation 2022/2023. For the purpose of open access, the author has applied a CC BY public copyright licence to any author accepted manuscript version arising from this submission.

References

1. Asif, M., Chen, L., Song, H., Yang, J., Frangi, A.F.: An automatic framework for endoscopic image restoration and enhancement. Appl. Intell. **51**(4), 1959–1971 (2021)
2. Azagra, P., et al.: Endomapper dataset of complete calibrated endoscopy procedures. arXiv preprint arXiv:2204.14240 (2022)

3. Barbed, O.L., Chadebecq, F., Morlana, J., Montiel, J.M.M., Murillo, A.C.: Super-point features in endoscopy. In: Manfredi, L., et al. (eds.) ISGIE GRAIL 2022. LNCS, vol. 13754, pp. 45–55. Springer, Cham (2022). https://doi.org/10.1007/978-3-031-21083-9_5

4. Borgli, H., et al.: Hyperkvasir, a comprehensive multi-class image and video dataset for gastrointestinal endoscopy. Sci. Data **7**(1), 1–14 (2020)

5. Chadebecq, F., Lovat, L.B., Stoyanov, D.: Artificial intelligence and automation in endoscopy and surgery. Nat. Rev. Gastroenterol. Hepatol. **20**(3), 171–182 (2023)

6. Chang, Y.L., Liu, Z.Y., Lee, K.Y., Hsu, W.: Free-form video inpainting with 3D gated convolution and temporal PatchGAN. In: Proceedings of the IEEE/CVF International Conference on Computer Vision, pp. 9066–9075 (2019)

7. Daher, R., Vasconcelos, F., Stoyanov, D.: A temporal learning approach to inpainting endoscopic specularities and its effect on image correspondence. arXiv preprint arXiv:2203.17013 (2022)

8. DeTone, D., Malisiewicz, T., Rabinovich, A.: Superpoint: self-supervised interest point detection and description. In: Proceedings of the IEEE Conference on Computer Vision and Pattern Recognition Workshops, pp. 224–236 (2018)

9. Diamantis, D.E., Gatoula, P., Iakovidis, D.K.: Endovae: generating endoscopic images with a variational autoencoder. In: 2022 IEEE 14th Image, Video, and Multidimensional Signal Processing Workshop (IVMSP), pp. 1–5. IEEE (2022)

10. Fischler, M.A., Bolles, R.C.: Random sample consensus: a paradigm for model fitting with applications to image analysis and automated cartography. Commun. ACM **24**(6), 381–395 (1981)

11. Funke, I., Bodenstedt, S., Riediger, C., Weitz, J., Speidel, S.: Generative adversarial networks for specular highlight removal in endoscopic images. In: Medical Imaging 2018: Image-Guided Procedures, Robotic Interventions, and Modeling, vol. 10576, pp. 8–16. SPIE (2018)

12. García-Vega, A., et al.: A novel hybrid endoscopic dataset for evaluating machine learning-based photometric image enhancement models. In: Pichardo Lagunas, O., Martínez-Miranda, J., Martínez Seis, B. (eds.) MICAI 2022. LNCS, vol. 13612, pp. 267–281. Springer, Cham (2022). https://doi.org/10.1007/978-3-031-19493-1_22

13. Hegenbart, S., Uhl, A., Vécsei, A.: Impact of endoscopic image degradations on LBP based features using one-class SVM for classification of celiac disease. In: 2011 7th International Symposium on Image and Signal Processing and Analysis (ISPA), pp. 715–720. IEEE (2011)

14. Mathew, S., Nadeem, S., Kaufman, A.: CLTS-GAN: color-lighting-texture-specular reflection augmentation for colonoscopy. In: Wang, L., Dou, Q., Fletcher, P.T., Speidel, S., Li, S. (eds.) MICCAI 2022. LNCS, vol. 13437, pp. 519–529. Springer, Cham (2022). https://doi.org/10.1007/978-3-031-16449-1_49

15. Mathew, S., Nadeem, S., Kumari, S., Kaufman, A.: Augmenting colonoscopy using extended and directional cyclegan for lossy image translation. In: Proceedings of the IEEE/CVF Conference on Computer Vision and Pattern Recognition, pp. 4696–4705 (2020)

16. Ozyoruk, K.B., et al.: Endoslam dataset and an unsupervised monocular visual odometry and depth estimation approach for endoscopic videos. Med. Image Anal. **71**, 102058 (2021)

17. Rivoir, D., et al.: Long-term temporally consistent unpaired video translation from simulated surgical 3D data. In: Proceedings of the IEEE/CVF International Conference on Computer Vision, pp. 3343–3353 (2021)

18. Schonberger, J.L., Frahm, J.M.: Structure-from-motion revisited. In: Proceedings of the IEEE Conference on Computer Vision and Pattern Recognition, pp. 4104–4113 (2016)
19. Schönberger, J.L., Zheng, E., Frahm, J.-M., Pollefeys, M.: Pixelwise view selection for unstructured multi-view stereo. In: Leibe, B., Matas, J., Sebe, N., Welling, M. (eds.) ECCV 2016. LNCS, vol. 9907, pp. 501–518. Springer, Cham (2016). https://doi.org/10.1007/978-3-319-46487-9_31
20. de Souza Jr, L.A., et al.: Assisting barrett's esophagus identification using endoscopic data augmentation based on generative adversarial networks. Comput. Biol. Med. **126**, 104029 (2020)
21. Xu, J., et al.: OfGAN: realistic rendition of synthetic colonoscopy videos. In: Martel, A.L., et al. (eds.) MICCAI 2020. LNCS, vol. 12263, pp. 732–741. Springer, Cham (2020). https://doi.org/10.1007/978-3-030-59716-0_70
22. Yamane, H., et al.: Automatic generation of polyp image using depth map for endoscope dataset. Procedia Comput. Sci. **192**, 2355–2364 (2021)
23. Zeng, Y., Fu, J., Chao, H.: Learning joint spatial-temporal transformations for video inpainting. In: Vedaldi, A., Bischof, H., Brox, T., Frahm, J.-M. (eds.) ECCV 2020. LNCS, vol. 12361, pp. 528–543. Springer, Cham (2020). https://doi.org/10.1007/978-3-030-58517-4_31
24. Zhu, J.Y., Park, T., Isola, P., Efros, A.A.: Unpaired image-to-image translation using cycle-consistent adversarial networks. In: 2017 IEEE International Conference on Computer Vision (ICCV) (2017)

Importance Weighted Variational Cardiac MRI Registration Using Transformer and Implicit Prior

Kangrong Xu[1,2,3], Qirui Huang[1,2,3], and Xuan Yang[1,2,3]([envelope]) [ORCID]

[1] Shenzhen University, Shenzhen 518060, Guangdong, China
yangxuan@szu.edu.cn
[2] Guangdong Provincial Key Laboratory of Popular High Performance Computers, Shenzhen, Guangdong, China
[3] Shenzhen Key Laboratory of Service Computing and Application, Shenzhen, Guangdong, China

Abstract. The variational registration model takes advantage of explaining uncertainties of registration results. However, most existing variational registration models are based on convolutional neural networks (CNNs), which cannot capture distant information in images. Besides, the evidence lower bound (ELBO) and the commonly used standard prior cannot close the gap between the real posterior and the variational posterior in the vanilla variational registration model. This paper proposes a network in a variational image registration model for cardiac motion estimation to effectively capture the spatial correspondence of long-distance images and solve the shortcomings of CNNs. Our proposed network comprises a Transformer with a T2T module and the cross attention between the moving and the fixed images. To close the gap between the real posterior and the variational posterior, the importance-weighted evidence lower bound (iwELBO) is introduced into the variational registration model with an implicit prior. The coefficients of a parametric transformation using multi-supports CSRBFs are latent variables in our variational registration model, which improve registration accuracy significantly. Experimental results show that the proposed method outperforms state-of-arts research on public cardiac datasets.

Keywords: Variational inference · Transformer · Cross attention · Compact support radial basis function (CSRBF)

1 Introduction

Cardiac motion estimation is vital in evaluating cardiac function, detecting heart diseases, and understanding cardiac biomechanics. Deformable image registration (DIR) is the critical technique of cardiac motion estimation. It minimizes the

---

Supplementary Information The online version contains supplementary material available at https://doi.org/10.1007/978-3-031-43999-5_55.

differences between the warped moving and fixed images to estimate a displacement vector field (DVF). Unsupervised deep-learning-based image registration has recently become mainstream due to the required non-annotation data [4,16] and rapid inference performance when the network is well-trained.

The probabilistic generative model shows potential in the unsupervised learning registration [11–13,17,18,22]. It allows the registration framework to be highly adaptable and can be applied to cases with a small amount of data and anatomical variability. Another advantage of the probabilistic generative registration model is that it can provide registration uncertainties [1], which plays a vital role in clinical application [14,21]. However, several issues exist with variational image registration approaches. The first is that traditional convolutions are limited in representing long-range relationships between image features. The second issue is a gap between the objective function ELBO and the log-likelihood of input image pairs, which deteriorates registration accuracy. Besides, non-parametric transformation is commonly used [5,20,27], which has the challenge of regularizing the DVF smooth and topology-preserving.

This paper proposed a novel variational image registration model to cope with the above issues by employing the Transformers with the cross-attention mechanism and introducing an importance-weighted ELBO (iwELBO) [7] with an implicit prior. Detailed contributions of our work include:

- A novel VAE network architecture is proposed, which employs the Transformer architecture to focus on cross-attention between the moving and fixed images. The predictive results of the transformation parameter distribution using our architecture are more accurate than traditional VAE architecture.
- We optimized the importance-weighted ELBO in the variational image registration model. We use approximated aggregated posterior as the prior to regularizing posterior. To our best knowledge, we are the first to combine the iwELBO and aggregated posterior to close the gap between the real and variational posterior.
- A parametric transformation based on multi-supports CSRBFs is embedded in our variational registration model. By imposing a sparse constraint, the coefficients of multi-CSRBFs are regularized to be sparse to select the optimal support for multi-support CSRBFs. The parametric transformation model improves registration accuracy significantly and makes it easy to regularize the smoothness of DVFs.

2   Proposed Method

2.1   Importance-Weighted Variational Image Registration Model

Given the moving and fixed images M, F, and n control points $\{p_i\}_{i=1}^n$, the parametric transformation based on multi-supports CSRBFs is $f_z(v) = v + \sum_{i=1}^n \sum_{k=1}^s z_{i,k} \phi(\frac{\|v-p_i\|_2}{c_k})$, where $\phi(\frac{\|\cdot\|}{c})$ is a CSRBF with support c; $\|v-p_i\|_2$ is the Euclidean distance between the pixel v and p_i. s different supports $\{c_k\}_{k=1}^s$

are provided for each CSRBF. $z = \{z_k\}_{k=1}^{s}$, $z_k = \{z_{k,i}\}_{i=1}^{n}$ is the latent variable whose distribution is required to be estimated. The parametric transformation can control deformations using different supports. By imposing sparse constraints, selecting the optimal support from these given supports is possible, leading to more flexible deformation results.

The variational registration model aims to estimate the posterior of $p(z|F, M)$. In the vanilla variational registration model, $q(z|F, M)$ is estimated to approximate $p(z|F, M)$ by optimizing $ELBO = \mathbb{E}_{z \sim q(z|F,M)}\left[\log \frac{p(F|z,M)p(z)}{q(z|F,M)}\right]$, where $p(F|z, M)$ is the probability of occurring F when the moving image M is deformed using a transformation f_z with latent variables z. It can be expressed as Boltzmann distribution $p(F|z, M) \propto e^{sim(F,M(f_z))}$ using a similarity metric sim. $p(z)$ is the prior of z.

The importance-weighted evidence lower bound (iwELBO) [7] is defined as,

$$iwELBO = \mathbb{E}_{z^1,..,z^K \sim q(z|F,M)}\left[\log \frac{1}{K}\sum_{k=1}^{K}\frac{p(F|z^k,M)p(z^k)}{q(z^k|F,M)}\right] \qquad (1)$$

where $z^1, \ldots, z^K$ are K-samples of latent variable z sampled from $q(z|F, M)$. It is assumed that $q(z|F, M) \sim \mathcal{N}(\boldsymbol{\mu}(F, M), \boldsymbol{\Sigma}(F, M))$, and $z^1, \ldots, z^K$ is sampled as $z^k = \boldsymbol{\mu}(F, M) + \boldsymbol{\Sigma}(F, M)\epsilon_k$, $\epsilon_k \sim \mathcal{N}(0, \boldsymbol{I})$ using the reparametrization trick.

Denote $w_k = \frac{p(F|z^k,M)p(z^k)}{q(z^k|F,M)}$, the gradient of iwELBO can be interpreted as normalized importance-weighted average gradients of each sample, which implies the sample with larger w_k contributes more to iwELBO. It is challenging to compute w_k directly due to the high dimensional latent variable z^k. We tackle this problem by a trick.

$$\arg\max iwELBO = \arg\max \mathbb{E}_{z^1,..,z^K \sim q(z|F,M)}\left[\log \frac{1}{K}\sum_{k=1}^{K}e^{\log w_k^{\frac{1}{\lambda}}}\right], \lambda > 0. \qquad (2)$$

Because $\log w_k \propto sim(F, M(f_{z^k})) + \log\frac{p(z^k)}{q(z^k|F,M)}$, our objective function is

$$\mathbb{E}_{z^1,..,z^K \sim q(z|F,M)}\left[\log \frac{1}{K}\sum_{k=1}^{K}e^{sim(F,M(f_{z^k}))+\frac{1}{\lambda}\log\frac{p(z^k)}{q(z^k|F,M)}}\right]. \qquad (3)$$

iwELBO is a tighter evidence lower bound of the log-likelihood of data; it approaches the log-likelihood $\log p(F|M)$ as $K \to \infty$. From the view of image registration, the iwELBO tends to converge to a transformation with the optimal w_k from K samples. When a large hyperparameter λ is used, $\frac{1}{\lambda}\log\frac{p(z^k|M)}{q(z^k|F,M)}$ is relative smaller compared with the similarity term. That implies the iwELBO prefers the sample z^k leading to the optimal similarity between the warped moving and fixed images, which is beneficial to push the network to predict more accurate z. Besides, Huang et al. [15] pointed out that when only one sample corresponding to a high loss is drawn to estimate the iwELBO, iwELBO tolerates this mistake due to the importance of weight. On the contrary, the sample with high loss will be highly penalized in traditional ELBO because the decoder treats the sample as real, observed data.

2.2   Aggregate Posterior as the Prior

A simple prior, such as the standard Normal in VAE, incurred over-regularization on the posterior and widened the gap between the variational posterior and the real posterior. Many researchers resolve this mismatch by proposing various priors [2,10,23–25]. Tomczak *et al.* stated that the optimal prior in VAE is the aggregated posterior of data. Takahashi *et al.* [23] introduced the density ratio trick to estimate this aggregated posterior implicitly. However, all these works are evaluated based on ELBO instead of iwELBO. We derive and approximate the optimal prior based on iwELBO, please refer to part 1 in the supplement.

The optimal prior should maximize the expectation of iwELBO:

$$\underset{p(z)}{\arg\max} \int p(F,M)\mathbb{E}_{z^1,..z^K \sim q(z|F,M)} \left[\log \frac{1}{K} \sum_{k=1}^{K} e^{sim(F,M(f_{z^k})) + \frac{1}{\lambda} \log \frac{p(z^k)}{q(z^k|F,M)}} \right] d(F,M) \tag{4}$$

It can be derived that the optimal prior $p^*(z)$ is approximated as the aggregated posterior $\mathbb{E}_{p(F,M)}q(z|F,M)$. Substituting the optimal prior $p^*(z)$ into Eq. (3), the objective function is rewritten as

$$\mathbb{E}_{z^1,..z^K \sim q(z|F,M)} \left[\log \frac{1}{K} \sum_{k=1}^{K} e^{sim(F,M(f_{z^k})) + \frac{1}{\lambda} \log \frac{p_0(z^k)}{q(z^k|F,M)} + \frac{1}{\lambda} \log \frac{p^*(z^k)}{p_0(z^k)}} \right]. \tag{5}$$

where $p_0(z)$ is a simple given prior. To estimate the density ratio $\log \frac{p^*(z)}{p_0(z)}$, a binary discriminator $T(\mathbf{z})$ is trained by maximizing [23],

$$\max \mathbb{E}_{p^*(z)}[\sigma(T(\mathbf{z}))] + \mathbb{E}_{p_0(z)}[\log(1 - \sigma(T(\mathbf{z})))] \tag{6}$$

where σ is the sigmoid function. The discriminator is a neural network composed of four fully connected layers, with the final layer outputting density ratio. A dropout layer is added before the output to prevent the discriminator network from overfitting. When $T(\mathbf{z})$ is well trained, $\log \frac{p^*(z^k)}{p_0(z^k)} \approx T(\mathbf{z}^k)$. The given prior $p_0(z)$ is defined as $\mathcal{N}(0, \mathbf{B}^{-1})$, where $\mathbf{B} = diag(B^1, \cdots, B^s)$, B^k is a $n \times n$ matirx with the entries $B_{ij}^k = \phi(\frac{\|p_i - p_j\|}{c_k})$, $k = 1, \ldots, s$. Then,

$$\log \frac{p_0(z^k)}{q(z^k|F,M)} = \log |\mathbf{\Sigma}(F,M)| + \log |\mathbf{B}| \\ - \frac{1}{2} \left[\mathbf{z}^{k^T} \mathbf{B} \mathbf{z}^k - (\mathbf{z}^k - \boldsymbol{\mu}(F,M))^T \mathbf{\Sigma}^{-1}(F,M)(\mathbf{z}^k - \boldsymbol{\mu}(F,M)) \right]. \tag{7}$$

where $\mathbf{z}^{k^T} \mathbf{B} \mathbf{z}^k$ is the bending energy of DVF using multi-supports CSRBFs aiming to regularize DVF smooth. The sampling size for k is 5; λ is 110000.

We optimize our network by iterating a two-step procedure. The encoder is updated using Eq. (7) by fixing the discriminator. Next, the discriminator is updated using Eq. (6) by fixing the encoder. Above two steps are performed alternatively.

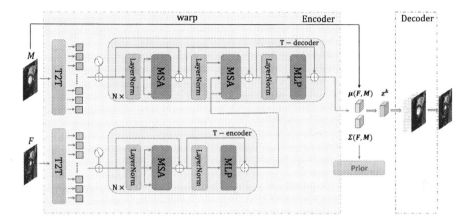

Fig. 1. The architecture of our network. The moving and fixed images M and F are preprocessed by T2T modules first and then input to our Transformer's T-encoder and T-decoder, respectively. Self-MSA and cross-MSA In our Transformer are marked by green and orange, respectively. (Color figure online)

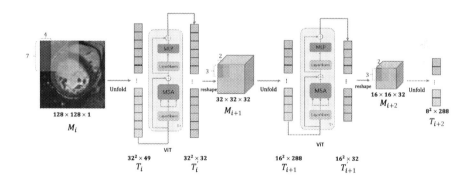

Fig. 2. T2T Module with three iterations.

2.3 Network

Our network architecture consists of an encoder and a decoder, as shown in Fig. 1. The encoder composes of T2T modules [26] and a Transformer to predict $\mu(F, M)), \Sigma(F, M)$. T2T modules preprocess M and F and then input to our Transformer's T-encoder and T-decoder, which pays self-attention and cross-attention of M and F, respectively. Outputs of the cross-MSA (marked by orange) are fused information of M and F, weighted by similarity. More details of the Transformer Network can be seen in part 2 of the supplementary materials. The inherent ability to capture the correlation between two images makes Transformers easier to extract effective features for image registration. The decoder of our network generates the warped moving image using the DVF obtained by transformation based on multi-supports CSRBFs. The number of Transformer blocks N is 3.

T2T modules tackle the partitioning issue in Transformer by modeling the local structure information iteratively. As shown in Fig. 2, overlapped patches are encoded by an unfold operation as a vector token T_i and re-encoded as T_i' using a ViT. T_i' is reshaped as M_{i+1} with the size of overlapped patches. The above process is repeated three times to obtain the final vector T_{i+2}, which is the input of our Transformer.

The total loss function of our network $\mathcal{L}$ is combing the iwELBO and sparse constraints on z as $\mathcal{L} = iwELBO - \|z\|_1$.

3   Experiments

Four public cardiac datasets are used to evaluate our method, including MIC-CAI2009 [19], York [3], ACDC [6], and M&Ms [8]. We combine MICCAI, York, and ACDC as a dataset denoted as ACDC+. Contours of the left ventricle (LV), the right ventricle (RV), or myocardium (MYO) at the end-diastolic (ED) phase or the end-systolic (ES) phase are provided by experts for different datasets. We take the ED and ES images as moving and fixed images, respectively. The training, validation, and testing slices are 1257, 130, 698 for ACDC+ and 1134, 266, 1030 for M&Ms. All images are cropped into the size of 128×128 containing the heart. Data augmentations such as flips and rotations are used. The LCC with the size of 9×9 is employed as the similarity metric. 64 global control points are evenly spaced on the 128×128 image, while 100 local control points are evenly spaced in an area of 64×64 in the center of the image that contains the heart. Our network is trained using PyTorch on a computer equipped with an Intel(R) Xeon(R) Silver 4210 CPU and Nvidia RTX 2080Ti GPU. The Adam optimizer with a learning rate of $5e^{-4}$ is employed. We use the estimated DVFs to map the contours of the moving image and compare the mapped contours with the contours of the fixed image using various metrics, including the Dice score, the average perpendicular distance (APD, in mm), and 95%-tile Hausdorff Distance (HD, in mm). Moreover, we count the number of anomalies to measure the topological property of the DVFs $|J_{f_z}| \le 0$ and calculate the bending energy (BE, $\times 10^{-4}$) of DVFs to measure their smoothness.

3.1   Results

To compare the performance of our method with unsupervised registration networks KrebsDiff [17], DalcaDiff [11], NetGI [13], VoxelMorph [4], and Transmorph [9]. KrebsDiff, DalcaDiff, and NetGI are networks of variational registration. Transmorph is a network embedding Transformers. VoxelMorph is a vanilla unsupervised registration network. Registration results of two datasets using different networks are listed in Table 1. Our network outperforms other networks regarding Dice, HD, and APD. NetGI is the most similar registration model to our method, achieving the smoothest DVFs, while DalcaDiff preserved the topology of DVF best due to the diffeomorphism deformation it used. The bending energy and topology-preserving of DVFs using our network are close to that of NetGI and DalcaDiff, which implies that the transformation model based on multi-supports CSRBFs is good at generating smooth and topology-preserving DVFs. Visualization of the registration results using different networks is shown in Fig. 3. The myocardium of the fixed image is marked by green, while the warped moving image is marked by blue. The overlap of the myocardium is marked by red. Here, registration results of the basal, middle, and apical slices are provided. Note that objects in apical slices are small, while our network matches the small myocardium better than other networks.

Table 1. Comparison of registration results on ACDC+ and M&Ms datasets using different networks. Data format: mean (standard deviations)

Dataset	Method	BE	$|J_{f_z}| \leq 0$	Dice	HD	APD
ACDC+	KrebsDiff [17]	27.71 (15.97)	2.01 (14.67)	0.840 (0.063)	5.66 (1.88)	2.21 (0.77)
	DalcaDiff [11]	165.60 (40.34)	**0.01 (0.04)**	0.849 (0.064)	5.78 (1.93)	2.09 (0.81)
	VoxelMorph [4]	149.26 (32.05)	26.17 (21.08)	0.845 (0.066)	5.84 (1.95)	2.15 (0.84)
	NetGI [13]	**12.68 (5.91)**	4.96 (14.45)	0.854 (0.057)	5.46 (1.87)	2.00 (0.67)
	TransMorph [9]	104.78 (27.16)	20.12 (27.55)	0.850 (0.064)	5.71 (1.89)	2.08 (0.82)
	Ours	24.09 (6.04)	0.84 (4.28)	**0.867 (0.049)**	**5.17** (1.57)	**1.87** (0.58)
M&Ms	KrebsDiff [17]	27.90 (15.97)	4.92 (10.09)	0.828 (0.054)	4.57 (2.24)	1.82 (0.93)
	DalcaDiff [11]	90.15 (55.86)	**0.01 (0.05)**	0.853 (0.050)	4.21 (2.07)	1.53 (0.82)
	VoxelMorph [4]	253.47 (59.40)	36.73 (25.40)	0.848 (0.059)	4.48 (2.34)	1.60 (0.92)
	NetGI [13]	**12.14 (4.47)**	0.18 (0.93)	0.847 (0.052)	4.32 (2.38)	1.63 (0.94)
	TransMorph [9]	169.36 (44.52)	33.08 (38.52)	0.860 (0.052)	4.13 (2.05)	1.48 (0.74)
	Ours	21.23 (5.64)	0.77 (1.56)	**0.869 (0.042)**	**3.84 (1.81)**	**1.41** (0.65)

3.2 Ablation Study

Ablation experiments are performed on the ACDC+ dataset to validate the influence of different components in our method. Table 2 lists evaluation results using the different combinations of components. Using ELBO as the objective function, the transformation based on multi-supports CSRBFs improves Dice 5%. Dice is improved 3% when the aggregated posterior is used as the prior. Whether the standard normal or the aggregated posterior as prior, the importance-weighted ELBO improves about 2–3% in Dice compared with ELBO. It is noted that when iwELBO is used, the aggregated posterior cannot improve registration compared

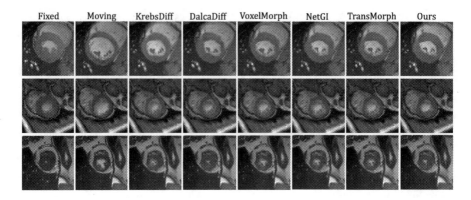

| Fixed | Moving | KrebsDiff | DalcaDiff | VoxelMorph | NetGI | TransMorph | Ours |

Fig. 3. Demonstration of registration results using different networks. (Color figure online)

with the standard Normal prior. The reason might be that the registration accuracy has a value close to its limit due to iwELBO; in this case, there is no space for the aggregated posterior as prior to improve registration accuracy further. Moreover, we find improvement in registration accuracy using iwELBO comes from the improvement of apical slices registration, details can be referred to in part 3 of the supplement.

Table 2. Comparison of the influence of different parts in our method on ACDC+ dataset. Data format: mean (stand deviation)

ELBO	iwELBO	Single support	Multi-supports	$\mathcal{N}(0, I)$ prior	Aggregated prior	Dice	BE	$|J_{f_z}| \leq 0$
✓		✓		✓		0.858 (0.053)	**7.48 (3.16)**	**0.36 (1.31)**
✓		✓			✓	0.861 (0.051)	8.48 (3.36)	0.50 (2.09)
✓			✓	✓		0.863 (0.053)	22.99 (5.88)	0.77 (3.52)
✓			✓		✓	0.865 (0.051)	21.16 (6.01)	0.62 (2.33)
	✓		✓	✓		0.866 (0.053)	20.63 (5.31)	0.69 (1.20)
	✓		✓		✓	**0.867 (0.049)**	24.09 (6.04)	0.84 (4.28)

Further, we replaced the encoder of our network using ViT. Since only the T-encoder existed in ViT, we concatenated the moving and fixed images as input of ViT. In this experiment, different loops in ViT and our Transformer, denoted as ViT-n and Ours-n (n is the number of loops), are employed to compare the performance of self-attention and cross-attention. Table 3 lists comparison results of ViT and our Transformer. It can be seen that cross-attention outperforms self-attention, and more loops are not beneficial in predicting the posterior parameters.

Table 3. Comparison of registration results on ACDC+ dataset using self-attention and cross-attention mechanisms, respectively. Data format: mean (standard deviations)

Method	BE	$|J_{f_z}| \leq 0$	Dice	HD	APD
ViT-3	26.28 (6.61)	0.68 (1.76)	0.865 (0.054)	5.23 (1.73)	**1.86 (0.63)**
ViT-6	19.58 (5.60)	0.70 (1.85)	0.862 (0.053)	5.30 (1.70)	1.91 (0.62)
Ours-3	24.09 (6.04)	0.84 (4.28)	**0.867 (0.049)**	**5.17 (1.57)**	1.87 (0.58)
Ours-6	**18.81 (5.26)**	**0.53 (1.32)**	0.862 (0.053)	5.33 (1.66)	1.92 (0.64)

4 Conclusion

In this paper, we proposed a novel variational registration model using Transformer to pay attention to cross-attention between images. The importance-weighted ELBO and the aggregated posterior as prior close the gap between the real posterior and the variational posterior. Our transformation using multi-supports CSRBFs generates flexible DVFs. Evaluation results on public cardiac datasets show that our method outperforms the state-of-art networks.

Acknowledgements. This paper is supported by the Shenzhen Fundamental Research Program (JCYJ20220531102407018).

References

1. Abdar, M., et al.: A review of uncertainty quantification in deep learning: techniques, applications and challenges. Inf. Fusion **76**, 243–297 (2021)
2. Akkari, N., Casenave, F., Daniel, T., Ryckelynck, D.: Data-targeted prior distribution for variational autoencoder. Fluids (2021)
3. Andreopoulos, A., Tsotsos, J.K.: Efficient and generalizable statistical models of shape and appearance for analysis of cardiac MRI. Med. Image Anal. **12**(3), 335–357 (2008)
4. Balakrishnan, G., Zhao, A., Sabuncu, M.R., Guttag, J., Dalca, A.V.: An unsupervised learning model for deformable medical image registration. In: Proceedings of the IEEE Conference on Computer Vision and Pattern Recognition, pp. 9252–9260 (2018)
5. Balakrishnan, G., Zhao, A., Sabuncu, M.R., Guttag, J., Dalca, A.V.: VoxelMorph: a learning framework for deformable medical image registration. IEEE Trans. Med. Imaging **38**(8), 1788–1800 (2019)
6. Bernard, O., et al.: Deep learning techniques for automatic MRI cardiac multi-structures segmentation and diagnosis: is the problem solved? IEEE Trans. Med. Imaging **37**(11), 2514–2525 (2018)
7. Burda, Y., Grosse, R., Salakhutdinov, R.: Importance weighted autoencoders. arXiv preprint arXiv:1509.00519 (2015)
8. Campello, V.M., et al.: Multi-centre, multi-vendor and multi-disease cardiac segmentation: the M&Ms challenge. IEEE Trans. Med. Imaging 9458279 (2021)
9. Chen, J., Frey, E.C., He, Y., Segars, W.P., Li, Y., Du, Y.: TransMorph: transformer for unsupervised medical image registration. arXiv preprint arXiv:2111.10480 (2021)

10. Connor, M., Canal, G.H., Rozell, C.J.: Variational autoencoder with learned latent structure. ArXiv abs/2006.10597 (2021)
11. Dalca, A.V., Balakrishnan, G., Guttag, J., Sabuncu, M.R.: Unsupervised learning for fast probabilistic diffeomorphic registration. In: Frangi, A.F., Schnabel, J.A., Davatzikos, C., Alberola-López, C., Fichtinger, G. (eds.) MICCAI 2018. LNCS, vol. 11070, pp. 729–738. Springer, Cham (2018). https://doi.org/10.1007/978-3-030-00928-1_82
12. Fan, J., Cao, X., Xue, Z., Yap, P.-T., Shen, D.: Adversarial similarity network for evaluating image alignment in deep learning based registration. In: Frangi, A.F., Schnabel, J.A., Davatzikos, C., Alberola-López, C., Fichtinger, G. (eds.) MICCAI 2018. LNCS, vol. 11070, pp. 739–746. Springer, Cham (2018). https://doi.org/10.1007/978-3-030-00928-1_83
13. Gan, Z., Sun, W., Liao, K., Yang, X.: Probabilistic modeling for image registration using radial basis functions: Application to cardiac motion estimation. IEEE Trans. Neural Netw. Learn. Syst. (2022)
14. Gong, X., Khaidem, L., Zhu, W., Zhang, B., Doermann, D.: Uncertainty learning towards unsupervised deformable medical image registration. In: Proceedings of the IEEE/CVF Winter Conference on Applications of Computer Vision, pp. 2484–2493 (2022)
15. Huang, C.W., Sankaran, K., Dhekane, E., Lacoste, A., Courville, A.: Hierarchical importance weighted autoencoders. In: International Conference on Machine Learning, pp. 2869–2878. PMLR (2019)
16. Kim, B., Kim, D.H., Park, S.H., Kim, J., Lee, J.G., Ye, J.C.: CycleMorph: cycle consistent unsupervised deformable image registration. Med. Image Anal. **71**, 102036 (2021)
17. Krebs, J., Mansi, T., Mailhé, B., Ayache, N., Delingette, H.: Unsupervised probabilistic deformation modeling for robust diffeomorphic registration. In: Stoyanov, D., et al. (eds.) DLMIA/ML-CDS -2018. LNCS, vol. 11045, pp. 101–109. Springer, Cham (2018). https://doi.org/10.1007/978-3-030-00889-5_12
18. Liu, R., Li, Z., Zhang, Y., Fan, X., Luo, Z.: Bi-level probabilistic feature learning for deformable image registration. In: Proceedings of the Twenty-Ninth International Conference on International Joint Conferences on Artificial Intelligence, pp. 723–730 (2021)
19. Radau, P., Lu, Y., Connelly, K., Paul, G., Dick, A., Wright, G.: Evaluation framework for algorithms segmenting short axis cardiac MRI. The MIDAS J.-Cardiac MR Left Ventricle Segment. Challenge **49** (2009)
20. Sandkühler, R., Andermatt, S., Bauman, G., Nyilas, S., Jud, C., Cattin, P.C.: Recurrent registration neural networks for deformable image registration. Adv. Neural Inf. Process. Syst. **32** (2019)
21. Sedghi, A., Kapur, T., Luo, J., Mousavi, P., Wells, W.M.: Probabilistic image registration via deep multi-class classification: characterizing uncertainty. In: Greenspan, H., et al. (eds.) CLIP/UNSURE -2019. LNCS, vol. 11840, pp. 12–22. Springer, Cham (2019). https://doi.org/10.1007/978-3-030-32689-0_2
22. Sheikhjafari, A., Noga, M., Punithakumar, K., Ray, N.: Unsupervised deformable image registration with fully connected generative neural network (2018)
23. Takahashi, H., Iwata, T., Yamanaka, Y., Yamada, M., Yagi, S.: Variational autoencoder with implicit optimal priors. In: AAAI (2019)
24. Tomczak, J.M., Welling, M.: VAE with a VampPrior. In: AISTATS (2018)
25. Xu, H., Chen, W., Lai, J., Li, Z., Zhao, Y., Pei, D.: On the necessity and effectiveness of learning the prior of variational auto-encoder. ArXiv abs/1905.13452 (2019)

26. Yuan, L., et al.: Tokens-to-token ViT: training vision transformers from scratch on ImageNet. In: Proceedings of the IEEE/CVF International Conference on Computer Vision, pp. 558–567 (2021)
27. Zhao, S., Dong, Y., Chang, E.I., Xu, Y., et al.: Recursive cascaded networks for unsupervised medical image registration. In: Proceedings of the IEEE/CVF International Conference on Computer Vision, pp. 10600–10610 (2019)

Make-A-Volume: Leveraging Latent Diffusion Models for Cross-Modality 3D Brain MRI Synthesis

Lingting Zhu[1], Zeyue Xue[1], Zhenchao Jin[1], Xian Liu[2], Jingzhen He[3(✉)], Ziwei Liu[4], and Lequan Yu[1(✉)]

[1] The University of Hong Kong, Hong Kong SAR, China
ltzhu99@connect.hku.hk, lqyu@hku.hk
[2] The Chinese University of Hong Kong, Hong Kong SAR, China
[3] Qilu Hospital of Shandong University, Jinan, China
hjzhhjzh@163.com
[4] S-Lab, Nanyang Technological University, Singapore, Singapore

Abstract. Cross-modality medical image synthesis is a critical topic and has the potential to facilitate numerous applications in the medical imaging field. Despite recent successes in deep-learning-based generative models, most current medical image synthesis methods rely on generative adversarial networks and suffer from notorious mode collapse and unstable training. Moreover, the 2D backbone-driven approaches would easily result in volumetric inconsistency, while 3D backbones are challenging and impractical due to the tremendous memory cost and training difficulty. In this paper, we introduce a new paradigm for volumetric medical data synthesis by leveraging 2D backbones and present a diffusion-based framework, **Make-A-Volume**, for cross-modality 3D medical image synthesis. To learn the cross-modality slice-wise mapping, we employ a latent diffusion model and learn a low-dimensional latent space, resulting in high computational efficiency. To enable the 3D image synthesis and mitigate volumetric inconsistency, we further insert a series of volumetric layers in the 2D slice-mapping model and fine-tune them with paired 3D data. This paradigm extends the 2D image diffusion model to a volumetric version with a slightly increasing number of parameters and computation, offering a principled solution for generic cross-modality 3D medical image synthesis. We showcase the effectiveness of our Make-A-Volume framework on an in-house SWI-MRA brain MRI dataset and a public T1-T2 brain MRI dataset. Experimental results demonstrate that our framework achieves superior synthesis results with volumetric consistency.

Keywords: Cross-modality medical image synthesis · Volumetric data · Latent diffusion model · Brain MRI

1 Introduction

Medical images are essential in diagnosing and monitoring various diseases and patient conditions. Different imaging modalities, such as computed tomography (CT) and magnetic resonance imaging (MRI), and different parametric images, such as T1 and T2 MRI, have been developed to provide clinicians with a comprehensive understanding of the patients from multiple perspectives [7]. However, in clinical practice, it is commonly difficult to obtain a complete set of multiple modality images for diagnosis and treatment due to various reasons, such as modality corruption, incorrect machine settings, allergies to specific contrast agents, and limited available time [5,10]. Therefore, cross-modality medical image synthesis is useful by allowing clinicians to acquire different characteristics across modalities and facilitating real-world applications in radiology and radiation oncology [28,32].

With the rise of deep learning, numerous studies have emerged and are dedicated to medical image synthesis [4,7,18]. Notably, generative adversarial networks (GANs) [8] based approaches have garnered significant attention in this area due to their success in image generation and image-to-image translation [11,33]. Moreover, GANs are also closely related to cross-modality medical image synthesis [2,10,32]. However, despite their efficacy, GANs are susceptible to mode collapse and unstable training, which can negatively impact the performance of the model and decrease the reliability in practice [1,17]. Recently, the advent of denoising diffusion probabilistic models (DDPMs) [9,24] has introduced a new scheme for high-quality generation, offering desirable features such as better distribution coverage and more stable training when compared to GAN-based counterparts. Benefiting from the better performance [6], diffusion-based models may be deemed much more reliable and dominant and recently researchers have made the first attempts to employ diffusion models for medical image synthesis [12–14,19].

Different from natural images, most medical images are volumetric. Previous studies employ 2D networks as backbones to synthesize slices of medical volumetric data due to their ease of training [18,32] and then stack 2D results for 3D synthesis. However, this fashion induces volumetric inconsistency, particularly along the z-axis when following the standard way of placing the coordinate system. Although training 3D models may avoid this issue, it is challenging and impractical due to the massive amount of volumetric data required, and the higher dimension of the data would result in costly memory requirements [3,16,26]. To sum up, balancing the trade-off between training and volumetric consistency remains an open question that requires further investigation.

In this paper, we propose **Make-A-Volume**, a diffusion-based pipeline for cross-modality 3D brain MRI synthesis. Inspired by recent works that factorize video generation into multiple stages [23,31], we introduce a new paradigm for volumetric medical data synthesis by leveraging 2D backbones to simultaneously facilitate high-fidelity cross-modality synthesis and mitigate volumetric inconsistency for medical data. Specifically, we employ a latent diffusion model (LDM) [20] to function as a slice-wise mapping that learns cross-modality translation in an image-to-image manner. Benefiting from the low-dimensional latent space of LDMs, the high memory requirements for training are mitigated. To enable the 3D image synthesis and enhance volumetric smoothness among medical slices, we further insert and fine-tune a series of volumetric layers to upgrade the slice-wise model to a volume-wise model. In summary, our contributions are three-fold: (1) We introduce a generic paradigm for 3D image synthesis with 2D backbones, which can mitigate volumetric inconsistency and training difficulty related to 3D backbones. (2) We propose an efficient latent diffusion-based framework for high-fidelity cross-modality 3D medical image synthesis. (3) We collected a large-scale high-quality dataset of paired susceptibility weighted imaging (SWI) and magnetic resonance angiography (MRA) brain images. Experiments on these in-house and public T1-T2 brain MRI datasets show the volumetric consistency and superior quantitative result of our framework.

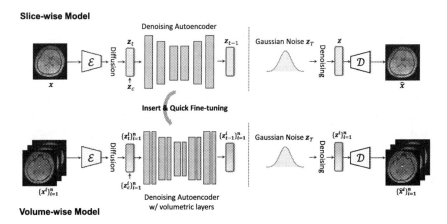

Fig. 1. Overview of our proposed two-stage Make-A-Volume framework. A latent diffusion model is used to predict the noises added to the image and synthesize independent slices from Gaussian noises. We insert volumetric layers and quickly fine-tune the model, which extends the slice-wise model to be a volume-wise model and enables synthesizing volumetric data from Gaussian noises.

2   Method

2.1   Preliminaries of DDPMs

In the diffusion process, DDPMs produce a series of noisy inputs $x_0, x_1, ..., x_T$, via sequentially adding Gaussian noises to the sample over a predefined number of timesteps T. Formally, given clean data samples which follow the real distribution $x_0 \sim q(x)$, the diffusion process can be written down with variances $\beta_1, ..., \beta_T$ as

$$q(x_t|x_{t-1}) = \mathcal{N}(x_t; \sqrt{1 - \beta_t}x_{t-1}, \beta_t\mathbf{I}). \tag{1}$$

Employing the property of DDPMs, the corrupted data x_t can be sampled easily from x_0 in a closed form:

$$q(x_t|x_0) = \mathcal{N}(x_t; \sqrt{\bar{\alpha}_t}x_0, (1-\bar{\alpha}_t)\mathbf{I}); \quad x_t = \sqrt{\bar{\alpha}_t}x_0 + \sqrt{1-\bar{\alpha}_t}\epsilon, \qquad (2)$$

where $\alpha_t = 1 - \beta_t$, $\bar{\alpha}_t = \prod_{s=1}^{t}\alpha_s$, and $\epsilon \sim \mathcal{N}(0,1)$ is the added noise.

In the reverse process, the model learns a Markov chain process to convert the Gaussian distribution into the real data distribution by predicting the parameterized Gaussian transition $p(x_{t-1}|x_t)$ with the learned model θ:

$$p_\theta(x_{t-1}|x_t) = \mathcal{N}(x_{t-1}; \mu_\theta(x_t, t), \sigma_t^2\mathbf{I}). \qquad (3)$$

In the model training, the model tries to predict the added noise ϵ with the simple mean squared error (MSE) loss:

$$L(\theta) = \mathbb{E}_{x_0 \sim q(x), \epsilon \sim \mathcal{N}(0,1), t}\left[\left\|\epsilon - \epsilon_\theta(\sqrt{\bar{\alpha}_t}x_0 + \sqrt{1-\bar{\alpha}_t}\epsilon, t)\right\|^2\right]. \qquad (4)$$

2.2   Slice-Wise Latent Diffusion Model

To improve the computational efficiency of DDPMs that learn data in pixel space, Rombach et al. [20] proposes training an autoencoder with a KL penalty or a vector quantization layer [15,27], and introduces the diffusion model to learn the latent distribution. Given calibrated source modality image x_c and target modality image x, we leverage a slice-wise latent diffusion model to learn the cross-modality translation. With the pretrained encoder $\mathcal{E}$, x_c and x are compressed into a spatially lower-dimensional latent space of reduced complexity, generating z_c and z. The diffusion and denoising processes are then implemented in the latent space and a U-Net [21] is trained to predict the noise in the latent space. The input consists of the concatenated z_c and z and the network learns the parameterized Gaussian transition $p_\theta(z_{t-1}|z_t, z_c) = \mathcal{N}(z_{t-1}; \mu_\theta(z_t, t, z_c), \sigma_t^2\mathbf{I})$. After learning the latent distribution, the slice-wise model can synthesize target latent $\hat{z}$ from Gaussian noise, given the source latent z_c. Finally, the decoder $\mathcal{D}$ restores the slice to the image space via $\hat{x} = \mathcal{D}(\hat{z})$.

2.3   From Slice-Wise Model to Volume-Wise Model

Figure 1 illustrates an overview of the Make-A-Volume framework. The first stage involves a latent diffusion model that learns the cross-modality translation in an image-to-image manner to synthesize independent slices from Gaussian noises. Then, to extend the slice-wise model to be a volume-wise model, we insert volumetric layers and quickly fine-tune the U-Net. As a result, the volume-wise model synthesizes volumetric data without inconsistency from Gaussian noises.

In the slice-wise model, distribution of the latent $z \in \mathbb{R}^{b_s \times c \times h \times w}$ is learned by the U-Net, where b_s, c, h, w are the batch size of slice, channels, height, and width dimensions respectively, and there is where little volume-awareness is introduced to the network. Since we target in synthesizing volumetric data and assume each volume consists of N slices, we can factorize the batch size of slices as $b_s = b_v n$, where B_v represents the batch size of volumes. Now, volumetric layers are injected and help the U-Net learn to latent feature $f \in \mathbb{R}^{(b_v \times n) \times c \times h \times w}$ with volumetric consistency. The volumetric layers are basic 1D convolutional layers and the i-th volumetric layer l_v^i takes in feature f and outputs f' as:

$$f' \leftarrow \text{Rearrange}(f, (b_v \times n)\ c\ h\ w \rightarrow (b_v \times h \times w)\ c\ n), \tag{5}$$

$$f' \leftarrow l_v^i(f'), \tag{6}$$

$$f' \leftarrow \text{rearrange}(f, (b_v \times h \times w)\ c\ n \rightarrow (b_v \times n)\ c\ h\ w). \tag{7}$$

Here, the 1D conv layers combined with the pretrained 2D conv layers, serve as pseudo 3D conv layers with little extra memory cost. We initialize the volumetric 1D convolution layers as Identity Functions for more stable training and we empirically find tuning is efficient. With the volume-aware network, the model learns volume data $\{x^i\}_{i=1}^n$, predicts $\{z^i\}_{i=1}^n$, and reconstruct $\{\hat{x}^i\}_{i=1}^n$. For diffusion model training, in the first stage, we randomly sample timestep t for each slice. However, when tuning the second stage, the U-Net with volumetric layers learns the relationship between different slices in one volume. As a result, fixing t for each volume data is necessary and we encourage the small t values to be sampled more frequently for easy training. In detail, we sample the timestep t with replacement from multinomial distribution, and the pre-normalized weight (used for computing probabilities after normalization) for timestep t equals $2T - t$, where T is the total number of timesteps. Therefore, we enable a seamless translation from the slice-wise model which processes slices individually, to a volume-wise model with better volumetric consistency.

3   Experiments

Datasets. The experiments were conducted on two brain MRI datasets: SWI-to-MRA (S2M) dataset and RIRE [30][1] T1-to-T2 dataset. To facilitate SWI-to-MRA brain MRI synthesis applications, we collected a high-quality SWI-to-MRA dataset. This dataset comprises paired SWI and MRA volume data of 111 patients that were acquired at Qilu Hospital of Shandong University using one 3.0T MRI scanner (*i.e.*, Verio from Siemens). The SWI scans have a voxel spacing of $0.3438 \times 0.3438 \times 0.8$ mm and the MRA scans have a voxel spacing of $0.8984 \times 0.8984 \times 2.0$ mm. While most public brain MRI datasets lack high-quality details along z-axis and therefore are weak to indicate volumetric inconsistency, this volume data provides a good way to illustrate the performances for volumetric synthesis due to the clear blood vessels. We also evaluate our method on the public RIRE dataset [30]. The RIRE dataset includes T1 and T2-weighted MRI volumes, and 17 volumes were used in the experiments.

Implementation Details. To summarize, for the S2M dataset, we randomly select 91 paired volumes for training and 20 paired volumes for inference; for the

---

[1] https://rire.insight-journal.org/index.html.

RIRE T1-to-T2 dataset, 14 volumes are randomly selected for training and 3 volumes are used for inference. All the volumes are resized to $256 \times 256 \times 100$ for S2M and $256 \times 256 \times 35$ for RIRE, where the last dimension represents the z-axis dimension, *i.e.*, the number of slices in one volume for 2D image-to-image setting. Our proposed method is built upon U-Net backbones. We use a pretrained KL autoencoder with a downsampling factor of $f = 4$. We train our model on an NVIDIA A100 80 GB GPU.

Table 1. Quantitative comparison on S2M and RIRE datasets.

Methods	S2M			RIRE [30]		
	MAE ↓	SSIM ↑	PSNR ↑	MAE ↓	SSIM ↑	PSNR ↑
Pix2pix [11]	8.175	0.739	25.663	16.812	0.538	20.106
Palette [22]	26.806	0.141	15.643	36.131	0.251	14.269
Pix2pix 3D [11]	6.234	0.765	28.395	11.369	0.650	22.854
CycleGAN 3D [33]	7.621	0.755	26.908	13.794	0.542	20.627
Ours 200 steps	**5.243**	**0.788**	**29.446**	**10.794**	**0.676**	**24.332**
Ours 1000 steps	**4.801**	**0.801**	**30.143**	**10.619**	**0.684**	**25.458**

Quantitative Results. We compare our pipeline to several baseline methods, including 2D-based methods: (1) Pix2pix [11], a solid baseline for image-to-image translation; (2) Palette [22], a diffusion-based method for 2D image translation; 3D-based methods: (3) a 3D version of Pix2pix, created by modifying the 2D backbone as a 3D backbone in the naive Pix2pix approach; and (4) a 3D version of CycleGAN [33]. Naive 3D diffusion-based models are not included due to the lack of efficient backbones and the matter of timesteps' sampling efficiency. We report the results in terms of mean absolute error (MAE), Structural Similarity Index (SSIM) [29], and peak signal-to-noise ratio (PSNR).

Table 1 presents a quantitative comparison of our method and baseline approaches on the S2M and RIRE datasets. Our method achieves better performance than the baselines in terms of various evaluation metrics. To accelerate the sampling of diffusion models, we implement DDIM [25] with 200 steps and report the results accordingly. It is worth noting that for the baseline approaches, the 3D version method (Pix2pix 3D) outperforms the corresponding 2D version (Pix2pix) at the cost of additional memory usage. For the Palette method, we implemented the 2D version but were unable to produce high-quality slices stably and failure cases dramatically affected the metrics results. Nonetheless, we included this method due to its great illustration of volumetric inconsistency.

Qualitative Results. Figure 2 presents a qualitative comparison of different methods, showcasing two axial slices of clear vessels. Our method synthesizes better images with more details, as shown in the qualitative results. The areas requiring special attention are highlighted with red arrows and red rectangles. It is worth noting that the synthesized axial slices not only depend on the source slice but also on the volume knowledge. For instance, for S2M case 1, the target slice shows a clear vessel cross-section that is based on the shape of the vessels in the volume. In Fig. 3, we provide coronal and sagittal views. For methods that rely on 2D generation, we synthesize individual slices and concatenate them to create volumes. It is clear to observe the volumetric inconsistency examining the coronal and sagittal views of these volumes. For instance, Palette synthesizes 2D slices unstably, where some good slices are synthesized but others are of poor quality. As a result, volumetric inconsistency severely impacts the performance of volumes. While 2D baselines inherently introduce inconsistency in the coronal and sagittal views, 3D baselines also generate poor results than ours, particularly in regard to blood vessels and ventricles.

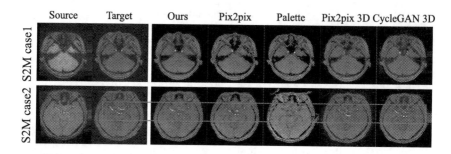

Fig. 2. Qualitative comparison. We compare our methods with baselines on two cases.

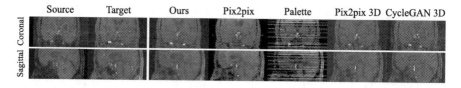

Fig. 3. Coronal view and sagittal view. To clearly indicate the volumetric consistency, we show a coronal view and a sagittal view of the volumes synthesized and the ground truth volumes.

Table 2. Ablation Quantitative Results.

Methods	S2M			RIRE [30]		
	MAE ↓	SSIM ↑	PSNR ↑	MAE ↓	SSIM ↑	PSNR ↑
w/o volumetric layers	5.128	0.792	29.894	10.925	0.667	24.623
w/ volumetric layers	4.801	0.801	30.143	10.619	0.684	25.458

Ablation Analysis. We conduct an ablation study to show the effectiveness of volumetric fine-tuning. Table 2 presents the quantitative results, demonstrating that our approach is able to increase the model's performance beyond that of the slice-wise model, without incurring significant extra training expenses. Figure 4 illustrates that fine-tuning volumetric layers helps to mitigate volumetric artifacts and produce clearer vessels, which is crucial for medical image synthesis.

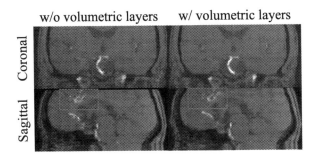

Fig. 4. Ablation qualitative results with coronal view and sagittal view.

4   Conclusion

In this paper, we propose Make-A-Volume, a diffusion-based framework for cross-modality 3D medical image synthesis. Leveraging latent diffusion models, our method achieves high performance and can serve as a strong baseline for multiple cross-modality medical image synthesis tasks. More importantly, we introduce a generic paradigm for volumetric data synthesis by utilizing 2D backbones and demonstrate that fine-tuning volumetric layers helps the two-stage model capture 3D information and synthesize better images with volumetric consistency. We collected an in-house SWI-to-MRA dataset with clear blood vessels to evaluate volumetric data quality. Experimental results on two brain MRI datasets demonstrate that our model achieves superior performance over existing baselines. Generating coherent 3D and 4D data is at an early stage in the diffusion models literature, we believe that by leveraging slice-wise models and extending them to 3D/4D models, more work can help achieve better volume synthesis with reasonable memory requirements. In the future, we will investigate more efficient approaches for more high-resolution volumetric data synthesis.

Acknowledgement. The work described in this paper was partially supported by grants from the Research Grants Council of the Hong Kong Special Administrative Region, China (T45-401/22-N), the National Natural Science Fund (62201483), HKU Seed Fund for Basic Research (202009185079 and 202111159073), RIE2020 Industry Alignment Fund - Industry Collaboration Projects (IAF-ICP) Funding Initiative, as well as cash and in-kind contribution from the industry partner(s).

References

1. Bau, D., et al.: Seeing what a GAN cannot generate. In: Proceedings of the IEEE/CVF International Conference on Computer Vision, pp. 4502–4511 (2019)
2. Ben-Cohen, A., et al.: Cross-modality synthesis from CT to pet using FCN and GAN networks for improved automated lesion detection. Eng. Appl. Artif. Intell. **78**, 186–194 (2019)
3. Chung, H., Ryu, D., McCann, M.T., Klasky, M.L., Ye, J.C.: Solving 3D inverse problems using pre-trained 2D diffusion models. arXiv preprint arXiv:2211.10655 (2022)
4. Dalmaz, O., Yurt, M., Çukur, T.: ResViT: residual vision transformers for multi-modal medical image synthesis. IEEE TMI **41**(10), 2598–2614 (2022)
5. Dar, S.U., Yurt, M., Karacan, L., Erdem, A., Erdem, E., Cukur, T.: Image synthesis in multi-contrast MRI with conditional generative adversarial networks. IEEE TMI **38**(10), 2375–2388 (2019)
6. Dhariwal, P., Nichol, A.: Diffusion models beat GANs on image synthesis. Adv. Neural. Inf. Process. Syst. **34**, 8780–8794 (2021)
7. Filippou, V., Tsoumpas, C.: Recent advances on the development of phantoms using 3d printing for imaging with CT, MRI, PET, SPECT, and ultrasound. Med. Phys. **45**(9), e740–e760 (2018)
8. Goodfellow, I., et al.: Generative adversarial networks. Commun. ACM **63**(11), 139–144 (2020)
9. Ho, J., Jain, A., Abbeel, P.: Denoising diffusion probabilistic models. Adv. Neural. Inf. Process. Syst. **33**, 6840–6851 (2020)
10. Hu, X., Shen, R., Luo, D., Tai, Y., Wang, C., Menze, B.H.: AutoGAN-synthesizer: neural architecture search for cross-modality MRI synthesis. In: Wang, L., Dou, Q., Fletcher, P.T., Speidel, S., Li, S. (eds.) MICCAI 2022, Part VI. LNCS, vol. 13436, pp. 397–409. Springer, Cham (2022). https://doi.org/10.1007/978-3-031-16446-0_38
11. Isola, P., Zhu, J.Y., Zhou, T., Efros, A.A.: Image-to-image translation with conditional adversarial networks. In: Proceedings of the IEEE Conference on Computer Vision and Pattern Recognition, pp. 1125–1134 (2017)
12. Kazerouni, A., et al.: Diffusion models for medical image analysis: a comprehensive survey. arXiv preprint arXiv:2211.07804 (2022)
13. Khader, F., et al.: Medical diffusion-denoising diffusion probabilistic models for 3D medical image generation. arXiv preprint arXiv:2211.03364 (2022)
14. Kim, B., Ye, J.C.: Diffusion deformable model for 4D temporal medical image generation. In: Wang, L., Dou, Q., Fletcher, P.T., Speidel, S., Li, S. (eds.) MICCAI 2022, Part I. LNCS, vol. 13431, pp. 539–548. Springer, Cham (2022). https://doi.org/10.1007/978-3-031-16431-6_51
15. Kingma, D.P., Welling, M.: Auto-encoding variational bayes. arXiv preprint arXiv:1312.6114 (2013)
16. Lee, S., Chung, H., Park, M., Park, J., Ryu, W.S., Ye, J.C.: Improving 3D imaging with pre-trained perpendicular 2D diffusion models. arXiv preprint arXiv:2303.08440 (2023)
17. Li, K., Malik, J.: On the implicit assumptions of GANs. arXiv preprint arXiv:1811.12402 (2018)
18. Nie, D., et al.: Medical image synthesis with deep convolutional adversarial networks. IEEE Trans. Biomed. Eng. **65**(12), 2720–2730 (2018)

19. Pinaya, W.H., et al.: Brain imaging generation with latent diffusion models. In: Mukhopadhyay, A., Oksuz, I., Engelhardt, S., Zhu, D., Yuan, Y. (eds.) DGM4MICCAI 2022. LNCS, vol. 13609, pp. 117–126. Springer, Cham (2022). https://doi.org/10.1007/978-3-031-18576-2_12

20. Rombach, R., Blattmann, A., Lorenz, D., Esser, P., Ommer, B.: High-resolution image synthesis with latent diffusion models. In: Proceedings of the IEEE/CVF Conference on Computer Vision and Pattern Recognition, pp. 10684–10695 (2022)

21. Ronneberger, O., Fischer, P., Brox, T.: U-net: convolutional networks for biomedical image segmentation. In: Navab, N., Hornegger, J., Wells, W.M., Frangi, A.F. (eds.) MICCAI 2015, Part III. LNCS, vol. 9351, pp. 234–241. Springer, Cham (2015). https://doi.org/10.1007/978-3-319-24574-4_28

22. Saharia, C., et al.: Palette: image-to-image diffusion models. In: ACM SIGGRAPH 2022 Conference Proceedings, pp. 1–10 (2022)

23. Singer, U., et al.: Make-a-video: text-to-video generation without text-video data. arXiv preprint arXiv:2209.14792 (2022)

24. Sohl-Dickstein, J., Weiss, E., Maheswaranathan, N., Ganguli, S.: Deep unsupervised learning using nonequilibrium thermodynamics. In: International Conference on Machine Learning, pp. 2256–2265. PMLR (2015)

25. Song, J., Meng, C., Ermon, S.: Denoising diffusion implicit models. arXiv preprint arXiv:2010.02502 (2020)

26. Uzunova, H., Ehrhardt, J., Handels, H.: Memory-efficient GAN-based domain translation of high resolution 3d medical images. Comput. Med. Imaging Graph. **86**, 101801 (2020)

27. Van Den Oord, A., Vinyals, O., et al.: Neural discrete representation learning. Adv. Neural Inf. Process. Syst. **30** (2017)

28. Wang, T., et al.: A review on medical imaging synthesis using deep learning and its clinical applications. J. Appl. Clin. Med. Phys. **22**(1), 11–36 (2021)

29. Wang, Z., Bovik, A.C., Sheikh, H.R., Simoncelli, E.P.: Image quality assessment: from error visibility to structural similarity. IEEE Trans. Image Process. **13**(4), 600–612 (2004)

30. West, J., et al.: Comparison and evaluation of retrospective intermodality brain image registration techniques. J. Comput. Assist. Tomogr. **21**(4), 554–568 (1997)

31. Wu, J.Z., et al.: Tune-a-video: One-shot tuning of image diffusion models for text-to-video generation. arXiv preprint arXiv:2212.11565 (2022)

32. Yu, B., Zhou, L., Wang, L., Shi, Y., Fripp, J., Bourgeat, P.: EA-GANs: edge-aware generative adversarial networks for cross-modality MR image synthesis. IEEE TMI **38**(7), 1750–1762 (2019)

33. Zhu, J.Y., Park, T., Isola, P., Efros, A.A.: Unpaired image-to-image translation using cycle-consistent adversarial networks. In: Proceedings of the IEEE International Conference on Computer Vision, pp. 2223–2232 (2017)

PIViT: Large Deformation Image Registration with Pyramid-Iterative Vision Transformer

Tai Ma, Xinru Dai, Suwei Zhang, and Ying Wen$^{(\boxtimes)}$

Shanghai Key Laboratory of Multidimensional Information Processing, School of
Communications and Electronic Engineering, East China Normal University,
Shanghai 200241, China
ywen@cs.ecnu.edu.cn

Abstract. Large deformation image registration is a challenging task in
medical image registration. Iterative registration and pyramid registration are two common CNN-based methods for the task. However, these
methods usually consume more parameters and time. Additionally, the
existing CNN-based registration methods mainly focus on local feature
extraction, limiting their ability to capture the long-distance correlation between image pairs. In this paper, we propose a fast and accurate learning-based algorithm, Pyramid-Iterative Vision Transformer
(PIViT), for 3D large deformation medical image registration. Our
method constructs a novel pyramid iterative composite structure to solve
large deformation problem by using low-scale iterative registration with a
Swin Transformer-based long-distance correlation decoder. Furthermore,
we exploit pyramid structure to supplement the detailed information of
the deformation field by using high-scale feature maps. Comprehensive
experimental results implemented on brain MRI and liver CT datasets
show that the proposed method is superior to the existing registration
methods in terms of registration accuracy, training time and parameters, especially of a significant advantage in running time. Our code is
available at https://github.com/Torbjorn1997/PIViT.

Keywords: Medical image registration · convolutional neural
networks · image processing

1 Introduction

Deformable image registration is one of the fundamental tasks in computer vision
and has been widely used in medical image processing. In recent years, deep
learning methods based on convolutional neural networks are widely applied in
deformable image registration. Balakrishnan et al. [3] proposed VoxelMorph with

Supplementary Information The online version contains supplementary material
available at https://doi.org/10.1007/978-3-031-43999-5_57.

a structure similar to Unet and further developed a diffeomorphism implementation of VoxelMorph [8]. Mok et al. [21] proposed SYMNet to achieve accurate diffeomorphic registration by exploiting the cycle consistency of registration. However, when there is a significant difference between the images, it is difficult to learn an accurate deformation field for alignment because large deformation image registration has a high degree of freedom in transformation. Typical registration methods utilize rigid or affine transformation with a low degree of freedom to provide initialized global transformation for large deformation, however, this requires the introduction of additional preprocessing to obtain the corresponding affine matrix [12] [23]. In order to solve the high degree of freedom of large deformation transformation, the end-to-end deformable image registration methods are mainly divided into two types: iterative registration (Fig. 1 (a)) and pyramid registration (Fig. 1 (b)). *(a)* Iterative registration achieves coarse-to-fine image registration by cascading several CNNs, which requires huge GPU memory during training. In addition, iterative registration methods learn separate image features in each iteration, which brings additional computational costs when repeatedly extracting features. Typical iterative registration methods include RCN [28] and LapIRN [22]. *(b)* Pyramid registration achieves coarse-to-fine registration within one iteration by warping feature maps. These methods successively learn feature maps and deformation fields from low to high resolution. Typical pyramid registration methods include Dual-PRNet [14] and NICE-Net [20]. However, current non-iterative registration methods still cannot well solve the image registration problem under the significant differences condition.

Inspired by the capabilities of Transformer in NLP, recent researchers have extended Transformer to computer vision tasks [11] [19] and acquired results that surpass CNNs' in many tasks [17] [27]. Many Transformer-based registration methods have also been proposed for image registration tasks, such as Trans-Morph [7], Swin-VoxelMorph [30] and XMorpher [26]. Compared with CNN-based methods, Transformer-based methods have achieved better registration results, which illustrates that the global receptive field of Transformer is helpful for image registration.

In this paper, we propose a novel Pyramid-Iterative Vision Transformer (PIViT) by combining Swin Transformer-based long-range correlation decoder and the proposed pyramid-iterative registration framework shown in Fig. 1 (c).

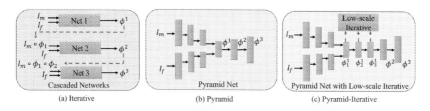

Fig. 1. Network architecture (a) iterative registration, (b) pyramid registration and (c) the proposed pyramid-iterative registration.

Our main contributions of this work are as: (1) We establish a pyramid-iterative registration framework to address large deformation image registration. The framework first extracts feature map pairs via a dual-stream weight-sharing encoder, then performs iterative registration on the low-scale feature space, and finally complements detail information and learns accurate deformation fields during pyramid decoding process. (2) We propose a Swin Transformer-based long-range correlation decoder, which exploits the global receptive field of Swin Transformer on low-scale feature maps to learn high accuracy large deformation fields while maintaining low parameters. (3) Compared with other popular registration methods, the proposed unsupervised end-to-end network is more lightweight and suitable for time-sensitive tasks.

Extensive experiments on 3D brain MRI and liver CT registration tasks demonstrate that PIViT achieves state-of-the-art performance in terms of accuracy but consumes less time and parameters.

2   The Proposed Method

In this section, we first propose a novel pyramid-iterative registration framework to solve large deformation image registration. The pyramid-iterative registration framework combines the advantages of iteration and pyramid registration framework to achieve fast and accurate registration. Then, we introduce a long-range correlation decoder based on Swin Transformer into the iterative registration stage of the proposed framework and utilize the global receptive field of the Swin Transformer to capture global correlations, thereby implementing high accurate and fast registration.

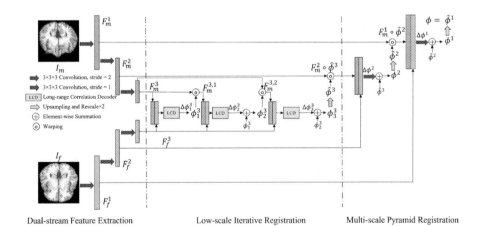

Fig. 2. Overview of the proposed PIViT. The number of pyramid levels N is set as 3 for illustration.

2.1   Pyramid-Iterative Registration Framework

As shown in Fig. 2, the proposed pyramid-iterative registration framework can be divided into three parts: dual-stream feature extraction, low-scale iterative registration and multi-scale pyramid registration.

Dual-Stream Feature Extraction: Similar to pyramid registration network, the proposed framework utilizes a weight-sharing feature encoder to construct feature pyramids for the fixed image I_f and the moving image I_m, respectively. At the ith step ($i \in [1 \cdots N]$), the feature maps of I_f and I_m are formulated as F_f^i and F_m^i, respectively. The weight-sharing feature encoder reuses the same network blocks to extract the feature maps F_f^i and F_m^i without adding parameters or complicating the training process while ensuring that F_f^i and F_m^i are in the same feature space.

Low-Scale Iterative Registration: The pyramid-iterative registration uses two different decoding modules at different scales. To capture large deformation, we adopt low-scale feature maps to obtain the coarse distribution of large deformation fields without considering the fine distribution in this paper. Therefore, at the last N^{th} level of feature pyramid, deformation field is predicted from F_f^N and F_m^N multiple times through an iterative structure. Similar to iterative-based registration methods, F_m^N is warped by the predicted deformation field ϕ_t^N, where t is the number of iterations. The warped $F_m^{N,t}$ and F_f^N are used for the next iteration. In the first iteration, the decoder obtains the initial deformation field ϕ_1^N, and in the subsequent iterations, the residual deformation field $\Delta\phi_t^N$ is obtained in each prediction and the updated overall deformation field ϕ_t^N is obtained. This procedure can be formulated as:

$$F_m^{N,t} = F_m^N \circ \phi_t^N, \phi_t^N = \begin{cases} \Delta\phi_t^N, t = 1, \\ \phi_{t-1}^N + \Delta\phi_t^N, t = 2, \cdots, T, \end{cases} \tag{1}$$

where T is the upper limit of iteration, $\circ$ denotes warping the feature map with deformation fields, and $+$ denotes element-wise summation of deformation fields.

Compared with other iterative registration methods, the advantage of iterating only at the Nth level is that there is no need to re-extract image features, thus the computational complexity and time consumption of our method can be greatly reduced. This can greatly accelerate the speed of model training and deformation field prediction, and better solve large deformation.

Multi-scale Pyramid Registration: After the implementation of low-scale iterative registration, the deformation field ϕ_T^N is rescaled by a factor of 2 and the rescaled flow $\widehat{\phi}^N$ is obtained. The subsequent process is the same as that of the pyramid registration method. At each level, warped feature $F_m^i \circ \widehat{\phi}^{i+1}$ and fixed feature F_f^i ($i = N-1, \cdots, 1$) are concatenated and the residual deformation field $\Delta\phi^i$ is predicted by 3D convolution. $\Delta\phi^i$ is used to update $\widehat{\phi}^{i+1}$ so as to obtain the deformation field ϕ^i corresponding to the ith layer. ϕ^i is rescaled by a factor of 2 and warps moving feature F_m^{i-1}. The purpose of introducing multi-scale pyramid registration is to supplement the lack of fine information caused by only using low-scale features in the iterative registration stage. This process is repeated at each level of the feature pyramid until the deformation field is rescaled to the original image resolution. Finally, the pyramid-iterative registration framework obtains the predicted global deformation field.

2.2   Long-Range Correlation Decoder

To capture large deformation at low-scale registration, the study of the decoder is very essential. Therefore, we propose a long-range correlation decoder (LCD) in the iterative registration phase. As shown in Fig. 3, the LCD consists of a Swin transformer-based block and two consecutive convolutions. The Swin transformer-based block models the long-range correlation between F_f^N and $F_m^N \circ \phi_{t-1}^N$ using the self-attention mechanism of the transformer, and then the residual flow field $\Delta\phi_t^N$ is obtained by the convolution block. In order to enhance the information interaction between non-overlapping windows, we adopt the shifted local window attention strategy of the Swin Transformer. The structure of the Swin Transformer-based block is shown in the red frame area in Fig. 3, which consists of shifted window-based self-attention modules (W-SA & SW-SA), followed by a 2-layer MLP. A LayerNorm (LN) layer is applied before each SA and MLP module, and a residual connection is applied after each module.

Current Transformer-based registration methods usually directly migrate the Transformer structure to the 3D image registration task, which leads to a large number of parameters and a remarkably long inference time. In contrast, the proposed PIViT models long-range correlations in low-scale iterative registration with LCD to warp corresponding voxels between feature maps to spatial neighborhoods, thus it is not necessary to use the Transformer on large feature maps at high scales. In addition, LCD also removes position embedding, only uses single-head self-attention and reduces the number of channels. These operations accelerate the speed of PIViT and significantly reduce parameters.

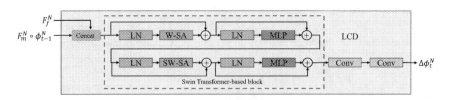

Fig. 3. An illustration of the structure of the proposed long-range correlation decoder (LCD). The red box indicates the Swin Transformer-based block. (Color figure online)

2.3   Loss Function

PIViT is an unsupervised end-to-end registration network. In this section, we design a loss function to train the proposed network. In the final stage of pyramid registration, PIViT obtains the deformation field ϕ between I_m and I_f and the warped image $I_w = I_m \circ \phi$ by using the differential operation based on the spatial transformer network [15]. In order to minimize the difference, we use the normalized cross-correlation (NCC) as a measure of the difference between the warped image I_w and fixed image I_f.

In order to ensure the continuity and smoothness of the deformation field ϕ in space, a regular term on its spatial gradient is introduced. The complete loss function is:

$$L_{I_f, I_m, \phi} = L_{sim} + \lambda L_{smooth} = -NCC(I_f, I_w) + \lambda \sum_{p \in \Omega} \|\nabla \phi(p)\|^2, \qquad (2)$$

where λ is the regularization hyperparameter.

3   Experiments

Data and Pre-processing: We evaluate the performance of PIViT on brain MRI datasets and liver CT datasets. In the experiments, we compare the proposed method with commonly used 3D convolutional registration methods Voxelmorph [3], Dual-PRNet [14], RCN [28], LKU-Net [16], TransMorph [7], Swin-VoxelMorph [30] and NICE-Net [20]. The accuracy of image registration is measured by Dice score [10]. We choose 2303 brain MRI scans from the ABIDE [9], ADHD [5] and ADNI [24] brain MRI datasets for training and LPBA [25] for testing. LPBA dataset contains 40 brain MRI scans with segmentation ground truth of 56 anatomical structures. For liver CT datasets, 1025 scans from MSD [2] and BFH [29] are selected for training and SLIVER [13], LiTS [6] and LSPIG [28] for testing. The images are all resampled to the size of 128×128×128. In order to better verify the effect of each method on large deformation image registration, we do not perform affine pre-alignment process. Atlas-based and scan-to-scan registrations are performed on brain and liver scans, respectively.

Implementation: We set λ to 1 for PIViT to guarantee the smoothness of the deformation field. Algorithm runtimes are computed on an NVIDIA GeForce RTX 3090 GPU and an Intel(R) Xeon(R) Silver 4210R CPU. We implement the model using Keras with a Tensorflow [1] backend and the ADAM [18] optimizer with a learning rate of $1e^{-4}$. The batch size is set as 1 and the networks are trained for 150,000 iterations.

Results: Table 1 shows the comparison of the proposed PIViT and other methods on 4 medical datasets. The Dice score, number of voxels with non-positive Jacobian determinants ($|J_s|_{\leq 0}$), GPU registration time (GRT), CPU registration time (CRT), network parameters and training time per iteration (TPI) of each method are presented.

Table 1. Comparison among VoxelMorph, VoxelMorph-diff, DualPRNet, RCN, Trans-Morph, Swin-VoxelMorph, LKU-Net, NICE-Net and the proposed PIViT on the LPBA, SLIVER, LiTs and LSPIG datasets. * indicates that the t-test p-value between PIViT and all other methods is less than 0.05.

Method	LPBA		SLIVER	LiTs	LSPIG	GRT ↓	CRT ↓	Params ↓	TPI ↓
	Dice ↑	$|J_s|_{\leq 0}$ ↓	Dice ↑	Dice ↑	Dice ↑				
VoxelMorph (CVPR 2018)	55.3 ± 6.3	17353.1	73.0 ± 6.7	67.3 ± 8.3	69.0 ± 8.8	**0.05 s**	0.73 s	**312.7K**	0.14 s
VoxelMorph-diff (MICCAI 2019)	60.1 ± 4.9	**0.0**	82.8 ± 6.2	80.1 ± 7.1	76.8 ± 6.5	0.08 s	0.99 s	319.7K	0.22 s
DualPRNet (MICCAI 2019)	58.3 ± 4.9	5446.9	79.1 ± 5.6	77.0 ± 6.5	71.5 ± 7.5	0.11 s	1.59 s	581.7K	0.28 s
RCN (ICCV 2019)	67.6 ± 2.6	6559.3	80.0 ± 6.6	73.7 ± 8.5	68.8 ± 7.3	0.35 s	2.20 s	938.7K	0.36 s
TransMorph (MIA 2022)	56.7 ± 5.6	24588.1	88.2 ± 4.1	78.7 ± 11.6	81.6 ± 5.3	0.10 s	1.99 s	45675.0K	0.86 s
Swin-VoxelMorph (MICCAI 2022)	58.1 ± 5.2	15857.6	77.1 ± 6.1	69.2 ± 10.5	70.6 ± 6.5	0.12 s	6.74 s	26573.9K	1.18 s
LKU-Net (MICCAI 2022)	58.9 ± 5.5	2215.7	81.4 ± 5.1	77.3 ± 7.4	73.9 ± 6.8	0.10 s	1.27 s	2037.4K	0.18 s
NICE-Net (MICCAI 2022)	68.1 ± 2.4	8061.6	87.1 ± 5.0	82.8 ± 6.8	79.6 ± 6.4	0.11 s	1.06 s	1033.7K	0.25 s
PIViT	**70.1 ± 1.4***	697.0	**90.9 ± 3.7***	**86.9 ± 5.3***	**84.7 ± 4.7**	0.07 s	**0.32 s**	420.8K	**0.14 s**

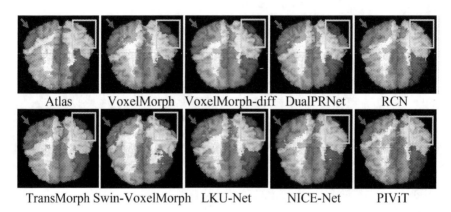

Atlas   VoxelMorph   VoxelMorph-diff   DualPRNet   RCN

TransMorph   Swin-VoxelMorph   LKU-Net   NICE-Net   PIViT

Fig. 4. Visualization of comparative registration results.

As shown in Table 1, VoxelMorph, VoxelMorph-diff, TransMorph, Swin-VoxelMorph and LKU-Net all get low Dice scores, indicating that these single-stream registration methods are difficult to solve large deformation. However, the iterative registration method RCN and the pyramid registration method NICE-Net obtain relatively good Dice scores, indicating that both iterative and pyramid registration methods can be useful to deal with large deformation. However, the Dice score of the proposed PIViT combining their advantages surpasses that of VoxelMorph by 14.8%, and the improvements compared to RCN and NiceNet also reach 2.5% and 2.0% on LPBA, respectively. On 3 liver datasets, compared with VoxelMorph, PIViT achieves 17.9%, 19.6% and 15.7% improvements, while compared with NICE-Net, PIViT achieves 3.8%, 4.1% and 5.1% improvements, respectively. The experiments indicate that the proposed PIViT implements large deformation fine registration better than other methods.

In addition to the superior Dice score, another advantage of the proposed PIViT is its fast and lightweight registration. Table 1 shows that the parameters, training and registration time of PIViT are close to that of VoxelMorph, far less than those of RCN and NICE-Net. These properties of PIViT make it easier to train and more suitable for time-sensitive tasks. Compared to other Transformer-based methods, the proposed PIViT has orders of magnitude optimization in parameters, which is because we only use the Transformer block at low scales, and LCD is tuned and optimized for 3D image registration tasks. What's more, although there is no additional constraint on the diffeomorphism of the deformation field, the deformation field obtained by PIViT has better diffeomorphism properties than those obtained by other methods except VoxelMorph-diff.

The visualization result of the experiment on the LPBA dataset is shown in Fig. 4. Obviously, most of the methods produce severe misregistration in the yellow regions of Fig. 4, due to the existence of large deformation. Compared with the registration results of RCN and NICE-Net, the proposed method achieves better alignment on the fine structure, which can be seen in the areas indicated by the red arrow in Fig. 4. It can be seen, since PIViT focuses on lightweight and fast registration of large deformations, its effectiveness on fine registration tasks is still somewhat weak.

Table 2. Dice score vs. different decoders and number of iterations.

t	Decoder Type											
	CNN				GRU				LCD			
	LPBA ↑	SLIVER ↑	Params ↓	TPI ↓	LPBA ↑	SLIVER ↑	Params ↓	TPI ↓	LPBA ↑	SLIVER ↑	Params ↓	TPI ↓
1(pyramid)	67.2	87.7	224.6K	0.12 s	68.3	88.4	413.7K	0.11 s	69.3	88.7	291.0K	0.12 s
2	68.2	88.7	278.7K	0.12 s	69.4	89.7	413.7K	0.12 s	69.8	90.4	355.4K	0.13 s
3	**68.9**	89.5	332.7K	0.12 s	69.5	90.2	413.7K	0.14 s	**70.1**	**90.9**	420.8K	0.14 s
4	68.8	**90.3**	386.7K	0.12 s	**70.0**	90.7	413.7K	0.14 s	70.0	90.8	486.1K	0.16 s
5	**68.9**	90.1	440.8K	0.13 s	69.9	**91.2**	413.7K	0.16 s	70.0	**90.9**	551.5K	0.19 s

Number of Iterations and Decoder Type: In this section, we explore how the number of iterations and decoder type of block in the long-range correlation decoder affect the registration performance. We select three different blocks, i.e. LCD, CNN and GRU [4], to perform iterative decoding and predict low-scale deformation fields. In order to verify the effectiveness of iterative registration and how the number of iterations affects the registration effect, we performed 1 to 5 iterations for each decoder.

The Dice scores corresponding to different decoders and number of iterations are shown in Table 2, t represents the time of iteration. Experiments are performed on LPBA and SLIVER datasets. Obviously, among the three decoders, LCD gets the best registration results in a limited number of iterations. When the number of iterations is 1, the proposed structure degenerates into pyramid registration, and when the number of iterations is greater than or equal to 2, the pyramid-iteration structure is used. Obviously, the registration accuracy is greatly improved compared with the pyramid structure when the number of iterations is 2. On this task, when the number of iterations reaches 3, the registration accuracy tends to be stable, which indicates that the large deformation has been basically captured. Compared with the GRU block commonly used in optical flow tasks, LCD requires fewer iterations to converge, which verifies that LCD can better capture long-distance correlation and learn accurate flow.

4   Conclusion

In this paper, we propose an unsupervised pyramid-iterative vision Transformer (PIViT) for large deformation image registration. PIViT is an iterative and pyramid composite framework to achieve fine registration of large deformable images by iterative registration of low-scale feature maps and pyramid feature supplementation on high-scale feature maps. Furthermore, in the iterative decoding stage, a Swin Transformer-based long-range correlation decoder is introduced to capture the long-distance dependencies between feature maps, which further improves the ability to handle large deformation. Experiments on brain MRI scans and liver CT scans demonstrate that our method can accurately register 3D large deformation medical images. Furthermore, our method has significant advantages in terms of parameters and time, which can make it more suitable for time-sensitive tasks.

Acknowledgments. This work was supported in part by the National Nature Science Foundation of China (62273150), Shanghai Natural Science Foundation (22ZR1421000), Shanghai Municipal Science and Technology Committee of Shanghai Outstanding Academic Leaders Plan (21XD1430600), the Science and Technology Commission of Shanghai Municipality (14DZ2260800).

References

1. Abadi, M., et al.: Tensorflow: a system for large-scale machine learning. In: 12th {USENIX} Symposium on Operating Systems Design and Implementation ({OSDI} 16), pp. 265–283 (2016)
2. Antonelli, M., et al.: The medical segmentation decathlon. arXiv preprint arXiv:2106.05735 (2021)
3. Balakrishnan, G., Zhao, A., Sabuncu, M.R., Guttag, J., Dalca, A.V.: An unsupervised learning model for deformable medical image registration. In: Proceedings of the IEEE Conference on Computer Vision and Pattern Recognition (CVPR) (2018)
4. Ballas, N., Yao, L., Pal, C., Courville, A.: Delving deeper into convolutional networks for learning video representations. arXiv preprint arXiv:1511.06432 (2015)
5. Bellec, P., Chu, C., Chouinard-Decorte, F., Benhajali, Y., Margulies, D.S., Craddock, R.C.: The neuro bureau ADHD-200 preprocessed repository. Neuroimage **144**, 275–286 (2017)
6. Bilic, P., et al.: The liver tumor segmentation benchmark (LITS). arXiv preprint arXiv:1901.04056 (2019)

7. Chen, J., Frey, E.C., He, Y., Segars, W.P., Li, Y., Du, Y.: TransMorph: transformer for unsupervised medical image registration. Med. Image Anal. **82**, 102615 (2022)

8. Dalca, A.V., Balakrishnan, G., Guttag, J., Sabuncu, M.R.: Unsupervised learning of probabilistic diffeomorphic registration for images and surfaces. Med. Image Anal. **57**, 226–236 (2019)

9. Di Martino, A., et al.: The autism brain imaging data exchange: towards a large-scale evaluation of the intrinsic brain architecture in autism. Mol. Psychiatry **19**(6), 659–667 (2014)

10. Dice, L.R.: Measures of the amount of ecologic association between species. Ecology **26**(3), 297–302 (1945)

11. Dosovitskiy, A., et al.: An image is worth 16x16 words: transformers for image recognition at scale. arXiv preprint arXiv:2010.11929 (2020)

12. He, Y., et al.: Geometric visual similarity learning in 3D medical image self-supervised pre-training (2023). https://doi.org/10.48550/ARXIV.2303.00874. https://arxiv.org/abs/2303.00874

13. Heimann, T., et al.: Comparison and evaluation of methods for liver segmentation from CT datasets. IEEE Trans. Med. Imaging **28**(8), 1251–1265 (2009)

14. Hu, X., Kang, M., Huang, W., Scott, M.R., Wiest, R., Reyes, M.: Dual-stream pyramid registration network. In: Shen, D., et al. (eds.) MICCAI 2019. LNCS, vol. 11765, pp. 382–390. Springer, Cham (2019). https://doi.org/10.1007/978-3-030-32245-8_43

15. Jaderberg, M., Simonyan, K., Zisserman, A., et al.: Spatial transformer networks. In: Advances in Neural Information Processing Systems, vol. 28 (2015)

16. Jia, X., Bartlett, J., Zhang, T., Lu, W., Qiu, Z., Duan, J.: U-Net vs TransFormer: is U-Net outdated in medical image registration? In: Lian, C., Cao, X., Rekik, I., Xu, X., Cui, Z. (eds.) MLMI 2022. LNCS, vol. 13583, pp. 151–160. Springer, Cham (2022). https://doi.org/10.1007/978-3-031-21014-3_16

17. Jiang, S., Campbell, D., Lu, Y., Li, H., Hartley, R.: Learning to estimate hidden motions with global motion aggregation. In: Proceedings of the IEEE/CVF International Conference on Computer Vision, pp. 9772–9781 (2021)

18. Kingma, D.P., Ba, J.: Adam: a method for stochastic optimization. arXiv preprint arXiv:1412.6980 (2014)

19. Liu, Z., et al.: Swin transformer: hierarchical vision transformer using shifted windows. In: Proceedings of the IEEE/CVF International Conference on Computer Vision, pp. 10012–10022 (2021)

20. Meng, M., Bi, L., Feng, D., Kim, J.: Non-iterative coarse-to-fine registration based on single-pass deep cumulative learning. In: Wang, L., Dou, Q., Fletcher, P.T., Speidel, S., Li, S. (eds.) MICCAI 2022. LNCS, vol. 13436, pp. 88–97. Springer, Cham (2022). https://doi.org/10.1007/978-3-031-16446-0_9

21. Mok, T.C., Chung, A.: Fast symmetric diffeomorphic image registration with convolutional neural networks. In: Proceedings of the IEEE/CVF Conference on Computer Vision and Pattern Recognition, pp. 4644–4653 (2020)

22. Mok, T.C.W., Chung, A.C.S.: Large deformation diffeomorphic image registration with Laplacian pyramid networks. In: Martel, L., et al. (eds.) MICCAI 2020. LNCS, vol. 12263, pp. 211–221. Springer, Cham (2020). https://doi.org/10.1007/978-3-030-59716-0_21

23. Mok, T.C., Chung, A.: Affine medical image registration with coarse-to-fine vision transformer. In: Proceedings of the IEEE/CVF Conference on Computer Vision and Pattern Recognition, pp. 20835–20844 (2022)

24. Mueller, S.G., et al.: Ways toward an early diagnosis in Alzheimer's disease: the Alzheimer's disease neuroimaging initiative (ADNI). Alzheimer's Dement. **1**(1), 55–66 (2005)

25. Shattuck, D.W., et al.: Construction of a 3D probabilistic atlas of human cortical structures. Neuroimage **39**(3), 1064–1080 (2008)

26. Shi, J., et al.: XMorpher: full transformer for deformable medical image registration via cross attention. In: Wang, L., Dou, Q., Fletcher, P.T., Speidel, S., Li, S. (eds.) MICCAI 2022. LNCS, vol. 1346, pp. 217–226. Springer, Cham (2022). https://doi.org/10.1007/978-3-031-16446-0_21

27. Xu, H., Zhang, J., Cai, J., Rezatofighi, H., Tao, D.: GMFlow: learning optical flow via global matching. In: Proceedings of the IEEE/CVF Conference on Computer Vision and Pattern Recognition, pp. 8121–8130 (2022)

28. Zhao, S., Dong, Y., Chang, E.I., Xu, Y., et al.: Recursive cascaded networks for unsupervised medical image registration. In: Proceedings of the IEEE/CVF International Conference on Computer Vision, pp. 10600–10610 (2019)

29. Zhao, S., Lau, T., Luo, J., Eric, I., Chang, C., Xu, Y.: Unsupervised 3D end-to-end medical image registration with volume Tweening network. IEEE J. Biomed. Health Inform. **24**(5), 1394–1404 (2019)

30. Zhu, Y., Lu, S.: Swin-VoxelMorph: a symmetric unsupervised learning model for deformable medical image registration using Swin transformer. In: Wang, L., Dou, Q., Fletcher, P.T., Speidel, S., Li, S. (eds.) Medical Image Computing and Computer Assisted Intervention - MICCAI 2022, vol. 13436, pp. 78–87. Springer, Cham (2022)

GSMorph: Gradient Surgery for Cine-MRI Cardiac Deformable Registration

Haoran Dou[1], Ning Bi[1], Luyi Han[2,3], Yuhao Huang[4,5,6], Ritse Mann[2,3], Xin Yang[4,5,6,7], Dong Ni[4,5,6], Nishant Ravikumar[1,8], Alejandro F. Frangi[1,8,9,10], and Yunzhi Huang[11(✉)]

[1] Centre for Computational Imaging and Simulation Technologies in Biomedicine (CISTIB), University of Leeds, Leeds, UK
[2] Department of Radiology and Nuclear Medicine, Radboud University Medical Centre, Nijmegen, The Netherlands
[3] Department of Radiology, Netherlands Cancer Institute, Amsterdam, The Netherlands
[4] National-Regional Key Technology Engineering Laboratory for Medical Ultrasound, School of Biomedical Engineering, Health Science Center, Shenzhen University, Shenzhen, China
[5] Medical Ultrasound Image Computing (MUSIC) Lab, Shenzhen University, Shenzhen, China
[6] Marshall Laboratory of Biomedical Engineering, Shenzhen University, Shenzhen, China
[7] Shenzhen RayShape Medical Technology Co., Ltd., Shenzhen, China
[8] Division of Informatics, Imaging and Data Science, Schools of Computer Science and Health Sciences, University of Manchester, Manchester, UK
[9] Medical Imaging Research Center (MIRC), Electrical Engineering and Cardiovascular Sciences Departments, KU Leuven, Leuven, Belgium
[10] Alan Turing Institute, London, UK
[11] Institute for AI in Medicine, School of Artificial Intelligence, Nanjing University of Information Science and Technology, Nanjing, China
yunzhi.huang.scu@gmail.com

Abstract. Deep learning-based deformable registration methods have been widely investigated in diverse medical applications. Learning-based deformable registration relies on weighted objective functions trading off registration accuracy and smoothness of the deformation field. Therefore, they inevitably require tuning the hyperparameter for optimal registration performance. Tuning the hyperparameters is highly computationally expensive and introduces undesired dependencies on domain knowledge. In this study, we construct a registration model based on the gradient surgery mechanism, named GSMorph, to achieve a hyperparameter-free balance on multiple losses. In GSMorph, we reformulate the optimization procedure by projecting the gradient of similarity loss orthogonally to the plane associated with the smoothness constraint, rather than additionally introducing a hyperparameter to balance these two competing

© The Author(s), under exclusive license to Springer Nature Switzerland AG 2023
H. Greenspan et al. (Eds.): MICCAI 2023, LNCS 14229, pp. 613–622, 2023.
https://doi.org/10.1007/978-3-031-43999-5_58

terms. Furthermore, our method is model-agnostic and can be merged into any deep registration network without introducing extra parameters or slowing down inference. In this study, We compared our method with state-of-the-art (SOTA) deformable registration approaches over two publicly available cardiac MRI datasets. GSMorph proves superior to five SOTA learning-based registration models and two conventional registration techniques, SyN and Demons, on both registration accuracy and smoothness.

Keywords: Medical image registration · Gradient surgery · Regularization

1   Introduction

Image registration is fundamental to many medical image analysis applications, e.g., motion tracking, atlas construction, and disease diagnosis [5]. Conventional registration methods usually require computationally expensive iterative optimization, making it inefficient in clinical practice [1,19]. Deep learning has recently been widely exploited in the registration domain due to its superior representation extraction capability and fast inference speed [2,7]. Deep-learning-based registration (DLR) formulates registration as a network learning process minimizing a composite objective function comprising one similarity loss to penalize the difference in the appearance of the image pair, and a regularization term to ensure the smoothness of deformation field. Typically, to balance the registration accuracy and smoothness of the deformation field, a hyperparameter is introduced in the objective function. However, performing hyperparameter tuning is labor-intensive, time-consuming, and *ad-hoc*; searching for the optimal parameter setting requires extensive ablation studies and hence training tens of models and establishing a reasonable parameter search space. Therefore, alleviating, even circumventing, hyperparameter search to accelerate development and deployment of DLR models remains challenging.

Recent advances [6,11,13] in DLR have primarily focused on network architecture design to boost registration performance. Few studies [9,16] investigated the potential in preventing hyperparameter searching by hypernetwork [8] and conditional learning [10]. Hoopes *et al.* [9] leveraged a hyper-network that takes the hyperparameter as input to generate the weight of the DLR network. Although effective, it introduces a large number of additional parameters to the basic DLR network, making the framework computationally expensive. In parallel, Mok *et al.* [16] proposed to learn the effect of the hyperparameter and condition it on the feature statistics (usually illustrated as *style* in computer vision [10]) to manipulate the smoothness of the deformation field in the inference phase. Both methods can avoid hyperparameter tuning while training the DLR model. However, they still require a reasonable sampling space and strategy of the hyperparameter, which can be empirically dependent.

Gradient surgery (GS) projects conflicting gradients of different losses during the optimization process of the model to mitigate gradient interference. This has proven useful in multi-task learning [20] and domain generalization [15]. Motivated by these studies, we propose utilizing the GS to moderate the discordance between the similarity loss and regularization loss. The proposed method can further avert searching the weight for balancing losses in training the DLR.

- We propose GSMorph, a gradient-surgery-based DLR model. Our method can circumvent tuning the hyperparameter in composite loss function with a gradient-level reformulation to reach the trade-off between registration accuracy and smoothness of the deformation field.
- Existing GS approaches have operated the parameters' gradients independently or integrally. We propose a layer-wise GS to group by the parameters for optimization to ensure the flexibility and robustness of the optimization process.
- Our method is model-agnostic and can be integrated into any DLR network without extra parameters or losing inference speed.

2   Methodology

Deformable image registration estimates the non-linear correspondence field ϕ between the moving, M, and fixed, F, images (Fig. 1). Such procedure is mathematically formulated as $\phi = f_\theta(F, M)$. For learning-based registration methods, f_θ (usually adopted by a neural network) takes the fixed and moving image pair as input and outputs the deformation field via the optimal parameters θ. Typically, θ can be updated using standard mini-batch gradient descent as follows:

$$\theta := \theta - \alpha \nabla_\theta \left(\mathcal{L}_{sim}(\theta; F, M \circ \phi) + \lambda \mathcal{L}_{Reg}(\theta; \phi) \right) \qquad (1)$$

where α is the learning rate; $\mathcal{L}_{sim}$ is the similarity loss to penalize differences in the appearance of the moving and fixed images (e.g., mean square error, mutual information or local negative cross-correlation); $\mathcal{L}_{reg}$ is the regularization loss to encourage the smoothness of the deformation field (this can be computed by the gradient of the deformation field); λ is the hyperparameter balancing the trade-off between $\mathcal{L}_{sim}$ and $\mathcal{L}_{reg}$ to achieve desired registration accuracy while preserving the smoothness of the deformation field in the meantime. However, hyperparameter tuning is time-consuming and highly experience-dependent, making it tough to reach the optimal solution.

Insight into the optimization procedure in Eq. 1, as registration accuracy and spatial smoothness are potentially controversial in model optimization, the two constraints might have different directions and strengths while going through the gradient descent. Based on this, we provide a geometric view to depict the gradient changes for θ based on the *gradient surgery* technique. The conflicting relationship between two controversial constraints can be geometrically projected as orthogonal vectors. Depending on the orthogonal relationship, merely updating the gradients of the similarity loss would automatically associate with the

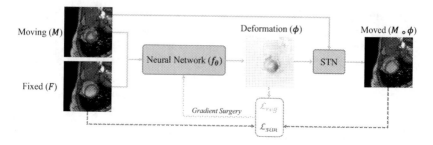

Fig. 1. Schematic illustration of our proposed GSMorph. GS modifies the gradients computed by similarity loss $\mathcal{L}_{sim}$ and regularization loss $\mathcal{L}_{reg}$, then updates the model's parameters θ.

updates of the regularization term. In this way, we avoid tuning the hyperparameter λ to optimize θ. The Eq. 1 can then be rewritten into a non-hyperparameter pattern:

$$\theta := \theta - \alpha \Phi(\nabla_\theta \mathcal{L}_{sim}(\theta; F, M \circ \phi)) \qquad (2)$$

where $\Phi(\cdot)$ is the operation of proposed GS method.

2.1 Layer-Wise Gradient Surgery

Figure 2 illustrates the two scenarios of gradients while optimizing the DLR network via vanilla gradient descent or gradient surgery. We first define that the gradient of similarity loss, g_{sim}, and that of regularization loss, g_{reg}, are conflicting when the angle between g_{sim} and g_{reg} is the obtuse angle, viz. $\langle g_{sim}, g_{reg} \rangle < 0$. In this study, we propose updating the parameters of neural networks by the original g_{sim} independently, when g_{sim} and g_{reg} are non-conflicting, representing g_{sim} has no incompatible component of the gradient along the direction of g_{reg}. Consequently, optimization with sole g_{sim} within a non-conflicting scenario can inherently facilitate the spatial smoothness of deformations.

Conversely, as shown in Fig. 2, conflicting gradients are the dominant reason associated with non-smooth deformations. Hence, deconflicting gradients in the optimization of the DLR network to ensure high registration accuracy, as well as smooth deformation, is the primary goal of our study. Following a simple and intuitive procedure, we project the g_{sim} onto the normal plane of the g_{reg}, where the projected similarity gradient g and g_{reg} are non-conflicting along each gradient's direction.

Existing studies [15,20] performed the GS in terms of independent parameters or the entire network. Despite the effectiveness, these can be either unstable or inflexible. Considering that a neural network usually extracts features through the collaboration of each parameter group in the convolution layers, we introduce a layer-wise GS to ensure its stability and flexibility. The parameter updating rule is detailed in the Algorithm 1. Specifically, in each gradient updating iteration, we first compute the gradients of two losses for the parameter group in

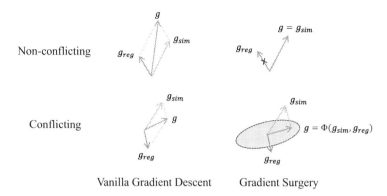

Fig. 2. Visualization of vanilla gradient descent and gradient surgery for non-conflicting and conflicting gradients. Regarding vanilla gradient descent, the gradient, g, is computed based on the average of g_{sim} and g_{reg}. Our GS-based approach projects the g_{sim} onto the normal vector of g_{reg} to prevent disagreements between the similarity loss and regularization loss. On the other hand, we only update the g_{sim} in non-conflicting scenarios.

each layer separately. Then, the conflicting relationship between the two gradients is calculated based on their inner production. Once the two gradients are non-conflicting, the gradients used to update its corresponding parameter group will be only the original gradients of similarity loss; on the contrary, the gradients will be the projected similarity gradients orthogonal to the gradients of regularization, which can be calculated as $g_{sim}^{i} - \frac{\langle g_{sim}^{i}, g_{reg}^{i} \rangle}{\|g_{reg}^{i}\|^{2}} g_{reg}^{i}$. After performing GS on all layer-wise parameter groups in the network, the final gradients will be used to update the model's parameters.

Algorithm 1. Gradient surgery

Require: Parameters θ_i in ith layer of the network; Number of layers in the network N.

1: $g_{sim} \leftarrow \nabla_\theta \mathcal{L}_{sim}$
2: $g_{reg} \leftarrow \nabla_\theta \mathcal{L}_{reg}$
3: **for** $i = 1 \rightarrow N$ **do**
4:     **if** $\langle g_{sim}^{i}, g_{reg}^{i} \rangle > 0$ **then**
5:         $g_i = g_{sim}^{i}$
6:     **else**
7:         $g_i = g_{sim}^{i} - \frac{\langle g_{sim}^{i}, g_{reg}^{i} \rangle}{\|g_{reg}^{i}\|^{2}} g_{reg}^{i}$
8:     **end if**
9:     $\Delta\theta_i = g_i$
10: **end for**
11: Update θ

2.2   Network Architecture

Our network architecture (seen in Fig. 1) is similar to VoxelMorph [7] that comprises naive U-Net [18] and spatial transform network (STN) [12]. The U-Net takes the moving and fixed image pair as input and outputs the deformation field, which is used to warp the moving image via STN. The U-Net consists of an encoder and a decoder with skip connections, which forward the features from each layer in the encoder to the corresponding layer in the decoder by concatenation to enhance the feature aggregation and prevent gradient vanishing. The number of feature maps in the encoder part of the network is 16, 32, 64, 128, and 256, increasing the number of features as their size shrinks, and vice versa in the decoder part. Each convolutional block in the encoder and decoder has two sequential convolutions with a kernel size of 3, followed by a batch normalization and a leaky rectified linear unit.

3   Experiments and Results

3.1   Datasets and Implementations

Datasets. In this study, we used two public cardiac cine-MRI datasets for investigation and comparison: ACDC [3] and M&M [4]. ACDC and M&M contain 100 and 249 subjects, respectively. We followed a proportion of 75%, 5%, and 20% to split each dataset for training, validation, and testing. We selected the image from the cine-MRI cardiac sequence at the End Systole (ES) time point of the cardiac cycle as the moving image, and that at the End Diastole (ED) as the fixed one. All images were cropped into the size of 128×128 centralized to the heart. We normalized the intensity of images into the range from 0 to 1 before inputting them into the model.

Implementation Details. We implemented our model in PyTorch [17], using a standard PC with an NVIDIA GTX 2080ti GPU. We trained the network through Adam optimizer [14] with a learning rate of 5e-3 and a batch size of 32 for 500 epochs. We also implemented and trained alternative methods for comparison with the same data and similar hyper-parameters for optimization. Our source code is available at https://github.com/wulalago/GSMorph.

3.2   Alternative Methods and Evaluation Criteria

To demonstrate the advantages of our proposed method in medical image registration, we compared it with two conventional deformable registration methods, i.e., **Demons** [19] and **SyN** [1], and a widely-used DLR model, **Voxel-Morph** [7]. These methods usually need laborious effort in hyperparameter tuning. Additionally, we reported the results of VoxelMorph trained with different λ (i.e., 0.1, 0.01, and 0.001, denoted as **VoxelMorph-l, VoxelMorph-m, VoxelMorph-s**). Meanwhile, we compared our approach to one alternative

DLR model based on the hyperparameter learning, i.e., **HyperMorph** [9]. This method only require additional validations in searching the optimal hyperparameter without necessarily tuning it from scratch. Finally, we reformulated two variations of GS based on our concept for further comparison. Specifically, **GS-Agr** [15] treats the gradient of each parameter independently. It updates the parameter with the gradient of similarity loss in the non-conflicting scenario, and a random gradient sampled from the Gaussian distribution when conflicting. While **GS-PCGrad** [20] uses the same GS strategy as ours, but with respect to the whole parameters of the entire network. The **Initial** represents the results without any deformation.

In this study, we used six criteria to evaluate the efficacy and efficiency of the investigated methods, including Dice score (Dice) and 95% Hausdorff distance (HD95) to validate the registration accuracy of the regions of interest, Mean square error (MSE) to evaluate the pixel-level appearance difference between the moved and fixed image-pairs, the percentage of pixels with negative Jacobian determinant (NJD) values to compare the smoothness and diffeomorphism of the deformation field, the number of parameters (Param) of the neural network and inference speed (Speed) to investigate the efficiency.

3.3   Results

As summarized in Table 1, our method could obtain the best MSE in the ACDC dataset and Dice in the M&M dataset while achieving comparable performance with the tuned VoxelMorph over other metrics. The Dice and HD95 reported in Table 1 were averaged over three anatomical regions of interest in the heart, i.e., Left ventricle, Myocardium, and Right Ventricle (LV, Myo, and RV). Consequently, the proposed model achieved superior registration accuracy and spatial regularization with faster inference speed than the two conventional registration methods. We also observed that our approach gained higher registration performance than HyperMorph in both datasets. Regarding the GS-based methods, GS-Agr totally collapsed, as the conflicting gradients accounted for most have

Table 1. Quantitative comparison of investigated methods on the testing datasets over ACDC and M&M.

Methods	Dataset							
	ACDC				M&M			
	Dice(%)	HD95(mm)	MSE(10^{-2})	NJD(%)	Dice(%)	HD95(mm)	MSE(10^{-2})	NJD(%)
Initial	61.81±8.68	4.40±1.33	1.58±0.52	–	61.03±10.16	4.79±1.82	1.90±1.08	–
Demons	85.38±3.52	1.67±0.75	0.46±0.21	1.31±0.59	75.66±10.30	17.79±6.23	0.71±0.61	1.84±1.19
SyN	79.28±8.23	2.24±1.28	0.65±0.21	0.30±0.27	81.97±9.36	**2.45±2.04**	0.84±0.12	0.49±0.47
VoxelMorph-s	86.69±2.17	1.30±0.24	0.39±0.14	2.01±0.96	77.12±9.36	3.43±2.18	**0.42±0.29**	3.45±2.33
VoxelMorph-m	**87.47±2.21**	**1.29±0.30**	0.42±0.15	0.67±0.48	79.93±8.57	2.91±1.98	0.48±0.32	1.31±1.10
VoxelMorph-l	82.12±4.30	1.87±0.64	0.59±0.18	**0.10±0.14**	77.18±8.69	2.81±1.60	0.74±0.43	**0.16±0.22**
HyperMorph	83.44±3.55	1.75±0.64	0.47±0.20	1.60±0.86	77.21±8.45	3.28±1.99	0.59±0.37	2.50±1.22
GS-Agr	63.40±9.15	4.20±1.35	1.33±0.43	0	63.41±9.85	4.50±1.77	1.55±0.86	<0.001
GS-PCGrad	84.59±3.53	1.62±0.53	0.51±0.16	0.11±0.17	80.67±8.18	2.48±1.67	0.59±0.36	0.41±0.44
GSMorph	87.45±2.27	1.34±0.40	**0.31±0.11**	0.87±0.52	**82.26±6.59**	2.66±1.93	0.49±0.27	0.98±0.84

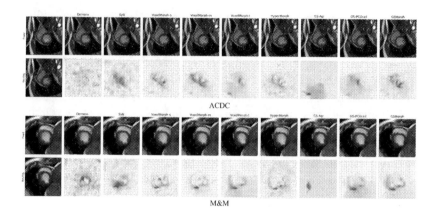

Fig. 3. Visual comparison of the registration results of the investigated methods for two representative test cases in ACDC and M&M datasets. The top rows are the fixed images and moved images from different methods; the bottom rows are the moving images and deformation fields. (We encourage you to zoom in for better visualization)

been replaced by random noise. On the other hand, GS-PCGrad only yielded an inadequate registration performance with an inclination of over-regularization. The comparison in the GS-based method shows the flexibility and robustness of our approach.

Figure 3 illustrates the sample cases of the warped images and the corresponding deformation fields from the compared methods. It can be observed our methods could obtain the moved images most similar to the fixed ones. Voxelmorph could achieve comparable results to us but still require a time-consuming hyperparameter tuning. Overall, the results of the comparisons in Table 1 and Fig. 3 indicate that our method performed the best among all the techniques that we implemented and examined, showing the effectiveness of our model in balancing the trade-off between registration accuracy and smoothness of deformations.

In Table 2, we have also reported the number of parameters and inference speed. We observed that DLR methods could obtain faster speed compared with conventional ones in general. As our proposed approach only modified the optimization procedure of the backbone network, it could maintain the original inference speed and the number of parameters. Conversely, HyperMorph intro-

Table 2. Number of parameters and inference speed of investigated methods on the testing datasets over ACDC and M&M.

Methods	Demons	SyN	VoxelMorph	HyperMorph	GSMorph
Params	–	–	1.96M	126M	1.96M
Speed	7.55±1.79	16.59±5.48	2.29±0.83	2.96±1.09	2.29±0.83

duced tremendous extra parameters and loss of inference speed as they adopted the secondary network to generate the conditions or weights of the main network architecture.

4   Conclusion

This work presents a gradient-surgery-based registration framework for medical images. To the best of our knowledge, this is the first study to employ gradient surgery to refine the optimization procedure in learning the deformation fields. In our GSMorph, the gradients from the similarity constraint were projected onto the plane orthogonal to those from the regularization term. In this way, merely updating the gradients in optimizing the registration accuracy would result in a joint updating of the gradients from the similarity and regularity constraints. Then, no additional regularization loss is required in the network optimization and no hyperparameter is further required to explicitly trade off between registration accuracy and spatial smoothness. Our model outperformed the conventional registration methods and the alternative DLR models. Finally, the proposed method is model-agnostic and can be integrated into any DLR network without introducing extra parameters or compromising the inference speed. We believe GSMorph will facilitate the development and deployment of DLR models and alleviate the influence of hyperparameters on performance.

Acknowledgement. This work was supported by the National Natural Science Foundation of China (62101365, 62171290, 62101343), Shenzhen-Hong Kong Joint Research Program (SGDX20201103095613036), Shenzhen Science and Technology Innovations Committee (20200812143441001), the startup foundation of Nanjing University of Information Science and Technology, the Ph.D. foundation for Innovation and Entrepreneurship in Jiangsu Province, the Royal Academy of Engineering (INSILEX CiET1819/19), Engineering and Physical Sciences Research Council UKRI Frontier Research Guarantee Programmes (INSILICO, EP/Y030494/1), and the Royal Society Exchange Programme CROSSLINK IES\NSFC\201380.

References

1. Avants, B.B., Epstein, C.L., Grossman, M., Gee, J.C.: Symmetric diffeomorphic image registration with cross-correlation: evaluating automated labeling of elderly and neurodegenerative brain. Med. Image Anal. **12**(1), 26–41 (2008)
2. Balakrishnan, G., Zhao, A., Sabuncu, M.R., Guttag, J., Dalca, A.V.: An unsupervised learning model for deformable medical image registration. In: Proceedings of the IEEE Conference on Computer Vision and Pattern Recognition, pp. 9252–9260 (2018)
3. Bernard, O., et al.: Deep learning techniques for automatic MRI cardiac multi-structures segmentation and diagnosis: is the problem solved? IEEE Trans. Med. Imaging **37**(11), 2514–2525 (2018)
4. Campello, V.M., et al.: Multi-centre, multi-vendor and multi-disease cardiac segmentation: the M&Ms challenge. IEEE Trans. Med. Imaging **40**(12), 3543–3554 (2021)

5. Chen, X., Diaz-Pinto, A., Ravikumar, N., Frangi, A.F.: Deep learning in medical image registration. Prog. Biomed. Eng. **3**(1), 012003 (2021)
6. Chen, X., Xia, Y., Ravikumar, N., Frangi, A.F.: A deep discontinuity-preserving image registration network. In: de Bruijne, M., et al. (eds.) MICCAI 2021. LNCS, vol. 12904, pp. 46–55. Springer, Cham (2021). https://doi.org/10.1007/978-3-030-87202-1_5
7. Dalca, A.V., Balakrishnan, G., Guttag, J., Sabuncu, M.R.: Unsupervised learning for fast probabilistic diffeomorphic registration. In: Frangi, A.F., Schnabel, J.A., Davatzikos, C., Alberola-López, C., Fichtinger, G. (eds.) MICCAI 2018. LNCS, vol. 11070, pp. 729–738. Springer, Cham (2018). https://doi.org/10.1007/978-3-030-00928-1_82
8. Ha, D., Dai, A., Le, Q.V.: Hypernetworks. arXiv preprint arXiv:1609.09106 (2016)
9. Hoopes, A., Hoffmann, M., Fischl, B., Guttag, J., Dalca, A.V.: HyperMorph: amortized hyperparameter learning for image registration. In: Feragen, A., Sommer, S., Schnabel, J., Nielsen, M. (eds.) IPMI 2021. LNCS, vol. 12729, pp. 3–17. Springer, Cham (2021). https://doi.org/10.1007/978-3-030-78191-0_1
10. Huang, X., Belongie, S.: Arbitrary style transfer in real-time with adaptive instance normalization. In: Proceedings of the IEEE International Conference on Computer Vision, pp. 1501–1510 (2017)
11. Huang, Y., Ahmad, S., Fan, J., Shen, D., Yap, P.T.: Difficulty-aware hierarchical convolutional neural networks for deformable registration of brain MR images. Med. Image Anal. **67**, 101817 (2021)
12. Jaderberg, M., Simonyan, K., Zisserman, A., et al.: Spatial transformer networks. In: Advances in Neural Information Processing Systems, vol. 28 (2015)
13. Jia, X., Thorley, A., Chen, W., Qiu, H., Shen, L., Styles, I.B., Chang, H.J., Leonardis, A., De Marvao, A., O'Regan, D.P., et al.: Learning a model-driven variational network for deformable image registration. IEEE Trans. Med. Imaging **41**(1), 199–212 (2021)
14. Kingma, D.P., Ba, J.: Adam: a method for stochastic optimization. arXiv preprint arXiv:1412.6980 (2014)
15. Mansilla, L., Echeveste, R., Milone, D.H., Ferrante, E.: Domain generalization via gradient surgery. In: Proceedings of the IEEE/CVF International Conference on Computer Vision, pp. 6630–6638 (2021)
16. Mok, T.C.W., Chung, A.C.S.: Conditional deformable image registration with convolutional neural network. In: de Bruijne, M., et al. (eds.) MICCAI 2021. LNCS, vol. 12904, pp. 35–45. Springer, Cham (2021). https://doi.org/10.1007/978-3-030-87202-1_4
17. Paszke, A., et al.: PyTorch: an imperative style, high-performance deep learning library. In: Advances in Neural Information Processing Systems, vol. 32 (2019)
18. Ronneberger, O., Fischer, P., Brox, T.: U-Net: convolutional networks for biomedical image segmentation. In: Navab, N., Hornegger, J., Wells, W.M., Frangi, A.F. (eds.) MICCAI 2015. LNCS, vol. 9351, pp. 234–241. Springer, Cham (2015). https://doi.org/10.1007/978-3-319-24574-4_28
19. Vercauteren, T., Pennec, X., Perchant, A., Ayache, N.: Diffeomorphic demons: efficient non-parametric image registration. Neuroimage **45**(1), S61–S72 (2009)
20. Yu, T., Kumar, S., Gupta, A., Levine, S., Hausman, K., Finn, C.: Gradient surgery for multi-task learning. In: Advances in Neural Information Processing Systems, vol. 33, pp. 5824–5836 (2020)

Progressively Coupling Network for Brain MRI Registration in Few-Shot Situation

Zuopeng Tan[1], Hengyu Zhang[1], Feng Tian[2,3], Lihe Zhang[1(✉)], Weibing Sun[2,3], and Huchuan Lu[1]

[1] School of Information and Communication Engineering, Dalian University of Technology, Dalian, China
{zuopengtan,skysthelimits}@mail.dlut.edu.cn,
{zhanglihe,lhchuan}@dlut.edu.cn
[2] Department of Urology, Affiliated Zhongshan Hospital of Dalian University, Dalian, China
[3] Key Laboratory of Microenvironment Regulation and Immunotherapy of Urinary Tumors in Liaoning Province, Dalian, China

Abstract. Segmentation-assisted registration models can leverage few available labels in exchange for large performance gains by their complementarity. Recent related works independently build the prediction branches of deformation field and segmentation label without any information interaction except for the joint supervision. They ignore underlying relationship between the two tasks, thereby failing to fully exploit their complementary nature. To this end, we propose a ProGressively Coupling Network (PGCNet) that relies on segmentation to regularize the correct projecting of registration. Our overall framework is a multitask learning paradigm in which features are extracted by one shared encoder and then separate prediction branches are built for segmentation and registration. In the prediction phase, we utilize the bidirectional deformation fields as bridges to warp the features of moving and fixed images to each other's segmentation branches, thereby progressively and interactively supplementing additional context information at multiple levels for their segmentation. By establishing the entangled correspondence, segmentation supervision can indirectly regularize registration stream to accurately project semantic layout for segmentation branches. In addition, we design the position correlation calculation for registration to easier capture the spatial correlation of the images from the shared features. Experimental results on public 3D brain MRI datasets show that our work performs favorably against the state-of-the-art methods.

1 Introduction

Deformable medical image registration has many applications in clinic, including but not limited to clinical case tracking, surgical navigation. In recent years, many advanced learning-based medical image registration methods [2,6,7,15] have been proposed. Due to the limitations of accessible labels, it is more common

H. Greenspan et al. (Eds.): MICCAI 2023, LNCS 14229, pp. 623–633, 2023.
https://doi.org/10.1007/978-3-031-43999-5_59

to employ unsupervised ways for optimization. Although unsupervised methods [5,12,14,17,19] can guide the registration network to optimize by maximizing the image-wise similarity between warped and fixed images, the lack of guidance from regions of interest (ROIs) makes the registration performance fall into a bottleneck. Moreover, medical image labeling usually requires professional medical staff, and it's laborious to get large-scale training labels.

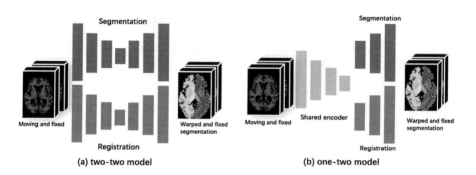

Fig. 1. Two common joint registration and segmentation frameworks. (a) Two-two model: two encoders and two decoders. (b) One-two model: one shared encoder and two separate decoders.

Integrating segmentation into registration can considerably compensate for the lack of image labels via their complementarity [11,23]. Specifically, warping segmentation result of moving image to align to that of fixed image by deformation field can provide additional supervision of ROIs for registration task. Meanwhile, as a way of data augmentation, the unlabelled images can participate in the segmentation optimization to prevent the overfitting of segmentation in few-shot situation [20,26].

Recently, several joint learning methods have been proposed. These methods can be generally divided into two categories in terms of model structure (see Fig. 1), *two-two model* (i.e., two encoders and two decoders) [10,11,23] and *one-two model* (i.e., one encoder and two decoders) [8,18,27]. The *two-two models* employ two completely independent models for segmentation and registration, and achieve mutual learning by joint loss. DeepAtlas [23] alternately trains registration and segmentation networks to achieve mutual improvement in brain and knee images. He et al. [10] feed the segmentation masks into the registration network to provide the internal texture information of the regions of interest (ROIs) so as to avoid incorrect deformation of internal areas.

In comparison, the *one-two models* are more in line with the general paradigm of multi-task learning, which can be regarded as a process of inductive bias [21], i.e., utilizing the shared encoder to induct the commonality of each task, and then using the respective decoder for preference prediction. They can effectively reduce the risk of overfitting and reduce the parameters of the network through the way of sharing parameters [3]. However, existing *one-two models* all focus on loss design to depict joint learning [8,18,27]. There is not any explicit feature interaction between two task streams. It is well known that sensible interactions help model capture extra key features or regularize networks to optimize in desired directions [22,24,25,28]. In addition, the semantic information of moving and fixed images are interchangeable and mutually exploitable. As it happens, the deformation field in registration is able to interconvert the context information of moving and fixed images and the converted features can be employed as additional contextual information for their segmentation.

Based on the framework of *one-two model*, we propose a progressively coupling network (PGCNet) as shown in Fig. 2. Specifically, progressive field estimators are designed to calculate multi-level deformation fields from coarse to fine. By warping the moving features and computing the correlation between warped and fixed features at each level, the deformation field is refined progressively. Different from common correlation calculation [13,14], we introduce learnable absolute position coordinates to provide spatial information for registration, which can better measure coordinate correspondence of voxels. In order to closely couple the registration and segmentation branches, we infuse the warped contexts into the segmentation branches. In this way, segmentation supervision sets an extra attention focusing for the registration decoder, which drives the deformation field to project semantic layout from moving image to fixed one. This semantic projection helps improve the registration accuracy. Meanwhile, the registration branch provides more sufficient semantic information for segmentation branch.

The main contributions can be summarized as follows: *1)* We propose a novel registration learning framework to establish the entangled relationship between registration and segmentation by progressively coupling the moving and fixed segmentation, which promotes both registration and segmentation performance in few-shot situation. *2)* To effectively measure the coordinate correspondence of voxels, we design the position correlation calculation, which provides spatial coordinate information for field estimator while measuring feature similarity. *3)* Experimental results on the general brain MRI datasets, OASIS and IXI, show that our method outperforms the state-of-the-art methods.

2   Method

Our model aims to learn registration and segmentation tasks simultaneously. Let M, F be moving and fixed images, respectively. We parameterize our model as a function with parameter θ, $f_\theta(M, F) = \phi, S_M, S_F$, where ϕ is the deformation field, S_M and S_F represent the segmentation results of M and F. Registration task completes the transformation from moving images to fixed ones, and segmentation task generates the segmentation maps of moving and fixed images.

2.1   Overall Network Structure

The overall pipeline of our method is shown in Fig. 2, which mainly consists of two parts, shared encoder and pyramid decoder. The parameters of the encoder

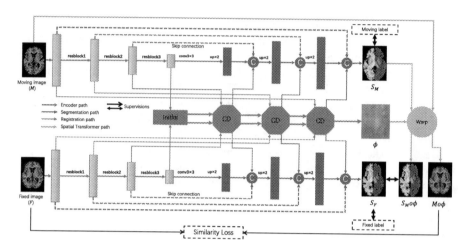

Fig. 2. The pipeline of progressively coupling network (PGCNet), which is an encoder-decoder structure. The red arrows denote the shared encoder path. The blue and purple arrows indicate the information flow of registration and segmentation, respectively, and the spatial transformation path is represented as brown. The structure of Coupling Decoder (the blue octagon) is shown in Fig. 3 (Color figure online).

(rose red block) and segmentation decoder (purple block) are shared. The shared encoder of our model is a 3D residual convolution network, which has 4 feature levels. Except for the input level, we use 3, 5, and 5 residual blocks at the other levels, respectively. A residual block contains two convolutional layers with pre-activation [9] structure and shortcut connection, to extract features, respectively. The numbers of channels at the four levels are 8, 16, 32, and 64, respectively, and each convolution layer is followed by Instance Normalization and Leaky-ReLU with a negative slope of 0.2 except for the last output. We first calculate the correlation of the deepest level features and then estimate the initial deformation field based on the correlation matrix. In the segmentation branch, we employ the commonly used skip connection structure, and concatenate the complementary features from the registration branch at each level, which are warped by bidirectional deformation fields.

2.2 Coupling Decoder

As shown in Fig. 3, there are four inputs in our coupling decoder: Field (φ), Inv-Field (φ^-), Moving Context (F_M) and Fixed Context (F_F). Among them, Field (φ) and Inv-Field (φ^-) are up-sampled by trilinear interpolation with the factor of 2 from the front level. Moving Context (F_M) and Fixed Context (F_F) are the context information from the corresponding level. We use φ to warp F_M and input the warped features $(F_M \circ \varphi)$ into the segmentation decoder of fixed image as additional complementary information (coupled features). In this way, the segmentation network obtains richer semantic information. The same operation is done in the segmentation branch of moving image as well. Relying on the projecting ability of the deformation field, we align and combine context information of moving and fixed images. Thus, registration and segmentation branches establish an entangled correspondence to constrain each other to learn a well generalized model. Generally speaking, if the fixed image is precisely segmented, the coupled features, which are projected from moving image to fixed one, usually are well matched. That is, the coupling process can constrain the deformation field to become accurate.

Position Correlation Calculation for Registration. In order to measure the similarity between each voxel and its neighbors, the warped ($F_M \circ \varphi$) and fixed (F_F) context features will conduct correlation calculation to obtain a cost volume. Note that the inverse correlation is calculated between the warped ($F_F \circ \varphi^-$) and moving (F_M) context features, which is not shown in Fig. 3. The spatial correlation between adjacent voxels in moving and fixed images is the key to determining the deformation field. According to the cost volume, we can judge the displacement between moving and fixed images at the current level. Different from existing works

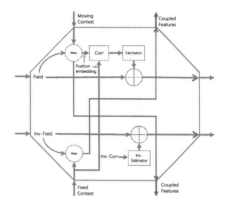

Fig. 3. Coupling decoder. Spatial transformer utilizes deformation field to warp context feature. Correlation calculation gets cost volume. Field estimation module predicts the increment of deformation field.

[13,14], we embed the position information into feature volumes before correlation calculation. Segmentation and registration tasks have different semantic expression requirements. The position coding can reduce the semantic confusion brought about by the shared encoder and enforce the decoder to understand spatial relationships of voxels. Let E be position coding map and r be the displacement radius, the correlation calculation is formulated as:

$$Corr(F_1, F_2, E) = \sum_{i,j,k}^{r} < (F_1 + E)_{x,y,z} \cdot (F_2 + E)_{x+i,y+j,z+k} >, \quad (1)$$

where $F_1 = F_M \circ \varphi$, $F_2 = F_F$. E is initialized as **0**. x, y and z represent the coordinate indexes of feature map in three directions, respectively. We move feature map point by point along x, y and z directions with radius r and do dot product to generate a correlation map. The dimension of correlation matrix (cost volume) is $(2r + 1)^3$, and we set $r = 1$ in this work. Then, Field and Inv-Field estimation modules utilize three residual blocks to compute the increment of forward and inverse displacement fields according to the correlation and inverse correlation matrices, respectively. The parameters of the two modules are shared.

2.3   Loss Function

The loss functions of our framework comprise three parts, which regularize registration, segmentation, and joint optimization, respectively.

The registration loss has similarity and smoothness penalty terms to align the moving and fixed images and ensure the smoothness of the deformation field. We use local normalized cross-correlation [2] with window size ω ($\omega = 9$) as similarity function. Let F and W represent the fixed and warped ($M \circ \phi$) scans. L_2 loss of deformation field gradient is set as regularization function. The registration loss can be defined as:

$$L_{reg} = LNCC_\omega(F, W) + \lambda_1 \sum \| \bigtriangledown \phi\|_2^2, \tag{2}$$

where λ_1 is a balance hyperparameter and is set as 1 in the experiments.

For segmentation, we use the combination of weighted cross entropy and Dice loss, which is formulated as:

$$L_{wce}(S, S^*) = -\frac{1}{NP} \sum_{n=0}^{N} \sum_{p=0}^{P} w_n S_{n,p}^* log S_{n,p}, \tag{3}$$

$$L_{dice}(S, S^*) = -\frac{1}{N} \sum_{n=0}^{N} \frac{\sum_{p=0}^{P} S_{n,p}^* \cdot S_{n,p}}{\sum_{p=0}^{P} S_{n,p}^* + \sum_{p=0}^{P} S_{n,p}}, \tag{4}$$

where N represents the total number of semantic classes, n indicates the channel of the corresponding category, and P is the number of voxels in each channel. S is the segmentation result and S^* is the voxel-wise manual segmentation label. In Eq. 3, w_n indicates the inverse proportion of voxels in category n to all voxels. For these two segmentation loss functions, their weights are fixed to the same value which means $L_{seg} = L_{wce} + L_{dice}$.

The joint loss function is defined as $L_{dice}(S_F, S_M \circ \phi)$. It regularizes registration network to provide additional training data for segmentation task. Meanwhile, registration network can pay more attention to the ROI regions.

3   Experiments and Results

3.1   Experimental Settings

Datasets and Preprocessing. We validate our method by using 414 T1-weighted brain MRI scans from OASIS [16] dataset and 576 T1-weighted brain MRI scans from IXI[1] dataset. These scans are preprocessed, which includes skull dissection, spatial normalization. All MRI scans are resampled with the same isotropic voxels of $1\,mm \times 1\,mm \times 1\,mm$, and we center crop the scans from OASIS dataset to $160 \times 192 \times 144$ and the scans from IXI dataset to $160 \times 160 \times 192$, which crops excess background to reduce the memory of GPU.

---

[1] IXI dataset is available in https://brain-development.org/ixi-dataset/.

The subcortical structures in OASIS and IXI are labeled as 35 and 44 categories by FreeSurfer for evaluation. We randomly select 5 scans as atlas from OASIS dataset, then the remaining scans are randomly divided into 255, 22, 132 for training, validation and testing. We do not use the labels of the training scans, only 5 labels of atlas are used to simulate a few-shot scenario. A total of 1,275, 110, 660 pairs of scans are used for training, validation and testing. Similarly, the IXI dataset is randomly split into 397, 58, 115, and 5 labeled scans are randomly taken as atlas for few-shot situation. The training, validation and testing include 1,985, 290, 575 pairs of scans, respectively.

Implementation. All trainings use Adam optimizer with learning rate $1e^{-4}$ and the batch size is set as 1. We first train the network by optimizing $L_{reg} + L_{seg}$ for 5,000 iterations and then by optimizing $L_{reg} + L_{seg} + L_{joint}$ for 80,000 iterations. All experiments are performed on 1 Nvidia RTX 3090 GPU.

Baseline Methods. We compare our method with a series of registration models. SYN [1] and LDDM [4] are two traditional registration algorithms. Voxelmorph [2] and transmorph [5] are the learning-based methods, which directly predict the deformation field. While LapIRN [17] and ULAE [19] are two progressive methods. We also evaluate these two methods under the semi-supervised setting with auxiliary segmentation labels. The semi-supervised loss function is the same as ours. DeepAtlas [23], UResNet [8], and PC-Reg [10] are joint learning methods, and we adopt the same training strategy as theirs.

3.2   Experimental Results

Registration Performance. As shown in Table 1, PGCNet achieves competitive registration accuracy, 87.72% DSC_{mean} and 1.59 $HD95$ on OASIS dataset and 78.08% DSC_{mean} and 3.19 $HD95$ on IXI dataset. For VoxelMorph and TransMorph, which directly estimate the deformation field, their DSC_{mean} are relatively low. Coarse-to-fine registration methods, LapIRN and ULAE, can meet the complex transformation needs between images through multiple deformations, but they still have poor performance due to the lack of ROIs supervision information. By introducing additional ROIs supervision information, their performance is improved by 1.1% and 1.8%, respectively. Their improvement is weak because of the limited semi-supervised information. Our method outperforms UResNet, DeepAtlas, and PC-Reg by 12.7%, 5.7%, and 1.7%, respectively, in joint learning models. The poor performance of UResNet can be attributed to its approach of concatenating moving and fixed images to extract features, while only predicting the moving segmentation result. Figure 4 presents visual comparison of various methods. It can be observed that our method exhibits lowest registration error in the third row.

Ablation Study. We conduct a series of ablation experiments on OASIS dataset to verify the effectiveness of each module, as shown in Table 2. We take the

Table 1. Quantitative results of different registration methods on OASIS and IXI datasets. Average Dice similarity coefficient (DSC_{mean}) and 95% percentile of the Hausdorff distance ($HD95$) are used to measure the registration accuracy of the models, and the average proportion of folding voxels in the deformation fields ($| J_\phi | < 0$) indicates the topological reversibility of the deformation field.

	OASIS			IXI		
	DSC_{mean} ↑	$HD95$ ↓	$| J_\phi | < 0$ ↓	DSC_{mean} ↑	$HD95$ ↓	$| J_\phi | < 0$ ↓
SYN [1]	76.08	2.22	0	67.21	3.76	0
LDDM [4]	76.07	2.23	0	66.94	3.90	0
VoxelMorph [2]	77.51	2.20	0.012	67.894	3.86	0.014
VoxelMorph-diff [7]	77.93	2.08	0	67.67	3.78	0
TransMorph [5]	79.94	2.04	0.010	69.58	3.72	0.014
LapIRN [17]	79.68	2.02	0.026	69.32	3.55	0.028
ULAE [19]	82.17	1.80	0.006	71.20	3.52	0.006
LapIRN-ROIs [17]	80.54	1.94	0.025	70.55	3.51	0.027
ULAE-ROIs [19]	83.62	1.70	0.006	72.47	3.42	0.007
UResNet [8]	77.83	2.16	0.020	67.76	3.70	0.019
DeepAtlas [23]	82.99	1.85	0.018	73.64	3.36	0.023
PC-Reg [10]	86.47	1.91	0.022	76.69	3.54	0.017
PGCNet(ours)	**87.72**	**1.59**	0.013	**78.08**	**3.19**	0.010

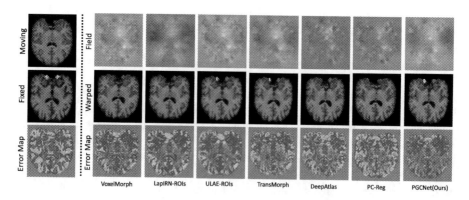

Fig. 4. Visualization results of different methods. In error maps, the redder color indicates the larger registration error, while the bluer color represents the smaller error.

Table 2. Ablation for registration.

JTrain	Couple	PEmbedding	RegDice
			82.51
√			86.29
√	√		87.03
√	√	√	**87.72**

Table 3. Ablation for segmentation.

Couple	JTrain	SegDice
		81.45
√		84.17
√	√	**86.15**

registration-only model as baseline, which has the same encoder and field estimation module as the final model, utilizing common correlation calculation. Joint training (JTrain) introduces semi-supervised loss for registration. And the coupling feature connection (Couple) improves the performance of registration by guiding the deformation field in advance to map semantic features. Position embedding (PEmbedding) assists the field estimator in understanding the correlation between voxels. In addition, we further analyse the factors that affect the performance of segmentation (see Table 3). Coupling feature connection provides more adequate semantic information, resulting in 3.3% improvement, while joint training provides additional trainable data, leading to 2.4% improvement.

4 Conclusion

We propose a progressively coupling network (PGCNet), which employs the deformation fields to couple the registration and segmentation branches. This is a novel mode of regularization for registration and segmentation, which promotes their performance in few-shot situation. In addition, position embedding provides additional spatial coordinate information for registration branch. The experimental results indicate that our work is superior to existing methods.

Acknowledgements. This work was supported by the National Natural Science Foundation of China #62276046, and the Liaoning Natural Science Foundation #2021-KF-12-10.

References

1. Avants, B.B., Epstein, C.L., Grossman, M., Gee, J.C.: Symmetric diffeomorphic image registration with cross-correlation: evaluating automated labeling of elderly and neurodegenerative brain. Med. Image Anal. **12**(1), 26–41 (2008)
2. Balakrishnan, G., Zhao, A., Sabuncu, M.R., Guttag, J., Dalca, A.V.: An unsupervised learning model for deformable medical image registration. In: IEEE Conference on Computer Vision and Pattern Recognition, pp. 9252–9260 (2018)
3. Baxter, J.: A Bayesian/information theoretic model of learning to learn via multiple task sampling. Mach. Learn. **28**(1), 7–39 (1997)
4. Beg, M.F., Miller, M.I., Trouve, A., Younes, L.: Computing large deformation metric mappings via geodesic flows. Int. J. Comput. Vision **61**(2), 139–157 (2005)

632   Z. Tan et al.

5. Chen, J., Frey, E.C., He, Y., Segars, W.P., Li, Y., Du, Y.: TransMorph: transformer for unsupervised medical image registration. Med. Image Anal. **82**, 102615 (2022)
6. Chen, X., Xia, Y., Ravikumar, N., Frangi, A.F.: A deep discontinuity-preserving image registration network. In: de Bruijne, M., et al. (eds.) MICCAI 2021. LNCS, vol. 12904, pp. 46–55. Springer, Cham (2021). https://doi.org/10.1007/978-3-030-87202-1_5
7. Dalca, A.V., Balakrishnan, G., Guttag, J., Sabuncu, M.R.: Unsupervised learning of probabilistic diffeomorphic registration for images and surfaces. Med. Image Anal. **57**, 226–236 (2019)
8. Estienne, T., et al.: U-ReSNet: ultimate coupling of registration and segmentation with deep nets. In: Shen, D., et al. (eds.) MICCAI 2019. LNCS, vol. 11766, pp. 310–319. Springer, Cham (2019). https://doi.org/10.1007/978-3-030-32248-9_35
9. He, K., Zhang, X., Ren, S., Sun, J.: Identity mappings in deep residual networks. In: Leibe, B., Matas, J., Sebe, N., Welling, M. (eds.) ECCV 2016. LNCS, vol. 9908, pp. 630–645. Springer, Cham (2016). https://doi.org/10.1007/978-3-319-46493-0_38
10. He, Y., et al.: Few-shot learning for deformable medical image registration with perception-correspondence decoupling and reverse teaching. IEEE J. Biomed. Health Inform. **26**(3), 1177–1187 (2022). https://doi.org/10.1109/JBHI.2021.3095409
11. He, Y., et al.: Deep complementary joint model for complex scene registration and few-shot segmentation on medical images. In: Vedaldi, A., Bischof, H., Brox, T., Frahm, J.-M. (eds.) ECCV 2020. LNCS, vol. 12363, pp. 770–786. Springer, Cham (2020). https://doi.org/10.1007/978-3-030-58523-5_45
12. Hu, B., Zhou, S., Xiong, Z., Wu, F.: Recursive decomposition network for deformable image registration. IEEE J. Biomed. Health Inform. **26**(10), 5130–5141 (2022)
13. Jonschkowski, R., Stone, A., Barron, J.T., Gordon, A., Konolige, K., Angelova, A.: What matters in unsupervised optical flow. In: Vedaldi, A., Bischof, H., Brox, T., Frahm, J.-M. (eds.) ECCV 2020. LNCS, vol. 12347, pp. 557–572. Springer, Cham (2020). https://doi.org/10.1007/978-3-030-58536-5_33
14. Kang, M., Hu, X., Huang, W., Scott, M.R., Reyes, M.: Dual-stream pyramid registration network. Med. Image Anal. **78**, 102379 (2022)
15. Lv, J., et al.: Joint progressive and coarse-to-fine registration of brain MRI via deformation field integration and non-rigid feature fusion. IEEE Trans. Med. Imaging **41**(10), 2788–2802 (2022). https://doi.org/10.1109/TMI.2022.3170879
16. Marcus, D.S., Wang, T.H., Parker, J., Csernansky, J.G., Morris, J.C., Buckner, R.L.: Open access series of imaging studies (OASIS): cross-sectional MRI data in young, middle aged, nondemented, and demented older adults. J. Cogn. Neurosci. **19**(9), 1498–1507 (2007)
17. Mok, T.C.W., Chung, A.C.S.: Large deformation diffeomorphic image registration with Laplacian pyramid networks. In: Martel, A.L., et al. (eds.) MICCAI 2020. LNCS, vol. 12263, pp. 211–221. Springer, Cham (2020). https://doi.org/10.1007/978-3-030-59716-0_21
18. Qin, C., et al.: Joint learning of motion estimation and segmentation for cardiac MR image sequences. In: Frangi, A.F., Schnabel, J.A., Davatzikos, C., Alberola-López, C., Fichtinger, G. (eds.) MICCAI 2018. LNCS, vol. 11071, pp. 472–480. Springer, Cham (2018). https://doi.org/10.1007/978-3-030-00934-2_53
19. Shu, Y., Wang, H., Xiao, B., Bi, X., Li, W.: Medical image registration based on uncoupled learning and accumulative enhancement. In: de Bruijne, M., et al. (eds.) MICCAI 2021. LNCS, vol. 12904, pp. 3–13. Springer, Cham (2021). https://doi.org/10.1007/978-3-030-87202-1_1

20. Vakalopoulou, M., et al.: AtlasNet: multi-atlas non-linear deep networks for medical image segmentation. In: Frangi, A.F., Schnabel, J.A., Davatzikos, C., Alberola-López, C., Fichtinger, G. (eds.) MICCAI 2018. LNCS, vol. 11073, pp. 658–666. Springer, Cham (2018). https://doi.org/10.1007/978-3-030-00937-3_75

21. Vandenhende, S., Georgoulis, S., Van Gansbeke, W., Proesmans, M., Dai, D., Van Gool, L.: Multi-task learning for dense prediction tasks: a survey. IEEE Trans. Pattern Anal. Mach. Intell. **44**(7), 3614–3633 (2021)

22. Xu, D., Ouyang, W., Wang, X., Sebe, N.: PAD-Net: multi-tasks guided prediction-and-distillation network for simultaneous depth estimation and scene parsing. In: IEEE Conference on Computer Vision and Pattern Recognition, pp. 675–684 (2018)

23. Xu, Z., Niethammer, M.: DeepAtlas: joint semi-supervised learning of image registration and segmentation. In: Shen, D., et al. (eds.) MICCAI 2019. LNCS, vol. 11765, pp. 420–429. Springer, Cham (2019). https://doi.org/10.1007/978-3-030-32245-8_47

24. Zhang, Z., Cui, Z., Xu, C., Jie, Z., Li, X., Yang, J.: Joint task-recursive learning for semantic segmentation and depth estimation. In: European Conference on Computer Vision, pp. 235–251 (2018)

25. Zhang, Z., Cui, Z., Xu, C., Yan, Y., Sebe, N., Yang, J.: Pattern-affinitive propagation across depth, surface normal and semantic segmentation. In: IEEE/CVF Conference on Computer Vision and Pattern Recognition, pp. 4106–4115 (2019)

26. Zhao, A., Balakrishnan, G., Durand, F., Guttag, J.V., Dalca, A.V.: Data augmentation using learned transformations for one-shot medical image segmentation. In: IEEE/CVF Conference on Computer Vision and Pattern Recognition, pp. 8543–8553 (2019)

27. Zhao, F., Wu, Z., Wang, L., Lin, W., Xia, S., Li, G.: A deep network for joint registration and parcellation of cortical surfaces. In: de Bruijne, M., et al. (eds.) MICCAI 2021. LNCS, vol. 12904, pp. 171–181. Springer, Cham (2021). https://doi.org/10.1007/978-3-030-87202-1_17

28. Zhao, X., Pang, Y., Zhang, L., Lu, H.: Joint learning of salient object detection, depth estimation and contour extraction. IEEE Trans. Image Process. **31**, 7350–7362 (2022)

Nonuniformly Spaced Control Points Based on Variational Cardiac Image Registration

Haosheng Su[1,2,3] and Xuan Yang[1,2,3](✉)

[1] Shenzhen University, Shenzhen 518060, Guangdong, China
yangxuan@szu.edu.cn
[2] Guangdong Provincial Key Laboratory of Popular High Performance Computers, Shenzhen, Guangdong, China
[3] Shenzhen Key Laboratory of Service Computing and Application, Shenzhen, Guangdong, China

Abstract. Non-uniformly spaced control points located on the interface of different objects are beneficial for constructing an accurate displacement field for image registration. However, extracting features of non-uniformly spaced control points in images is challenging for convolutional neural networks (CNNs). We extend a probabilistic image registration model using uniformed-spaced control points by employing non-uniformly-spaced control points. We construct a network to extract the image and spatial features of non-uniformly-spaced control points. Moreover, a variational Bayesian (VB) model using a factorized prior is employed to estimate the distribution of latent variables. In theory, we analyze the KL divergence between the posterior and the two separated priors. We found that the factorized prior has the advantage of decreasing the KL divergence, but too more factorized priors, such as the standard normal, might deteriorate registration accuracy. Moreover, we analyze the relationship between the uncertainty of the displacement field and the spatial distribution of control points. Experimental results on four public datasets show that our network outperforms the state-of-arts registration networks and can provide registration uncertainty.

Keywords: Cardiac image registration · Variational Bayesian · Non-uniformly-spaced control points

1 Introduction

Image registration is critical for estimating cardiac motion, aiming to estimate the displacements between cardiac anatomical tissues at different time points. Generative models focus on data distribution and tend to model the underlying

---

Supplementary Information The online version contains supplementary material available at https://doi.org/10.1007/978-3-031-43999-5_60.

patterns or data distribution. They make it possible to train the network using fewer data, improve robustness when data is missing, and, most importantly, allow quantifying the uncertainty associated with the output [12,22]. Variational Bayesian (VB) [16] is commonly used in generative models. In recent years, unsupervised registration methods based on Variational Bayesian (VB) have been proposed, including point-set-based and intensity-based [2–4,9,14,15,28]. Point-set-based methods extract critical points from two images and simultaneously estimate the probabilistic correspondence and spatial transformation between two point sets [7,18,21,26,27,29]. In these methods, the points and their correspondence are random variables, and the transformation parameters are latent variables. Intensity-based methods estimate the distribution of parameters of transformation [8,11,13,17,19,20]. These methods extract image features and pay attention to image correspondence. This paper focuses on intensity-based VB methods.

Gan *et al.* [11] proposed a probabilistic image registration method based on a parametric transformation model. They pointed out that the spatial locations of control points influence registration accuracy and delicately located control points can improve registration results. On the other hand, most existing priors either each dimension with identical independent distribution [17,19,20] or all dimensions obey a global distribution [8,11]. Priors with identical independent distributions are too simple to constrain the variational posterior; on the contrary, the global distribution might enforce each dimension of the variational posterior to correlate too much.

To address the above issues, we propose a probabilistic model based on variational Bayesian using non-uniformly spaced control points for cardiac image registration. Details of our contributions include:

- Employing nonuniformly spaced control points in a variational Bayesian image registration model improves registration accuracy. The control points are spaced on the contours of objects, and their intensity and spatial features are extracted using a network. We addressed the inherent disorder challenge in the control-points-based image registration model using CNNs, which can locate control points freely instead of only on grids. Additionally, our approach is not sensitive to location errors of control points.
- The global prior is partitioned into several independent priors, which correspond to different control points. We analyzed the KL divergence between the variational posterior and the factorized prior in theory and found that properly factorized priors can close the gap with the variational posterior and increase the evidence of a lower bound in the VB model.
- Our approach can provide more available information about registration uncertainty. Our uncertainty maps concentrate on the boundaries of objects instead of spreading over everywhere. It is favorable in real applications, where surgeons only pay attention to regions of interest.

2   Method

2.1   Posterior Estimation and Prior in Variational Registration Model

Given the source image S and the target image T, the goal of image registration is estimating the spatial transformation $f_z : R^d \to R^d$ between S and T. The VB model used a variational posterior $q(z|T, S)$ to approximate the intractable registration posterior $p(z|T, S)$ by maximizing the evidence lower bound (ELBO) $\mathcal{L}(T, S)$ of log-likelihood $p(T, S)$,

$$\mathcal{L}(T, S) = E_{q(z|T,S)} \log p(T|z, S) - KL[q(z|T, S)||p(z)] \qquad (1)$$

In Eq. (1), the first term is expected to be significant to ensure registration accuracy. It can be expressed by the similarity between two images, such as the Boltzmann distribution with a parameter λ, $p(T|z, S) \propto exp(-\lambda(1 - sim(T, S(f_z))))$, where sim is the similarity measure. By using the Monte Carlo method, the first term in $\mathcal{L}(T, S)$ can be approximated. The second term quantifies the amount of information the model absorbed through learning. It is a measure of the complexity of the model that is expected to be small.

The parametric transformation using compact radial basis functions(CSRBFs) with control points $\{p_i\}_{i=1}^n$ to interpolate the dense DVF in our model as $f_z(u) = u + \sum_{i=1}^n z_i \psi(\frac{\|u-p_i\|_2}{r})$, where $z = \{z_i\}_{i=1}^n$ is the latent variable, u is a pixel. ψ is the CSRBF with support r. The value of r is obtained by the distance between control points [11]. The VB-based image registration aims to estimate the variational posterior $q(z|T, S)$ to approximate registration posterior $p(z|T, S)$. We employed multivariate normal distribution as the variational posterior $q(z|T, S) = \mathcal{N}(\mu, Diag(\sigma^2))$, where $\mu = [\mu_1, \ldots, \mu_n]^T$, $\sigma^2 = [\sigma_1^2, \ldots, \sigma_n^2]^T$. (μ_k, σ_k^2) is the distribution parameter of the kth element of the latent variable z.

Control points $\{p_i\}_{i=1}^n$ influence the DVF greatly. The network NetGI proposed by Gan $et\ al.$ [11] spaced the global control points (GPs) and local control points (LPs) uniformly, which cannot describe the anatomical contours of cardiac tissues. When the control points are located in the boundary of objects, one advantage is that it is easier to extract significant features; the other advantage is that these points can dominate the DVF discriminately, which might control the DVF more delicately than uniformly spaced control points. In this paper, we employed the farthest point sampling (FPS) [10] to sample control points on the contours of LV, RV, and MYO. All these non-uniformly spaced control points (NuPs) can roughly reflect the shape of the objects, as illustrated in Fig. 1.

Since GPs and NuPs are used, the latent variable can be represented as $z = \begin{bmatrix} z_u \\ z_{ru} \end{bmatrix}$, where z_u and z_{ru} correspond to GPs and NuPs, respectively. Distribution parameters of z_u and z_{ru}, denoted as (μ_u, σ_u^2) and $(\mu_{ru}, \sigma_{ru}^2)$, respectively, are estimated by a VAE. Since NuPs locate disorderly, it is challenging for CNNs to extract corresponding features. A specially designed VAE network NuNet

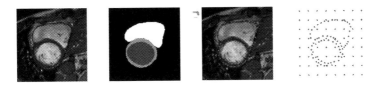

Fig. 1. Spatial distribution of control points. From left to right: the image, the mask of the image, the contour of objects, and uniformly spaced global control points and non-uniformly spaced local control points.

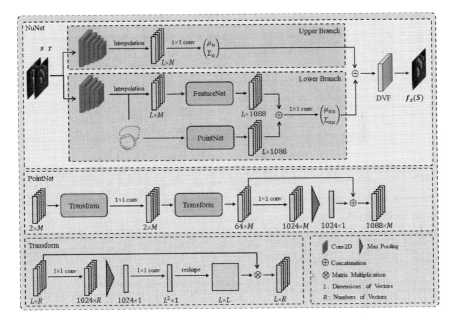

Fig. 2. Architecture of our network NuNet.

deals with this issue, as shown in Fig. 2. Our NuNet comprises an encoder and a decoder. The encoder of our NuNet contains two branches aiming to predict $(\boldsymbol{\mu}_u, \boldsymbol{\sigma}_u^2)$ and $(\boldsymbol{\mu}_{nu}, \boldsymbol{\sigma}_{nu}^2)$, respectively. The upper branch is for GPs with several convolutional layers. To obtain more representative features, the interpolation operation was employed in the feature maps. The lower branch is for NuPs. Since NuPs are disordered and diverse from each other for different image pairs. We embedded the PointNet (PN) architecture proposed by Qi *et al.* [23] in our NuNet. PointNet aims to extract the geometry features of a set of points without a specific order. In PointNet, two transform blocks are used to align points and features. The FeatureNet (FN) is similar to the PointNet, while the second transform is deleted because only feature matching is required. The decoder contains a CSRBF layer and an interpolation layer, where the CSRBF layer constructs a DVF using the sampled z, and the interpolation layer warps the source image S.

Gan *et al.* [11] proposed a normal distribution $p(z) = \mathcal{N}(\boldsymbol{\mu}, \boldsymbol{B}^{-1})$ as the global prior $p(z)$. We partition the global prior as $p(z) = p(z_u)p(z_{nu})$, where $p(z_u) = \mathcal{N}(0, \boldsymbol{B}_u^{-1})$ and $p(z_{nu}) = \mathcal{N}(0, \boldsymbol{B}_{nu}^{-1})$ correspond to the uniformly and non-uniformly spaced control points, respectively. We prove that the factorized prior results in a small KL divergence between the prior and the variational posterior with a high probability, which is favorable in increasing the ELBO in the VB model. Details can be referred to in the supplement. The conclusion is especially applicable to the control-points-based image registration model; it is favorable to make different control points have different priors. That implies we can regularize the variational posterior finely and control the DVF delicately. The extreme case of the prior factorization leads to the standard normal prior $\mathcal{N}(0, \boldsymbol{I})$. However, the standard normal prior is not conducive to estimate reasonable DVF because it makes control points independent of each other. It is contrary to the idea of CSBRF-based transformation, that is, control points that are close influence each other to the DVF.

2.2 Registration Uncertainty

Our registration network predicts the variance of latent variables, which corresponds to the deviation of parameters of elastic transformation. It is a kind of data uncertainty. We estimate the uncertainty of DVF using its variance. The displacement of pixel u is $d(u) = \sum_{i=1}^{n} z_i \psi(\frac{\|u - p_i\|_2}{r})$. Since z_i is independent to each other, the variance of $d(u)$ is $Var(d_u) = \sum_{p_i \in A_u} \sigma_i^2 \psi(\frac{\|u-p_i\|}{r})^2$, where A_u is the local region centered at u with radius r. We found that σ_i^2 of NuPs is larger than that of GPs in a statistical sense. The reason is that NuPs locate at the boundaries of objects, where large displacements occur in these areas for cardiac motion. However, cardiac motion varies subject to subject, resulting in the different displacements of points located at the boundaries of LV or RV. On the contrary, GPs distribute uniformly and locate mainly in the background with small motion in general. Correspondingly, the displacement variances of GPs are relatively small compared with that of NuPs. Moreover, when the pixel u is close to NuPs, $\psi(\frac{\|u-p_i\|}{r})^2$ is relatively large. Then, it can be concluded that the region where the NuPs are gathered generally has significant variances, such as the corner of the RV and thin myocardium. On the contrary, the regions with sparse control points, such as the background, usually have low uncertainty.

3 Experiments

3.1 Datasets and Implement Details

Four public datasets are used to evaluate our NuNet in experiments, including the York dataset [1], MICCAI2009 challenge dataset [24], ACDC dataset [5], and M&Ms dataset [6]. We combine the York, MICCAI2009, and ACDC as a hybrid dataset. There are 1060, 160, and 486 image pairs for training, validation, and testing in the hybrid dataset, respectively, while 1134, 266, and 859 image pairs in the M&Ms dataset, respectively. Image slices at the end-diastolic (ED) phase and the end-systolic (ES) in one cardiac cycle are the source and target images, respectively. All images are cropped as 128×128 containing the heart in the center of the image. The local correlation coefficient between two images is used as the similarity measure. The Dice score, bending energy (BE), the average perpendicular distance (APD, in mm), and the number of nonpositive Jacobian determinants ($|J_{f_z}| \leq 0$) are used to evaluate the performance.

Table 1. Evaluation of registration results for all networks on the hybrid dataset and M&Ms dataset. Data format: mean (standard deviations).

Dataset	Method	Dice	APD	BE	$|J_{f_z}| \leq 0$
Hybrid	KrebsDiff	0.835 (0.062)	2.30 (0.79)	28.88 (15.23)	10.47 (18.11)
	DalcaDiff	0.847 (0.059)	2.09 (0.70)	157.97 (104.15)	0.25 (0.57)
	VoxelMorph	0.842 (0.066)	2.20 (0.84)	157.34 (96.98)	55.81 (46.87)
	CycleMorph	0.841 (0.079)	2.25 (1.09)	304.49 (72.32)	65.85 (40.56)
	NetGI	0.843 (0.068)	2.19 (0.87)	3.94 (1.69)	0.00 (0.00)
	Ours+mNuPs+mNuPs	0.855 (0.060)	2.01 (0.70)	6.13 (3.70)	2.80 (5.55)
	Ours+mNuPs+pNuPs	0.852 (0.060)	2.04 (0.73)	5.73 (3.10)	3.37 (10.26)
	Ours+pNuPs+pNuPs	0.850 (0.064)	2.10 (0.80)	4.85 (2.86)	1.14 (2.53)
M&Ms	KrebsDiff	0.836 (0.054)	1.84 (1.00)	36.65 (34.96)	8.00 (16.89)
	DalcaDiff	0.851 (0.052)	1.64 (0.95)	183 (126)	0.42 (1.81)
	VoxelMorph	0.844 (0.058)	1.71 (0.96)	194 (102)	105 (110)
	CycleMorph	0.852 (0.058)	1.64 (0.99)	519 (137)	95 (55)
	NetGI	0.847 (0.054)	1.69 (0.80)	4.74 (1.71)	1.17 (9.20)
	Ours+mNuPs+mNuPs	0.861 (0.052)	1.54 (0.78)	2.19 (0.76)	1.19 (7.47)
	Ours+mNuPs+pNuPs	0.859 (0.049)	1.57 (0.79)	9.03 (4.50)	6.06 (14.01)
	Ours+pNuPs+pNuPs	0.858 (0.050)	1.57 (0.78)	9.44 (4.45)	6.15 (15.20)

3.2 Registration Results

To compare the performance of our proposed approach, five deep learning networks, KrebsDiff [17], DalcaDiff [8], VoxelMorph [4], CycleMorph [14] and NetGI [11] are employed. Two strategies are used to extract NuPs to train and test our network, including extracting NuPs from contours of masks provided by the dataset (mNuPs) and extracting NuPs from contours of predicted results (pNuPs) using a trained U-Net [25]. mNuPs are located precisely on the contours, while pNuPs are the ones with location errors due to network performance. We denote "Ours+training NuPs+ testing NuPs" as our approach. For example, "Ours+mNuPs+pNuPs" means we use mNuPs to train our network and pNuPs to predict the registration results.

Registration results of different networks on two datasets are listed in Table 1. Whether the predictive NuPs are employed for training or testing, our NuNet outperforms other networks regarding Dice and APD for two datasets. It implies

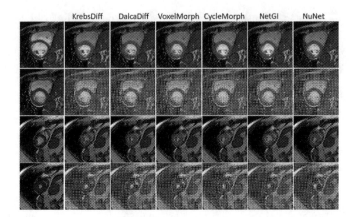

Fig. 3. Demonstration of registration results using different networks. The first column is the source images (odd rows) and the target images (even rows). The green, blue, and red colors mark the mask of the target image, the warped mask of the source image, and the overlap region of the two masks. The warped grids illustrate the estimated DVFs using different networks. (Color figure online)

that our approach is not sensitive to the location error of NuPs. NetGI had better performances on BE and the number of negative Jacobian determinants. The reason is that the influence between non-uniformly spaced control points varies in different regions, which makes it challenging to control the smoothness of DVF. Besides, the factorized prior regularizes the distribution of latent variables less, leading to more flexible DVFs. As shown in Fig. 3, our NuNet matched the contours of the myocardium and the right ventricle more accurately. Both NetGI and our NuNet achieve smoother DVFs compared with other networks.

3.3 Uncertainty

We provide uncertainty using different hyperparameters λ for NetGI and our NuNet, as shown in Fig. 4. An image pair is input to trained networks to predict the distribution parameters of the latent variable z. Next, z is sampled 500 times to construct DVFs, and the displacement vector's magnitude deviation is used as the uncertainty of a DVF. In Fig. 4, it is observed that the uncertainty estimated by our approach concentrates on the boundaries of objects, while NetGI diffuses the uncertainty around the heart region. The reason is that our NuPs locate on the boundaries of objects, while NetGI spreads local control points uniformly in the heart region. Our approach focuses on uncertainty in specific regions, which provides more valuable uncertainty information.

3.4 Ablation Study

To verify the effectiveness of the different modules of our network, we employ different variants of our network to conduct an ablation study. "GPs", "LPs",

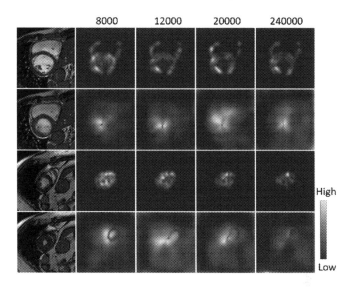

Fig. 4. Uncertainty maps using our network and NetGI, respectively, under different registration accuracy. The first column is the source images (odd rows) and the target images (even rows). Other columns: the odd and even rows illustrate uncertainty using NuNet and NetGI, respectively.

Table 2. Results of ablation experiments on two datasets.

No.	GPs	LPs	FN	PN	Σ_P	Σ_f	I	Hybrid Dice	Hybrid APD	M&Ms Dice	M&Ms APD
1	✓						✓	0.813 (0.085)	2.66 (1.17)	0.807 (0.066)	2.09 (0.97)
2		✓					✓	0.802 (0.089)	2.71 (1.10)	0.807 (0.068)	2.09 (1.17)
3		✓	✓				✓	0.821 (0.082)	2.46 (1.05)	0.825 (0.067)	1.94 (1.19)
4	✓	✓	✓				✓	0.824 (0.078)	2.48 (1.05)	0.832 (0.058)	1.85 (0.89)
5	✓	✓	✓			✓		0.851 (0.063)	2.07 (0.78)	0.859 (0.052)	1.57 (0.86)
6	✓	✓	✓	✓	✓			0.827 (0.077)	2.45 (1.03)	0.830 (0.060)	1.85 (0.87)
7	✓	✓	✓	✓		✓		0.855 (0.060)	2.01 (0.70)	0.861 (0.052)	1.54 (0.78)
8	✓	✓	✓	✓			✓	0.833 (0.072)	2.28 (0.97)	0.844 (0.064)	1.75 (1.07)

"FN" and "PN" represent global control points, local control points, FeatureNet, and PointNet, respectively. Three priors with different covariance matrices are compared. The experimental ablation results are listed in Table 2. All results are average evaluations on two testing datasets using mNuPs. From the first two rows, it can be seen that the upper branch is also vital for registration, even if the background deforms slightly between the two images. By focusing on the global and the local simultaneously, the performance can be improved further, as listed in the fourth row. By comparing the results of the second and third rows, it can be concluded that the FeatureNet embedded in our lower branch can address the disorder issue of intensity features of NuPs. Besides, it is observed from the fifth and seventh rows that the spatial features boosted the performance of our NuNet. Results of the last three rows indicate that our factorized prior generates complex deformation but has little influence to BE and the number of nonpositive Jacobian determinants.

4   Conclusion

This paper addressed the issue of non-uniformly spaced control points in the VB-based image registration model for cardiac motion estimation. We employed the FPS algorithm to sample control points from the contour of the heart. The PointNetis embedded in our network to learn the intensity and spatial features. We found that the factorized prior leads to small KL divergence and is beneficial to produce more flexible DVFs. Experimental results on four datasets show that our proposed approach achieves optimal performance compared to state-of-art networks. The uncertainty estimated by our network focuses on important regions and provides more information about uncertainty in applications.

Acknowledgements. This work is supported by the Shenzhen Fundamental Research Program (JCYJ20220531102407018).

References

1. Andreopoulos, A., Tsotsos, J.K.: Efficient and generalizable statistical models of shape and appearance for analysis of cardiac MRI. Med. Image Anal. **12**(3), 335–357 (2008)
2. Arar, M., Ginger, Y., Danon, D., Bermano, A.H., Cohen-Or, D.: Unsupervised multi-modal image registration via geometry preserving image-to-image translation. In: Proceedings of the IEEE/CVF Conference on Computer Vision and Pattern Recognition, pp. 13410–13419 (2020)
3. Balakrishnan, G., Zhao, A., Sabuncu, M.R., Guttag, J., Dalca, A.V.: An unsupervised learning model for deformable medical image registration. In: Proceedings of the IEEE Conference on Computer Vision and Pattern Recognition, pp. 9252–9260 (2018)
4. Balakrishnan, G., Zhao, A., Sabuncu, M.R., Guttag, J., Dalca, A.V.: VoxelMorph: a learning framework for deformable medical image registration. IEEE Trans. Med. Imaging **38**(8), 1788–1800 (2019)
5. Bernard, O., Lalande, A., et al.: Deep learning techniques for automatic MRI cardiac multi-structures segmentation and diagnosis: is the problem solved? IEEE Trans. Med. Imaging **37**(11), 2514–2525 (2018)
6. Campello, V.M., et al.: Multi-centre, multi-vendor and multi-disease cardiac segmentation: the M&Ms challenge. IEEE Trans. Med. Imaging 9458279 (2021)
7. Cao, H., Wang, H., Zhang, N., Yang, Y., Zhou, Z.: Robust probability model based on variational bayes for point set registration. Knowl.-Based Syst. **241**, 108182 (2022)
8. Dalca, A.V., Balakrishnan, G., Guttag, J., Sabuncu, M.R.: Unsupervised learning for fast probabilistic diffeomorphic registration. In: Frangi, A.F., Schnabel, J.A., Davatzikos, C., Alberola-López, C., Fichtinger, G. (eds.) MICCAI 2018. LNCS, vol. 11070, pp. 729–738. Springer, Cham (2018). https://doi.org/10.1007/978-3-030-00928-1_82

9. De Vos, B.D., Berendsen, F.F., Viergever, M.A., Sokooti, H., Staring, M., Išgum, I.: A deep learning framework for unsupervised affine and deformable image registration. Med. Image Anal. **52**, 128–143 (2019)

10. Eldar, Y., Lindenbaum, M., Porat, M., Zeevi, Y.Y.: The farthest point strategy for progressive image sampling. IEEE Trans. Image Process. **6**(9), 1305–1315 (1997)

11. Gan, Z., Sun, W., Liao, K., Yang, X.: Probabilistic modeling for image registration using radial basis functions: application to cardiac motion estimation. IEEE Trans. Neural Netw. Learn. Syst. (2022)

12. Gawlikowski, J., et al.: A survey of uncertainty in deep neural networks. arXiv preprint arXiv:2107.03342 (2021)

13. Grzech, D., et al.: A variational Bayesian method for similarity learning in non-rigid image registration. In: Proceedings of the IEEE/CVF Conference on Computer Vision and Pattern Recognition, pp. 119–128 (2022)

14. Kim, B., Kim, D.H., Park, S.H., Kim, J., Lee, J.G., Ye, J.C.: CycleMorph: cycle consistent unsupervised deformable image registration. Med. Image Anal. **71**, 102036 (2021)

15. Kim, B., Kim, J., Lee, J.-G., Kim, D.H., Park, S.H., Ye, J.C.: Unsupervised deformable image registration using cycle-consistent CNN. In: Shen, S., et al. (eds.) MICCAI 2019. LNCS, vol. 11769, pp. 166–174. Springer, Cham (2019). https://doi.org/10.1007/978-3-030-32226-7_19

16. Kingma, D.P., Welling, M.: Auto-encoding variational bayes. arXiv preprint arXiv:1312.6114 (2013)

17. Krebs, J., Delingette, H., Mailhé, B., Ayache, N., Mansi, T.: Learning a probabilistic model for diffeomorphic registration. IEEE Trans. Med. Imaging **38**(9), 2165–2176 (2019)

18. Liang, L., et al.: Dual-features student-T distribution mixture model based remote sensing image registration. IEEE Geosci. Remote Sens. Lett. **19**, 1–5 (2021)

19. Liu, L., Hu, X., Zhu, L., Heng, P.-A.: Probabilistic multilayer regularization network for unsupervised 3D brain image registration. In: Shen, D., et al. (eds.) MICCAI 2019. LNCS, vol. 11765, pp. 346–354. Springer, Cham (2019). https://doi.org/10.1007/978-3-030-32245-8_39

20. Liu, R., Li, Z., Zhang, Y., Fan, X., Luo, Z.: Bi-level probabilistic feature learning for deformable image registration. In: Proceedings of the Twenty-Ninth International Conference on International Joint Conferences on Artificial Intelligence, pp. 723–730 (2021)

21. Liu, Y., et al.: A remote sensing image registration algorithm based on multiple constraints and a variational Bayesian framework. Remote Sens. Lett. **12**(3), 296–305 (2021)

22. Nenoff, L., et al.: Deformable image registration uncertainty for inter-fractional dose accumulation of lung cancer proton therapy. Radiother. Oncol. **147**, 178–185 (2020)

23. Qi, C.R., Su, H., Mo, K., Guibas, L.J.: PointNet: deep learning on point sets for 3D classification and segmentation. In: Proceedings of the IEEE Conference on Computer Vision and Pattern Recognition, pp. 652–660 (2017)

24. Radau, P., Lu, Y., Connelly, K., Paul, G., Dick, A., Wright, G.: Evaluation framework for algorithms segmenting short axis cardiac MRI. MIDAS J.-Cardiac MR Left Ventricle Segment. Challenge **49** (2009)

25. Ronneberger, O., Fischer, P., Brox, T.: U-net: convolutional networks for biomedical image segmentation. In: Navab, N., Hornegger, J., Wells, W.M., Frangi, A.F. (eds.) MICCAI 2015. LNCS, vol. 9351, pp. 234–241. Springer, Cham (2015). https://doi.org/10.1007/978-3-319-24574-4_28

26. Schultz, S., Krüger, J., Handels, H., Ehrhardt, J.: Bayesian inference for uncertainty quantification in point-based deformable image registration. In: Medical Imaging 2019: Image Processing, vol. 10949, pp. 459–466. SPIE (2019)
27. Zhang, A., Min, Z., Zhang, Z., Meng, M.Q.H.: Generalized point set registration with fuzzy correspondences based on variational Bayesian inference. IEEE Trans. Fuzzy Syst. **30**, 1529–1540 (2022)
28. Zhao, S., Dong, Y., Chang, E.I., Xu, Y., et al.: Recursive cascaded networks for unsupervised medical image registration. In: Proceedings of the IEEE/CVF International Conference on Computer Vision, pp. 10600–10610 (2019)
29. Zhou, J., et al.: Robust variational Bayesian point set registration. In: Proceedings of the IEEE/CVF International Conference on Computer Vision, pp. 9905–9914 (2019)

Implicit Neural Representations for Joint Decomposition and Registration of Gene Expression Images in the Marmoset Brain

Michal Byra[1,2]([✉]), Charissa Poon[1], Tomomi Shimogori[3], and Henrik Skibbe[1]

[1] Brain Image Analysis Unit, RIKEN Center for Brain Science, Wako, Japan
michal.byra@riken.jp
[2] Institute of Fundamental Technological Research, Polish Academy of Sciences, Warsaw, Poland
[3] Laboratory for Molecular Mechanisms of Brain Development, RIKEN Center for Brain Science, Wako, Japan

Abstract. We propose a novel image registration method based on implicit neural representations that addresses the challenging problem of registering a pair of brain images with similar anatomical structures, but where one image contains additional features or artifacts that are not present in the other image. To demonstrate its effectiveness, we use 2D microscopy *in situ* hybridization gene expression images of the marmoset brain. Accurately quantifying gene expression requires image registration to a brain template, which is difficult due to the diversity of patterns causing variations in visible anatomical brain structures. Our approach uses implicit networks in combination with an image exclusion loss to jointly perform the registration and decompose the image into a support and residual image. The support image aligns well with the template, while the residual image captures individual image characteristics that diverge from the template. In experiments, our method provided excellent results and outperformed other registration techniques.

Keywords: brain · deep learning · gene expression · implicit neural representations · registration

1 Introduction

Image registration is a crucial prerequisite for image comparison, data integration, and group studies in contemporary medical and neuroscience research. In research and clinical settings, pairs of images often show similar anatomical structures but may contain additional features or artifacts, such as specific staining, electrodes, or lesions, that are not present in the other image. This difficulty of finding corresponding structures for automatically aligning images complicates image registration. In this work, we address the challenging problem

Supplementary Information The online version contains supplementary material available at https://doi.org/10.1007/978-3-031-43999-5_61.

of the gene expression image registration in the marmoset brain. Brain atlases of gene expression, created using images of brain tissue processed through *in situ* hybridization (ISH), offer single-cell resolution of spatial gene expression patterns across the entire brain [2,7]. However, accurately quantifying gene expression requires brain image registration to spatially align ISH images to a common atlas space. The diversity of gene expression patterns in ISH images causes variations in visible anatomical brain structures with respect to the template image. ISH microscopy images are also susceptible to tissue processing artifacts, resulting in non-specific staining and tissue deformations.

Traditional pair-wise image registration methods use optimization algorithms to find the deformation field that maximizes the similarity between a pair of images. While several deep learning methods based on convolutional neural networks (CNNs) have been proposed for calculating the deformation field between two images [3], such models typically require large training sets and may suffer from generalization issues when applied to images presenting texture patterns that diverge from the training data. Therefore, classic algorithms, such as Advanced Normalization Tools (ANTs) [1], are still preferred as off-the-shelf tools for image registration in neuroscience due to scarce experimental data and the diversity of data acquisition protocols and registration tasks. Recently, implicit neural representations (INRs) have been utilized for image registration in MRI and CT [14,16], offering a hybrid approach that connects modern deep learning techniques with per-case optimization as used in classical approaches. INRs are defined on continuous coordinate spaces, making them suitable for registration of images that differ in geometry.

In this work, we propose a novel INR-based framework well-suited to address the challenging problem of gene expression brain image registration. We associate the registration problem with an image decomposition task. We utilize implicit neural networks to decompose the ISH image into two separate images: a support image and a residual image. The support image corresponds to the part of the ISH image that is well-aligned with the registration template image in respect to the texture. On the contrary, the residual image presents features of the ISH image, such as artifacts or texture patterns (e.g. gene expression), which presumably undermine the registration procedure. The support image is used to improve the deformation field calculations. We also introduce an exclusion loss to encourage clearer separation of the support and residual images. The usefulness of the proposed method is demonstrated using 2D ISH gene expression images of the marmoset brain.

2   Methods

2.1   Registration with Implicit Networks

The goal of the pairwise image registration is to determine a spatial transformation that maximizes the similarity between the moving image M and the target fixed template image F. INRs serve as a continuous, coordinate based approximation of the deformation field obtained through a fully connected neural network. In this study, as the backbone for our method, we utilized the standard approach to registration with INRs, as described in [14,16]. We used a single implicit deformation network D to map 2D spatial coordinates $\bar{x} \in [-1,1]^2$ of the moving image M to a displacement vector $\Delta\bar{x} \in \mathbb{R}^2$. Next, the transformation field was determined as $\Phi(\bar{x}) = \bar{x} + \Delta\bar{x}$ and the bilinear interpolation algorithm was applied to obtain the corresponding moved image $T_\Phi(M)$.

To train the deformation network, the following loss function based on correlation coefficients was applied to assess the similarity between the moved image $T_\Phi(M)$ and the fixed template image F:

$$\mathcal{L}_{cc}(F, T_\Phi(M)) = \frac{1}{2N} \sum_{\bar{x}} \Big(\text{NCC}(F, T_\Phi(M)) + \text{LNCC}(F, T_\Phi(M)) \Big), \quad (1)$$

where NCC and LNCC stand for the normalized cross-correlation and local normalized cross-correlation based loss functions averaged over the entire image domain consisting of N elements. NCC was used to stabilize the training of the network, while LNCC ensured good local registration results. Additionally, following the standard approach to INR based registration, we regularized the deformation field based on the Jacobian matrix determinant $|J_{\Phi(\bar{x})}|$ using following equation [16]:

$$\mathcal{L}_{reg}(\Phi(\bar{x})) = \frac{1}{N} \sum_{\bar{x}} |1 - |J_{\Phi(\bar{x})}||. \quad (2)$$

2.2   Registration Guided Image Decomposition

Our aim is to improve the registration performance associated with the implicit deformation network D. The proposed framework is presented in Fig. 1. We assume that the moving image M can be decomposed with separate implicit networks, S and R, into two images: the support image M_S and the residual image M_R. Ideally, the support image should correspond to the part of the moving image that contributes to the registration performance. On the contrary, we expect the residual image to include image artifacts and texture patterns (e.g. ISH gene expression patterns) that diverge from the fixed template image and undermine the registration procedure. We impose the following condition based on the mean squared error loss function for the decomposition of the moving image:

$$\mathcal{L}_{rec}(M, M_S + M_R) = \frac{1}{N} \sum_{\bar{x}} (M - M_S - M_R)^2, \quad (3)$$

stating that the support M_S and residual M_R images should sum up to the moving image M. To ensure that the support image M_S contributes to the registration with respect to the fixed image F, we utilize the cross-correlation based loss function $\mathcal{L}_{cc}(F, T_\Phi(M_S))$ (Eq. 1), where $T_\Phi(M_S)$ stands for the transformed support image M_S. Therefore, the deformation network is trained to provide the transformation field $\Phi(x)$ both for the moving image and the support image

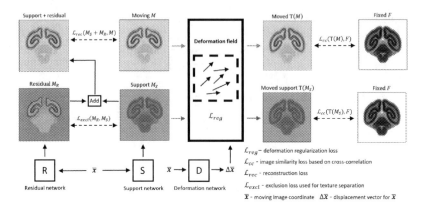

Fig. 1. We use implicit networks S and R to decompose the moving image into the support and residual images. The moving and support images are jointly registered to the fixed template image, which guides the image decomposition procedure to generate a support image that is well-aligned to the fixed image with respect to the texture. The residual image includes the remaining moving image contents that do not contribute to the registration, such as local gene expression patterns or image artifacts.

using two cross-correlation based loss functions. This way the training of the deformation network is guided to provide a more detailed transformation field for the contents of the moving image that actually correspond to the fixed template image. Moving image texture patterns that do not correspond to the fixed image have lower impact on the training of the deformation network.

In practice, it might be beneficial, following INR based methods for obstruction and rain removal, to additionally constrain the image decomposition procedure to obtain more clearly separated support M_S and residual M_R images [10]. For this, we utilize the following exclusion loss to encourage the gradient structure of the implicit networks S and R to be decorrelated [4]:

$$\mathcal{L}_{excl}(M_S, M_R) = \frac{1}{N} \sum_{\bar{x}} \sum_{i,j} |\Gamma(J_S(\bar{x}), J_R(\bar{x}))| \qquad (4)$$

$\Gamma(J_S(\bar{x}), J_R(\bar{x})) = \tanh(J_S(\bar{x})) \otimes \tanh(J_R(\bar{x}))$, $\otimes$ indicates element-wise multiplication and indices i, j go over all elements of the matrix Γ.

In our framework, we jointly optimize all three implicit networks (D, S and R) using the following composite loss function:

$$\begin{aligned} Loss = \alpha_1 \mathcal{L}_{cc}(F, T_\Phi(M)) + \alpha_2 \mathcal{L}_{cc}(F, T_\Phi(M_S)) + \alpha_3 \mathcal{L}_{reg}(\Phi(\bar{x})) \\ + \alpha_4 \mathcal{L}_{rec}(M, M_S + M_R) + \alpha_5 \mathcal{L}_{excl}(M_S, M_R). \end{aligned} \qquad (5)$$

The first row of Eq. 5 can be perceived as a standard registration loss, while the second row stands for a regularized image reconstruction loss.

2.3   Evaluation

We designed the proposed method with the aim to address the problem of ISH gene expression image registration. For the evaluation, we used neonate marmoset brain ISH images collected at the Laboratory for Molecular Mechanisms of Brain Development, RIKEN Center for Brain Science, Wako, Japan (geneatlas.brainminds.jp) [6,12]. We prepared manual annotations for 2D images from 50 gene expression datasets. Atlas template images were created using ANTs [1], based on semi-automatically aligned sets of 2D ISH images from 1942 gene expression datasets. ISH images used to generate the template were converted to gray-scale to meet ANTs requirements and better highlight brain tissue interfaces.

Performance of the proposed approach was compared to the SynthMorph network and the ANTs SyN registration algorithm based on mutual information metric, as these two methods do not require pre-training and can serve as off-the-shelf registration tools for neuroscience [1,5]. We conducted an ablation study to assess the effectiveness of the proposed representation decomposition approach with and without the exclusion loss. Registration methods were evaluated quantitatively based on Dice scores using manual 2D segmentations prepared for the following five brain structures ranging in size and shape complexity: aqueduct (AQ, 95 masks), hippocampus area (HA, 570 masks), dorsal lateral geniculate (DLG, 370 masks), inferior colliculus (IC, 70 masks) and visual cortex area (VCA, 68 masks). Segmentations were outlined both for the template and ISH 2D images, resulting in 1114 image pairs corresponding to the same brain regions. We also calculated the percentage of the non-positive Jacobian determinant values to assess the deformation field folding. Moreover, we determined the structural similarity index (SSIM) between the moved images and the template fixed images.

2.4   Implementation

We utilized sinusoidal representation networks to determine the implicit representations [13]. Each network contained five fully connected hidden layers with 256 neurons. We used the Fourier mapping with six frequencies to encode the input coordinates [15]. The coordinates and the encoded coordinates were additionally concatenated within the middle layer of the network. Weights of the networks were initialized following the original paper except for the last linear layer of the deformation network D, for which we uniformly sampled the weights from $[-0.0001, 0.0001]$ interval to ensure small deformations at initial epochs. Additional details about the network architecture can be found in the supplementary materials. Networks were trained for 1000 epochs using AdamW optimizer with learning rate of 0.0001 on a server equipped with several NVIDIA A100 GPUs [8]. ISH images of size 360×420 were downsampled to 256×256. Each epoch corresponded to a batch of all image pixel coordinates [13]. After some initial experiments, we set the composite loss function weights (Eq. 5) to

$\alpha_1 = \alpha_2 = \alpha_3 = \alpha_5 = 1$ and $\alpha_4 = 100$, partially following the previous studies on INRs [10,14,16]. The window size for the LNCC loss was set to [32, 32]. Our PyTorch implementation of the proposed INR based registration method is available at https://github.com/BrainImageAnalysis/ImpRegDec.

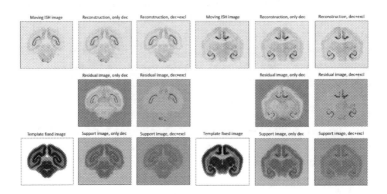

Fig. 2. Illustration of the moving image decomposition obtained with the proposed method (dec). Incorporation of the exclusion loss (excl) resulted in clearer separation of the gene expression texture patterns in the residual images.

3   Results

3.1   Qualitative Results

Support and residual images generated with the proposed method are shown in Fig. 2. The support images retain the main style and content of the fixed template image, while the residual images include the remaining image contents, along with gene expression patterns not present in the template image. Utilization of the exclusion loss resulted in a clearer and more visually plausible separation between the support and residual images, particularly for gene expression patterns. Figure 3 further highlights the usefulness of the proposed registration guided image decomposition technique. First, our method can be applied to extract microscopy image artifacts, and therefore mitigate their impact on the registration. Second, the proposed method is general and can also be applied to register an ISH gene expression image to a Nissl image. In this case, the color distribution of the support image corresponds to that of a Nissl image, while the residual image presents the local contents of the gene expression image. We also used the proposed method to register an ISH brain image to another ISH image with a different gene expression. For this example, the residual image highlighted the gene expression patterns of the moving image, while the support image showed the gene expression patterns of the fixed image.

Figure 4 visually compares the registration performance of the proposed technique, equipped with the exclusion loss, to ANTs. We found that the proposed method provided good results both in respect to the image registration and the transformation of the manual segmentations.

3.2   Quantitative Results

Table 1 shows Dice scores obtained for the selected marmoset brain regions. Registration techniques based on INRs outperformed the other methods on four out of five brain regions. ANTs achieved better registration results for only one structure, the VCA, which was the largest among the annotated brain regions

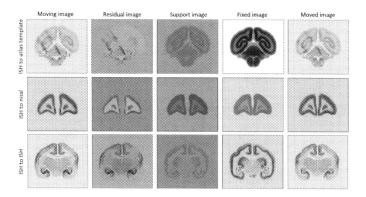

Fig. 3. Proposed technique can be useful for the extraction of microscope image artifacts (e.g. diagonal lines in the first row of images). It can also be applied to register ISH brain images to Nissl images or other ISH images. For such cases the support image presents image style of the fixed image, while the residual image includes local image patterns of the moving image.

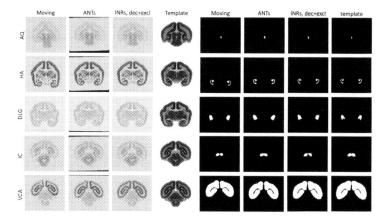

Fig. 4. Comparison of the proposed registration technique based on implicit networks and ANTs. AQ, HA, DLG, IC and VCA indicate the aqueduct, hippocampus area, dorsal lateral geniculate, inferior colliculus and visual cortex area, respectively.

Table 1. Dice scores (mean ± std) determined for the aqueduct (AQ), hippocampus area (HA), dorsal lateral geniculate (DLG), inferior colliculus (IC) and visual cortex are (VCA). Best results are shown in bold. dec and excl stand for the proposed image decomposition technique and the exclusion loss.

Method	AQ ↑	HA ↑	DLG ↑	IC ↑	VCA ↑
None	0.497 ± 0.194	0.311 ± 0.132	0.612 ± 0.141	0.742 ± 0.169	0.848 ± 0.051
ANTs SyN	0.673 ± 0.102	0.644 ± 0.141	0.757 ± 0.130	0.831 ± 0.133	**0.941 ± 0.015**
SynthMorph	0.625 ± 0.129	0.503 ± 0.190	0.719 ± 0.146	0.798 ± 0.157	0.922 ± 0.034
INRs	0.734 ± 0.071	0.657 ± 0.127	0.756 ± 0.130	0.804 ± 0.191	0.922 ± 0.022
INRs, dec	0.748 ± 0.067	0.662 ± 0.138	**0.767 ± 0.125**	0.839 ± 0.142	0.916 ± 0.033
INRs, dec+excl	**0.749 ± 0.063**	**0.665 ± 0.134**	0.766 ± 0.128	**0.845 ± 0.143**	0.920 ± 0.017

Table 2. Structural similarity index (SSIM) and the percentage of the non-positive Jacobian determinant values (mean ± std) calculated for the investigated registration methods. Best results are shown in bold. dec and excl indicate the proposed image decomposition technique and the exclusion loss, respectively.

Method	SSIM ↑	$|J_\Phi| \leq 0$ [%] ↓
None	0.619 ± 0.046	–
ANTs	0.656 ± 0.059	**<0.001**
SynthMorph	0.683 ± 0.039	**<0.001**
INRs	0.713 ± 0.052	0.353 ± 0.459
INRs, dec	0.725 ± 0.054	0.359 ± 0.415
INRs, dec+excl	**0.727 ± 0.054**	0.429 ± 0.460

and already similar in unregistered images with an initial Dice score of 0.848. Additionally, the Dice score for the VCA was high and comparable across all investigated registration methods. Our approach achieved significantly better Dice scores compared to the standard INRs for AQ, HA, DLG and IC (t-test's p-values < 0.05). Furthermore, incorporating the exclusion loss slightly improved the Dice scores for three structures.

SSIM values in Table 2 show that the registration based on implicit networks provided the most structurally similar results to the template images. With respect to the SSIM metric, our method significantly outperformed other approaches (t-test's p-values < 0.05). ANTs and SynthMorph provided smoother deformation fields compared to the implicit networks, with significantly lower percentage of folding (t-test's p-values < 0.05). However, the percentage of the folding obtained for the implicit networks was small and acceptable, as defined by folds in 0.5% of all pixels [11]. The main disadvantage of the proposed approach was the relatively long optimization time of about 90 s for a single pairwise registration, resulting from the requirement to jointly train three implicit networks.

4   Conclusion

Our approach based on implicit networks and registration-guided image decomposition has demonstrated excellent performance for the challenging task of registering ISH gene expression images of the marmoset brain. The results show that our approach outperformed pairwise registration methods based on ANTs and SynthMorph CNN, highlighting the potential of INRs as versatile off-the-shelf tools for image registration. Moreover, the proposed registration-guided image decomposition mechanism not only improved the registration performance, but also could be used to effectively separate the patterns that diverge from the target fixed image. In the future, we plan to investigate the possibility of using image decomposition for simultaneous registration and pattern segmentation, and methods to speed up the training [9]. We also plan to extend our technique to 3D and test it on medical images that include pathologies.

Acknowledgement. The authors do not have any conflicts of interest. This work was supported by the program for Brain Mapping by Integrated Neurotechnologies for Disease Studies (Brain/MINDS) from the Japan Agency for Medical Research and Development AMED (JP15dm0207001) and the Japan Society for the Promotion of Science (JSPS, Fellowship PE21032).

References

1. Avants, B.B., Tustison, N.J., Song, G., Cook, P.A., Klein, A., Gee, J.C.: A reproducible evaluation of ants similarity metric performance in brain image registration. Neuroimage **54**(3), 2033–2044 (2011)
2. Corrales, M., et al.: A single-cell transcriptomic atlas of complete insect nervous systems across multiple life stages. Neural Dev. **17**(1), 8 (2022)
3. Fu, Y., Lei, Y., Wang, T., Curran, W.J., Liu, T., Yang, X.: Deep learning in medical image registration: a review. Phys. Med. Biol. **65**(20), 20TR01 (2020)
4. Gandelsman, Y., Shocher, A., Irani, M.: "Double-dip": unsupervised image decomposition via coupled deep-image-priors. In: Proceedings of the IEEE/CVF Conference on Computer Vision and Pattern Recognition, pp. 11026–11035 (2019)
5. Hoffmann, M., Billot, B., Greve, D.N., Iglesias, J.E., Fischl, B., Dalca, A.V.: SynthMorph: learning contrast-invariant registration without acquired images. IEEE Trans. Med. Imaging **41**(3), 543–558 (2021)
6. Kita, Y., et al.: Cellular-resolution gene expression profiling in the neonatal marmoset brain reveals dynamic species-and region-specific differences. Proc. Natl. Acad. Sci. **118**(18), e2020125118 (2021)
7. Lein, E.S., et al.: Genome-wide atlas of gene expression in the adult mouse brain. Nature **445**(7124), 168–176 (2007)
8. Loshchilov, I., Hutter, F.: Decoupled weight decay regularization. arXiv preprint arXiv:1711.05101 (2017)
9. Mehta, I., Gharbi, M., Barnes, C., Shechtman, E., Ramamoorthi, R., Chandraker, M.: Modulated periodic activations for generalizable local functional representations. In: Proceedings of the IEEE/CVF International Conference on Computer Vision, pp. 14214–14223 (2021)

10. Nam, S., Brubaker, M.A., Brown, M.S.: Neural image representations for multi-image fusion and layer separation. In: Avidan, S., Brostow, G., Cissé, M., Farinella, G.M., Hassner, T. (eds.) ECCV 2022, Part VII. LNCS, vol. 13667, pp. 216–232. Springer, Cham (2022). https://doi.org/10.1007/978-3-031-20071-7_13
11. Qiu, H., Qin, C., Schuh, A., Hammernik, K., Rueckert, D.: Learning diffeomorphic and modality-invariant registration using B-splines. In: Medical Imaging with Deep Learning (2021)
12. Shimogori, T., et al.: Digital gene atlas of neonate common marmoset brain. Neurosci. Res. **128**, 1–13 (2018)
13. Sitzmann, V., Martel, J., Bergman, A., Lindell, D., Wetzstein, G.: Implicit neural representations with periodic activation functions. Adv. Neural. Inf. Process. Syst. **33**, 7462–7473 (2020)
14. Sun, S., Han, K., Kong, D., You, C., Xie, X.: MIRNF: medical image registration via neural fields. arXiv preprint arXiv:2206.03111 (2022)
15. Tancik, M., et al.: Fourier features let networks learn high frequency functions in low dimensional domains. Adv. Neural. Inf. Process. Syst. **33**, 7537–7547 (2020)
16. Wolterink, J.M., Zwienenberg, J.C., Brune, C.: Implicit neural representations for deformable image registration. In: International Conference on Medical Imaging with Deep Learning, pp. 1349–1359. PMLR (2022)

FSDiffReg: Feature-Wise and Score-Wise Diffusion-Guided Unsupervised Deformable Image Registration for Cardiac Images

Yi Qin and Xiaomeng Li[✉]

The Hong Kong University of Science and Technology, Kowloon, Hong Kong SAR, China
eexmli@ust.hk

Abstract. Unsupervised deformable image registration is one of the challenging tasks in medical imaging. Obtaining a high-quality deformation field while preserving deformation topology remains demanding amid a series of deep-learning-based solutions. Meanwhile, the diffusion model's latent feature space shows potential in modeling the deformation semantics. To fully exploit the diffusion model's ability to guide the registration task, we present two modules: Feature-wise Diffusion-Guided Module (FDG) and Score-wise Diffusion-Guided Module (SDG). Specifically, FDG uses the diffusion model's multi-scale semantic features to guide the generation of the deformation field. SDG uses the diffusion score to guide the optimization process for preserving deformation topology with barely any additional computation. Experiment results on the 3D medical cardiac image registration task validate our model's ability to provide refined deformation fields with preserved topology effectively. Code is available at: https://github.com/xmed-lab/FSDiffReg.git.

Keywords: Deformable Image Registration · Score-based Generative Model

1 Introduction

Deformable image registration is the process of accurately estimating non-rigid voxel correspondences, such as the deformation field, between the same anatomical structure of a moving and fixed image pair. Fast, accurate, and realistic image registration algorithms are essential to improving the efficiency and accuracy of clinical practices. By observing dynamic changes, such as lesions, physicians can more comprehensively design treatment plans for patients [8,11]. When images during surgery align with preoperative ones, surgeons can locate instruments better and improve surgical prognosis [1]. As reported in [12], cardiac image registration is especially vital in improving heart chamber analysis accuracy,

Supplementary Information The online version contains supplementary material available at https://doi.org/10.1007/978-3-031-43999-5_62.

correcting cardiac imaging errors, and guiding cardiac surgeries. Thus, several studies have explored classical [2,16] and deep-learning-based [3,10,14,20] registration methods over the years.

Classical registration methods [16] used hand-crafted features to align images by solving computational-expensive optimization problems. Recently, researchers explored the deep-learning-based unsupervised deformable image registration [3,14,19,20] to address the computational burden while reducing the need for accurate ground truth in the registration task. VoxelMorph [3], as the baseline, took moving and fixed image pairs as the input and maximized image pair similarity to train a registration network. To achieve higher accuracy, most unsupervised methods adopted a cascaded network with several sub-networks or an iterative refinement strategy [6,10,14,20]. These strategies made the training procedure complicated and computational resources demanding. Meanwhile, to obtain smoother and more realistic deformation fields, i.e., topology preservation, many existing works introduced explicit diffeomorphic constraints [7,17,19] or additional calculations on cycle consistency [14]. For example, CycleMorph [14] utilized the bidirectional registration consistency to preserve the topology during training. VoxelMorph-Diff [7] adopted velocity field-based deformation field and new diffeomorphic estimation. SYMNet [19] used symmetric deformation field estimation to achieve the goal. However, these schemes did not fully exploit the inherent network features, thereby overlooking these features' ability for better topology preservation.

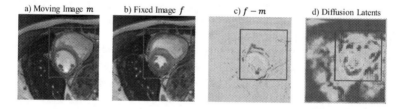

Fig. 1. $f - m$ intuitively indicates where significant deformation occurs, as the red-boxed area shows. We calculated the per voxel energy score in the diffusion model's latent obtained in [13] as [10] suggests to identify the areas where complex deformation is likely to happen. The result indicates the same area, which was not explicitly utilized in prior work [13]. (Color figure online)

Recently, Kim et al. [13] first proposed a diffusion model [9], which is simpler to train than other generative models yet rich in semantics, for the registration task. They used the latent feature from the diffusion model's score function, i.e., the gradient field of a distribution's log-likelihood function [22], as one of the registration network's inputs for a better registration result. However, this method only used the final diffusion score as an image level guidance, which *ignored diffusion model's rich task-specific semantics in the feature levels*, as proven in [4,18,23]. This resulted in the latent semantics of the diffusion model not being able to directly guide the features learned at the hidden layers of the registration network. As a result, the informativeness of these features for image registration was reduced. Moreover, this method only preserved deformation topology by simply using the diffusion score as the input, thereby *ignoring the informative details about areas where significant deformations occur*; see Fig. 1d for unexploited informative semantics. Therefore, the registration network was unable to explicitly prioritize hard-to-register areas, thereby limiting its effectiveness in preserving the deformation topology.

To address these issues, we present two novel modules, namely **F**eature-wise **D**iffusion-**G**uided Module (**FDG**) and **S**core-wise **D**iffusion-**G**uided Module (**SDG**) in the registration network. FDG introduces a direct feature-wise diffusion guidance technique for generating deformation fields by utilizing cross-attention to integrate the intermediate features of the diffusion model into the hidden layer of the registration network's decoder. Furthermore, we embed the feature-wise guidance into multiple layers of the registration network and produce the feature-level deformation fields in multiple scales. Finally, after obtaining deformation fields at multiple scales, we upsample and average them to generate the full-resolution deformation field for registration. Our SDG introduces explicit score-wise diffusion guidance for deformation topology preservation by reweighing the similarity-based unsupervised registration loss based on the diffusion score. Through this reweighing scheme, direct attention is given during the optimization process to ensure the preservation of the deformation topology. Our main contribution can be summarized as follows:

- We propose a novel feature-wise diffusion-guided module (FDG), which utilizes multi-scale intermediate features from the diffusion model to effectively guide the registration network in generating deformation fields.
- We also propose a score-wise diffusion-guided module (SDG), which leverages the diffusion model's score function to guide deformation topology preservation during the optimization process without incurring any additional computational burden.
- Experimental results on the cardiac dataset validated the effectiveness of our proposed method.

2   Method

2.1   Baseline Registration Model

Figure 2a shows the overview of our proposed method. We first sample a perturbed noisy image x_t from the fixed target image f following the same scheme in [9], which can be formulated as Eq. 1:

$$x_t = \sqrt{\alpha_t}f + \sqrt{1 - \alpha_t}\epsilon,$$
$$\text{where } \alpha_t = \prod_{s=1}^{t}(1 - \beta_s), \epsilon \sim \mathcal{N}(0, I) \tag{1}$$

where $0 < \beta_s < 1$ is the variance of the noise, t is the noise level. Then we perform the registration training task. Given an input x_{in} consisting of a fixed reference image f, a moving unaligned image m, and the perturbed noisy image x_t, we feed this input $x_{in} = \{f, m, x_t\}$ into the registration network's shared encoder E_β, followed by the registration decoder R_θ. Then, the registration decoder R_θ outputs a deformation field ϕ, guided by our *Feature-wise Diffusion-Guided* module G_σ. Afterward, we feed m and ϕ into the spatial transformation layer (STL) to generate the warped image $m(\phi)$. Finally, by optimizing the similarity-based loss function $L_{scoreNCC}$ guided by our *Score-wise Diffusion-Guided* module, we can obtain the final registration model.

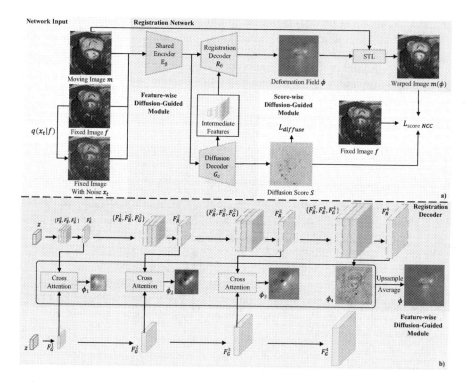

Fig. 2. a) The workflow of FSDiffReg. b) The illustration of the guidance process of the Feature-wise Diffusion-Guided Module.

2.2 Feature-Wise Diffusion-Guided Module

The main component of the Feature-wise Diffusion-Guided module (FDG) is an auxiliary denoising diffusion decoder G_σ. The workflow of FDG is shown in Fig. 2b. Given the input $x_{in} = \{f, m, x_t\}$, the UNet shared encoder E_β extracts

the representation z. z is then fed into the diffusion decoder G and the regis-
tration decoder R to get intermediate feature map pairs $F_i = \{(F_G^i, F_R^i)\}, i = 1, ..., N$ from the i-th layer of the decoder. Of note, we generate the registration
decoder's feature map by incorporating the guidance from the diffusion decoder,
which can be formulated as Eq. 2:

$$F_R^i = r_i(\text{concat}(F_R^{i-1}, F_E^i, F_G^i)), \text{ where } i = 1, ..., N \tag{2}$$

where r_i is the i-th layer of the registration decoder, and F_E^i is the skip connection
of features from the shared encoder layer at the same depth.

After obtaining the feature map pairs, our FDG module estimates the i-th
feature level deformation field ϕ_i from the feature map pair (F_G^i, F_R^i) using linear
cross attention [21], which can be defined as Eq. 3:

$$\phi_i = \text{Conv}(\text{softmax}(F_R^i(\text{GroupNorm}(F_G^i)^T \cdot F_G^i)) + F_R^i) \tag{3}$$

After obtaining all feature-level deformation fields from the shallowest layer
to the deepest layer, we generate the final deformation field ϕ by enlarging
and averaging all feature-level deformation fields. This is a commonly adopted
method for multi-scale deformation field merging so as to merge features which
attend different scales and granularity. The final ϕ is then fed into the spatial
transformation layer with the moving image m to generate the registered image
$m(\phi)$.

2.3 Score-Wise Diffusion-Guided Module

Given the representation z encoded by the shared encoder $z = E_\beta(x_{in})$, the
diffusion decoder G_σ outputs a diffusion score estimation $S = G_\sigma(z)$. Then,
the Score-wise Diffusion-Guided Module (SDG) uses this score to reweigh the
similarity-based normalized cross-correlation loss function, formulated as Eq. 4:

$$L_{scoreNCC}(m, f, S) = (\frac{1}{1 + e^{-S}})^\gamma \odot -(m(\phi) \otimes f) \tag{4}$$

where $m(\phi)$ is the warped moving image, $\odot$ defines the Hadamard product, and
$\otimes$ defines the local normalized cross-correlation function. γ is a hyperparameter
to amplify the reweighing effect.

By this means, SDG utilizes the diffusion score to explicitly indicate the hard-
to-register areas, i.e., areas where deformation topology is hard to preserve, then
assigning higher weights in the loss function for greater attention, and vice versa
for easier-to-register areas. Therefore, the information on deformation topology
is effectively incorporated into the optimization process without additional constraints by the SDG module.

2.4   Overall Training and Inference

Loss Function. Our network predicts the deformable fields at the feature level and then outputs the registered image. The total loss function of our method is defined as Eq. 5:

$$L_{total} = L_{diffusion}(x_{in}, t) + \lambda L_{scoreNCC}(m, f, S) + \lambda_\phi \sum ||\nabla_\phi||^2 \qquad (5)$$

$$L_{diffusion}(x_{in}, t) = \mathbb{E}_z ||G_\sigma(E_\beta(x_{in}, t)) - \epsilon||_2^2 \text{ ,where } \epsilon \sim \mathcal{N}(0, I) \qquad (6)$$

where $L_{diffusion}$ is the auxiliary loss function for training the diffusion decoder G_σ (Eq. 6), and t is the noise level of x_t, following the method in [9]. Our proposed $L_{scoreNCC}$ encourages maximizing the similarity between the registered and reference images while preserving the deformation topology. $\sum ||\nabla_\phi||^2$ is the conventional smoothness penalty on the deformation field. λ and λ_ϕ are hyperparameters, and we empirically set them to 20 in our experiments.

Inference. In the inference stage, we perform image registration in the same style as [13]. Instead of the perturbed image x_t, we input the original reference image f into the network, and the total network input becomes $x_{in} = \{f, m, f\}$. Given this network input x_{in}, our network first generates the deformation field ϕ between the moving image m and the reference image f and produces the registered moving image $m(\phi)$ by feeding the moving image m and the deformation field ϕ into the spatial transformation layer (STL). The registered moving image is the final output of our network.

3   Experiments and Results

Dataset and Preprocessing. Following the previous work [13], we used the publicly available 3D cardiac MR dataset ACDC [5] for experiments. The dataset includes 100 4D temporal cardiac MRI data with corresponding segmentation maps. We selected the 3D image at the end of the diastolic stage as the fixed image and the image at the end of the systolic stage as the moving image. We resampled all scans to the voxel spacing of $1.5 \times 1.5 \times 3.15$ mm, then cropped them to the voxel size of $128 \times 128 \times 32$. We normalized the intensity of all images to $[-1, 1]$. The training set contains 90 image pairs, while the remaining 10 pairs form the test set. The abovementioned preprocessing steps were performed in accordance with the approach described in prior work [13] to ensure a fair comparison.

Moving Image   Fixed Image   Ours   DiffuseMorph   VoxelMorph

Fig. 3. The visualization of registration results. As the red-boxed area shows, our deformation field and the corresponding image are more refined in the area where larger deformation happens. (Color figure online)

Implementation Details. The proposed framework was implemented using the PyTorch library, version 1.12.0. Following [13], we used DDPM UNet's 3D encoder as our shared encoder and DDPM UNet's 3D decoder as our diffusion decoder. For the registration part, instead of a complete 3D UNet in [13], we only used DDPM UNet's 3D decoder as our registration decoder to generate the deformation field. During the diffusion task, we gradually increased the noise schedule from 10^{-6} to 10^{-2} over 2000 timesteps. We utilized an Nvidia RTX3090 GPU and the Adam optimization algorithm [15] to train the model with $\lambda = 20$, $\lambda_\phi = 20$, $\gamma = 1$, batch size $\mathbb{B} = 1$, a learning rate of 2×10^{-4}, and a maximum of 700 epochs.

Evaluation Metrics. We employed three evaluation metrics, i.e., DICE, $|J| \leq 0(\%)$, and $SD(|J|)$ to measure the image registration performance, following existing registration methods [3,13,14]. DICE measures the spatial overlap of anatomical segmentation maps between the warped moving image and the fixed reference image. A higher Dice score indicates better alignment between the warped moving image and the fixed reference image, thus reflecting an improved registration quality. $|J| \leq 0(\%)$ indicates the percentage of non-positive values in the Jacobian determinant of the registration field. This metric indicates the percentage of voxels that lacks a one-to-one registration mapping relation, causing unrealistic deformations and roughness. $SD(|J|)$ refers to the standard deviation of the Jacobian determinant of the registration field. A lower standard deviation indicates that the registration field is relatively smooth and consistent across the image.

Compare with the State-of-the-Art Methods. Table 1 shows the comparison of our method with existing state-of-the-art methods including Voxel-Morph [3], VoxelMorph-Diff [7], and DiffuseMorph [13] on the same training and testing dataset. We produced baseline results using the recommended hyperparameters in their paper. The result shows that our proposed method outperforms existing baseline methods by a substantial margin (Wilcoxon signed-rank test, $p < 0.005$) (Also see Fig. 3). Furthermore, our method aligned better in areas where larger deformation happened, such as myocardium (myo).

Table 1. Image registration results with standard deviation in parenthesis on the 3D cardiac dataset. "LV", "Myo", "RV" refers to Left Ventricle, Myocardium, and Right Ventricle, respectively. "Overall" refers to the averaged registration result of the left blood pool, myocardium, left ventricle, right ventricle, and these total region, following [13]. ↑: the higher, the better results. ↓: the lower, the better results.

Method	DICE ↑				$|J| \leq 0(\%) \downarrow$	$SD(|J|) \downarrow$
	LV	Myo	RV	Overall		
Initial	0.585(0.074)	0.357(0.120)	0.741(0.069)	0.655(0.188)	–	–
VM [3]	0.770(0.086)	0.679(0.129)	0.816(0.065)	0.799(0.110)	0.079(0.058)	0.183
VM-Diff [7]	0.755(0.092)	0.659(0.137)	0.815(0.066)	0.789(0.117)	0.083(0.063)	0.182
DiffuseMorph [13]	0.783(0.086)	0.678(0.148)	0.821(0.067)	0.805(0.114)	0.061(0.038)	0.178
Ours	**0.809(0.077)**	**0.724(0.119)**	**0.827(0.061)**	**0.823(0.096)**	**0.054(0.026)**	**0.176**

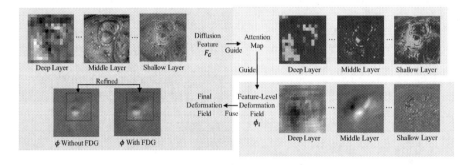

Fig. 4. The visualization example of the effectiveness of FDG. The deformation field shows more details and is more distinct when guided by the FDG, thus improving registration accuracy, as the area in the red box shows. (Color figure online)

Ablation Study. To validate the effectiveness of our proposed learning strategies, including the Feature-wise Diffusion-Guided module(FDG) and Score-wise Diffusion-Guided module(SDG), we conducted ablative experiments, as shown in Table 2. The network without FDG also uses the denoising diffusion decoder but generates the deformation field from the encoded feature directly, and without SDG means that we optimize the network using the vanilla NCC loss. By integrating multi-scale intermediate latent diffusion features into generating deformation fields, we can see that the network's performance increased by 1%. By deploying the reweighing loss, the Jacobian metric decreased by 60.5%. The result achieved a balance when all components were deployed. These results demonstrated that our proposed components could effectively guide the deformation field generation by using multi-scale diffusion features (Also see Fig. 4). Optimization guided by diffusion score led to better preservation of deformation topology. It is worth noticing that the results without FDG or SDG showed only marginal improvement over baseline results, indicating the importance of feature-level deformation field generation and the reweighing scheme. The ablative study of hyperparameter λ is illustrated in Supp. Fig. 1.

Table 2. Ablation study on FDG and SDG.

FDG	SDG	DICE↑	$|J| \leq 0(\%)$ ↓
		0.811(0.098)	0.114(0.076)
√		0.818(0.102)	0.062(0.038)
	√	0.810(0.188)	0.045(0.025)
√	√	0.823(0.096)	0.054(0.026)

Table 3. Ablation Study on hyperparameter γ in SDG.

γ	0.5	1	2
DICE↑	0.817(0.100)	0.823(0.096)	0.816(0.104)
$|J| \leq 0(\%)$ ↓	0.069(0.029)	0.054(0.026)	0.042(0.023)

Analysis of γ. Furthermore, to validate SDG's effectiveness on topology preservation, we conducted another ablative study on SDG's hyperparameter γ, as Table 3 shows. Increased γ indicates a more substantial reweighing effect. The results showed that by adding stronger reweighing influence, we could obtain deformation fields with better topology preservation almost without compromising accuracy.

4   Conclusion

This work proposes two novel modules for unsupervised deformable image registration: the Feature-wise Diffusion-Guided module (FDG) and the Score-wise Diffusion-Guided module (SDG). Among these modules, FDG can effectively guide the deformation field generation by utilizing the multi-scale intermediate diffusion features. SDG demonstrates its ability to guide the optimization process for better deformation topology preservation using the diffusion score. Extensive experiments show that the proposed framework brings impressive improvements over all baselines. The proposed work models the non-linear deformation semantics using the diffusion model. Therefore, it is sound to generalize to other registration tasks and images, which may be one of the future research directions.

Acknowledgements. This work was supported by the Hong Kong Innovation and Technology Fund under Project ITS/030/21 & PRP/041/22FX, as well as by Foshan HKUST Projects under Grants FSUST21-HKUST10E and FSUST21-HKUST11E.

References

1. Alam, F., Rahman, S.U., Ullah, S., Gulati, K.: Medical image registration in image guided surgery: issues, challenges and research opportunities. Biocybern. Biomed. Eng. **38**(1), 71–89 (2018)
2. Avants, B.B., Epstein, C.L., Grossman, M., Gee, J.C.: Symmetric diffeomorphic image registration with cross-correlation: evaluating automated labeling of elderly and neurodegenerative brain. Med. Image Anal. **12**(1), 26–41 (2008)
3. Balakrishnan, G., Zhao, A., Sabuncu, M.R., Guttag, J., Dalca, A.V.: VoxelMorph: a learning framework for deformable medical image registration. IEEE Trans. Med. Imaging **38**(8), 1788–1800 (2019)

4. Baranchuk, D., Voynov, A., Rubachev, I., Khrulkov, V., Babenko, A.: Label-efficient semantic segmentation with diffusion models. In: International Conference on Learning Representations (2022). https://openreview.net/forum?id=SlxSY2UZQT

5. Bernard, O., et al.: Deep learning techniques for automatic MRI cardiac multi-structures segmentation and diagnosis: is the problem solved? IEEE Trans. Med. Imaging **37**(11), 2514–2525 (2018)

6. Che, T., et al.: AMNet: adaptive multi-level network for deformable registration of 3D brain MR images. Med. Image Anal. **85**, 102740 (2023)

7. Dalca, A.V., Balakrishnan, G., Guttag, J., Sabuncu, M.R.: Unsupervised learning of probabilistic diffeomorphic registration for images and surfaces. Med. Image Anal. **57**, 226–236 (2019)

8. Giger, M.L., Karssemeijer, N., Schnabel, J.A.: Breast image analysis for risk assessment, detection, diagnosis, and treatment of cancer. Annu. Rev. Biomed. Eng. **15**, 327–357 (2013)

9. Ho, J., Jain, A., Abbeel, P.: Denoising diffusion probabilistic models. In: Advances in Neural Information Processing Systems, vol. 33, pp. 6840–6851 (2020)

10. Huang, Y., Ahmad, S., Fan, J., Shen, D., Yap, P.T.: Difficulty-aware hierarchical convolutional neural networks for deformable registration of brain MR images. Med. Image Anal. **67**, 101817 (2021)

11. Jain, M., Rai, C., Jain, J., Gambhir, D.: Amalgamation of machine learning and slice-by-slice registration of MRI for early prognosis of cognitive decline. Int. J. Adv. Comput. Sci. Appl. **12**(1) (2021)

12. Khalil, A., Ng, S.C., Liew, Y.M., Lai, K.W.: An overview on image registration techniques for cardiac diagnosis and treatment. Cardiol. Res. Pract. **2018** (2018)

13. Kim, B., Han, I., Ye, J.C.: DiffuseMorph: unsupervised deformable image registration using diffusion model. In: Avidan, S., Brostow, G., Cissé, M., Farinella, G.M., Hassner, T. (eds.) ECCV 2022. LNCS, vol. 13691, pp. 347–364. Springer, Cham (2022). https://doi.org/10.1007/978-3-031-19821-2_20

14. Kim, B., Kim, D.H., Park, S.H., Kim, J., Lee, J.G., Ye, J.C.: CycleMorph: cycle consistent unsupervised deformable image registration. Med. Image Anal. **71**, 102036 (2021)

15. Kingma, D.P., Ba, J.: Adam: a method for stochastic optimization. arXiv preprint arXiv:1412.6980 (2014)

16. Klein, S., Staring, M., Murphy, K., Viergever, M.A., Pluim, J.P.: Elastix: a toolbox for intensity-based medical image registration. IEEE Trans. Med. Imaging **29**(1), 196–205 (2009)

17. Krebs, J., Mansi, T., Mailhé, B., Ayache, N., Delingette, H.: Unsupervised probabilistic deformation modeling for robust diffeomorphic registration. In: Stoyanov, D., et al. (eds.) DLMIA/ML-CDS 2018. LNCS, vol. 11045, pp. 101–109. Springer, Cham (2018). https://doi.org/10.1007/978-3-030-00889-5_12

18. Kwon, M., Jeong, J., Uh, Y.: Diffusion models already have a semantic latent space. In: The Eleventh International Conference on Learning Representations (2023). https://openreview.net/forum?id=pd1P2eUBVfq

19. Mok, T.C., Chung, A.: Fast symmetric diffeomorphic image registration with convolutional neural networks. In: Proceedings of the IEEE/CVF Conference on Computer Vision and Pattern Recognition, pp. 4644–4653 (2020)

20. Mok, T.C.W., Chung, A.C.S.: Large deformation diffeomorphic image registration with Laplacian pyramid networks. In: Martel, A.L., et al. (eds.) MICCAI 2020. LNCS, vol. 12263, pp. 211–221. Springer, Cham (2020). https://doi.org/10.1007/978-3-030-59716-0_21

21. Shen, Z., Zhang, M., Zhao, H., Yi, S., Li, H.: Efficient attention: attention with linear complexities. In: Proceedings of the IEEE/CVF Winter Conference on Applications of Computer Vision, pp. 3531–3539 (2021)
22. Song, Y., Sohl-Dickstein, J., Kingma, D.P., Kumar, A., Ermon, S., Poole, B.: Score-based generative modeling through stochastic differential equations. arXiv preprint arXiv:2011.13456 (2020)
23. Tumanyan, N., Geyer, M., Bagon, S., Dekel, T.: Plug-and-play diffusion features for text-driven image-to-image translation. arXiv preprint arXiv:2211.12572 (2022)

A Denoised Mean Teacher for Domain Adaptive Point Cloud Registration

Alexander Bigalke[(✉)] [iD] and Mattias P. Heinrich[iD]

Institute of Medical Informatics, University of Lübeck, Lübeck, Germany
{alexander.bigalke,mattias.heinrich}@uni-luebeck.de

Abstract. Point cloud-based medical registration promises increased computational efficiency, robustness to intensity shifts, and anonymity preservation but is limited by the inefficacy of unsupervised learning with similarity metrics. Supervised training on synthetic deformations is an alternative but, in turn, suffers from the domain gap to the real domain. In this work, we aim to tackle this gap through domain adaptation. Self-training with the Mean Teacher is an established approach to this problem but is impaired by the inherent noise of the pseudo labels from the teacher. As a remedy, we present a denoised teacher-student paradigm for point cloud registration, comprising two complementary denoising strategies. First, we propose to filter pseudo labels based on the Chamfer distances of teacher and student registrations, thus preventing detrimental supervision by the teacher. Second, we make the teacher dynamically synthesize novel training pairs with noise-free labels by warping its moving inputs with the predicted deformations. Evaluation is performed for inhale-to-exhale registration of lung vessel trees on the public PVT dataset under two domain shifts. Our method surpasses the baseline Mean Teacher by 13.5/62.8%, consistently outperforms diverse competitors, and sets a new state-of-the-art accuracy (TRE = 2.31 mm). Code is available at https://github.com/multimodallearning/denoised_mt_pcd_reg.

Keywords: Point cloud registration · Domain adaptation · Mean Teacher

1 Introduction

Recent deep learning-based registration methods have shown great potential in solving medical image registration problems [10,14]. Most of these methods perform the registration based on the raw volumetric intensity images, e.g. [1,5,6,18,31]. By contrast, only a few recent works [13,21] operate on sparse, purely geometric point clouds extracted from the images, even though this representation promises multiple potential benefits, including computational efficiency, robustness against intensity shifts in the image domain, and anonymity

Supplementary Information The online version contains supplementary material available at https://doi.org/10.1007/978-3-031-43999-5_63.

H. Greenspan et al. (Eds.): MICCAI 2023, LNCS 14229, pp. 666–676, 2023.
https://doi.org/10.1007/978-3-031-43999-5_63

preservation. The latter, for instance, can facilitate public data access and federated learning, as exemplified by a recently released point cloud dataset of lung vessels [21] whose underlying CT scans are not publicly accessible. On the other hand, the sparsity of point clouds and the absence of intensity information make the registration problem more challenging. In particular, unsupervised learning with similarity metrics – as established for dense image registration [5,18] – was shown ineffective for deformable point cloud registration [21], as confirmed by our experiments. Since manual annotations for supervised learning are prohibitively costly, an alternative consists of training on synthetic deformations with known displacements [21], as known from dense registration [7,26]. The inevitable domain gap between synthetic and real deformations, however, involves the risk of suboptimal performance on real data. In this work, we aim to bridge this gap through domain adaptation (DA).

DA has widely been studied for classification and segmentation tasks [12], with popular techniques ranging from adversarial feature [11,25] or output [24] alignment to self-supervised feature learning [22]. However, these methods are insufficient for the specific characteristics of the registration problem, involving a more complex output space and requiring the detection of local correspondences. Instead, recent works adapted the Mean Teacher paradigm [23], previously established for domain adaptive classification [9] and segmentation [19], to the registration problem [3,16,29]. The basic idea is to supervise the learning student model with displacement fields (pseudo labels) provided by a teacher model, whose weights represent the exponential moving average of the student's weights. A significant limitation of this method, however, is the inevitable noise in the pseudo labels, potentially misguiding the adaptation process. Prior works addressed this problem by refining the pseudo labels [16] or weighting them according to model uncertainty, estimated through Monte Carlo dropout [29,30]. However, even refined pseudo labels remain inaccurate, and the proposed refinement strategy [16] assumes piecewise rigid motions of 3D objects and does not apply to complex deformations in medical applications. And weighting pseudo labels according to teacher uncertainty [29,30] does not explicitly consider the quality of the actual registrations, completely ignores the quality and certainty of the current student predictions, and can, therefore, not prevent detrimental supervision of the student through inferior teacher predictions.

Contributions. We introduce two complementary strategies to denoise the Mean Teacher for domain adaptive point cloud registration, addressing the above limitations (see Fig. 1). Both strategies are based on our understanding of an optimal student-teacher relationship. First, if the student's solution to a problem is superior to that of the teacher, good teachers should not insist on their solution but accept the student's approach. To implement this, inspired by a recent technique to filter pseudo labels for human pose estimation [2], we propose to assess the quality of both the teacher and student registrations with the Chamfer distance and to provide only those registrations of the teacher as supervision to the student that are more accurate. This approach differs from previous uncertainty-based methods [29,30] in two decisive aspects: 1) It explicitly assesses the quality

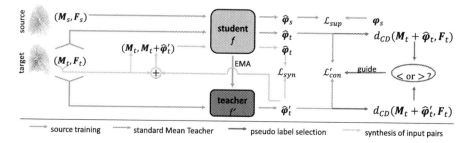

Fig. 1. Overview of our denoised Mean Teacher for domain adaptive registration. We overcome noisy supervision by the teacher with a novel pseudo label selection strategy and the synthesis of new training pairs with precisely known displacements.

of final registrations, using a model-free and objective measure with little computational overhead compared to multiple forward passes in Monte Carlo dropout. 2) The selection process considers both teacher and student predictions and can thus prevent detrimental supervision by the teacher. Our second strategy follows the intuition that good teachers should not pose problems to which they do not know the solution. Instead, they should come up with novel tasks with precisely known solutions. Consequently, we propose a completely novel teacher paradigm, where predicted deformations by the teacher are used to synthesize new training pairs for the student, consisting of the original moving inputs and their warps. These input pairs come with precise noise-free displacement labels and significantly differ from static hand-crafted synthetic deformations [21]. 1) The deformations are based on a real data pair that the teacher aims to align. 2) The deformations are dynamic and become more realistic as the teacher improves. Finally, we unify both strategies in a joint framework for domain adaptive point cloud registration. It is compatible with arbitrary geometric registration models, stable to train, and involves only a few hyper-parameters. We experimentally evaluate the method for inhale-to-exhale registration of lung vessel point clouds on the public PVT dataset [21], demonstrating substantial improvements over diverse competing methods and state-of-the-art performance.

2 Methods

2.1 Problem Setup and Standard Mean Teacher

In point cloud registration, we are given fixed and moving point clouds $F \in \mathbb{R}^{N_F \times 3}$, $M \in \mathbb{R}^{N_M \times 3}$ and aim to predict a displacement vector field $\varphi \in \mathbb{R}^{N_M \times 3}$ that spatially aligns M to F as $M + \varphi$. We address the task in a domain adaptation setting with training data comprising a labeled source dataset $\mathcal{S}$ of triplets (M_s, F_s, φ_s) and a shifted unlabeled target dataset $\mathcal{T}$ of tuples (M_t, F_t). While the formulation of our method is agnostic to the specific domain shift between $\mathcal{S}$ and $\mathcal{T}$, in this work, we generate the source samples on the fly as random synthetic deformations of the target clouds using a fixed hand-crafted deformation function $def : \mathbb{R}^{N \times 3} \to \mathbb{R}^{N \times 3}$, i.e. source triplets are given as $(def(\boldsymbol{F}_t), \boldsymbol{F}_t, \boldsymbol{F}_t - def(\boldsymbol{F}_t))$ or $(def(\boldsymbol{M}_t), \boldsymbol{M}_t, \boldsymbol{M}_t - def(\boldsymbol{M}_t))$. Note that def preserves point correspondences enabling ground truth computation through point-wise subtraction. Given the training data, we aim to learn a function f that predicts deformation vector fields as $\hat{\varphi} = f(\boldsymbol{M}, \boldsymbol{F})$ with optimal performance in the target domain.

Baseline Mean Teacher. To solve the problem, the standard Mean Teacher framework [3,9,23] employs two identical networks, denoted as the student f and teacher f', with parameters $\boldsymbol{\theta}$ and $\boldsymbol{\theta}'$. While the student's weights $\boldsymbol{\theta}$ are optimized through gradient descent, the teacher's weights correspond to the exponential moving average (EMA) of the student and are updated as $\boldsymbol{\theta}'_i = \alpha \boldsymbol{\theta}'_{i-1} + (1-\alpha)\boldsymbol{\theta}_i$ at iteration i with momentum α. Meanwhile, the student is trained by minimizing

$$\mathcal{L}(\boldsymbol{\theta}) = \lambda_1 \underbrace{\|f(\boldsymbol{M}_s, \boldsymbol{F}_s) - \varphi_s\|_2^2}_{\mathcal{L}_{\text{sup}}} + \lambda_2 \underbrace{\|f(\boldsymbol{M}_t, \boldsymbol{F}_t) - f'(\boldsymbol{M}_t, \boldsymbol{F}_t)\|_2^2}_{\mathcal{L}_{\text{con}}} \qquad (1)$$

consisting of the supervised loss $\mathcal{L}_{\text{sup}}$ on source data and the consistency loss $\mathcal{L}_{\text{con}}$ on target data, weighted by λ_1 and λ_2. $\mathcal{L}_{\text{con}}$ guides the learning of the student in the target domain with pseudo-supervision from the teacher, which, as a temporal ensemble, is expected to be superior to the student. Nonetheless, predictions by the teacher can still be noisy and inaccurate, limiting the efficacy of the adaptation process.

2.2   Chamfer Distance-Based Filtering of Pseudo Labels

In a worst-case scenario, the student might predict an accurate displacement field $\hat{\varphi}_t$, which is strongly penalized by the consistency loss due to an inaccurate teacher prediction $\hat{\varphi}'_t$. To prevent such detrimental supervision, we aim to select only those teacher predictions for supervision that are superior to the corresponding student predictions, which, however, is complicated by the absence of ground truth. We, therefore, propose to assess the quality of student and teacher registrations by measuring the similarity/distance between fixed and warped moving clouds, with higher similarities/lower distances indicating more accurate registrations. Among existing similarity measures, we opt for the symmetric Chamfer distance [28], which computes the distance between two point clouds $\boldsymbol{X}, \boldsymbol{Y}$ as

$$d_{\text{CD}}(\boldsymbol{X}, \boldsymbol{Y}) = \sum_{\boldsymbol{x} \in \boldsymbol{X}} \min_{\boldsymbol{y} \in \boldsymbol{Y}} \|\boldsymbol{x} - \boldsymbol{y}\|_2^2 + \sum_{\boldsymbol{y} \in \boldsymbol{Y}} \min_{\boldsymbol{x} \in \boldsymbol{X}} \|\boldsymbol{x} - \boldsymbol{y}\|_2^2 \qquad (2)$$

While we experimentally found the Chamfer distance insufficient as a direct loss function – presumably due to sparse differentiability and susceptibility to local minima, we still observed a strong correlation between Chamfer distance and actual registration error, making it a suitable choice for our purposes. We also explored other measures (Laplacian curvature [28], Gaussian MMD [8]), which proved slightly inferior (Supp., Table 2). Formally, we thus measure the quality of the student prediction $\hat{\varphi}_t = f(\boldsymbol{M}_t, \boldsymbol{F}_t)$ as $d_{\mathrm{CD}}(\boldsymbol{M}_t + \hat{\varphi}_t, \boldsymbol{F}_t)$ and analogously for the teacher prediction $\hat{\varphi}'_t$. We then define our indicator function

$$I(\hat{\varphi}_t, \hat{\varphi}'_t) = \begin{cases} 1 & d_{\mathrm{CD}}(\boldsymbol{M}_t + \hat{\varphi}'_t, \boldsymbol{F}_t) < d_{\mathrm{CD}}(\boldsymbol{M}_t + \hat{\varphi}_t, \boldsymbol{F}_t) \\ 0 & \text{else} \end{cases} \tag{3}$$

and reformulate the consistency loss in Eq. 1 as

$$\mathcal{L}'_{\mathrm{con}} = I(\hat{\varphi}_t, \hat{\varphi}'_t) \cdot \|f(\boldsymbol{M}_t, \boldsymbol{F}_t) - f'(\boldsymbol{M}_t, \boldsymbol{F}_t)\|_2^2 \tag{4}$$

2.3   Synthesizing Inputs with Noise-Free Supervision

While the above filtering strategy mitigates detrimental supervision, the selected pseudo labels are still inaccurate. Therefore, we complement the strategy with a novel teacher paradigm, where the teacher dynamically synthesizes new training pairs with precisely known displacements for supervision. Specifically, given a teacher prediction $\hat{\varphi}'_t = f'(\boldsymbol{M}_t, \boldsymbol{F}_t)$, we do not only use it to supervise the student on the same input pair but also generate a new input sample $(\boldsymbol{M}_t, \boldsymbol{M}_t + \hat{\varphi}'_t)$ by warping $\boldsymbol{M}_t$ with $\hat{\varphi}'_t$. The underlying displacement field is naturally precisely known, enabling noise-free training of the student by minimizing

$$\mathcal{L}_{\mathrm{syn}} = \|f(\boldsymbol{M}_t, \boldsymbol{M}_t + \hat{\varphi}'_t) - \hat{\varphi}'_t\|_2^2 \tag{5}$$

To our knowledge, there is no prior work with a similarly "generative" teacher model. Altogether, we train the student network by minimizing the loss

$$\mathcal{L}(\boldsymbol{\theta}) = \lambda_1 \mathcal{L}_{\mathrm{sup}} + \lambda_2 \mathcal{L}'_{\mathrm{con}} + \lambda_3 \mathcal{L}_{\mathrm{syn}} \tag{6}$$

Technical Details. The synthesized input pairs $(\boldsymbol{M}_t, \boldsymbol{M}_t + \hat{\varphi}'_t)$ exhibit exact point correspondence, i.e., for each point in $\boldsymbol{M}_t$ exists a corresponding point in $\boldsymbol{M}_t + \hat{\varphi}'_t$. That is usually not the case for real data pairs and thus introduces another domain shift, which prevented proper convergence in our initial experiments. To overcome the problem, we exploit that the original point clouds in a dataset, denoted as $\boldsymbol{M}_{t*}$, usually comprise more points than the subsampled clouds $\boldsymbol{M}_t$ that are fed to the network. Given predicted displacements $\hat{\varphi}'_t$ for $\boldsymbol{M}_t$, we interpolate the displacement vectors to $\boldsymbol{M}_{t*}$ with an isotropic Gaussian kernel, yielding $\hat{\varphi}'_{t*}$. The final input pair is then obtained by sampling disjoint point subsets from $(\boldsymbol{M}_{t*}, \boldsymbol{M}_{t*} + \hat{\varphi}'_{t*})$, excluding one-to-one correspondences.

3   Experiments

3.1   Experimental Setup

Datasets. We evaluate our method for inhale-to-exhale registration of lung vessel point clouds on the public PVT dataset [21] (https://github.com/uncbiag/

robot, License: CC BY-NC-SA 3.0). The dataset comprises 1,010 such data pairs, which were extracted from lung CT scans as part of the IRB-approved COPDGene study (NCT00608764). Ten of these scan pairs are cases from the Dirlab-COPDGene dataset [4] and thus annotated with 300 landmark correspondences. We use these cases as the test set and split the remaining unlabeled pairs into 800 cases for training and 200 for validation (on synthetic deformations only). The original point clouds in the dataset have a very high resolution (~100k points), making the processing with deep networks computationally costly. Therefore, we extract distinctive keypoints by local density estimation followed by non-maximum suppression. We extract two sets of such keypoints for each cloud: one with the ~8k most distinctive points for inference, and another with ~16k points, from which we randomly sample subsets during training for increased variability (see Sect. 2.3, technical details). Finally, we pre-align each pair by matching the mean and standard deviation of the coordinates.

Implementation Details. The registration network f is implemented as the default 4-scale architecture of PointPWC-Net [28], operating on 8192 points per cloud. Following [28], we implement $\mathcal{L}_{sup}$, $\mathcal{L}'_{con}$, and $\mathcal{L}_{syn}$ as multi-scale losses. Optimization is performed with the Adam optimizer. We first pre-train the network on source data (batch size 4) for 160 epochs and subsequently minimize the joint loss (Eq. 6) for 140 epochs, both with a constant learning rate of 0.001, which requires up to 11 GB and 13/23 h on an RTX2080. For joint optimization, we use mixed batches of 4 source and 4 target samples, set $\lambda_1 = \lambda_2 = \lambda_3 = 10$, and the EMA-parameter to $\alpha = 0.996$. While the original PVT data pairs represent the target domain in all experiments, we consider two variants of the function def to synthesize source data pairs: a realistic task-specific 2-scale random field similar to [21] and a simple rigid transformation. This enables us to evaluate our method under two differently severe domain shifts. Since real validation data are unavailable, hyper-parameters of all compared methods were tuned in a synthetic adaptation scenario, with the rigid deformations in the source and the 2-scale random field deformations in the target domain. For further implementation details, we refer to our public code.

Comparison Methods. 1) The source-only model is exclusively trained on source data without DA. 2) We adopt the standard Mean Teacher [3]. 3) An uncertainty-aware Mean Teacher (UA-MT), similar to [29,30]. 4) As proposed in [28], we performed purely unsupervised training on target data with a Chamfer loss. However, consistent with the findings in [21], this approach could not converge for complex geometric lung structures. Instead, we use the Chamfer loss on target data as an additional loss to complement supervised source training. 5) We guide the learning on target data with the cycle-consistency method from [17]. 6) As a classical algorithm, we adapt sLBP [13]. 7) We collect the results of two current SOTA methods, S-Robot and D-Robot, from [21], which combine deep networks (Point U-Net, PointPWC-Net), trained on synthetic deformations, with optimal transport modules. Note, however, that the experimental setup in

672   A. Bigalke and M. P. Heinrich

Table 1. Quantitative results on the PVT dataset, reported as mean TRE and 25/75% percentiles in mm and SDlogJ. † indicates a deviating experimental setup (Sect. 3.1).

Method	def = 2-scale rnd. field				def = rigid			
	TRE	25%	75%	SDlogJ	TRE	25%	75%	SDlogJ
initial	23.32	13.22	31.61	–	23.32	13.22	31.61	–
pre-align	12.83	8.25	16.68	–	12.83	8.25	16.68	–
sLBP [13]	3.62	1.24	3.29	0.038	3.62	1.24	3.29	0.038
S-Robot† [21]	5.48	2.86	7.14	N/A	N/A	N/A	N/A	N/A
D-Robot† [21]	2.86	1.25	3.11	N/A	N/A	N/A	N/A	N/A
source-only	4.50	1.62	5.49	0.034	7.62	3.12	11.15	0.019
Chamfer loss [28]	3.96	1.47	4.43	0.036	4.18	1.54	5.27	0.043
cycle-consistency [17]	3.93	1.48	4.36	0.035	6.47	2.43	9.08	0.029
Mean Teacher [3]	2.67	1.33	3.12	**0.028**	6.40	2.42	9.50	**0.013**
UA-MT [29]	2.58	1.28	3.04	0.029	5.71	1.83	8.77	0.015
ours w/o $\mathcal{L}_{\text{syn}}$	2.49	1.23	2.88	0.030	2.57	1.22	2.93	0.027
ours w/o $\mathcal{L}'_{\text{con}}$	2.96	1.27	3.29	0.035	3.00	1.21	3.39	0.034
ours	**2.31**	**1.16**	**2.66**	0.034	**2.38**	**1.12**	**2.66**	0.033

[21] slightly differs from our setting in terms of more input points (60k vs. 8k) and additional input features (vessel radii), thus accessing more information.

Metrics. We interpolate the predicted displacements from the moving input cloud to the annotated moving landmarks with an isotropic Gaussian kernel ($\sigma = 5$ mm) and measure the target registration error (TRE) with respect to the fixed landmarks. To assess the smoothness of the predictions, we interpolate the sparse displacement fields to the underlying image grid and measure the standard deviation of the logarithm of the Jacobian determinant (SDlogJ).

3.2   Results

Quantitative results are shown in Table 1 and reveal the following insights: 1) The source-only model benefits from realistic synthetic deformations in the source domain, yielding a 40.9% lower TRE. 2) The standard Mean Teacher proves effective under the weaker domain shift (-40.7% TRE compared to source-only) but only achieves a slight improvement of 16.0% in the more challenging scenario, where pseudo labels by the teacher are naturally noisier, in turn limiting the efficacy of the adaptation process. 3) Our proposed strategy to filter pseudo labels (ours w/o $\mathcal{L}_{\text{syn}}$) improves the standard teacher and its uncertainty-aware extension, particularly notable under the more severe domain shift ($-59.8/-55.0\%$ TRE). 4) Synthesizing novel data pairs with the teacher (ours w/o $\mathcal{L}'_{\text{con}}$) alone is

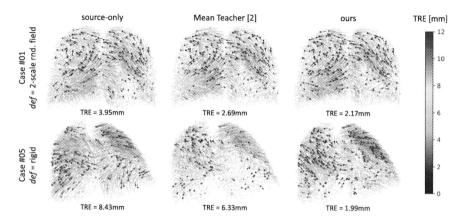

Fig. 2. Qualitative results on two sample cases of the PVT dataset. Predicted displacement fields are shown in gray. Colored dots and lines represent moving landmarks and their interpolated flow, with colors encoding the TRE (clamped to 12 mm). (Color figure online)

slightly inferior to the standard teacher for realistic deformations in the source domain but substantially superior for simple rigid transformations. 5) Combining our two strategies yields further considerable improvements to TREs of 2.31 and 2.38 mm, demonstrating their complementarity. Thus, our method improves the standard Mean Teacher by 13.5/62.8%, outperforms all competitors by statistically significant margins ($p < 0.001$ in a Wilcoxon signed-rank test), and sets a new state-of-the-art accuracy. Remarkably, our method achieves almost the same accuracy for simple rigid transformations in the source domain as for complex, realistic deformations. Thus, it eliminates the need for designing task-specific deformation models, which requires strong domain knowledge. Qualitative results are presented in Fig. 2 and Supp., Fig. 3, demonstrating accurate and smooth deformation fields by our method, as confirmed by the SDlogJ in Table 1, which takes small values for all methods.

4 Conclusion

Our work addressed domain adaptive point cloud registration to bridge the gap between synthetic source and real target deformations. Starting from the established Mean Teacher paradigm, we presented two novel strategies to tackle the noise of pseudo labels from the teacher model, which is a persistent, significant limitation of the method. Specifically, we 1) proposed to prevent detrimental supervision through the teacher by filtering pseudo labels according to Chamfer distances of student and teacher registrations and 2) introduced a novel teacher-student paradigm, where the teacher synthesizes novel training data pairs with perfect noise-free displacement labels. Our experiments for lung vessel registration on the PVT dataset demonstrated the efficacy of our method under two scenarios, outperforming the standard Mean Teacher by up to 62.8% and setting a new state-of-the-art accuracy (TRE = 2.31 mm). As such, our method even favorably compares to popular image-based deep learning methods (VoxelMorph [1] and LapIRN [18], e.g., achieve TREs of 7.98 and 4.99 mm on the original DIR-Lab CT images) but lags behind conventional image-based optimization methods [20] with 0.83 mm TRE. But while the latter require run times of several minutes to process the dense intensity scans with 30M+ voxels, our method processes sparse, purely geometric point clouds with 8k points only, enabling anonymity-preservation and extremely fast inference within 0.2 s. In this light, we see two significant potential impacts of our work: First, our method could generally advance purely geometric keypoint-based medical registration, previously limited by the inefficacy of unsupervised learning with similarity metrics. In particular, medical point cloud registration, currently primarily focusing on lung anatomies, still needs to be investigated for other anatomical structures (abdomen, brain) in future work, which might benefit from our generic approach. Second, our method is conceptionally transferable to dense image registration (e.g., intensity-based similarity metrics [15,27] can replace the Chamfer distance). In this context, it appears of great interest to revisit learning from synthetic deformations [7] within a DA setting or to combine our method with unsupervised learning under metric supervision.

Acknowledgement. We gratefully acknowledge the financial support by the Federal Ministry for Economic Affairs and Climate Action of Germany (FKZ: 01MK20012B) and by the Federal Ministry for Education and Research of Germany (FKZ: 01KL2008).

References

1. Balakrishnan, G., Zhao, A., Sabuncu, M.R., Guttag, J., Dalca, A.V.: VoxelMorph: a learning framework for deformable medical image registration. IEEE Trans. Med. Imaging **38**(8), 1788–1800 (2019)
2. Bigalke, A., Hansen, L., Diesel, J., Hennigs, C., Rostalski, P., Heinrich, M.P.: Anatomy-guided domain adaptation for 3D in-bed human pose estimation. arXiv preprint arXiv:2211.12193 (2022)
3. Bigalke, A., Hansen, L., Heinrich, M.P.: Adapting the mean teacher for keypoint-based lung registration under geometric domain shifts. In: Wang, L., Dou, Q., Fletcher, P.T., Speidel, S., Li, S. (eds.) MICCAI 2022. LNCS, vol. 13436, pp. 280–290. Springer, Cham (2022). https://doi.org/10.1007/978-3-031-16446-0_27
4. Castillo, R., et al.: A reference dataset for deformable image registration spatial accuracy evaluation using the COPDgene study archive. Phys. Med. Biol. **58**(9), 2861 (2013)
5. Chen, J., Frey, E.C., He, Y., Segars, W.P., Li, Y., Du, Y.: TransMorph: transformer for unsupervised medical image registration. Med. Image Anal. **82**, 102615 (2022)
6. De Vos, B.D., Berendsen, F.F., Viergever, M.A., Sokooti, H., Staring, M., Išgum, I.: A deep learning framework for unsupervised affine and deformable image registration. Med. Image Anal. **52**, 128–143 (2019)
7. Eppenhof, K.A., Pluim, J.P.: Pulmonary CT registration through supervised learning with convolutional neural networks. IEEE Trans. Med. Imaging **38**(5), 1097–1105 (2018)

8. Feydy, J.: Geometric data analysis, beyond convolutions. Ph.D. thesis, Université Paris-Saclay Gif-sur-Yvette, France (2020)
9. French, G., Mackiewicz, M., Fisher, M.: Self-ensembling for visual domain adaptation. In: International Conference on Learning Representations (2018)
10. Fu, Y., Lei, Y., Wang, T., Curran, W.J., Liu, T., Yang, X.: Deep learning in medical image registration: a review. Phys. Med. Biol. **65**(20), 20TR01 (2020)
11. Ganin, Y., Lempitsky, V.: Unsupervised domain adaptation by backpropagation. In: International Conference on Machine Learning, pp. 1180–1189. PMLR (2015)
12. Guan, H., Liu, M.: Domain adaptation for medical image analysis: a survey. IEEE Trans. Biomed. Eng. **69**, 1173–1185 (2021)
13. Hansen, L., Heinrich, M.P.: Deep learning based geometric registration for medical images: how accurate can we get without visual features? In: Feragen, A., Sommer, S., Schnabel, J., Nielsen, M. (eds.) IPMI 2021. LNCS, vol. 12729, pp. 18–30. Springer, Cham (2021). https://doi.org/10.1007/978-3-030-78191-0_2
14. Haskins, G., Kruger, U., Yan, P.: Deep learning in medical image registration: a survey. Mach. Vis. Appl. **31**(1), 1–18 (2020)
15. Heinrich, M.P., et al.: Mind: modality independent neighbourhood descriptor for multi-modal deformable registration. Med. Image Anal. **16**(7), 1423–1435 (2012)
16. Jin, Z., Lei, Y., Akhtar, N., Li, H., Hayat, M.: Deformation and correspondence aware unsupervised synthetic-to-real scene flow estimation for point clouds. In: Proceedings of the IEEE/CVF Conference on Computer Vision and Pattern Recognition, pp. 7233–7243 (2022)
17. Mittal, H., Okorn, B., Held, D.: Just go with the flow: self-supervised scene flow estimation. In: Proceedings of the IEEE/CVF Conference on Computer Vision and Pattern Recognition, pp. 11177–11185 (2020)
18. Mok, T.C.W., Chung, A.C.S.: Large deformation diffeomorphic image registration with Laplacian pyramid networks. In: Martel, A.L., et al. (eds.) MICCAI 2020. LNCS, vol. 12263, pp. 211–221. Springer, Cham (2020). https://doi.org/10.1007/978-3-030-59716-0_21
19. Perone, C.S., Ballester, P., Barros, R.C., Cohen-Adad, J.: Unsupervised domain adaptation for medical imaging segmentation with self-ensembling. Neuroimage **194**, 1–11 (2019)
20. Rühaak, J., et al.: Estimation of large motion in lung CT by integrating regularized keypoint correspondences into dense deformable registration. IEEE Trans. Med. Imaging **36**(8), 1746–1757 (2017)
21. Shen, Z., et al.: Accurate point cloud registration with robust optimal transport. In: Advances in Neural Information Processing Systems, vol. 34, pp. 5373–5389 (2021)
22. Sun, Y., Tzeng, E., Darrell, T., Efros, A.A.: Unsupervised domain adaptation through self-supervision. arXiv preprint arXiv:1909.11825 (2019)
23. Tarvainen, A., Valpola, H.: Mean teachers are better role models: weight-averaged consistency targets improve semi-supervised deep learning results. In: Advances in Neural Information Processing Systems, vol. 30 (2017)
24. Tsai, Y.H., Hung, W.C., Schulter, S., Sohn, K., Yang, M.H., Chandraker, M.: Learning to adapt structured output space for semantic segmentation. In: Proceedings of the IEEE Conference on Computer Vision and Pattern Recognition, pp. 7472–7481 (2018)
25. Tzeng, E., Hoffman, J., Saenko, K., Darrell, T.: Adversarial discriminative domain adaptation. In: Proceedings of the IEEE Conference on Computer Vision and Pattern Recognition, pp. 7167–7176 (2017)

26. Uzunova, H., Wilms, M., Handels, H., Ehrhardt, J.: Training CNNs for image registration from few samples with model-based data augmentation. In: Descoteaux, M., Maier-Hein, L., Franz, A., Jannin, P., Collins, D.L., Duchesne, S. (eds.) MICCAI 2017. LNCS, vol. 10433, pp. 223–231. Springer, Cham (2017). https://doi.org/10.1007/978-3-319-66182-7_26

27. de Vos, B.D., van der Velden, B.H., Sander, J., Gilhuijs, K.G., Staring, M., Išgum, I.: Mutual information for unsupervised deep learning image registration. In: Medical Imaging 2020: Image Processing, vol. 11313, pp. 155–161. SPIE (2020)

28. Wu, W., Wang, Z.Y., Li, Z., Liu, W., Fuxin, L.: PointPWC-Net: cost volume on point clouds for (self-)supervised scene flow estimation. In: Vedaldi, A., Bischof, H., Brox, T., Frahm, J.-M. (eds.) ECCV 2020. LNCS, vol. 12350, pp. 88–107. Springer, Cham (2020). https://doi.org/10.1007/978-3-030-58558-7_6

29. Xu, Z., et al.: Double-uncertainty guided spatial and temporal consistency regularization weighting for learning-based abdominal registration. In: Wang, L., Dou, Q., Fletcher, P.T., Speidel, S., Li, S. (eds.) MICCAI 2022. LNCS, vol. 13436, pp. 14–24. Springer, Cham (2022). https://doi.org/10.1007/978-3-031-16446-0_2

30. Yu, L., Wang, S., Li, X., Fu, C.-W., Heng, P.-A.: Uncertainty-aware self-ensembling model for semi-supervised 3D left atrium segmentation. In: Shen, D., et al. (eds.) MICCAI 2019. LNCS, vol. 11765, pp. 605–613. Springer, Cham (2019). https://doi.org/10.1007/978-3-030-32245-8_67

31. Zhao, S., Dong, Y., Chang, E.I., Xu, Y., et al.: Recursive cascaded networks for unsupervised medical image registration. In: Proceedings of the IEEE/CVF International Conference on Computer Vision, pp. 10600–10610 (2019)

Unsupervised 3D Registration Through Optimization-Guided Cyclical Self-training

Alexander Bigalke[1]([✉]), Lasse Hansen[2], Tony C. W. Mok[3],
and Mattias P. Heinrich[1]

[1] Institute of Medical Informatics, University of Lübeck, Lübeck, Germany
{alexander.bigalke,mattias.heinrich}@uni-luebeck.de
[2] EchoScout GmbH, Lübeck, Germany
[3] DAMO Academy, Alibaba Group, Hangzhou, China

Abstract. State-of-the-art deep learning-based registration methods employ three different learning strategies: supervised learning, which requires costly manual annotations, unsupervised learning, which heavily relies on hand-crafted similarity metrics designed by domain experts, or learning from synthetic data, which introduces a domain shift. To overcome the limitations of these strategies, we propose a novel self-supervised learning paradigm for unsupervised registration, relying on self-training. Our idea is based on two key insights. Feature-based differentiable optimizers 1) perform reasonable registration even from random features and 2) stabilize the training of the preceding feature extraction network on noisy labels. Consequently, we propose cyclical self-training, where pseudo labels are initialized as the displacement fields inferred from random features and cyclically updated based on more and more expressive features from the learning feature extractor, yielding a self-reinforcement effect. We evaluate the method for abdomen and lung registration, consistently surpassing metric-based supervision and outperforming diverse state-of-the-art competitors. Source code is available at https://github.com/multimodallearning/reg-cyclical-self-train.

Keywords: Registration · Unsupervised learning · Self-training

1 Introduction

Medical image registration is a fundamental task in medical imaging with applications ranging from multi-modal data fusion to temporal data analysis. In recent years, deep learning has advanced learning-based registration methods [11], which achieve competitive performances at low runtimes and thus constitute a promising alternative to accurate but slow classical optimization methods. A decisive factor in successfully training deep learning-based methods is the choice of a suitable strategy to supervise the learning process. In the literature, there exist three different learning strategies. The first is supervised learning

Supplementary Information The online version contains supplementary material available at https://doi.org/10.1007/978-3-031-43999-5_64.

based on manual annotations such as landmark correspondences [9] or semantic labels [16]. However, manual annotations are costly and may introduce a label bias [2]. Alternatively, a second strategy employs synthetic deformation fields to generate image pairs with precisely known displacement fields [7]. However, this introduces a domain gap between synthetic training and real test pairs, limiting the performance at inference time. Elaborated deformation techniques can reduce the gap but require strong domain knowledge, are tailored to specific problems, and do not generalize across tasks. The third widely used training strategy is unsupervised metric-based learning, maximizing a similarity metric between fixed and warped moving images, e.g. implemented in [2,17]. Popular metrics include normalized cross-correlation [19] and MIND [13]. However, the success of this strategy strongly depends on the specific hand-crafted metric, and the performance of the trained deep learning models is often inferior to a classical optimization-based counterpart. Considering the deficiencies of the above training techniques, in this work, we introduce a novel learning strategy for unsupervised registration based on the concept of self-training.

Self-training is a widespread training strategy for semi-supervised learning [24] and domain adaptation [29]. The core idea is to pre-train a network on available labeled data and subsequently apply the model to the unlabeled data to generate so-called pseudo labels. Afterwards, one alternates between re-training the model on the union of labeled and pseudo-labeled data and updating the pseudo labels with the current model. This general concept was successfully adapted to diverse tasks and settings, with methods in medical context primarily focusing on segmentation [8,18]. These methods resort to a special form of self-training, the Mean Teacher paradigm [22], where pseudo labels are continuously provided by a teacher model, representing a temporal ensemble of the learning network. A persistent problem of classical and Mean Teacher-based self-training is the inherent noise of the pseudo labels, which can severely hamper the learning process. As a remedy, some works aim to filter reliable pseudo labels based on model uncertainty [28]. Only recently, the Mean Teacher was adapted to the registration problem, tackling domain adaptation [3] or complementing metric-based supervision for adaptive regularization weighting [25]. Contrary to these methods, we introduce self-training for registration in a fully unsupervised setting, with pseudo labels as the single source of supervision.

Contributions. We introduce a novel learning paradigm for unsupervised registration by adapting the concept of self-training to the problem. This involves two principal challenges. First, labeled data for the pre-training stage is unavailable, raising the question of how to generate initial pseudo labels. Second, as a general problem in self-training, the negative impact of noise in the pseudo labels needs to be mitigated. In our pursuit to overcome these challenges, we made two decisive observations (see Fig. 2) when exploring a combination of deep learning-based feature extraction with differentiable optimization algorithms for the displacement prediction, such as [9,20]. First, we found that feature-based optimizers predict reasonable displacement fields and improve the initial registration even when applied to the output of random feature networks (orange line in

Fig. 2). We attribute this feature to the inductive bias of deep neural networks, which extract somewhat useful features even with random weights [4]. These predicted displacements thus constitute meaningful initial pseudo labels, solving the first problem and leaving us with the second problem to overcome the noise in the labels. In this context, we made the second observation that the intrinsic regularizing capacity of the optimizers stabilizes the learning from noisy labels. Specifically, training the feature extractor on our initial pseudo labels yielded registrations surpassing the accuracy of the noisy labels used for training (green, red, purple, brown, and magenta lines in Fig. 2). Consequently, we propose a cyclical self-training scheme, alternating between training the feature extractor and updating the pseudo labels. As such, our novel learning paradigm does not require costly manual annotations, prevents the domain shift of synthetic deformations, and is independent of hand-crafted similarity metrics. Moreover, our method significantly differs from previous uncertainty-based pseudo label filtering strategies since it implicitly overcomes the negative impact of noisy labels by combining deep feature learning with regularizing differentiable optimization. We evaluate the method for CT abdomen registration and keypoint-based lung registration, demonstrating substantial improvements over diverse state-of-the-art comparison methods.

2 Methods

2.1 Problem Setup

Given a data pair (F, M) of a fixed and a moving image as input, registration aims at finding a displacement field φ that spatially aligns M to F. We address the task in an unsupervised setting, where training data $\mathcal{T} = \{(F_i, M_i)\}_{i=1}^{|\mathcal{T}|}$ consists of $|\mathcal{T}|$ unlabeled data pairs. Given the training data, we aim to learn a function f with parameters θ_f, (partially) represented by a deep network, which predicts displacement fields as $\hat{\varphi} = f(F, M; \theta_f)$.

2.2 Cyclical Self-training

We propose to solve the above problem with a cyclical self-training strategy visualized in Fig. 1. While existing self-training methods assume the availability of some labeled data, annotations are unavailable in our unsupervised setting. To overcome this issue and generate an initial set of pseudo labels for the first stage of self-training, we parameterize the function f as the combination of a deep neural network g for feature extraction with a non-learnable but differentiable feature-based optimization algorithm h for displacement prediction, i.e.

$$f(F, M; \theta_f) = h(g(F, M; \theta_g)) \tag{1}$$

The approach is based on our empirical observation that a suitable optimization algorithm h can predict reasonable initial displacement fields $\hat{\varphi}^{(0)}$ from random features provided by a network $g^{(0)}$ with random initialization $\theta_g^{(0)}$, which is in

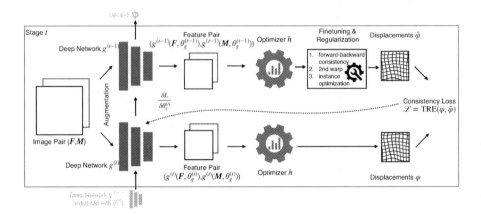

Fig. 1. Overview of the proposed cyclical self-training paradigm for unsupervised registration. The underlying registration pipeline comprises a deep network for feature extraction g and a differentiable optimizer h to predict the displacements. At stage t, we supervise the training of the network $g^{(t)}$ with pseudo labels generated based on the features from the network $g^{(t-1)}$ from the previous stage. For optimal feature learning, the pseudo displacements from the optimizer are further refined and regularized.

line with recent studies on the inductive bias of CNNs [4]. We leverage these predicted displacements as pseudo labels to supervise the first stage of self-training, where the parameters of the feature extractor with different initialization $\boldsymbol{\theta}_g^{(1)}$ are optimized by minimizing the loss

$$\mathcal{L}(\boldsymbol{\theta}_g^{(1)}; \mathcal{T}) = \frac{1}{|\mathcal{T}|} \sum_i \text{TRE}\left(h\left(g\left(\boldsymbol{F}_i, \boldsymbol{M}_i; \boldsymbol{\theta}_g^{(1)} \right) \right), \hat{\boldsymbol{\varphi}}_i^{(0)} \right) \qquad (2)$$

with $\text{TRE}(\hat{\boldsymbol{\varphi}}_i^{(1)}, \hat{\boldsymbol{\varphi}}_i^{(0)})$ denoting the mean over the element-wise target registration error between the displacement fields $\hat{\boldsymbol{\varphi}}_i^{(1)}$ and $\hat{\boldsymbol{\varphi}}_i^{(0)}$.

A critical problem of this basic setup is that the network might overfit the initial pseudo labels and learn to reproduce random features. Therefore, in the spirit of recent techniques from contrastive learning [6], we propose to improve the efficacy of feature learning by incorporating asymmetries into the learning and pseudo label streams at two levels. First, we apply different random augmentations to the input pairs in both streams. Second, we augment the pseudo label stream with additional (non-differentiable) fine-tuning and regularization steps after the optimizer to improve the pseudo displacement fields (see Sec. 2.3 for details). As demonstrated in our ablation experiments (Fig. 2, Table 1), both strategies improve feature learning and strengthen the self-improvement effect.

Once the first stage of self-training has converged, we repeat the process T times. Specifically, at stage t, we generate refined pseudo labels with the trained network $g^{(t-1)}$ from the previous stage, initialize the learning network $g^{(t)}$ with the weights from $g^{(t-1)}$ and perform a warm restart on the learning rate to escape potential local minima from the previous stage.

2.3   Registration Framework

Our proposed self-training scheme is a flexible, modular framework, agnostic to the input modality and the specific implementation of feature extractor g and optimizer h. This section describes our specific design choices for g and h for image and point cloud registration, with the former being our main focus.

Image Registration. To extract features from 3D input volumes, we implement g a standard 3D CNN with six convolution layers with kernel sizes $3 \times 3 \times 3$ and 32, 64, or 128 channels. Each convolution is followed by BatchNorm and ReLU, and every second convolution contains a stride of 2, yielding a downsampling factor of 8. The outputs for both images are mapped to 16-dimensional features using a $1 \times 1 \times 1$ convolution and fed into a correlation layer [21] that captures 125 discrete displacements.

As the optimizer, we adapt the coupled convex optimization for learning-based 3D registration from [20], which, given fixed and moving features, infers a displacement field that minimizes a combined objective of smoothness and feature dissimilarity. Our proposed refinement strategy in the pseudo label stream comprises three ingredients. 1) Forward-backward consistency additionally computes the reverse displacement field (F to M) and then iteratively minimizes the discrepancy between both fields. 2) For a second warp, the moving image is warped with the inferred displacement field before repeating all previous steps. 3) Iterative instance optimization finetunes the final displacement field with Adam by jointly minimizing regularization cost and feature dissimilarity. For the latter, we use the CNN features after the second convolution block and map them with a $1 \times 1 \times 1$ convolution to 16 channels. We apply the same refinement steps at test time. Moreover, we propose to leverage the difference between network-predicted and finetuned displacements to estimate the difficulty of the training samples. Consequently, we apply a weighted batch sampling at training that increases the probability of using less difficult registration pairs with a higher agreement between both fields. We rank all training pairs and use a sigmoid function with arguments ranging linearly from -5 to 5 for the weighted random sampler.

Point Cloud Registration. For point cloud registration, we implement the feature extractor as a graph CNN and rely on sparse loopy belief propagation for differentiable optimization, as introduced in [9].

3   Experiments

3.1   Experimental Setup

Datasets. We conduct our main experiments for inter-patient abdomen CT registration using the corresponding dataset of the Learn2Reg (L2R) Challenge[1]

---

[1] https://learn2reg.grand-challenge.org/.

[15]. The dataset contains 30 abdominal 3D CT scans of different patients with 13 manually labeled anatomical structures of strongly varying sizes. The original image data and labels are from [26]. As part of L2R, they were affinely pre-registered into a canonical space and resampled to identical voxel resolutions (2 mm) and spatial dimensions ($192 \times 160 \times 256\,vx$). Following the data split of L2R, we use 20 scans (190 pairs) for training and the remaining 10 scans (45 pairs) for evaluation. Hence, data split and preprocessing are consistent with compared previous works [9,27]. As metrics, we report the mean Dice overlap (DSC) between the semantic labels and the standard deviation of the logarithmic Jacobian determinant (SDlogJ).

We perform a second experiment for inhale-to-exhale lung CT registration on the DIR-Lab COPDGene dataset[2] [5], which comprises 10 such scan pairs. For each pair, 300 expert-annotated landmark correspondences are available for evaluation. We pre-process all scans in multiple steps: 1) resampling to $1.75 \times 1.00 \times 1.25\,mm$ for exhale and $1.75 \times 1.25 \times 1.75\,mm$ for inhale, 2) cropping with fixed-size bounding boxes ($192 \times 192 \times 208\,vx$), centered around automatically generated lung masks, 3) affine pre-registration, aligning the lung masks. Since we focus on keypoint-based registration of the lung CTs, we follow [9] and extract distinctive keypoints from the CTs using the Förstner algorithm with non-maximum suppression, yielding around 1k points in the fixed and 2k points in the moving cloud. In our experiments, we perform 5-fold cross-validation, with each fold comprising eight data pairs for training and two for testing. We report the target registration error (TRE) at the landmarks and the SDlogJ as metrics.

Implementation Details. We implement all methods in Pytorch and optimize network parameters with the Adam optimizer. For abdomen registration, we train for $T = 8$ stages, each stage comprising 1000 iterations with a batch size of 2. The learning rate follows a cosine annealing warm restart schedule, decaying from 10^{-3} to 10^{-5} at each stage. Hyper-parameters were set based on the DSC on three cases from the training set. For lung registration, the model converged after $T = 5$ stages of 60 epochs with batch size 4, with an initial learning rate of 0.001, decreased by a factor of 10 at epochs 40 and 52. Here, hyper-parameters were adopted from [9]. For both datasets, training requires 90-100 min and 8 GB on an RTX2080, and input augmentations consist of random affine transformations.

3.2   Results

Abdomen. First, we analyze our method in several ablation experiments. In Fig. 2, we visualize the performance of our method on a subset of classes over several cycles of self-training. We observe consistent improvements over the stages, particularly pronounced at early stages while the performance converges later on. This highlights the self-reinforcing effect achieved through alternating pseudo label updates and network training. In the upper part of Table 1, we verify

---

[2] https://med.emory.edu/departments/radiation-oncology/research-laboratories/ deformable-image-registration/downloads-and-reference-data/copdgene.html.

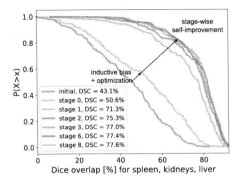

Fig. 2. "Opposite" cumulative distribution of Dice overlaps for Abdomen CT registration after different stages of self-training.

Table 1. Ablation study for abdomen CT registration.

Method	DSC	SDlogJ
prealign	25.9	-
w/o input augm	48.8	.129
w/o PL refinement	48.8	.200
w/o weighted sampling	50.1	.147
ours	**51.1**	.146
1 warp w/o Adam	38.6	**.061**
1 warp w/ Adam	49.6	.119
2 warps w/o Adam	41.1	.088
2 warps w/Adam (ours)	**51.1**	.146

the efficacy of incorporating asymmetries (input augmentations, finetuning of pseudo labels) into both streams and weighted sampling. The results confirm the importance of each component to reach optimal performance. In the lower part of Table 1, we evaluate our final model under different test configurations, highlighting the improvements through a second warp and Adam finetuning.

Next, we compare our method to a comprehensive set of state-of-the-art unsupervised methods, including classical algorithms [1,10,14] and deep learning-based approaches, trained with MIND [2,12]/NCC [17] supervision or contrastive learning [27]. The results are collected from [9,27]. Moreover, we train our own registration framework with metric-based supervision (MIND [13], NCC [19]) to directly verify the advantage of our self-training strategy. Results are shown in Table 2, Fig. 3, and Supp., Fig. 1. Our method substantially outperforms all comparison methods in terms of DSC (statistical significance is confirmed by a

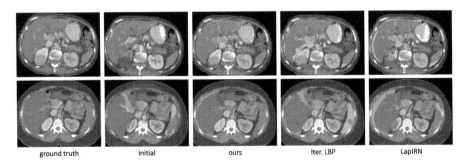

Fig. 3. Qualitative results of selected methods on two cases of the Abdomen CT dataset (axial view). We show overlays of the warped segmentation labels with the fixed scan: liver ■, stomach ■, left kidney ■, right kidney ▢, spleen ■, gall bladder ▢, esophagus ■, pancreas ■, aorta ■, inferior vena cava ■, portal vein ■, left ■/right ■ adrenal gland.

Table 2. Results for unsupervised abdomen CT registration.

Method	Dice [%]	SDlogJ	Time [s]
pre-aligned	25.9	-	
Adam [10]	36.6	**.080**	1.6
Iter. LBP [10]	40.1	.093	0.6
ANTs (SyN) [1]	28.4	N/A	74.3
DEEDS [14]	46.5	N/A	45.4
VoxelMorph [2]	35.4	.134	**0.2**
PDD [12]	41.5	.129	1.4
LapIRN [17]	42.4	.089	3.8
SAME [27]	49.8	N/A	1.2
MIND sup	47.7	.237	1.2
NCC sup	48.1	.299	1.2
ours	**51.1**	.146	1.2

Table 3. Results for lung CT registration on the COPD dataset.

Method	TRE [mm]	SDlogJ
initial (pre-aligned)	11.99	-
VoxelMorph [2]	7.98	N/A
LapIRN [17]	4.99	N/A
PDD [12]	2.16	N/A
rigid deform	2.98	.037
rnd. field deform	3.19	.035
metric sup. [23]	6.79	.042
landmark sup. [9]	2.27	.036
ours	**1.93**	**.033**

Wilcoxon-signed rank test with p<0.0001 for all competitors with public code, which excludes SAME) and sets a new state-of-the-art accuracy of 51.1% DSC. This highlights the advantages of our new learning paradigm over previous unsupervised strategies. Meanwhile, the smoothness of predicted displacement fields (SDlogJ) is comparable with most unsupervised deep learning-based methods [2,12] and superior to MIND- and NCC-supervision.

Lung. For point cloud-based lung registration, we compare our cyclical self-training strategy to three alternative learning strategies: supervision with manually annotated landmark correspondences as in [9], metric-based supervision with Chamfer distance and local Laplacian penalties as in [23], and training on synthetic rigid/random field deformations. All strategies are implemented for the same baseline registration model from [9]. Moreover, we report the performance of three unsupervised image-based deep learning methods [2,12,17] trained with MIND supervision. Results are shown in Table 3, demonstrating the superiority of our self-training strategy over all competing learning strategies and the reported image-based SOTA methods. Qualitative results of the experiment are shown in Supp., Fig. 2, demonstrating accurate and smooth displacements, as also confirmed by low values of SDlogJ.

4   Conclusion

We introduced a novel cyclical self-training paradigm for unsupervised registration. To this end, we developed a modular registration pipeline of a deep feature extraction network coupled with a differentiable optimizer, stabilizing learning from noisy pseudo labels through regularization and iterative, cyclical refinement. That way, our method avoids pitfalls of popular metric supervision (NCC, MIND), which relies on shallow features or image intensities and is prone to noise and local minima. By contrast, our supervision through optimization-refined and -regularized pseudo labels promotes learning task-specific features that are more robust to noise, and our cyclical learning strategy gradually improves the expressiveness of features to avoid local minima. In our experiments, we demonstrated the efficacy and flexibility of our approach, which outperformed the competing state-of-the-art methods and learning strategies for dense image-based abdomen and point cloud-based lung registration. In summary, we did not only present the first fully unsupervised self-training scheme but also a new perspective on unsupervised learning-based registration. In particular, we consider our strategy complementary to existing techniques (metric-based and contrastive learning), opening up the potential for combined training schemes in future work.

Acknowledgement. We gratefully acknowledge the financial support by the Federal Ministry for Economic Affairs and Climate Action of Germany (FKZ: 01MK20012B) and by the Federal Ministry for Education and Research of Germany (FKZ: 01KL2008).

References

1. Avants, B.B., Epstein, C.L., Grossman, M., Gee, J.C.: Symmetric diffeomorphic image registration with cross-correlation: evaluating automated labeling of elderly and neurodegenerative brain. Med. Image Anal. **12**(1), 26–41 (2008)
2. Balakrishnan, G., Zhao, A., Sabuncu, M.R., Guttag, J., Dalca, A.V.: VoxelMorph: a learning framework for deformable medical image registration. IEEE Trans. Med. Imaging **38**(8), 1788–1800 (2019)
3. Bigalke, A., Hansen, L., Heinrich, M.P.: Adapting the mean teacher for keypoint-based lung registration under geometric domain shifts. In: Wang, L., Dou, Q., Fletcher, P.T., Speidel, S., Li, S. (eds.) Medical Image Computing and Computer Assisted Intervention – MICCAI 2022: 25th International Conference, Singapore, September 18–22, 2022, Proceedings, Part VI, pp. 280–290. Springer, Cham (2022). https://doi.org/10.1007/978-3-031-16446-0_27
4. Cao, Y.H., Wu, J.: A random CNN sees objects: one inductive bias of CNN and its applications. In: Proceedings Of The AAAI Conference On Artificial Intelligence. vol. 36, pp. 194–202 (2022)
5. Castillo, R., et al.: A reference dataset for deformable image registration spatial accuracy evaluation using the copdgene study archive. Phys. Med. Bio. **58**(9), 2861 (2013)
6. Chen, X., He, K.: Exploring simple siamese representation learning. In: Proceedings of the IEEE/CVF conference on computer vision and pattern recognition. pp. 15750–15758 (2021)
7. Eppenhof, K.A., Pluim, J.P.: Pulmonary CT registration through supervised learning with convolutional neural networks. IEEE Trans. Med. Imaging **38**(5), 1097–1105 (2018)
8. Hang, W., et al.: Local and global structure-aware entropy regularized mean teacher model for 3D left atrium segmentation. In: Martel, A.L., et al. (eds.) Medical Image Computing and Computer Assisted Intervention – MICCAI 2020: 23rd International Conference, Lima, Peru, October 4–8, 2020, Proceedings, Part I, pp. 562–571. Springer, Cham (2020). https://doi.org/10.1007/978-3-030-59710-8_55

9. Hansen, L., Heinrich, M.P.: Deep learning based geometric registration for medical images: how accurate can we get without visual features? In: Feragen, A., Sommer, S., Schnabel, J., Nielsen, M. (eds.) Information Processing in Medical Imaging: 27th International Conference, IPMI 2021, Virtual Event, June 28–June 30, 2021, Proceedings, pp. 18–30. Springer, Cham (2021). https://doi.org/10.1007/978-3-030-78191-0_2

10. Hansen, L., Heinrich, M.P.: Revisiting iterative highly efficient optimisation schemes in medical image registration. In: de Bruijne, M., et al. (eds.) Medical Image Computing and Computer Assisted Intervention – MICCAI 2021: 24th International Conference, Strasbourg, France, September 27–October 1, 2021, Proceedings, Part IV, pp. 203–212. Springer, Cham (2021). https://doi.org/10.1007/978-3-030-87202-1_20

11. Haskins, G., Kruger, U., Yan, P.: Deep learning in medical image registration: a survey. Mach. Vision Appl. **31**(1), 1–18 (2020)

12. Heinrich, M.P.: Closing the gap between deep and conventional image registration using probabilistic dense displacement networks. In: Shen, D., et al. (eds.) Medical Image Computing and Computer Assisted Intervention – MICCAI 2019: 22nd International Conference, Shenzhen, China, October 13–17, 2019, Proceedings, Part VI, pp. 50–58. Springer, Cham (2019). https://doi.org/10.1007/978-3-030-32226-7_6

13. Heinrich, M.P., et al.: Mind: modality independent neighbourhood descriptor for multi-modal deformable registration. Med. Image Anal. **16**(7), 1423–1435 (2012)

14. Heinrich, M.P., Jenkinson, M., Brady, M., Schnabel, J.A.: MRF-based deformable registration and ventilation estimation of lung CT. IEEE Trans. Med. Imaging **32**(7), 1239–1248 (2013)

15. Hering, A., et al.: Learn2reg: comprehensive multi-task medical image registration challenge, dataset and evaluation in the era of deep learning. IEEE Trans. Med. Imaging **42**, 697–712 (2022)

16. Hu, Y., et al.: Weakly-supervised convolutional neural networks for multimodal image registration. Med. Image Anal. **49**, 1–13 (2018)

17. Mok, T.C.W., Chung, A.C.S.: Large deformation diffeomorphic image registration with laplacian pyramid networks. In: Martel, A.L., et al. (eds.) Medical Image Computing and Computer Assisted Intervention – MICCAI 2020: 23rd International Conference, Lima, Peru, October 4–8, 2020, Proceedings, Part III, pp. 211–221. Springer, Cham (2020). https://doi.org/10.1007/978-3-030-59716-0_21

18. Perone, C.S., Ballester, P., Barros, R.C., Cohen-Adad, J.: Unsupervised domain adaptation for medical imaging segmentation with self-ensembling. NeuroImage **194**, 1–11 (2019)

19. Sarvaiya, J.N., Patnaik, S., Bombaywala, S.: Image registration by template matching using normalized cross-correlation. In: 2009 International Conference on Advances in Computing, Control, and Telecommunication Technologies, pp. 819–822. IEEE (2009)

20. Siebert, H., Heinrich, M.P.: Learn to fuse input features for large-deformation registration with differentiable convex-discrete optimisation. In: Hering, A., Schnabel, J., Zhang, M., Ferrante, E., Heinrich, M., Rueckert, D. (eds.) Biomedical Image Registration: 10th International Workshop, WBIR 2022, Munich, Germany, July 10–12, 2022, Proceedings, pp. 119–123. Springer, Cham (2022). https://doi.org/10.1007/978-3-031-11203-4_13

21. Sun, D., Yang, X., Liu, M.Y., Kautz, J.: PWC-Net: CNNs for optical flow using pyramid, warping, and cost volume. In: Proceedings of the IEEE Conference on Computer Vision and Pattern Recognition, pp. 8934–8943 (2018)

22. Tarvainen, A., Valpola, H.: Mean teachers are better role models: weight-averaged consistency targets improve semi-supervised deep learning results. Adv. Neural Inf. Process. Syst. **30** (2017)

23. Wu, W., Wang, Z.Y., Li, Z., Liu, W., Fuxin, L.: PointPWC-Net: cost volume on point clouds for (self-)supervised scene flow estimation. In: Vedaldi, A., Bischof, H., Brox, T., Frahm, J.-M. (eds.) Computer Vision – ECCV 2020: 16th European Conference, Glasgow, UK, August 23–28, 2020, Proceedings, Part V, pp. 88–107. Springer, Cham (2020). https://doi.org/10.1007/978-3-030-58558-7_6

24. Xie, Q., Luong, M.T., Hovy, E., Le, Q.V.: Self-training with noisy student improves imagenet classification. In: Proceedings of the IEEE/CVF conference on computer vision and pattern recognition. pp. 10687–10698 (2020)

25. Xu, Z., et al.: Double-uncertainty guided spatial and temporal consistency regularization weighting for learning-based abdominal registration. In: Wang, L., Dou, Q., Fletcher, P.T., Speidel, S., Li, S. (eds.) Medical Image Computing and Computer Assisted Intervention – MICCAI 2022: 25th International Conference, Singapore, September 18–22, 2022, Proceedings, Part VI, pp. 14–24. Springer, Cham (2022). https://doi.org/10.1007/978-3-031-16446-0_2

26. Xu, Z., et al.: Evaluation of six registration methods for the human abdomen on clinically acquired CT. IEEE Trans. Biomed. Eng. **63**(8), 1563–1572 (2016)

27. Yan, K., et al.: Sam: self-supervised learning of pixel-wise anatomical embeddings in radiological images. IEEE Trans. Med. Imaging **41**(10), 2658–2669 (2022)

28. Yu, L., Wang, S., Li, X., Fu, C.-W., Heng, P.-A.: Uncertainty-aware self-ensembling model for semi-supervised 3d left atrium segmentation. In: Shen, D., et al. (eds.) Medical Image Computing and Computer Assisted Intervention – MICCAI 2019: 22nd International Conference, Shenzhen, China, October 13–17, 2019, Proceedings, Part II, pp. 605–613. Springer, Cham (2019). https://doi.org/10.1007/978-3-030-32245-8_67

29. Zou, Y., Yu, Z., Kumar, B., Wang, J.: Unsupervised domain adaptation for semantic segmentation via class-balanced self-training. In: Proceedings of the European conference on computer vision (ECCV). pp. 289–305 (2018)

Inverse Consistency by Construction for Multistep Deep Registration

Hastings Greer[1]([✉]), Lin Tian[1], Francois-Xavier Vialard[2], Roland Kwitt[3],
Sylvain Bouix[4], Raul San Jose Estepar[5], Richard Rushmore[6],
and Marc Niethammer[1]

[1] University of North Carolina at Chapel Hill, Chapel Hill, USA
tgreer@cs.unc.edu
[2] University Paris-Est, Créteil, France
[3] University of Salzburg, Salzburg, Austria
[4] ÉTS Montréal, Montreal, Canada
[5] Brigham and Women's Hospital, Boston, USA
[6] Boston University, Boston, USA

Abstract. Inverse consistency is a desirable property for image registration. We propose a simple technique to make a neural registration network inverse consistent by construction, as a consequence of its structure, as long as it parameterizes its output transform by a Lie group. We extend this technique to multi-step neural registration by composing many such networks in a way that preserves inverse consistency. This multi-step approach also allows for inverse-consistent coarse to fine registration. We evaluate our technique on synthetic 2-D data and four 3-D medical image registration tasks and obtain excellent registration accuracy while assuring inverse consistency.

Keywords: Registration · Deep Learning

1 Introduction

Image registration, or finding the correspondence between a pair of images, is a fundamental task in medical image computing. One desirable property for registration algorithms is inverse consistency – the property that the transform found registering image A onto image B, composed with the transform found by registering image B onto image A, yields the identity map. Inverse consistency is useful for several reasons. Practically, it is convenient to have a single transform and its inverse associating two images instead of two transforms of unknown relationship. For within-subject registration, inverse consistency is often a natural assumption as long as images are consistent with each other, e.g., did not undergo surgical removal of tissue. For time series analysis, inverse consistency prevents bias [19]. We propose a novel deep network structure that registers images in multiple steps in a way that is *inverse-consistent by construction*. Our approach is flexible and allows different transform types for different steps (Fig. 1).

H. Greenspan et al. (Eds.): MICCAI 2023, LNCS 14229, pp. 688–698, 2023.
https://doi.org/10.1007/978-3-031-43999-5_65

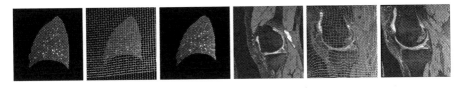

Fig. 1. Cases registered by ConstrICON from DirLab and OAI

2   Related Work

Inverse consistency in deep image registration approaches is commonly promoted via a penalty [7,15,22,27] on the inverse consistency error. Extensive work also exists on optimization-based *exactly* inverse consistent image registration. For example, by using a symmetric image similarity measure and an inverse consistency loss on the transformations [5] or by performing robust inverse consistent rigid registrations with respect to a middle space [19]. ANTs SyN [2] is an approach to inverse consistent deformable registration, but by default is part of a multi-step affine then SyN pipeline which is not as a whole inverse consistent.

Mok *et al.* [14] introduce a deep-learning framework that is exactly inverse consistent. They take advantage of the fact that a stationary velocity field (SVF) transform representation allows for fast inversion of a transform by integrating the negated velocity field. Thus, by calling their network twice, the second time with the inputs reversed, they can construct a transform $\Phi^{AB} = \exp(N_\theta[I^A, I^B]) \circ \exp(-N_\theta[I^B, I^A])$. This registration network is inverse-consistent by construction, but only supports one step. *Our approach will provide a general inverse consistent multi-step framework.*

Iglesias *et al.* [10] introduce a two-step deep registration framework for brain registration that is inverse consistent by construction. First, they independently segment each image with a U-Net into 97 anatomical regions. The centroids of these regions and the corresponding regions of an atlas are then used to obtain an affine transformation to the atlas. This is inverse consistent. Second, each brain image is resampled to the atlas space followed by an SVF-based transformation, where the velocity field is obtained by two calls to their velocity field network: $\exp(N_\theta[I^A, I^B] - N_\theta[I^B, I^A])$. This symmetrization retains inverse consistency and is conceptually similar to our approach. *However, their approach, unlike ours, does not directly extend to N steps and is not trained end to end.*

There is extensive literature on deep multi-step approaches. The core idea is to conduct the registration in multiple steps with the warped image produced by the previous step being the input to the latter step. Thus, the original input image pairs can be registered progressively. AVSM [22] achieves this by reusing the same neural network at each step. Other works in the literature [7,13,23] setup different neural networks at each step. In addition, these steps are often conducted in a coarse-to-fine manner. Namely, the neural network at the current step registers the input images at a coarse resolution, interpolates the output deformation field to a finer resolution, composes the interpolated deformation field with the composed transformation from previous steps, warps the moving image at the finer resolution, and passes the warped image and target image to the neural network at next step. Greer *et al.* and Tian *et al.* [7,23] define an abstract `TwoStep` operator to represent the process described above. However, this `TwoStep` operation does not guarantee inverse consistency between the *composed* forward transformation and the *composed* backward transformation. *To address this issue, we propose a novel operator for multi-step registration to obtain inverse consistent registration by construction.*

Definitions and Notation. We use subscripted capital letters, e.g., N_θ, to represent neural networks that return arrays of numbers, and capital Greek letters Φ, Ψ, and Ξ to represent registration neural networks, i.e., neural networks that return transforms. A transform is a function $\mathbb{R}^D \to \mathbb{R}^D$ with D denoting the dimension of the images we are registering. N_θ^{AB} is shorthand for N_θ called on the images I^A and I^B, likewise Φ^{AB} is shorthand for $\Phi[I^A, I^B]$. A deep registration network outputs a transform such that $I^A \circ \Xi^{AB} \sim I^B$. For a Lie group G and associated algebra $\mathfrak{g}$, exp is the (Lie-)exponential map from $\mathfrak{g} \to G$ [6,11].

3   Lie-Group Based Inverse Consistent Registration

To design a registration algorithm, one must pick a class of transforms that the algorithm will return. Many types of transforms that are useful for practical medical registration problems happen to also be Lie groups. We describe a procedure for designing a neural network that outputs a member of a specified Lie group in an inverse consistent manner and provide several examples.

Recall that a Lie group G is always associated with a Lie algebra $\mathfrak{g}$. Create a neural network N_θ (of arbitrary design) with two input images and an output that can be considered an element of $\mathfrak{g}$.

A registration network Φ defined to act as follows on two images

$$\Phi[I^A, I^B] := \exp(g(I^A, I^B)), \quad g(I^A, I^B) := N_\theta[I^A, I^B] - N_\theta[I^B, I^A] \quad (1)$$

is inverse consistent, because $g(I^A, I^B) = -g(I^B, I^A)$ by construction. We explore how this applies to several Lie groups.

Rigid Registration. The Lie algebra of rigid rotations is skew-symmetric matrices. N_θ outputs a skew-symmetric matrix R and a vector t, so that

$$N_\theta^{AB} = \begin{bmatrix} R & t \\ 0 & 1 \end{bmatrix}, \Phi_{(\text{rigid})}[I^A, I^B](x) := \exp(N_\theta[I^A, I^B] - N_\theta[I^B, I^A]) \begin{bmatrix} x \\ 1 \end{bmatrix}, \quad (2)$$

where $\Phi_{(\text{rigid})}$ will output a rigid transformation in an inverse consistent manner. Here, the exponential map is just the matrix exponent.

Affine Registration. Relaxing R to be an arbitrary matrix instead of a skew-symmetric matrix in the above construction produces a network that performs inverse consistent affine registration.

Nonparametric Vector Field Registration. In the case of the group of diffeomorphisms, the corresponding Lie algebra[1] is the space of vector fields. If N_θ outputs a vector field, implemented as a grid of vectors which are linearly interpolated, then, by using scaling and squaring [1,3] to implement the Lie exponent, we have

$$\Phi_{(\mathrm{svf})}[I^A, I^B](x) := \exp(N_\theta[I^A, I^B] - N_\theta[I^B, I^A])(x), \tag{3}$$

which is an inverse consistent nonparametric registration network. This is equivalent to the standard SVF technique for image registration, with a velocity field represented as a grid of vectors equal to $N_\theta[I^A, I^B] - N_\theta[I^B, I^A]$.

MLP Registration. An ongoing research question is how to represent the output transform as a multi-layer perceptron (MLP) applied to coordinates. One approach is to reshape the vector of outputs of a ConvNet so that the vector represents the weight matrices defining an MLP (with D inputs and D outputs). This MLP is then a member of the Lie algebra of vector-valued functions, and the exponential map to the group of diffeomorphisms can be computed by solving the following differential equation to $t = 1$ using an integrator such as fourth-order Runge-Kutta. Again, by defining the velocity field to flip signs when the input image order is flipped, we obtain an inverse consistent transformation:

$$v(z) = N_\theta^{AB}(z) - N_\theta^{BA}(z), \frac{\partial}{\partial t}\Phi^{AB}(x,t) = v(\Phi^{AB}(x,t)),$$

$$\Phi^{AB}(x,0) = x, \quad \Phi^{AB}(x) = \Phi^{AB}(x,1). \tag{4}$$

4   Multi-step Registration

The standard approach to composing two registration networks is to register the moving image to the fixed image, warp the moving image and then register the warped moving image to the fixed image again and compose the transforms. This is formalized in [7,23] as TwoStep, i.e.,

$$\mathtt{TwoStep}\{\Phi, \Psi\} := \Phi[I^A, I^B] \circ \Psi[I^A \circ \Phi[I^A, I^B], I^B]. \tag{5}$$

Unfortunately, TwoStep *[7,23] is not always inverse consistent even with inverse consistent arguments. First, although* $\Psi[\tilde{I}^A, I^B]$ *is the inverse of* $\Psi[I^B, \tilde{I}^A]$, *it does not necessarily have any relationship with* $\Psi[\tilde{I}^B, I^A]$ *which is the term that appears when swapping the inputs to* TwoStep. *Second, composing* TwoStep $\{\Phi, \Psi\}[I^A, I^B] \circ$ TwoStep$\{\Phi, \Psi\}[I^B, I^A]$, *results in* $\Phi \circ \Psi \circ \Phi^{-1} \circ \sim \Psi^{-1}$. *The inverses are interleaved so that even if they were exact, they can not cancel.*

Our contribution is an operator, TwoStepConsistent, that is inverse consistent if its components are inverse consistent. We assume that our component

---

[1] Although in infinite dimensions, the name Lie algebra does not apply, in our case we only need the notions of the exponential map and tangent space at identity to preserve the inverse consistency property.

networks Φ and Ψ are inverse consistent, and that Φ returns a transform that we can explicitly find the square root of, such that $\sqrt{\Phi^{AB}} \circ \sqrt{\Phi^{AB}} = \Phi^{AB}$. Note that for transforms defined by $\Phi^{AB} = \exp(g)$, $\sqrt{\Phi^{AB}} = \exp(g/2)$. Since each network is inverse consistent, we have access to the inverses of the transforms they return. We begin with the relationship that Φ will be trained to fulfill

$$I^A \circ \Phi[I^A, I^B] \sim I^B, \ \hat{I}^A := I^A \circ \sqrt{\Phi[I^A, I^B]} \sim \hat{I}^B := I^B \circ \sqrt{\Phi[I^B, I^A]}, \quad (6)$$

and apply Ψ to register $\hat{I}^A$ and $\hat{I}^B$

$$I^A \circ \sqrt{\Phi[I^A, I^B]} \circ \Psi[\hat{I}^A, \hat{I}^B] \simeq I^B \circ \sqrt{\Phi[I^B, I^A]}, \quad (7)$$

$$I^A \circ \sqrt{\Phi[I^A, I^B]} \circ \Psi[\hat{I}^A, \hat{I}^B] \circ \sqrt{\Phi[I^A, I^B]} \simeq I^B. \quad (8)$$

We isolate the transform in the left half of Eq. (8) as our new operator, i.e.,

$$\texttt{TwoStepConsistent}\{\Phi, \Psi\}[I^A, I^B] := \sqrt{\Phi[I^A, I^B]} \circ \Psi[\hat{I}^A, \hat{I}^B] \circ \sqrt{\Phi[I^A, I^B]}. \quad (9)$$

In fact, we can verify that

$$\texttt{TwoStepConsistent}\{\Phi, \Psi\}[I^A, I^B] \circ \texttt{TwoStepConsistent}\{\Phi, \Psi\}[I^B, I^A] \quad (10)$$

$$= \sqrt{\Phi} \circ \Psi \circ \sqrt{\Phi} \circ \sqrt{\Phi^{-1}} \circ \Psi^{-1} \circ \sqrt{\Phi^{-1}} = \text{id}. \quad (11)$$

Notably, This Procedure Extends to N-Step Registration. With the operator of Eq. (9), a registration network composed from an arbitrary number of steps may be made inverse consistent. This is because $\texttt{TwoStepConsistent}\{\cdot, \cdot\}$ is a valid *second* argument to $\texttt{TwoStepConsistent}$. For instance, a three-step network can be constructed as $\texttt{TwoStepConsistent}\{\Phi, \texttt{TwoStepConsistent}\{\Psi, \Xi\}\}$.

5   Synthetic Experiments

Inverse Consistent Rigid, Affine, Nonparametric, and MLP Registration. We train networks on MNIST 5s using the methods in Sects. 3 and 4, demonstrating that the resulting networks are inverse-consistent. Our $\texttt{TwoStepConsistent}$ (TSC) operator can be used on any combination of the networks defined in Sect. 3. For demonstrations, we join an MLP registration network to a vector field registration network, and join two affine networks to two vector field networks. Figure 2 shows successful inverse-consistent sample registrations.

Affine Registration Convergence. In addition to being inverse consistent, our method accelerates convergence and stability of affine registration, compared to directly predicting the matrix of an affine transform. Here, we disentangle whether this happens for any approach that parameterizes an affine transform by taking the exponent of a matrix, or whether this acceleration is

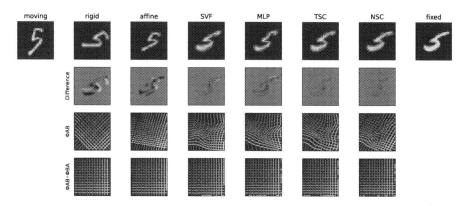

Fig. 2. We train single-step rigid, affine, vector field parameterized SVF, and neural deformation field (MLP) networks, as well as a two-step registration network (TSC) composed of a neural deformation field step followed by a vector field parameterized SVF step and a 4 step network (NSC) composed of two affine steps and two SVF steps. We observe excellent registration results indicated by the small differences (second row) after applying the estimated transformation Φ^{AB} (third row). Composing with the inverse produces results very close to the identity map (last row) as desired.

unique to our inverse consistent method. We also claim that multi-step registration is important for registration accuracy and convergence time and that an *inverse consistent* multi-step operator, TwoStepConsistent, is thus beneficial.

To justify these claims, we investigate training for affine registration on the synthetic *Hollow Triangles and Circles* dataset from [23] while varying the method used to obtain a matrix from the registration network and the type of multi-step registration used. To obtain an affine matrix, we either directly use the neural network output N_θ^{AB}, use $\exp(N_\theta^{AB})$, or, as suggested in Sect. 3, use $\exp(N_\theta^{AB} - N_\theta^{BA})$. We either register in one step, use the TwoStep operator from [7,23], or use our new TwoStepConsistent operator. This results in 9 training configurations, which we run 65 times each.

We observe that parameterizing an affine registration using the $\exp(N_\theta^{AB} - N_\theta^{BA})$ construction speeds up the first few epochs of training and gets even faster when combined with any multi-step method. In Fig. 3, note that in the top-left corner of the first plot, the green loss curves (corresponding to models using $N_\theta^{AB} - N_\theta^{BA}$) are roughly vertical, while the other loss curves are roughly horizontal, eventually bending down. After this initial lead, these green curves also converges to a better final loss. Further, all methods that use the $N_\theta^{AB} - N_\theta^{BA}$ construction train reliably, while other methods sometimes fail to converge (Fig. 3, right plot). This has a dramatic effect on the practicality of a method since training on 3-D data can take multiple days on expensive hardware.

Finally, as expected, the *only* two approaches that are inverse consistent are the single-step inverse consistent by construction network, and the network using two inverse-consistent by construction subnetworks, joined by the TwoStepConsistent operator. (Fig. 3, middle, dotted and solid green).

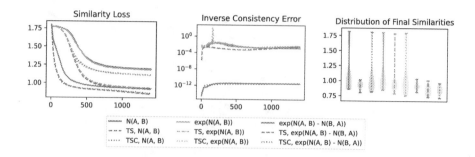

Fig. 3. We vary the network used to perform affine registration and the method for composing steps on the Hollow Triangles and Circles dataset. Average loss curves and distribution of final losses are shown for 65 training runs per experimental configuration. Our `TwoStepConsistent` approach performs best overall. It shows fast convergence, high accuracy (indicated by a low similarity loss, left), is highly inverse consistent (middle), and trains reliably (indicated by the tight violin plot on the right). (Color figure online)

6   Evaluation on 3-D Medical Datasets

We evaluate on several datasets, where we can compare to earlier registration approaches. We use the network $\Phi := \text{TSC}\{\Psi_1, \text{TSC}\{\Psi_2, \text{TSC}\{\Xi_1, \Xi_2\}\}\}$ with Ξ_i inverse-consistent SVF networks backed by U-Nets and Ψ_i inverse-consistent affine networks backed by ConvNets[2]. We rely on local normalized cross-correlation as our similarity measure, with $\sigma = 5vx$, and regularize the SVF networks by the sum of the bending energies of their velocity fields, with $\lambda = 5$. We train end to end, minimizing $-\text{LNCC}(I^A \circ \Phi[I^A, I^B], I^B) + \lambda \mathcal{L}_{\text{reg}}$ for 100,000 iterations ($\sim$2 days on 4 NVIDIA A6000s) with Adam optimization and a learning rate of $1e$-4. In all cases, we normalize images to the range $(0, 1)$. We evaluate registration accuracy with and without instance optimization [23,26]. Without instance optimization, registration takes $\sim$0.23 s on an NVIDIA RTX A6000 on the HCP [24] dataset. With instance optimization, registration takes $\sim$43 s.

6.1   Datasets

COPDGene/Dirlab Lung CT. We follow the data selection and preprocessing of [23]. We train on 999 inhale/exhale pairs from COPDGene [18], resampled to 2mm spacing at $[175 \times 175 \times 175]$, masked with lung segmentations, clipped to [-1000, 0] Hounsfield units, and scaled to $(0, 1)$. We evaluate landmark error (MTRE) on the ten inhale/exhale pairs of the Dirlab challenge dataset [4][3].

---

[2] Specifically, `networks.tallUNet2` and `networks.ConvolutionalMatrixNet` from the library icon_registration version 1.1.1 on `pypi`.

[3] https://tinyurl.com/msk56ss5.

OAI Knee MRI. We train and test on the split published with [22], with 2532 training examples and 301 test pairs from the Osteoarthritis Initiative (OAI) [16][4]. We evaluate using the mean Dice score of femoral and tibial cartilage. To compare directly to [7,22,23] we train and evaluate at $[80 \times 192 \times 192]$.

HCP Brain MRI. We train on 1076 brain-extracted T1w images from the HCP dataset [24] and test on a sample of 100 pairs between 36 images via mean Dice over 28 midbrain structures [20,21]. We train and execute the network at $[130 \times 155 \times 130]$, then compute the Dice score at full resolution.

OASIS Brain MRI. We use the OASIS-1 [12] data preprocessed by [9]. This dataset contains images of 414 subjects. Following the data split in [14], we train on 255 images and test on 153 images[5]. The images in the dataset are of size $[160 \times 192 \times 224]$, and we crop the center of the image according to the preprocessing in [14], leading to a size of $[160 \times 144 \times 192]$. During training, we sample image pairs randomly from the train set. For evaluation, we randomly pick 5 cases as the fixed images and register all the remaining 148 cases to the 5 cases, resulting in 740 image pairs overall.

6.2 Comparisons

We use publicly-available pretrained weights and code for ANTs [2], PTVReg [25], GradICON [23], SynthMorph [8], SymNet [14], and EasyReg [10]. SymNet, GradICON, and PTVReg are run on the datasets associated with their original publication. SynthMorph, which we evaluate on HCP, was originally trained and evaluated on HCP-A and OASIS. EasyReg was trained on HCP [24] and ADNI [17]. Our ConstrICON method outperforms the state of the art on HCP, OAI, and OASIS registration but underperforms on the DirLab data. Since we use shared hyperparameters between these datasets, which are not tuned to a specific task, we assert that this performance level will likely generalize to new datasets. We find that our method is more inverse consistent than existing inverse consistent by construction methods SymNet and SyNOnly with higher accuracy, and more inverse consistent than inverse-consistent-by-penalty GradICON (Table 1).

---

[4] https://nda.nih.gov/oai.

[5] Due to changes in the OASIS-1 data, our test set slightly differs from [14]. We evaluate all methods using our testing protocol so that results are consistent.

696   H. Greer et al.

Table 1. Results on 3-D medical registration. $\%|J|$ indicates the percentage of voxels with negative Jacobian. $\|\Phi^{AB} \circ \Phi^{BA} - \mathrm{id}\|$ indicates the mean deviation in voxels from inverse consistency. Instance optimization is denoted by *io*. Our ConstrICON approach shows excellent registration performance while being highly inverse consistent.

HCP

Approach	DICE	$\%|J|$	$\|\Phi^{AB} \circ \Phi^{BA} - \mathrm{id}\|$
ANTs SyNOnly [2]	75.8	0	0.0350
ANTs SyN	77.2	0	1.30
ConstrICON	79.3	3.81e-6	0.000386
ConstrICON + io	80.1	0	0.00345
GradICON [23]	78.6	0.00120	0.309
GradICON + io	80.2	0.000352	0.123
SynthMorph [8] brain	78.4	0.364	–
SynthMorph shape	79.7	0.298	–

OASIS

Approach	DICE	$\%|J|$	$\|\Phi^{AB} \circ \Phi^{BA} - \mathrm{id}\|$
ConstrICON	79.7	9.73e-5	0.00776
SymNet [14]	79.1	0.00487	0.0595
EasyReg [10]	77.2	–	0.181

DirLab

Approach	MTRE	$\%|J|$	$\|\Phi^{AB} \circ \Phi^{BA} - \mathrm{id}\|$
ANTs SyN [2]	1.79	0	4.23
PTVReg [25]	0.96	0.60	6.13
ConstrICON	1.61	6.6e-6	0.00518
ConstrICON + io	1.32	3.02e-6	0.00306
GradICON [23]	1.26	0.0003	0.575
GradICON + io	0.96	0.0002	0.161

OAI

Approach	DICE	$\%|J|$	$\|\Phi^{AB} \circ \Phi^{BA} - \mathrm{id}\|$
ANTs SyN	65.7	0	5.32
GradICON	70.4	0.0261	1.84
GradICON + io	71.2	0.0042	0.504
ConstrICON	70.7	2.41e-7	0.0459
ConstrICON + io	71.5	3.36e-7	0.0505

7   Conclusion

The fundamental impact of this work is as a recipe for constructing a broad class of exactly inverse consistent, multi-step registration algorithms. We also are pleased to present registration results on four medically relevant datasets that are competitive with the current state of the art, and in particular are more accurate than existing inverse-consistent-by-construction neural approaches.

Acknowledgements. This research was supported by NIH awards RF1MH126732, 1R01-AR072013, 1R01-HL149877, 1R01 EB028283, R41-MH118845, R01MH112748, 5R21LM013670, R01NS125307, 2-R41-MH118845; by the Austrian Science Fund: FWF P31799-N38; and by the Land Salzburg (WISS 2025): 20102-F1901166-KZP, 20204-WISS/225/197-2019. The work expresses the views of the authors and not of the funding agencies. The authors have no conflicts of interest.

References

1. Arsigny, V., Commowick, O., Pennec, X., Ayache, N.: A Log-Euclidean framework for statistics on diffeomorphisms. In: Larsen, R., Nielsen, M., Sporring, J. (eds.) MICCAI 2006. LNCS, vol. 4190, pp. 924–931. Springer, Heidelberg (2006). https://doi.org/10.1007/11866565_113
2. Avants, B.B., Epstein, C.L., Grossman, M., Gee, J.C.: Symmetric diffeomorphic image registration with cross-correlation: evaluating automated labeling of elderly and neurodegenerative brain. Media **12**(1), 26–41 (2008)
3. Balakrishnan, G., Zhao, A., Sabuncu, M.R., Guttag, J., Dalca, A.V.: VoxelMorph: a learning framework for deformable medical image registration. TMI **38**(8), 1788–1800 (2019)
4. Castillo, R., et al.: A reference dataset for deformable image registration spatial accuracy evaluation using the COPDgene study archive. Phys. Med. Biol. **58**(9), 2861 (2013)

5. Christensen, G.E., Johnson, H.J.: Consistent image registration. TMI **20**(7), 568–582 (2001)
6. Eade, E.: Lie groups for 2D and 3D transformations (2013). http://ethaneade.com/lie.pdf. Revised Dec 117, 118
7. Greer, H., Kwitt, R., Vialard, F.X., Niethammer, M.: ICON: learning regular maps through inverse consistency. In: ICCV (2021)
8. Hoffmann, M., Billot, B., Greve, D.N., Iglesias, J.E., Fischl, B., Dalca, A.V.: SynthMorph: learning contrast-invariant registration without acquired images. TMI **41**(3), 543–558 (2022)
9. Hoopes, A., Hoffmann, M., Greve, D.N., Fischl, B., Guttag, J., Dalca, A.V.: Learning the effect of registration hyperparameters with HyperMorph. arXiv preprint arXiv:2203.16680 (2022)
10. Iglesias, J.E.: EasyReg: a ready-to-use deep learning tool for symmetric affine and nonlinear brain MRI registration (2023)
11. Lie, S.: Theorie der transformationsgruppen i. Math. Ann. **16**, 441–528 (1880)
12. Marcus, D.S., Wang, T.H., Parker, J., Csernansky, J.G., Morris, J.C., Buckner, R.L.: Open access series of imaging studies (OASIS): cross-sectional MRI data in young, middle aged, nondemented, and demented older adults. J. Cogn. Neurosci. **19**(9), 1498–1507 (2007)
13. Mok, T.C.W., Chung, A.C.S.: Large deformation diffeomorphic image registration with Laplacian pyramid networks. In: Martel, A.L., et al. (eds.) MICCAI 2020. LNCS, vol. 12263, pp. 211–221. Springer, Cham (2020). https://doi.org/10.1007/978-3-030-59716-0_21
14. Mok, T.C., Chung, A.C.: Fast symmetric diffeomorphic image registration with convolutional neural networks. In: Proceedings of the IEEE/CVF Conference on Computer Vision and Pattern Recognition (CVPR) (2020)
15. Nazib, A., Fookes, C., Salvado, O., Perrin, D.: A multiple decoder CNN for inverse consistent 3D image registration. In: 2021 IEEE 18th International Symposium on Biomedical Imaging (ISBI), pp. 904–907 (2021). https://doi.org/10.1109/ISBI48211.2021.9433911
16. Nevitt, M.C., Felson, D.T., Lester, G.: The osteoarthritis initiative. Protocol Cohort Study **1** (2006)
17. Petersen, R., et al.: Alzheimer's disease neuroimaging initiative (ADNI): clinical characterization. Neurology **74**(3), 201–209 (2010). https://doi.org/10.1212/WNL.0b013e3181cb3e25
18. Regan, E.A., et al.: Genetic epidemiology of COPD (COPDGene) study design. COPD: J. Chronic Obstr. Pulm. Dis. **7**(1), 32–43 (2011)
19. Reuter, M., Rosas, H.D., Fischl, B.: Highly accurate inverse consistent registration: a robust approach. NeuroImage **53**(4), 1181–1196 (2010). https://doi.org/10.1016/j.neuroimage.2010.07.020. https://www.sciencedirect.com/science/article/pii/S1053811910009717
20. Rushmore, R.J., et al.: Anatomically curated segmentation of human subcortical structures in high resolution magnetic resonance imaging: an open science approach. Front. Neuroanat. **16** (2022)
21. Rushmore, R.J., et al.: HOA-2/SubcorticalParcellations: release-50-subjects-1.1.0 (2022). https://doi.org/10.5281/zenodo.7080547
22. Shen, Z., Han, X., Xu, Z., Niethammer, M.: Networks for joint affine and non-parametric image registration. In: CVPR (2019)
23. Tian, L., et al.: GradICON: approximate diffeomorphisms via gradient inverse consistency (2022). https://doi.org/10.48550/ARXIV.2206.05897

24. Van Essen, D.C., et al.: The human connectome project: a data acquisition perspective. Neuroimage **62**(4), 2222–2231 (2012)
25. Vishnevskiy, V., Gass, T., Szekely, G., Tanner, C., Goksel, O.: Isotropic total variation regularization of displacements in parametric image registration. TMI **36**(2), 385–395 (2017)
26. Wang, D., et al.: PLOSL: population learning followed by one shot learning pulmonary image registration using tissue volume preserving and vesselness constraints. Media **79**, 102434 (2022)
27. Zhang, J.: Inverse-consistent deep networks for unsupervised deformable image registration. CoRR abs/1809.03443 (2018). http://arxiv.org/abs/1809.03443

X2Vision: 3D CT Reconstruction from Biplanar X-Rays with Deep Structure Prior

Alexandre Cafaro[1,2,3]($\boxtimes$), Quentin Spinat[1], Amaury Leroy[1,2,3],
Pauline Maury[2], Alexandre Munoz[3], Guillaume Beldjoudi[3], Charlotte Robert[2],
Eric Deutsch[2], Vincent Grégoire[3], Vincent Lepetit[4], and Nikos Paragios[1]

[1] TheraPanacea, Paris, France
`a.cafaro@therapanacea.eu`
[2] Gustave Roussy, Inserm 1030, Paris-Saclay University, Villejuif, France
[3] Department of Radiation Oncology, Centre Léon Bérard, Lyon, France
[4] LIGM, Ecole des Ponts, Univ Gustave Eiffel, CNRS, Paris, France

Abstract. We propose an unsupervised deep learning method to reconstruct a 3D tomographic image from biplanar X-rays, to reduce the number of required projections, the patient dose, and the acquisition time. To address this ill-posed problem, we introduce prior knowledge of anatomic structures by training a generative model on 3D CTs of head and neck. We optimize the latent vectors of the generative model to recover a volume that both integrates this prior knowledge and ensures consistency between the reconstructed image and input projections. Our method outperforms recent methods in terms of reconstruction error while being faster and less radiating than current clinical workflow. We evaluate our method in a clinical configuration for radiotherapy.

Keywords: Image reconstruction · Inverse problem · Sparse sampling · Deep generative model · CT

1 Introduction

Tomographic imaging estimates body density using hundreds of X-ray projections, but it's slow and harmful to patients. Acquisition time may be too high for certain applications, and each projection adds dose to the patient. A quick, low-cost 3D estimation of internal structures using only bi-planar X-rays can revolutionize radiology, benefiting dental imaging, orthopedics, neurology, and more. This can improve image-guided therapies and preoperative planning, especially for radiotherapy, which requires precise patient positioning with minimal radiation exposure.

However, this task is an ill-posed inverse problem: X-ray measurements are the result of attenuation integration across the body, which makes them very

Supplementary Information The online version contains supplementary material available at https://doi.org/10.1007/978-3-031-43999-5_66.

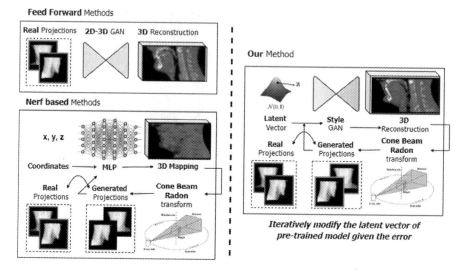

Fig. 1. Current methods vs our method. Feed-forward methods do not manage to predict a detailed and matching tomographic volume from a few projections. Iterative methods based on neural radiance fields lack prior for good reconstruction. By learning an embedding for the possible volumes, we can recover an accurate volume from very few projections with an optimization based on a Bayesian formulation.

ambiguous. Traditional reconstruction methods require hundreds of projections to get sufficient constraints on the internal structures. With very few projections, it is very difficult to disentangle the structures for even coarse 3D estimation. In other words, many 3D volumes may have generated such projections *a priori*.

Classical analytical and iterative methods [8] fail when very few projections are available. Several works have attempted to largely decrease the number of projections needed for an accurate volumetric reconstruction. Some deep learning methods [7,12,24,25,30] predict directly a 3D volume in a forward way from very few projections. The volume is however not guaranteed to be consistent with the projections and it is not clear which solution is retrieved. Other recent methods have adapted NeRFs [20] to tomographic reconstruction [23,31]. These non-learning methods show good results when the number of input projections remains higher than a dozen but fail when very few projections are provided, as our experiments in Sect. 3.3 show.

As illustrated in Fig. 1, to be able to reconstruct a volume accurately given as low as two projections only, we first learn a prior on the volume. To do this, we leverage the potential of generative models to learn a low-dimensional manifold of the target body part. Given projections, we find by a Bayesian formulation the intermediate latent vectors conditioning the generative model that minimize the error between synthesized projections of our reconstruction and these input projections. Our work builds on Hong et al. [10]'s 3D style-based generative model, which we extend via a more complex network and training framework.

Compared to other 3D GANs, it is proven to provide the best disentanglement of the feature space related to semantic features [2].

By contrast with feed-forward methods, our approach does not require paired projections-reconstructions, which are very tedious to acquire, and it can be used with different numbers of projections and different projection geometries without retraining. Compared to NeRF-based methods, our method exploits prior knowledge from many patients to require only two projections. We evaluate our method on reconstructing cancer patients' head-and-neck CTs, which involves intricate and complicated structures. We perform several experiments to compare our method with a feed-forward-based method [30] and a recent NeRF-based method [23], which are the previous state-of-the-art methods for the very few or few projections cases, respectively.

We show that our method allows to retrieve results with the finest reconstructions and better matching structures, for a variety of number of projections. To summarize, our contributions are two-fold: (i) A new paradigm for 3D reconstruction with biplanar X-rays: instead of learning to invert the measurements, we leverage a 3D style-based generative model to learn deep image priors of anatomic structures and optimize over the latent space to match the input projections; (ii) A novel unsupervised method, fast and robust to sampling ratio, source energy, angles and geometry of projections, all of which making it general for downstream applications and imaging systems.

2   Method

Figure 2 gives an overview of the pipeline we propose. We first learn the low-dimensional manifold of CT volumes of a target body region. At inference, we estimate the Maximum A Posteriori (MAP) volume on this manifold given very few projections: we find the latent vectors that minimize the error between the synthetic projections from the corresponding volume on the manifold and the real ones. In this section, we formalize the problem, describe how we learn the manifold, and detail how we optimize the latent vectors.

2.1   Problem Formulation

Given a small set of projections $\{I_i\}_i$, possibly as few as two, we would like to reconstruct the 3D tomographic volume v that generates these projections. This is a hard ill-posed problem, and to solve it, we need prior knowledge about the possible volumes. To do this, we look for the maximum *a posteriori* (MAP) estimate given the projections $\{I_i\}_i$:

$$v^* = \operatorname*{argmax}_{v} p(v|\{I_i\}_i) = \operatorname*{argmax}_{v} p(v)p(\{I_i\}_i|v) = \operatorname*{argmin}_{v} \sum_i \mathcal{L}(v, I_i) + R(v) \,. \tag{1}$$

Term $\mathcal{L}(v, I_i)$ is a log-likelihood. We take it as:

$$\mathcal{L}(v, I_i) = \lambda_2 \|A_i \circ v - I_i\|_2 + \lambda_p \mathcal{L}_p(A_i \circ v, I_i) \,, \tag{2}$$

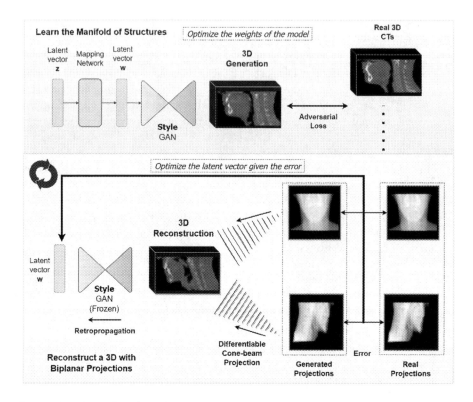

Fig. 2. Our pipeline. We first learn the low-dimensional manifold of 3D structures using a generative model. Then, given projections, we find the latent vectors that minimize the error between the projections of our generation and the input projections.

where A_i is an operator that projects volume v under view i. We provide more details about operator A in Sect. 2.3. $\mathcal{L}_p$ is the perceptual loss [13] between projection of v and the observed projection I_i. Term $R(v)$ is a regularization term. It is crucial as it is the term that embodies prior knowledge about the volume to reconstruct. As discussed in the introduction, we rely on a generative model, which we describe in the next section. Then, we describe how exactly we use this generative model for regularization term $R(v)$ and how this changes our optimization problem.

2.2   Manifold Learning

To regularize the domain space of solutions, we leverage a style-based generative model to learn deep priors of anatomic structures. Our model relies on Style-GAN2 [15] that we extend in 3D by changing the 2D convolutions into 3D ones as done in 3DStyleGAN [10] except that we start from the StyleGAN2 architecture.

Our generator G generates a volume v given a latent vector $\mathbf{w}$ and Gaussian noise vectors $\mathbf{n} = \{\mathbf{n}_j\}_j$: $v = G(\mathbf{w}, \mathbf{n})$. Latent vector $\mathbf{w} \in \mathcal{N}(\mathbf{w}|\mu, \sigma)$ is computed from an initial latent vector $\mathbf{z} \in \mathcal{N}(0, I)$ mapped using a learned network m: $\mathbf{w} = m(\mathbf{z})$. $\mathbf{w}$ controls the global structure of the predicted volumes at different scales by its components $\mathbf{w}_i$, while the noise vectors $\mathbf{n}$ allow more fine-grained details. The mean μ and standard deviation σ of the mapped latent space can be computed by mapping over initial latent space $\mathcal{N}(0, I)$ after training. The mapping network learns to disentangle the initial latent space relatively to semantic features which is crucial for the inverse problem. We train this model using the non-saturating logistic loss [5] and path length regularization [15]. For the discriminator, we use the non-saturating logistic loss with R1 regularization [19]. We implement adaptive discriminator augmentation from StyleGAN-ADA [14] to improve learning of the model's manifold with limited medical imaging data.

2.3   Reconstruction from Biplanar Projections

Since our generative model provides a volume v as a function of vectors $\mathbf{w}$ and $\mathbf{n}$, we can reparameterize our optimization from Eq. (1) into:

$$\mathbf{w}^*, \mathbf{n}^* = \underset{\mathbf{w}, \mathbf{n}}{\operatorname{argmin}} \sum_i \mathcal{L}(G(\mathbf{w}, \mathbf{n}), I_i) + R(\mathbf{w}, \mathbf{n}) . \tag{3}$$

Note that by contrast with [18] for example, we optimize on the noise vectors $\mathbf{n}$ as well: as we discovered in our early experiments, the $\mathbf{n}$ are also useful to embed high-resolution details. We take our regularization term $R(\mathbf{w}, \mathbf{n})$ as:

$$R(\mathbf{w}, \mathbf{n}) = \lambda_w \mathcal{L}_w(\mathbf{w}) + \lambda_c \mathcal{L}_c(\mathbf{w}) + \lambda_n \mathcal{L}_n(\mathbf{n}) . \tag{4}$$

Term $\mathcal{L}_w(\mathbf{w}) = -\sum_k \log \mathcal{N}(\mathbf{w}_k|\mu, \sigma)$ ensures that $\mathbf{w}$ lies on the same distribution as during training. $\mathcal{N}(\cdot|\mu, \sigma)$ represents the density of the standard normal distribution of mean μ and standard deviation σ.

Term $\mathcal{L}_c(\mathbf{w}) = -\sum_{i,j} \log \mathcal{M}(\theta_{i,j}|0, \kappa)$ encourages the $\mathbf{w}_i$ vectors to be collinear so to keep the generation of coarse-to-fine structures coherent. $\mathcal{M}(\cdot; \mu, \kappa)$ is the density of the Von Mises distribution of mean μ and scale κ, which we take fixed, and $\theta_{i,j} = \arccos(\frac{\mathbf{w}_i \cdot \mathbf{w}_j}{\|\mathbf{w}_i\| \|\mathbf{w}_j\|})$ is the angle between vectors $\mathbf{w}_i$ and $\mathbf{w}_j$.

Term $\mathcal{L}_n(\mathbf{n}) = -\sum_j \log \mathcal{N}(\mathbf{n}_j|\mathbf{0}, I)$ ensures that the $\mathbf{n}_j$ lie on the same distribution as during training, $i.e.$, a multivariate standard normal distribution. The λ_* are fixed weights.

Projection Operator. In practice, we take operator A as a 3D cone beam projection that simulates X-ray attenuation across the patient, adapted from [21,27]. We model a realistic X-ray attenuation as a ray tracing projection using material and spectrum awareness:

$$\mathcal{I}_{\text{atten}} = \sum_E \mathcal{I}_0 e^{-\sum_m \mu(m,E)t_m} , \tag{5}$$

with $\mu(m, E)$ the linear attenuation coefficient of material m at energy state E that is known [11], t_m the material thickness, $\mathcal{I}_0$ the intensity of the source X-ray.

For materials, we consider the bones and tissues that we separate by threshold on electron density. A inverts the attenuation intensities $\mathcal{I}_{\text{atten}}$ to generate an X-ray along few directions successively. We make A differentiable using [21] to allow end-to-end optimization for reconstruction.

3   Experiments and Results

3.1   Dataset and Preprocessing

Manifold Learning. We trained our model with a large dataset of 3500 CTs of patients with head-and-neck cancer, more exactly 2297 patients from the publicly available The Cancer Imaging Archive (TCIA) [1,6,16,17,28,32] and 1203 from private internal data, after obtention of ethical approbations. We split this data into 3000 cases for training, 250 for validation, and 250 for testing. We focused CT scans on the head and neck region above shoulders, with a resolution of $80 \times 96 \times 112$, and centered on the mouth after automatic segmentation using a pre-trained U-Net [22]. The CTs were preprocessed by min-max normalization after clipping between -1024 and 2000 Hounsfield Units (HU).

3D Reconstruction. To evaluate our approach, we used an external private cohort of 80 patients who had undergone radiotherapy for head-and-neck cancer, with their consent. Planning CT scans were obtained for dose preparation, and CBCT scans were obtained at each treatment fraction for positioning with full gantry acquisition. As can be seen in Fig. 3 and the supplementary material, all these cases are challenging as there are large changes between the original CT scan and the CBCT scans. We identified these cases automatically by comparing the CBCTs with the planning CTs. To compare our reconstruction in the calibrated HU space, we registered the planning CTs on the CBCTs by deformable registration with MRF minimization [4]. We hence obtained 3D volumes as virtual CTs we considered as ground truths for our reconstructions after normalization. From these volumes, we generated projections using the projection module described in Sect. 2.3.

3.2   Implementation Details

Manifold Learning. We used Pytorch to implement our model, based on Style-GAN2 [15]. It has a starting base layer of $256 \times 5 \times 6 \times 7$ and includes four upsamplings with 3D convolutions and filter maps of $256, 128, 64, 32$. We also used 8 fully-convolutional layers with dimension 512 and an input latent vector of dimension 512, with tanh function as output activation. To optimize our model, we used lazy regularization [15] and style mixing [15], and added a 0.2 probability for generating images without Gaussian noise to focus on embedding the most information. We augmented the discriminator with vertical and depth-oriented flips, rotation, scaling, motion blur and Gaussian noise at a probability of 0.2. Our training used mixed precision on a single GPU Nvidia Geforce GTX

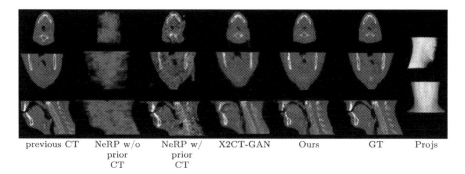

previous CT   NeRP w/o   NeRP w/   X2CT-GAN   Ours   GT   Projs
                  prior   prior
                   CT   CT

Fig. 3. Visual comparison of 3D reconstruction from biplanar projections by our model and baselines. Without a previous CT volume, NeRP fails by lack of constraints. When initialized with an earlier CT (left), NeRP tends to create artefacts to match the projections rather than really change the anatomy. Our method produces better matching structures than X2CT-GAN, almost matching the CT volume deformed on the CBCT volume (GT, right).

3090 with a batch size of 6, and we optimized the generator, discriminator, and mapping networks using Adam at learning rates $6e-5$ and $1e-5$ to avoid mode collapse and unstable training. After training for 4 weeks, we achieved stabilization of the Fréchet Inception Distance (FID) [9] and Multi-scale Structural Similarity (MS-SSIM) [29] on the validation set.

3D Reconstruction. For the reconstruction, we performed the optimization on GPU V100 PCI-E using Adam, with learning rate of $1e-3$. By grid search on the validation set, we selected the best weights that well balance between structure and fine-grained details, $\lambda_2 = 10$, $\lambda_p = 0.1$, $\lambda_w = 0.1$, $\lambda_c = 0.05$, $\lambda_n = 10$. We perform 100 optimization steps starting from the mean of the mapped latent space, which takes 25 s, enabling clinical use.

3.3   Results and Discussion

Manifold Learning. We tested our model's ability to learn the low-dimensional manifold. We used FID [9] to measure the distance between the distribution of generated volumes and real volumes, and MS-SSIM [29] to evaluate volumes' diversity and quality. We obtained a 3D FID of 46 and a MS-SSIM of 0.92. For reference, compared to 3DStyleGAN [10], our model achieved half their FID score on another brain MRI dataset, with comparable MS-SSIM. This may be due to a more complex architecture, discriminator augmentation, or simpler anatomy.

Baselines. We compared our method against the main feed-forward method X2CT-GAN [30] and the neural radiance fields with prior image embedding method NeRP [23] meant for modest sparsely-sampled reconstruction. Recent methods like [24] and [12] were excluded because they provide only minor

Table 1. Metrics for our method and baselines, for reconstruction from 1 to 8 cone beam projections. Standard deviations are provided in parentheses.

Method	1 Projection		2 Projections	
	PSNR (dB)↑	SSIM↑	PSNR (dB)↑	SSIM↑
NeRP (w/o prior volume)	14.8 (±2.7)	0.12 (±0.10)	18.4 (±3.8)	0.17 (±0.10)
NeRP (w/ prior volume)	22.5 (±3.2)	0.29 (±0.07)	23.5 (±3.5)	0.30 (±0.06)
X2CT-GAN	20.7 (±2.4)	0.57 (±0.07)	21.8 (±2.5)	0.72 (±0.08)
Ours	**23.2** (±**2.8**)	**0.79** (±**0.09**)	**25.8** (±**3.2**)	**0.85** (±**0.10**)
	4 Projections		8 Projections	
NeRP (w/o prior volume)	19.9 (±2.6)	0.21 (±0.04)	20.0 (±2.5)	0.23 (±0.05)
NeRP (w/ prior volume)	24.2 (±2.7)	0.32 (±0.05)	24.9 (±4.9)	0.34 (±0.08)
Ours	**28.2** (±**3.5**)	**0.89** (±**0.10**)	**30.1** (±**3.9**)	**0.92** (±**0.11**)

improvements compared to X2CT-GAN [30] and have similar constraints to feed-forward methods. Additionally, no public implementation is available. [26] uses a flow-based generative model, but the results are of lower quality compared to GANs and similar to X2CT-GAN [30].

3D Reconstruction. To evaluate our method's performance with biplanar projections, we focused on positioning imaging for radiotherapy. Figure 3 compares our reconstruction with those of the baselines from biplanar projections. Our method achieves better fitting of the patient structure, including bones, tissues, and air separations, almost matching the real CT volume. X2CT-GAN [30] produced realistic structures, but failed to match the actual structures as it does not enforce consistency with the projections. In some clinical procedures, an earlier CT volume of the patient may be available and can be used as an additional input for NeRP [23]. Without a previous CT volume, NeRP lacks the necessary prior to accurately solve the ill-posed problem. Even when initialised with a previous CT volume, NeRP often fails to converge to the correct volume and introduces many artifacts when few projections are used. In contrast, our method is more versatile and produces better results.

We used quantitative metrics (PSNR and SSIM) to evaluate reconstruction error and human perception, respectively. Table 1 shows these metrics for our method and baselines with 1 to 8 cone beam projections. Deviation from projections, as in X2CT-GAN, leads to inaccurate reconstruction. However, relying solely on projection consistency is inadequate for this ill-posed problem. NeRP matches projections but cannot reconstruct the volume correctly. Our approach balances between instant and iterative methods by providing a reconstruction in 25 s with 100 optimization steps, while ensuring maximal consistency. In contrast, NeRP requires 7 min, and X2CT-GAN produces structures instantly but unmatching. Clinical CBCT acquisition and reconstruction by FDK [3] take about 1–2 min and 10 s respectively. Our approach significantly reduces clinical time and radiation dose by using instant biplanar projections, making it promising for fast 3D visualization towards complex positioning.

4 Conclusion, Limitations, and Future Work

We proposed a new unsupervised method for 3D reconstruction from biplanar X-rays using a deep generative model to learn the structure manifold and retrieve the maximum a posteriori volume with the projections, leading to state-of-the-art reconstruction. Our approach is fast, robust, and applicable to various human body parts, making it suitable for many clinical applications, including positioning and visualization with reduced radiation.

Future hardware improvements may increase resolution, and our approach could benefit from other generative models like latent diffusion models. This approach may provide coarse reconstructions for patients with rare abnormalities, as most learning methods, but a larger dataset or developing a prior including tissue abnormalities could improve robustness.

References

1. Beichel, R.R., et al.: Data from QIN-HEADNECK (2015)
2. Ellis, S., et al.: Evaluation of 3D GANs for Lung Tissue Modelling in Pulmonary CT. arXiv (2022)
3. Feldkamp, L.A., Davis, L.C., Kress, J.W.: Practical cone-beam algorithm. J. Opt. Soc. Am. A-Opt. Image Sci. Vis. (1984)
4. Glocker, B., Komodakis, N., Tziritas, G., Navab, N., Paragios, N.: Dense image registration through MRFs and efficient linear programming. Med. Image Anal. **12**(6), 731–741 (2008)
5. Goodfellow, I., et al.: Generative adversarial networks. Commun. ACM **63**(11) (2020)
6. Grossberg, A., et al.: Anderson Cancer Center Head and Neck Quantitative Imaging Working Group. HNSCC (2020)
7. Henzler, P., Rasche, V., Ropinski, T., Ritschel, T.: Single-image tomography: 3D volumes from 2D cranial X-rays. In: Computer Graphics Forum (2018)
8. Herman, G.T.: Fundamentals of Computerized Tomography: Image Reconstruction from Projections. Springer, London (2009). https://doi.org/10.1007/978-1-84628-723-7
9. Heusel, M., Ramsauer, H., Unterthiner, T., Nessler, B., Hochreiter, S.: GANs trained by a two time-scale update rule converge to a local nash equilibrium. In: NeurIPS (2017)
10. Hong, S., et al.: 3D-StyleGAN: a style-based generative adversarial network for generative modeling of three-dimensional medical images. In: Engelhardt, S., et al. (eds.) DGM4MICCAI/DALI -2021. LNCS, vol. 13003, pp. 24–34. Springer, Cham (2021). https://doi.org/10.1007/978-3-030-88210-5_3
11. Hubbell, J.H.: Tables of X-Ray Mass Attenuation Coefficients 1 keV to 20 MeV for Elements Z=1 to 92 and 48 Additional Substance of Dosimetric Interest. NISTIR 5632 (1995)

12. Jiang, Y.: MFCT-GAN: multi-information network to reconstruct CT volumes for security screening. J. Intell. Manuf. Spec. Equip. **3**(1), 17–30 (2022)
13. Johnson, J., Alahi, A., Fei-Fei, L.: Perceptual losses for real-time style transfer and super-resolution. In: Leibe, B., Matas, J., Sebe, N., Welling, M. (eds.) ECCV 2016. LNCS, vol. 9906, pp. 694–711. Springer, Cham (2016). https://doi.org/10.1007/978-3-319-46475-6_43
14. Karras, T., Aittala, M., Hellsten, J., Laine, S., Lehtinen, J., Aila, T.: Training generative adversarial networks with limited data. In: NeurIPS (2020)
15. Karras, T., Laine, S., Aittala, M., Hellsten, J., Lehtinen, J., Aila, T.: Analyzing and improving the image quality of stylegan. In: CVPR (2020)
16. Kinahan, P., Muzi, M., Bialecki, B., Coombs, L.: Data from the ACRIN 6685 Trial HNSCC-FDG-PET/CT (2020)
17. Kwan, J.Y.Y., et al.: Data from Radiomic Biomarkers to Refine Risk Models for Distant Metastasis in Oropharyngeal Carcinoma (2019)
18. Marinescu, R.V., Moyer, D., Golland, P.: Bayesian Image Reconstruction Using Deep Generative Models. arXiv (2020)
19. Mescheder, L., Geiger, A., Nowozin, S.: Which training methods for GANs do actually converge? In: ICML (2018)
20. Mildenhall, B., Srinivasan, P.P., Tancik, M., Barron, J.T., Ramamoorthi, R., Ng, R.: NeRF: representing scenes as neural radiance fields for view synthesis. Commun. ACM **65**(1) (2021)
21. Peng, C., Liao, H., Wong, G., Luo, J., Zhou, S.K., Chellappa, R.: XraySyn: realistic view synthesis from a single radiograph through CT priors. In: AAAI (2021)
22. Ronneberger, O., Fischer, P., Brox, T.: U-Net: convolutional networks for biomedical image segmentation. In: Navab, N., Hornegger, J., Wells, W.M., Frangi, A.F. (eds.) MICCAI 2015. LNCS, vol. 9351, pp. 234–241. Springer, Cham (2015). https://doi.org/10.1007/978-3-319-24574-4_28
23. Shen, L., Pauly, J., Xing, L.: NeRP: implicit neural representation learning with prior embedding for sparsely sampled image reconstruction. IEEE Trans. Neural Netw. (2022)
24. Shen, L., Zhao, W., Capaldi, D., Pauly, J., Xing, L.: A geometry-informed deep learning framework for ultra-sparse 3D tomographic image reconstruction. Comput. Biol. Med. **148**, 105710 (2022)
25. Shen, L., Zhao, W., Xing, L.: Patient-specific reconstruction of volumetric computed tomography images from a single projection view via deep learning. Nature **3**(11), 880–888 (2019)
26. Shibata, H., et al.: On the simulation of ultra-sparse-view and ultra-low-dose computed tomography with maximum a posteriori reconstruction using a progressive flow-based deep generative model. Tomography **8**(5), 2129–2152 (2022)
27. Unberath, M., et al.: DeepDRR – a catalyst for machine learning in fluoroscopy-guided procedures. In: Frangi, A.F., Schnabel, J.A., Davatzikos, C., Alberola-López, C., Fichtinger, G. (eds.) MICCAI 2018. LNCS, vol. 11073, pp. 98–106. Springer, Cham (2018). https://doi.org/10.1007/978-3-030-00937-3_12
28. Vallières, M., et al.: Data from Head-Neck-PET-CT (2020)
29. Wang, Z., Simoncelli, E.P., Bovik, A.C.: Multiscale structural similarity for image quality assessment. In: The Thrity-Seventh Asilomar Conference on Signals, Systems & Computers (2003)
30. Ying, X., Guo, H., Ma, K., Wu, J., Weng, Z., Zheng, Y.: X2CT-GAN: reconstructing CT from biplanar X-rays with generative adversarial networks. In: CVPR (2019)

31. Zha, R., Zhang, Y., Li, H.: NAF: neural attenuation fields for sparse-view CBCT reconstruction. In: Wang, L., Dou, Q., Fletcher, P.T., Speidel, S., Li, S. (eds.) MICCAI 2022. LNCS, vol. 13436, pp. 442–452. Springer, Cham (2022). https://doi.org/10.1007/978-3-031-16446-0_42
32. Zuley, M.L., et al.: The Cancer Genome Atlas Head-Neck Squamous Cell Carcinoma Collection TCGA-HNSC) (2015)

Fast Reconstruction for Deep Learning PET Head Motion Correction

Tianyi Zeng[1], Jiazhen Zhang[2], Eléonore V. Lieffrig[1], Zhuotong Cai[1],
Fuyao Chen[2], Chenyu You[3], Mika Naganawa[1], Yihuan Lu[5],
and John A. Onofrey[1,2,4(✉)]

[1] Department of Radiology and Biomedical Imaging, Yale University, New Haven, CT, USA
{tianyi.zeng,jiazhen.zhang,eleonore.lieffrig,zhuotong.cai,
mika.naganawa,fuyao.chen,chenyu.you,john.onofrey}@yale.edu
[2] Department of Biomedical Engineering, Yale University, New Haven, CT, USA
[3] Department of Electrical Engineering, Yale University, New Haven, CT, USA
[4] Department of Urology, Yale University, New Haven, CT, USA
[5] United Imaging Healthcare, Shanghai, China
yihuan.lu@united-imaging.com

Abstract. Head motion correction is an essential component of brain PET imaging, in which even motion of small magnitude can greatly degrade image quality and introduce artifacts. Building upon previous work, we propose a new head motion correction framework taking fast reconstructions as input. The main characteristics of the proposed method are: (i) the adoption of a high-resolution short-frame fast reconstruction workflow; (ii) the development of a novel encoder for PET data representation extraction; and (iii) the implementation of data augmentation techniques. Ablation studies are conducted to assess the individual contributions of each of these design choices. Furthermore, multi-subject studies are conducted on an ^{18}F-FPEB dataset, and the method performance is qualitatively and quantitatively evaluated by MOLAR reconstruction study and corresponding brain Region of Interest (ROI) Standard Uptake Values (SUV) evaluation. Additionally, we also compared our method with a conventional intensity-based registration method. Our results demonstrate that the proposed method outperforms other methods on all subjects, and can accurately estimate motion for subjects out of the training set. All code is publicly available on GitHub: https://github.com/OnofreyLab/dl-hmc_fast_recon_miccai2023.

Keywords: Deep Learning · PET fast reconstruction · Data-driven motion correction · Brain PET

1 Introduction

Positron emission tomography (PET) has been widely used in human brain imaging, thanks to the availability of a vast array of specific radiotracers. These

---

Supplementary Information The online version contains supplementary material available at https://doi.org/10.1007/978-3-031-43999-5_67.

H. Greenspan et al. (Eds.): MICCAI 2023, LNCS 14229, pp. 710–719, 2023.
https://doi.org/10.1007/978-3-031-43999-5_67

compounds allow for studying various neurotransmitters and receptor dynamics for different brain targets [11]. Brain PET images are commonly used to diagnose and monitor neurodegenerative diseases, such as Alzheimer's disease, Parkinson's disease, epilepsy, and certain types of brain tumors [3]. Head motion in PET imaging reduces brain image resolution, lowers tracer distribution estimation, and introduces attenuation correction (AC) mismatch artifacts [12]. Consequently, the capability to monitor and correct head motion is of utmost importance in brain PET studies.

The first step of PET head motion correction is motion tracking. When head motion information is acquired, either frame-based motion correction or event-by-event (EBE) motion correction methods can be applied in the reconstruction workflow to derive motion-free PET images. EBE motion correction provides better results for real-time motion tracking compared to frame-based methods, as the latter does not allow for correction of motion that occurs within each dynamic frame [1]. Currently, there are two main categories of head motion tracking methods, hardware-based motion tracking (HMT) and data-driven methods. For HMT, head motion is obtained from external devices. Generally, HMT systems offer accurate tracking results with high time resolution. Marker-based HMT such as Polaris Vicra (NDI, Canada) use light-reflecting markers on the patient's head and track the markers for motion correction [6]. However, Vicra is not routinely used in the clinic, as setup and calibration of the tracking device can be complicated and attaching markers to each patient increases the logistical burden of the scan. In response, some researchers began to use markerless motion tracking systems for brain PET [4,13]. These methods typically rely on the use of cameras and computer vision algorithms to detect and analyze the movement of a person's head in real-time, but these methods still require additional hardware setup. In data-driven motion tracking methods, head motion is estimated from PET reconstructions or raw data. With the development of commercial PET systems and technological advancements such as time of flight (TOF), data-driven head PET motion tracking has shown promising results in reducing motion artifacts and improving image quality. For instance, [12] developed a novel data-driven head motion detection method based on the centroid of distribution (COD) of PET 3D point cloud image (PCI). Image registration methods that seek to align two or more images offer a data-driven solution for correcting head motion. Intensity-based registration methods have been used to track head motion using good-quality PET reconstruction frames to achieve stable performance [14]. However, because of the dynamic change in PET images, current registration-based methods need to split the data into several discrete time frames, e.g., 5 min. Therefore, they will introduce a cumulative error when dealing with inter-frame motion. Finally, inspired by the development of deep learning-based registration methods, a deep learning head motion correction (DL-HMC) network using Vicra as ground truth was proposed [15]. This study achieved accurate motion tracking on single subject testing data, but showed less accurate motion predictions for multi-subject motion studies. Meanwhile, the input images were low-resolution PCIs without TOF and had large voxel spacing, which can negatively affect motion tracking accuracy.

In this study, we proposed a new method to perform deep learning-based brain PET motion prediction across multiple subjects by utilizing high-resolution one-second fast reconstruction images (FRIs) with TOF. A novel encoder and data augmentation strategy was also applied to improve model performance. Ablation studies were conducted to assess the individual contributions of key method components. Multi-subject studies were conducted on a dataset of 20 subject and its results were quantitatively and qualitatively evaluated by MOLAR reconstruction studies and corresponding brain Region of Interest (ROI) Standard Uptake Values (SUV) evaluation.

2 Methods

2.1 Data

We identified 20 ^{18}F-FPEB studies from a database of brain PET scans acquired on a conventional mCT scanner (Siemens, Germany) at the Yale PET center. All subjects are healthy controls and the mean activity injected was 3.90 ± 1.02 mCi. The mean translational and rotational motions across time and scans are 3.75 ± 6.88 mm and $3.30 \pm 8.77°$, respectively. PET list-mode data and Vicra motion tracking information are available for each subject, as well as T1-weighted MR images and MR-space to PET-space transformation matrices. We consider data acquired between 60 and 90 min post injection (30 min total).

2.2 Fast Reconstruction Images

To overcome the challenges when using low-quality, noisy PCI for motion correction, ultra-fast reconstruction techniques [14] that generate one-second dynamic fast reconstruction images (FRIs), can be utilized as input for deep learning motion correction methods. Leveraging the availability of CPU-parallel reconstruction platforms [5], we develop a reconstruction package for one-second FRI. Our proposed method employs a TOF-based projector and utilizes pure maximum-likelihood expectation maximization (MLEM) for reconstruction [10]. Attenuation correction (AC) and scatter correction are turned off to avoid AC mismatch and expensive computation. Normalization correction, random correction and decay correction are applied and the iteration number was set to two. Standard Uptake Value (SUV) calculation was also conducted to normalize the activities between different subjects. The final reconstructed image dimension is $150 \times 150 \times 111$, with voxels spaced at $2.04 \times 2.04 \times 2.00$ mm^3. For comparison purposes, we also computed the same resolution PCI by TOF back-projection of the PET list-mode data along the line-of-response (LOR) with normalization for scanner sensitivity. Both FRI and PCI were resized to $96 \times 96 \times 64$, with voxel spacing of $3.18 \times 3.18 \times 3.45$ mm^3. As shown in Fig. 1, the resized FRI and PCI from the same time pairs are displayed. Due to the PET corrections, the FRI quality and noise level is superior to the PCI, particularly in areas outside of the head.

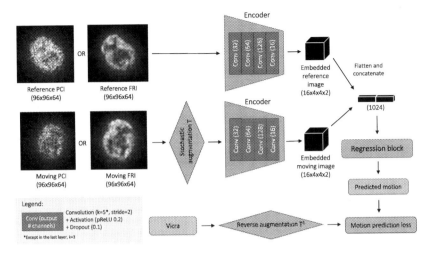

Fig. 1. The proposed network architecture and different network inputs comparison. The proposed network takes two one-second FRIs as input to encoder block with shared weights. Stochastic motion T from the data augmentation block is applied to the moving reconstruction, and the reverse augmentation T^{-1} is applied on corresponding Vicra data to match the augmented FRI pairs. A regression block then estimates the rigid motion transformation parameters. PCI input is shown on the left for visual comparison.

2.3   Motion Correction Framework

Network Architecture. We propose a modified version of the DL-HMC framework to learn rigid head motion in a supervised manner [15] (Fig. 1). Our proposed method uses two FRIs I_{ref} and I_{mov} from two different time points t_{ref} and t_{mov} to predict the relative rigid motion transformation between the reference and moving time points with respect to the Vicra gold-standard. Our encoder consists of 3 convolution layers with kernel size of 5^3 and an intermediate convolution layer with convolution size of 3^3. PReLU activation layers follow each convolution. In addition, we add dropout in the regression layers with rate 0.3. In this new architecture, the embedding space consists of feature maps of size $16 \times 4 \times 4 \times 2$, which preserves the spatial information in the FRI. No padding is applied to the images or feature maps. The extracted features are fed into the fully connected regression block to predict the six translation and rotation components of the relative rigid motion.

Data Augmentation. To improve the performance and generalizability of our network, we use a task-specific data augmentation strategy to expose it to more varied and diverse training data. As a rule of thumb, translations of 2–5 mm and rotations of $2°$–$3°$ are common and larger magnitudes are expected without restraint or if non-customized supports are used [9]. Due to our sampling strategy during model training, statistically, most of the sampled pairs will have small relative motion. However, during the inference, the relative motion between the moving frames at late time points and the reference frame at the beginning will be large due to the accumulation of motion. Therefore, the model may not be able to make accurate predictions when facing large relative motions. To take this problem into account, we perform data augmentation by simulating an additional relative motion that can be concatenated with the true relative motion. To be specific, the synthetic translation and rotation are uniformly sampled in the range of $[-10, 10]$ mm and $[-5, 5]°$, respectively. The randomly simulated motion T will be applied to the moving frame to generate a synthetic moving frame $T \circ I_{\text{ref}}$ and be concatenated with the real relative motion to acquire the synthetic relative motion between the reference and the synthetic moving frame. The synthetic moving frame and the synthetic relative motion will be used for training to increase the data variability.

Network Training and Inference. To train the network, we randomly sampled image pairs $(t_{\text{ref}}, t_{\text{mov}})$ under the condition $(t_{\text{ref}} < t_{\text{mov}})$. The network was optimized by minimizing the mean square error (MSE) between the predicted motion estimate and Vicra parameters. More specifically, the prediction error for a given pair of reference and moving clouds is defined as $\mathcal{L}(\hat{\theta}, \theta) = \|\hat{\theta} - \theta\|^2$ with $\theta = [t_x, t_y, t_z, r_x, r_y, r_z]$ the Vicra information for the three translational and three rotational parameters (t_x, t_y, t_z) and (r_x, r_y, r_z), respectively, and $\hat{\theta}$ the network prediction. After training the model, we perform motion tracking inference by setting the image from first time point $t_{\text{ref}} = 3{,}600$ (60 min post-injection) as the reference image and predict the motion from this reference image to all subsequent one-second image frames in the next 30 min (1,800 one-second time points).

3   Results

We performed quantitative and qualitative experiments to validate our approach. We evaluated motion correction performance by comparing our proposed method to DL-HMC [15], intensity-based registration using the BioImage Suite (BIS) software package [7] using the one-second FRIs, and ablation studies to demonstrate the effectiveness of our design choices. We qualitatively assessed motion correction performance by reconstructing the PET images with motion tracking result and comparing to DL-HMC and the Vicra gold-standard reconstruction results. We split our dataset of 20 subjects into distinct subsets for training and testing with 14 and 6 subjects, respectively. From the training cohort, we randomly sampled 10% of the time frames to be used as a validation set. Training the network required 6,000 epochs for convergence using a mini-batch size of 64. Adam optimization was used with initial learning rate set to 5e-4, γ set to 0.98, and exponential decay with a step size of 150. All computations were performed on a server with Intel Xeon Gold 5218 processors, 256 GB RAM, and an NVIDIA Quadro RTX 8000 GPU (48 GB RAM). The network was implemented in Python (v3.9) using PyTorch (v1.13.1) and MONAI (v1.0.1).

Table 1. Quantitative motion correction results. We compared our proposed approach using fast reconstruction image (FRI) as input with DL-HMC and with using point cloud image (PCI) as input. An ablation study quantifies the effect of stochastic data augmentation (DA) and to standard intensity-based registration (BIS). Reported values are MSE (mean $\pm$ SD) comparing motion estimates to Vicra gold-standard.

Method	Val. loss	Test Set		
		Total loss	Translation (mm)	Rotation (°)
DL-HMC PCI	0.60 ± 1.18	0.72 ± 0.18	1.05 ± 0.75	0.38 ± 0.43
DL-HMC FRI	0.33 ± 0.58	0.87 ± 0.42	1.14 ± 1.03	0.59 ± 1.03
Proposed PCI	0.35 ± 0.28	0.32 ± 0.22	$\mathbf{0.28 \pm 0.26}$	0.35 ± 0.31
Proposed FRI	$\mathbf{0.18 \pm 0.12}$	$\mathbf{0.30 \pm 0.19}$	0.35 ± 0.31	0.25 ± 0.19
Proposed w/o DA	0.20 ± 0.24	0.41 ± 0.25	0.61 ± 0.49	$\mathbf{0.21 \pm 0.20}$
BIS	N/A	8.40 ± 12.34	2.48 ± 2.35	1.15 ± 1.05

Quantitative Evaluation. For quantitative comparisons of motion tracking, we compare MSE loss in the validation set and test set. We calculate MSE for the 6 parameter rigid motion as well as the translation and rotational components separately. To verify feasibility of traditional intensity-based registration method on FRIs, we use BIS with a multi-resolution hierarchical representation (3 levels) and minimize the sum of squared differences (SSD) similarity metric (Fig. S1 shows a BIS result on a example testing subject). Compared to Vicra gold-standard, BIS fails to predict the motion. We evaluate the following motion prediction methods (Table 1): (i) DL-HMC with PCI as input (DL-HMC PCI); (ii) DL-HMC with FRI as input (DL-HMC FRI); (iii) proposed network with PCI as input (Proposed PCI); (iv) proposed network with FRI as input (Proposed FRI); and (v) proposed method with FRI but without the data augmentation module (Proposed w/o DA); Results demonstrate that the proposed network with FRI input provides the best motion tracking performance in both validation and testing data. We also observes that using FRI yields a lower loss for the proposed network, indicating that high image quality enhanced the motion correction performance. For testing translation results, Proposed PCI outperforms Proposed FRI and has similar total motion loss, which indicates that the proposed network can still estimate motion on testing subjects even with noisy input. Figure 2 shows motion prediction results for different variations of the proposed method in a single test subject. These results show that the proposed FRI method is more similar to Vicra than the other methods, especially for translation in the x and z directions. However, the Proposed FRI method exhibits higher variance than other methods, which may be a result of the data augmentation distribution. Overall, our experiments demonstrate that the strategies in proposed FRI method enhance the motion tracking performance of the network.

Qualitative Reconstruction Evaluation. After inference, the 6 rigid degrees of freedom transformation estimated from the network were used to reconstruct

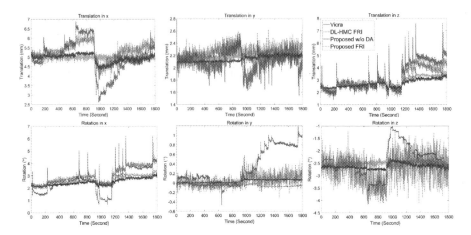

Fig. 2. Motion information comparison for testing subject inference. Columns show rigid transformation parameters from Vicra (Orange), DL-HMC FRI (yellow), proposed network without data augmentation (purple), and proposed FRI method (blue). (Color figure online)

the PET images using Motion-compensation OSEM List-mode Algorithm for Resolution-Recovery Reconstruction (MOLAR) algorithm [5]. We applied PET head motion correction using the Proposed FRI model and compared with Vicra and no motion correction (NMC) reconstruction results. Figure 3 shows the reconstruction results for the same testing subject in quantitative evaluation. Based on the tracer distribution of ^{18}F-PEB, we selected some frames from reconstructed images to illustrate the proposed FRI motion correction performance. In general, the Proposed FRI results in qualitatively enhanced anatomical interpretation of PET images. In Fig. 3, NMC reconstruction has motion blurring on margins of the brain, while Proposed FRI reconstruction shows well-defined gyrus and sulcus comparable to Vicra reconstruction. ^{18}F-FPEB tracer is a metabotropic glutamate 5 receptor antagonist with moderate to high accumulation in multiple brain regions such as insula/caudate nucleus, thalamus and temporal lobe. Thus, we compared visualization of these regions among NMC, Proposed FRI and Vicra reconstructions (Fig. 3). Specifically, our Proposed FRI method yields clear delineation of the insula/caudate nucleus, thalamus, and temporal lobe nearly indistinguishable from Vicra reconstructed images.

In addition, brain region of interest (ROI) analyses were also performed for quantitative use. Each subject's MR image was segmented into 74 regions using FreeSurfer software [2]. These regions were then merged into twelve large grey matter (GM) ROIs. For all testing subjects, the SUV difference of 12 ROIs from DL-HMC, Proposed PCI, and Proposed FRI reconstruction were calculated for comparison with Vicra reference (Fig. 3 right). The proposed FRI method yields the lowest absolute difference (0.8%) from Vicra images in SUV, while the absolute SUV difference for DL-HMC FRI is 1.5% and for NMC is 1.9%.

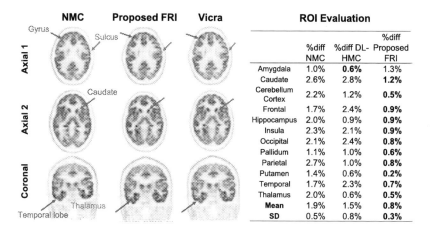

Fig. 3. PET image reconstruction and ROI evaluation results. MOLAR reconstructed images using Vicra gold-standard motion correction, the proposed FRI motion correction, and no motion correction (NMC). The table on the right shows quantitative SUV difference values with respect to Vicra.

The table from the figure:

	%diff NMC	%diff DL-HMC	%diff Proposed FRI
Amygdala	1.0%	0.6%	1.3%
Caudate	2.6%	2.8%	1.2%
Cerebellum Cortex	2.2%	1.2%	0.5%
Frontal	1.7%	2.4%	0.9%
Hippocampus	2.0%	0.9%	0.9%
Insula	2.3%	2.1%	0.9%
Occipital	2.1%	2.4%	0.8%
Pallidum	1.1%	1.0%	0.6%
Parietal	2.7%	1.0%	0.8%
Putamen	1.4%	0.6%	0.2%
Temporal	1.7%	2.3%	0.7%
Thalamus	2.0%	0.6%	0.5%
Mean	1.9%	1.5%	0.8%
SD	0.5%	0.8%	0.3%

Specifically, results of proposed FRI method are closest to Vicra results in regions such as thalamus, temporal lobe, insula (absolute activity difference are 0.5%, 0.7%, 0.9%, respectively), which are the target areas of ^{18}F-FPEB tracer with highest empirical tracer accumulation. The reconstruction and ROI evaluation results indicate that the proposed FRI method holds potential to improve clinical applicability through amelioration of PET motion correction accuracy.

4   Discussion and Conclusion

In this work, we propose a new head motion correction approach using fast reconstructions as input. The proposed method outperforms other methods in a multi-subject cohort, and ablation studies demonstrate the effectiveness of our strategies. We apply our proposed FRI motion correction to get motion-free reconstruction using MOLAR. The proposed FRI method achieves good image quality and similar ROI evaluation results compared to Vicra gold-standard HMT. In this study, we showed that conventional intensity-based registration fails at performing motion tracking on FRI data. This is likely due to the PET dynamic changes and non-optimal registration parameters. Compared with previous deep learning motion correction [15], the training speed and GPU memory usage of the proposed method are much better thanks to the proposed shallower encoder architecture and our efficient training and testing strategies. Though HMT method such as Vicra achieves good accuracy and time resolution for PET head motion tracking, two common types of Vicra failure may occur: slipping and wobbling. Our method would be robust enough to compensate for the

Vicra failure. Because of the limited Vicra data, in the future, we will develop semi-supervised deep learning methods for PET head motion correction. Our study used TOF PET data because it can yield high signal to noise ratio (SNR) for both FRI and PCI due to the better location identification of photons, thus the one-second FRI still retains some essential brain structures. Limitations of this work include partial limited tracking time and low time resolution compared to HMT methods mentioned in Sect. 1. In the future, with the development of PET techniques such as depth-of-interaction [8], higher resolution and sensitivity PET will be available. Such PET will give data-driven PET motion correction a revolutionary opportunity to have more accurate tracking and higher time resolution. We plan to apply the proposed method to other datasets, developing a generalized model for multi-tracer and multi-scanner PET data.

Acknowledgments. Research reported in this publication was supported by the National Institute Of Biomedical Imaging And Bioengineering (NIBIB) of the National Institutes of Health (NIH) under Award Number R21 EB028954. The content is solely the responsibility of the authors and does not necessarily represent the official views of the NIH.

References

1. Carson, R.E., Barker, W.C., Liow, J.S., Johnson, C.A.: Design of a motion-compensation OSEM list-mode algorithm for resolution-recovery reconstruction for the HRRT. In: 2003 IEEE Nuclear Science Symposium. Conference Record (IEEE Cat. No. 03CH37515), vol. 5, pp. 3281–3285. IEEE (2003)
2. Fischl, B.: Freesurfer. Neuroimage **62**(2), 774–781 (2012)
3. Hoffman, J.M., et al.: FDG PET imaging in patients with pathologically verified dementia. J. Nucl. Med. **41**(11), 1920–1928 (2000)
4. Iwao, Y., Akamatsu, G., Tashima, H., Takahashi, M., Yamaya, T.: Marker-less and calibration-less motion correction method for brain pet. Radiol. Phys. Technol. **15**(2), 125–134 (2022)
5. Jin, X., et al.: List-mode reconstruction for the biograph MCT with physics modeling and event-by-event motion correction. Phys. Med. Biol. **58**(16), 5567 (2013)
6. Jin, X., Mulnix, T., Sandiego, C.M., Carson, R.E.: Evaluation of frame-based and event-by-event motion-correction methods for awake monkey brain pet imaging. J. Nucl. Med. **55**(2), 287–293 (2014)
7. Joshi, A., et al.: Unified framework for development, deployment and robust testing of neuroimaging algorithms. Neuroinformatics **9**(1), 69–84 (2011)
8. Kuang, Z., et al.: Design and performance of SIAT APET: a uniform high-resolution small animal pet scanner using dual-ended readout detectors. Phys. Med. Biol. **65**(23), 235013 (2020)
9. Kyme, A.Z., Fulton, R.R.: Motion estimation and correction in SPECT, PET and CT. Phys. Med. Biol. **66**(18), 18TR02 (2021)
10. Lange, K., Carson, R., et al.: EM reconstruction algorithms for emission and transmission tomography. J. Comput. Assist. Tomogr. **8**(2), 306–16 (1984)
11. Phelps, M.E., Mazziotta, J.C.: Positron emission tomography: human brain function and biochemistry. Science **228**(4701), 799–809 (1985)

12. Revilla, E.M., et al.: Adaptive data-driven motion detection and optimized correction for brain pet. Neuroimage **252**, 119031 (2022)
13. Slipsager, J.M., et al.: Markerless motion tracking and correction for PET, MRI, and simultaneous PET/MRI. PLoS ONE **14**(4), e0215524 (2019)
14. Spangler-Bickell, M.G., Deller, T.W., Bettinardi, V., Jansen, F.: Ultra-fast list-mode reconstruction of short pet frames and example applications. J. Nucl. Med. **62**(2), 287–292 (2021)
15. Zeng, T., et al.: Supervised deep learning for head motion correction in pet. In: Wang, L., Dou, Q., Fletcher, P.T., Speidel, S., Li, S. (eds.) MICCAI 2022. LNCS, vol. 13434, pp. 194–203. Springer, Cham (2022). https://doi.org/10.1007/978-3-031-16440-8_19

An Unsupervised Multispectral Image Registration Network for Skin Diseases

Songhui Diao[1,2], Wenxue Zhou[2,3], Chenchen Qin[2], Jun Liao[2],
Junzhou Huang[4], Wenming Yang[3], and Jianhua Yao[2(✉)]

[1] Shenzhen Institute of Advanced Technology, Chinese Academy of Sciences,
Shenzhen, China
[2] Tencent AI Lab, Shenzhen, China
jianhuayao@tencent.com
[3] Shenzhen International Graduate School, Tsinghua University, Shenzhen, China
[4] University of Texas at Arlington, Arlington, USA

Abstract. Multispectral imaging has a broad, promising and advantageous application prospect in the diagnosis of skin diseases. However, there are inherent deviations such as rigid or non-rigid deformation among multispectral images (MSI), which makes accurate and robust registration algorithms desirable to extract reliable multispectral features. Existing registration algorithms are susceptible to significant and nonlinear amplitude differences and geometric distortions among MSI, resulting in an unsatisfactory estimation of the registration field (RF). In this study, we propose an end-to-end multispectral image registration (MSIR) network with unsupervised learning for human skin disease diagnosis. First, we propose a basic adjacent-band pair registration (ABPR) model to obtain the corresponding RFs through simultaneously modeling a series of image pairs from adjacent bands. Second, we introduce a multispectral attention module (MAM) for extraction and adaptive weight allocation of the high-level pathological features of multiple MSI pairs. Third, we design a registration field refinement module (RFRM) to rectify and reconstruct a general RF solution. Fourth, we propose an unsupervised center-toward registration loss function, combining a similarity loss for features in the frequency domain and a smoothness loss for RF. In addition, we built a MSI dataset of multi-type skin diseases and conducted extensive experiments. The results show that our method not only outperforms state-of-the-art methods on MSI registration task, but also contributes to the subsequent task of benign and malignant disease classification.

Keywords: Image registration · Multispectral image · Unsupervised registration · Registration field refinement

---

S. Diao and W. Zhou—These authors contributed equally to this work.

Supplementary Information The online version contains supplementary material available at https://doi.org/10.1007/978-3-031-43999-5_68.

1   Introduction

Skin disease is common in clinic, which is characterized by the complexity of pathological morphology and etiology, as well as the diversity of disease types and locations. Multispectral imaging (MSI) has the characteristics of non-tissue contact puncture, no radiation and no need for exogenous contrast agents. The imaging mechanism characterizes specific, correlated and complementary tissue features, which makes it having a broad, promising and advantageous application prospect in the diagnosis of skin diseases [1]. Correspondingly, multispectral imaging also has some shortcomings. On one hand, wavelength and focal length vary with frequency, resulting in non-rigid deformation such as scaling deviation among MSI. On the other hand, the motion of imaging device or patient may introduce further deviation among images. Consequently, MSI registration, that is, the identification and mapping of the same or similar structure or content at the pixel level, is a fundamental and critical process for subsequent tasks such as image fusion, pathological analysis, disease identification and diagnosis.

The difficulties for MSI registration are twofold. Firstly, conventional registration method is to register two images [2–5], while group image registration (GIR) is the joint registration of a group of related images. Current GIR research focuses on time-series MRI. MSI contains multiple images with significant and nonlinear amplitude differences and geometric distortions, not only making the pair-wise image registration not applicable, but also bringing great challenges to GIR due to the inability to take advantage of image intensity or structural similarity. Secondly, the type and location of diseases both affect the light reflection coefficient of skin tissue, making it challenging to find a general registration field (RF) for GIR.

In the field of image registration, many inspiring methods based on traditional or deep learning techniques have been developed and applied to computer vision tasks in medical imaging, remote sensing, etc. Whereas, the traditional methods [6,7] are not suitable for MSI dataset with significant non-rigid deformation, gray jump, noise and other factors, which will lead to low efficiency and poor accuracy. The supervised deep learning methods [8,9] have the limitation of relying on the groundtruth of RF which is difficult to obtain in medical images. For the unsupervised deep learning method, G. Balakrishnan et al. [10] proposed a VoxelMorph framework for deformable and pairwise brain image registration based on image intensity. Y. Ye et al. [11] presented a MU-Net framework, which stacks several DNN models on multiple scales to generate a coarse-to-fine registration pipeline. L. Meng et al. [12] proposed an DSIM network for MSI registration, which utilized pyramid structure similarity loss to optimize the network and regress the homography parameters. Although the existing algorithms can achieve relatively accurate registration for images with weak or repeated texture, they are susceptible to significant and nonlinear amplitude differences and geometric distortions among MSI, and generally have the disadvantages of poor accuracy, low robustness and low efficiency in realizing group image registration, leading to unsatisfactory RF results.

To address the aforementioned issues, we propose an end-to-end multispectral image registration (MSIR) network with unsupervised learning for multiple types of human skin diseases, which improves the capability of CNN architecture to learn the cross-band transformation relationship among pathological features, so as to obtain an efficient and robust RF solution. First, we design a basic adjacent-band pair registration (ABPR) model, which simultaneously models a series of image pairs from adjacent bands based on CNN, and makes full use of the feature transformation relationship between images to obtain their corresponding RFs. Second, we introduce a multispectral attention module (MAM), which is used to achieve extraction and adaptive weight allocation for the high-level pathological features. Third, we design a registration field refinement module (RFRM) to obtain a general RF solution of MSI through rectifying and reconstructing the RFs learned from all adjacent-band MSI pairs. Fourth, we propose an unsupervised center-toward registration loss function, combining a similarity loss for features in the frequency domain and a smoothness loss for RF. We perform extensive experiments on a MSI dataset of multi-type skin diseases. The evaluation results demonstrate that our method not only outperforms prior arts on MSI registration task, but also contributes to the subsequent task of benign and malignant disease classification.

2 Methodology

As shown in Fig. 1, the proposed MSIR framework consists of four main components: (1) an ABPR model to extract pixel-wise representations of corresponding features from a pair of images synchronously. (2) a MAM to reassign the weight of the high-level features. (3) a RFRM to yield a general RF of MSI. (4) a center-toward registration loss function to optimize the effect of GIR.

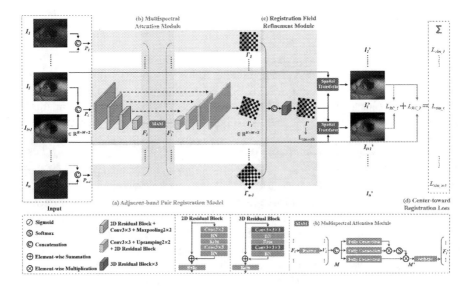

Fig. 1. The overall framework of our proposed MSIR network for multiple types of human skin diseases.

2.1 Adjacent-Band Pair Registration (ABPR) Model

The ABPR model is based on Unet [13] cascaded with 2D residual blocks [14]. In the feature encoder represented by the gray part, 4 convolution blocks with size of 3×3 and stride of 2 are used for down-sampling, and feature maps with channel number of 16, 32, 32 and 32 are obtained. Correspondingly, in the decoder represented by the yellow part, the convolution blocks contain interpolation operations for up-sampling.

The input to ABPR model is the concatenation of a pair of MSI images in adjacent bands $P_i \in R^{H \times W \times 2}$ (images I_i and I_{i+1}, $1 \leq i \leq n-1$. n is the number of bands). The module constructs a function $\Gamma_i = F_\rho(P_i)$ to synchronously extracts the mutual transformation relationship. F represents the registration function fit by the designed Unet architecture. ρ refers to the weights and bias parameters of the kernels of the convolutional layers. $\Gamma_i \in R^{H \times W \times 2}$ is the RF between the given P_i, where 2 denotes two channels along the x-axis and y-axis.

Since P_i in multi-band imaging has unequal contribution to the solution of the final RF, we introduce a multispectral attention module (MAM) into ABPR model, which can not only facilitate the extraction of mutual complementary non-local features between MSI, but also guide the model to pay more attention to features from specific spectra with high impact.

Specifically, we carry out global average pooling operation on the high-level features F_i of the encoder for image pair P_i, and flatten F_i into a feature vector V_i. Then we obtain a feature map $M \in R^{(n-1) \times H' \times W' \times C'}$ by concatenating the accumulated $n-1$ feature vectors in column. The formula for reallocating attention weights is defined as [15]:

$$M' = softmax(\frac{MW_Q MW_K^T}{\sqrt{d_k}})MW_v \tag{1}$$

where W_Q, W_K and W_V are weights of the fully connection layers. d_k represents the dimension of V_i. Next, we reshape M' to obtain the updated high-level features F_i' and continue the subsequent decoding calculation in ABPR model.

2.2 Registration Field Refinement Module (RFRM)

In order to make full use of the potential complementarity of corresponding features between cross-band images, we design a RFRM to obtain a general RF of MSI through rectifying and reconstructing a series of RFs learned from multiple adjacent-band image pairs, which is more conducive to further improving the accuracy and generalization of the MSI registration network.

First, RFRM concatenates all $\Gamma_i \in R^{H \times W \times 2}$ learned from $n-1$ adjacent-band image pairs. The obtained RF$\in R^{(n-1) \times H \times W \times 2}$ is then refined through three 3D residual blocks, and the reliable $\Gamma \in R^{H \times W \times 2}$ is finally generated. T_Γ means a coordinate mapping that is parameterized by Γ and its spatial transformation. That is, for each pixel $p \in I_i$, there is a Γ such that $I_i(p)$ and $T_\Gamma(I_i)(p)$ are two corresponding points in adjacent bands.

2.3   Center-Toward Registration Loss Function

The ultimate goal of image registration is to obtain a RF from MSI with significant amplitude difference and geometric distortion, so that the perceived images corrected by the RF have the best similarity with each other. The pixel similarity in spatial domain among MSI is prone to large difference due to spectral intensity, while the feature similarity in the frequency domain is more stable. In order to optimize the adaptive center-toward registration of a group of images simultaneously, we propose an unsupervised loss function, including a similarity loss for features in the frequency domain and a smoothness loss for RF. The specific scheme is described in detail as follows:

(1) The first is a similarity loss function based on the residual complexity of features in the frequency domain, which is used to penalize differences in appearance and optimize the registration effect under different lighting conditions. The residual complexity loss function is defined as [16]:

$$L_{RC}(I, I') = \frac{1}{m} \sum_{j=1}^{m} \log[(DCT(I - I'))^2/\alpha + 1] \tag{2}$$

where m is the number of pixels of the images I and I'. DCT denotes discrete cosine transform whose weight is regulated by a hyperparameter α with an empirical value of 0.05. It is worth noting that the similarity loss consists of two components. For the image with band i, we first use the fused Γ and its spatial transformation to obtain the warped image $I'_i = T_\Gamma(I_i)$. Then we evaluate its similarity to the reference image with the adjacent band I_{i+1} and the warped image $I'_{i+1} = T_\Gamma(I_{i+1})$, which not only ensures that the transformed images do not deviate from the original spatial distribution, but also realizes center-toward registration of a group of images synchronously and uniformly. Then the total similarity loss can be obtained through adding the residual complexity results from $n - 1$ image pairs. The formula is defined as:

$$L_{sim} = \frac{1}{n-1} \sum_{i=1}^{n-1} [L_{RC}(I'_i, I_{i+1}) + L_{RC}(I'_i, I'_{i+1})] \tag{3}$$

(2) The second is an auxiliary loss function that constrains the smoothness of RF and penalizes the local spatial variation. L_{smooth} is constructed based on Γ_i through differentiable operation on spatial gradient, and the formula is as follows [10]:

$$L_{smooth} = \frac{1}{n-1} \sum_{i=1}^{n-1} \|\nabla(\Gamma_i)\|^2 \tag{4}$$

where ∇ is the gradient operator calculated along the x-axis and y-axis. Then the loss function of our module L_{total} is the weighted sum of L_{sim} and L_{smooth}, which is defined as:

$$L_{total} = L_{sim} + \lambda L_{smooth} \tag{5}$$

where λ is a hyperparameter used to balance the similarity and smoothness of RF, and the empirical value 4 is taken, which is consistent with the Voxelmorph [10] method. All these components are unified in the MSIR network and trained end-to-end.

3 Experiments

3.1 Dataset and Implementation Details

We validated the proposed MSIR network on an in-house MSI dataset containing 85 cases with multi-type skin diseases collected from our partner hospital from November 2021 to March 2022, including 36 cases with benign diseases (keloid, fibrosarcoma, cyst, lipoma, hemangioma and nevus) and 49 cases with malignant diseases (eczematoid paget disease, squamous cell carcinoma, malignant melanoma and basal cell carcinoma). Specifically, each case consists of 22 scans with wavelengths ranging from 405 nm to 1650 nm and a uniform size of 640×512. For each case, a clinician manually annotated four corresponding landmarks for registration on each scan, and their positions were further reviewed by an experienced medical expert. These landmarks are used to evaluate the registration accuracy. The detailed information for the training set and test set is shown in Table 1. We conducted a 4-fold cross-validation in training set to select models and hyperparameters.

Table 1. Characteristics for in-house dataset.

Characteristic	Entry	Training Set	Test Set
Patient Demographics	No. of cases	59	26
	No. of scans	1298	572
Prevalence No. of scans (percentage) with	Benign	528(28.24%)	264(14.12%)
	Malignant	770(41.18%)	308(16.47%)

3.2 Quantitative and Qualitative Evaluation

In this paper, we adopt three evaluation indexes to assess the registration performance of different methods, including normalized mutual information (NMI), registration feature error (RFE) and target registration point error (TRE) [17].

Table 2 shows the quantitative comparisons of the registration performance among our MSIR network and three state-of-the-art competitive methods, including Voxelmorph [10], DSIM [12] framework and MU-Net [11]. The mean value of the initial TRE is 8.77 pixels. It can be observed that our method achieves the superior performance. Specifically, it improves the best NMI to 0.547, RFE to 1.166 and TRE to 4.984 pixels, which significantly outperforms the other competitors.

Table 2. Quantitative comparisons of different methods.

Method	Input	NMI	RFE	TRE
Voxelmorph [10]	Multidimensional Image	0.498(±0.057)	1.909(±0.246)	5.734(±3.265)
DSIM [12]	Multispectral Image	0.479(±0.064)	2.122(±0.393)	5.763(±3.560)
MU-Net [11]	Multispectral Image	0.513(±0.027)	1.851(±0.891)	5.692(±3.206)
MSIR (Ours)	Multispectral Image	**0.547(±0.029)**	**1.166(±0.642)**	**4.984(±3.471)**

Table 3. Quantitative results of ablation study on components and augmented dataset.

NCC Loss	Our Loss	MAM	Augmented Dataset	NMI	RFE	TRE
✓	✗	✗	✗	0.523(±0.051)	1.331(±0.819)	5.227(±4.441)
✗	✓	✗	✗	0.539(±0.046)	1.168(±0.674)	5.085(±3.905)
✓	✗	✓	✗	0.526(±0.078)	1.245(±0.795)	4.935(±6.422)
✗	✓	✓	✗	**0.547(±0.029)**	**1.166(±0.642)**	**4.984(±3.471)**
✗	✓	✓	✓	**0.576(±0.024)**	**1.109(±0.547)**	**4.736(±3.353)**

We further conduct ablation study to verify the contributions of individual components, as shown in Table 3. NCC refers to the normalization cross correlation, which describes the relevance and similarity of targets. Compared with NCC loss, the proposed loss function significantly improves the registration performance (0.142 reduction in TRE, 0.163 reduction in RFE and 0.016 increase in NMI). The introduction of MAM further improves the registration performance. On this basis, we introduce an augmented dataset to expand the training set to 50 times of its original size by performing affine transformation operations on MSI, such as translation, rotation, scaling, clipping, oblique cutting, etc. The adoption of the augmented dataset makes the indexes continuously optimized. Quantitative results validate the effectiveness of MSIR network and augmented dataset in improving registration performance.

Next, we conduct ablation experiments to explore the registration performance based on images with different subsets of bands, as illustrated in Table 4. VIS and NIR represent the visible bands (405 nm–780 nm) and the near infrared bands (780 nm–1650 nm), respectively. ALL represents the whole bands (405 nm–1650 nm). The results demonstrate that our method can achieve remarkable registration improvement for images of different bands.

To verify the generalization of the proposed method, we test the model using a synthetic dataset (with |5| degrees of rotation, |0.02| of scaling, |6| and |8| pixels of translation along the x-axis and y-axis). The visualization results are shown in Fig. 2. It can be seen that MSIR method can achieve accurate registration not only for the MSI of the raw dataset, but also for that of the synthetic dataset with more complex transformation that are challenging for registration.

In addition, in order to verify the effect of this registration method to subsequent tasks, we further conduct a classification task for benign and malignant diseases based on the established MSI dataset. Table 5 shows the quantitative

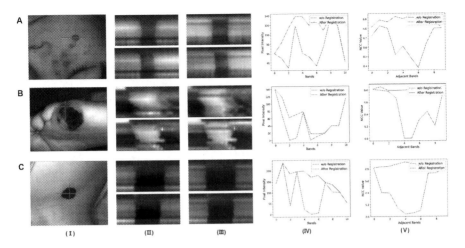

Fig. 2. Visual comparisons of test results on raw dataset and synthetic dataset. A, B and C represent three patient cases respectively. Column (I) shows the example of the original image at $650nm$, in which the boxes marked with red crosses indicate the position of zooming for column (II) to column (III). Column (II) and column (III) are the qualitative results of the raw dataset processed without and with our method, respectively. The first row represents the sectioning of the red cross on the x-axis, and the second row represents that on the y-axis. Column (IV) and column (V) are the quantitative results of two indicators in the red cross region of the synthetic dataset, namely, the pixel intensity among bands and the normalized correlation coefficient (NCC) between adjacent bands. The blue and red lines represent the results without and with registration, respectively.

comparisons of different classifiers using single-band and all-bands images. Four classifiers are compared, namely KNN [18], SVM [19], Resnet18 [20] and Inception V3 [21]. Columns 2 and 3 represent the worst and best classification accuracy based on single-band images, which come from the wavelengths of 450 nm and 525 nm, respectively. The fourth column is the classification results based on the original MSI dataset without registration, and the last column shows that based on the images processed by MSIR network. It can be seen that the MSI contains more abundant information than the single-band image, which is more conducive to the subsequent analysis. More importantly, due to the contribution of MSIR network for image registration, the classification accuracy on

Table 4. Quantitative mean($\pm$std) results of ablation experiments on bands.

Bands	w/o Registration			MSIR (Ours)		
	NMI	RFE	TRE	NMI	RFE	TRE
VIS	0.295($\pm$0.103)	1.505($\pm$1.197)	8.625($\pm$7.055)	0.517($\pm$0.089)	1.287($\pm$0.788)	4.632($\pm$3.947)
NIR	0.227($\pm$0.101)	1.324($\pm$1.241)	12.448($\pm$15.093)	0.576($\pm$0.048)	1.013($\pm$0.718)	5.595($\pm$5.406)
ALL	0.271($\pm$0.097)	1.323($\pm$0.904)	10.070($\pm$9.544)	0.547($\pm$0.029)	1.166($\pm$0.642)	4.984($\pm$3.471)

Table 5. Quantitative comparisons of different classifiers on band and registration operation.

Classifier	Accuracy			
	Single Band		All Bands	
	Worst (450 nm)	Best (525 nm)	w/o Registration	MSIR (Ours)
KNN [18]	0.524(±0.017)	0.567(±0.079)	0.607(±0.063)	**0.644(±0.058)**
SVM [19]	0.539(±0.064)	0.583(±0.072)	0.615(±0.084)	**0.672(±0.049)**
Resnet18 [20]	0.540(±0.072)	0.602(±0.051)	0.647(±0.051)	**0.754(±0.037)**
Inception V3 [21]	0.551(±0.061)	0.606(±0.046)	0.659(±0.022)	**0.762(±0.018)**

MSI dataset has been significantly improved, which verifies the necessity and effectiveness of MSIR network.

4 Conclusion

In this study, an efficient and robust framework for multispectral image registration is proposed and validated on a self-established dataset of multiple types of skin diseases, which holds great potentials for the further analysis, such as the classification of benign and malignant diseases. We intend to release the MSI dataset in future. The quantitative results of experiments demonstrate the superiority of our method over the current state-of-the-art methods.

References

1. Spreinat, A., Selvaggio, G., Erpenbeck, L., Kruss, S.: Multispectral near infrared absorption imaging for histology of skin cancer. J. Biophotonics **13**(1), e201960080 (2020)
2. Mok, T.C.W., Chung, A.C.S.: Large deformation diffeomorphic image registration with Laplacian pyramid networks. In: Martel, A.L., et al. (eds.) MICCAI 2020. LNCS, vol. 12263, pp. 211–221. Springer, Cham (2020). https://doi.org/10.1007/978-3-030-59716-0_21
3. Shu, Y., Wang, H., Xiao, B., Bi, X., Li, W.: Medical image registration based on uncoupled learning and accumulative enhancement. In: de Bruijne, M., et al. (eds.) MICCAI 2021. LNCS, vol. 12904, pp. 3–13. Springer, Cham (2021). https://doi.org/10.1007/978-3-030-87202-1_1
4. Meng, M., Bi, L., Feng, D., Kim, J.: Non-iterative coarse-to-fine registration based on single-pass deep cumulative learning. In: Wang, L., Dou, Q., Fletcher, P.T., Speidel, S., Li, S. (eds.) MICCAI 2022. LNCS, vol. 13436, pp. 88–97. Springer, Cham (2022). https://doi.org/10.1007/978-3-031-16446-0_9
5. Chen, J., Frey, E.C., He, Y., Segars, W.P., Li, Y., Du, Y.: Transmorph: transformer for unsupervised medical image registration. Med. Image Anal. **82**, 102615 (2022)
6. Ma, W., et al.: Remote sensing image registration with modified sift and enhanced feature matching. IEEE Geosci. Remote Sens. Lett. **14**(1), 3–7 (2016)

7. Cao, X., et al.: Deformable image registration based on similarity-steered CNN regression. In: Descoteaux, M., Maier-Hein, L., Franz, A., Jannin, P., Collins, D.L., Duchesne, S. (eds.) MICCAI 2017. LNCS, vol. 10433, pp. 300–308. Springer, Cham (2017). https://doi.org/10.1007/978-3-319-66182-7_35

8. Blendowski, M., Hansen, L., Heinrich, M.P.: Weakly-supervised learning of multi-modal features for regularised iterative descent in 3D image registration. Med. Image Anal. **67**, 101822 (2021)

9. Guo, H., Kruger, M., Xu, S., Wood, B.J., Yan, P.: Deep adaptive registration of multi-modal prostate images. Comput. Med. Imaging Graph. **84**, 101769 (2020)

10. Balakrishnan, G., Zhao, A., Sabuncu, M.R., Guttag, J., Dalca, A.V.: Voxelmorph: a learning framework for deformable medical image registration. IEEE Trans. Med. Imaging **38**(8), 1788–1800 (2019)

11. Ye, Y., Tang, T., Zhu, B., Yang, C., Li, B., Hao, S.: A multiscale framework with unsupervised learning for remote sensing image registration. IEEE Geosci. Remote Sens. Lett. **60**, 1–15 (2022)

12. Meng, L., et al.: Investigation and evaluation of algorithms for unmanned aerial vehicle multispectral image registration. Int. J. Appl. Earth Obs. Geoinf. **102**, 102403 (2021)

13. Ronneberger, O., Fischer, P., Brox, T.: U-Net: convolutional networks for biomedical image segmentation. In: Navab, N., Hornegger, J., Wells, W.M., Frangi, A.F. (eds.) MICCAI 2015. LNCS, vol. 9351, pp. 234–241. Springer, Cham (2015). https://doi.org/10.1007/978-3-319-24574-4_28

14. He, K., Zhang, X., Ren, S., Sun, J.: Deep residual learning for image recognition. In: Proceedings of the IEEE Conference on Computer Vision and Pattern Recognition, pp. 770–778 (2016)

15. Vaswani, A., et al.: Attention is all you need. In: Advances in Neural Information Processing Systems, vol. 30 (2017)

16. Myronenko, A., Song, X.: Intensity-based image registration by minimizing residual complexity. IEEE Trans. Med. Imaging **29**(11), 1882–1891 (2010)

17. Johnson, J., Alahi, A., Fei-Fei, L.: Perceptual losses for real-time style transfer and super-resolution. In: Leibe, B., Matas, J., Sebe, N., Welling, M. (eds.) ECCV 2016. LNCS, vol. 9906, pp. 694–711. Springer, Cham (2016). https://doi.org/10.1007/978-3-319-46475-6_43

18. Uddin, S., Haque, I., Lu, H., Moni, M.A., Gide, E.: Comparative performance analysis of k-nearest neighbour (KNN) algorithm and its different variants for disease prediction. Sci. Rep. **12**(1), 1–11 (2022)

19. Therese, M.J., Devi, A., Kavya, G.: Melanoma detection on skin lesion images using k-means algorithm and SVM classifier. In: Handbook of Deep Learning in Biomedical Engineering and Health Informatics, pp. 227–251. Apple Academic Press (2021)

20. Vasu, K., et al.: Effective classification of colon cancer using Resnet-18 in comparison with squeezenet. J. Pharm. Negat. Results 1413–1421 (2022)

21. Khamparia, A., Singh, P.K., Rani, P., Samanta, D., Khanna, A., Bhushan, B.: An internet of health things-driven deep learning framework for detection and classification of skin cancer using transfer learning. Trans. Emerg. Telecommun. Technol. **32**(7), e3963 (2021)

CortexMorph: Fast Cortical Thickness Estimation via Diffeomorphic Registration Using VoxelMorph

Richard McKinley$^{(\boxtimes)}$ and Christian Rummel

Support Center for Advanced Neuroimaging (SCAN), University Institute of
Diagnostic and Interventional Neuroradiology, Inselspital, Bern University Hospital,
Bern, Switzerland
richard.mckinley@insel.ch

Abstract. The thickness of the cortical band is linked to various neurological and psychiatric conditions, and is often estimated through surface-
based methods such as Freesurfer in MRI studies. The DiReCT method,
which calculates cortical thickness using a diffeomorphic deformation of
the gray-white matter interface towards the pial surface, offers an alternative to surface-based methods. Recent studies using a synthetic cortical
thickness phantom have demonstrated that the combination of DiReCT
and deep-learning-based segmentation is more sensitive to subvoxel cortical thinning than Freesurfer.

While anatomical segmentation of a T1-weighted image now takes seconds, existing implementations of DiReCT rely on iterative image registration methods which can take up to an hour per volume. On the other
hand, learning-based deformable image registration methods like Voxel-
Morph have been shown to be faster than classical methods while improving registration accuracy. This paper proposes CortexMorph, a new method
that employs unsupervised deep learning to directly regress the deformation
field needed for DiReCT. By combining CortexMorph with a deep-learning-
based segmentation model, it is possible to estimate region-wise thickness in
seconds from a T1-weighted image, while maintaining the ability to detect
cortical atrophy. We validate this claim on the OASIS-3 dataset and the synthetic cortical thickness phantom of Rusak et al.

Keywords: MRI · Morphometry · cortical thickness · Unsupervised
image registration · Deep learning

1 Introduction

Cortical thickness (CTh) is a crucial biomarker of various neurological and psychiatric disorders, making it a primary focus in neuroimaging research. The
cortex, a thin ribbon of grey matter at the outer surface of the cerebrum, plays
a vital role in cognitive, sensory, and motor functions, and its thickness has

---

Supplementary Information The online version contains supplementary material
available at https://doi.org/10.1007/978-3-031-43999-5_69.

been linked to a wide range of neurological and psychiatric conditions, including Alzheimer's disease, multiple sclerosis, schizophrenia, and depression, among others. Structural magnetic resonance imaging (MRI) is the primary modality used to investigate CTh, and numerous computational methods have been developed to estimate this thickness on the sub-millimeter scale. Among these, surface-based methods like Freesurfer [5,6] have been widely used, but they are computationally intensive, making them less feasible for clinical applications. Optimizations based on Deep Learning have brought the running time for a modified Freesurfer pipeline down to one hour. [7] The DiReCT method [4] offers an alternative to surface-based morphometry methods, calculating CTh via a diffeomorphic deformation of the gray-white matter interface (GWI) towards the pial surface (the outer edge of the cortical band). The ANTs package of neuroimaging tools provides an implementation of DiReCT via the function `KellyKapowski`: for readablility we refer below to `KellyKapowski` with its default parameters as ANTs-DiReCT. The ANTs cortical thickness pipeline uses ANTs-DiReCT together with a three-class segmentation (grey matter, white matter, cerebrospinal fluid) provided by the Atropos segmentation method, taking between 4 and 15 h depending on the settings and available hardware [1,15]. A more recent version of ANTs provides a deep-learning based alternative to Atropos, giving comparable results to ANTs but accelerating the overall pipeline to approximately one hour, such that now the running time is dominated by the time needed to run ANTs-DiReCT [16]. Meanwhile, Rebsamen et al. have shown that applying DiReCT to the output of a deep-learning-based segmentation model trained on Freesurfer segmentations (rather than Atropos) yields a CTh method which agrees strongly with Freesurfer, while having improved repeatability on repeated scans [12]. Subsequently, a digital phantom using GAN-generated scans with simulated cortical atrophy showed that the method of Rebsamen et al. is more sensitive to cortical thinning than Freesurfer [13].

The long running time of methods for determining CTh remains a barrier to application in clinical routine: a running time of one hour, while a substantial improvement over Freesurfer and ANTs cortical thickness, is still far beyond the real-time processing desirable for on-demand cortical morphometry in clinical applications. In terms of both the speed and performance, VoxelMorph and related models are known to outperform classical deformable registration methods, suggesting that a DiReCT-style CTh algorithm based on unsupervised registration models may enable faster CTh estimation. [2,3,18] In this paper, we demonstrate that a VoxelMorph style model can be trained to produce a diffeomorphism taking the GWI to the pial surface, and that this model can be used to perform DiReCT-style CTh estimation in seconds. We trained the model on 320 segmentations derived from the IXI and ADNI datasets, and demonstrate excellent agreement with ANTs-DiReCT on the OASIS-3 dataset. Our model also shows improved performance on the digital CTh phantom of Rusak et al. [13].

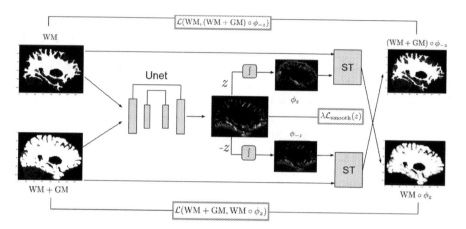

Fig. 1. End-to-end unsupervised architecture for DiReCT: velocity field z is regressed from WM and WM+GM segmentations, using a Unet. This velocity field is then integrated by seven scaling and squaring layers ($\int$) to yield forward and reverse deformation fields ϕ_z and ϕ_{-z}, which are used to deform the input images in spatial transformer (ST) blocks. Components of the loss function are marked in orange.

2 Methods

2.1 DiReCT Cortical Thickness Estimation

The estimation of CTh using the DiReCT method [4] proceeds as follows: first a (partial volume) segmentation of the cortical white matter (WM) and cortical grey matter (GM) is obtained. Second, a forward deformation field ϕ mapping the white-matter (WM) image towards the WM+GM image is computed. This forward deformation field should be a diffeomorphism, in order that the deformation field is invertible and the topology of the inferred pial surface is the same as the GWI. Third, the diffeormorphism is inverted to obtain the reverse the deformation field, taking the pial surface towards the GWI. Finally, the CTh is determined by computing the magnitude of the reverse field at the GWI: specifically, at each voxel of WM adjacent to the GM. In ANTs-DiReCT, the forward transform (from WM to WM+GM) is calculated by a modified greedy algorithm, in which the WM surface is propagated iteratively in the direction of the surface normal until it reaches the outer GM surface or a predefined spatial prior maximum is reached. The approximate inverse field is then determined by numerical means using kernel based splines (as implemented in ITK).

The absence of a reliable gold-standard ground truth for CTh makes comparisons between methods difficult. This situation has recently been improved by the publication of a synthetic cortical atrophy phantom: a dataset generated using a GAN conditioned on subvoxel segmentations, consisting of 20 synthetic subjects with 19 induced sub-voxel atrophy levels per subject (ten evenly spaced atrophy levels from 0 to 0.1 mm, and a further nine evenly spaced atrophy levels from 0.1 mm to 1 mm). [13] The purpose of this digital phantom is to explore the ability of CTh algorithms to resolve subtle changes of CTh. The paper of Rusak et al. analyzed the performance of several CTh methods on this dataset, finding that the DL+DiReCT method [12] (which combines a deep network trained on Freesurfer annotations with ANTs-DiReCT) was the most sensitive to cortical atrophy and had the best agreement with the synthetically induced thinning.

2.2   CortexMorph: VoxelMorph for DiReCT

The original VoxelMorph architecture, introduced in [3], utilized a Unet architecture to directly regress a displacement field from a fixed brain image and a moving brain image. Application of a spatial transform layer allows the moving image to be transformed to the space of the fixed image, and compared using a differentiable similarity metric such as mean squared error or cross-correlation. Since the spatial transformation is also a differentiable operation, the network can be trained end-to-end. Later adaptations of the concept employed a regression of a stationary *velocity field*, with the deformation field being calculated via an integration layer: the principal advantage of this formulation is that integrating through a velocity field yields a *diffeomorphism.* [2] Since diffeomorphic registration is required in the DiReCT method, we adopt this velocity-field form of VoxelMorph for our purposes.

The setup of our VoxelMorph architecture, CortexMorph, is detailed in Fig. 1. The two inputs to the network are a partial volume segmentation of white matter (WM), and a partial volume segmentation of grey matter plus white matter (WM+GM). These are fed as entries into a Unet, the output of which is a velocity field z, which is then integrated using 7 steps of scaling and squaring to yield a displacement field ϕ_z. This displacement field is then applied to the WM image to yield the deformed white matter volume $WM \circ \phi_z$. By integrating $-z$ we obtain the reverse deformation field ϕ_{-z}, which is applied to the WM+GM image to obtain a deformed volume $(WM + GM) \circ \phi_{-z}$. This simplifies the DiReCT method substantially: instead of needing to perform a numerical inversion of the deformation field, the reverse deformation field can be calculated directly. The deformed volumes are then compared using a loss function $\mathcal{L}$ to their non-deformed counterparts: both directions of deformation are weighted equally in the final objective function. To encourage smoothness, a discrete approximation of the squared gradient magnitude of the velocity field $\mathcal{L}_{\text{smooth}}$ is added to the loss as a regularizer. [2] As a result, our loss has the following form

$$\mathcal{L}(WM, (WM + GM) \circ \phi_{-z}) + \mathcal{L}(WM + GM, WM \circ \phi_z) + \lambda \mathcal{L}_{\text{smooth}}(z) \quad (1)$$

2.3   Data and WM/GM Segmentation

Training data and validation for our VoxelMorph model was derived from two publicly available sources: images from 200 randomly selected elderly individuals from the ADNI dataset [10] and images from 200 randomly selected healthy adults from the IXI dataset (https://brain-development.org/ixi-dataset). From each of these datasets, 160 images were randomly chosen to serve as training data, yielding in total 320 training cases and 80 validation cases. For testing our pipeline, we use two sources different from the training/validation data: the well-known OASIS-3 dataset (2,643 scans of 1,038 subjects, acquired over >10 years on three different Siemens scanners), and the CTh phantom of Rusak et al. [13,14]

For WM/GM segmentation, we employed the DeepSCAN model [11,12], which is available as part of DL+DiReCT (https://github.com/SCAN-NRAD/DL-DiReCT), since this is already known to give high-quality CTh results when combined with ANTs-DiReCT. This model takes as input a T1-weighted image, performs resampling and skull-stripping if necessary (provided by HD-BET [9]) and produces a partial volume segmentation P_w of the white matter and P_g of the cortex (the necessary inputs to the DiReCT algorithm) with 1mm isovoxel resolution. It also produces a cortical parcellation in the same space (necessary to calculate region-wise CTh measures). We applied this model to the training data, validation data, and the 400 synthetic MRI cases of the CTh phantom, both to produce ANTs-DiReCT CTh measurements and also as an input to our VoxelMorph models.

2.4   Training and Model Selection

Our network was implemented and trained in Pytorch (1.13.1). We utilized a standard Unet (derived from the nnUnet framework [8]) with 3 pooling steps and a feature depth of 24 features at each resolution. The spatial transformer/squaring and scaling layers/gradient magnitude loss were incorporated from the official VoxelMorph repository. For the loss function $\mathcal{L}$ we tested both L1 loss and mean squared error (MSE). We tested values of the smoothness parameter lambda between 0 and 0.05. The models were trained with the Adam optimizer, with a fixed learning rate of 10^{-3} and weight decay 10^{-5}. Patches of size 128^3 were used as training data in batches of size 2.

The training regime was fully unsupervised with respect to cortical thickness: neither the deformation fields yielded by ANTs-DiReCT nor the CTh results computed from those deformation fields were used in the objective function. Since we are interested in replacing the iterative implementation of DiReCT with a deep learning counterpart, we used the 80 validation examples for model selection, selecting the model which showed best agreement in mean global CTh with the results of ANTs-DiReCT. The metric for agreement chosen is intraclass correlation coefficient, specifically ICC(2,1) (the proportion of variation explained by the individual in a random effects model, assuming equal means of the two CTh measurement techniques), since this method is sensitive to both absolute agreement and relative consistency of the measured quantity. ICC was calculated using the python package `Pingouin` [17].

2.5   Testing

The VoxelMorph model which agreed best with ANTs-DiReCT on the validation set was applied to segmentations of the OASIS-3 dataset, to confirm whether model selection on a small set of validation data would induce good agreement with ANTs-DiReCT on a much larger test set (metric, ICC(2,1)) and to the synthetic CTh phantom of Rusak et al., to determine whether the VoxelMorph model is able to distinguish subvoxel changes in CTh (metric, coefficient of determination (R^2)).

3   Results

The best performing model on the validation set (in terms of agreement with DiReCT) was the model trained with MSE loss and a λ of 0.02. When used to measure mean global CTh, this model scored an ICC(2,1) of 0.91 (95% confidence interval [0.9, 0.92]) versus the mean global CTh yielded by ANTs-DiReCT on the OASIS-3 dataset. For comparison, on the same dataset the ICC between Freesurfer and the ANTs-DiReCT method was 0.50 ([95% confidence interval -0.08, 0.8]). A breakdown of the ICC by cortical subregion can be seen in Fig. 2: these range from good agreement (entorhinal right, ICC = 0.87) to poor (caudalanteriorcingulate right, ICC = 0.26), depending on the region. However, ICC(2,1) is a measure of absolute agreement, as well as correlation: all regional Pearson correlation coefficients lie in a range [0.64–0.90] (see supplementary material for a region-wise plot of the Pearson correlation coefficients).

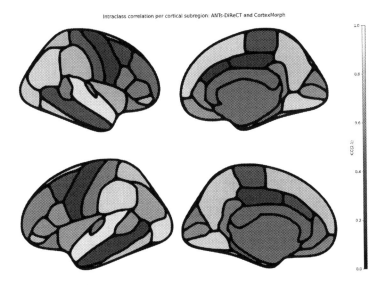

Fig. 2. Region-wise performance of CortexMorph: ICC(2,1) of mean region-wise cortical thickness between CortexMorph and ANTs-DiReCT, using the segmentations generated by DeepSCAN on the OASIS-3 dataset.

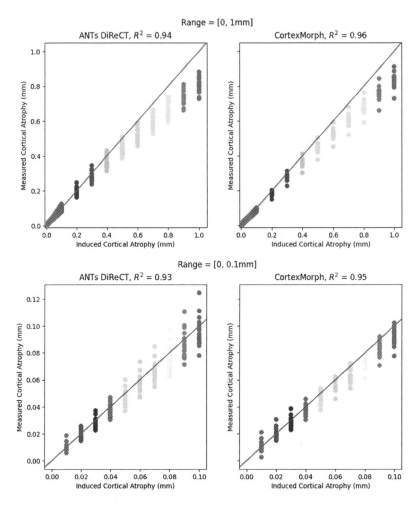

Fig. 3. Performance of ANTs-DiReCT and CortexMorph on the CTh phantom of Rusak et al., based on segmentations derived from DeepSCAN. Above: performance on the whole synthetic dataset, comprising twenty synthetic individuals, each with a baseline scan and 19 'follow-up' images with induced levels of uniform cortical atrophy. Measured atrophy is defined as the difference between the mean CTh as measured on the synthetic baseline scan and the mean CTh measured on the synthetic follow-up, averaged across the whole cortex. Below: The same data, but focused only on the range [0-0.1mm] of induced atrophy. R^2 denotes the coefficient of determination between the induced and measured atrophy levels.

Performance of this model on the CTh digital phantom can be seen in Fig. 3: agreement with the induced level of atrophy is high (metric: Coefficient of Determination between the induced and the measured level of atrophy, across all 20 synthetic subjects) in both the wide range of atrophy (up to 1mm) and the fine-grained narrower range of atrophy (up to 0.1mm), suggesting that the VoxelMorph model is able to resolve small changes in CTh.

Calculating regional CTh took between 2.5 s and 6.4 s per subject (mean, 4.3 s, standard deviation 0.71 s) (Nvidia A6000 GP, Intel Xeon(R) W-11955M CPU).

4   Conclusion

Our experiments suggest that the classical, iterative approach to cortical thickness estimation by diffeomorphic registration can be replaced with a VoxelMorph network, with ~ 800 fold reduction in the time needed to calculate CTh from a partial volume segmentation of the cortical grey and white matter. Since such segmentations can also be obtained in a small number of seconds using a CNN or other deep neural network, we have demonstrated for the first time reliable CTh estimation running on a timeframe of seconds. This level of acceleration offers increased feasibility to evaluate CTh in the clinical setting. It would also enable the application of ensemble methods to provide multiple thickness measures for an individual: given an ensemble of, say, 15 segmentation methods, a plausible distribution of CTh values could be reported for each cortical subregion within one minute: this would allow better determination of the presence of cortical atrophy in an individual than is provided by point estimates. We are currently investigating the prospect of leveraging the velocity field to enable fast calculation of other morphometric labels such as grey-white matter contrast and cortical curvature: these too could be calculated with error bars via ensembling.

This work allows the fast calculation of diffeomorphisms for DiReCT on the GPU. We did not consider the possibility of directly implementing/accelerating the classical DiReCT algorithm on a GPU in this work. Elements of the ANTs-DiReCT pipeline implement multithreading, yielding for example a 20 min runtime with 4 threads: however, since some parts of the pipeline cannot be parallelized it is unlikely that iterative methods can approach the speed of direct regression by CNN.

Given the lack of a gold standard ground truth for CTh, it is necessary when studying a new definition of CTh to compare to an existing silver standard method: this would typically be Freesurfer, but recent results suggest that this may not be the optimal method when studying small differences in CTh. [13] We have focused on comparison to the DL+DiReCT method for this study, since the results of this model on the CTh phantom are already reported and represent the state-of-the-art. For this reason, it made sense to use the outputs of the underlying CNN as inputs to our pipeline. However, the method we describe is general and could be applied to any highly performing segmentation method. Similarly, while we performed model selection to optimize agreement with the CTh values produced by Rebsamen et al., this optimization could easily be tuned to instead optimize agreement with Freesurfer. Alternatively, we could abandon agreement and instead select models based on consistency (given by a different variant of ICC) or Pearson correlation with a baseline model: this could lead to models which deviate from the baseline model but are better able to capture differences between patients or cohorts.

Acknowledgements. This work was supported by a Freenovation grant from the Novartis Forschungsstiftung, and by the Swiss National Science Foundation (SNSF) under grant number 204593 (ScanOMetrics).

References

1. Avants, B.B., Tustison, N.J., Wu, J., Cook, P.A., Gee, J.C.: An open source multivariate framework for n-tissue segmentation with evaluation on public data. Neuroinformatics **9**(4), 381–400 (2011). https://doi.org/10.1007/s12021-011-9109-y
2. Balakrishnan, G., Zhao, A., Sabuncu, M.R., Guttag, J., Dalca, A.V.: VoxelMorph: a learning framework for deformable medical image registration. IEEE Trans. Med. Imaging **38**(8), 1788–1800 (2019). https://doi.org/10.1109/TMI.2019.2897538. arXiv:1809.05231
3. Dalca, A.V., Balakrishnan, G., Guttag, J., Sabuncu, M.R.: Unsupervised learning of probabilistic diffeomorphic registration for images and surfaces. Med. Image Anal. **57**, 226–236 (2019). https://doi.org/10.1016/j.media.2019.07.006
4. Das, S.R., Avants, B.B., Grossman, M., Gee, J.C.: Registration based cortical thickness measurement. Neuroimage **45**(3), 867–879 (2009). https://doi.org/10.1016/j.neuroimage.2008.12.016
5. Fischl, B.: FreeSurfer. Neuroimage **62**(2), 774–781 (2012). https://doi.org/10.1016/j.neuroimage.2012.01.021
6. Fischl, B., Dale, A.M.: Measuring the thickness of the human cerebral cortex from magnetic resonance images. Proc. Natl. Acad. Sci. **97**(20), 11050–11055 (2000). https://doi.org/10.1073/pnas.200033797
7. Henschel, L., Conjeti, S., Estrada, S., Diers, K., Fischl, B., Reuter, M.: FastSurfer - a fast and accurate deep learning based neuroimaging pipeline. NeuroImage **219**, 117012 (2020). https://doi.org/10.1016/j.neuroimage.2020.117012. https://www.sciencedirect.com/science/article/pii/S1053811920304985
8. Isensee, F., Jaeger, P.F., Kohl, S.A.A., Petersen, J., Maier-Hein, K.H.: nnU-Net: a self-configuring method for deep learning-based biomedical image segmentation. Nat. Methods **18**(2), 203–211 (2021)
9. Isensee, F., et al.: Automated brain extraction of multisequence MRI using artificial neural networks. Hum. Brain Mapp. **40**(17), 4952–4964 (2019)
10. Jack Jr., C.R., et al.: The Alzheimer's disease neuroimaging initiative (ADNI): MRI methods. J. Magn. Reson. Imaging **27**(4), 685–691 (2008). https://doi.org/10.1002/jmri.21049. https://onlinelibrary.wiley.com/doi/abs/10.1002/jmri.21049
11. McKinley, R., Rebsamen, M., Meier, R., Reyes, M., Rummel, C., Wiest, R.: Few-shot brain segmentation from weakly labeled data with deep heteroscedastic multi-task networks. arXiv preprint arXiv:1904.02436 (2019). https://arxiv.org/abs/1904.02436
12. Rebsamen, M., Rummel, C., Reyes, M., Wiest, R., McKinley, R.: Direct cortical thickness estimation using deep learning-based anatomy segmentation and cortex parcellation. Hum. Brain Mapp. (2020). https://doi.org/10.1002/hbm.25159. https://onlinelibrary.wiley.com/doi/abs/10.1002/hbm.25159
13. Rusak, F., et al.: Quantifiable brain atrophy synthesis for benchmarking thickness estimation of cortical methods. Med. Image Anal. **82**, 102576 (2022)

14. Rusak, F., et al.: Synthetic brain MRI dataset for testing of cortical thickness estimation methods. v1. https://doi.org/10.25919/4ycc-fc11. https://data.csiro.au/collection/csiro:53241v1
15. Tustison, N.J., et al.: The ANTs cortical thickness processing pipeline. In: Medical Imaging 2013: Biomedical Applications in Molecular, Structural, and Functional Imaging, vol. 8672, pp. 126–129. SPIE (2013). https://doi.org/10.1117/12.2007128. https://www.spiedigitallibrary.org/conference-proceedings-of-spie/8672/86720K/The-ANTs-cortical-thickness-processing-pipeline/10.1117/12.2007128.full
16. Tustison, N.J., et al.: The ANTsX ecosystem for quantitative biological and medical imaging. Sci. Rep. **11**(1), 9068 (2021)
17. Vallat, R.: Pingouin: statistics in python. J. Open Source Softw. **3**(31), 1026 (2018). https://doi.org/10.21105/joss.01026
18. Zou, J., Gao, B., Song, Y., Qin, J.: A review of deep learning-based deformable medical image registration. Front. Oncol. **12**, 1047215 (2022). https://doi.org/10.3389/fonc.2022.1047215. https://www.frontiersin.org/articles/10.3389/fonc.2022.1047215

ModeT: Learning Deformable Image Registration via Motion Decomposition Transformer

Haiqiao Wang, Dong Ni, and Yi Wang[✉]

Smart Medical Imaging, Learning and Engineering (SMILE) Lab, Medical UltraSound Image Computing (MUSIC) Lab, School of Biomedical Engineering, Shenzhen University Medical School, Shenzhen University, Shenzhen, China
onewang@szu.edu.cn

Abstract. The Transformer structures have been widely used in computer vision and have recently made an impact in the area of medical image registration. However, the use of Transformer in most registration networks is straightforward. These networks often merely use the attention mechanism to boost the feature learning as the segmentation networks do, but do not sufficiently design to be adapted for the registration task. In this paper, we propose a novel motion decomposition Transformer (ModeT) to explicitly model multiple motion modalities by fully exploiting the intrinsic capability of the Transformer structure for deformation estimation. The proposed ModeT naturally transforms the multihead neighborhood attention relationship into the multi-coordinate relationship to model multiple motion modes. Then the competitive weighting module (CWM) fuses multiple deformation sub-fields to generate the resulting deformation field. Extensive experiments on two public brain magnetic resonance imaging (MRI) datasets show that our method outperforms current state-of-the-art registration networks and Transformers, demonstrating the potential of our ModeT for the challenging nonrigid deformation estimation problem. *The benchmarks and our code are publicly available at* https://github.com/ZAX130/SmileCode.

Keywords: Deformable image registration · Motion decomposition · Transformer · Attention · Pyramid structure

1 Introduction

Deformable image registration has always been an important focus in the society of medical imaging, which is essential for the preoperative planning, intraoperative information fusion, disease diagnosis and follow-ups [10,23]. The deformable registration is to solve the non-rigid deformation field to warp the moving image, so that the warped image can be anatomically similar to the fixed image. Let $I_f, I_m \in \mathbb{R}^{H \times W \times L}$ be the fixed and moving images (H, W, L denote image size), in the deep-learning-based registration paradigm, it is often necessary to employ a spatial transformer network (STN) [13] to apply the estimated sampling grid $G \in \mathbb{R}^{H \times W \times L \times 3}$ to the moving image, where G is obtained by adding the regular grid and the deformation field. For any position $p \in \mathbb{R}^3$ in the sampling

H. Greenspan et al. (Eds.): MICCAI 2023, LNCS 14229, pp. 740–749, 2023.
https://doi.org/10.1007/978-3-031-43999-5_70

grid, $G(p)$ represents the corresponding relation, which means that the voxel at position p in the fixed image corresponds to the voxel at position $G(p)$ in the moving image. That is to say, image registration can be understood as finding the corresponding voxels between the moving and fixed images, and converting this into the relative positional relationship between voxels, which is very similar to the calculation method of Transformer [8].

Transformers have been successfully used in the society of computer vision and have recently made an impact in the field of medical image computing [11,17]. In medical image registration, there are also several related studies that employ Transformers to enhance network structures to obtain better registration performance, such as Transmorph [5], Swin-VoxelMorph [26], Vit-V-Net [6], etc. The use of Transformer in these networks, however, often merely leverages the self-attention mechanism in Transformers to boost the feature learning (the same as the segmentation tasks do), but does not sufficiently design for the registration tasks. Some other methods use cross-attention to model the corresponding relationship between moving and fixed images, such as Attention-Reg [22] and Xmorpher [21]. The cross-attention Transformer (CAT) module is used in the bottom layer of Attention-Reg [22] and each layer in Xmorpher [21] to establish the relationship between the features of moving and fixed images. However, the usage of Transformer in [21,22] is still limited to improving the feature learning, with no additional consideration given to the relationship between the attention mechanism and the deformation estimation. Furthermore, due to the large network structure of [21], only small windows can be created for similarity calculation, which may result in performance degradation. Few studies consider the relationship between attention and deformation estimation, such as Coordinate Translator [18] and Deformer [4]. Deformer [4] uses the calculation mode of multiplication of attention map and Value matrix in Transformer to weight the predicted basis to generate the deformation field, but its attention map calculation is only the concatenation and projection of moving and fixed feature maps, without using similarity calculation part. Coordinate Translator [18] calculates the matching score of the fixed feature map and the moving feature map. Then the computed scores are employed to re-weight the deformation field. However, for feature maps with coarse-level resolution, a voxel often has multiple possibilities of different motion modes [25], which is not considered in [18]. Traditional methods have explored multiple modes of deformations, e.g., probabilistic registration [12], to improve the performance.

In this study, we propose a novel motion decomposition Transformer (ModeT) to explicitly model multiple motion modalities by fully exploiting the intrinsic capability of the Transformer structure for deformation estimation. Experiments on two public brain magnetic resonance imaging (MRI) datasets demonstrate our method outcompetes several cutting-edge registration networks and Transformers. The main contributions of our work are summarized as follows:

- We propose to leverage the Transformer structure to naturally model the correspondence between images and convert it into the deformation field,

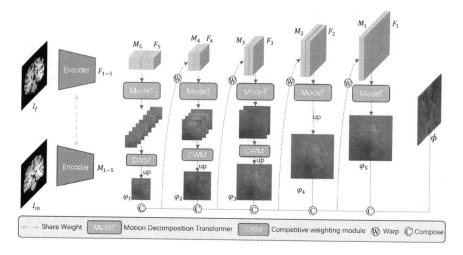

Fig. 1. Illustration of the proposed deformable registration network. The encoder takes the fixed image I_f and moving image I_m as input to extract hierarchical features F_1-F_5 and M_1-M_5. The motion decomposition Transformer (ModeT) is used to generate multiple deformation sub-fields and the competitive weighting module (CWM) fuses them. Finally the decoding pyramid outputs the total deformation field ϕ.

thus explicitly separating the two tasks of feature extraction and deformation estimation in deep-learning-based registration networks in which to make the registration procedure more sensible.

- The proposed ModeT makes full use of the multi-head neighborhood attention mechanism to efficiently model multiple motion modalities, and then the competitive weighting module (CWM) fuses multiple deformation sub-fields in a competitive way, which can improve the interpretability and consistency of the resulting deformation field.
- The pyramid structure is employed for feature extraction and deformation propagation, and is beneficial to reduce the scope of attention calculation required for each level.

2   Method

2.1   Network Overview

The proposed deformable registration network is illustrated in Fig. 1. We employ a pyramidal registration structure, which has the advantage of reducing the scope of attention calculation required at each decoding level and therefore alleviating the computational consumption. Given the fixed image I_f and moving image I_m as input, the encoder extracts hierarchical features using a 5-layer convolutional block, which doubles the number of channels in each layer. This generates two sets of feature maps F_1, F_2, F_3, F_4, F_5 and M_1, M_2, M_3, M_4, M_5. The feature maps M_5 and F_5 are sent into the ModeT to generate multiple deformation sub-fields, and then the generated deformation sub-fields are input into the CWM to obtain the fused deformation field φ_1 of the coarsest decoding layer as the initial of the total deformation field ϕ. The moving feature map M_4 is deformed using ϕ, and the deformed moving feature map is fed into the ModeT along with F_4 to generate multiple sub-fields, which are input into the CWM to get φ_2. Then φ_2 is compounded with previous total deformation field to generate the updated ϕ. The feature maps M_3 and F_3 go through the similar operations. As the decoding feature maps become finer, the number of motion modes at position p decreases, along with the number of attention heads we need to model. At the F_2/M_2 and F_1/M_1 levels, we no longer generate multiple deformation sub-fields, i.e., the number of attention heads in ModeT is 1. Finally, the obtained total deformation field ϕ is used to warp I_m to obtain the registered image.

To guide the network training, the normalized cross correlation $\mathcal{L}_{\mathrm{ncc}}$ [19] and the deformation regularization $\mathcal{L}_{\mathrm{reg}}$ [3] is used:

$$\mathcal{L}_{\mathrm{train}} = \mathcal{L}_{\mathrm{ncc}}(I_f, I_m \circ \phi) + \lambda \mathcal{L}_{\mathrm{reg}}(\phi), \qquad (1)$$

where $\circ$ is the warping operation, and λ is the weight of the regularization term.

2.2 Motion Decomposition Transformer (ModeT)

In deep-learning-based registration networks, a position p in the low-resolution feature map contains semantic information of a large area in the original image and therefore may often have multiple possibilities of different motion modalities. To model these possibilities, we employ a multi-head neighborhood attention mechanism to decompose different motion modalities at low-resolution level. The illustration of the motion decomposition is shown in Fig. 2.

Let $F, M \in \mathbb{R}^{c \times h \times w \times l}$ stand for the fixed and moving feature maps from a specific level of the hierarchical encoder, where h, w, l denote feature map size and c is the channel number. The feature maps F and M go through linear projection $(proj)$ and LayerNorm (LN) [2] to get Q $(query)$ and K (key):

$$Q = LN(proj(F)), \quad K = LN(proj(M)),$$
$$Q = \{Q^{(1)}, Q^{(2)}, \dots, Q^{(S)}\}, \qquad (2)$$
$$K = \{K^{(1)}, K^{(2)}, \dots, K^{(S)}\},$$

where the projection operation is shared weight, and the weight initialization is sampled from $N(0, 1e^{-5})$, the bias is initialized to 0. The Q and K are then divided according to channels, and S represents the number of divided heads.

We then calculate the neighborhood attention map. We use $c(p)$ to denote the neighborhood of voxel p. For a neighborhood of size $n \times n \times n$, $\|c(p)\| = n^3$. The neighborhood attention map of multiple heads is obtained by:

$$NA(p, s) = softmax(Q_p^{(s)} \cdot K_{c(p)}^{(s)T} + B^{(s)}), \qquad (3)$$

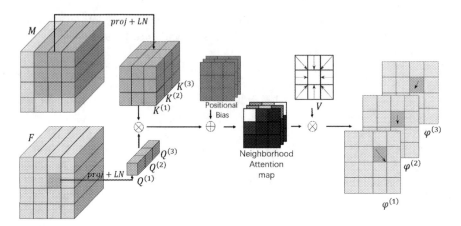

Fig. 2. Illustration of the proposed motion decomposition Transformer, which employs the multi-head neighborhood attention mechanism to decompose different motion modalities. ($S = 3$ in this illustration)

where $B \in \mathbb{R}^{S \times n \times n \times n}$ is a learnable relative positional bias, initialized to all zeros. We pad the moving feature map with zeros to calculate boundary voxels because the registration task sometimes requires voxels outside the field-of-view to be warped. Equation (3) shows how the neighborhood attention is computed for the s-th head at position p, so that the semantic information of voxels on low resolution can be decomposed to compute similarity one by one, in preparation for modeling different motion modalities. Moreover, the neighborhood attention operation narrows the scope of attention calculation to reduce the computational effort, which is very friendly to volumetric processing.

The next step is to obtain the multiple sub-fields at this level by computing the regular displacement field weighted via the neighborhood attention map:

$$\varphi_p^{(s)} = NA(p, s)V, \tag{4}$$

where $\varphi^{(s)} \in \mathbb{R}^{h \times w \times l \times 3}$, $V \in \mathbb{R}^{n \times n \times n}$, and V (*value*) represents the relative position coordinates for the neighborhood centroid, which is not learned so that the multi-head attention relationship can be naturally transformed into a multi-coordinate relationship. With the above steps, we obtain a series of deformation sub-fields for this level:

$$\varphi^{(1)}, \varphi^{(2)}, \dots, \varphi^{(S)} \tag{5}$$

2.3 Competitive Weighting Module (CWM)

Multiple low-resolution deformation fields need to be reasonably fused when deforming a high-resolution feature map. As shown in Fig. 3, we first upsample these deformation sub-fields, then convolve them in three layers to get the score of each sub-field, and use softmax to compete the motion modality for each

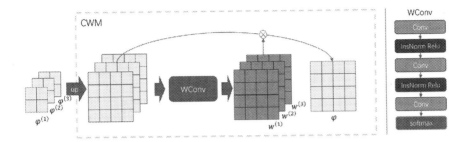

Fig. 3. Illustration of the proposed competitive weighting module (CWM).

voxel. The convolution uses $3 \times 3 \times 3$ convolution rather than direct projection because deformation fields often require correlation of adjacent displacements to determine if they are reasonable. We formulate above competitive weighting operation to obtain the deformation field φ at this level as follows:

$$
\begin{aligned}
w^{(1)}, w^{(2)}, \ldots, w^{(S)} &= WConv(cat(\varphi^{(1)}, \varphi^{(2)}, \ldots, \varphi^{(S)})), \\
\varphi &= w^{(1)}\varphi^{(1)} + w^{(2)}\varphi^{(2)} +, \ldots, + w^{(S)}\varphi^{(S)},
\end{aligned}
\tag{6}
$$

where $w^{(s)} \in \mathbb{R}^{h \times w \times l}$, and $\varphi^{(s)}$ has already been upsampled. $WConv$ represents the ConvBlock used to calculate weights, as shown in the right part of Fig. 3.

3 Experiments

Datasets. Experiments were carried on two public brain MRI datasets, including LPBA [20] and Mindboggle [16]. For LPBA, each MRI volume contains 54 manually labeled region-of-interests (ROIs). All volumes in LPBA were rigidly pre-aligned to mni305. 30 volumes (30×29 pairs) were employed for training and 10 volumes (10×9 pairs) were used for testing. For Mindboggle, each volume contains 62 manually labeled ROIs. All volumes in Mindboggle were affinely aligned to mni152. 42 volumes (42×41 pairs from the NKI-RS-22 and NKI-TRT-20 subsets) were employed for training, and 20 volumes from OASIS-TRT-20 (20×19 pairs) were used for testing. All volumes were pre-processed by min-max normalization, and skull-stripping using FreeSurfer [9]. The final size of each volume was $160 \times 192 \times 160$ after a center-cropping operation.

Evaluation Metrics. To quantitatively evaluate the registration performance, Dice score (DSC) [7] was calculated as the primary similarity metric to evaluate the degree of overlap between corresponding regions. In addition, the average symmetric surface distance (ASSD) [24] was evaluated, which can reflect the similarity of the region contours. The quality of the predicted deformation ϕ was assessed by the percentage of voxels with non-positive Jacobian determinant (i.e., folded voxels). All above metrics were calculated in 3D. A better registration shall have larger DSC, and smaller ASSD and Jacobian.

Table 1. The numerical results of different registration methods on two datasets.

	Mindboggle (62 ROIs)			LPBA (54 ROIs)		
	DSC (%)	ASSD	$\%|J_\phi| \leq 0$	DSC (%)	ASSD	$\%|J_\phi| \leq 0$
SyN [1]	56.7 ± 1.5	1.38 ± 0.09	$< 0.00001\%$	70.1 ± 6.2	1.72 ± 0.12	$< 0.0004\%$
VM [3]	56.0 ± 1.6	1.49 ± 0.11	$< 1\%$	64.3 ± 3.2	2.03 ± 0.21	$< 0.7\%$
TM [5]	60.7 ± 1.5	1.35 ± 0.10	$< 0.9\%$	67.0 ± 3.0	1.90 ± 0.20	$< 0.6\%$
I2G [18]	59.8 ± 1.3	1.30 ± 0.07	$< 0.03\%$	71.0 ± 1.4	1.64 ± 0.10	$< 0.01\%$
PR++ [14]	61.1 ± 1.4	1.34 ± 0.10	$< 0.5\%$	69.5 ± 2.2	1.76 ± 0.17	$< 0.2\%$
XM [21]	53.6 ± 1.5	1.46 ± 0.09	$< 1\%$	66.3 ± 2.0	1.92 ± 0.15	$< 0.1\%$
DMR [4]	60.6 ± 1.4	1.34 ± 0.09	$< 0.7\%$	69.2 ± 2.4	1.79 ± 0.18	$< 0.4\%$
Ours	$\mathbf{62.8 \pm 1.2}$	$\mathbf{1.22 \pm 0.07}$	$< 0.03\%$	$\mathbf{72.1 \pm 1.4}$	$\mathbf{1.58 \pm 0.11}$	$< 0.007\%$

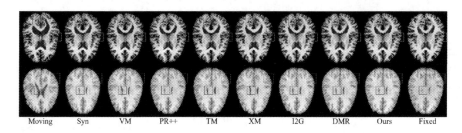

| Moving | Syn | VM | PR++ | TM | XM | I2G | DMR | Ours | Fixed |

Fig. 4. Visualized registration results from different methods on Mindboggle (top row) and LPBA (bottom row).

Implementation Details. Our method was implemented with PyTorch, using a GPU of NVIDIA Tesla V100 with 32GB memory. The regularization term λ and neighborhood size n were set as 1 and 3. For the encoder part, we used the same convolution structure as [18]. In the pyramid decoder, from coarse to fine, the number of attention heads were set as $8, 4, 2, 1, 1$, respectively. We used 6 channels for each attention head. The Adam optimizer [15] with a learning rate decay strategy was employed as follows:

$$lr_m = lr_{init} \cdot (1 - \frac{m-1}{M})^{0.9}, m = 1, 2, ..., M \tag{7}$$

where lr_m represents the learning rate of m-th epoch and $lr_{init} = 1e - 4$ represents the learning rate of initial epoch. We set the batch size as 1, M as 30 for training. In the inference phase, our method averagely took 0.56 second and 9GB memory to register a volume pair of size $160 \times 192 \times 160$.

Comparison Methods. We compared our method with several state-of-the-art registration methods: (1) SyN [1]: a classical traditional approach, using the *SyNOnly* setting in ANTS. (2) VoxelMorph(VM) [3]: a popular single-stage registration network. (3) TransMorph(TM) [5]: a single-stage registration network

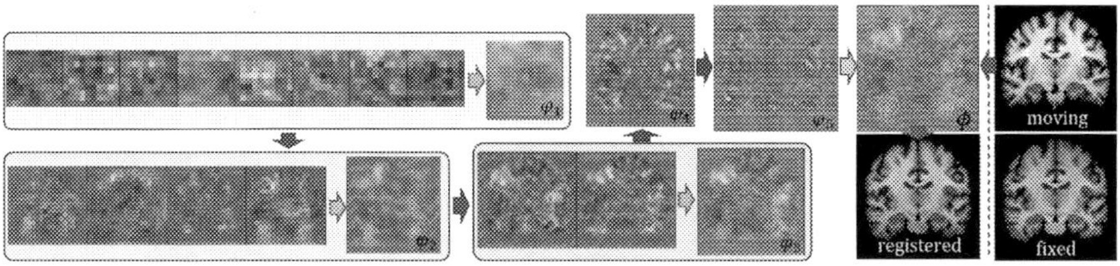

Fig. 5. Visualization of the generated multi-level deformation fields (φ_1-φ_5) to register one image pair. At low-resolution levels, multiple deformation sub-fields are decomposed to effectively model different motion modalities.

with SwinTransformer enhanced encoder. (4) `PR++` [14]: a pyramid registration network using 3D correlation layer. (5) `XMorpher(XM)` [21]: a registration network using CAT modules for each level of encoder and decoder. (6) `Im2grid(I2G)` [18]: a pyramid network using a coordinate translator. (7) `DMR` [4]: a registration network using a Deformer and a multi-resolution refinement module.

Quantitative and Qualitative Analysis. The numerical results of different methods on datasets Mindboggle and LPBA are reported in Table 1. It can be observed that our method consistently attained the best registration accuracy with respect to DSC and ASSD metrics. For the DSC results, our method surpassed the second-best networks by 1.7% and 1.1% on Mindboggle and LPBA, respectively. We further investigated the statistical significance of our method over comparison methods on DSC and ASSD metrics, by conducting the paired and two-sided Wilcoxon signed-rank test. The null hypotheses for all pairs (our method $v.s.$ other method) were not accepted at the 0.05 level. As a result, our method can be regarded as significantly better than all comparison methods on DSC and ASSD metrics. Table 1 also lists the percentage of voxels with non-positive Jacobian determinant ($\%|J_\phi| \leq 0$). Our method achieved satisfactory performance, which was the best among all deep-learning-based networks.

Figure 4 visualizes the registered images from different methods on two datasets. Our method generated more accurate registered images, and internal structures can be consistently preserved using our method. Figure 5 takes the registration of one image pair as an example to show the multi-level deformation fields generated by our method. Our ModeT effectively modeled multiple motion modalities and our CWM fused them together at low-resolution levels. The final deformation field ϕ accurately warped the moving image to registered with the fixed image.

4   Conclusion

We present a motion decomposition Transformer (ModeT) to naturally model the correspondence between images and convert this into the deformation field, which improves the interpretability of the deep-learning-based registration network. The proposed ModeT employs the multi-head neighborhood attention mechanism to identify various motion patterns of a voxel in the low-resolution feature map. Then with the help of competitive weighting module and pyramid structure, the motion modes contained in a voxel can be gradually fused and determined in the coarse-to-fine pyramid decoder. The experimental results have proven the superior performance of the proposed method. In our future study, we attempt to implement our ModeT in a more efficient way, and also investigate more effective fusion strategy to combine the displacement field from multiple attention heads.

Acknowledgements. This work was supported in part by the National Natural Science Foundation of China under Grants 62071305, 61701312, 81971631 and 62171290, in part by the Guangdong Basic and Applied Basic Research Foundation under Grant 2022A1515011241, and in part by the Shenzhen Science and Technology Program (No. SGDX 20201103095613036).

References

1. Avants, B., Epstein, C., Grossman, M., Gee, J.: Symmetric diffeomorphic image registration with cross-correlation: evaluating automated labeling of elderly and neurodegenerative brain. Med. Image Anal. **12**(1), 26–41 (2008)
2. Ba, J.L., Kiros, J.R., Hinton, G.E.: Layer normalization. arXiv preprint arXiv:1607.06450 (2016)
3. Balakrishnan, G., Zhao, A., Sabuncu, M.R., Guttag, J., Dalca, A.V.: VoxelMorph: a learning framework for deformable medical image registration. IEEE Trans. Med. Imaging **38**(8), 1788–1800 (2019)
4. Chen, J., et al.: Deformer: towards displacement field learning for unsupervised medical image registration. In: Wang, L., Dou, Q., Fletcher, P.T., Speidel, S., Li, S. (eds.) MICCAI 2022. LNCS, vol. 13436, pp. 141–151. Springer, Cham (2022). https://doi.org/10.1007/978-3-031-16446-0_14
5. Chen, J., Frey, E.C., He, Y., Segars, W.P., Li, Y., Du, Y.: Transmorph: transformer for unsupervised medical image registration. Med. Image Anal. **82**, 102615 (2022)
6. Chen, J., He, Y., Frey, E.C., Li, Y., Du, Y.: ViT-V-Net: vision transformer for unsupervised volumetric medical image registration. arXiv preprint arXiv:2104.06468 (2021)
7. Dice, L.R.: Measures of the amount of ecologic association between species. Ecology **26**(3), 297–302 (1945)
8. Dosovitskiy, A., et al.: An image is worth 16x16 words: transformers for image recognition at scale. In: International Conference on Learning Representations (ICLR) (2021)
9. Fischl, B.: FreeSurfer. NeuroImage **62**(2), 774–781 (2012)
10. Fu, Y., Lei, Y., Wang, T., Curran, W.J., Liu, T., Yang, X.: Deep learning in medical image registration: a review. Phys. Med. Biol. **65**(20), 20TR01 (2020)
11. He, K., et al.: Transformers in medical image analysis. Intell. Med. **3**(1), 59–78 (2023)
12. Heinrich, M.P., Simpson, I.J., Papież, B.W., Brady, S.M., Schnabel, J.A.: Deformable image registration by combining uncertainty estimates from supervoxel belief propagation. Med. Image Anal. **27**, 57–71 (2016)

13. Jaderberg, M., Simonyan, K., Zisserman, A., et al.: Spatial transformer networks. In: Advances in Neural Information Processing Systems, pp. 2017–2025 (2015)
14. Kang, M., Hu, X., Huang, W., Scott, M.R., Reyes, M.: Dual-stream pyramid registration network. Med. Image Anal. **78**, 102379 (2022)
15. Kingma, D.P., Ba, J.: Adam: a method for stochastic optimization. arXiv preprint arXiv:1412.6980 (2014)
16. Klein, A., Tourville, J.: 101 labeled brain images and a consistent human cortical labeling protocol. Front. Neurosci. **6**, 171 (2012)
17. Li, J., Chen, J., Tang, Y., Wang, C., Landman, B.A., Zhou, S.K.: Transforming medical imaging with transformers? A comparative review of key properties, current progresses, and future perspectives. Med. Image Anal. **85**, 102762 (2023)
18. Liu, Y., Zuo, L., Han, S., Xue, Y., Prince, J.L., Carass, A.: Coordinate translator for learning deformable medical image registration. In: Li, X., Lv, J., Huo, Y., Dong, B., Leahy, R.M., Li, Q. (eds.) MMMI 2022. LNCS, vol. 13594, pp. 98–109. Springer, Cham (2022). https://doi.org/10.1007/978-3-031-18814-5_10
19. Rao, Y.R., Prathapani, N., Nagabhooshanam, E.: Application of normalized cross correlation to image registration. Int. J. Res. Eng. Technol. **3**(5), 12–16 (2014)
20. Shattuck, D.W., et al.: Construction of a 3D probabilistic atlas of human cortical structures. Neuroimage **39**(3), 1064–1080 (2008)
21. Shi, J., et al.: Xmorpher: full transformer for deformable medical image registration via cross attention. In: Wang, L., Dou, Q., Fletcher, P.T., Speidel, S., Li, S. (eds.) MICCAI 2022. LNCS, vol. 13436, pp. 217–226. Springer, Cham (2022). https://doi.org/10.1007/978-3-031-16446-0_21
22. Song, X., et al.: Cross-modal attention for MRI and ultrasound volume registration. In: de Bruijne, M., et al. (eds.) MICCAI 2021. LNCS, vol. 12904, pp. 66–75. Springer, Cham (2021). https://doi.org/10.1007/978-3-030-87202-1_7
23. Sotiras, A., Davatzikos, C., Paragios, N.: Deformable medical image registration: a survey. IEEE Trans. Med. Imaging **32**(7), 1153–1190 (2013)
24. Taha, A.A., Hanbury, A.: Metrics for evaluating 3D medical image segmentation: analysis, selection, and tool. BMC Med. Imaging **15**(1), 1–28 (2015)
25. Zheng, J.Q., Wang, Z., Huang, B., Lim, N.H., Papiez, B.W.: Residual aligner network. arXiv preprint arXiv:2203.04290 (2022)
26. Zhu, Y., Lu, S.: Swin-VoxelMorph: a symmetric unsupervised learning model for deformable medical image registration using swin transformer. In: Wang, L., Dou, Q., Fletcher, P.T., Speidel, S., Li, S. (eds.) MICCAI 2022. LNCS, vol. 13436, pp. 78–87. Springer, Cham (2022). https://doi.org/10.1007/978-3-031-16446-0_8

Non-iterative Coarse-to-Fine Transformer Networks for Joint Affine and Deformable Image Registration

Mingyuan Meng[1,2] (ID), Lei Bi[2(✉)] (ID), Michael Fulham[1,3] (ID), Dagan Feng[1,4] (ID), and Jinman Kim[1] (ID)

[1] School of Computer Science, The University of Sydney, Sydney, Australia
[2] Institute of Translational Medicine, Shanghai Jiao Tong University, Shanghai, China
lei.bi@sjtu.edu.cn
[3] Department of Molecular Imaging, Royal Prince Alfred Hospital, Sydney, Australia
[4] Med-X Research Institute, Shanghai Jiao Tong University, Shanghai, China

Abstract. Image registration is a fundamental requirement for medical image analysis. Deep registration methods based on deep learning have been widely recognized for their capabilities to perform fast end-to-end registration. Many deep registration methods achieved state-of-the-art performance by performing coarse-to-fine registration, where multiple registration steps were iterated with cascaded networks. Recently, Non-Iterative Coarse-to-finE (NICE) registration methods have been proposed to perform coarse-to-fine registration in a single network and showed advantages in both registration accuracy and runtime. However, existing NICE registration methods mainly focus on deformable registration, while affine registration, a common prerequisite, is still reliant on time-consuming traditional optimization-based methods or extra affine registration networks. In addition, existing NICE registration methods are limited by the intrinsic locality of convolution operations. Transformers may address this limitation for their capabilities to capture long-range dependency, but the benefits of using transformers for NICE registration have not been explored. In this study, we propose a Non-Iterative Coarse-to-finE Transformer network (NICE-Trans) for image registration. Our NICE-Trans is the first deep registration method that (i) performs joint affine and deformable coarse-to-fine registration within a single network, and (ii) embeds transformers into a NICE registration framework to model long-range relevance between images. Extensive experiments with seven public datasets show that our NICE-Trans outperforms state-of-the-art registration methods on both registration accuracy and runtime.

Keywords: Image Registration · Coarse-to-fine Registration · Transformer

Supplementary Information The online version contains supplementary material available at https://doi.org/10.1007/978-3-031-43999-5_71.

1 Introduction

Image registration is a fundamental requirement for medical image analysis and has been an active research focus for decades [1]. It aims to find a spatial transformation between a pair of fixed and moving images, through which the moving image can be warped to spatially align with the fixed image. Similar to natural image registration [2], medical image registration usually requires affine registration to eliminate rigid misalignments and then performs additional deformable registration to address non-rigid deformations. Traditional methods usually formulate medical image registration as a time-consuming iterative optimization problem [3, 4]. Recently, deep registration methods based on deep learning have been widely adopted to perform end-to-end registration [5, 6]. Deep registration methods learn a mapping from image pairs to spatial transformations based on training data in an unsupervised manner, which have shown advantages in registration accuracy and computational efficiency [7–18].

Many deep registration methods perform coarse-to-fine registration to improve registration accuracy, where the registration is decoupled into multiple coarse-to-fine registration steps that are iteratively performed by using multiple cascaded networks [10–13] or repeatedly running a single network for multiple iterations [14, 15]. Mok et al. [13] proposed a Laplacian pyramid Image Registration Network (LapIRN), where multiple networks at different pyramid levels were cascaded. Shu et al. [14] proposed to use a single network (ULAE-net) to perform coarse-to-fine registration with multiple iterations. These methods perform iterative coarse-to-fine registration and extract image features repeatedly in each iteration, which inevitably increases computational loads and prolongs the registration runtime. Recently, Non-Iterative Coarse-to-finE (NICE) registration methods have been proposed to perform coarse-to-fine registration with a single network in a single iteration [16–18]. For example, we previously proposed a NICE registration network (NICE-Net) [18, 19], where multiple coarse-to-fine registration steps are performed with a single network in a single iteration. These NICE registration methods show advantages in both registration accuracy and runtime on the benchmark task of intra-patient brain MRI registration. Nevertheless, we identified that existing NICE registration methods still have two main limitations.

Firstly, existing NICE registration methods merely focus on deformable coarse-to-fine registration, while affine registration, a common prerequisite, is still reliant on traditional registration methods [16, 18] or extra affine registration networks [17]. Using traditional registration methods incurs time-consuming iterative optimization, while cascading extra networks consumes additional computational resources (e.g., extra GPU memory and runtime). Secondly, existing NICE registration methods are based on Convolution Neural Networks (CNN) and thus are limited by the intrinsic locality (i.e., limited receptive field) of convolution operations. Transformers have been widely adopted in many medical applications for their capabilities to capture long-range dependency [20]. Recently, transformers have also been shown to improve registration with conventional Voxelmorph [7]-like architecture [21–23]. However, the benefits of using transformers for NICE registration have not been explored.

In this study, we propose a Non-Iterative Coarse-to-finE Transformer network (NICE-Trans) for joint affine and deformable registration. Our technical contributions are two folds: (i) We extend the existing NICE registration framework to affine registration, where multiple steps of both affine and deformable coarse-to-fine registration are performed with a single network in a single iteration. (ii) We explore the benefits of transformers for NICE registration, where Swin Transformer [24] is embedded into the NICE-Trans to model long-range relevance between fixed and moving images. This is the first deep registration method that integrates previously separated affine and deformable coarse-to-fine registration into a single network, and this is also the first deep registration method that exploits transformers for NICE registration. Extensive experiments with seven public datasets show that our NICE-Trans outperforms state-of-the-art registration methods on both registration accuracy and runtime.

2 Method

Image registration aims to find a spatial transformation ϕ that warps a moving image I_m to a fixed image I_f, so that the warped image $I_{m\circ\phi} = I_m \circ \phi$ is spatially aligned with the I_f. In this study, we assume the I_m and I_f are two single-channel, grayscale volumes defined in a 3D spatial domain $\Omega \subset \mathbb{R}^3$, which is consistent with common medical image registration studies [7–18]. The ϕ is parameterized as a displacement field, and we parametrized the image registration problem as a function $\mathcal{R}_\theta(I_f, I_m) = \phi$ using NICE-Trans. As shown in Fig. 1, our NICE-Trans consists of an intra-image feature learning encoder and an inter-image relevance modeling decoder (refer to Sect. 2.1). Multiple steps of affine and deformable registration are performed within a single network iteration (refer to Sect. 2.2). The θ is a set of learnable parameters that are optimized through unsupervised learning (refer to Sect. 2.3).

2.1 Non-iterative Coarse-to-fine Transformer Networks (NICE-Trans)

The architecture of the proposed NICE-Trans is presented in Fig. 1, which consists of a dual-path encoder to learn image features from I_m and I_f separately and a single-path decoder to model the spatial relevance between I_m and I_f. Skip connections are used at multiple scales to propagate features from the encoder to the decoder. Here, we assume the NICE-Trans performs L_a and L_d steps of affine and deformable registration, resulting in a total of $L = L_a + L_d$ steps of coarse-to-fine registration.

The encoder has two identical, weight-shared paths P_m and P_f that take I_m and I_f as input, respectively. Each path consists of L successive Conv modules with $2 \times 2 \times 2$ max pooling applied between two adjacent modules, which produces two L-level feature pyramids $F_m \in \{F_m^1, F_m^2, \ldots, F_m^L\}$ and $F_f \in \{F_f^1, F_f^2, \ldots, F_f^L\}$, where the F_f^i and F_m^i are the output of the ith Conv module in the P_f and P_m. Each Conv module consists of two $3 \times 3 \times 3$ convolutional layers followed by LeakyReLU activation with parameter 0.2. This dual-path design can learn uncoupled image features of I_m and I_f, which enables the NICE-Trans to reuse the learned features at multiple registration steps, thereby discarding the requirement for repeated feature learning.

The decoder consists of $L-1$ SwinTrans modules and a Conv module, with a patch expanding layer [23] applied between two adjacent modules to double the feature resolution and halve the feature dimension. Each SwinTrans module consists of one 1

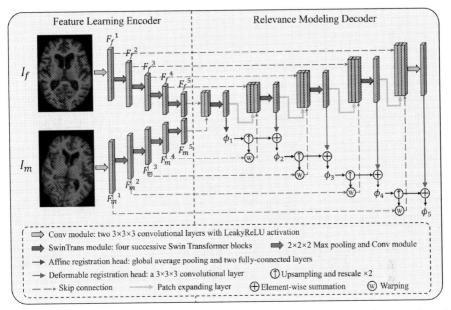

Fig. 1. The Architecture of our NICE-Trans. The affine and deformable registration steps, L_a and L_d, are set as 1 and 4 for illustration.

$\times 1 \times 1$ convolutional layer for feature dimension reduction and four successive Swin Transformer blocks [24] including layer normalization, Window/Shifted Window-based Multi-head Self-Attention (W/SW-MSA), Multilayer Perceptron (MLP), and residual connections. The output of each decoder module is fed into an affine or deformable registration head that maps the input features into a displacement field, which produces L displacement fields $\phi_i \in \{\phi_1, \phi_2, \ldots, \phi_L\}$ for L steps of coarse-to-fine registration (detailed in Sect. 2.2). The output of each patch expanding layer is concatenated with F_f^i and $F_m^i \circ \phi_{i-1}$, which is then fed into its later decoder module. The decoder performs finer registration after each decoder module, where the ϕ_L is the final output ϕ. Detailed architecture settings (e.g., feature dimensions, head numbers of self-attention) are presented in the supplementary materials.

Our NICE-Trans differs from the existing NICE-Net [18] mainly in two aspects: (i) our NICE-Trans integrates affine and deformable registration into a unified network, and (ii) our NICE-Trans leverages Swin Transformer to model long-range spatial relevance between I_m and I_f. In addition, the existing NICE-Net extracts features from the intermediately warped image at each registration step, while our NICE-Trans directly warps the F_m to avoid this process and achieves similar performance.

2.2 Joint Affine and Deformable Registration

The output features of the first L_a decoder modules are fed into L_a affine registration heads, where the features are mapped to a 3×4 affine matrix through global average pooling and two fully-connected layers, which are then sampled as a dense displacement field. After the first L_a steps of affine registration, the output features of the last L_d decoder modules are fed into L_d deformable registration heads, where the features are directly mapped to a dense displacement field via a $3 \times 3 \times 3$ convolutional layer.

At the beginning of coarse-to-fine registration, the ϕ_1 is the output of the first registration head. Then, the ϕ_1 is upsampled ($\times 2$) and voxel-wisely added to the output of the second registration head to derive ϕ_2. This process is repeated until the ϕ_L is derived, which realizes joint affine and deformable coarse-to-fine registration. In our experiments, we set L_a and L_d as 1 and 4 (illustrated in Fig. 1) as this setting achieved the best validation results (refer to the supplementary materials). Figure 2 exemplifies a registration result of the NICE-Trans with five steps of coarse-to-fine registration.

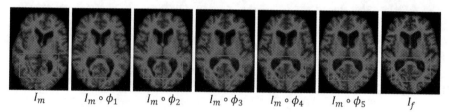

$$I_m \qquad I_m \circ \phi_1 \qquad I_m \circ \phi_2 \qquad I_m \circ \phi_3 \qquad I_m \circ \phi_4 \qquad I_m \circ \phi_5 \qquad I_f$$

Fig. 2. Registration results of the NICE-Trans with $L_a = 1$ and $L_d = 4$. From left to right are the moving image, the images warped by 5 registration steps, and the fixed image.

2.3 Unsupervised Learning

The learnable parameters θ are optimized using an unsupervised loss $\mathcal{L}$ that does not require labels. The $\mathcal{L}$ is defined as $\mathcal{L} = \mathcal{L}_{sim} + \sigma \mathcal{L}_{reg}$, where the $\mathcal{L}_{sim}$ is an image similarity term that penalizes the differences between the warped image $I_{m \circ \phi}$ and the fixed image I_f, the $\mathcal{L}_{reg}$ is a regularization term that encourages smooth and invertible transformations ϕ, and the σ is a regularization parameter.

We adopt negative local normalized cross-correlation (NCC) as the $\mathcal{L}_{sim}$, which is a widely used similarity metric in image registration methods [7–10, 12–18]. For the $\mathcal{L}_{reg}$, we impose a diffusion regularizer on the ϕ to encourage its smoothness and also adopt a Jacobian Determinant (JD) loss [25] to enhance its invertibility. As the ϕ is not invertible at voxel p where the Jacobian determinant is negative ($|J\phi(p)| \leq 0$) [26], the JD loss explicitly penalizes the negative Jacobian determinants of ϕ. Finally, the $\mathcal{L}_{reg}$ is defined as $\mathcal{L}_{reg} = \sum_{p \in \Omega} ||\nabla \phi(p)||^2 + \lambda JD(\phi)$, where the λ is a regularization parameter balancing registration accuracy and transformation invertibility.

3   Experimental Setup

3.1   Dataset and Preprocessing

We evaluated the proposed NICE-Trans on the task of inter-patient brain MRI registration, which is a common benchmark task in medical image registration studies [7–9, 12–18]. We followed the dataset settings in [18]: 2,656 brain MRI images acquired from four public datasets (ADNI [27], ABIDE [28], ADHD [29], and IXI [30]) were used for training; two public brain MRI datasets with anatomical segmentation (Mindboggle [31] and Buckner [32]) were used for validation and testing. The Mindboggle dataset contains 100 MRI images and were randomly split into 50/50 images for validation/testing. The Buckner dataset contains 40 MRI images and were used for testing only. In addition to the original settings of [18], we adopted an additional public brain MRI dataset (LPBA [33]) for testing, which contains 40 MRI images.

We performed brain extraction and intensity normalization for each MRI image with FreeSurfer [32]. Each image was placed at the same position via Center of Mass (CoM) initialization [34], and then was cropped into $144 \times 192 \times 160$ voxels.

3.2 Implementation Details

We implemented our NICE-Trans using PyTorch on a NVIDIA Titan V GPU with 12 GB memory. We used an ADAM optimizer with a learning rate of 0.0001 and a batch size of 1 to train the NICE-Trans for 100,000 iterations. At each iteration, two images were randomly picked from the training data as the fixed and moving images. A total of 100 image pairs, randomly picked from the validation data, were used to monitor the training process and to optimize hyper-parameters. We set σ as 1 to ensure that the $\mathcal{L}_{sim}$ and $\sigma \mathcal{L}_{reg}$ have close values, while the λ was set as 10^{-4} to ensure that the percentage of voxels with negative Jacobian determinants is less than 0.05% (refer to the supplementary materials for detailed regularization analysis). Our code will be available in https://github.com/ MungoMeng/Registration-NICE-Trans.

3.3 Comparison Methods

Our NICE-Trans was compared with nine image registration methods, including two traditional methods and seven deep registration methods. The compared traditional methods are SyN [3] and NiftyReg [4]. For these methods, we used cross-correlation as the similarity measure and adopted FLIRT [35] for affine registration. The compared deep registration methods are VoxelMorph (VM) [7], Diffeomorphic VoxelMorph (DifVM) [8], TransMorph [21], Swin-VoxelMorph (Swin-VM) [22], LapIRN [13], ULAE-net [14], and NICE-Net [18]. The VM and DifVM are two commonly benchmarked registration methods in the literature [12–18, 21–23]. The TransMorph and Swin-VM are two state-of-the-art methods that embed Swin Transformer into VM-like architecture. The LapIRN, ULAE-net, and NICE-Net are three state-of-the-art coarse-to-fine registration methods. For the compared deep registration methods, we adopted NCC as the similarity loss and followed [17, 36] to cascade a CNN-based registration network (AffineNet) for affine registration.

3.4 Experimental Settings

We compared the NICE-Net to the nine comparison methods for subject-to-subject registration. For testing, we randomly picked 100 image pairs from each of the Mindboggle, Buckner, and LPBA testing sets. We used standard evaluation metrics for medical image registration [7–18]. The registration accuracy was evaluated using the Dice similarity coefficients (DSC) of segmentation labels, while the smoothness and invertibility of spatial transformations were evaluated using the percentage of Negative Jacobian Determinants (NJD). Generally, a higher DSC and a lower NJD indicate better registration performance. A two-sided P value less than 0.05 is considered to indicate a statistically significant difference between two DSCs.

We also performed an ablation study to explore the benefits of transformers. We built a baseline method that has the same architecture as the NICE-Trans but only uses Conv modules. After that, we embedded Swin Transformer into the baseline method, where SwinTrans modules replaced the Conv modules in the encoder (Trans-Encoder), decoder (Trans-Decoder), or both (Trans-All).

4 Results and Discussion

Table 1 presents the registration performance of our NICE-Trans and all comparison methods. The registration accuracy of all methods degraded by 1–3% in DSC when affine registration was not performed, which demonstrates the importance of affine registration. However, using FLIRT or AffineNet for affine registration incurred extra computational loads and increased the registration runtime. Our NICE-Trans performed joint affine and deformable registration, which enabled it to realize affine registration with negligible additional runtime. Moreover, we suggest that integrating affine and deformable registration into a single network also brings convenience for network training. Training two separate affine and deformable registration networks will prolong the whole training time, while joint training will consume more GPU memory. As for registration accuracy, the TransMorph and Swin-VM achieved higher DSCs than the conventional VM and DifVM, but still cannot outperform the existing CNN-based coarse-to-fine registration methods (LapIRN, ULAE-net, and NICE-Net). Our NICE-Trans leverages Swin Transformer to perform coarse-to-fine registration, which enabled it to achieve the highest DSCs among all methods. This means that our NICE-Trans also has advantages on registration accuracy. We present a qualitative comparison in the supplementary materials, which shows that the registration result produced by our NICE-Trans is more consistent with the fixed image. In addition, there usually exists a trade-off between DSC and NJD as imposing constraints on the spatial transformations limits their flexibility, which results in degraded registration accuracy [13, 18]. For example, compared with VM, the DifVM with diffeomorphic constraints achieved better NJDs and worse DSCs. Nevertheless, our NICE-Trans achieved both the best DSCs and NJDs. We suggest that, if we set λ as 0 to maximize the registration accuracy with the cost of transformation invertibility, our NICE-Trans can achieve higher DSCs and outperform the comparison methods by a larger margin (refer to the regularization analysis in the supplementary materials).

Table 2 shows the results of our ablation study. Swin Transformer improved the registration performance when embedded into the decoder, but had limited benefits in the encoder. This suggests that Swin Transformer can benefit registration in modeling inter-image spatial relevance while having limited benefits in learning intra-image representations. This finding is intuitive as image registration aims to find spatial relevance between images, instead of finding the internal relevance within an image. Under

Table 1. Registration performance of our NICE-Trans and all comparison methods.

Method		Mindboggle		Buckner		LPBA		Runtime (s)	
		DSC	NJD (%)	DSC	NJD (%)	DSC	NJD (%)	CPU	GPU
Before registration		0.269*	/	0.330*	/	0.536*	/	/	/
FLIRT (affine only)		0.347*	/	0.406*	/	0.626*	/	58.2	/
SyN (no affine)		0.535*	0.25	0.566*	0.28	0.674*	0.12	3688	/
NiftyReg (no affine)		0.558*	0.32	0.601*	0.35	0.690*	0.15	165	/
FLIRT +	SyN	0.548*	0.26	0.577*	0.25	0.692*	0.09	3746	/
	NiftyReg	0.567*	0.34	0.610*	0.30	0.705*	0.13	223	/
AffineNet (affine only)		0.341*	/	0.400*	/	0.611*	/	1.12	0.118
VM (no affine)		0.518*	2.63	0.558*	2.37	0.663*	1.21	3.85	0.395
DifVM (no affine)		0.502*	0.042	0.548*	0.032	0.671*	0.005	3.92	0.446
TransMorph (no affine)		0.545*	2.25	0.585*	2.15	0.682*	1.27	3.90	0.432
Swin-VM (no affine)		0.542*	0.022	0.589*	0.017	0.684*	0.004	5.82	0.550
LapIRN (no affine)		0.563*	0.046	0.599*	0.039	0.688*	0.006	6.52	0.624
ULAE-net (no affine)		0.579*	2.08	0.611*	2.00	0.695*	1.06	7.21	0.730
NICE-Net (no affine)		0.580*	0.048	0.611*	0.034	0.696*	0.004	4.17	0.423
AffineNet +	VM	0.548*	2.54	0.580*	2.24	0.682*	1.17	4.97	0.513
	DifVM	0.526*	0.048	0.565*	0.027	0.686*	0.005	5.04	0.564
	TransMorph	0.568*	2.14	0.604*	2.18	0.694*	1.10	5.02	0.550
	Swin-VM	0.563*	0.024	0.607*	0.021	0.696*	0.003	6.94	0.668
	LapIRN	0.581*	0.042	0.611*	0.036	0.699*	0.006	7.64	0.742
	ULAE-net	0.595*	2.12	0.625*	1.92	0.705*	0.97	8.33	0.848
	NICE-Net	0.596*	0.034	0.624*	0.026	0.705*	0.004	5.29	0.541
NICE-Trans (affine only)		0.353*	/	0.410*	/	0.618*	/	1.04	0.105
NICE-Trans (no affine)		0.594*	0.018	0.622*	0.016	0.704*	0.003	4.52	0.480
NICE-Trans (ours)		**0.612**	**0.016**	**0.636**	**0.015**	**0.715**	**0.002**	4.69	**0.486**

Bold: the best DSC and NJD in each testing dataset and the shortest runtime of completing both affine and deformable registration. *: $P<0.05$, in comparison to NICE-Trans (ours).

this aim, embedding transformers in the decoder helps to capture long-range relevance between images and improves registration performance. We noticed that previous studies gained improvements by embedding Swin Transformer in the encoder [21] or leveraging a full transformer network [22]. This is attributed to the fact that they used a VM-like architecture that entangles image representation learning and spatial relevance modeling throughout the whole network. Our NICE-Trans decouples these two parts and provides further insight on using transformers for registration: leveraging transformers to learn intra-image relevance might not be beneficial but merely incurs extra computational loads.

Table 2. Results of our ablation study.

Method	Mindboggle		Buckner		LPBA		Runtime (s)	
	DSC	NJD (%)	DSC	NJD (%)	DSC	NJD (%)	CPU	GPU
Baseline	0.600	0.028	0.627	0.025	0.706	0.003	4.25	0.438
Trans-Encoder	0.600	0.024	0.625	0.018	0.705	0.003	4.82	0.488
Trans-Decoder (ours)	**0.612**	0.016	**0.636**	**0.015**	**0.715**	**0.002**	4.69	0.486
Trans-All	**0.612**	**0.015**	0.634	0.017	0.714	**0.002**	5.48	0.557

Bold: the best DSC and NJD in each testing dataset.

It should be acknowledged that there are a few limitations in our study. First, the experiment (Table 1) demonstrated that our NICE-Trans can well address the inherent misalignments among inter-patient brain MRI images, but the sensitivity of affine registration to different degrees of misalignments is still awaiting further exploration. Second, in this study, we evaluated the NICE-Trans on the benchmark task of inter-patient brain MRI registration, while we believe that our NICE-Trans also could apply to other image registration applications (e.g., brain tumor registration [37]).

5   Conclusion

We have outlined a Non-Iterative Coarse-to-finE Transformer network (NICE-Trans) for medical image registration. Unlike the existing image registration methods, our NICE-Trans performs joint affine and deformable coarse-to-fine registration with a single network in a single iteration. The experimental results show that our NICE-Trans can outperform the state-of-the-art coarse-to-fine or transformer-based deep registration methods on both registration accuracy and runtime. Our study also suggests that transformers benefit registration in modeling inter-image spatial relevance while having limited benefits in learning intra-image representations.

Acknowledgement. This work was supported by Australian Research Council (ARC) under Grant DP200103748.

References

1. Sotiras, A., Davatzikos, C., Paragios, N.: Deformable medical image registration: a survey. IEEE Trans. Med. Imaging **32**(7), 1153–1190 (2013)
2. Meng, M., Liu, S.: High-quality panorama stitching based on asymmetric bidirectional optical flow. In: International Conference on Computational Intelligence and Applications (ICCIA), pp. 118–122 (2020)
3. Avants, B.B., Epstein, C.L., Grossman, M., Gee, J.C.: Symmetric diffeomorphic image registration with cross-correlation: evaluating automated labeling of elderly and neurodegenerative brain. Med. Image Anal. **12**(1), 26–41 (2008)
4. Modat, M., et al.: Fast free-form deformation using graphics processing units. Comput. Meth. Programs Biomed. **98**(3), 278–284 (2010)
5. Haskins, G., Kruger, U., Yan, P.: Deep learning in medical image registration: a survey. Mach. Vis. Appl. **31**, 8 (2020)
6. Xiao, H., et al.: A review of deep learning-based three-dimensional medical image registration methods. Quant. Imaging Med. Surg. **11**(12), 4895–4916 (2021)
7. Balakrishnan, G., et al.: Voxelmorph: a learning framework for deformable medical image registration. IEEE Trans. Med. Imaging **38**(8), 1788–1800 (2019)
8. Dalca, A.V., et al.: Unsupervised learning of probabilistic diffeomorphic registration for images and surfaces. Med. Image Anal. **57**, 226–236 (2019)
9. Meng, M., et al.: Enhancing medical image registration via appearance adjustment networks. Neuroimage **259**, 119444 (2022)
10. De Vos, B.D., et al.: A deep learning framework for unsupervised affine and deformable image registration. Med. Image Anal. **52**, 128–143 (2019)

11. Hering, A., van Ginneken, B., Heldmann, S.: mlVIRNET: multilevel variational image registration network. In: Shen, D., et al. (eds.) MICCAI 2019. LNCS, vol. 11769, pp. 257–265. Springer, Cham (2019)

12. Zhao, S., et al.: Recursive cascaded networks for unsupervised medical image registration. In: IEEE International Conference on Computer Vision, pp. 10600–10610 (2019)

13. Mok, T.C.W., Chung, A.C.S.: Large deformation diffeomorphic image registration with Laplacian pyramid networks. In: Martel, A.L., et al. (eds.) MICCAI 2020. LNCS, vol. 12263, pp. 211–221. Springer, Cham (2020)

14. Shu, Y., et al.: Medical image registration based on uncoupled learning and accumulative enhancement. In: deBruijne, M., et al. (eds.) MICCAI 2021. LNCS, vol. 12904, pp. 3–13. Springer, Cham (2021)

15. Hu, B., Zhou, S., Xiong, Z., Wu, F.: Recursive decomposition network for deformable image registration. IEEE J. Biomed. Health Inform. 26(10), 5130–5141 (2022)

16. Kang, M., et al.: Dual-stream pyramid registration network. Med. Image Anal. 78, 102379 (2022)

17. Lv, J., et al.: Joint progressive and coarse-to-fine registration of brain MRI via deformation field integration and non-rigid feature fusion. IEEE Trans. Med. Imaging 41(10), 2788–2802 (2022)

18. Meng, M., Bi, L., Feng, D., Kim, J.: Non-iterative coarse-to-fine registration based on single-pass deep cumulative learning. In: Wang, L., et al. (eds.) MICCAI 2022. LNCS, vol. 13436, pp. 88–97. Springer, Cham (2022)

19. Meng, M., Bi, L., Feng, D. Kim, J.: Brain Tumor Sequence Registration with Non-iterative Coarse-to-fine Networks and Dual Deep Supervision. arXiv preprint arXiv:2211.07876 (2022)

20. Dosovitskiy, A., et al.: An image is worth 16×16 words: transformers for image recognition at scale. In: International Conference on Learning Representations (2021)

21. Chen, J., et al.: Transmorph: transformer for unsupervised medical image registration. Med. Image Anal. 82, 102615 (2022)

22. Zhu, Y., Lu, S.: Swin-voxelmorph: a symmetric unsupervised learning model for deformable medical image registration using swin transformer. In: Wang, L., et al. (eds.) MICCAI 2022. LNCS, vol. 13436, pp. 78–87. Springer, Cham (2022)

23. Shi, J., et al.: Xmorpher: full transformer for deformable medical image registration via cross attention. In: Wang, L., et al. (eds.) MICCAI 2022. LNCS, vol. 13436, pp. 217–226. Springer, Cham (2022)

24. Liu, Z., et al.: Swin transformer: hierarchical vision transformer using shifted windows. In: IEEE/CVF International Conference on Computer Vision, pp. 10012–10022 (2021)

25. Kuang, D., Schmah, T.: Faim–a convnet method for unsupervised 3d medical image registration. In: Suk, H.I., et al. (eds.) MLMI 2019. LNCS, vol. 11861, pp. 646–654. Springer, Cham (2019)

26. Ashburner, J.: A fast diffeomorphic image registration algorithm. Neuroimage 38(1), 95–113 (2007)

27. Mueller, S.G., et al.: Ways toward an early diagnosis in Alzheimer's disease: the Alzheimer's Disease Neuroimaging Initiative (ADNI). Alzheimers Dement. 1(1), 55–66 (2005)

28. Martino, D., et al.: The autism brain imaging data exchange: towards a large-scale evaluation of the intrinsic brain architecture in autism. Mol. Psychiatry 19(6), 659–667 (2014)

29. ADHD-200 consortium.: the ADHD-200 consortium: a model to advance the translational potential of neuroimaging in clinical neuroscience. Front. Syst. Neurosci. 6, 62 (2012)

30. The Information eXtraction from Images (IXI) dataset. https://brain-development.org/ixi-dataset/. Accessed 31 Oct 2022

31. Klein, A., Tourville, J.: 101 labeled brain images and a consistent human cortical labeling protocol. Front. Neurosci. 6, 171 (2012)

32. Fischl, B.: FreeSurfer. Neuroimage **62**(2), 774–781 (2012)
33. Shattuck, D.W., et al.: Construction of a 3D probabilistic atlas of human cortical structures. Neuroimage **39**(3), 1064–1080 (2008)
34. McCormick, M., et al.: ITK: enabling reproducible research and open science. Front. Neuroinform. **8**, 13 (2014)
35. Jenkinson, M., Smith, S.: A global optimisation method for robust affine registration of brain images. Med. Image Anal. **5**(2), 143–156 (2001)
36. Zhao, S., et al.: Unsupervised 3D end-to-end medical image registration with volume tweening network. IEEE J. Biomed. Health Inform. **24**(5), 1394–1404 (2019)
37. Baheti, B., et al.: The brain tumor sequence registration challenge: establishing correspondence between pre-operative and follow-up MRI scans of diffuse glioma patients. arXiv preprint arXiv:2112.06979 (2021)

DISA: DIfferentiable Similarity Approximation for Universal Multimodal Registration

Matteo Ronchetti[1,2]([✉]), Wolfgang Wein[1], Nassir Navab[2], Oliver Zettinig[1], and Raphael Prevost[1]

[1] ImFusion GmbH, Munich, Germany
ronchetti@imfusion.com
[2] Computer Aided Medical Procedures (CAMP), Technische Universität München, Munich, Germany

Abstract. Multimodal image registration is a challenging but essential step for numerous image-guided procedures. Most registration algorithms rely on the computation of complex, frequently non-differentiable similarity metrics to deal with the appearance discrepancy of anatomical structures between imaging modalities. Recent Machine Learning based approaches are limited to specific anatomy-modality combinations and do not generalize to new settings. We propose a generic framework for creating expressive cross-modal descriptors that enable fast deformable global registration. We achieve this by approximating existing metrics with a dot-product in the feature space of a small convolutional neural network (CNN) which is inherently differentiable can be trained without registered data. Our method is several orders of magnitude faster than local patch-based metrics and can be directly applied in clinical settings by replacing the similarity measure with the proposed one. Experiments on three different datasets demonstrate that our approach generalizes well beyond the training data, yielding a broad capture range even on unseen anatomies and modality pairs, without the need for specialized retraining. We make our training code and data publicly available.

Keywords: Image Registration · Multimodal · Metric Learning · Differentiable · Deformable Registration

1 Introduction

Multimodal imaging has become increasingly popular in healthcare due to its ability to provide complementary anatomical and functional information. However, to fully exploit its benefits, it is crucial to perform accurate and robust registration of images acquired from different modalities. Multimodal image registration is a challenging task due to differences in image appearance, acquisition

Supplementary Information The online version contains supplementary material available at https://doi.org/10.1007/978-3-031-43999-5_72.

protocols, and physical properties of the modalities. This holds in particular if ultrasound (US) is involved, and has not been satisfactorily solved so far.

While simple similarity measures directly based on the images' intensities such as sum of absolute (L1) or squared (L2) differences and normalized cross-correlation (NCC) [16] work well in monomodal settings, a more sophisticated approach is needed when intensities cannot be directly correlated. Historically, a breakthrough in CT-MRI registration was achieved by Viola and Wells, who proposed Mutual Information [19]. Essentially, it abstracts the problem to the statistical concept of information theory and optimizes image-wide alignment statistics. Broken down to patch level and inspired by ultrasound physics, the Linear Correlation of Linear Combination (LC^2) measure has shown to work well for US to MRI or CT registration [2,22]. While dealing well with US specifics, it is not differentiable and expensive to compute.

As an alternative to directly assessing similarity on the original images, various groups have proposed to first compute intermediate representations, and then align these with conventional L1 or L2 metrics [5,20]. A prominent example is the Modality-Independent Neighbourhood Descriptor (MIND) [5], which is based on image self-similarity and has with minor adaptations (denoted MIND-SSC for self-similarity context) also been applied to US problems [7]. Most recently, it has been shown that using 2D confidence maps-based weighting and adaptive normalization may further improve registration accuracy [21]. Yet, such feature descriptors are not expressive enough to cope with complex US artifacts and exhibit many local optima, therefore requiring closer initialization.

More recently, multimodal registration has been approached using various Machine Learning (ML) techniques. Some of these methods involve the utilization of Convolutional Neural Networks (CNN) to extract segmentation volumes from the source data, transforming the problem into the registration of label maps [13,24]. Although these methods have demonstrated promising results, they are anatomy-specific and require the identification and labeling of structures that are visible in both modalities. Other approaches are trained using ground truth registrations to directly predict the pose [9,12] or to establish keypoint correspondences [1,11]. However, these methods are not generalizable to different anatomies or modalities. Moreover, the paucity of precise and unambiguous ground truth registration, particularly in abdominal MR-US registration, exacerbates the overfitting problem, restricting generalization even within the same modality and anatomy. It has furthermore been proposed in the past to utilize CNNs as a replacement for a similarly metric. In [3,17], the two images being registered are resampled into the same grid in each optimizer iteration, concatenated and fed into a network for similarity evaluation. While such a measure can directly be integrated into existing registration methods, it still suffers from similar limitations in terms of runtime performance and modality dependance.

In contrast, we propose in this work to use a small CNN to approximate an expensive similarity metric with a straightforward dot product in its feature space. Crucially, our method does not necessitate to evaluate the CNN at every optimizer iteration. This approach combines ML and classical multimodal image registration techniques in a novel way, avoiding the common limitations of ML approaches: ground truth registration is not required, it is differentiable and computationally efficient, and generalizes well across anatomies and imaging modalities.

2   Approach

We formulate image registration as an optimization problem of a similarity metric s between the moving image M and the fixed image F with respect to the parameters α of a spatial transformation $T_\alpha : \Omega \to \Omega$. Most multi-modal similarity metrics are defined as weighted sums of local similarities computed on patches. Denoting $M \circ T_\alpha$ the deformed image, the optimization target can be expressed in the following way:

$$f(\alpha) = \sum_{p \in \Omega} w(p) \, s(F[p], M \circ T_\alpha[p]), \tag{1}$$

where $w(p)$ is the weight assigned to the point p, $s(\cdot, \cdot)$ defines a local similarity and the $[\cdot]$ operator extracts a patch (or a pixel) at a given spatial location. This definition encompasses SSD but also other more elaborate metrics like LC^2 or MIND. The function w is typically used to reduce the impact of patches with ambiguous content (e.g. with uniform intensities), or can be chosen to encode prior information on the target application.

The core idea of our method is to approximate the similarity metric $s(P_1, P_2)$ of two image patches with a dot product $\langle \phi(P_1), \phi(P_2) \rangle$ where $\phi(\cdot)$ is a function that extracts a feature vector, for instance in $\mathbb{R}^{16}$, from its input patch. When ϕ is a fully convolutional neural network (CNN), we can simply feed it the entire volume in order to pre-compute the feature vectors of every voxel with a single forward pass. The registration objective (Eq. 1) is then approximated as

$$f(\alpha) \approx \sum_{p \in \Omega} w(p) \, \langle \phi(F)[p], \phi(M) \circ T_\alpha[p] \rangle, \tag{2}$$

thus converting the original problem into a registration of pre-computed feature maps using a simple and differentiable dot product similarity. This approximation is based on the assumption that the CNN is approximately equivariant to the transformation, i.e. $\phi(M \circ T_\alpha)[p] \approx \phi(M) \circ T_\alpha[p]$. Our experiments show that this assumption (implicitly made also by other descriptors like MIND) does not present any practical impediment. Our method exhibits a large capture range and can converge over a wide range of rotations and deformations.

Advantages. In contrast to many existing methods, our approach doesn't require any ground truth registration and can be trained using patches from unregistered pairs of images. This is particularly important for multi-modal deformable registration as ground truths are harder to define, especially on ultrasound. The simplicity of our training objective allows the use of a CNN with a limited number of parameters and a small receptive field. This means

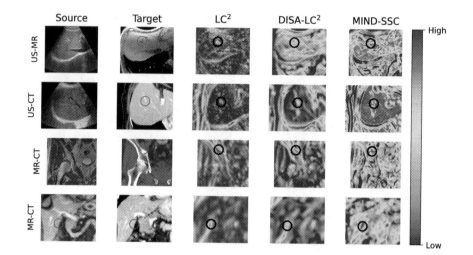

Fig. 1. Similarity maps across different modalities and anatomies. Each heatmap shows the similarity of the marked point on the source image to every point in the target image. Our method (DISA-LC^2) approximates LC^2 well in a fraction of the computation time and produces less ambiguous heatmaps than MIND.

that the CNN has a negligible computational cost and can generalize well across anatomies and modalities: a single network can be used for all types of images and does not need to be retrained for a new task. Furthermore, the objective function (Eq. 2) can be easily differentiated without backpropagating the gradient through the CNN. This permits efficient gradient-based optimization, even when the original metric is either non-differentiable or costly to differentiate. Finally, we quantize the feature vectors to 8-bit precision further increasing the computational speed of registration without impacting accuracy.

3   Method

We train our model to approximate the three-dimensional LC^2 similarity, as it showed good performance on a number of tasks, including ultrasound [2,22]. The LC^2 similarity quantifies whether a target patch can be approximated by a linear combination of the intensities and the gradient magnitude of the source patch. In order to reduce the sensitivity on the scale, our target is actually the average LC^2 over different radiuses of 3, 5, and 7. In order to be consistent with the original implementation of LC^2 we use the same weighting function w based on local patch variance. Note that the network will be trained only once, on a fixed dataset that is fully independent of the datasets that will be used in the evaluation (see Sect. 4).

Dataset. Our neural network is trained using patches from the "Gold Atlas - Male Pelvis - Gentle Radiotherapy" [14] dataset, which is comprised of 18 patients each with a CT, MR T1, and MR T2 volumes. We resample each volume to a spacing of 2 mm and normalize the voxel intensities to have zero mean and standard variation of one. Since our approach is unsupervised, we don't make use of the provided registration but leave the volumes in their standard DICOM orientation. As LC^2 requires the usage of gradient magnitude in one of the modalities, we randomly pick it from either CT or MR.

We would like to report that, initially, we also made use of a proprietary dataset including US volumes. However, as our investigation progressed, we observed that the incorporation of US data did not significantly contribute to the generalization capabilities of our model. Consequently, for the purpose of ensuring reproducibility, all evaluations presented in this paper exclusively pertain to the model trained solely on the public MR-CT dataset.

Patch Sampling from Unregistered Datasets. For each pair of volumes (M, F) we repeat the following procedure 5000 times: (1) Select a patch from M with probability proportional to its weight w; (2) Compute the similarity with all the patches of F; (3) Uniformly sample $t \in [0, 1]$; (4) Pick the patch of F with similarity score closest to t. Running this procedure on our training data results in a total of 510000 pairs of patches.

Architecture and Training. We use the same feed-forward 3D CNN to process all data modalities. The proposed model is composed of residual blocks [4], LeakyReLU activations [10] and uses BlurPool [25] for downsampling, resulting in a total striding factor of 4. We do not use any normalization layer, as this resulted in a reduction in performance. The output of the model is 16-channels volume with the norm of each voxel descriptor clipped at 1. The architecture consists of ten layers and a total of 90,752 parameters, making it notably smaller than many commonly utilized neural networks.

Augmentation on the training data is used to make the model as robust as possible while leaving the target similarity unchanged. In particular, we apply the same random rotation to both patches, randomly change the sign and apply random linear transformation on the intensity values. We train our model for 35 epochs using the L2 loss and batch size of 256. The training converges to an average patch-wise L2 error of 0.0076 on the training set and 0.0083 on the validation set. The total training time on an NVIDIA RTX4090 GPU is 5 h, and inference on a 256^3 volume takes 70 ms. We make the training code and preprocessed data openly available online[1].

4   Experiments and Results

We present an evaluation of our approach across tasks involving diverse modalities and anatomies. Notably, the experimental data utilized in our analysis differs significantly from our model's training data in terms of both anatomical structures and combination of modalities. To assess the effectiveness of our method,

---

[1] https://github.com/ImFusionGmbH/DISA-universal-multimodal-registration.

Table 1. Results on registration of brain US-MR data from the RESECT Challenge. FRE is the average of fiducial errors in millimeters across all cases, while FRE25, FRE50, and FRE75 refer to the 25th, 50th, and 75th percentiles.

Method	Mode	Avg. FRE	FRE25	FRE50	FRE75
MIND-SSC	Rigid	5.05	1.69	2.20	3.31
MIND-SSC	Affine	2.01	1.44	1.84	2.29
LC^2	Rigid	1.71	1.31	1.56	1.72
LC^2	Affine	1.73	1.32	1.67	1.89
DISA-LC^2	Rigid	1.82	1.37	1.65	1.80
DISA-LC^2	Affine	1.74	1.33	1.58	1.73

Table 2. Results on the Abdomen MR-CT task of the Learn2Reg challenge 2021. The best results and the ones not significantly different from them are in bold.

Method	Stride	DSC25	DSC50	DSC75	HD95
MIND-SSC	4	42.3%	70.9%	84.9%	26.4 mm
MIND-SSC	2	49.8%	70.9%	84.9%	24.8 mm
MIND-SSC	1	48.8%	70.9%	84.9%	24.5 mm
DISA-LC^2	4	61.4%	72.7%	85.2%	23.6 mm
DISA-LC^2	2	**61.5%**	73.2%	85.5%	22.8 mm
DISA-LC^2	1	**61.5%**	**74.0%**	**85.5%**	**22.6 mm**

we compare it against LC^2, which is the metric we approximate, and MIND-SSC [7]. In all experiments, we use a Wilcoxon signed-rank test with p-value 10^{-2} to establish the significance of our results.

As will be demonstrated in the next subsections, our method is capable of achieving comparable levels of accuracy as LC^2 while retaining the speed and flexibility of MIND-SSC. In particular, on abdominal US registration (Sect. 4.3) our method obtains a significantly larger capture range, opening new possibilities for tackling this challenging problem.

4.1   Affine Registration of Brain US-MR

In this experiment, we evaluate the performance of different methods for estimating affine registration of the REtroSpective Evaluation of Cerebral Tumors (RESECT) MICCAI challenge dataset [23]. This dataset consists of 22 pairs of pre-operative brain MRs and intra-operative ultrasound volumes. The initial pose of the ultrasound volumes exhibits an orientation close to the ground truth but can contain a significant translation shift. For both MIND-SSC and DISA-LC^2, we resample the input volumes to 0.4 mm spacing and use the BFGS [18] optimizer with 500 random initializations within a range of $\pm 10°$ and ± 25 mm.

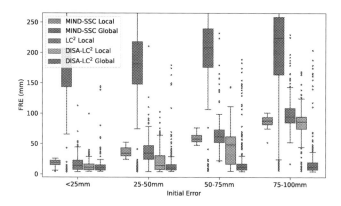

Fig. 2. Boxplot of fiducial registration errors for the different methods on deformable registration of abdominal US-CT and US-MR.

We report the obtained Fiducial Registration Errors (FRE) in Table 1. DISA-LC^2 is significantly better than MIND-SSC while the difference with LC^2 is not significant. In conclusion, our experiments demonstrate that the proposed DISA-LC^2, combined with a simple optimization strategy, is capable of achieving equivalent performance to manually tuned LC^2.

4.2   Deformable Registration of Abdominal MR-CT

Our second application is the Abdomen MR-CT task of the Learn2Reg challenge 2021 [8]. The dataset comprises 8 sets of MR and CT volumes, both depicting the abdominal region of a single patient and exhibiting notable deformations. We estimate dense deformation fields using the methodology outlined in [6] (without inverse consistency) which first estimates a discrete displacement using explicit search and then iteratively enforces global smoothness. Segmentation maps of anatomical structures are used to measure the quality of the registration. In particular, we compute the 25th, 50th, and 75th quantile of the Dice Similarity Coefficient (DSC) and the 95th quantile of the Hausdorff distance (HD95) between the registered label maps. We compare MIND-SCC and DISA-LC^2 used with different strides and followed by a downsampling operation that brings the spacing of the descriptors volumes to 8 mm. The hyperparameters of the registration algorithm have been manually optimized for each approach. Table 2 shows that our method obtains significantly better results than MIND-SCC on the DSC metrics while being not significantly better on HD95.

4.3   Deformable Registration of Abdominal US-CT and US-MR

As the most challenging experiment, we finally use our method to achieve deformable registration of abdominal 3D freehand US to a CT or MR volume.

We are using a heterogeneous dataset of 27 cases, comprising liver cancer patients and healthy volunteers, different ultrasound machines, as well as optical

Table 3. Results on deformable registration of abdominal US-CT and US-MR. A case is considered "converged" if the FRE after registration is less than 15 mm. The best results and the ones not significantly different from them are highlighted in bold. (*)Time and evaluations for Global LC^2 are estimated by extrapolation.

Similarity	Search	Converged cases w.r.t. initialization error				Time (s)	Num. eval.
		0–25 mm	25–50 mm	50–75 mm	75–100 mm		
MIND-SSC	Local	23.6%	0.0%	0.0%	0.0%	0.4	17
LC^2	Local	54.1%	14.0%	0.0%	0.0%	1.9	98
DISA-LC^2	Local	**70.3%**	52.0%	21.1%	5.8%	0.9	70
MIND-SSC	Global	17.9%	14.6%	5.3%	12.0%	1.3	26370
LC^2	Global		N/A			948.0*	38740*
DISA-LC^2	Global	**75.5%**	**73.2%**	**65.0%**	**64.0%**	1.8	29250

vs. electro-magnetic external tracking, and sub-costal vs. inter-costal scanning of the liver. All 3D ultrasound data sets are accurately calibrated, with overall system errors in the range of commercial ultrasound fusion options. Between 4 and 9 landmark pairs (vessel bifurcations, liver gland borders, gall bladder, kidney) were manually annotated by an expert. In order to measure the capture range, we start the registration from 50 random rigid poses around the ground truth and calculate the Fiducial Registration Error (FRE) after optimization. For local optimization, LC^2 is used in conjunction with BOBYQA [15] as in the original paper [22], while MIND-SCC and DISA-LC^2 are instead used with BFGS. Due to an excessive computation time, we don't do global optimization with LC^2 while with other methods we use BFGS with 500 random initializations within a range of $\pm40°$ and ±150 mm. We use six parameters to define the rigid pose and two parameters to describe the deformation caused by the ultrasound probe pressure.

From the results shown in Table 3 and Fig. 2, it can be noticed that the proposed method obtains a significantly larger capture range than MIND-SCC and LC^2 while being more than 300 times faster per evaluation than LC^2 (the times reported in the table include not just the optimization but also descriptor extraction). The differentiability of our objective function allows our method to converge in fewer iterations than derivative-free methods like BOBYQA. Furthermore, the evaluation speed of our objective function allows us to exhaustively search the solution space, escaping local minima and converging to the correct solution with pose and deformation parameters at once, in less than two seconds.

Note that this registration problem is much more challenging than the prior two due to difficult ultrasonic visibility in the abdomen, strong deformations, and ambiguous matches of liver vasculature. Therefore, to the best of our knowledge, these results present a significant leap towards reliable and fully automatic fusion, doing away with cumbersome manual landmark placements.

5   Conclusion

We have discovered that a complex patch-based similarity metric can be approximated with feature vectors from a CNN with particularly small architecture, using the same model for any modality. The training is unsupervised and merely requires unregistered data. After features are extracted from the volumes, the actual registration comprises a simple iterative dot-product computation, allowing for global and derivative-based optimization. This novel combination of classical image processing and machine learning elevates multi-modal registration to a new level of performance, generality, but also algorithm simplicity.

We demonstrate the efficiency of our method on three different use cases with increasing complexity. In the most challenging scenario, it is possible to perform global optimization within seconds of both pose and deformation parameters, without any organ-specific distinction or successive increase of parameter sizes.

While we specifically focused on developing an unsupervised and generic method, a sensible extension would be to specialize our method by including global information, such as segmentation maps, into the approximated measure or by making use of ground-truth registration during training. Finally, the cross-modality feature descriptors produced by our model could be exploited by future research for tasks different from registration such as modality synthesis or segmentation.

References

1. Esteban, J., Grimm, M., Unberath, M., Zahnd, G., Navab, N.: Towards fully automatic X-ray to CT registration. In: Shen, D., et al. (eds.) MICCAI 2019. LNCS, vol. 11769, pp. 631–639. Springer, Cham (2019). https://doi.org/10.1007/978-3-030-32226-7_70
2. Fuerst, B., Wein, W., Müller, M., Navab, N.: Automatic ultrasound-MRI registration for neurosurgery using the 2D and 3D LC2 metric. Med. Image Anal. **18**(8), 1312–1319 (2014)
3. Haskins, G., et al.: Learning deep similarity metric for 3D MR-TRUS image registration. Int. J. Comput. Assist. Radiol. Surg. **14**, 417–425 (2019)
4. He, K., Zhang, X., Ren, S., Sun, J.: Deep residual learning for image recognition. In: Proceedings of the IEEE Conference on Computer Vision and Pattern Recognition, pp. 770–778 (2016)
5. Heinrich, M.P., et al.: Mind: modality independent neighbourhood descriptor for multi-modal deformable registration. Med. Image Anal. **16**(7), 1423–1435 (2012)
6. Heinrich, M.P., Papież, B.W., Schnabel, J.A., Handels, H.: Non-parametric discrete registration with convex optimisation. In: Ourselin, S., Modat, M. (eds.) WBIR 2014. LNCS, vol. 8545, pp. 51–61. Springer, Cham (2014). https://doi.org/10.1007/978-3-319-08554-8_6
7. Heinrich, M.P., Jenkinson, M., Papież, B.W., Brady, S.M., Schnabel, J.A.: Towards realtime multimodal fusion for image-guided interventions using self-similarities. In: Mori, K., Sakuma, I., Sato, Y., Barillot, C., Navab, N. (eds.) MICCAI 2013. LNCS, vol. 8149, pp. 187–194. Springer, Heidelberg (2013). https://doi.org/10.1007/978-3-642-40811-3_24

8. Hering, A., et al.: Learn2reg: comprehensive multi-task medical image registration challenge, dataset and evaluation in the era of deep learning. IEEE Trans. Med. Imaging **42**, 697–712 (2022)

9. Horstmann, T., Zettinig, O., Wein, W., Prevost, R.: Orientation estimation of abdominal ultrasound images with multi-hypotheses networks. In: Medical Imaging with Deep Learning (2022)

10. Maas, A.L., Hannun, A.Y., Ng, A.Y.: Rectifier nonlinearities improve neural network acoustic models. In: Proceedings of the ICML, vol. 30, p. 3. Citeseer (2013)

11. Markova, V., Ronchetti, M., Wein, W., Zettinig, O., Prevost, R.: Global multi-modal 2D/3D registration via local descriptors learning. In: Wang, L., Dou, Q., Fletcher, P.T., Speidel, S., Li, S. (eds.) MICCAI 2022. LNCS, pp. 269–279. Springer, Cham (2022). https://doi.org/10.1007/978-3-031-16446-0_26

12. Montaña-Brown, N., et al.: Towards multi-modal self-supervised video and ultrasound pose estimation for laparoscopic liver surgery. In: Aylward, S., Noble, J.A., Hu, Y., Lee, S.L., Baum, Z., Min, Z. (eds.) ASMUS 2022. LNCS, vol. 13565, pp. 183–192. Springer, Cham (2022). https://doi.org/10.1007/978-3-031-16902-1_18

13. Müller, M., et al.: Deriving anatomical context from 4D ultrasound. In: 4th Bi-annual Eurographics Workshop on Visual Computing for Biology and Medicine (2014)

14. Nyholm, T., et al.: Gold atlas - male pelvis - gentle radiotherapy (2017)

15. Powell, M.J.: The Bobyqa algorithm for bound constrained optimization without derivatives. Cambridge NA Report NA2009/06, vol. 26. University of Cambridge, Cambridge (2009)

16. Roche, A., Malandain, G., Ayache, N.: Unifying maximum likelihood approaches in medical image registration. Int. J. Imaging Syst. Technol. **11**(1), 71–80 (2000)

17. Sedghi, A., et al.: Semi-supervised deep metrics for image registration. arXiv preprint arXiv:1804.01565 (2018)

18. Skajaa, A.: Limited memory BFGS for nonsmooth optimization. Master's thesis, Courant Institute of Mathematical Science, New York University (2010)

19. Viola, P., Wells, W.M.: Alignment by maximization of mutual information. In: Proceedings of IEEE International Conference on Computer Vision, pp. 16–23. IEEE (1995)

20. Wachinger, C., Navab, N.: Entropy and Laplacian images: structural representations for multi-modal registration. Med. Image Anal. **16**(1), 1–17 (2012)

21. Wang, Y., et al.: Multimodal registration of ultrasound and MR images using weighted self-similarity structure vector. Comput. Biol. Med. **155**, 106661 (2023)

22. Wein, W., Brunke, S., Khamene, A., Callstrom, M.R., Navab, N.: Automatic CT-ultrasound registration for diagnostic imaging and image-guided intervention. Med. Image Anal. **12**(5), 577–585 (2008)

23. Xiao, Y., Fortin, M., Unsgård, G., Rivaz, H., Reinertsen, I.: Retrospective evaluation of cerebral tumors (resect): a clinical database of pre-operative MRI and intra-operative ultrasound in low-grade glioma surgeries. Med. Phys. **44**(7), 3875–3882 (2017)

24. Zeng, Q., et al.: Learning-based US-MR liver image registration with spatial priors. In: Wang, L., Dou, Q., Fletcher, P.T., Speidel, S., Li, S. (eds.) MICCAI 2022. LNCS, vol. 13436, pp. 174–184. Springer, Cham (2022). https://doi.org/10.1007/978-3-031-16446-0_17

25. Zhang, R.: Making convolutional networks shift-invariant again. In: ICML (2019)

StructuRegNet: Structure-Guided Multimodal 2D-3D Registration

Amaury Leroy[1,2,3]([✉]), Alexandre Cafaro[1,2,3], Grégoire Gessain[4],
Anne Champagnac[5], Vincent Grégoire[3], Eric Deutsch[2], Vincent Lepetit[6],
and Nikos Paragios[1]

[1] Therapanacea, Paris, France
a.leroy@therapanacea.eu
[2] Gustave Roussy, Inserm 1030, Paris-Saclay University, Villejuif, France
[3] Department of Radiation Oncology, Centre Léon Bérard, Lyon, France
[4] Department of Pathology, Gustave Roussy, Villejuif, France
[5] Department of Pathology, Centre Léon Bérard, Lyon, France
[6] Ecole des Ponts ParisTech, Marne-la-Vallée, France

Abstract. Multimodal 2D-3D co-registration is a challenging problem with numerous clinical applications, including improved diagnosis, radiation therapy, or interventional radiology. In this paper, we present StructuRegNet, a deep-learning framework that addresses this problem with three novel contributions. First, we combine a 2D-3D deformable registration network with an adversarial modality translation module, allowing each block to benefit from the signal of the other. Second, we solve the initialization challenge for 2D-3D registration by leveraging tissue structure through cascaded rigid areas guidance and distance field regularization. Third, StructuRegNet handles out-of-plane deformation without requiring any 3D reconstruction thanks to a recursive plane selection. We evaluate the quantitative performance of StructuRegNet for head and neck cancer between 3D CT scans and 2D histopathological slides, enabling pixel-wise mapping of low-quality radiologic imaging to gold-standard tumor extent and bringing biological insights toward homogenized clinical guidelines. Additionally, our method can be used in radiation therapy by mapping 3D planning CT into the 2D MR frame of the treatment day for accurate positioning and dose delivery. Our framework demonstrates superior results to traditional methods for both applications. It is versatile to different locations or magnitudes of deformation and can serve as a backbone for any relevant clinical context.

Keywords: Multimodal · Registration · 2D-3D · Histopathology · Radiology

1 Introduction

2D-3D registration refers to the highly challenging process of aligning an input 2D image to its corresponding slice inside a given 3D volume [4]. It has received growing

Supplementary Information The online version contains supplementary material available at https://doi.org/10.1007/978-3-031-43999-5_73.

attention in medical imaging due to the various contexts where it applies, like image fusion between 2D real-time acquisitions and either pre-operative 3D images for guided interventions or reference planning volumes for patient positioning in radiation therapy (RT). Another important task is the volumetric reconstruction of a sequence of misaligned slices *ex vivo*, enabling multimodal comparison toward improved diagnosis. In this respect, overlaying 3D radiology and 2D histology could significantly enhance radiologists' understanding of the links between tissue characteristics and radiologic signals [9]. Indeed, MRI or CT scans are the baseline source of information for cancer treatment but fail to provide an accurate assessment of disease proliferation, leading to high variability in tumor detection [5,13,17]. On the other hand, high-resolution digitized histopathology, called Whole Slide Imaging (WSI), provides cell-level information on the tumor environment from the surgically resected specimens. However, the registration process is substantially difficult due to the visual characteristics, resolution scale, and dimensional differences between the two modalities. In addition, histological preparation involves tissue fixation and slicing, leading to severe collapse and out-of-plane deformations. (Semi-)automated methods have been developed to avoid time-consuming and biased manual mapping, including protocols with 3D mold or landmarks [10,22], volume reconstruction to perform 3D registration [2,18,19,23], or optimization algorithms for direct multimodal comparison [3,15]. More recently, deep learning (DL) has been introduced but is limited to 2D/2D and requires prior plane selection [20]. On the other hand, successful DL methods have been proposed to address the 2D/3D mapping problem for other medical modalities [6,8,16,21]. However, given the extreme deformation that the tissue undergoes during the histological process, additional guidance is needed. One promising solution is to rely on rigid structures that are supposedly more robust during the preparation. Structural information to guide image registration has been studied with the help of segmentations into the training loop [11], or by learning new image representations for refined mapping [12].

In this paper, we propose to leverage the structural features of tissue and more particularly the rigid areas to guide the registration process with two distinct contributions: (1) a cascaded rigid alignment driven by stiff regions and coupled with recursive plane selection, and (2) an improved 2D/3D deformable motion model with distance field regularization to handle out-of-plane deformation. To our knowledge, no previous study proposed 2D/3D registration combined with structure awareness. We also use the CycleGAN for image translation and direct monomodal signal comparison [25]. Like [14,24], we combine registration with modality translation and integrate the two aforementioned components. We demonstrate superior quantitative results for Head and Neck (H&N) 3D CT and 2D WSIs than traditional approaches failing due to the histological constraints. In addition, we show that StructuRegNet performs better than the state-of-the-art model from [14] on 3D CT/2D MR for the pelvis in RT.

2   Methods

2.1   Histology-to-CT Modality Translation

Our three-step structure-aware pipeline is thoroughly detailed in Fig. 1. For clarity, we focus on the radiology-histology application but the pipeline is versatile

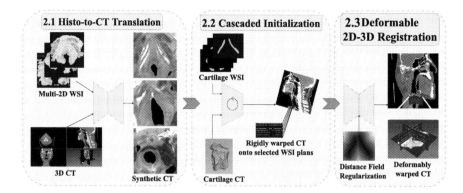

Fig. 1. Overview of StructuRegNet, where the WSIs are first translated into synthetic CTs (2.1) before getting matched with the most similar CT slices thanks to a cascaded structure-aware plane selection (2.2). Finally, the rigid transformation is refined through a deformable network to handle out-of-plane distortions (2.3).

to any 2D/3D setting. The modality transfer is a 2D image-to-image translation problem defined as follows: Given a sequence of n slices $H = \{h_1, ..., h_n\}$ and a volume considered as a full stack of m axial slices $CT = \{ct_1, ..., ct_m\}$, we build a CycleGAN with two generators and two discriminators $G_{H \to CT}, G_{CT \to H}, D_H$ and D_{CT}. With a symmetric situation for $G_{CT \to H}$, $G_{H \to CT}$ outputs a synthetic CT image, which is then processed by D_{CT} along with randomly sampled original input slices with an associated adversarial loss L_{adv}. The cyclical pattern lies in the similarity between the original images and the reconstructed samples $G_{CT \to H} \circ G_{H \to CT}(h_i)$ through a pixel-wise cycle loss L_{cyc}. Finally, we employ two additional metrics: an identity loss L_{Id} to encourage modality-specific feature representation when considering h_i being the input for $G_{CT \to H}$ with an expected identity synthesis; and a structure consistency MIND loss from [7] to ensure style transfer without content alteration. These losses are the classical implementations for CycleGAN and are detailed in the supplementary material.

2.2 Recursive Cascaded Plane Selection

We replace the volume reconstruction step with a recursive dual model. We first rotate and translate the 3D CT to match the stack of slices H, which is crucial as an initialization step to help the 2D-3D network to focus on small out-of-plane deformations and avoid local minima. Then, we perform a precise plane selection and solve the spacing gap by adjusting the z-position of each slice to its most similar CT section. These two steps are performed iteratively until convergence, with a recursive algorithm to reduce computational cost (Fig. 2).

For rigid initialization, the hypothesis is that the histological specimen is cut with an unknown spacing and angle, but the latter is supposed constant between WSIs. A rigid alignment is thus sufficient to reorient moving CT onto fixed H. Based on a theoretical axial slice sequence $Z = (Z_1, ..., Z_m)$, we define

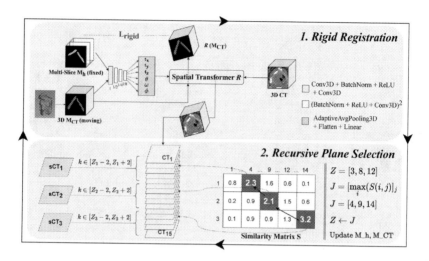

Fig. 2. Cascaded alignment through rigid structure-aware warping followed by recursive plane selection. The deformed CT from 1. is the input for 2., along with sCT and slice sequence Z. The updated Z from 2. is applied to M_h while the rigid deformation is applied to M_{CT} so that new inputs can feed 1. again as iterative refining.

H as a sparse 3D volume the same size as CT, filled in with h_i at $z = Z_i$ and zeros elsewhere (the same applies for the corresponding sCT from the previous module). Because soft tissues undergo too large out-of-plane deformations, we leverage the rigid structures which are supposed not to be distorted or shrunk during the histological process. We extract their segmentation masks M_{ct}, M_h for both modalities (see preprocessing in Sect. 3), concatenate and fed them into an encoder followed by a fully connected layer that outputs six transformation parameters (3 rotations, 3 translations). A differentiable spatial transform R finally warps M_{ct} for similarity optimization with M_h. Similarly to [14], we adopt a loss L_{rigid} masked on empty slices to avoid the introduction of noise at slices within the gradient where no data is provided, and directly train on the Dice Similarity Coefficient (DSC) between rigid areas:

$$L_{\text{rigid}}(M_h, R(M_h, M_{ct})) = \sum_{i \in [1,m]} \text{DSC}(M_{h_{Z_i}}, R(M_h, M_{ct})(M_{ct_{Z_i}})) . \quad (1)$$

Additionally, R also warps CT without gradient backpropagation and is the input with sCT for plane selection. We then introduce a sequence alignment problem, the objective being to update the slice sequence Z of sCT by mapping it to a corresponding sequence J of 2D images from CT. We define S a similarity matrix, where $S(i, j)$ is the total similarity (measured with MI) when mapping $sct_{Z_1}, ..., sct_{Z_i}$ with $ct_{J_1}, ..., ct_{J_j}$ for $i \in [1, n], j \in [1, m]$:

$$S(i,j) = \begin{cases} MI(sct_{Z_i}, ct_{J_j}) & \text{if } i = 1 \\ \max_k S(i-1, k) + MI(sct_{Z_i}, ct_{J_j}) & \text{else} \end{cases} , \quad (2)$$

which means that each row of S will be filled by computing the sum of the MI for the corresponding column j and the maximum similarity from the last row. Like any dynamic programming method, we want to find the optimal sequence J by following the backward path of S building. To do so, we retrieve the new index j that yielded the maximized similarity for each step $J = [\max_i(S(i,j))]_j$, and we update $Z \leftarrow J$ accordingly. In addition, the J sequence cannot be too different from Z as it would induce overlap between ordered WSIs. We thus constrained the possible matching values with $k \in [Z_i - 2, Z_i + 2]$. Based on these rigid registration and plane selection blocks, we build a cascaded module to iteratively refine the alignment where the intermediate warping becomes the new input. We defined the number of iterations as a hyperparameter to reach a good balance between computational time and similarity maximization. This dual model is crucial for initialization but does not take into account out-of-plane deformations and a perfect alignment is not accessible yet. The deformable framework bridges this gap by focusing on irregular displacements caused by tissue manipulation and refining the rigid warping.

2.3 Deformable 2D-3D Registration

Given one fixed multi-slice sCT' and a moving rigidly warped $R(CT)$ from the previous module, we adopt an architecture close to Voxelmorph [1]. Still, the rigidly warped $CT' = R(CT)$ and the plane-adjusted sparse sCT' are fed through two different encoders for independent feature extraction. The architecture is depicted in Fig. 3. Both latent representations are element-wise subtracted. A decoder is connected to both encoders and generates a displacement field Φ the same size as input images but with (x, y, z)-channels corresponding to the displacement in each spatial coordinate. A differentiable sampler D warps CT' in a deformable setting, which is then compared to sCT' through a masked Normalized Cross-Correlation (NCC) loss L_{defo}:

$$L_{\text{defo}}(sCT', D(sCT', CT')) = \sum_{i \in [1,m]} \text{NCC}(sCT'_{Z_i}, D(sCT', CT')(CT'_{Z_i})). \quad (3)$$

Finally, we add two sources of regularization. Soft tissues away from bones and cartilage are more subject to shrinkage or disruption, so we harness the information from the cartilage segmentation mask of CT to generate a distance transform map Δ defined as $\Delta(\mathbf{v}) = \min_{\mathbf{m} \in M_{CT}} ||\mathbf{v} - \mathbf{m}||_2$. It maps each voxel $\mathbf{v}$ of CT to its distance with the closest point $\mathbf{m}$ to the rigid area M_{CT}. We can then control the displacement field, with close tissue being more highly constrained than isolated areas: $\Phi' = \Phi \odot (\Delta + \epsilon)$, where $\odot$ is the Hadamard product and ϵ is a hyperparameter matrix allowing small displacement even for cartilage areas for which distance transform is null. A second regularization takes the form of a loss $L_{\text{regu}}(\Phi') = \sum_{\mathbf{v} \in \mathbb{R}^3} ||\nabla \Phi'(\mathbf{v})||^2$ on the volume to constrain spatial gradients and thus encourage smooth deformation, which is essential for empty slices which are excluded from L_{defo}. The total loss is a weighted sum of L_{defo} and L_{regu}.

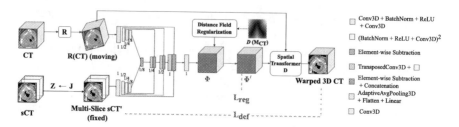

Fig. 3. Deformable 2D-3D registration pipeline, made of two encoders and a shared decoder, with regularization applied on the displacement field Φ thanks to the distance map from CT.

3   Experiments

Dataset and Preprocessing. Our clinical dataset consists of 108 patients for whom were acquired both a pre-operative H&N CT scan and 4 to 11 WSIs after laryngectomy (with a total amount of 849 WSIs). The theoretical spacing between each slice is 5 mm, and the typical pixel size before downsampling is $100K \times 100K$. Two expert radiation oncologists on CT delineated both the thyroid and cricoid cartilages for structure awareness and the Gross Tumor Volume (GTV) for clinical validation, while two expert pathologists did the same on WSIs. They then meet and agreed to place 6 landmarks for each slice at important locations (not used for training). We ended up with images of size 256×256 ($\times 64$ for 3D CT) of 1 mm isotropic grid space. We split the dataset patient-wise into three groups for training (64), validation (20), and testing (24). To demonstrate the performance of our model on another application, we also retrieved the datasets from [14] for pelvis 3D CT/2D MR. It is made of 451 pairs between CT and TrueFISP sequences, and 217 other pairs between CT and T2 sequences. We guided the registration thanks to the rigid left/right femoral heads and computed similarity metrics on the 7 additional organs at risk (anal canal, bladder, rectum, penile bulb, seminal vesicle, and prostate). All masks were provided by the authors and were originally segmented by internal experts.

Hyperparameters. We drew our code from CycleGAN and Voxelmorph implementations with modifications explained above, and we thank the authors of MSV-RegSynNet for making their code and data available to us [1,14,25]. A detailed description of architectures and hyperparameters can be found in the supplementary material. We implemented our model with Pytorch1.13 framework and trained for 600 (800 for MR/CT) epochs with a batch size of 8 (4 for MR/CT) patients parallelized over 4 NVIDIA GTX 1080 Tis.

Evaluation. We benchmarked our method against three baselines: First, to assess the benefit of modality translation over the multimodal loss, we re-used the original 3D VoxelMorph model with MIND as a multimodal metric for optimization. We also modified this approach by masking the loss function to account for the 2D-3D setting. Next, we implemented the modality translation-based MSV-

RegSyn-Net and modified it for our application to measure the importance of joint structure-aware initialization and regularization. Finally, to differentiate the latter contributions, we tested two ablation studies: without the cascaded rigid mapping or without the distance field control. According to the MR/CT application in RT, we compared our model against the state-of-the-art results of MSV-RegSynNet which were computed on the same dataset.

4 Results

4.1 Modality Translation

Three samples from the test set are displayed in Fig. 1. From a qualitative perspective, the densities of the different tissues are well reconstructed, with rigid structures like cartilage being lighter than soft tissues or tumors. The general shape of the larynx also complies with the original radiologic images. We achieve a mean Structural Similarity (SSIM) index of 0.76/1 between both modalities, demonstrating the strong synthesis capabilities of our network compared to MSV-RegSynNet and our ablative study without initialization process, with an SSIM of 0.72 (respectively 0.69). Therefore, the cascaded rigid initialization is crucial and helps the modality translation module in getting more similar pairs of images for eased synthesis on the next pass.

4.2 Registration

We present visual results in Fig. 4. The initialization enables an accurate plane selection as proved by the similar shape of cartilages in (b). Even for some severe

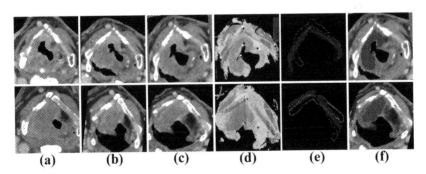

Fig. 4. Visual results for two samples. (a) Original CT, (b) warped CT after rigid initialization and plane selection, (c) warped CT after deformable registration, (d) original histology with landmarks from pathologist (black) and warped projected landmarks from radiologists (yellow), (e) overlaid cartilage masks after registration of histology (filled blue) and radiology (red for our method, yellow for MSV-RegSynNet), (f) Overlaid contours between warped CT (GTV, red) and WSI (true tumor extent, blue). (Color figure online)

Table 1. Mean and stand. dev. registration performance in terms of Dice Score (%), Hausdorff Distance (mm) and Landmark Error (mm). Inference runtime is in seconds.

Method	Dice	Hausdorff	Landmark	Runtime
VoxelMorph 3D	68.4 ± 0.6	8.53 ± 0.32	6.71 ± 0.16	**1.3**
VoxelMorph 2/3D	71.9 ± 1.7	7.19 ± 0.24	5.99 ± 0.22	1.4
MSV-RegSynNet	76.3 ± 1.4	6.88 ± 0.28	4.98 ± 0.15	2.1
Ours (no init)	77.9 ± 1.9	6.91 ± 0.19	4.73 ± 0.31	2.1
Ours (no regu)	85.1 ± 0.8	4.23 ± 0.27	3.71 ± 0.19	2.8
Ours	**86.9 ± 1.3**	**3.81 ± 0.20**	**3.28 ± 0.16**	2.9
3D CT/2D MR	0.35T TrueFISP → 3D CT		1.5T T2 → 3D CT	
	Dice	Hausdorff	Dice	Hausdorff
MSV-RegSynNet	84.6 ± 0.9	7.25 ± 0.05	86.1 ± 1.0	5.84 ± 0.15
Ours	**84.8 ± 1.1**	**7.12 ± 0.08**	**87.9 ± 1.2**	**5.21 ± 0.09**

difficulties inherent to the histological process like a cut larynx, the model successfully maps both cartilage and soft tissue without completely tearing the CT image thanks to regularization (c-d-e). For quantitative assessment, we computed the DSC as well as the Hausdorff Distance between cartilages, and the average distance between characteristic landmarks disposed before registration(Table 1). Our method outperforms all baselines, proving the necessity of a singular approach to handle the specific case of histology. The popular Voxelmorph framework fails, and the 2D-3D adaptation demonstrates the value of the masked loss function. The superior performance of MSV-RegSynNet advocates for a modality translation-based method compared to a direct multimodal similarity criterion. In addition, the ablation studies prove the benefit of the distance field regularization and more importantly the cascaded initialization. Concerning the GPU runtime, with a 3-step cascade for initialization, the inference remains in a similar time scale to baseline methods and performs mapping in less than $3s$. We also compared against MSV-RegSynNet on its own validation dataset for generalization assessment: we yielded comparable results for the first cohort and significantly better ones for the second, which proves that StructuRegNet behaves well on other modalities and that the structure awareness is an essential asset for better registration, as pelvis is a location where organs are moving. Visuals of registration results are displayed in the supplementary material. Eventually, an important clinical endpoint of our study is to compare the GTV delineated on CT with gold-standard tumor extent after co-registration to highlight systematic errors and better understand the biological environment from the radiologic signals. We show in (f) that the GTV delineated on CT overestimates the true tumor extent of around 31%, but does not always encompass the tumor with a proportion of histological tumor contained within the CT contour of 0.86. The typical error cases are the inclusion of cartilage or edema, which highlights the limitations and variability of radiology-based examinations, leading to increased toxicity or untreated areas in RT.

5   Discussion and Conclusion

We introduced a novel framework for 2D/3D multimodal registration. StructuRegNet leverages the structure of tissues to guide the registration through both initial plane selection and deformable regularization; it combines adversarial training for modality translation with a 2D-3D mapping setting and does not require any protocol for 3D reconstruction. It is worth noticing that even if the annotation of cartilage was manual, automating this process is not a bottleneck as the difference in contrast between soft tissue and stiff areas is clear enough to leverage any image processing tool for this task. Finally, it is entirely versatile as we designed our experiments for CT-WSI but any 3D radiological images are suitable. We achieve superior results than state-of-the-art methods in DL-based registration in a similar time scale, allowing precise mapping of both modalities and a better understanding of the tumor microenvironment. The main limitation lies in the handling of organs without any rigid areas like the prostate. Future work also includes a study with biomarkers from immunohistochemistry mapped onto radiology to go beyond binary tumor masks and move toward virtual biopsy.

References

1. Balakrishnan, G., Zhao, A., Sabuncu, M.R., Guttag, J., Dalca, A.V.: VoxelMorph: a learning framework for deformable medical image registration. IEEE Trans. Med. Imaging **38**(8), 1788–1800 (2019). https://doi.org/10.1109/TMI.2019.2897538
2. Caldas-Magalhaes, J., et al.: The accuracy of target delineation in laryngeal and hypopharyngeal cancer. Acta Oncologica **54**(8), 1181–1187 (2015). https://doi.org/10.3109/0284186X.2015.1006401
3. Chappelow, J., et al.: Elastic registration of multimodal prostate MRI and histology via multiattribute combined mutual information. Med. Phys. **38**(4), 2005–2018 (2011). https://doi.org/10.1118/1.3560879
4. Ferrante, E., Paragios, N.: Slice-to-volume medical image registration: a survey. Med. Image Anal. **39**, 101–123 (2017). https://doi.org/10.1016/j.media.2017.04.010
5. Geets, X., et al.: Inter-observer variability in the delineation of pharyngo-laryngeal tumor, parotid glands and cervical spinal cord: comparison between CT-scan and MRI. Radiother. Oncol.: J. Eur. Soc. Ther. Radiol. Oncol. **77**(1), 25–31 (2005). https://doi.org/10.1016/j.radonc.2005.04.010
6. Guo, H., Xu, X., Xu, S., Wood, B.J., Yan, P.: End-to-end ultrasound frame to volume registration (2021)
7. Heinrich, M.P., et al.: MIND: modality independent neighbourhood descriptor for multi-modal deformable registration. Med. Image Anal. **16**(7), 1423–1435 (2012). https://doi.org/10.1016/j.media.2012.05.008
8. Jaganathan, S., Wang, J., Borsdorf, A., Shetty, K., Maier, A.: Deep iterative 2D/3D registration. arXiv:2107.10004 [cs, eess], vol. 12904, pp. 383–392 (2021). https://doi.org/10.1007/978-3-030-87202-1_37

9. Jager, E.A., et al.: Interobserver variation among pathologists for delineation of tumor on H&E-sections of laryngeal and hypopharyngeal carcinoma. How good is the gold standard? Acta Oncologica **55**(3), 391–395 (2016). https://doi.org/10.3109/0284186X.2015.1049661

10. Kimm, S.Y., et al.: Methods for registration of magnetic resonance images of ex vivo prostate specimens with histology. J. Magn. Reson. Imaging **36**(1), 206–212 (2012)

11. Kuckertz, S., Papenberg, N., Honegger, J., Morgas, T., Haas, B., Heldmann, S.: Learning deformable image registration with structure guidance constraints for adaptive radiotherapy. In: Špiclin, Ž, McClelland, J., Kybic, J., Goksel, O. (eds.) WBIR 2020. LNCS, vol. 12120, pp. 44–53. Springer, Cham (2020). https://doi.org/10.1007/978-3-030-50120-4_5

12. Lee, M.C.H., Oktay, O., Schuh, A., Schaap, M., Glocker, B.: Image-and-spatial transformer networks for structure-guided image registration (2019). https://doi.org/10.48550/arXiv.1907.09200

13. Leroy, A., et al.: MO-0476 statistical discrepancies in GTV delineation for H&N cancer across expert centers. Radiother. Oncol. **170**, S426–S427 (2022). https://doi.org/10.1016/S0167-8140(22)02370-2

14. Leroy, A., et al.: End-to-end multi-slice-to-volume concurrent registration and multimodal generation. In: Wang, L., Dou, Q., Fletcher, P.T., Speidel, S., Li, S. (eds.) MICCAI 2022. LNCS, pp. 152–162. Springer, Cham (2022). https://doi.org/10.1007/978-3-031-16446-0_15

15. Li, L., et al.: Co-registration of ex vivo surgical histopathology and in vivo T2 weighted MRI of the prostate via multi-scale spectral embedding representation. Sci. Rep. **7**(1), 8717 (2017). https://doi.org/10.1038/s41598-017-08969-w

16. Markova, V., Ronchetti, M., Wein, W., Zettinig, O., Prevost, R.: Global multimodal 2D/3D registration via local descriptors learning (2022)

17. Njeh, C.F.: Tumor delineation: the weakest link in the search for accuracy in radiotherapy. J. Med. Phys./Assoc. Med. Physicists India **33**(4), 136–140 (2008). https://doi.org/10.4103/0971-6203.44472

18. Ohnishi, T., et al.: Deformable image registration between pathological images and MR image via an optical macro image. Pathol. Res. Pract. **212**(10), 927–936 (2016). https://doi.org/10.1016/j.prp.2016.07.018

19. Rusu, M., et al.: Registration of presurgical MRI and histopathology images from radical prostatectomy via RAPSODI. Med. Phys. **47**(9), 4177–4188 (2020)

20. Shao, W., et al.: ProsRegNet: a deep learning framework for registration of MRI and histopathology images of the prostate. arXiv:2012.00991 [eess] (2020)

21. Tian, L., Lee, Y.Z., Estépar, R.S.J., Niethammer, M.: LiftReg: limited angle 2D/3D deformable registration (2023)

22. Ward, A.D., et al.: Prostate: registration of digital histopathologic images to in vivo MR images acquired by using endorectal receive coil. Radiology **263**(3), 856–864 (2012). https://doi.org/10.1148/radiol.12102294

23. Xiao, G., et al.: Determining histology-MRI slice correspondences for defining MRI-based disease signatures of prostate cancer. Comput. Med. Imaging Graph. **35**(7), 568–578 (2011). https://doi.org/10.1016/j.compmedimag.2010.12.003

24. Xu, Z., et al.: Adversarial uni- and multi-modal stream networks for multimodal image registration. arXiv:2007.02790 [cs, eess] (2020)

25. Zhu, J.Y., Park, T., Isola, P., Efros, A.A.: Unpaired image-to-image translation using cycle-consistent adversarial networks. arXiv:1703.10593 [cs] (2020)

X-Ray to CT Rigid Registration Using Scene Coordinate Regression

Pragyan Shrestha[1([✉])], Chun Xie[1], Hidehiko Shishido[1], Yuichi Yoshii[2], and Itaru Kitahara[1]

[1] University of Tsukuba, Tsukuba, Ibaraki, Japan
shrestha.pragyan@image.iit.tsukuba.ac.jp,
{xiechun,shishid,kitahara}@ccs.tsukuba.ac.jp
[2] Tokyo Medical University, Ami, Ibaraki, Japan
yyoshii@tokyo-med.ac.jp

Abstract. Intraoperative fluoroscopy is a frequently used modality in minimally invasive orthopedic surgeries. Aligning the intraoperatively acquired X-ray image with the preoperatively acquired 3D model of a computed tomography (CT) scan reduces the mental burden on surgeons induced by the overlapping anatomical structures in the acquired images. This paper proposes a fully automatic registration method that is robust to extreme viewpoints and does not require manual annotation of landmark points during training. It is based on a fully convolutional neural network (CNN) that regresses the scene coordinates for a given X-ray image. The scene coordinates are defined as the intersection of the back-projected rays from a pixel toward the 3D model. Training data for a patient-specific model were generated through a realistic simulation of a C-arm device using preoperative CT scans. In contrast, intraoperative registration was achieved by solving the perspective-n-point (PnP) problem with a random sample and consensus (RANSAC) algorithm. Experiments were conducted using a pelvic CT dataset that included several real fluoroscopic (X-ray) images with ground truth annotations. The proposed method achieved an average mean target registration error (mTRE) of $3.79+/1.67$ mm in the 50th percentile of the simulated test dataset and projected mTRE of $9.65+/-4.07$ mm in the 50th percentile of real fluoroscopic images for pelvis registration. The code is available at https://github.com/Pragyanstha/SCR-Registration.

Keywords: Registration · X-Ray Image · Scene Coordinates

1 Introduction

Image-guided navigation plays a crucial role in modern surgical procedures. In the field of orthopedics, many surgical procedures such as total hip arthro-

Supplementary Information The online version contains supplementary material available at https://doi.org/10.1007/978-3-031-43999-5_74.

plasty, total knee arthroplasty, and pedicle screw injections utilize intraoperative fluoroscopy for surgical navigation [2, 4, 11]. Due to overlapping anatomical structures in X-ray images, it is often difficult to correctly identify and reason the 3D structure from solely the image. Therefore, registering an intraoperatively acquired X-ray image to the preoperatively acquired CT scan is crucial in performing such procedures [13, 15, 18, 19]. The standard procedure for acquiring highly accurate registration involves embedding fiducial markers into the patient and acquiring a preoperative CT scan to obtain 2D-3D correspondences [6, 10, 17]. Inserting fiducial markers onto the body involves extra surgical costs and might not be viable for minimally invasive surgeries. To circumvent such issues with the feature-based method, an intensity-based optimization scheme for registration has been extensively studied [1, 9]. Since the objective function is highly nonlinear for optimizing pose parameters, a good initialization is necessary for the method to converge in a global minimum. Therefore, it is usually accompanied by initial coarse registration using manual alignment of the 3D model to the image, interrupting the surgical flow. On the other hand, learning-based methods have proved to be efficient in solving the registration task. Existing learning-based methods can be broadly categorized into landmark estimation and direct pose regression. Landmark estimation methods aim to solve for pose using correspondences between 3D landmark annotations and its estimated 2D projection points [3, 5, 7], while methods based on pose regression estimate the global camera pose in a single inference [12]. Pose regressors are known to overfit training data and generalize poorly to unseen images [14]. This makes the landmark estimation methods stand out in terms of registration quality and generalization. However, there exist two main issues with landmark estimation methods: 1) Annotation cost of a sufficiently large number of landmarks in the CT image. 2) Failure to solve for the pose in extreme views where projected landmarks are not visible or the number of visible landmarks is small.

This paper addresses these issues by introducing scene coordinates [16] to establish dense 2D-3D correspondences. Specifically, the proposed method regresses the scene coordinates of the CT-scan model from corresponding X-ray images. A rigid transformation that aligns the CT-scan model to the image is then calculated by solving the Perspective-n-point (PnP) problem with the Random sample and consensus (RANSAC) algorithm.

2 Method

2.1 Problem Formulation

The problem of 2D-3D registration can be formulated a finding the rigid transformation that transforms the 3D model defined in the anatomical or world coordinate system into the camera coordinate system. Specifically, given a CT-scan volume $V_{CT}(x_w)$ where x_w is defined in the world coordinate system, the registration problem is concerned with finding $T_w^c = [R|t]$ such that the following holds.

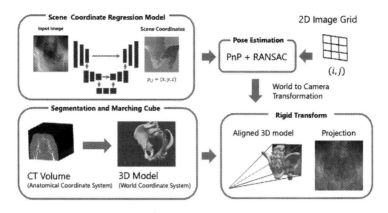

Fig. 1. An overview of the proposed method. Scene coordinates are regressed using a U-Net architecture given an X-ray image. With the obtained dense correspondences, PnP with RANSAC is run to get the transformation matrix that aligns the projection of the 3D model with the X-ray image in the camera coordinate system.

$$I = \Re\{V_{CT}(T_w^{c^{-1}} x_c); K\} \tag{1}$$

where $\Re\{\cdot\}$ is the X-ray transform that can be applied to volumes in the camera coordinate system given an intrinsic matrix K and I, the target X-ray image.

2.2 Registration

The proposed registration pipeline overview is shown in Fig. 1. The proposed method comprises the following four parts: first, the scene coordinates were regressed using a single-view X-ray image as input to the U-Net model; second, the PnP + RANSAC algorithm is used to solve for the pose of the captured X-ray system; third, the CT-scan volume was segmented to obtain a 3D model of the bone regions; and fourth, the computed rigid transformation from world coordinates to camera coordinates is used to generate projection overlay images.

Scene Coordinates. The scene coordinates are defined as the points of intersection between the camera's back-projected rays and the 3D model in a world coordinate system (i.e., only the first intersection and the last intersections are considered). The same concept was adapted for X-ray images and their underlying 3D models obtained from the CT-scans. Specifically, given an arbitrary point x_{ij} in the image plane, the scene coordinates X_{ij} satisfy the following conditions.

$$X_{ij} = [R^T| - t](dK^T x_{ij}) \tag{2}$$

where R and t are the rotation matrix and translation vector that maps points in the world coordinate system to the camera coordinate system, K is the intrinsic matrix, d is the depth, as seen from the camera, of the point X on the 3D model.

Uncertainty Estimation. The task of scene coordinate regression is to estimate these X_{ij} for every pixel ij, given an X-ray image I. However, the existence of X_{ij} is not guaranteed for all pixels because back-projected rays may not intersect the 3D model. One of the many ways to address such a case is to prepare a mask (i.e., 1 if the bone area, 0 otherwise) in advance so that only the pixels that lie inside the mask are estimated. As this approach requires an explicit method for estimating the mask image, an alternative approach was adopted in this study. Instead of estimating a single X_{ij}, the mean and variance of the scene coordinates are estimated. The non-intersecting scene coordinates were identified by applying thresholding to the estimated variance (i.e., points with high variance were considered non-existent scene coordinates and were filtered out). This approach assumes that the observed scene coordinates are corrupted with a zero mean, non-zero and non-constant variance, and isotropic Gaussian noise.

$$X_{ij} \sim N(u(I, x_{ij}), \sigma(I, x_{ij})) \tag{3}$$

where $u(I, x_{ij})$ and $\sigma(I, x_{ij})$ are the functions that produce the mean and standard deviation of the scene coordinates, respectively. This work represents these functions using a fully convolutional neural network.

Loss Function. A U-Net architecture was used to estimate the mean and standard deviation of the scene coordinates at every pixel in a given image. The loss function for the intersecting scene coordinates is derived from the maximum likelihood estimates using the likelihood X_{ij}. This can be expressed as follows:

$$Loss_{intersecting} = (\frac{(X_{ij} - u(I, x_{ij}))}{\sigma(I, x_{ij})})^2 + 2\log(\sigma(I, x_{ij})) \tag{4}$$

Because it is desirable to have a high variance for non-existent scene coordinates, the loss function for non-existent coordinates is designed as follows:

$$Loss_{non-existent} = \frac{1}{\sigma(I, x_{ij})} \tag{5}$$

2D-3D Registration. An iterative PnP implementation from OpenCV was run using RANSAC with maximum iteration of 1000 and reprojection error of 10px and 20px for the simulated and real X-ray images respectively. An example of a successful registration is shown in the left part of Fig. 2 below.

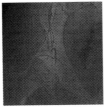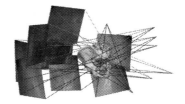

Fig. 2. An example of successful registration with the proposed method (left two images) and Randomly picked data samples in the test set (right). The X-ray image and model's gradient projection overlay (middle) and the model's pose in the camera coordinates system (left). The origin of the view frustum is the X-ray source position and the simulated X-ray images are placed in the detector plane for visualization (right).

3 Experiments and Results

3.1 Dataset

A dataset containing six annotated CT scans, each with several registered real X-ray images from [7] was used to properly evaluate the proposed method. The annotations included 14 landmarks and 7 segmentation labels. The CT scans were of the pelvic bones of cadaveric specimens. Because there were only a few real X-ray images, simulated X-ray images were generated from each CT-scans to train and test the model. In particular, DeepDRR [3] was used to simulate a Siemens Cios Fusion Mobile C-arm imaging device. Similar to [3], the left anterior oblique/right anterior oblique (LAO/RAO) views were sampled at angles of [45, 45] degrees with 1-degree intervals. A random offset was applied in each direction. The offset vector was sampled from a normal distribution with a mean of zero and standard deviations of 90 mm in the lateral direction, and 30 mm in the other two directions. Images intentionally included partially visible structures. A selection of randomly sampled images is displayed on the right side of Fig. 2. Ground-truth scene coordinates for each image were obtained by rendering the depth map of the 3D models and converting them to 3D coordinates using Eq. 2. In total, 8100 simulated X-rays were generated for each specimen. Among these, 5184 images were randomly assigned to the training set, 1296 to the validation set, and the remaining 1620 to the test set.

3.2 Implementation Details

The input image and output scene coordinates had a size of 512×512 pixels. The U-Net model had eight output channels, which consisted of three channels for scene coordinates and one channel for standard deviation, multiplied by two for the entry and exit points. Patient-specific models were trained individually for each dataset. Adam optimizer with a constant learning rate of 0.0001 and a batch size of 16 was used. Online data augmentation which includes random inversion, color jitter, and random erasing was applied. The scene coordinates were filtered using a log variance threshold of 0.0 for simulated images and -2.0 for real X-ray images.

3.3 Baselines and Evaluation Metrics

The proposed method was compared with two other baseline methods: PoseNet [8] and DFLNet [7]. PoseNet was implemented using ResNet-50 as the backbone for the feature extractor and was trained using geometric loss. DFLNet uses the same architecture as the proposed method however the last layer regresses 14 heatmaps of the landmarks instead of scene coordinates. Note that the original study's segmentation layer and the gradient-based optimization phase were omitted for architectural comparison. Each baseline was trained in a patient-specific manner following the proposed method. The mean target registration error (mTRE) and Gross Failure Rate (GFR), were used as the evaluation metrics for comparison with the baselines. mTRE is defined in 6, where X_k is the position of the ground truth landmark $\hat{X}_k$ after applying the predicted transformation. The GFR is the ratio of failed cases, defined as the registration results with an mTRE greater than 10 mm. Because we could only obtain the projection of the ground truth landmarks and not the ground truth transformation matrix for the real X-ray images, the projected mTRE (proj. mTRE) was used for the evaluation. It is similar to mTRE except the X_k and $\hat{X}_k$ represent the projected coordinates of the landmarks, in the detector plane (i.e., the pixel coordinates are scaled according to the detector size to match the units).

$$\text{mTRE} = \frac{1}{N} \sum_{k=1}^{k=N} \|X_k - \hat{X}_k\|_2 \tag{6}$$

3.4 Registration Results

Simulated X-Ray Images. Table 1 shows the mTRE for the 25th, 50th, and 95th percentiles of the total test sample size and the GFR. The proposed method could retain a GFR below 20% for most of the specimens, whereas PoseNet and DFLNet failed to register with more than 20% GFR in most cases. This is because the network in PoseNet cannot reason about the spatial structure or its local relation to the image patches. For DFLNet, this is inevitable because of the visibility issue of the landmark points, mostly located in the pubic region of the pelvis. Comparing the mTRE of each specimen with that of each method, the proposed method achieved an mTRE of 7.98 mm even in the 95th percentile of Specimen 2. DFLNet achieved the lowest mTRE of 0.98 mm in the 25th percentile of Specimen 4. This illustrates the highly accurate registration of landmark estimation methods. However, with extreme or partial views such as the one shown in Fig. 3, the method cannot estimate the correct pose parameter because of incorrect landmark localization or insufficiently visible landmarks. Please refer to the supplemental material for the registration overlay results of the different specimens using the proposed method.

Real X-Ray Images. Table 2 lists the mTRE values calculated for the projected image points (abbreviated as proj. mTRE) for PoseNet and the proposed method, respectively. DFLNet did not adapt to real X-ray images, therefore, it was omitted from the table. Because our dataset consisted mostly of images with partially visible hips, only a few landmarks were visible in each image. This causes the DFLNet to overfit to the partially visible landmark distribution, whereas our proposed model mitigates this issue by learning the general structure (i.e., every surface point that is visible). The proposed method estimates good transformations (that is proj. mTRE approximately 10 mm in the 50th percentile). In contrast, the proj. mTRE for PoseNet is significantly higher. This suggests that PoseNet overfitted the training data despite applying domain randomization. This result agrees with previous reports [14] addressing this issue. A visualization of the overlays is presented in the Supplemental Material.

Table 1. The mean target registration errors each in 25th, 50th and 95th percentile of the simulated test dataset. All models are trained individually on the 6 specimens shown below. The proposed method outperforms other methods regarding 50th percentile mTRE and GFR in most specimens.

Specimen	PoseNet				DFLNet				Ours			
	mTRE[mm]↓			GFR[%]↓	mTRE[mm]↓			GFR[%]↓	mTRE[mm]↓			GFR[%]↓
	25th	50th	95th		25th	50th	95th		25th	50th	95th	
#1	5.75	8.37	24.81	38.53	3.36	212.59	680.58	62.27	1.37	**2.50**	9.80	**4.87**
#2	6.97	10.23	25.95	51.63	1.98	7.04	656.41	46.15	1.15	**2.14**	7.98	**2.54**
#3	5.42	7.86	23.34	35.35	1.03	**2.51**	583.63	28.65	1.67	3.05	12.25	**8.56**
#4	4.67	6.46	16.91	18.77	0.98	**2.30**	558.20	23.59	1.76	3.38	19.19	**12.54**
#5	4.81	6.52	18.98	22.43	1.51	**4.28**	767.56	37.47	3.09	5.30	17.32	**18.85**
#6	4.06	**5.85**	18.42	**22.69**	2.26	139.96	15321.19	58.72	3.80	6.37	18.25	23.18
mean	5.28	7.55	21.40	31.57	1.85	61.45	3094.60	42.81	2.14	**3.79**	14.13	**11.76**
std	1.02	1.62	3.77	12.58	0.90	91.88	5990.24	15.76	1.06	1.67	4.75	8.05

Table 2. The projected mean target registration errors for real X-ray images. The proposed method achieved significantly low registration errors compared to PoseNet, implying that it generalizes well to unseen data and domains.

Specimen	Number of Images	PoseNet			Ours		
		proj. mTRE [mm]↓			proj. mTRE [mm]↓		
		25th	50th	95th	25th	50th	95th
#1	111	43.64	49.11	64.23	5.45	**8.02**	55.87
#2	24	19.42	27.18	43.68	2.74	**3.32**	6.48
#3	104	31.18	38.97	66.06	7.60	**11.85**	162.83
#4	24	35.52	38.37	57.07	11.12	**15.52**	92.34
#5	48	38.97	46.60	69.06	6.09	**9.07**	21.91
#6	55	34.51	37.13	47.72	7.18	**10.14**	20.78
mean		33.87	39.56	57.97	6.70	**9.65**	60.04
std		8.25	7.77	10.37	2.77	4.07	59.15

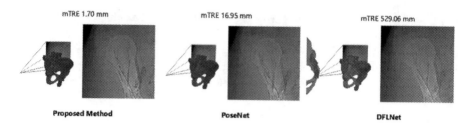

Fig. 3. An example case illustrating an extreme partial viewpoint. The proposed method successfully registers the image with 1.70 mm mTRE, while PoseNet struggles with 16.95 mm mTRE. Since there is an insufficient number (less than 4) of visible landmarks, the DFLNet hallucinates landmarks providing incorrect 2D-3D correspondences, leading to a large mTRE.

4   Limitations

As the proposed method was designed to give initial estimates of the pose parameters, a further refinement step using an intensity-based optimization method would be required to obtain clinically relevant registration accuracy. Although the proposed method provided a good initial estimate, the average runtime for the entire pipeline was 1.75 s which was approximately two orders of magnitude greater than that of PoseNet, which had an average runtime of 0.06 s. This is because RANSAC must determine a good pose from a dense set of correspondences. This issue can be addressed by heuristically selecting a good variance threshold per image that filters out bad correspondences.

5   Conclusion

This paper presented a scene coordinate regression-based approach for the X-ray to CT-scan model registration problem. Experiments with simulated and real X-ray images showed that the proposed method performed well even under partially visible structures and extreme view angles, compared with direct pose regression and landmark estimation methods. Testing the model trained solely on simulated X-ray images, on real X-ray images did not result in catastrophic failure. Instead, the results were positive for instantiating further refinement steps.

Acknowledgement. This work was partially supported by a grant from JSPS KAK-ENHI grant number JP23K08618. This study (in part) used the computational resources for Cygnus provided by the Multidisciplinary Cooperative Research Program at the Center for Computational Sciences, University of Tsukuba, Japan.

References

1. Aouadi, S., Sarry, L.: Accurate and precise 2D–3D registration based on X-ray intensity. Comput. Vis. Image Underst. **110**(1), 134–151 (2008)
2. Belei, P., et al.: Fluoroscopic navigation system for hip surface replacement. Comput. Aided Surg. **12**(3), 160–167 (2007)
3. Bier, B., et al.: X-ray-transform invariant anatomical landmark detection for pelvic trauma surgery (2018)
4. Bradley, M.P., Benson, J.R., Muir, J.M.: Accuracy of acetabular component positioning using computer-assisted navigation in direct anterior total hip arthroplasty. Cureus **11**(4), e4478 (2019)
5. Esteban, J., Grimm, M., Unberath, M., Zahnd, G., Navab, N.: Towards fully automatic X-ray to CT registration. In: Shen, D., et al. (eds.) MICCAI 2019. LNCS, vol. 11769, pp. 631–639. Springer, Cham (2019). https://doi.org/10.1007/978-3-030-32226-7_70
6. George, A.K., Sonmez, M., Lederman, R.J., Faranesh, A.Z.: Robust automatic rigid registration of MRI and X-ray using external fiducial markers for XFM-guided interventional procedures. Med. Phys. **38**(1), 125–141 (2011)
7. Grupp, R.B., et al.: Automatic annotation of hip anatomy in fluoroscopy for robust and efficient 2D/3D registration. Int. J. Comput. Assist. Radiol. Surg. **15**(5), 759–769 (2020)
8. Kendall, A., Grimes, M., Cipolla, R.: PoseNet: a convolutional network for real-time 6-DOF camera relocalization. In: 2015 IEEE International Conference on Computer Vision (ICCV), pp. 2938–2946 (2015)
9. Livyatan, H., Yaniv, Z., Joskowicz, L.: Gradient-based 2-D/3-D rigid registration of fluoroscopic X-ray to CT. IEEE Trans. Med. Imaging **22**(11), 1395–1406 (2003)
10. Maurer, C.R., Jr., Fitzpatrick, J.M., Wang, M.Y., Galloway, R.L., Jr., Maciunas, R.J., Allen, G.S.: Registration of head volume images using implantable fiducial markers. IEEE Trans. Med. Imaging **16**(4), 447–462 (1997)
11. Merloz, P., et al.: Fluoroscopy-based navigation system in spine surgery. Proc. Inst. Mech. Eng. H **221**(7), 813–820 (2007)
12. Miao, S., Jane Wang, Z., Liao, R.: Real-time 2D/3D registration via CNN regression (2015)
13. Reichert, J.C., Hofer, A., Matziolis, G., Wassilew, G.I.: Intraoperative fluoroscopy allows the reliable assessment of deformity correction during periacetabular osteotomy. J. Clin. Med. Res. **11**(16) (2022)
14. Sattler, T., Zhou, Q., Pollefeys, M., Leal-Taixé, L.: Understanding the limitations of CNN-Based absolute camera pose regression. In: 2019 IEEE/CVF Conference on Computer Vision and Pattern Recognition (CVPR), pp. 3297–3307 (2019)
15. Selles, C.A., Beerekamp, M.S.H., Leenhouts, P.A., Segers, M.J.M., Goslings, J.C., Schep, N.W.L.: EF3X study group: the value of intraoperative 3-dimensional fluoroscopy in the treatment of distal radius fractures: a randomized clinical trial. J. Hand Surg. Am. **45**(3), 189–195 (2020)
16. Shotton, J., Glocker, B., Zach, C., Izadi, S., Criminisi, A., Fitzgibbon, A.: Scene coordinate regression forests for camera relocalization in RGB-D images. In: 2013 IEEE Conference on Computer Vision and Pattern Recognition, pp. 2930–2937 (2013)
17. Tang, T.S.Y., Ellis, R.E., Fichtinger, G.: Fiducial registration from a single X-ray image: a new technique for fluoroscopic guidance and radiotherapy. In: Delp, S.L., DiGoia, A.M., Jaramaz, B. (eds.) MICCAI 2000. LNCS, vol. 1935, pp. 502–511. Springer, Heidelberg (2000). https://doi.org/10.1007/978-3-540-40899-4_51

18. Woerner, M., et al.: Visual intraoperative estimation of cup and stem position is not reliable in minimally invasive hip arthroplasty. Acta Orthop. **87**(3), 225–230 (2016)

19. Wylie, J.D., Ross, J.A., Erickson, J.A., Anderson, M.B., Peters, C.L.: Operative fluoroscopic correction is reliable and correlates with postoperative radiographic correction in periacetabular osteotomy. Clin. Orthop. Relat. Res. **475**(4), 1100–1106 (2017)

Author Index

H. Greenspan et al. (Eds.): MICCAI 2023, LNCS 14229, pp. 791–795, 2023.
https://doi.org/10.1007/978-3-031-43999-5

Printed in the United States
by Baker & Taylor Publisher Services